American Mathematical Society
Colloquium Publications
Volume 57

Opera de Cribro

John Friedlander
Henryk Iwaniec

AMS American Mathematical Society

2010 *Mathematics Subject Classification.* Primary 11N35, 11N36; Secondary 11N05, 11N13, 11N32, 11N37, 11J71, 11E25.

For additional information and updates on this book, visit
www.ams.org/bookpages/coll-57

Library of Congress Cataloging-in-Publication Data

Friedlander, J. B. (John B.)
Opera de cribro / John Friedlander, Henryk Iwaniec.
p. cm. — (Colloquium publications ; v. 57)
In English.
Includes bibliographical references and index.
ISBN 978-0-8218-4970-5 (alk. paper)| Softcover ISBN 978-1-4704-8168-1
1. Sieves (Mathematics) I. Iwaniec, Henryk. II. Title.

QA246.F75 2010
512.7′3–dc22 2009046518

Printed in the United States of America.
Reprinted by the American Mathematical Society, 2025.

∞ The paper used in this book is acid-free and falls within the guidelines
established to ensure permanence and durability.
Visit the AMS home page at http://www.ams.org/

11 10 9 8 7 6 5 4 3 2 30 29 28 27 26 25

Contents

Preface

Libretto

What possesses us to write a book on sieve methods? Hopefully we have something to say that is worth listening to and, if we didn't write it, maybe a little something would be lost.

Many young people are drawn to mathematics and frequently what is most attractive to them are problems which are related to the most elementary objects, the integers, the squares, and the primes. The sieve method, itself an elementary idea two thousand years old, offers, even to the high school teenager, a natural approach within her own compass. Many of our colleagues can relate to this experience, as do we. Just as the student during his development acquires more sophisticated tools and tastes, so has the subject of sieve methods itself; and, whereas it is the elementary ideas that offer the initial attraction, it is the ability to infuse these with the more advanced tools that brings joy in later life. One of our goals is to transmit some of these joys to new investigators.

Eratosthenes, about 300 BC, is the first to be mentioned. His idea, illustrated in Chapter 1, is essentially an algorithm for tabulating primes. Following a very long gap, the subject was taken up by A. Legendre who gave a formula (Section 1.1) for $\pi(x)$. This was the beginning of the principle of inclusion-exclusion. Legendre's formula, however, is of very limited practical value and the ideas which were needed to turn it into a useful instrument were initiated by V. Brun in 1915. He developed his methods over the following decade before leaving the topic. We discuss his ideas in Chapter 6. Such a late placement should alert the reader that we do not follow a chronological order of sieve developments in this book.

The combinatorial complexity of Brun's most advanced works inhibited later researchers so that the subject did not receive the huge initial impetus that the power of his ideas merited.

Brun's results were improved considerably by A. Buchstab in the late 1930's using an iterative scheme, the basis of which has remained valuable to this day and which is discussed in several chapters, especially Chapter 11.

A new impetus came to the subject with the ideas of A. Selberg in the late 1940's (discussed in Chapter 7). His upper-bound sieve, which, following Selberg, we call the Λ^2-sieve, is, in particular, very elegant in comparison with the earlier Brun sieves. It is also considerably stronger in some respects, although subsequent developments have shown that the strongest results are obtained by keeping both the Λ^2-sieve and the combinatorial sieve in mind.

More recent progress has involved many new ideas and many names. Not only has the theory of the sieve apparatus grown but it has developed enhancements which permit the implementation of sophisticated results from harmonic analysis, from exponential sums, from arithmetic geometry, and from automorphic forms. The reachable targets have expanded, not only from almost-primes to primes, but also to questions of greater variety, for example to solutions of diophantine equations. We hope the contents of the book will speak for themselves.

Let's take a quick run-through of the chapters. Although we follow roughly the order of development of the subject there are significant detours. Thus, for example, Bombieri who appears in Chapter 3 is thought to be somewhat younger than Eratosthenes who appears in Chapter 4. The reader we hope has sufficient mathematical maturity not to be troubled by this.

In Chapter 1, following the introduction to exclusion-inclusion and the formula of Eratosthenes-Legendre, we give a number of examples of arithmetic problems which can be attacked by the sieve. Although there are many we do not cover, those presented here suffice to show the versatility of sieve ideas.

In Chapter 2 we present, still following the historical order, ideas (due to Dirichlet and Tchebyshev) which are of an elementary nature and which will be used in sieve arguments. These are methods for evaluating sums of arithmetic functions of convolution type and are based on changing the order of summation in the spirit of the Dirichlet hyperbola method. A second purpose is the evaluation of sums and products over prime numbers to which we shall need frequent reference. Further results of this nature are postponed to Appendix A.

In Chapter 3 we develop Bombieri's sieve which, despite the name, we consider to be the transition between the preceding ideas of Tchebyshev and the sieve ideas to follow. Consequently, although historically out of place, it seems to fit here naturally. An understanding of these ideas should help the reader to appreciate the magic of positivity in the sieves that are to follow. More importantly, the results of this section offer the best way to view the parity phenomenon which lies at the heart of the limitations of classical sieve theory.

We are now ready to start with the sieve. However, before entering the mainstream, we take the opportunity in Chapter 4, for historical reasons, to show the extent to which the original sieve of Eratosthenes and Legendre can be developed by today's techniques. Surprisingly, these straightforward ideas can even lead to asymptotic formulae under favourable conditions.

In Chapter 5 we introduce many of the basic sieve principles, terminology and notation. To some extent, we have already out of necessity been doing this, even in Chapter 1. However, at the risk of a little repetition, we want to gather these in a single location for ease of reference. From the start, a serious impediment to an understanding of the sieve has been the difficulty in coming up with a satisfactory notation. The literature is littered with awkward choices. It is especially difficult to be consistent throughout the presentation of a full-length book. We think the reader is better served by the adoption of a notation which sacrifices uniqueness in order to be more suggestive. Consequently, one finds slightly modified names for similar objects. In such cases, the context helps to clarify the issue. The material in the final few sections of Chapter 5 is of a more advanced nature but still deals with general sieve principles so is appropriate to include before we get to specific sieve procedures. Moreover, it is beneficial to see it inserted now, while the notation is fresh, making its exposition more transparent.

In Chapter 6 we finally come to the real beginning of our story. Brun gave up on the goal of finding asymptotic formulae and was willing to settle for upper and lower bounds. This allowed him, by positivity, to simply throw away many of the pieces of the problem which he could not handle. The fact that the few pieces remaining were enough to give useful results still today seems a miracle. Even the simplest version of Brun's sieve, which we present in the first section, produces remarkable results. Although less than perfect, such results seem hopeless to derive without the sieve by any modern means. In the following few sections we refine the original constructions of Brun, not as he did, but following along his lines and equipped with a good deal of hindsight. In particular, we take advantage of an important observation due to Buchstab which, by iteration, produces a sequence of successively stronger sieves. This idea will lead us later to the construction of sieves which are optimal in the most important cases. Before we come to these strong but complicated results, we give in Section 6.5 a simpler treatment, nevertheless, already obtaining theorems of fundamental nature. The resulting upper and lower bounds have the right order of magnitude and are sufficient for many, possibly most, of the sieve applications in the general literature. In the final two sections we illustrate some of these applications. In particular, the application in the last section gives a glimpse of the techniques we shall use to go beyond the natural limitations of the sieve. Along the lines of the earlier Section 6.8, which is needed to prepare for this, one learns that it is useful to understand how the sieve mechanism works and not merely the final statements of the theorems.

Returning to the theorem in Section 6.5, an important feature of the result is the fact that the upper and lower bounds for the sifting function approach each other, and at more than exponential speed, as the sifting range gets smaller. This feature is of such importance throughout the subject that it merits the name by which it has become known, "Fundamental Lemma", a term which was coined by Halberstam and Richert.

Chapter 7 is devoted to Selberg's sieve. His upper-bound sieve, prompted by his earlier work on the zeta-function, introduced ideas fundamentally different from those of Brun. We discuss various features of the Selberg sieve at length. It is distinguished not only by the power of the results but also by the elegance of the arguments. The estimates are particularly strong when the sieve dimension is large. This power is exploited brilliantly in the work of Goldston, Pintz and Yildirim on small gaps between consecutive primes. Their argument is one case where the Selberg sieve seems to be essential. In the final sections of the chapter we show how these ideas work.

Given any upper-bound sieve, one can obtain a lower-bound sieve (and vice-versa) by Buchstab's recurrence formula. This possibility is particularly convenient in the case of the Selberg upper bound which, unlike the situation with Brun, has no natural lower bound counterpart. Ankeny and Onishi were the first to perform this procedure. The resulting estimates, upper and lower, can be further improved by iterations of the operation. Actually, it requires infinitely many such iterations to obtain the best results in all ranges. We do not present these in the book beyond the first iteration. In the case of the linear sieve, a combination of Selberg's upper-bound sieve with the construction of a combinatorial sieve of Brun's type led Jurkat and Richert to the essentially best possible upper and lower bounds for the sifting function in all ranges. These are the same results as could have been obtained by an infinite number of iterations. We shall also obtain the same results as Jurkat and

Richert (in Chapter 11), but by a more direct construction, using the combinatorial sieve.

An important problem in the theory of the sieve is concerned with the "sifting limit", that is, the maximal possible range for which one can get a positive lower bound in a general setting. If κ denotes the dimension of the sieve problem it has been conjectured that this sifting limit parameter should be 2κ (actually we shall see that it is sometimes better). The results that we shall develop later show that this indeed is true for $\kappa = 1$ and $\kappa = \frac{1}{2}$. For κ very large the results in this respect were rather far from the truth until Selberg introduced a new combinatorial device into his upper-bound sieve, giving bounds for the sifting limits which approach 2κ as $\kappa \to \infty$. We present his arguments (slightly modified) in Section 7.6.

There are interesting problems which require the removal of elements from many residue classes per modulus. The combinatorial sieve proves to be quite effective if the number of unwanted classes is small or at least bounded. On the other hand, Selberg's upper-bound sieve shows to advantage here in case the number of classes is large or even increases with the modulus. For example, it is capable of bounding correctly the numbers of squares by removing half of the classes for each prime modulus. Here, the concept of sieve dimension is meaningless and the Selberg sieve benefits from being insensitive to the sieve dimension. In Chapter 8 we give a number of applications of Selberg's sieve in such circumstances.

There are, however, very different techniques for studying these "large sieve" problems. In Chapter 9 we develop some of these techniques, discussing in Section 9.5 the extent to which they are related. Linnik originated the large sieve theory. Over time, various alternative approaches and general formulations have been developed, reaching the point that the subject is no longer recognizable as a sieve. Nevertheless, the main statement retains the name "large sieve inequality". The most frequently used, among its many forms, is an inequality for trigonometrical polynomials at well-spaced points, which we establish in Theorem 9.1. The main idea behind results of this type is the duality principle for linear operators on a vector space.

An important example of well-spaced points is provided by the Farey fractions. This example leads to estimates for character sums, both additive and multiplicative. The results are so strong that they compete with the consequences of the Riemann Hypothesis. Indeed, the large sieve inequalities for multiplicative characters lead to bounds for the error term in the Prime Number Theorem in arithmetic progressions (Bombieri–Vinogradov theorem) which have served as an unconditional substitute for the Riemann Hypothesis in many applications. This will be evident in our frequent use of this theorem throughout. The Farey points are just the roots of linear congruences. Another example of well-spaced points is given by the roots of quadratic congruences and the large sieve produces powerful consequences; see Propositions 20.5 and 20.7. A beautiful aspect of the large sieve inequality is that it holds for character sums with general complex coefficients. This offers applications inaccessible even to the Riemann Hypothesis. In the later sections of Chapter 9 we derive several estimates for bilinear forms in arithmetic progressions which, due to this generality, will be of great use to us later.

By this time, we feel it is appropriate to say a little more about the sieve weights. Although introduced simply as tools to serve our purpose, these surrogates for the Möbius function have an inner structure which is of interest on its own. Revealing such properties can also provide a greater understanding of the sieve mechanism. In Chapter 10 we probe this "molecular" structure.

We have already seen in Chapter 6 the motivation for and construction of the beta-sieve weights and their application to the Fundamental Lemma. Although extremely powerful when one is not sifting very far, the results there suffer in quality when the sifting range is large, which is the case in the most advanced applications. This is not the fault of the choice of the weights but rather because we used there some crude bounds which could be quickly executed and are sufficient for the Fundamental Lemma. In Chapter 11 we reconsider the beta-sieve and work a lot harder to determine the optimal value of β and, using that, the optimal upper and lower bounds for the sifting function in all ranges. The analysis is very delicate. It depends on the theory of differential-difference equations which we postpone to Appendix B and, in particular, on a careful treatment of a sensitive convergence question. We conclude the chapter with some applications of the results for the fractional dimensions $0 < \kappa < 1$. The most important cases of $\kappa = 1$ and $\kappa = \frac{1}{2}$ are given individual additional treatment in subsequent chapters.

In Chapters 12 and 13 we concentrate attention on the linear sieve, that is, the sieve of dimension $\kappa = 1$. Because the upper and lower bounds are best possible with respect to the main terms, as we illustrate in Section 12.3 using a construction of Selberg, it makes sense to invest the effort to obtain sharpened secondary terms which didn't merit attention in the general development in Chapter 11. In particular, we expend considerable energy to get the best possible error term in the main terms by using differential-difference equations especially designed for that purpose. More importantly, we pay renewed attention to the remainder. We manage to give the remainder in the shape of a bilinear form. This opens the possibility to apply various techniques of harmonic analysis, such as given in Chapter 13, to obtain estimates superior to those following from the traditional trivial treatment. Consequently, the results are stronger. The flexibility of ranges in our bilinear remainder allows us to choose the parameters to fit the results of the harmonic analysis in the best way. This new input from harmonic analysis to sieve theory makes the latter less elementary but the rewards are great.

Chapter 13 is entirely devoted to the illustration of this input from harmonic analysis for a variety of sifting sequences. First we consider the sequence of integers in an arithmetic progression, use Fourier expansion of the remainder terms, apply bounds for Kloosterman type sums, and conclude an upper bound for $\pi(x; q, a)$ which is quite strong, uniformly in large q. We then consider the sequence of integers in a short interval and use similar analysis, but with exponential sums suited to this sequence, getting similar improvements. Finally, we show how this analysis can be combined with Tchebyshev's method rather than the sieve to show the existence, in the sequence, of integers having a large prime power factor. Here we have only considered two of the simplest sequences. But harmonic analysis is very versatile. A lot more will follow.

The beta-sieve with $\kappa = \frac{1}{2}$ is called the semi-linear sieve and several of its applications are featured in Chapter 14. In this case the sifting limit is $\beta = 1$ which is the best that one could hope for in any sieve. This means that one can perform sieving in the maximal range conceivable, right up to the level of distribution. The lower and upper bounds given by the sifted sums are optimal. The example which demonstrates this optimality is attractive from an arithmetic point of view. Indeed, we show the upper bound becomes an asymptotic for some interesting sequences, such as the sequence of values of a binary quadratic form. The lower bound is also very interesting, even when the exponent of distribution is as small as $\frac{1}{2}$, in which case it just misses in general to give the desired output. Moreover, in many cases it can, with additional arguments, be made to do so. For example, when the sifting sequence consists of shifted primes we do have exponent of distribution $\frac{1}{2}$, due to the Bombieri–Vinogradov theorem, and we conclude that shifted primes are representable, for example by sums of two squares, infinitely often.

Another interesting instance which involves a double application of the semi-linear sieve arises in the study of primes represented by polynomials over varieties. We give in Theorem 14.10 an example wherein one can derive bounds of the correct order of magnitude provided the level of distribution is sufficiently good. The level needed is not yet known to hold but is weaker than the Elliott–Halberstam conjecture. In addition to its applications to prime number problems the semi-linear sieve proves to be a powerful tool for the study of diophantine problems. We show how this works for cubic surfaces. The arguments exemplify once again that one can accomplish more with an imaginative transformation of sieve results as compared to a direct application of the theorems.

The versatility of the sieve shows itself not only in its adaptation to different mathematical techniques but also in its applicability to differing arithmetic goals. In Chapter 15 we consider a few problems which at first glance may seem quite outside the compass of the sieve.

In the next Chapter 16 we switch from practical aspects to a question of great theoretical importance, the parity problem of the linear sieve. As mentioned earlier, this phenomenon is most clearly visible when examined in the context of the Bombieri sieve and so we follow closely his treatment. This requires considerable generalization of the results of Chapter 3 so as to provide a wider class of test functions. Such general results, when confronted with Selberg-type examples, reveal that one cannot produce primes within the framework of the classical sieve. Actually, the parity phenomenon reveals even further limitations.

Chapter 17 deals with combinatorial identities for primes and related arithmetical functions. The subject received its impetus from Vinogradov who, like Brun, took his inspiration from the Eratosthenes sieve. Although not, strictly speaking, a part of sieve methods the ideas are closely related and the results play an important supporting role in modern sieve theory. The objective of these identities is to create from sums over primes, by replacement, sums of special linear forms and bilinear forms. These identities are truly of a combinatorial nature, as opposed to identities such as Poisson summation, obtained by analytic transformations. It is amazing how such simple transformations, obtained by merely adding and subtracting terms, can be so powerful that they lead to asymptotic formulas for sums over primes in many sequences. The point is that, by decomposing in this fashion, one may apply different types of analytic arguments to different pieces, none of which would be appropriate for all. It is interesting to compare at this point Brun's decomposition with that of Vinogradov. The latter has the advantage, when it works, of producing asymptotic formulas. Brun's decomposition, which is also into special linear forms and bilinear forms, has been cleverly chosen so that all the bilinear forms appear with the same sign. Therefore, they can be discarded by positivity and this offers the possibility of successful bounds (but not asymptotic formulae) even in cases when some of the parts are inaccessible to analysis of any kind.

Brun's decomposition could be regarded as a combinatorial identity, à la Vinogradov, were we not to discard the relevant bilinear forms. In fact, we can and do reclaim some of these terms, getting results beyond the classical framework. Therefore, one can consider modern sieve theory as standing between the two. Vinogradov's original decomposition has over the years been replaced by much simpler variants and in this chapter we give a panorama of identities of this type. Similar in principle, they nevertheless have different features which can make one or the other more convenient in specific cases.

In Chapter 18 we develop some further combinatorial identities, designed for application to specific problems in Chapters 19 and 21. These identities have the added feature that they contain pieces which we shall be unable to properly evaluate, but which fortunately have zero measure in the decomposition so that, for these, a crude sieve upper bound will suffice. The fact that we can put aside these zero measure parts makes the other parts just barely fit to the analysis. Without this feature we would fail. In fact we are, in this situation, very close to the combinatorial identity end of the spectrum but we succeed only by moving the tiniest little bit in the direction of the sieve. We should like to point out that the Fundamental Lemma is essential in this sort of "asymptotic identity". We state the main results of this chapter as formulae for sums over primes in terms of special linear forms, bilinear forms and small terms. To complete the job in any given case one does not need to look back into the machinery but only to estimate the forms. One may consider these identities as an axiomatization of the asymptotic sieve for primes. The linear form part is classical, while the bilinear form part is the new axiom by means of which one breaks the parity barrier. In Theorem 18.6 one sees clearly what is missing in the Bombieri sieve in the case $k = 1$, in other words, the reason it failed for primes. The treatment of the bilinear forms in the asymptotic sieve for primes, no matter what methods are used, ends up with sums over primes or the Möbius function, but without restrictions, and so the theory of L-functions applies. Therefore, the asymptotic sieve for primes produces primes in special sequences because it reduces the question to primes in regular sequences.

In Chapter 19 we apply one of the asymptotic identities of the previous chapter to establish the equidistribution of roots of a quadratic polynomial to prime moduli. What still needs to be done is to verify the axioms, and that is the bulk of the problem. To this end we apply harmonic analysis, this time the spectral theory of automorphic forms which is stronger and more appropriate to this problem than is classical Fourier analysis. This application shows clearly that modern sieve theory does not live on its own. It draws strength from external sources and its ability to adapt to do so is the essence of its current success and the key to its future growth.

In Chapter 20 we give another application of the asymptotic sieve for primes, this time to primes $p = \ell^2 + m^2$ with ℓ in an arbitrary but relatively dense set, such as, for example, the primes. Along the way we need to establish a large sieve inequality for "quadratic Farey points". We shall need this type of inequality again.

In Chapter 21 we present some results concerning the multiplicative structure of polynomials in two variables. These polynomial sequences are rather sparse and sieve theory has only begun to scratch the surface. In the case of the polynomials X^2+Y^4 and X^3+2Y^3 one is able to produce primes. In our third example X^2+Y^6

only the first difficult step has been successfully treated, namely the divisor function τ_3. Even for these few successes, the technical details are far too lengthy to include in this book. The sieve aspects for X^2+Y^4 have already been treated in Chapter 18 and we briefly discuss here those for X^3+2Y^3. We also point out some interesting results established in the course of the proofs.

One may say that the theory of the sieve can be divided into two basic components. The first of these is the search for sieve weights which give results as strong as possible for the main term without causing havoc for the remainder. The main ideas used are partly combinatorial and partly analytic. Any set of sieve weights reduces the problem of finding, say primes, to the problem of finding a very good level of distribution for the congruence sums (and bilinear forms). It is in relation to this second component that an infusion of ideas from many parts of mathematics has proved to be particularly fruitful. In Chapter 22 we demonstrate a number of such achievements. A great challenge is provided by sparse sequences and we concentrate on these, for example, sums of two biquadrates. The problem becomes more acute for a general polynomial in several variables and for varieties of high dimension. Here, it is not even obvious in which order to count the points, less so even to guess which varieties should possess primes and, if so, how many of them. We highlight very recent work of Bourgain, Gamburd and Sarnak which captures much of the essence of these issues and solves some of the problems.

In Chapter 23 we apply the linear sieve to estimate from below the number of primes in a short interval. As in many other places, we do not attempt to present the most advanced results. The point we wish to make is that one can go further by combining the sieve with analytic tools than did the earlier results on this problem which applied zeta-function techniques alone. The extra advantage coming from the sieve is the added flexibility it brings to the theory of Dirichlet polynomials. This flexibility is sufficiently great that we could entirely dispense with the zero-density estimates. As did Hoheisel, we still need a good zero-free region, which is implicitly responsible for allowing us to break the parity barrier of the linear sieve.

In the ensuing Chapter 24 we address the problem of the least prime in an arithmetic progression. This is analogous to the Hoheisel problem but, as is well known, it is much more difficult and requires sophisticated new ideas. Our goal is to prove Linnik's theorem using ideas from sieve theory and as little else as we can get away with. Especially, we succeeded in avoiding any use of the repulsion phenomenon of Deuring-Heilbronn and the log-free zero-density theorems for L-functions which are centrepieces of the previous proofs. That we could do so is largely due to combinatorial arrangements in the sieve decomposition. As in all proofs, the arguments split into two cases. In our sieve treatment the case when the exceptional character $\chi \bmod q$ exists is particularly attractive. Rather than considering the sequence of integers in the progression $a \bmod q$ we apply the sieve to that sequence weighted by $1 * \chi$. In this illusory circumstance our sequence is rather lacunary. One would think that our task would be harder but in fact it is easier. The point is that, by weighting in this fashion, we have implicitly performed a preliminary sieve (of relatively small level) which has killed so many prime divisors that what is left to do in the second step is sieve by a very sparse set of primes. Hence, that becomes a problem of sieve dimension zero. So, no surprise that we succeeded. The arguments involving the exceptional character hold within a quite general framework and are of a pleasing sieve-theoretic flavor so we examine them further in the final section of the chapter.

Despite our best efforts, in the most famous problems it is still impossible to produce primes. Right from Brun's early work, there have been results producing "almost-primes" and an interest in reducing as far as possible the number of prime factors. Any lower bound for $S(\mathcal{A}, z)$ of the right order gives plenty of these. However, as first noticed by Kuhn, one can do better in reducing the number of factors by attaching to the sequence certain weights of a combinatorial nature which are negative at unwanted elements. There has been considerable development in the art of choosing these weights and the most powerful choices are due to G. Greaves. We use in Chapter 25 a simpler earlier construction due to Richert, which is still quite powerful. We strengthen it further by implementing within it the new shape of the linear sieve having bilinear remainder. Consequently, using harmonic analysis, we succeeded in many cases to get results which surpass not only the earlier ones but, more importantly, even those which are conjectured to be the limit for such a weighted sieve in the traditional setting.

The book ends with two appendices. The first one delivers standard but much-needed asymptotics for sums of multiplicative functions and related results. In some of these asymptotics the main terms involve functions which are most conveniently described as continuous solutions to certain differential-difference equations. There is a large theory concerning such equations and such equations occur in many places in sieve theory and elsewhere. In the second appendix we present what we need.

We had a number of goals in writing this book. Foremost among them, we would like to encourage young people to study the subject. In doing so it would be very helpful, in case they do not already have the relevant background, to have at hand some books on analytic number theory, such as [**9**], [**28**], [**109**], [**129**], [**132**]. An inexperienced reader may find it helpful to also consult some classic texts in sieve theory, such as the first few chapters of [**79**] or of [**76**]. For experienced readers, we also recommend the article of Selberg [**145**] as a companion to the more theoretical aspects of our book.

We do intend the book also for more experienced readers, especially from other parts of number theory and mathematics, whether to master the subject or to learn those parts they need for application. We made an effort to make the book a handy reference for theorems and techniques. Experience shows, however, that frequently the theorems available are not precisely in the form that is wanted for a particular application. So, it is important to be in command of the ideas behind the theorems in order to perform the necessary adjustments, sometimes small and sometimes not so small.

A word of warning. A lot of ground is covered and the shortest route is not always chosen. The book is intentionally not written in a linear order. Some topics require tools developed later, some topics are discussed before earlier discoveries and some topics are more difficult than others which are treated later. Don't get discouraged! Although we hope that very few readers will choose to take a minimal route, an exceedingly bare-bones introduction to the subject can be attempted by means of Sections 1.1–1.3, Sections 5.1–5.8, Sections 6.1–6.5 and 6.9, and finally, Sections 7.1, 7.4–7.6 and 7.8–7.12.

On the other hand, we did not intend to make the book encyclopaedic. The subject of sieve methods has undergone tremendous growth in the past few decades, especially when it comes to applications, many of them highly interesting. Already on their own, Erdös and his collaborators are responsible for dozens of these. Reluctant choices needed to be made, including omission of some personal favourites of the authors. Inevitably, there are instruments which we do not allow to perform in our opera. Nevertheless, we hope that there will be readers who find plenty of interesting points and will devote their time to the further development of the subject. Not least, we hope we have succeeded in conveying to many an appreciation of the music that is the sieve.

* * *

Acknowledgments. Thanks are due to Barbara Miller, Lucile Lo, Ida Bulat, Jemima Merisca and Allison Lee for typing various parts of the manuscript and to Luann Cole for directing the book through the production process. We are grateful to Sergei Gelfand for his unflagging interest over the (approximately five year) period we enjoyed working on the book.

JF was supported in part by NSERC Grant A5123 and, during 2003–2005, by a Killam Research Fellowship. HI was supported in part by NSF Grants DMS-03-01168 and DMS-08-02246 and the University of Toronto during numerous visits. Both authors are grateful to the Banff International Research Station for a Research in Teams Award in August 2007, during which, in beautiful surroundings, some of our favourite tunes were composed.

CHAPTER 1

Sieve Questions

Overture

1.1. Inclusion-Exclusion

The sieve begins with Eratosthenes. Let x be a positive number and

$$\mathcal{A} = \{n \leqslant x\},$$

the set of positive integers up to x. We are going to select the primes in this set.

For purposes of illustration a good choice is $x = 30$. Thus we begin by listing these integers

1	2	3	4	5	6	7	8	9	10	11	12	13	14	15
16	17	18	19	20	21	22	23	24	25	26	27	28	29	30

and from them we are going to delete those that are composite. A composite number is characterized as one having a prime divisor smaller than itself. The first prime number is two, so we circle it and our first step is to cross out those which are larger than two and are multiples of two. Our list above now appears as follows.

1	②	3	~~4~~	5	~~6~~	7	~~8~~	9	~~10~~	11	~~12~~	13	~~14~~	15
~~16~~	17	~~18~~	19	~~20~~	21	~~22~~	23	~~24~~	25	~~26~~	27	~~28~~	29	~~30~~

The first number after two which is not crossed out is the next prime, three, so we circle it and cross out all of its multiples. This leaves us in the following position:

1	②	③	~~4~~	5	~~6~~	7	~~8~~	~~9~~	~~10~~	11	~~12~~	13	~~14~~	~~15~~
~~16~~	17	~~18~~	19	~~20~~	~~21~~	~~22~~	23	~~24~~	25	~~26~~	~~27~~	~~28~~	29	~~30~~

Note that there are now some numbers (the multiples of six) which have been crossed out twice. Again we progress to the next uncrossed number, five, which is necessarily the next prime number, circle it and cross out the multiples of that one. Our diagram now looks like this:

1	②	③	~~4~~	⑤	~~6~~	7	~~8~~	~~9~~	~~10~~	11	~~12~~	13	~~14~~	~~15~~
~~16~~	17	~~18~~	19	~~20~~	~~21~~	~~22~~	23	~~24~~	~~25~~	~~26~~	~~27~~	~~28~~	29	~~30~~

Here again we find more numbers, specifically the multiples of ten and of fifteen, which have been crossed out twice. Moreover, we have even come to a number, thirty, which has been crossed out as a multiple of each of the three primes, two, three and five.

We are now ready to proceed to the multiples of the next prime, seven. However, before doing so one can save a good deal of trouble by noticing that every one of the remaining numbers on our list, apart from the integer one, is in fact a prime number. This will turn out to be a quite general observation, a consequence of the fact that every composite integer must be divisible by some prime number which is no larger than its square root. In our case we started with numbers not exceeding thirty and hence we only need to delete multiples of primes $p \leqslant \sqrt{30}$. Since five is the largest such prime we can gain no more by continuing further.

We have here constructed an algorithm which produces the table of all primes up to x. But we are also interested in counting these primes. To this end, we follow the above procedure and count in each step the multiples which were crossed out (it is convenient to also include in this count the prime which was circled). Because some numbers are crossed out several times, we must rectify the count by adding back or subtracting such numbers, according to the multiplicity of times this has happened.

Let's sum up what we have done. On the one hand, totaling up the results of the count of our inclusion–exclusion, we began with $[x]$ integers (in our case $x = 30$), for each prime $p \leqslant \sqrt{x}$ we subtracted out $[x/p]$ multiples of p then for each pair of distinct primes $p_1 < p_2 \leqslant \sqrt{x}$ we added back in the $[x/p_1p_2]$ multiples of p_1p_2, and so on. In all, we are left with the final count

$$[x] - \sum_{p \leqslant \sqrt{x}} \left[\frac{x}{p}\right] + \sum\sum_{p_1 < p_2 \leqslant \sqrt{x}} \left[\frac{x}{p_1\, p_2}\right] - \sum\sum\sum_{p_1 < p_2 < p_3 \leqslant \sqrt{x}} \left[\frac{x}{p_1\, p_2\, p_3}\right] + - \cdots .$$

On the other hand, this was after all just the count for the number of integers not crossed out and these integers are just the primes less than or equal to x, other than those which were circled (hence less than or equal to $\sqrt{x}$), together with the integer one.

Equating the two we obtain the formula (attributed to Legendre):

$$\pi(x) - \pi(\sqrt{x}) + 1 = \sum_{\substack{d \\ p|d \Rightarrow p \leqslant \sqrt{x}}} \mu(d) \left[\frac{x}{d}\right].$$

Here, in the usual notation, $\pi(x)$ is the prime counting function

$$\pi(x) = \sum_{p \leqslant x} 1,$$

and, here and almost always, the letter p will denote a prime. As usual, the Möbius function $\mu(d)$ is $(-1)^\nu$ when d is the product of $\nu \geqslant 0$ distinct primes and is zero if d has a multiple prime factor. This function provides a concise way of expressing the right-hand side of the formula.

As it happens, $\pi(x)$ will turn out to be considerably larger than $\sqrt{x}$. Hence the left-hand side of the Legendre formula is approximately $\pi(x)$ and, in order to estimate $\pi(x)$, we want to develop further the right-hand side.

The obvious starting point for any estimation of the right-hand side is the replacement everywhere of the awkward function $[t]$, the integral part of t, by the simpler function t. This gives an error or "remainder term" of $\{t\}$, the fractional part. More precisely, we have

$$\pi(x) - \pi(\sqrt{x}) + 1 = x \sum_{d} \frac{\mu(d)}{d} + R = x \prod_{p \leqslant \sqrt{x}} \left(1 - \frac{1}{p}\right) + R$$

where d runs over the products of distinct primes which are less than or equal to $\sqrt{x}$ and where the "remainder" $R = R(x)$ is

$$R = -\sum_{d} \mu(d) \left\{\frac{x}{d}\right\}.$$

At first glance, the best we can expect to do is to use the trivial bound $\{t\} < 1$ which leads us to bound the remainder by

$$|R| \leqslant \sum_{d} 1 = 2^{\pi(\sqrt{x})},$$

which is absolutely enormous, much larger even than the number of integers $[x]$ with which we began. Of course, we have been particularly careless here, for example, fruitlessly sifting out multiples of d even for many integers d which exceed x. Hence, the above bound can certainly be improved somewhat. But not enough! In reality R is genuinely large. In fact, using old ideas due to Tchebyshev and to Mertens (see (2.21)), one knows that

$$\prod_{p \leqslant \sqrt{x}} \left(1 - \frac{1}{p}\right) \sim \frac{2e^{-\gamma}}{\log x},$$

so what we have been expecting to be our main term is actually wrong. Since, by the Prime Number Theorem,

$$\pi(x) \sim \frac{x}{\log x},$$

we see that the quantity R we have been referring to as the remainder has order of magnitude just as large as the main term.

1.2. Some Generality

At the moment we are in the rather depressing position of having a method which fails to give us good estimates for the number $\pi(x)$ of primes up to x, but worse yet, the only reason we even know that it must inevitably fail is because other techniques, coming from analytic number theory, succeed (in proving the Prime Number Theorem), thereby telling us so. What possibility remains for the value of the sieve is its capacity for generalization, giving some information in cases where the finer analytic machinery is lacking. Therefore, to consider the situation more generally is not merely a useful enterprise; it is the sieve's only excuse for being.

We thus consider a finite sequence of non-negative real numbers

$$\mathcal{A} = (a_n), \ n \leqslant x,$$

and a general set $\mathcal{P}$ of primes. For notational purposes it is convenient to introduce the product

$$P(z) = \prod_{\substack{p \in \mathcal{P} \\ p < z}} p.$$

Our goal will be to estimate the "sifting function"

$$S(\mathcal{A}, z) = \sum_{\substack{n \leqslant x \\ (n, P(z))=1}} a_n .$$

We shall proceed as in our original example, but in slightly different notation. We need the most basic property of the Möbius function

$$\sum_{d|n} \mu(d) = \begin{cases} 1 & \text{if } n = 1, \\ 0 & \text{if } n > 1. \end{cases} \tag{1.1}$$

We also use the simple observation that δ divides both a and b if and only if δ divides (a, b), that is, the set of common divisors of two positive integers is just the same as the set of divisors of their greatest common divisor. Hence,

$$\sum_{\substack{d|n \\ d|P(z)}} \mu(d) = \begin{cases} 1 & \text{if } (n, P(z)) = 1, \\ 0 & \text{if } (n, P(z)) > 1. \end{cases}$$

Inserting this coprimality detector and then interchanging the order of summation, we obtain

$$\begin{aligned} S(\mathcal{A}, z) &= \sum_n a_n \sum_{d|(n,P(z))} \mu(d) = \sum_n a_n \sum_{\substack{d|n \\ d|P(z)}} \mu(d) \\ &= \sum_{d|P(z)} \mu(d) \sum_{n \equiv 0 \, (\mathrm{mod}\, d)} a_n = \sum_{d|P(z)} \mu(d) A_d(x), \end{aligned}$$

say. This is just once again, but in more general garb, the Legendre formula and here, as before, we need information about the "congruence sums"

$$A_d(x) = \sum_{\substack{n \leqslant x \\ n \equiv 0 \, (\mathrm{mod}\, d)}} a_n$$

which give the mass of the subsequence running over the multiples of d, in other words, over the subsequence $\mathcal{A}_d = (a_m)$, $m \equiv 0 \, (\mathrm{mod}\, d)$, and which, in our beginning example, was just $[x/d]$. Specifically, we need a useful approximation formula. We assume we can write this sum in the form

$$A_d(x) = g(d)X + r_d(x) , \tag{1.2}$$

where X is a handy approximation to

$$A(x) = A_1(x) = \sum_{n \leqslant x} a_n ,$$

the total mass of our sequence, where $g(d)$, the "density function", has several nice properties ($g(d) = 1/d$ in our example) and $r_d(x)$ is a "remainder term" which is small, at least on average over d (this was $r_d(x) = -\{x/d\} \ll 1$ in our example). In other words, we need to count integers before we can count primes, and how well we count the first determines how well we count the second.

On insertion of our approximation formula (1.2) the sifting function becomes

$$S(\mathcal{A}, z) = X \sum_{d|P(z)} \mu(d) g(d) + \sum_{d|P(z)} \mu(d) r_d(x) .$$

The function $g(d)$ acts like a probability, approximating the fraction of the total mass which resides in the multiples of d. (It is useful to remember $g(d) = 1/d$ as being the prototype for such a function.) Because of this it is natural that we assume that $g(1) = 1$ and that, for each $d > 1$, we have $0 \leqslant g(d) < 1$. This last inequality will be needed in some places and is also natural in our situation. Indeed, if for some $d > 1$ we had $g(d) = 1$, then virtually everything would be a multiple of d. There would not be much point in looking for primes in such a sequence. We shall also assume that g is a multiplicative function, that is, whenever $(d_1, d_2) = 1$ we have

$$g(d_1 d_2) = g(d_1)g(d_2).$$

The essence of this is that we are assuming that divisibility by two relatively prime integers are independent events. In practice, this is true only to a quite limited extent and this fact is in large measure responsible for the failure of the sieve to achieve more than it has.

Very often we only use (1.2) for d squarefree, but sometimes it helps to assume (1.2) for all d. Because $\mathcal{A}_d$ is a subsequence of $\mathcal{A}$ it is natural to assume that

$$g(d_1) \leqslant g(d_2) \tag{1.3}$$

if $d_2 | d_1$. In particular, $g(p^\ell)$ is non-increasing in ℓ for any p.

1.3. Some Examples

We consider some examples. In many of the most basic examples the sequence $\mathcal{A}$ is just the characteristic function of an interesting set of integers. In such a case we shall frequently not bother to distinguish between the function and the set on which it is supported.

Example 1.1. We begin with a slight extension of our original example to the set of integers in an interval. Thus, we consider

$$\mathcal{A} = \{m \mid x - y < m \leqslant x\}, \qquad \mathcal{P} = \{\text{all primes}\},$$

$$A_d(x) = \left[\frac{x}{d}\right] - \left[\frac{x-y}{d}\right], \qquad X = y,$$

$$g(d) = \frac{1}{d}, \quad r_d(x) = -\left\{\frac{x}{d}\right\} + \left\{\frac{x-y}{d}\right\}, \quad |r_d(x)| \leqslant 1.$$

Example 1.2. Now, for a little more variety, consider

$$\mathcal{A} = \{m^2 + 1 \leqslant x\}, \quad \mathcal{P} = \{p;\ p \not\equiv 3 \pmod 4\},$$

$$A(x) = \left[\sqrt{x-1}\,\right], \qquad X = \sqrt{x}\,,$$

$$g(p) = \begin{cases} 2/p, & p \equiv 1 \pmod 4, \\ 1/2, & p = 2, \end{cases} \qquad |r_d(x)| \leqslant 2^{\nu(d)},$$

this last estimate following from the bound $|r_p(x)| \leqslant 2$ and the Chinese Remainder Theorem. Here, there is no need to sieve by the primes congruent to three modulo four since none of the integers in our set is divisible by such a prime. Equivalently, we could achieve the same results sifting by the set of all primes and simply setting $g(p) = 0$ for the additional primes. In this example, were we able to get a positive lower bound for $S(\mathcal{A}, \sqrt{x})$ we would be producing primes of the form m^2+1. A proof that there are infinitely many such primes would settle an outstanding problem in the subject.

EXAMPLE 1.3. Moving to another famous conjecture, we consider the following:

$$\mathcal{A} = \{m(m+2) \leqslant x\}, \qquad \mathcal{P} = \{\text{all primes}\},$$
$$g(p) = \begin{cases} 2/p\,, & p \text{ odd}, \\ 1/2\,, & p = 2, \end{cases} \qquad |r_d(x)| \leqslant 2^{\nu(d)}.$$

Here, were we to give a positive lower bound for $S\left(\mathcal{A}, x^{1/4}\right)$ we would be producing integers $m(m+2)$ where both factors are prime and differ by two. The twin prime conjecture predicts that there are infinitely many such pairs.

EXAMPLE 1.4. As an alternative approach via the sieve to attack the twin prime conjecture, we consider the sequence:

$$\mathcal{A} = \{p-2;\ p \leqslant x\}, \qquad \mathcal{P} = \{\text{odd primes}\},$$
$$A_d(x) = \pi(x; d, 2), \qquad X = \pi(x), \qquad g(d) = \frac{1}{\varphi(d)},$$

where $\pi(x; d, a)$ is the number of primes up to x which are congruent to a modulo d and where $\varphi(d)$, the Euler function, counts the number of reduced residue classes modulo d. This sequence offers some advantages over the previous one for studying the twin prime problem and it gives stronger results in that direction, although this was not so in the earliest results. The most obvious advantage is that we are starting from the beginning with the knowledge that one of our two numbers, namely p, is a prime. On the other hand, the remainder term is more complicated, namely $r_d(x) = \pi(x; d, 2) - \pi(x)/\varphi(d)$, and it is much more difficult to bound it successfully. In the current state of knowledge, a reasonably good bound can only be given on average over d; the most powerful bound of this type being the Bombieri–Vinogradov theorem which we shall prove in Section 9.18. Again in this example, if we were to be successful in giving a positive lower bound, this time for $S(\mathcal{A}, \sqrt{x})$, then we would be producing twin primes.

There are considerable generalizations to all of the above examples. One may take a polynomial with integer coefficients, say in one variable (although not necessarily so), and consider $\mathcal{A}$ to be the sequence of its values as the variable runs through the integers in a segment, or the primes in a segment, or the primes in a segment of an arithmetic progression. It is possible to give many other cases wherein well–known problems concerning primes, for instance the Goldbach conjecture that every even integer exceeding 2 is the sum of two primes, can be phrased in such a manner as to follow from sufficiently strong sieve-theoretic estimates. Formulating them this way is, however, by far the easier part of the problem; producing successful estimates is a very stern challenge indeed!

Not always will the target of the sieve be a set of primes. Perhaps the simplest case is the following.

EXAMPLE 1.5. Let

$$\mathcal{A} = \{m \leqslant x\}, \quad \mathcal{P} = \{p;\ p \mid q\},$$

where q is a given positive integer. Now the target is the set of integers $m \leqslant x$ with $(m, q) = 1$. By the inclusion-exclusion argument in Section 1.1 the number of such integers is

$$\sum_{d|q} \mu(d)\left[\frac{x}{d}\right] = \frac{\varphi(q)}{q}x + R$$

where

$$R = -\sum_{d|q} \mu(d)\left\{\frac{x}{d}\right\}$$

so $|R| \leqslant \tau(q)$. Note that, in the event that x is an integral multiple of q, the remainder terms all vanish so we have an exact formula. We shall never be so lucky again.

The precision in the above example comes from the fact that the sifting set of primes is fixed. If instead we allow q to grow the problem again becomes difficult. In two very basic situations we may take q to be the product of the primes $p < z$ in which case we have simply rephrased our original problem or we may take q to be the product of primes $p > z$ in which case we are counting integers without large prime factors. We shall touch on the asymptotics for these examples much later, in Section 12.2.

Another important sequence that appears as the target of a sieve is the set of squarefree numbers. In this scenario, rather than sieving by a set of primes we sift by squares of primes. For this purpose the basic formula is

$$\sum_{d^2|n} \mu(d) = \begin{cases} 1 & \text{if } n \text{ is squarefree,} \\ 0 & \text{otherwise.} \end{cases}$$

Sieving them from the set of all positive integers is easy, so we consider a somewhat more ambitious problem.

EXAMPLE 1.6. Let

$$\mathcal{A} = \{m^2 + 1 \leqslant x\}\ , \qquad \mathcal{P} = \{p \equiv 1 \,(\operatorname{mod} 4)\}\ .$$

Then, by the corresponding Legendre formula we have

$$\left|\{m^2 + 1 \leqslant x;\ m^2 + 1 \text{ squarefree}\}\right| = \sum_{d|P(x)} \mu(d) A_{d^2}(x)\ .$$

Splitting into classes $m \equiv \nu \,(\operatorname{mod} d^2)$ we find that

$$A_{d^2}(x) = \frac{\rho(d^2)}{d^2}\sqrt{x} + O\big(\rho(d^2)\big)$$

where $\rho(q)$ is the number of solutions to the congruence $\nu^2 + 1 \equiv 0 \,(\operatorname{mod} q)$.

We have $\rho(d^2) = \rho(d) = \tau(d)$ for $d \mid P(x)$, so the above approximation is only good for $d \leqslant x^{\frac{1}{4}}$. For larger d the remainder term is too large. In reality $\mathcal{A}_{d^2}$ is frequently empty. To treat the contribution from larger d, say $d > D$, we change the role of the variables. We write

$$m^2 + 1 = d^2 k \leqslant x\ ,$$

and we estimate the number of solutions in m, d for every given $k \leqslant K = xD^{-2}$. For fixed k this reduces to the counting of units in the real quadratic field $\mathbb{Q}(\sqrt{k})$.

Since the units (solutions to Pell's equation) grow exponentially, the number of these in the relevant range is $O(\log x)$. Therefore,

$$\sum_{d|P(x)} \mu(d)A_{d^2}(x) = \sum_{\substack{d|P(x)\\ d\leqslant D}} \mu(d)\left\{\frac{\tau(d)}{d^2}\sqrt{x} + O\big(\tau(d)\big)\right\} + O(K\log x)$$

$$= \sqrt{x}\prod_{p\in\mathcal{P}}\Big(1-\frac{2}{p^2}\Big) + O\Big(\Big(D+\frac{x}{D^2}\Big)\log x\Big) .$$

Choosing $D = x^{\frac{1}{3}}$ we conclude that

$$\Big|\big\{m^2+1\leqslant x;\ m^2+1 \text{ squarefree}\big\}\Big| = cx^{\frac{1}{2}} + O\big(x^{\frac{1}{3}}\log x\big)$$

where c is a positive constant given by

$$c = \prod_{p\equiv 1(\mathrm{mod}\,4)}\Big(1-\frac{2}{p^2}\Big) .$$

We remark that the more general asymptotic formula for ℓ−free values of degree ℓ polynomials was established by G. Ricci [**139**] in 1933.

In our next example the target is the set of integers which can be written as the sum of two squares. The result is cleaner if we restrict to odd integers a^2+b^2 with a, b relatively prime.

EXAMPLE 1.7. Here we take

$$\mathcal{A} = \big\{n\leqslant x;\ n\equiv 1 \pmod 4\big\},$$
$$\mathcal{P} = \big\{p;\ p\equiv 3 \pmod 4\big\},$$

getting

$$S(\mathcal{A},\mathcal{P},\sqrt{x}) = \Big|\big\{n\leqslant x;\ n = a^2+b^2 \text{ odd with } (a,b)=1\big\}\Big| .$$

This problem (of estimating $S(\mathcal{A},\mathcal{P},z)$) is intermediate in difficulty between the squarefree numbers and the primes. Whereas the first was easy enough to do right here, the second we shall solve in Chapter 14, but only with techniques that are well advanced.

As we have seen in the last few examples, there are many variations on our original sieve format. An important one deals with the situation where we want to sift out many residue classes for each prime in our set $\mathcal{P}$. In essence, Example 1.3 deals with two residue classes for each odd prime. Now we present a problem in which we wish to remove a great many more.

EXAMPLE 1.8. Let

$$\mathcal{A} = \big\{n\leqslant x\big\},$$
$$\mathcal{P} = \text{the set of all primes},$$
$$\Omega_p = \Big\{\omega\,(\mathrm{mod}\,p);\ \omega = 0 \text{ or } \Big(\frac{\omega}{p}\Big) = 1\Big\} .$$

Note that the number of classes to be removed $\omega(p) = \frac{1}{2}(p+1)$, if $p>2$, is very large. The problem is to estimate

$$S(\mathcal{A},\mathcal{P},\Omega) = \Big|\big\{n\leqslant x;\ n\,(\mathrm{mod}\,p)\notin\Omega_p \text{ for each } p\leqslant\sqrt{x}\big\}\Big| .$$

This counts the integers $n \leqslant x$ which are quadratic non-residues for all primes $p \leqslant \sqrt{x}$. This is the problem which gave rise to the large sieve. We shall develop the large sieve theory in Chapter 9 and apply it to obtain the upper bound

$$S(\mathcal{A}, \mathcal{P}, \Omega) \ll \sqrt{x}\ .$$

In slightly more general form this idea will be used to estimate the least quadratic non-residue.

Finally, we just mention:

EXAMPLE 1.9. Let

$$F = F_1 \dots F_r$$

be the product of irreducible polynomials with integer coefficients and take $\mathcal{A}$ to be the sequence of values $F(n)$ or $F(p)$. In this case, if $r > 1$ there is no chance to find primes, but it might be possible to find integers with r prime factors, and it is interesting to see how close one can come to this goal. Here, in general, it is not so simple to describe precisely the density function g.

1.4. A Model of a Sifting Sequence for a Given Density

In this section we are going to create a sequence $\mathcal{B} = \big(b_n\big)$ which satisfies the sieve axioms for a given density function g of dimension $\kappa = 1$; see Section 5.5. Naturally, we assume that $g(d)$ is multiplicative with

$$0 \leqslant g(p) < 1 \tag{1.4}$$

and

$$g(p^\alpha) \geqslant g(p^{\alpha+1}) \geqslant 0\ , \tag{1.5}$$

for any prime p and any $\alpha \geqslant 0$. Given such a function g, we define the companion function h, which is multiplicative with

$$h(p^\alpha) = \frac{g(p^\alpha) - g(p^{\alpha+1})}{1 - g(p)}\ . \tag{1.6}$$

We call h the relative density function. Note that if $g(d)$ is completely multiplicative then $h(d) = g(d)$. If $g(d)$ is supported on squarefree numbers, then so is $h(d)$ and

$$h(p) = \frac{g(p)}{1 - g(p)}\ , \tag{1.7}$$

hence the name "relative density function".

For simplicity we assume that the function

$$f(n) = h(n)n \tag{1.8}$$

satisfies the conditions of Lemma A.15, that is, (A.78), (A.82) and (A.83). Hence we obtain

$$\sum_{\substack{n \leqslant x \\ n \equiv 0 (\mathrm{mod}\, d)}} f(n) = \hat{f}(d) x + O\Big(\tilde{f}(d)(x/d)^\theta\Big)\ , \tag{1.9}$$

where $\hat{f}$ is given by (A.91) and $\tilde{f}$ by (A.90).

In our case, the leading factor is

$$\hat{f}(d) = \prod_{p^\nu \| d} \left(1 - \frac{1}{p}\right)\Big(\sum_{\alpha \geqslant \nu} h(p^\alpha)\Big)$$
$$= \prod_{p^\nu \| d} \left(1 - \frac{1}{p}\right)\big(1 - g(p)\big)^{-1} g(p^\nu) = g(d) H^{-1} ,$$

with

$$H = \prod_p \big(1 - g(p)\big)\Big(1 - \frac{1}{p}\Big)^{-1} . \tag{1.10}$$

Now, define the sequence $\mathcal{B} = \big(b_n\big)$ by

$$b_n = Hf(n) = Hh(n)n . \tag{1.11}$$

Then, for any d we have

$$\mathcal{B}_d(x) = \sum_{\substack{n \leqslant x \\ n \equiv 0 (d)}} b_n = g(d)x + r_d(\mathcal{B}, x) , \tag{1.12}$$

where

$$r_d(\mathcal{B}, x) \ll \tilde{f}(d)(x/d)^\theta \tag{1.13}$$

with

$$\tilde{f}(d) = \sum_{abc^2 = d} f(a) b^{-\varepsilon} c^{\varepsilon} \tag{1.14}$$

and any $\varepsilon > 0$. Hence,

$$\sum_{d \leqslant D} \tilde{f}(d) d^{-\theta} = \sum_{abc^2 \leqslant D} f(a) b^{-\varepsilon} c^{\varepsilon} (abc^2)^{-\theta}$$
$$\ll D^{1-\theta} \sum_{bc^2 \leqslant D} b^{-\varepsilon} c^{\varepsilon} (bc^2)^{-1} \ll D^{1-\theta} .$$

This shows that the sequence $\mathcal{B}$ has a remainder which satisfies the bound

$$\sum_{d \leqslant D} |r_d(\mathcal{B}, x)| \ll x^\theta D^{1-\theta} \tag{1.15}$$

with some $0 < \theta < 1$. This is satisfactory for $D \leqslant x(\log x)^{-A}$, with A sufficiently large in terms of θ.

Such a model can be useful in the situation where we consider an arithmetic sequence $\mathcal{A} = \big(a_n\big)$ with density function $g(d)$ and compare it with its model $x^{-1}A(x)\mathcal{B}$ (normalized for the occasion). In the difference sequence $\mathcal{C} = \big(c_n\big)$, $c_n = a_n - x^{-1}A(x)b_n$, the main terms cancel and we are left only with the remainder terms. We shall use this device in Chapter 18. We conclude the above considerations with the following result.

Proposition 1.1. *Let $\mathcal{A} = (a_n)$ be a sequence of non-negative numbers. Put*

$$A_d(x) = \sum_{\substack{n \leqslant x \\ n \equiv 0 (d)}} a_n = g(d)A(x) + r_d(x) , \tag{1.16}$$

where $g(d)$ *satisfies* (1.4), (1.5), *and the conditions* (A.78), (A.82), (A.83). *Put*

$$R(x;D) = \sup_{y \leqslant x} \sum_{d \leqslant D} |r_d(y)| \ . \tag{1.17}$$

Then, for $\mathcal{C} = \big(c_n\big)$ *with* $c_n = a_n - x^{-1}A(x)Hh(n)n$, *we have*

$$\sum_{d \leqslant D} \big|C_d(x)\big| \ll R(x;D) + A(x)\big(D/x\big)^{1-\theta} \tag{1.18}$$

and

$$\sum_{d \leqslant D} \big|C'_d(x)\big| \ll R(x;D)\log x + A(x)\big(D/x\big)^{1-\theta} \ , \tag{1.19}$$

where

$$C_d(x) = \sum_{\substack{n \leqslant x \\ n \equiv 0(d)}} c_n \ , \qquad C'_d(x) = \sum_{\substack{n \leqslant x \\ n \equiv 0(d)}} c_n \log n \ .$$

REMARK. Recall that the notations $A = O(B)$, $A \ll B$, and $B \gg A$ all mean that $|A| \leqslant cB$ where c is a positive constant. Thus A can take complex values whereas B is real and non-negative. In those cases where c is not absolute but depends on one or more parameters this dependence will be spelled out, except occasionally when it is clear from the context or simply too repetitious to do so.

CHAPTER 2

Elementary Considerations on Arithmetic Functions

First Notes

Sieve methods are to a great extent elementary in nature, yet they can require quite sophisticated analysis, especially when one wishes to estimate the relevant sums asymptotically. Already before Legendre, Euler had given some of the basic theorems in the elementary theory (including a variant of the Eratosthenes sieve [**42**]) and also had laid the groundwork for the introduction of analysis into number theory by seeing arithmetic functions through their generating series (power series for additive functions and zeta-functions for multiplicative functions). His Opera Omnia contain them all. But it was left to Dirichlet to firmly plant the elementary and analytic roots into the subject.

In this chapter we bring to the reader's attention selected arguments from the theory of arithmetic functions which will often appear in the background. To our tastes, some of these arguments have a flavour of sieve ideas. We shall need much more than these however and many of the heavier proofs of the results we require are postponed to Appendix A on sums of multiplicative functions.

2.1. Dirichlet Convolution and Switching Divisors

Let $\tau(n)$ denote the number of positive divisors of n; in other words, it is the n-th coefficient in the Dirichlet series

$$\zeta^2(s) = \sum_1^\infty \frac{\tau(n)}{n^s} \, .$$

Although the size of $\tau(n)$ is rather irregular, it is easy to obtain the asymptotic formula for the sum

$$\sum_{n\leqslant x} \tau(n) = x\log x + O(x) \, , \tag{2.1}$$

showing that $\tau(n)$ is about $\log n$ on average. The arguments go as follows:

$$\sum_{n\leqslant x} \tau(n) = \sum_{d\leqslant x} \left[\frac{x}{d}\right] = \sum_{d\leqslant x} \left\{\frac{x}{d} + O(1)\right\}$$
$$= x\log x + O(x) \, .$$

Because the number of terms is x and most of the error terms have the same sign, the total error term $O(x)$ cannot be improved with such crude bounds.

The function $\tau(n)$ can also be viewed as the number of positive solutions of the equation $cd = n$, and using the symmetry between c and d one can arrange the sum (2.1) as follows:

$$2 \sum_{d \leqslant \sqrt{x}} \left[\frac{x}{d}\right] - \left[\sqrt{x}\right]^2 .$$

In this expression we have far fewer terms, leading to the more precise formula

$$\sum_{n \leqslant x} \tau(n) = x\big(\log x + 2\gamma - 1\big) + O(\sqrt{x}) . \tag{2.2}$$

The above arguments can be visualized as counting the positive integral points under the hyperbola $cd = x$. The idea of switching divisors in this fashion is due to Dirichlet and so we call it the Dirichlet hyperbola trick. This will be used many times in more subtle situations and in fact we have already used it in Example 1.6.

Recall that for two arithmetic functions f, g the Dirichlet convolution $f * g$ is given by

$$(f * g)(n) = \sum_{d|n} f(d) g\Big(\frac{n}{d}\Big) . \tag{2.3}$$

Denoting the generating series by

$$F(s) = \sum_1^\infty \frac{f(n)}{n^s} , \qquad G(s) = \sum_1^\infty \frac{g(n)}{n^s} ,$$

we have

$$F(s)G(s) = \sum_1^\infty \frac{(f * g)(n)}{n^s} .$$

In the case of the divisor function $\tau(n)$ we considered the convolution of a constant function with itself. The Dirichlet hyperbola trick applies rather generally to convolutions $f * g$ provided both f and g are nice functions. Another example is the function $r(n)$, the number of representations of n as the sum of two squares. We have

$$r(n) = 4 \sum_{d|n} \chi_4(d) , \tag{2.4}$$

where χ_4 is the character of conductor four. Arguing as with $\tau(n)$ we obtain

$$\sum_{n \leqslant x} r(n) = 4 \sum_{d \leqslant \sqrt{x}} \chi_4(d) \left[\frac{x}{d}\right] + 4 \sum_{c \leqslant \sqrt{x}} \Bigg(\sum_{d \leqslant x/c} \chi_4(d) \Bigg) + O\left(\sqrt{x}\right) .$$

The sum of $\chi_4(d)$ is bounded, giving a contribution $O\left(\sqrt{x}\right)$. Replacing $[x/d]$ by x/d, we get an extra error term $O\left(\sqrt{x}\right)$. Then, extending the summation over $d \leqslant \sqrt{x}$ in the main term to all d, we find that

$$\sum_{d \leqslant \sqrt{x}} \frac{\chi_4(d)}{d} = L(1, \chi_4) + O\Big(\frac{1}{\sqrt{x}}\Big) .$$

Hence, we conclude that

$$\sum_{n \leqslant x} r(n) = \pi x + O\left(\sqrt{x}\right) \tag{2.5}$$

because $L(1,\chi_4)=\pi/4$. In fact, the left side of (2.5) can be visualized geometrically as the number of integral points in the disc of radius $\sqrt{x}$ and (2.5) can alternatively be deduced by the Gauss method of packing the disc with unit squares. One can also view the Gauss circle problem as an additive convolution problem and treat it accordingly by switching the summands (that is the squares) rather than the divisors.

To go a little deeper in this direction we consider the sum

$$\begin{aligned}\sum_{n\leqslant x}\tau(n^2+1) &= 2\sum_{d\leqslant x}\sum_{\substack{n\leqslant x\\ n^2+1\equiv 0(d)}} 1-\sum\sum_{\substack{c,d\leqslant x\\ cd=n^2+1}} 1+O(x)\\ &=\sum_{d\leqslant x}\Big\{\frac{2\rho(d)}{d}x-\frac{\rho(d)}{d}\sqrt{dx}+O\big(\rho(d)\big)\Big\}+O(x)\\ &=2x\sum_{d\leqslant x}\frac{\rho(d)}{d}+O(x)\ ,\end{aligned}$$

which follows by the bounds

$$\sum_{d\leqslant x}\rho(d)\ll x\ ,\quad \sum_{d\leqslant x}\frac{\rho(d)}{\sqrt{d}}\ll\sqrt{x}\ .$$

Note that the generating series for $\rho(d)$ is

$$\sum_{1}^{\infty}\frac{\rho(d)}{d^s}=\zeta(2s)^{-1}\zeta(s)L(s,\chi_4)\ ,$$

whence

$$\rho(d)=\sum_{a^2bc=d}\mu(a)\chi_4(c)\ .$$

Using this convolution formula we can derive elementarily that

$$\sum_{d\leqslant x}\frac{\rho(d)}{d}=\frac{L(1,\chi_4)}{\zeta(2)}\log x+O(1)\ .$$

Hence, we conclude that

$$\sum_{n\leqslant x}\tau(n^2+1)=\frac{3}{\pi}x\log x+O(x)\ . \tag{2.6}$$

EXERCISE 2.1. Prove elementarily by induction that, for $r\geqslant 2$ and $x\geqslant 2$,

$$\sum_{n\leqslant x}\tau_r(n)=\frac{1}{(r-1)!}\,x\,(\log x)^{r-1}+O\big(x(\log x)^{r-2}\big)\ . \tag{2.7}$$

Hence, derive the formula

$$\sum_{n\leqslant x}\tau^2(n)=\frac{1}{\pi^2}\,x\,(\log x)^3+O\big(x(\log x)^2\big)\ . \tag{2.8}$$

Hint. Use the expression

$$\tau^2(n)=\sum_{\substack{abcd=n\\ (c,d)=1}}1=\sum_{m^2|n}\mu(m)\tau_4(n/m^2)\ .$$

2.2. Tchebyshev's Prime Sprigs

Arithmetic functions form a ring with ordinary addition and with Dirichlet convolution as multiplication. The identity of this ring is

$$\theta(n) = \begin{cases} 1 & \text{if } n = 1 , \\ 0 & \text{else} . \end{cases} \tag{2.9}$$

The constant function 1 is generated by $\zeta(s)$ while its inverse, which is generated by

$$\zeta(s)^{-1} = \prod_p \Big(1 - \frac{1}{p^s}\Big) = \sum_1^\infty \frac{\mu(n)}{n^s} ,$$

is the Möbius function $\mu(n)$. Its most basic property (1.1), when rephrased in terms of generating functions, is merely the identity $\zeta(s)\zeta(s)^{-1} = 1$.

Similarly, the equivalence between $\zeta(s)F(s) = G(s)$ and $F(s) = \zeta(s)^{-1}G(s)$ translates into the Möbius inversion formula

$$\sum_{d|n} f\big(\frac{n}{d}\big) = g(n) \Longleftrightarrow \sum_{d|n} \mu(d) g\big(\frac{n}{d}\big) = f(n) , \tag{2.10}$$

each holding for all n if and only if the other does.

Many interesting functions arise as convolutions and this fact plays a role in sieve theory. The simplest such functions are $\tau = 1 * 1$ and $r = 1 * \chi_4$ which were dealt with above. Perhaps the most basic for studying primes is the convolution $\log = 1 * \Lambda$ which defines the von Mangoldt function $\Lambda = \mu * \log$, that is,

$$\Lambda(n) = \sum_{d|n} \mu(d) \log \frac{n}{d} = -\sum_{d|n} \mu(d) \log d . \tag{2.11}$$

By unique factorization we have

$$\sum_{p^k \| n} \log p^k = \log n ,$$

so $1 * \Lambda = \log$ means

$$\Lambda(n) = \begin{cases} \log p & \text{if } n = p^k,\ k \geqslant 1 , \\ 0 & \text{else} . \end{cases} \tag{2.12}$$

This can also be verified by examining the Euler product

$$\begin{aligned} -\frac{\zeta'}{\zeta}(s) &= -\frac{d}{ds} \log \zeta(s) = -\sum_p \frac{d}{ds} \log\Big(1 - \frac{1}{p^s}\Big)^{-1} \\ &= \sum_p \sum_k p^{-ks} \log p . \end{aligned}$$

We call an integer $n > 1$ primary if it is a power of a prime.

It was P. Tchebyschev who first recognized the importance of, and exploited, the formula (2.11) for counting primes. The von Mangoldt function is non-negative and its support is just the primary numbers. It turns out for various reasons to be more convenient than the characteristic function of the primes.

Therefore, in addition to $\pi(x)$ we also consider

$$\psi(x) = \sum_{n \leqslant x} \Lambda(n) \ . \tag{2.13}$$

Summing the identity

$$\sum_{d|n} \Lambda(d) = \log n \tag{2.14}$$

over $n \leqslant x$, we derive

$$\sum_{d \leqslant x} \Lambda(d) \Big[\frac{x}{d}\Big] = \sum_{m \leqslant x} \psi\Big(\frac{x}{m}\Big) = x \log x - x + O(\log x) \ , \tag{2.15}$$

where the right side follows from

$$\sum_{n \leqslant x} \log n = \int_1^x (\log y) \ dy + O(\log x) \ .$$

Subtracting from (2.15) twice the same formula for $x/2$ we get the alternating series

$$\psi(x) - \psi\Big(\frac{x}{2}\Big) + \psi\Big(\frac{x}{3}\Big) - \psi\Big(\frac{x}{4}\Big) + \cdots = x \log 2 + O(\log x) \ . \tag{2.16}$$

Hence, dropping all but two terms we deduce by positivity,

$$x \log 2 + O(\log x) < \psi(x) < \psi\Big(\frac{x}{2}\Big) + x \log 2 + O(\log x) \ .$$

Adding the last inequality for $x, \frac{x}{2}, \frac{x}{4}, \dots$ we see that the terms $\psi(\frac{x}{2}), \psi(\frac{x}{4}), \dots$ cancel out and we are left with an upper bound for $\psi(x)$.

Summarizing these results we obtain the following estimates of Tchebyshev.

PROPOSITION 2.1. *For $x \geqslant 2$ we have*

$$x \log 2 + O(\log x) < \psi(x) < x \log 4 + O\big((\log x)^2\big) \ . \tag{2.17}$$

Now knowing that $\psi(x) \ll x$, we derive from (2.15), by replacing $[\frac{x}{d}]$ with $\frac{x}{d}$, that

$$\sum_{d \leqslant x} \frac{\Lambda(d)}{d} = \log x + O(1) \ . \tag{2.18}$$

The contribution from higher powers of primes is bounded, so we also obtain

$$\sum_{p \leqslant x} \frac{\log p}{p} = \log x + O(1) \ . \tag{2.19}$$

By partial summation, this yields

$$\sum_{p \leqslant x} \frac{1}{p} = \log\log x + \text{ constant } + O\Big(\frac{1}{\log x}\Big) \ . \tag{2.20}$$

Furthermore, we get

$$\begin{aligned} \prod_{p \leqslant x} \Big(1 - \frac{1}{p}\Big) &= \exp\Big(\sum_{p \leqslant x} \log\Big(1 - \frac{1}{p}\Big)\Big) \\ &= \exp\Big(-\sum_{p \leqslant x} \frac{1}{p} + \text{ constant } + O\Big(\frac{1}{\log x}\Big)\Big) \end{aligned}$$

and, applying (2.20), we conclude the following result of F. Mertens.

PROPOSITION 2.2. *For $x \geqslant 2$ we have*

$$\prod_{p \leqslant x} \Big(1 - \frac{1}{p}\Big) = \frac{e^{-\gamma}}{\log x} \Big\{1 + O\Big(\frac{1}{\log x}\Big)\Big\} \tag{2.21}$$

with some real constant γ.

Actually, Mertens [**127**] showed the above formula with $\gamma = 0.577\ldots$ being the Euler constant (for further remarks see the end of Section 12.2).

As we have seen, Tchebyshev's arguments work nicely for finding primes in the sequence of all integers. However, they do not extend to find primes in more general sequences $\mathcal{A} = (a_n)$ of the type considered in Chapter 1. If instead we ask only for numbers having a large primary factor then the method does extend considerably, depending on our ability to estimate the remainder terms $r_d(x)$. This begins with an evaluation of the sum

$$S(x,y) = \sum_{n \leqslant x} a_n \Big(\sum_{\substack{d|n \\ d>y}} \Lambda(d)\Big) . \tag{2.22}$$

Ignoring the restriction $d > y$, the convolution completes to $\log n$ which is a nice function, growing sufficiently slowly that

$$A'(x) = \sum_{n \leqslant x} a_n \log n = A(x)(\log x + O(1)) , \tag{2.23}$$

for most natural sequences $\mathcal{A}$ that one encounters.

Rearranging the order of summation in (2.22), we obtain

$$S(x,y) = A'(x) - \sum_{d \leqslant y} \Lambda(d) A_d(x) . \tag{2.24}$$

If we succeed in showing that $S(x,y)$ is positive then we can infer that there are $a_n > 0$ with $n \leqslant x$ having a primary divisor exceeding y. How large y can be taken depends on how many of the congruence sums $A_d(x)$ we can handle.

Inserting the approximations (1.2), we obtain

$$S(x,y) = A'(x) - A(x) \sum_{d \leqslant y} g(d)\Lambda(d) - \sum_{d \leqslant y} \Lambda(d) r_d(x) .$$

In many natural cases the first sum can be evaluated as

$$\sum_{d \leqslant y} g(d)\Lambda(d) = \log y + O(1) ,$$

see for example (2.18) for $g(d) = 1/d$.

Suppose also that the remainder terms satisfy

$$\sum_{d \leqslant y} \Lambda(d) r_d(x) \ll A(x) . \tag{2.25}$$

PROPOSITION 2.3. *If* (2.25) *holds for $y = x^\theta$ with $0 < \theta < 1$, then*

$$S(x,y) = (1-\theta)A(x)(\log x + O(1)) . \tag{2.26}$$

This shows that, if x is sufficiently large, there are $a_n > 0$ with $n \leqslant x$ having a primary factor larger than x^θ (also, if $\theta > 1/2$, that there are $a_n > 0$ with $n \leqslant x$ having no primary factor larger than x^θ).

If we only require a lower bound for $S(x, y)$ then we can afford to give up some percentage of the asymptotic (2.26) and this opens the possibility to choose $y = x^\vartheta$ with ϑ somewhat larger than the "exponent of distribution" θ. We shall illustrate this procedure in Section 13.5 for quite general sequences $\mathcal{A} = (a_n)$ and we give one application for the sequence of integers in a very short interval.

One expects that higher powers of primes contribute a negligible amount and this is in practice verifiable in cases where the sequence $\mathcal{A}$ is not too thin. We can then deduce the existence of large prime factors. What we require for this purpose is a crude bound such as

$$\sum_{n \leqslant x} a_n \Big(\sum_{p^2 | n} \log p \Big) \ll A(x) . \tag{2.27}$$

CHAPTER 3

Bombieri's Sieve

Prelude

3.1. Heuristics for Estimating Sums over Primes

In Chapter 1 we used Eratosthenes' idea to count primes and we failed. However, we were more successful in Chapter 2 with Tchebyshev's ideas, although we did not get an asymptotic formula for $\psi(x)$. In this section we intend to explore this idea further, and in a general context.

Let $\mathcal{A} = \big(a_n\big)$ be a sequence of non-negative reals. We study the sum

$$S(x) = \sum_{n \leqslant x} a_n \Lambda(n)\ . \tag{3.1}$$

Using the expression

$$\Lambda(n) = -\sum_{d|n} \mu(d) \log d$$

we arrange $S(x)$ as follows:

$$S(x) = \sum_{d \leqslant x} -\mu(d)(\log d) A_d(x)\ .$$

Replacing $A_d(x)$ by the approximation (1.2) we get

$$S(x) = A(x)H(x) + R(x) \tag{3.2}$$

where $H(x)$ is the main term factor

$$H(x) = \sum_{d \leqslant x} -\mu(d) g(d) \log d \tag{3.3}$$

and $R(x)$ is the remainder

$$R(x) = \sum_{d \leqslant x} -\mu(d)(\log d) r_d(x)\ . \tag{3.4}$$

In the main term factor we assume that $g(d)$ is sufficiently regular (see (3.45) below) so that the sum $H(x)$ is well approximated by the completed series

$$H = \sum_{1}^{\infty} -\mu(d) g(d) \log d\ , \tag{3.5}$$

which by Proposition 3.4 can also be written as the infinite product

$$H = \prod_p \big(1 - g(p)\big)\Big(1 - \frac{1}{p}\Big)^{-1}\ . \tag{3.6}$$

Ignoring the remainder $R(x)$, we arrive at the asymptotic formula

$$S(x) \sim HA(x) \ . \tag{3.7}$$

Unlike the heuristic result from the Eratosthenes sieve, we believe that the formula (3.7) gives the correct answer.

EXAMPLES. For $x \to \infty$ we have

$$\sum_{m \leqslant x} \Lambda(m) \sim x \ , \tag{3.8}$$

$$\sum_{m \leqslant x} \Lambda(m^2+1) \sim Hx \ , \qquad H = \prod_p \Big(1 - \frac{1+\chi_4(p)}{p}\Big)\Big(1-\frac{1}{p}\Big)^{-1} \ , \tag{3.9}$$

$$\sum_{m \leqslant x} \Lambda(m)\Lambda(m+2) \sim Hx \ , \qquad H = 2\prod_{p>2} \Big(1 - \frac{1}{(p-1)^2}\Big) \ , \tag{3.10}$$

$$\sum\sum_{a^2+b^4 \leqslant x} \Lambda(a^2+b^4) \sim Hx^{\frac{3}{4}} \ , \qquad H = \frac{\sqrt{2}\,\Gamma^2(1/4)}{3\pi^{3/2}} \ . \tag{3.11}$$

Actually, in this last case H is also given by the integral

$$H = \frac{4}{\pi}\int_0^1 (1-t^4)^{\frac{1}{2}}\, dt \ . \tag{3.12}$$

The asymptotic formula (3.8) is known to be true and this is called the Prime Number Theorem. The formulas (3.9), (3.10) were predicted by Hardy and Littlewood [**80**] using heuristic arguments from the circle method. Finally, (3.11) is now a theorem; see Chapter 21. The reason that these asymptotics are reliable is that the terms with $d > D$, which we cannot handle, are twisted by the Möbius function $\mu(d)$ and this is expected to change sign randomly, causing cancellation. By contrast the Möbius function, when truncated as in the sieve of Eratosthenes (or any other combinatorial sieve, see Chapter 6), loses its randomness.

For a rigorous proof of (3.7) we need to show that the remainder $R(x)$ is small. For most sequences of interest there is a "level of distribution" $D = D(x) < x$ such that

$$R(x, D) = \sum_{d \leqslant D} |r_d(x)| \tag{3.13}$$

is small, and so may be neglected.

There remains the problem of dealing with large d, that is the sum

$$- \sum_{D < d \leqslant x} \mu(d)(\log d) r_d(x) \ .$$

We do expect this sum to be small, not because the individual terms $r_d(x)$ are small, but due to the cancellation which results from the sign changes of the Möbius function $\mu(d)$ and the variation of $r_d(x)$. However, these two effects should not conspire against each other. Hence it is dangerous to insert the absolute values as in $R(x, D)$.

If we reinterpret the sum

$$-\sum_{D<d\leqslant x} \mu(d)(\log d)A_d(x)$$

using the Dirichlet hyperbola trick we are led to consider the contribution of the sums

$$-\sum_{D<d\leqslant x/m} \mu(d)(\log d)a_{dm} ,$$

for any $m \leqslant xD^{-1}$. Since m is relatively small we cannot lose by treating these sums separately for each individual m. In particular, for $m = 1$ we need to estimate

$$-\sum_{D<d\leqslant x} \mu(d)(\log d)a_d .$$

We cannot handle such sums at this stage; one main punchline of this book will emerge toward the end when we succeed in doing so in many cases.

3.2. Almost-Primes

E. Bombieri [**8**] showed that the ideas of the previous section can work for a modified problem wherein one considers "almost-primes" rather than primes.

Define $\Lambda_k = \mu * (\log)^k$ for any integer $k \geqslant 1$, that is,

$$\Lambda_k(n) = \sum_{d|n} \mu(d)\Big(\log \frac{n}{d}\Big)^k . \tag{3.14}$$

That these functions generalize the von Mangoldt function is clear since $\Lambda_1 = \Lambda$. It turns out that the support of Λ_k is on integers having at most k distinct prime factors. This and a number of other nice properties follow from the recurrence formula

$$\Lambda_{k+1} = \Lambda_k \cdot L + \Lambda_k * \Lambda \tag{3.15}$$

which is easily proved by induction on k. Here, L denotes the arithmetic function $L(n) = \log n$.

We have

$$0 \leqslant \Lambda_k(n) \leqslant \big(\log n\big)^k , \tag{3.16}$$

the first inequality following by induction on k and using (3.15), while the second holds since, by Möbius inversion $1 * \Lambda_k = L^k$, that is,

$$\sum_{d|n} \Lambda_k(d) = \big(\log n\big)^k . \tag{3.17}$$

For $k = 2$ we have

$$\Lambda_2(n) = \Lambda(n)\log n + \sum_{d|n} \Lambda(d)\Lambda\Big(\frac{n}{d}\Big) , \tag{3.18}$$

a function which first came to prominence when Selberg gave an elementary proof of the asymptotic formula

$$\sum_{n\leqslant x} \Lambda_2(n) \sim 2x\log x , \tag{3.19}$$

which in turn played a fundamental role in the first few elementary proofs of the Prime Number Theorem

$$\sum_{n\leqslant x} \Lambda(n) \sim x \ , \tag{3.20}$$

proofs which yielded the result with a rather weak error term. The generalization to larger k was subsequently introduced and exploited by Bombieri to considerably sharpen that error term in the elementary proof.

From (3.20) it is evident that half of the mass of the sum (3.19) comes from integers having an even number of prime factors and half from those with an odd number. It turns out that the same dichotomy holds as well for the sum

$$\sum_{n\leqslant x} \Lambda_k(n) \sim kx\big(\log x\big)^{k-1} \ , \tag{3.21}$$

for each $k \geqslant 3$.

3.3. Multi-dimensional von Mangoldt Functions

We intend to study sums over a very wide class of functions weighted on almost-primes. The functions Λ_k introduced in the previous section generate a space too small for what we need. As an intermediate step we consider a function of von Mangoldt type for vectors

$$\Lambda_{(k)} = \Lambda_{k_1} * \cdots * \Lambda_{k_r} \ , \tag{3.22}$$

where $(k) = (k_1, \ldots, k_r)$ is a vector of positive integers. Note that $\Lambda_{(k)}$ is supported on positive integers having at most

$$|k| = k_1 + \cdots + k_r \tag{3.23}$$

distinct prime divisors. Note also that the vector von Mangoldt functions can be bounded in terms of the corresponding scalar ones, specifically:

PROPOSITION 3.1. *We have*

$$0 \leqslant \Lambda_{(k)} \leqslant \Lambda_{|k|} \ . \tag{3.24}$$

PROOF. This follows by induction on r provided we show that

$$\Lambda_{h+k} \geqslant \Lambda_h * \Lambda_k \tag{3.25}$$

for all h, $k \geqslant 1$. The latter is true for $h = k = 1$ since, by (3.15), $\Lambda_2 = \Lambda L + \Lambda * \Lambda$.

If we assume (3.25) holds for given h, k, then, by the recurrence formula (3.15),

$$\begin{aligned}
\Lambda_{h+k+1} &= \Lambda_{h+k}L + \Lambda * \Lambda_{h+k} \\
&\geqslant (\Lambda_h * \Lambda_k)L + \Lambda * (\Lambda_h * \Lambda_k) \\
&\geqslant (\Lambda_h L + \Lambda * \Lambda_h) * \Lambda_k \\
&= \Lambda_{h+1} * \Lambda_k \ .
\end{aligned}$$

This gives (3.25) by induction on h and k. □

We define the vector Möbius function by

$$\mu_{(k)} = \mu * \Lambda_{(k)} \ . \tag{3.26}$$

Note that $\mu_{(0)} = \mu$.

PROPOSITION 3.2. *We have*

$$|\mu_{(k)}| \leqslant L^{|k|} . \tag{3.27}$$

PROOF. First note that

$$|\mu_{(k)}| \leqslant |\mu| * \Lambda_{(k)} \leqslant 1 * \Lambda_{(k)} . \tag{3.28}$$

We write

$$\Lambda_{(k)} = \Lambda_{(k')} * \Lambda_\ell \tag{3.29}$$

where (k') is a vector of dimension $r-1$. Then

$$1 * \Lambda_{(k)} = 1 * \mu * L^\ell * \Lambda_{(k')} = L^\ell * \Lambda_{(k')} \leqslant L^\ell \big(1 * \Lambda_{(k')}\big)$$

so, from (3.16) and (3.28), the proposition follows by induction on r. □

Note that by (3.29) and (3.26), using Möbius inversion we get $\Lambda_{(k)} = \mu_{(k)} * 1 = \mu * \Lambda_{(k')} * \Lambda_\ell * 1$, so

$$\Lambda_{(k)} = \mu_{(k')} * L^\ell . \tag{3.30}$$

3.4. Asymptotic Formula for $S_{(k)}(x)$

We now wish to study the general sum

$$S_{(k)}(x) = \sum_{n \leqslant x} a_n \Lambda_{(k)}(n) . \tag{3.31}$$

It is expected that the dichotomy to which we alluded in connection with $S_k(x)$ holds here as well, except for very peculiar sequences, and this is indeed the essence of the parity problem which we discuss in Chapter 16.

Applying (3.30) we arrange the sum (3.31) as

$$S_{(k)}(x) = \sum\sum_{dm \leqslant x} \mu_{(k')}(d) a_{dm} \big(\log m\big)^\ell . \tag{3.32}$$

For the partial sum, say with $d \leqslant D$,

$$S_{(k)}(x, D) = \sum\sum_{\substack{dm \leqslant x \\ d \leqslant D}} \mu_{(k')}(d) a_{dm} \big(\log m\big)^\ell . \tag{3.33}$$

We are going to proceed as we did before for $k=1$ in Section 3.1. For $D < d \leqslant x$ we are still not able to exploit any cancellation from $\mu_{(k')}(d)$, so we estimate this part trivially by

$$T_{(k)}(x, M) = \big(\log M\big)^\ell \big(\log D\big)^{|k'|} \sum_{m \leqslant M} A_m(x) , \tag{3.34}$$

where $M = xD^{-1}$. If $D = D(x)$ is very close to x, then $\log M$ is small compared to $\log x$ and the factor $\log M / \log x$ is saved ℓ times. It turns out that this saving suffices if $\ell \geqslant 2$ and just fails if $\ell = 1$ (one has to recover at least the factor $\log x$ which is lost in the removal of the generalized Möbius function). Furthermore, we use the trivial estimate $A_m(x) \leqslant g(m)A(x) + |r_m(x)|$, getting

$$T_{(k)}(x, M) \leqslant \big(A(x)G(M) + R(x, M)\big) \big(\log M\big)^\ell \big(\log D\big)^{|k'|} , \tag{3.35}$$

where

$$G(M)=\sum_{m\leqslant M} g(m)\ , \tag{3.36}$$

$$R(x,M)=\sum_{d\leqslant M} |r_d(x)|\ . \tag{3.37}$$

Now we proceed to $S_{(k)}(x,D)$. We relax the factor $(\log m)^\ell$ by the following device:

$$\big(\log m\big)^\ell=\int_1^m d\big(\log t\big)^\ell\ ,$$

obtaining (we apologize for the double use of the symbol d)

$$\begin{aligned}S_{(k)}(x,D)&=\sum_{d\leqslant D}\mu_{(k')}(d)\int_1^{x/d}\Big(\sum_{t<m\leqslant x/d} a_{dm}\Big)d\big(\log t\big)^\ell\\&=\sum_{d\leqslant D}\mu_{(k')}(d)\int_d^x\big(A_d(x)-A_d(t)\big)d\Big(\log\frac{t}{d}\Big)^\ell\\&=\int_1^x\sum_{d\leqslant\min(t,D)}\mu_{(k')}(d)\,(A_d(x)-A_d(t))\,d\Big(\log\frac{t}{d}\Big)^\ell\ .\end{aligned}$$

Writing $A_d(x)-A_d(t)=g(d)\big(A(x)-A(t)\big)+r_d(x)-r_d(t)$ and estimating the contribution of the remainder terms trivially by

$$R_{(k)}(x,D)=\int_1^x\big(R(x,D)+R(t,D)\big)d\big(\log t\big)^\ell\big(\log D\big)^{|k'|}\ , \tag{3.38}$$

we find that

$$S_{(k)}(x,D)=\int_1^x\big(A(x)-A(t)\big)\sum_{d\leqslant\min(t,D)}\mu_{(k')}(d)g(d)d\Big(\log\frac{t}{d}\Big)^\ell+\theta R_k(D,x)$$

with $|\theta|\leqslant 1$. Next we simplify the main term by enlarging $\min(t,D)$ to t. We get

$$H_{(k)}(x)=\int_1^x\big(A(x)-A(t)\big)\sum_{d\leqslant t}\mu_{(k')}(d)g(d)d\Big(\log\frac{t}{d}\Big)^\ell\ .$$

After an interchange of the integration with the summation, this becomes

$$H_{(k)}(x)=\sum_{n\leqslant x} a_n\,h_{(k)}(n)\ , \tag{3.39}$$

where

$$h_{(k)}(y)=\sum_{d<y}\mu_{(k')}(d)g(d)\Big(\log\frac{y}{d}\Big)^\ell\ . \tag{3.40}$$

We estimate the excess of the main term resulting from the above enlargement by

$$\begin{aligned}&\int_D^x\big(A(x)-A(t)\big)\Big(\sum_{D<d\leqslant t} g(d)\Big)d\Big(\log\frac{t}{D}\Big)^\ell\big(\log x\big)^{|k'|}\\&\leqslant A(x)\sum_{D<d\leqslant x} g(d)\Big(\log\frac{x}{D}\Big)^\ell\big(\log x\big)^{|k'|}\\&=A(x)\big(G(x)-G(D)\big)\big(\log M\big)^\ell\big(\log x\big)^{|k'|}\ .\end{aligned}$$

Gathering the above estimates, we arrive at:

PROPOSITION 3.3. *For any* $(k) = \big(\ell, (k')\big)$ *and* $\sqrt{x} < D \leqslant x$ *we have*
(3.41)
$$| \, S_{(k)}(x) - H_{(k)}(x) \, | \leqslant A(x)\Big(G(x) - G(D) + G\big(\frac{x}{D}\big)\Big)\Big(\log \frac{x}{D}\Big)^{\ell}\big(\log x\big)^{|k'|} + 2R_{(k)}(D, x)$$
where the main term $H_{(k)}(x)$ *is given by* (3.39), (3.40), *the remainder* $R_{(k)}(D, x)$ *is given by* (3.37), (3.38), *and* $G(y)$ *by* (3.36).

Suppose the density function $g(d)$ satisfies the following (quite crude) bound:
$$\sum_{y<d\leqslant x} g(d) \ll \log \frac{x}{y} \ , \tag{3.42}$$
for all y with $1 \leqslant y \leqslant x/2$. Suppose also that, for some D with $\sqrt{x} < D \leqslant x/2$, the remainder $R(t, D)$ satisfies
$$R(t, D) \ll A(x)\big(\log x\big)^{-2} \ , \tag{3.43}$$
for all $1 \leqslant t \leqslant x$. Then, (3.41) yields the following formula:
$$S_{(k)}(x) = H_{(k)}(x) + O\bigg(A(x)\Big[\Big(\log \frac{x}{D}\Big)^{\ell+1} + \big(\log x\big)^{\ell-2}\Big]\big(\log x\big)^{|k'|}\bigg) \ . \tag{3.44}$$

To evaluate the main term (3.39) more explicitly we need a condition on the density function $g(d)$ somewhat stronger than (3.42).

PROPOSITION 3.4. *Suppose that* $g(d)$ *is a multiplicative function with* $0 \leqslant g(d) < 1$ *such that*
$$\sum_{p\leqslant y} g(p) \log p = \log y + c_g + O\big((\log 2y)^{-A}\big) \tag{3.45}$$
for all A, *with an implied constant depending on* g *and* A, *and also*
$$g(d) \leqslant \tau_r(d) d^{-1} \ , \tag{3.46}$$
for some $r \geqslant 1$. *Then, for* $(k) = \big(\ell, (k')\big)$ *with* $\ell \geqslant 2$, *we have*
$$\sum_{d\leqslant y} \mu_{(k')}(d) g(d)\Big(\log \frac{y}{d}\Big)^{\ell} = \frac{(k)!}{(|k| - 1)!} H \big(\log y\big)^{|k|-1} + O\Big(\big(\log y\big)^{|k|-2}\Big) \ , \tag{3.47}$$
where H *is a positive constant given by the product*
$$H = \prod_p \big(1 - g(p)\big)\Big(1 - \frac{1}{p}\Big)^{-1} \ . \tag{3.48}$$

PROOF. This will be given in a general setting in Section A.7. □

From Proposition 3.4 we have
$$h_{(k)}(n) = \frac{(k)!}{(|k| - 1)!} H \big(\log n\big)^{|k|-1} + O\Big(\big(\log n\big)^{|k|-2}\Big) \tag{3.49}$$
and
$$H_{(k)}(x) = \frac{(k)!}{(|k| - 1)!} H \sum_{n\leqslant x} a_n \big(\log n\big)^{|k|-1} + O\Big(A(x)\big(\log x\big)^{|k|-2}\Big) \ . \tag{3.50}$$
Introducing (3.49) to (3.50) we conclude the following result.

THEOREM 3.5. *Suppose a sequence $\mathcal{A} = (a_n)$ of real non-negative numbers a_n satisfies* (1.2) *with a density function $g(d)$ satisfying* (3.45). *Suppose that the remainders $R(t, D)$ satisfy* (3.43) *for some $D = D(x)$ with $\sqrt{x} \leqslant D \leqslant x/2$, and all $1 \leqslant t \leqslant x$. Then, for $(k) \neq (1, \dots, 1)$ we have*

$$\begin{aligned}\sum_{n \leqslant x} a_n \Lambda_{(k)}(n) = \frac{(k)!}{(|k|-1)!} H \sum_{n \leqslant x} a_n (\log n)^{|k|-1} \\ + O\Big(A(x)\Big(\log \frac{x}{D}\Big)^3 (\log x)^{|k|-2}\Big) .\end{aligned} \tag{3.51}$$

Assuming one can take

$$D = D(x) = x \; \exp\big(-(\log x)^{\frac{1}{4}}\big) \; ,$$

the above error term simplifies to $O\big(A(x)(\log x)^{|k|-\frac{5}{4}}\big)$, while the main term is of order $A(x)(\log x)^{|k|-1}$. For $k = 2$ we get

$$\sum_{n \leqslant x} a_n \Lambda_2(n) = 2H \sum_{n \leqslant x} a_n \log n + O\big(A(x)(\log x)^{\frac{3}{4}}\big) \; . \tag{3.52}$$

We should emphasize that Bombieri [**8**] (see also [**50**]) proves that the asymptotic formula in (3.51) holds, of course with weaker error term, under the milder assumption that the remainders are acceptable with $D = D(x) = x^{1-\varepsilon}$ for every $\varepsilon > 0$. The proof of this fortissimo version would have required quite a bit more groundwork, for example, by postponing the issue until we could appeal to the Fundamental Lemma of Chapter 6.

On the other hand, the assumption that the remainders are acceptable with $D = D(x) = x^{\theta}$ for a given fixed $\theta < 1$ is not sufficient to give the asymptotic formula for $S_k(x)$ for any k, no matter how large. This was revealed by the counterexamples provided by Ford [**43**], which become increasingly sophisticated as θ approaches one.

REMARK. Our conditions (3.45) and (3.46) for the density function are satisfied by virtually any example which occurs in practice. The error term in (3.45) is as strong as in the Prime Number Theorem so, strictly speaking, our proof of Theorem 3.5 uses the PNT. Such a strong assumption was introduced only to simplify the computation of the main term given in Section A.7. In earlier versions of the Theorem (see [**50**]) the proof is given under a much milder version of (3.45).

3.5. Application to Additive Convolutions

As a quick illustration of Theorem 3.5 we consider the sum

$$S_{(k)}(x) = \sum\sum_{\ell + m \leqslant x} \alpha_\ell \, \beta_m \, \Lambda_{(k)}(\ell + m) \tag{3.53}$$

where α_ℓ, β_m are quite general numbers $0 \leqslant \alpha_\ell, \beta_m \leqslant 1$. This is the sum (3.31) for the sequence $\mathcal{A} = (a_n)$ with

$$a_n = \sum_{\ell + m = n} \alpha_\ell \, \beta_m \; .$$

We cannot treat very sparse sequences so we assume that

$$A(x)=\sum\sum_{\ell+m\leqslant x}\alpha_\ell\,\beta_m\gg x^2(\log x)^{-A}\ , \tag{3.54}$$

where A is any fixed number. For simplicity we assume that β_m is supported on primes, which eases considerably the technical details, and that α_ℓ is supported on even numbers.

However, the following Siegel–Walfisz type condition, although mild, is essential to our argument. We assume:

$$\sum_{\substack{m\leqslant y\\ m\equiv a\,(\mathrm{mod}\ q)}}\beta_m-\frac{1}{\varphi(q)}\sum_{m\leqslant y}\beta_m\ll x(\log x)^{-2A-8}\ , \tag{3.55}$$

for any $q\geqslant 1$, $(a,q)=1$ and $2\leqslant y\leqslant x$.

Under these conditions on the sequences we deduce:

THEOREM 3.6. *For* $(k)\neq(1,\ldots,1)$ *we have*

$$S_{(k)}(x)\sim\frac{(k)!}{\big(|k|-1\big)!}\,H\,A(x)\big(\log x\big)^{|k|-1} \tag{3.56}$$

where H *is the twin prime constant*

$$H=\prod_{p>2}\Big(1-\frac{1}{(p-1)^2}\Big)\ . \tag{3.57}$$

PROOF. For d odd we have

$$\begin{aligned}A_d(x)&=\sum\sum_{\substack{\ell+m\leqslant x\\ \ell+m\equiv 0\,(\mathrm{mod}\ d)}}\alpha_\ell\,\beta_m\\ &=\sum\sum_{\substack{\ell+m\leqslant x\\ (\ell,m)=1\\ \ell+m\equiv 0\,(\mathrm{mod}\ d)}}\alpha_\ell\,\beta_m+O(x\log\log x)\ .\end{aligned}$$

Hence,

$$A_d(x)=\frac{A(x)}{\varphi(d)}+r_d(x)$$

with

$$\sum_{d\leqslant D}|r_d(x)|\ll\big(D+x(\log x)^{-2A-8}\big)^{\frac12}x^{\frac32}(\log x)^2\ ,$$

by Theorem 22.6. This satisfies the condition (3.43) so Theorem 3.5 is applicable with $g(d)=1/\varphi(d)$ for d odd and $g(d)=0$ for d even and with $D=x(\log x)^{-2A-8}$. Moreover, we have

$$\sum_{n\leqslant x}a_n\big(\log n\big)^{|k|-1}\sim A(x)\big(\log x\big)^{|k|-1}$$

by (3.54), giving the result. □

CHAPTER 4

Sieve of Eratosthenes-Legendre

Cadenza

Despite our earlier pessimistic comments about the usefulness of the sieve in the form given by Eratosthenes and Legendre, it can in certain circumstances be made to give worthwhile estimates, even asymptotic formulas. Before entering into the more advanced areas of the sieve we wish to show how this comes about.

We begin with a general sequence $\mathcal{A} = \big(a_n\big)$ of non-negative numbers and want to evaluate the sifting function

$$S(\mathcal{A}, z) = \sum_{\substack{n \leqslant x \\ (n, P(z))=1}} a_n \tag{4.1}$$

where $P(z)$ is the product of the primes, $p < z$, from a certain set $\mathcal{P}$. Applying the Möbius formula we write

$$S(\mathcal{A}, z) = \sum_{d \mid P(z)} \mu(d) A_d(x) \tag{4.2}$$

where

$$A_d(x) = \sum_{\substack{n \leqslant x \\ n \equiv 0 (d)}} a_n \ .$$

Note that if $d > x$ this sum is void so there is no need to keep these terms in (4.2). This apparently trivial observation reduces tremendously the number of terms and, if the set $\mathcal{P}$ of primes is sufficiently thin, this truncation of Möbius is enough to make a difference.

Introducing the approximation (1.2) for each $A_d(x)$, $d \leqslant x$, we obtain

$$S(\mathcal{A}, z) = XG(x, z) + R \tag{4.3}$$

where

$$G(x, z) = \sum_{\substack{d \mid P(z) \\ d \leqslant x}} \mu(d) g(d) \ , \tag{4.4}$$

and

$$R = \sum_{\substack{d \mid P(z) \\ d \leqslant x}} \mu(d) r_d(x) \ . \tag{4.5}$$

First we treat the remainder. As in Bombieri's sieve in the previous chapter, we cannot take advantage of any cancellation in terms due to the sign changes of the Möbius function (even though we expect these take place) and so we simply have to estimate the larger sum

$$R(x,z)=\sum_{\substack{d|P(z)\\ d\leqslant x}} \left|r_d(x)\right| .$$

To simplify our presentation we assume a bound for the individual terms,

$$\left|r_d(x)\right| \leqslant g(d)d , \tag{4.6}$$

for every $d\leqslant x$, $d\mid P(z)$. This assumption is natural in some important cases. For example, if $a_n=1$ we have $g(d)=1/d$ and $\left|r_d(x)\right|\leqslant 1$. By (4.6) we have

$$R(x,z)\leqslant \sum_{\substack{d|P(z)\\ d\leqslant x}} g(d)d . \tag{4.7}$$

We assume that our density function g satisfies

$$\sum_{\substack{w\leqslant p<y\\ p\in P}} g(p)\log p \leqslant \kappa \log\frac{y}{w}+b \tag{4.8}$$

for all $2\leqslant w<y$, with constants $\kappa\geqslant 0$, $b\geqslant 1$.

Elementary arguments of a Tchebychev flavour, which will be presented in Section A.1, allow us to bound (4.7). By (A.8) we have

$$R(x,z)\leqslant \frac{3(\kappa+b)x}{\log x}\prod_{p|P(z)}\big(1+g(p)\big) . \tag{4.9}$$

The condition (4.8) implies that

$$\prod_{p|P(z)}\big(1+g(p)\big) \ll \big(\log z\big)^{\kappa} . \tag{4.10}$$

Therefore, the remainder is bounded by $O\big(x(\log x)^{\kappa-1}\big)$. On the other hand, by heuristic reasoning one expects that $S(\mathcal{A},z)$ should be of order $X(\log z)^{-\kappa}$. Therefore, our bound (4.9) for the remainder is useful only if $\kappa<1/2$. Interestingly, if we make this assumption then the main term $G(x,z)$ is of order of magnitude $(\log z)^{-\kappa}$ and this can be established by elementary methods.

To do so, we require information about the density function more precise than that given in (4.8). We assume

$$0\leqslant g(p)<1 , \quad p\in\mathcal{P} , \tag{4.11}$$

$$\sum_{\substack{p\leqslant y\\ p\in\mathcal{P}}} g(p)\log p=\kappa\log y+O(1) , \tag{4.12}$$

and

$$\sum_{p\in\mathcal{P}} g(p)^2\log p<\infty . \tag{4.13}$$

PROPOSITION 4.1. *Let $z=x^{1/s}$ with $s\geqslant 1$. Assuming (4.12) with $\kappa<\frac{1}{2}$ and also (4.13) we have*

$$G(x,z)=V(z)\left\{h(s)+O\big((\log x)^{2\kappa-1}\big)\right\} , \tag{4.14}$$

where the implied constant depends on κ and s. Here

$$V(z) = \prod_{p|P(z)} (1 - g(p)) \ . \tag{4.15}$$

The function $h(s)$ is the continuous solution to the differential-difference equation

$$\begin{aligned} s^\kappa h(s) &= \frac{e^{\gamma\kappa}}{\Gamma(1-\kappa)} \ , \qquad & 0 < s \leqslant 1 \ , \\ sh'(s) &= -\kappa h(s) + \kappa h(s-1) \ , \quad & \text{if } s > 1 \ . \end{aligned} \tag{4.16}$$

Proof. This is a special case of Theorem A.7 for the function $\mu(d)g(d)$. □

Combining Proposition 4.1 with (4.9) and (4.10) we deduce:

Theorem 4.2. *Let $\mathcal{A} = \big(a_n\big)$ be a sequence of non-negative reals and $\mathcal{P}$ a set of primes. Assume that the density function g satisfies (4.12) with $\kappa < 1/2$ and (4.13), and that the remainder terms satisfy (4.6) for every $d \leqslant x$. Let $s \geqslant 1$. Then, for $z = x^{1/s}$ we have*

$$S(\mathcal{A}, z) = XV(z) \left\{h(s) + O\left((\log x)^{2\kappa-1}\right)\right\} + O\left(x(\log x)^{\kappa-1}\right) \ , \tag{4.17}$$

where the implied constants depend on κ and s.

Note that, if $\mathcal{A}$ is dense in the sense that $X \gg x$, the error terms are small and we have the asymptotic formula for $S(\mathcal{A}, z)$. For $z = x$ we have $s = 1$, $h(1) = e^{\gamma\kappa}/\Gamma(1-\kappa)$, and

$$G(x, x) \sim \frac{e^{\gamma\kappa}}{\Gamma(1-\kappa)} V(x) \ . \tag{4.18}$$

This fails for $\kappa = 1/2$ where the true asymptotic is twice that given by (4.18); see Chapter 14. It is interesting that, for $\kappa < 1/2$, an asymptotic formula exists at all, since for $\kappa = 1$ the corresponding sum with $g(d) = 1/d$ is much smaller, certainly

$$\sum_{d \leqslant x} \frac{\mu(d)}{d} \ll \big(\log x\big)^{-2007} \ .$$

Just how much smaller, we would surely like to know!

Probably there is also no asymptotic for $1/2 < \kappa < 1$. Note that the asymptotic for $\kappa < 1/2$,

$$G(x, z) \sim V(z)\, h(s) \ , \tag{4.19}$$

does not give the result $G(x, z) \sim V(z)$ which might have been expected on probabilistic grounds. These outcomes do match as $s \to \infty$ because $h(s) \to 1$. In fact, $h(s) - 1$ changes sign in every interval of length one (see Corollary B.9) so the probabilistic expectation is right at least once a day (like a broken clock).

CHAPTER 5

Sieve Principles and Terminology

Tuning Up

5.1. Introduction

In this chapter we gather together a number of very basic ideas which occur repeatedly in the development of sieve methods and direct attention to some of the terminology we use to describe these in further developments. Some of them will already be familiar from the earlier chapters but we want to make them easily accessible for the reader.

It is difficult to come up with sieve terminology which is completely satisfactory so that sometimes one symbol will do more than one job. For example,

$$\mathcal{A} = \big(a_n\big) \tag{5.1}$$

is a sequence of non-negative reals (the sifting sequence), but it is sometimes restricted to $n \leqslant x$, and occasionally is confused with the set of numbers on which it is supported; see the examples in Section 1.3. We find it more informative to have this slight imprecision as opposed to introducing excess baggage.

Conversely, sometimes the same concept may have slightly different names in different places. Thus, an important formula may bear a new number in a subsequent chapter. As another example, we shall often encounter functions of several variables and frequently suppress in the notation reference to those which do not play a role in the argument. We believe that this shortcut makes things more descriptive rather than less.

5.2. Cast of Characters

The basic input to any sieve mechanism comes via the subsequences $\mathcal{A}_d$ consisting of those a_n with $n \equiv 0 \,(\mathrm{mod}\ d)$ and, in particular, from estimates for the congruence sums

$$\big|\mathcal{A}_d\big| = \sum_{n \equiv 0\,(\mathrm{mod}\ d)} a_n\ . \tag{5.2}$$

Sometimes, particularly if $\mathcal{A}$ is restricted to $n \leqslant x$, we denote this by

$$A_d(x) = \sum_{\substack{n \leqslant x \\ n \equiv 0\,(\mathrm{mod}\ d)}} a_n\ . \tag{5.3}$$

Expecting that divisibility by d is a regularly occurring event we should be able to write

$$A_d(x) = g(d)A(x) + r_d(x) \tag{5.4}$$

where

$$|\mathcal{A}| = A(x) = A_1(x) = \sum_{n \leqslant x} a_n \tag{5.5}$$

which we envision as an approximation formula so that $r_d(x)$ is a relatively small remainder term. The form of the main term $g(d)A(x)$ separates the variable d from the sequence $\mathcal{A}$. Here $g(d)$ is a multiplicative function, expressing the belief that divisibility by distinct primes are independent events. This belief is true only to a limited extent, a fact which can be blamed for the inability of sieve methods to work better than they do. Naturally, we call $g(d)$ the "density function" and the "local densities" $g(p)$ satisfy

$$0 \leqslant g(p) < 1 \tag{5.6}$$

for every prime p.

As we have seen in the examples of Section 1.3, it is frequently convenient to have a smooth approximation to $A(x)$, which we label as X, and to replace (5.4) by

$$A_d(x) = g(d)X + r_d\ , \tag{5.7}$$

or for brevity, especially when the restriction $n \leqslant x$ is hidden,

$$|\mathcal{A}_d| = g(d)X + r_d\ . \tag{5.8}$$

In practice X will be sufficiently close to $A(x)$ that (5.7) (or (5.8)) will be equivalent to (5.4) with the same density function $g(d)$ and with the remainder terms r_d only slightly different from $r_d(x)$. Occasionally, we should like to study the dependence of the remainder terms on the sequence $\mathcal{A}$, or on the cut-off point x, in which case we may write $r_d(\mathcal{A})$, or once again $r_d(x)$, in place of r_d.

In some sieve problems we do not need to sieve by all the primes but only by a subset $\mathcal{P}$, which we call the sifting set of primes; see Example 1.7, say. Naturally, we do not need in $\mathcal{P}$ any prime p for which all a_n with $n \equiv 0 \,(\mathrm{mod}\ p)$ vanish, that is, such that $|\mathcal{A}_p| = 0$, because such a p does nothing in the sifting process. For example, if a_n is supported on the polynomial values $n = 4m^2 + 1$, then we can restrict $\mathcal{P}$ to primes $p \equiv 1 (\mathrm{mod}\ 4)$. Therefore, the approximations (5.4), (5.7), (5.8) are needed only for moduli d which are composed of primes from the set $\mathcal{P}$ and very often only on squarefree moduli. These facts are sometimes omitted in the range of summation for notational simplicity but should be clear from the context.

For a given $\mathcal{A}$ there is at most one natural choice of the multiplicative function g which produces an adequate approximation (5.4). Since $g(p)$ is employed only at primes $p \in \mathcal{P}$ we can (for notational convenience) arbitrarily revise the natural values of $g(p)$ at $p \notin \mathcal{P}$; we may set

$$g(p) = 0 \quad \text{if } p \notin \mathcal{P}. \tag{5.9}$$

The modified remainder terms $r_d(x)$ with d having a prime divisor outside the sifting set could be very poor but it doesn't matter because these have no effect on the results.

Usually, the sifting set $\mathcal{P}$ is a naturally defined infinite set but we need to truncate it. An instrumental role in the sifting process is played by the sifting function

$$S(\mathcal{A}, \mathcal{P}, z) = \sum_{(n, P(z))=1} a_n \tag{5.10}$$

where

$$P(z) = \prod_{\substack{p \in \mathcal{P} \\ p < z}} p \ . \tag{5.11}$$

In (5.10) we refer to z as the "sifting level" and to the set of primes dividing $P(z)$ (or simply to the number $P(z)$) as "the sifting range". Almost always, when the sifting set of primes is understood from the context, we simply write $S(\mathcal{A}, z)$ in place of $S(\mathcal{A}, \mathcal{P}, z)$.

Having in mind the approximation (5.7), we may expect by probabilistic considerations that $S(\mathcal{A}, z)$ looks like $XV(z)$ where

$$V(z) = \prod_{\substack{p \in \mathcal{P} \\ p < z}} (1 - g(p)) \ . \tag{5.12}$$

We shall see that this expectation points to the truth, but not precisely. Note that, already in the notation $V(z)$, we have abandoned any mention of its dependence on the set $\mathcal{P}$.

The remainder of this section contains remarks of a more technical nature which might be omitted on a first reading.

We should point out that sometimes in practice one has an approximation to $A_d(x)$ somewhat different than (5.4) in which the dependence on d and on $\mathcal{A}$ is not so simple to separate. A typical case occurs when the main term comes from the residue of a relevant generating Dirichlet series

$$Z_d(s) = \sum_{n \equiv 0 (\text{mod } d)} a_n n^{-s}$$

having a multiple pole at $s = 1$. In this situation the main term is

$$\operatorname*{res}_{s=1} s^{-1} Z_d(s) x^s$$

and may be a linear combination (over j) of terms

$$g(d) x \big(c_j + h_j(d)\big) \big(\log(x/d)\big)^j$$

where the c_j are constants and the h_j are additive functions. The presence, especially of $\log d$, ruins the multiplicativity. Nevertheless, the damage can be contained and the corresponding theory rather fully developed. Vaguely speaking, the logarithmic factor $(\log d)^j$ in the main term has the analytic effect of increasing the sieve dimension by j. In this book we do not venture into this generality, apart from one application in Section 15.4. Fortunately, when such variants of the main term obstruct our path we are always able to pass through by ad-hoc techniques, such as by use of the monotonicity principles described in Section 5.7.

Another variation of the approximation (5.4) which sometimes occurs in practice is

$$A_d(x) = \sum_{n \leqslant x} g_n(d) a_n + r_d(x) \ , \tag{5.13}$$

where the $g_n(d)$ are multiplicative functions, as is $g(d)$, but not necessarily the same ones for every n. However, the dependence on n is rather mild, essentially a harmless perturbation. Typically, one encounters

$$g_n(d) = \begin{cases} g(d) & \text{if } (n,d)=1, \\ 0 & \text{otherwise.} \end{cases} \tag{5.14}$$

In other words, (5.13) becomes

$$A_d(x) = g(d) \sum_{\substack{n \leqslant x \\ (n,d)=1}} a_n + r_d(x) \ . \tag{5.15}$$

Here, the restriction $(n,d)=1$ is very friendly so that we may be able to relax it if provided with a little extra information about the sequence $\mathcal{A}=(a_n)$, say

$$\sum_{\substack{n \leqslant x \\ (n,d)=1}} a_n = g^*(d) \sum_{n \leqslant x} a_n + r_d^*(x) \ , \tag{5.16}$$

where $g^*(d)$ is a multiplicative function having very little variation and $r_d^*(x)$ is somewhat smaller than $A(x)$, say

$$r_d^*(x) \ll A(x)(\log x)^{-2\kappa-1} \ . \tag{5.17}$$

In such a scenario we have

$$A_d(x) = g(d)g^*(d)A(x) + r_d(x) + g(d)r_d^*(x) \ . \tag{5.18}$$

This has the shape of (5.4) with the density function $g(d)$ replaced by $g(d)g^*(d)$ and the remainder term $r_d(x)$ replaced by $r_d(x)+g(d)r_d^*(x)$. The resulting modification of the density function does not make any essential change in the sieve arguments while the additional remainder term is negligible by (5.17). Although the slight extra requirement (5.16) is easy to verify in practice, one can live without it, as we show in Section 5.7.

5.3. Sifting Weights

The coprimality condition $\big(n,P(z)\big)=1$ in the sifting function $S(\mathcal{A},z)$ can be detected exactly by the expression

$$\sum_{\substack{d|n \\ d|P(z)}} \mu(d) = \begin{cases} 1 & \text{if } \big(n,P(z)\big)=1, \\ 0 & \text{if } \big(n,P(z)\big)>1. \end{cases}$$

This decomposition as a convolution allows us to express $S(\mathcal{A},z)$ in terms of the congruence sums $|\mathcal{A}_d|$.

Recall the problem we had in Chapter 1 with the sieve of Eratosthenes, caused by the huge number of terms in the remainder

$$\sum_{d|P(z)} \mu(d) r_d(x) \ .$$

Even with the more careful treatment of Chapter 4 we can only obtain results by severely limiting the range of the sifting set of primes.

The essence of sieve methods is the replacement of the Möbius function $\mu(d)$ by a function $\Lambda=(\lambda_d)$ which to some extent mimics the Möbius function but has finite support

$$\Lambda = (\lambda_d), \qquad d|P(z) \ , \ \ d < D \ ,$$

making it more successful for estimation. This idea, which we have so far neglected, will play a central role in the following chapters. We refer to (λ_d) as "sieve weights (or sifting weights) of level D". An important role will also be played by the "sifting variable"

$$s = \frac{\log D}{\log z}$$

which measures on a logarithmic scale the size of the desired sifting range relative to the available sifting level.

The price for the change to weights with finite support is that we no longer have the exact formula for the sifting function

$$S(\mathcal{A}, z) = \sum_{d|P(z)} \mu(d)|\mathcal{A}_d|$$

but rather obtain the "sifted sum"

$$S^\Lambda(\mathcal{A}, z) = \sum_d \lambda_d |\mathcal{A}_d| = \sum_n a_n \theta_n \ , \tag{5.19}$$

where

$$\theta = 1 * \lambda, \quad \text{that is} \quad \theta_n = \sum_{d|n} \lambda_d \ .$$

Although Λ cannot be chosen so that S^Λ is extremely close to S (unless z is very small), it can be chosen to have a number of attractive features. Most significantly, it can be chosen to give upper and lower bounds

$$S^-(\mathcal{A}, z) \leqslant S(\mathcal{A}, z) \leqslant S^+(\mathcal{A}, z) \ , \tag{5.20}$$

where we use the notation S^+, S^- to articulate the sifted sums S^Λ for weights $\Lambda^+ = (\lambda_d^+)$, $\Lambda^- = (\lambda_d^-)$. To achieve (5.20) we merely require that

$$\theta_n^- \leqslant \theta_n^0 \leqslant \theta_n^+ \tag{5.21}$$

for all n, where θ^0 is the Möbius convolution (1.1) and $\theta^- = 1 * \lambda^-$, $\theta^+ = 1 * \lambda^+$.

A weight system having these properties will be called a lower-bound sieve, upper-bound sieve, respectively. There are many possibilities to construct such weights and a substantial part of the book describes such constructions. Of course, in the case of an upper-bound (respectively, lower-bound) sieve, we should like to choose the weights (λ_d^+) so as to make S^+ as small as possible (respectively, (λ_d^-) to make S^- as large as possible).

5.4. Main Terms and Remainders

On introducing the approximation (5.8) to the sifted sum $S^\Lambda(\mathcal{A}, z)$, (5.19) is transformed into

$$S^\Lambda(\mathcal{A}, z) = XV(D, z) + R^\Lambda(D, z) \tag{5.22}$$

where

$$V(D, z) = V^\Lambda(D, z) = \sum_{d|P(z)} \lambda_d \, g(d) \tag{5.23}$$

and

$$R^\Lambda(D, z) = \sum_{d|P(z)} \lambda_d \, r_d \ . \tag{5.24}$$

In the main term $XV(D,z)$ the sequence $\mathcal{A}$ has disappeared from view so the evaluation of $V(D,z)$ is a primary objective and, at least from the viewpoint of the main term, the choice of weights λ_d should be the same for all sequences $\mathcal{A}$ having the same density function $g(d)$. The notation $V(D,z)$ reminds us that this sum is expected to approximate, at least to some extent, the product $V(z)$ defined in (5.12). Sometimes we shall denote the sum (5.23) instead by G, especially in the context of Selberg's sieve where its resemblance to $V(z)$ is not so clear.

We call $R^{\Lambda}(D,z)$ the "remainder", or the remainder sum of level D, in contrast to the individual "remainder terms" r_d which constitute it. Unlike the main term, this remainder still depends on the sequence $\mathcal{A}$ and so, in many places in the following chapters, this dependence will be displayed in the notation, for example, as $R^{\Lambda}(\mathcal{A},D,z)$ or $R^{\Lambda}(\mathcal{A},D)$ or perhaps just $R^{\Lambda}(x,D)$. However, we do not want to lose sight of the level of the weights so we almost never drop D from the notation for the remainder.

In the case of the upper-bound, lower-bound sieves (λ_d^+) and (λ_d^-) we denote the corresponding sifted sums by $S^{\pm}(\mathcal{A},z)$, their main term sums by

$$V^+(D,z) = \sum_{\substack{d|P(z)\\ d<D}} \lambda_d^+ \, g(d)\ , \quad V^-(D,z) = \sum_{\substack{d|P(z)\\ d<D}} \lambda_d^- \, g(d)\ , \tag{5.25}$$

and their remainders by

$$R^+(D,z) = \sum_{\substack{d|P(z)\\ d<D}} \lambda_d^+ \, r_d\ , \quad R^-(D,z) = \sum_{\substack{d|P(z)\\ d<D}} \lambda_d^- \, r_d\ . \tag{5.26}$$

Here, of course, we could simply drop the restriction $d<D$ because it follows from the support of the sieve weights, however it sometimes makes for easier reading if a little redundancy reminds us where we are.

The primary goal for the sums $V^{\pm}(D,z)$ is that they give the expected order of magnitude $V^{\pm}(D,z) \asymp V(z)$. The success of a particular sieve in achieving this will be measured in terms of the size of the sifting variable $s = \log D/\log z$, namely that it be not too small. In the case of a good lower-bound sieve Λ, there will be a value $s=\beta$ beyond which this expected bound starts to occur, which we call the "sifting limit" for the sieve Λ.

Since the sieve weights λ_d are in some sense surrogates for the Möbius function it might seem reasonable to expect some cancellation in the remainder sums. Unfortunately, in many cases we won't be able to demonstrate this, so cannot take advantage of any such cancellation. In such circumstances we may as well replace the remainder sums (5.26) by

$$\sum_d \bigl|\lambda_d \, r_d\bigr|\ . \tag{5.27}$$

For most of the sieves we shall consider, the weights satisfy

$$\bigl|\lambda_d\bigr| \leqslant 1 \quad \text{for all } d. \tag{5.28}$$

In this case our remainders are bounded by $\bigl|R^{\pm}(D,z)\bigr| \leqslant R(D,z)$, where

$$R(D,z) = R(\mathcal{A},D,z) = \sum_{\substack{d|P(z)\\ d<D}} |r_d|\ . \tag{5.29}$$

In some situations (see Chapter 7) the weights satisfy only a weaker bound

$$|\lambda_d| \leqslant \tau_k(d) \tag{5.30}$$

and then we have to consider

$$R_k(D,z) = \sum_{\substack{d|P(z)\\ d<D}} \tau_k(d)|r_d| \,. \tag{5.31}$$

These nice bounds for the size of the sieve weights come as a bonus. When we try to find sieve weights which give excellent results for the main term, the bounds (5.28), (5.30) will just fall in our laps. This is quite fortunate since it is somewhat daunting to consider the prospect of trying to choose weights which simultaneously optimize the two sums in (5.22) (but see Subsection 7.2.1).

Despite these bounds, one should not rush to insert absolute values as in (5.27) because sometimes there will be a chance to exploit a cancellation in the sum over the remainder terms r_d with the effect of admitting larger D, hence larger z for the same value of the sifting variable s. In subsequent chapters we shall develop various shapes for the remainder which have special features of this nature. In many of the instances where we can take advantage of cancellation in the remainder sums it is because the structure of the sifting weights allows us to transform the sum (5.26) into double sums of the form

$$R(M,N,z) = \sum\sum_{\substack{mn|P(z)\\ m\leqslant M, n\leqslant N}} \alpha_m \, \beta_n \, r_{mn} \tag{5.32}$$

where $|\alpha_m| \leqslant \tau_k(m)$, $|\beta_n| \leqslant \tau_k(n)$ and $MN = D$.

Of course, with any form of the remainder, what we require is that it be negligible in comparison with the main term. This amounts to the assumption

$$R^{\Lambda}(D,z) \ll A(x)\big(\log x\big)^{-B} \tag{5.33}$$

for some sufficiently large positive constant B, and similarly for $R^{\pm}(D,z)$, $R_k(D,z)$ and $R(M,N,z)$. Equivalently, we may request (5.33) to hold with $A(x)$ replaced by X.

Clearly, the larger the value of D which is admissible in (5.33), the better. We call any such admissible D, "the level of distribution". Although somewhat dependent on Λ, this is primarily a property of the sequence $\mathcal{A}$, so we say that $\mathcal{A}$ has level of distribution D. If the bound in (5.33) holds for the sum (5.29), we say that D is the level of absolute distribution. In practice, $D = x$ represents a rough limit to what one can hope to attain as the level of distribution of the sequence $\mathcal{A}$, whereas the frequently smaller quantity $A(x)$ represents a rough limit to what one can hope to attain as the level of absolute distribution.

When the bound in (5.33) holds for $R(M,N,z)$ we say $\mathcal{A}$ has level of bilinear distribution MN. In some cases the coefficients in the bilinear remainder are special, say $\beta_n = 1$, and we are able to take advantage of this in the estimations to get a higher level of distribution, in this case a couple M, N, called the level of "special bilinear" distribution. Examples of these bilinear levels are plentiful in several chapters, beginning in Chapter 13.

Very often we are able to verify (5.33) with $D = D(x) = x^{\alpha-\varepsilon}$ for some $0 < \alpha \leqslant 1$ and every $\varepsilon > 0$, the implied constant depending on ε. In this case we say $\mathcal{A}$ has exponent of distribution α. In Example 1.1 it is obvious that $\mathcal{A}$ has exponent of distribution $\alpha = 1$. More generally, we see that if $A(x) \gg x^\alpha$ and $|r_d| \leqslant g(d)d$ as in Chapter 4, or even if $r_d \ll x^\varepsilon$, then $\mathcal{A}$ has exponent of distribution α. For a much less immediate instance where we do not know the individual remainder terms are small, consider Example 1.4, where, nevertheless, $\mathcal{A}$ has been shown to have exponent of distribution $\alpha = \frac{1}{2}$ by the celebrated Bombieri–Vinogradov theorem; see Theorem 9.18.

If we apply an upper-bound sieve to a sequence $\mathcal{A}$ having any exponent of distribution $\alpha > 0$, then we obtain an upper bound of the expected order for the sifting function. If we apply a lower-bound sieve with sifting limit β to a sequence having exponent of distribution α, then we can conclude that $\mathcal{A}$ has some support on integers having no prime factor in $\mathcal{P}$ that is less than $x^{\alpha/\beta-\varepsilon}$.

5.5. The Sieve Dimension

In order to be able to examine the main terms $V(D,z)$ in the sieve estimates we are going to assume some regularity in the distribution of g over primes. A reasonable condition, based on many examples, would be

$$\sum_{p\leqslant z} g(p)\log p = \kappa \log z + O(1) \tag{5.34}$$

where $\kappa \geqslant 0$ is a constant. Comparing this with Mertens' formula (2.19) one can interpret (5.34) as saying that $g(p)p$ is κ on average. By partial summation (5.34) yields

$$\sum_{p\leqslant z} g(p) = \kappa \log\log z + \alpha + O((\log z)^{-1}) \tag{5.35}$$

where α is a certain constant, hence, by another formula of Mertens (2.21) we have

$$\prod_{p\leqslant z}(1-g(p)) = e^{-\gamma\kappa} H (\log z)^{-\kappa}\left\{1 + O((\log z)^{-1})\right\} \tag{5.36}$$

where H is the positive constant given by the convergent product

$$H = \prod_{p} (1-g(p))\left(1-\frac{1}{p}\right)^{-\kappa}. \tag{5.37}$$

Much of sieve theory requires weaker conditions and some methods require nothing of $g(p)$ at all. In this book we shall be working most of the time with a one-sided inequality of the type:

$$\prod_{w\leqslant p<z} (1-g(p))^{-1} \leqslant K\left(\frac{\log z}{\log w}\right)^{\kappa} \tag{5.38}$$

or, in other words, in the notation of (5.12),

$$\frac{V(w)}{V(z)} \leqslant K\Big(\frac{\log z}{\log w}\Big)^{\kappa}$$

for any $z > w \geqslant 2$, where K is a constant, $K > 1$. Note that this inequality implies

$$g(p) \leqslant 1 - \frac{1}{K}.$$

Here, K is quite close to one in our minds and is sometimes replaced by a function like

$$K = 1 + \frac{L}{\log w}\,, \tag{5.39}$$

where $L \geqslant 1$ is a constant.

The important parameter κ depends on both the density function g and the sifting set of primes $\mathcal{P}$. For a sequence $\mathcal{A}$ with density function g it encapsulates in a single number the size of the sifting set of primes and the susceptibility of elements of $\mathcal{A}$ to being divisible by those primes. Since it is given by a weighted average of the local densities $g(p)$ it would make a lot of sense to refer to it as the "global density". But, we are already using the word "density" a lot and, besides, this has already had quite a number of names in the literature so we shall take one of those and call κ the sieve dimension. In the most central case we have $\kappa = 1$ and we refer to this as a linear sieve problem.

Note that in (5.38) any number larger than κ also serves. If (5.38) holds with κ, but with no smaller number, then we expect that $V(z)$ has size

$$V(z) \asymp (\log z)^{-\kappa}\,. \tag{5.40}$$

There are some advantages to postulating the one-sided inequality (5.38) in preference to the asymptotic (5.36). First of all, the sieve results for dimension κ will be automatically valid for any smaller dimension. In particular, we can remove any inconvenient prime from the sifting range without losing results. These adjustments must not be performed on a grand scale (or else the resulting bounds, although true, will be rather weak), but rather occasionally to clear technical obstacles. One of the involved problems could be with parameters upon which the sequence $\mathcal{A}$ depends and which distort its asymptotic behaviour, yet by making slight modifications one may achieve an adequate uniformity without altering the condition (5.38).

In practice we can establish by elementary means the following:

$$\prod_{w \leqslant p < z} (1 - g(p))^{-1} \leqslant \left(1 + \frac{L}{\log w}\right)\left(\frac{\log z}{\log w}\right)^{\kappa} \tag{5.41}$$

for any $z > w \geqslant 2$ where L is a positive constant. Note that this inequality implies

$$g(p) \leqslant \frac{L}{L + \log p}$$

and it yields (5.38) with the constant $K = 1 + L/\log 2$. If this K is unacceptably large one can put aside a few primes from the sifting range, say all $p < y$ for some y, and deduce (5.38) with $K = 1 + L/\log y$ from (5.41). Thus (5.38) can be assumed to hold with K close to 1 provided $\mathcal{P}$ contains no small primes. For small primes one can apply exact sifting by means of Möbius inversion (Legendre formula) or else apply some other available sieve.

Another way to secure (5.38) with K close to 1 goes by enlarging slightly the dimension κ. Indeed, the right-hand side of (5.41) is bounded by

$$\left\{1 + \frac{L}{\log y}\left(\frac{\log y}{\log z}\right)^{\varepsilon}\right\}\left(\frac{\log z}{\log w}\right)^{\kappa+\varepsilon}$$

provided $\mathcal{P}$ contains no primes $p < y$. Hence, we have (5.38) with

$$K = 1 + L(\log y)^{\varepsilon-1}(\log z)^{-\varepsilon} \tag{5.42}$$

and $\kappa + \varepsilon$ in place of κ. Since z does not change through the applied arguments and it is usually large, the constant K given by (5.42) is fine, even for $y = 2$ (no preliminary sifting is needed). In particular, if we increase the dimension by $\varepsilon = 1$ we have (5.38) with

$$K = 1 + L(\log z)^{-1}. \tag{5.43}$$

Occasionally, we require the approximation (5.4) also at integers d which are not squarefree. This necessitates some assumption about the multiplicative function $g(d)$ at prime powers. Such an assumption is quite minor in practice, for example, it could take the form

$$\sum_p \sum_{\alpha \geqslant 2} g(p^\alpha) \ll 1 . \tag{5.44}$$

5.6. Sums of the Local Densities

For numerous applications, we derive from (5.38) the following handy estimates.

LEMMA 5.1. *Let h be a continuous, non-negative and non-decreasing function on the segment $[y, z]$. Let g be a multiplicative function with $0 \leqslant g(p) < 1$ such that (5.38) holds for all w with $y \leqslant w \leqslant z$. Then we have the following inequalities:*

$$\sum_{y \leqslant p < z} g(p)h(p)V(p) \leqslant -KV(z) \int_y^z h(w)d\left(\frac{\log z}{\log w}\right)^\kappa + (K-1)h(z)V(z), \tag{5.45}$$

$$\sum_{y \leqslant p < z} g(p)h(p)V(p) \leqslant -V(z) \int_y^z h(w)d\left(\frac{\log z}{\log w}\right)^\kappa + (1-\frac{1}{K})h(z)V(y), \tag{5.46}$$

and

$$\sum_{y \leqslant p < z} g(p)h(p) \leqslant \kappa \int_y^z \frac{h(w)}{w \log w} dw + h(z)\log K. \tag{5.47}$$

By the remarks at the end of the Section 5.5 we can take K to be of the form (5.43) at the expense of increasing the dimension by one so that (5.47) implies

$$\sum_{y \leqslant p < z} g(p)h(p) \leqslant (\kappa + 1) \int_y^z \frac{h(w)}{w \log w} dw + \frac{Lh(z)}{\log z}. \tag{5.48}$$

PROOF. By partial summation the first sum is equal to

$$G(y)h(y) + \int_y^z G(w)dh(w) ,$$

where

$$G(w) = \sum_{w \leqslant p < z} g(p)V(p) = V(w) - V(z) < \left[K\left(\frac{\log z}{\log w}\right)^\kappa - 1\right]V(z) ,$$

by (5.38). Applying this bound and then integrating by parts, we obtain the first inequality. By a slightly different arrangement,

$$G(w) = K^{-1}V(w) - V(z) + \left(1 - K^{-1}\right)V(w) ,$$

so we have

$$G(w) < \left[\left(\frac{\log z}{\log w} \right)^{\kappa} - 1 \right] V(z) + (1 - \frac{1}{K}) V(y) \ .$$

Hence we obtain the second inequality.

To obtain the final part of the lemma we begin by applying the simple inequality $x \leqslant \log(1-x)^{-1}$, from which we find that

$$\sum_{w \leqslant p < z} g(p) \leqslant \log \frac{V(w)}{V(z)} \ .$$

Hence, by (5.38),

$$\sum_{w \leqslant p < z} g(p) \leqslant \kappa \log \left(\frac{\log z}{\log w} \right) + \log K.$$

Therefore, using partial summation we obtain

$$\begin{aligned} \sum_{w \leqslant p < z} g(p) h(p) &= h(y) \sum_{y \leqslant p < z} g(p) + \int_y^z \Big(\sum_{w \leqslant p < z} g(p) \Big) h'(w) dw \\ &\leqslant h(y) \left(\kappa \log \left(\frac{\log z}{\log y} \right) + \log K \right) \\ &\quad + \int_y^z \left(\kappa \log \left(\frac{\log z}{\log w} \right) + \log K \right) h'(w) dw, \end{aligned}$$

from which the result follows. □

5.7. Sieve-Twisted Sums of the Density Function

The main term $V^{\Lambda}(D, z)$ is a sum of the density function twisted by the sieve weights and many of the transformations of $V^{\Lambda}(D, z)$ which we shall perform hold in a more general setting. Let $\Lambda = (\lambda_d)$ be a sequence of real numbers supported on a finite set of squarefree numbers. Let $g(d)$ be a multiplicative function. We consider sums of the type

$$G = \sum_d \lambda_d \, g(d) \ . \tag{5.49}$$

For our purposes we may assume that $g(d)$ is supported on squarefree numbers and

$$0 \leqslant g(p) < 1 \tag{5.50}$$

at every prime p. We may also assume that $g(p)$ is zero for all but finitely many p to avoid having to write restrictions on the summation variables. We shall refer to any such g as a density function.

Associated with the sequence Λ is the convolution sequence $\theta = 1 * \lambda$, that is,

$$\theta_n = \sum_{d|n} \lambda_d \ . \tag{5.51}$$

By Möbius inversion $\lambda = \mu * \theta$, that is,

$$\lambda_d = \sum_{mn = d} \mu(m) \theta_n \ . \tag{5.52}$$

Hence, the sum G can be expressed in terms of the convolution sequence θ as follows:

$$G = \sum_m \sum_n \mu(m)\, g(mn)\, \theta_n \ .$$

By the multiplicativity of g this gives

$$G = \sum_n \theta_n\, g(n) \prod_{p \nmid n} \bigl(1 - g(p)\bigr) \ . \tag{5.53}$$

Let $h(n)$ be the multiplicative function supported on squarefree numbers with

$$h(p) = g(p)/\bigl(1 - g(p)\bigr) \ , \quad \text{so} \quad g(p) = h(p)/\bigl(1 + h(p)\bigr) \ . \tag{5.54}$$

We shall refer to h as a relative density function. We also define $j(d) = h(d)/g(d)$ which is the multiplicative function with local values

$$j(p) = \bigl(1 - g(p)\bigr)^{-1} = 1 + h(p) \ , \tag{5.55}$$

whence $j = 1 * h$. Now, (5.53) can be written as

$$G = VG^* \tag{5.56}$$

where

$$V = \prod_p \bigl(1 - g(p)\bigr) \tag{5.57}$$

and

$$G^* = \sum_n \theta_n\, h(n) \ . \tag{5.58}$$

In sieve theory we shall be dealing with sequences $\Lambda = \Lambda^{\pm} = \bigl(\lambda_d^{\pm}\bigr)$ supported on divisors d of $P(z)$. From the normalization condition $\lambda_1 = \lambda_1^{\pm} = 1$ it follows that

$$\theta_n^{\pm} = 1 \quad \text{for } (n, P(z)) = 1 \ . \tag{5.59}$$

Moreover,

$$\theta_n^{+} \geqslant 0 \quad \text{for all } (n, P(z)) > 1 \ , \tag{5.60}$$

$$\theta_n^{-} \leqslant 0 \quad \text{for all } (n, P(z)) > 1 \ . \tag{5.61}$$

Then, by (5.58) we have $G^* \geqslant 1$ if (5.60) holds and $G^* \leqslant 1$ if (5.61) holds. Therefore, by (5.56) we get

$$\begin{aligned} G &\geqslant V \quad \text{if (5.60) holds,} \\ G &\leqslant V \quad \text{if (5.61) holds.} \end{aligned}$$

The formula (5.53) can be generalized to sums over numbers co-prime with a given number $q \geqslant 1$ (simply re-define g by setting $g(p) = 0$ if $p \mid q$):

$$\begin{aligned} \sum_{(d,q)=1} \lambda_d\, g(d) &= \sum_{(n,q)=1} \theta_n\, g(n) \prod_{p \nmid nq} \bigl(1 - g(p)\bigr) \\ &= \prod_{p \mid q} \bigl(1 - g(p)\bigr)^{-1} \sum_{(n,q)=1} \theta_n\, g(n) \prod_{p \nmid n} \bigl(1 - g(p)\bigr) \ . \end{aligned} \tag{5.62}$$

If we drop the condition $(n, q) = 1$ in the last sum we obtain the following principles of monotonicity of sieves.

MONOTONICITY PRINCIPLE I. *Assuming* (5.59) *and* (5.60) *we have*

$$\sum_{(d,q)=1} \lambda_d^+ g(d) \leqslant j(q) \sum_d \lambda_d^+ g(d) . \tag{5.63}$$

Assuming (5.59) *and* (5.61) *we have*

$$\sum_{(d,q)=1} \lambda_d^- g(d) \geqslant j(q) \sum_d \lambda_d^- g(d) . \tag{5.64}$$

By inclusion-exclusion we deduce from the above monotonicity:

COROLLARY 5.2. *Assuming* (5.59) *and* (5.60) *we have*

$$-\sum_{d\equiv 0(\mathrm{mod}\, p)} \lambda_d^+ g(d) \leqslant h(p) \sum_d \lambda_d^+ g(d) . \tag{5.65}$$

Assuming (5.59) *and* (5.61) *we have*

$$-\sum_{d\equiv 0(\mathrm{mod}\, p)} \lambda_d^- g(d) \geqslant h(p) \sum_d \lambda_d^- g(d) . \tag{5.66}$$

We next develop similar inequalities in the opposite direction. Transferring the first product on the right side of (5.62) to the other side we get

$$\prod_{p_1 | q} (1 - g(p_1)) \sum_{(d,q)=1} \lambda_d g(d) = \sum_{(n,q)=1} \theta_n g(n) \prod_{p_1 \nmid n} (1 - g(p_1)) .$$

Subtracting this equation for $q = p$ prime from that for $q = 1$ we obtain

$$\begin{aligned} g(p) \sum_{(d,p)=1} \lambda_d g(d) + \sum_{d\equiv 0(\mathrm{mod}\, p)} \lambda_d g(d) &= \sum_{n\equiv 0(\mathrm{mod}\, p)} \theta_n g(n) \prod_{p_1 \nmid n} (1 - g(p_1)) \\ &= \frac{g(p)}{g'(p)} \sum_{n\equiv 0(\mathrm{mod}\, p)} \theta_n g'(n) \prod_{p_1 \nmid n} (1 - g'(p_1)) , \end{aligned}$$

where g' is any multiplicative function which agrees with g at all primes except for p and where $0 \leqslant g'(p) < 1$. Let $\Lambda = (\lambda_d)$ be an upper-bound sieve, so $\theta = 1 * \lambda \geqslant 0$. Now, ignoring the congruence $n \equiv 0(p)$ we arrive at

$$g(p) \prod_{p_1 \neq p} (1 - g(p_1)) + \sum_{d\equiv 0(\mathrm{mod}\, p)} \lambda_d g(d) \leqslant \frac{g(p)}{g'(p)} \sum_d \lambda_d g'(d) - \frac{g(p)}{g'(p)} \prod_{p_1} (1 - g(p_1)) .$$

This inequality also holds for a lower-bound sieve but in the reverse direction. Hence, we conclude

COROLLARY 5.3. *Let* $\Lambda^\pm = (\lambda_d^\pm)$ *be an upper-bound and a lower-bound sieve and* $g(d)$ *a density function. Then we have*

$$\sum_{d\equiv 0(\mathrm{mod}\, p)} \lambda_d^+ g(d) \leqslant \frac{g(p)}{g'(p)} \sum_d \lambda_d^+ g'(d) - \frac{h(p)}{g'(p)} V , \tag{5.67}$$

$$\sum_{d\equiv 0(\mathrm{mod}\, p)} \lambda_d^- g(d) \geqslant \frac{g(p)}{g'(p)} \sum_d \lambda_d^- g'(d) - \frac{h(p)}{g'(p)} V , \tag{5.68}$$

where g' *is the modified density function agreeing with* g *at all primes except for* p, *and* V *is the product* (5.57),

Note that, if the density function g satisfies (5.38) then so does the modified density function g', but with the constant $K' = K(1-g(p))/(1-g'(p))$ in place of K. An interesting choice is $g'(p) = \frac{1}{2}(1+g(p))$ which gives $K' = 2K$.

The inequality (5.67) is not as sharp as (5.65), yet it is sufficient for applications, for example in the argument leading to the bound (15.106). The sum on the left-hand side of (5.67) is expected to be negative. Unfortunately, that is not always the case. For the optimal Selberg sieve weights it is true, see (7.51). One can also show it is true for Brun's sieve if the sifting range is relatively small, or when the sum $\sum \lambda_d^+ g'(d)$ is well-approximated by the product $V' = V(1-g'(p))/(1-g(p))$. In this situation the right side of (5.67) is approximately equal to $-h(p)V$ and so is the right side of (5.65).

Sometimes we encounter the sums (5.49) twisted by some weights $w(d)$. If the weights are additive and non-negative, that is, if they have the form

$$w(d) = \sum_{p|d} w(p)\ , \quad \text{with } w(p) \geqslant 0, \tag{5.69}$$

then the above inequalities yield respectively:

COROLLARY 5.4. *We have*

$$-\sum_d \lambda_d^+ g(d)\, w(d) \leqslant \sigma \sum_d \lambda_d^+ g(d)\ , \tag{5.70}$$

$$-\sum_d \lambda_d^- g(d)\, w(d) \geqslant \sigma \sum_d \lambda_d^- g(d)\, , \tag{5.71}$$

where

$$\sigma = \sum_p h(p)\, w(p)\ . \tag{5.72}$$

We shall need this result in Section 15.4.

Typically, the main term in the sequence $\mathcal{A} = (a_n)$ arises from a simple pole of a relevant L-function. However, as already remarked in Section 5.2, it occasionally comes from a double pole or even one of higher multiplicity. In this case the leading term is not a multiplicative function but rather a combination of logarithms, multiplicative functions and additive functions of type (5.69), the last of these resulting from the logarithmic derivative of the L-function in question. In such a case we also need the results in the following form.

COROLLARY 5.5. *Let $\Lambda^{\pm} = (\lambda_d^{\pm})$ be upper-bound, lower-bound sieves of level D. Then*

$$-\sum_d \lambda_d^+ g(d)\log d \leqslant \Big(\sum_{p\leqslant D} h(p)\log p\Big) \sum_d \lambda_d^+ g(d)\ , \tag{5.73}$$

$$-\sum_d \lambda_d^- g(d)\log d \geqslant \Big(\sum_{p\leqslant D} h(p)\log p\Big) \sum_d \lambda_d^- g(d)\ . \tag{5.74}$$

PROOF. Take $w(p) = \log p$ for $p \leqslant D$. □

Sometimes the sieve weights λ_d are designed with a specific density function $g(d)$ in mind. However, when applying these in different places some small changes may occur in $g(d)$. Such changes can be accepted with little loss by the following inequality.

MONOTONICITY PRINCIPLE II. *Let $\Lambda^+ = (\lambda_d^+)$ be an upper-bound sieve of level D. Put*

$$G = \sum_d \lambda_d^+ g(d) , \qquad G' = \sum_d \lambda_d^+ g'(d) ,$$

where $g(d)$, $g'(d)$ are density functions. If $g'(d) \geqslant g(d)$ then,

$$G' \geqslant G \prod_p \Big(\frac{1 - g'(p)}{1 - g(p)}\Big) . \tag{5.75}$$

PROOF. This follows immediately by (5.56) and (5.58). □

The monotonicity principles can be used to nicely handle sieve problems for which the congruence sums $A_d(x)$ satisfy an approximation of type (5.15) without having the extra requirement (5.16). In this scenario the formula (5.22) for the sifted sum becomes

$$S^\Lambda(\mathcal{A}, z) = \sum_{n \leqslant x} a_n \Big(\sum_{\substack{d|P(z)\\(d,n)=1}} \lambda_d\, g(d)\Big) + R(D, z) . \tag{5.76}$$

By (5.63) and (5.64) this yields

$$S^+(\mathcal{A}, z) \leqslant V^+(D, z) X_g + R^+(D, z) \tag{5.77}$$

and

$$S^-(\mathcal{A}, z) \geqslant V^-(D, z) X_g + R^-(D, z) , \tag{5.78}$$

where X_g differs slightly from X by twisting a_n with $j(n)$, precisely

$$X_g = \sum_{n \leqslant x} a_n \prod_{p|n} (1 - g(p))^{-1} , \tag{5.79}$$

and $V^\pm(D, z)$, $R^\pm(D, z)$ are given by the same settings (5.25), (5.26).

Note that we have estimated here the sifted sums $S^\pm(\mathcal{A}, z)$ rather than the sifting function $S(\mathcal{A}, z)$, and this gives upper and lower bounds for $S(\mathcal{A}, z)$ as well. The use of the monotonicity property in this case has cost us almost nothing in the quality of the bounds. The point is that the only change being made was in dropping the coprimality condition $(n, d) = 1$ at terms which were weighted by θ_n. This weight can be thought of as being supported on integers n having no small prime factors so the coprimality condition makes no real difference in practice.

5.8. Composition of Sieves

Let $\Lambda' = (\lambda')$ and $\Lambda'' = (\lambda'')$ be two sieves (by which we mean that $\theta' = 1 * \lambda'$ and $\theta'' = 1 * \lambda''$ do not change sign for $d > 1$), of level D', D'' respectively. Then $\Lambda = (\lambda)$, given by

$$\lambda_d = \sum_{[d_1,d_2]=d} \lambda'_{d_1} \lambda''_{d_2} ,$$

is a sieve of level $D'D''$ which we call the "composition" of the two sieves. We have

$$\theta = 1 * \lambda = \theta'\theta'',$$

so naturally we write $\Lambda = \Lambda'\Lambda''$. Obviously, the composition of an upper-bound sieve with a lower-bound sieve yields a new lower-bound sieve, whereas the composition of two upper-bound sieves or of two lower-bound sieves yields a new upper-bound sieve.

Let g be a density function and let G, G', G'' be the corresponding sums of g weighted respectively by λ, λ', λ'':

$$G = \sum_d \lambda_d g(d)\,, \quad G' = \sum_d \lambda'_d g(d)\,, \quad G'' = \sum_d \lambda''_d g(d)\,.$$

By (5.56), (5.58) we have

$$G = V \sum_n h(n)\, \theta_n = V \sum_n h(n)\, \theta'_n \theta''_n \tag{5.80}$$

where V is the product (5.57) and $h(n)$ is the relative density function; see (5.54).

Recall that P is the sifting range (a finite product of distinct primes from a set $\mathcal{P}$, not necessarily all primes $p < z$ as in (5.11)) and our sieves are supported on the divisors of P. Also the product (5.57) runs over primes in the sifting range. While the sieves $\Lambda = (\lambda)$, $\Lambda' = (\lambda')$, $\Lambda'' = (\lambda'')$ are supported on divisors $d < D$, $d < D'$, $d < D''$ respectively, the convolutions $\theta = 1 * \lambda$, $\theta' = 1 * \lambda'$, $\theta'' = 1 * \lambda''$ do not obey these support restrictions. For this reason one may prefer to see G in terms of the original sieve weights so we provide the following computations:

$$\begin{aligned}
G &= \sum\sum\sum_{abc|P} \lambda'_{ac} \lambda''_{bc} g(abc) \\
&= \sum_{c|P} g(c) \sum_d \mu(d) g^2(d) \Big(\sum_a \lambda'_{acd} g(a)\Big)\Big(\sum_b \lambda''_{bcd} g(b)\Big) \\
&= \sum_m (g * \mu g^2)(m) \Big(\sum_a \lambda'_{am} g(a)\Big)\Big(\sum_b \lambda''_{bm} g(b)\Big) \\
&= \sum_m \frac{g(m)}{j(m)} \Big(\sum_a \lambda'_{am} g(a)\Big)\Big(\sum_b \lambda''_{bm} g(b)\Big)\,.
\end{aligned}$$

We write this in the following form:

$$G = \sum_m \frac{1}{h(m)} \Big(\sum_{d_1 \equiv 0 \,(\mathrm{mod}\, m)} \lambda'_{d_1} g(d_1)\Big)\Big(\sum_{d_2 \equiv 0 \,(\mathrm{mod}\, m)} \lambda''_{d_2} g(d_2)\Big)\,. \tag{5.81}$$

After we have specialized the constructions of the sieves Λ', Λ'', we shall, in Sections 6.3, 7.6 and 7.7, develop further formulas for sieve compositions.

5.9. Reduced Composition of Sieve-Twisted Sums

Let Λ', Λ'' be sequences of real numbers supported on finite sets of squarefree numbers and g', g'' be multiplicative functions satisfying (5.50). In sieve theory we sometimes encounter sums of type

$$G' * G'' = \sum\sum_{(d_1,d_2)=1} \lambda'_{d_1} \lambda''_{d_2}\, g'(d_1)\, g''(d_2)\,. \tag{5.82}$$

Without the coprimality restriction we would just have the product $G'G''$.

A model of the reduced composition $G' * G''$ occurs when two different sieves are applied to two sequences which are linearly dependent. For example, these arise when we consider the problem of estimating the number of solutions to the equation

$ap_1 - bp_2 = c$ in prime numbers p_1, $p_2 \leqslant x$ for fixed positive integers a, b, c. Such a composition will appear in an essential way in Section 14.7 in the proof of the Hyperbolic Prime Number Theorem.

LEMMA 5.6. *We have*

$$G' * G'' = \sum\sum_{(b_1,b_2)=1} \theta'_{b_1} \theta''_{b_2} g'(b_1) g''(b_2) \prod_{p \nmid b_1 b_2} (1 - g'(p) - g''(p)) \ . \tag{5.83}$$

PROOF. Using (5.52) we get

$$\begin{aligned} G' * G'' &= \sum\sum\sum\sum_{(a_1 b_1, a_2 b_2)=1} \mu(a_1 a_2) \theta'_{b_1} \theta''_{b_2} g'(a_1 b_1) g''(a_2 b_2) \\ &= \sum\sum_{(b_1,b_2)=1} \theta'_{b_1} \theta''_{b_2} g'(b_1) g''(b_2) W(b_1 b_2) \end{aligned}$$

where

$$\begin{aligned} W(b) &= \sum_{(a_1 a_2, b)=1} \mu(a_1 a_2) g'(a_1) g''(a_2) \\ &= \sum_{(a_1,b)=1} \mu(a_1) g'(a_1) \prod_{p \nmid a_1 b} (1 - g''(p)) \\ &= \prod_{p \nmid b} (1 - g''(p)) \sum_{(a_1,b)=1} \mu(a_1) g'(a_1) \prod_{p | a_1} (1 - g''(p))^{-1} \\ &= \prod_{p \nmid b} (1 - g''(p)) \prod_{p \nmid b} \Big(1 - \frac{g'(p)}{1 - g''(p)}\Big) = \prod_{p \nmid b} (1 - g'(p) - g''(p)) \ . \end{aligned}$$

This completes the proof. □

Note that at every p we have

$$1 - g' - g'' = (1 - g')(1 - g'')(1 - h'h'') \ ,$$

where (see (5.54))

$$h' = g'(1 - g')^{-1}, \quad h'' = g''(1 - g'')^{-1} \ .$$

Since $|1 - h'h''| \leqslant 1 + h'h''$, we find

$$|W(b)| \leqslant \prod_{p \nmid b} (1 - g'(p)) (1 - g''(p)) (1 + h'h''(p)) \ .$$

Hence,

$$|W(b)| \leqslant CV'V'' \prod_{p | b} (1 - g'(p))^{-1} (1 - g''(p))^{-1} \ , \tag{5.84}$$

where

$$C = \prod_p (1 + h'h''(p)) \tag{5.85}$$

and

$$V' = \prod_p (1 - g'(p)) \ , \quad V'' = \prod_p (1 - g''(p)) \ . \tag{5.86}$$

Inserting (5.84) into (5.83) and dropping the condition $(b_1, b_2) = 1$ we obtain

COROLLARY 5.7. *We have*

$$|G' * G''| \leqslant CV'V''G^*_{12}G^*_{21} \tag{5.87}$$

where

$$G^*_{12} = \sum_b |\theta'_b| h_{12}(d) \ , \quad G^*_{21} = \sum_b |\theta''_b| h_{21}(d) \ , \tag{5.88}$$

and $h_{12}(d)$, $h_{21}(d)$ are the multiplicative functions supported on squarefree numbers such that

$$\begin{aligned} h_{12}(p) &= g'(p)\big(1-g'(p)\big)^{-1}\big(1-g''(p)\big)^{-1} \ , \\ h_{21}(p) &= g''(p)\big(1-g''(p)\big)^{-1}\big(1-g'(p)\big)^{-1} \ . \end{aligned} \tag{5.89}$$

Now assume that Λ', Λ'' are such that $\theta' = 1 * \lambda' \geqslant 0$ and $\theta'' = 1 * \lambda'' \geqslant 0$, hence the absolute values in (5.88) are redundant. Define the corresponding multiplicative functions g_{12}, g_{21} by the formula (5.54), so we have (at primes)

$$g_{12} = h_{12}\big(1+h_{12}\big)^{-1} \ , \quad g_{21} = h_{21}\big(1+h_{21}\big)^{-1} \ . \tag{5.90}$$

Using (5.89), these are (at primes)

$$g_{12} = g'\big(1-g''(1-g')\big)^{-1} \ , \quad g_{21} = g''\big(1-g'(1-g'')\big)^{-1} \ . \tag{5.91}$$

By (5.56) we get

$$\begin{aligned} V_{12}G^*_{12} &= G_{12} = \sum_d \lambda'_d \, g_{12}(d) \ , \\ V_{21}G^*_{21} &= G_{21} = \sum_d \lambda''_d \, g_{21}(d) \ , \end{aligned} \tag{5.92}$$

where

$$V_{12} = \prod_p \big(1 - g_{12}(p)\big) \ , \qquad V_{21} = \prod_p \big(1 - g_{21}(p)\big) \ . \tag{5.93}$$

Inserting (5.92) into (5.88) we deduce

PROPOSITION 5.8. *Let Λ', Λ'' be sequences of real numbers supported on finite sets of squarefree numbers and let g', g'' be density functions. Suppose that $\theta' = 1 * \lambda' \geqslant 0$ and $\theta'' = 1 * \lambda'' \geqslant 0$. We have*

$$|G' * G''| \leqslant BCG_{12}G_{21} \ , \tag{5.94}$$

where

$$\begin{aligned} B = \frac{V'V''}{V_{12}V_{21}} &= \prod_p \frac{\big(1-g'(p)\big)\big(1-g''(p)\big)}{\big(1-g_{12}(p)\big)\big(1-g_{21}(p)\big)} \\ &= \prod_p \big(1 + g'g''(p) + h'h''(p)\big) \end{aligned} \tag{5.95}$$

and C is given by (5.85). We also have $B \leqslant C^2$.

5.10. An Example of a Reduced Composition

Now suppose the density functions g', g'' satisfy the linear sieve conditions, in which case so do the functions g_{12}, g_{21}. In Chapter 11 we shall find upper-bound sieve weights Λ', Λ'' of level D', D'' respectively, for which we accomplish the following estimates:

$$G_{12} \leqslant \big(2e^{\gamma} + o(1)\big)V_{12}\ , \qquad G_{21} \leqslant \big(2e^{\gamma} + o(1)\big)V_{21}\ , \tag{5.96}$$

where V_{12}, V_{21} are the products (5.93) restricted by $p < D'$, $p < D''$, respectively. By (5.92) we obtain the corresponding bound $2e^{2\gamma} + o(1)$ both for G^*_{12} and for G^*_{21}. Hence, (5.87) gives

$$|G' * G''| \leqslant 4CV'V''\big(e^{2\gamma} + o(1)\big) \tag{5.97}$$

where

$$V' = \prod_{p<D'} \big(1 - g'(p)\big) = \big(e^{-\gamma} + o(1)\big)H'\big(\log D'\big)^{-1}\ ,$$

$$V'' = \prod_{p<D''} \big(1 - g''(p)\big) = \big(e^{-\gamma} + o(1)\big)H''\big(\log D''\big)^{-1}\ ,$$

by Mertens' formula, and H', H'' are the products

$$\begin{aligned} H' &= \prod_p \big(1 - g'(p)\big)\Big(1 - \frac{1}{p}\Big)^{-1}\ ,\\ H'' &= \prod_p \big(1 - g''(p)\big)\Big(1 - \frac{1}{p}\Big)^{-1}\ . \end{aligned} \tag{5.98}$$

Inserting these asymptotic formulas in (5.97) we conclude the following result.

THEOREM 5.9. *Let Λ', Λ'' be the upper-bound linear sieve weights of level D', D'' respectively, and let g', g'' be density functions satisfying the linear sieve conditions. Then we have*

$$|G' * G''| \leqslant \big(4 + o(1)\big)\, CH'H''\big(\log D'\big)^{-1}\big(\log D''\big)^{-1}\ , \tag{5.99}$$

where C and H', H'' are given by the infinite products (5.85), (5.98) *respectively, and $o(1)$ means a quantity tending to zero as both D' and D'' tend to infinity.*

CHAPTER 6

Brun's Sieve – The Big Bang

6.1. Brun's Pure Sieve

The sieve of Eratosthenes lay virtually ignored for some two thousand years. Despite the few touches applied by Legendre, the modern subject of sieve methods really begins with Viggo Brun. (The results of Chapter 4, despite the name, are much more recent.) Recall that the Eratosthenes sieve in the form of Legendre's formula suffers seriously from having a huge number of remainder terms which make an error beyond our control. The first successful sieve $\Lambda = (\lambda_d)$, Brun's pure sieve, was achieved by restricting the Möbius function $\mu(d)$ to numbers d having a limited number of prime divisors. Subsequently, Brun discovered some considerably more refined sieve weights which imposed certain restrictions on the size of the prime divisors of d and not just the number of them. Although this first sieve does not give results as strong as the later ones, it was sufficient for Brun [**16**] to prove the first striking application of the theory, to the twin prime problem. In this section we present the Brun pure sieve, not only for historical reasons, but because it is a very fine place to start showing the first ideas and prepare the ground for the far more complicated arguments of the combinatorial sieve.

As in the previous chapter we begin with a sequence $\mathcal{A} = (a_n)$ of non-negative reals and sift by a set of primes $p \in \mathcal{P}$, $p < z$. We denote by $P = P(z)$ the product of these primes. If we terminate the inclusion-exclusion process of Chapter 1 after r steps we have included too much if r is odd and get the upper bound

$$S(\mathcal{A}, z) \leqslant \sum_{\substack{d|P(z) \\ \omega(d)<r}} \mu(d)|\mathcal{A}_d| \ , \tag{6.1}$$

or have excluded too much if r is even and get the lower bound

$$S(\mathcal{A}, z) \geqslant \sum_{\substack{d|P(z) \\ \omega(d)<r}} \mu(d)|\mathcal{A}_d| \ . \tag{6.2}$$

Actually, we can find the missing terms from the identity

$$S(\mathcal{A}, z) = \sum_{\substack{d|P(z) \\ \omega(d)<r}} \mu(d)|\mathcal{A}_d| + (-1)^r \sum_{\substack{d|P(z) \\ \omega(d)=r}} S(\mathcal{A}_d, p(d)) \tag{6.3}$$

where $p(d)$ denotes the least prime divisor of d. This identity is obtained by the repeated application r times of the very useful Buchstab formula

$$S(\mathcal{A}, z) = |\mathcal{A}| - \sum_{p|P(z)} S(\mathcal{A}_p, p) \ , \tag{6.4}$$

which in turn is easily seen by grouping the elements sifted out in accordance with the first prime that removes them. Hence, one derives (6.1) and (6.2) by dropping the terms $S(\mathcal{A}_d, p(d))$ in (6.3), permissible since these are non-negative.

In a reader who has familiarity with the complexity of the multiplicative structure of the integers the formula (6.3) may generate a disbelief of being of any possible use. Recall that the number of integers $d \leqslant x$ having exactly k prime divisors is asymptotically $c_k x(\log\log x)^{k-1}(\log x)^{-1}$, a quantity of different orders of magnitude for distinct k. Hence, there is no possibility to obtain any cancellation in the terms $\mu(d)|\mathcal{A}_d|$ in (6.3) when r is fixed. Brun had the great imagination and courage to proceed along this path despite this natural objection. He noticed that as r increased without bound the sum began to collapse to the expected order.

The general inequalities (6.1), (6.2) show that the sequence $\Lambda = (\lambda_d)$ with

$$\lambda_d = \begin{cases} \mu(d) & \text{if } \omega(d) < r, \\ 0 & \text{if } \omega(d) \geqslant r, \end{cases} \tag{6.5}$$

makes an upper-bound sieve if r is odd or a lower-bound sieve if r is even, that is, the corresponding inequalities (5.21) hold for the summatory functions $\theta = 1 * \Lambda$. One can also see this property directly from the identity

$$\sum_{\substack{d|n \\ \omega(d)<r}} \mu(d) = -(-1)^r \binom{\omega(n)-1}{r-1}, \tag{6.6}$$

which follows by applying (6.3) for the sequence consisting of a single non-zero element at n.

Similar identities and inequalities are satisfied by the expected asymptotic main term $XV(z)$. Recall that

$$V(z) = \sum_{d|P(z)} \mu(d)g(d) = \prod_{p|P(z)} (1 - g(p)).$$

The Buchstab formula becomes

$$V(z) = 1 - \sum_{p|P(z)} g(p)V(p(d)). \tag{6.7}$$

After r iterations we arrive at

$$V(z) = \sum_{\substack{d|P(z) \\ \omega(d)<r}} \mu(d)g(d) + (-1)^r \sum_{\substack{d|P(z) \\ \omega(d)=r}} g(d)V(p(d)). \tag{6.8}$$

Inserting the approximation (1.2) for each $|\mathcal{A}_d|$ in (6.3) and comparing the resulting main term with (6.8) we obtain

$$\begin{aligned} S(\mathcal{A}, z) = XV(z) &+ \sum_{\substack{d|P(z) \\ \omega(d)<r}} \mu(d)r_d(\mathcal{A}) \\ &+ (-1)^r \sum_{\substack{d|P(z) \\ \omega(d)=r}} \{S(\mathcal{A}_d, p(d)) - g(d)XV(p(d))\}. \end{aligned}$$

Applying the trivial estimates

$$0 \leqslant S(\mathcal{A}_d, p(d)) \leqslant |\mathcal{A}_d| = g(d)X + r_d(\mathcal{A})$$

and

$$0 \leqslant g(d)XV(p(d)) \leqslant g(d)X \ ,$$

we get

$$S(\mathcal{A}, z) = XV + \theta X G_r + \theta R_r \tag{6.9}$$

where $V = V(z)$, $|\theta| \leqslant 1$, and

$$G_r = \sum_{\substack{d|P(z)\\ \omega(d)=r}} g(d), \quad R_r = \sum_{\substack{d|P(z)\\ \omega(d)\leqslant r}} |r_d(\mathcal{A})| \ . \tag{6.10}$$

For $r = 1$ we have

$$G = \sum_{p|P(z)} g(p) \leqslant \sum_{p|P(z)} -\log(1 - g(p)) = -\log V.$$

Hence, for any $r \geqslant 1$ we deduce by $r! \geqslant e(r/e)^r$ that

$$G_r \leqslant \frac{G^r}{r!} \leqslant \frac{1}{e}\left(\frac{eG}{r}\right)^r \leqslant \frac{1}{e}\left(\frac{e}{r}|\log V|\right)^r .$$

To make this bound small we choose r sufficiently large. Clearly, we must take $r \geqslant c\big|\log V\big|$, where $c > e$. Let c be the number which solves the equation

$$\left(\frac{c}{e}\right)^c = e, \qquad c = 3.591\ldots \quad . \tag{6.11}$$

This special constant c, probably a transcendental number, appears naturally in various aspects of the combinatorial sieve. Note that for $b \geqslant c$ we have $b^b \geqslant e^{2b-c+1}$. We apply this inequality for $b = r/|\log V|$ getting $G_r \leqslant e^{-r-1}V^{1-c}$. We choose $r = [s]$ where $s \geqslant 1 - c\log V$, finding that

$$G_r \leqslant e^{-s}V^{1-c}. \tag{6.12}$$

Since we are only sifting by the primes of the range $\mathcal{P}$ which are $\leqslant z$, thus every $d|P(z)$ with $\omega(d) \leqslant r$ satisfies $d \leqslant z^r \leqslant z^s$. Hence R_r is bounded by

$$R(\mathcal{A}, D) = \sum_{\substack{d|P(z)\\ d<D}} |r_d(\mathcal{A})| \ , \tag{6.13}$$

with $D = z^s$. By (6.9), (6.12) and (6.13) we obtain

THEOREM 6.1. *Suppose that $\mathcal{A}$ is a sequence of non-negative reals and $\mathcal{P}$ a set of primes. Let $z \geqslant 2$ and $D = z^s$ with $s \geqslant 1 + c|\log V(z)|$. Then, we have*

$$S(\mathcal{A}, z) = XV(z)\{1 + \theta e^{-s}V(z)^{-c}\} + \theta R(\mathcal{A}, D) \tag{6.14}$$

with $|\theta| \leqslant 1$.

REMARKS. In practice $V(z) \gg (\log z)^{-\kappa}$ so our condition is satisfied for $s \geqslant c\kappa \log\log z + \eta(z)$ where $\eta(z) \to \infty$. Ignoring the remainder $R(\mathcal{A}, D)$ we would obtain $S(\mathcal{A}, z) \sim XV(z)$. It is remarkable that such simple arguments yield the asymptotic formula for the sifting function in such a wide range. No wonder that this result might have inspired Brun to the further pursuit of an expansion in the sifting range. We shall undertake this task in Section 6.5 but, in the meanwhile we provide in this section some estimates, more complete yet simple enough so as to dispel the early myth that an understanding of the combinatorial sieve represents a formidable venture.

Suppose the remainder terms satisfy

$$|r_d(\mathcal{A})| \leqslant g(d)d \quad \text{if } d \mid P(z) \ . \tag{6.15}$$

In this case we return to (6.9) and (6.10). The remainder R_r can be estimated as follows:

$$\begin{aligned} R_r &\leqslant \sum_{\substack{d|P(z)\\ \omega(d)\leqslant r}} g(d)d \leqslant \sum_{\substack{d|P(z)\\ \omega(d)\leqslant r}} g(d)z^{\omega(d)} \\ &= \sum_{0\leqslant \ell\leqslant r} G_\ell z^\ell \leqslant \sum_{0\leqslant \ell\leqslant r} \frac{(zG)^\ell}{\ell!} \leqslant A^r e^{zG/A} \end{aligned}$$

for any real numbers $A \geqslant 1$. We take $A = \max(1, zG/r)$, getting

$$R_r \leqslant (ezG/r)^r + e^r \ .$$

Recall that we have $G_r \leqslant e^{-1}(eG/r)^r$. Hence (6.9) gives

$$|S(\mathcal{A}, z) - XV(z)| \leqslant \left(\frac{eG}{r}\right)^r \left(X + z^r\right) + e^r$$

for any integer $r \geqslant 1$. Recall that $G \leqslant |\log V|$. We choose $r = [\log X/\log z]$, so the above bound yields the following result.

COROLLARY 6.2. *Suppose the remainder terms $r_d(\mathcal{A})$ satisfy* (6.15). *Then*

$$|S(\mathcal{A}, z) - XV(z)| \leqslant 2V(z)X(\log X)^{-1} + X^{\frac{3}{4}} \tag{6.16}$$

for $X \geqslant 4$ and z in the range

$$4 \leqslant z \leqslant X^{1/c\log(V^{-1}\log X)} \ . \tag{6.17}$$

Note that $1 \leqslant V^{-1} \leqslant K(\log X)^\kappa$ by (5.38). Hence, (6.16) holds in the range

$$4 \leqslant z \leqslant X^{1/c(\kappa+1)\log(K\log X)} \ . \tag{6.18}$$

Let us apply (6.16) for the sequence of numbers $F(m)$, $1 \leqslant m \leqslant x$, where F is the product of k distinct irreducible polynomials over $\mathbb{Z}$ with positive leading coefficients and having no fixed prime divisor; see Example 1.9 of Chapter 1. This will be a sieve of dimension k. Suppose the sifting range $\mathcal{P}$ is the set of all primes $p < z$. Put

$$\pi_F(x, z) = |\{1 \leqslant m \leqslant x : (F(m), P(z)) = 1\}|. \tag{6.19}$$

By (6.18) we obtain

$$\pi_F(x, z) \asymp x(\log z)^{-k} \tag{6.20}$$

provided $\log z \ll (\log x)(\log\log x)^{-1}$. Hence

COROLLARY 6.3. *The number of integers $1 \leqslant m \leqslant x$ for which each of the irreducible factors of $F(m)$ is a prime satisfies*

$$\pi_F(x) \ll x\left(\frac{\log\log x}{\log x}\right)^k . \tag{6.21}$$

In particular, for $F(m) = m(m-2)$ we get this upper bound with $k = 2$, hence for the number $\pi_2(x)$ of twin primes p, $p-2$ with $p \leqslant x$ which was the target of Brun's original study. He couldn't establish a lower bound for $\pi_2(x)$ (neither have we!), but nevertheless put his upper bound into striking form, stating that the series of reciprocals of twin primes converges. This limit, "Brun's constant", has been computed by Shanks and Wrench [**148**] and Brent [**14**]:

$$\sum_{p,\,p-2 \text{ primes}} \left(\frac{1}{p} + \frac{1}{p-2}\right) = 1.9021602393\ldots\ . \tag{6.22}$$

6.2. Setting up a Sieve by Iterations

We shall refer to a sequence (λ_d) of sieve weights as a "combinatorial" sieve if λ_d takes only the values $\mu(d)$ and zero. In the previous section we considered a combinatorial sieve which takes the Möbius function on integers having a number of prime divisors limited by a certain fixed (although large) parameter r. The parity of r alone determines whether this is a lower-bound or an upper-bound sieve. If r is chosen properly the expected asymptotic formula derived from such a pure sieve is valid in a range much larger than that allowed by the Eratosthenes-Legendre sieve.

Subsequently, Brun [**17**], [**18**] extended his results to still wider ranges by choosing the sequences $\Lambda^+ = (\lambda_d^+)$ and $\Lambda^- = (\lambda_d^-)$ to be the Möbius function truncated to sets of the type

$$\begin{aligned} \mathcal{D}^+ &= \{d = p_1 \ldots p_\ell : p_m < y_m \quad \text{for } m \text{ odd}\}\ , \\ \mathcal{D}^- &= \{d = p_1 \ldots p_\ell : p_m < y_m \quad \text{for } m \text{ even}\}\ , \end{aligned} \tag{6.23}$$

where d is written as the product of distinct primes enumerated in decreasing order,

$$d = p_1 \ldots p_\ell \quad \text{with } p_1 > \cdots > p_\ell. \tag{6.24}$$

By convention both sets $\mathcal{D}^+$ and $\mathcal{D}^-$ contain $d = 1$. Note that these sets are divisor-closed in the sense that if d belongs to the set then so does every divisor of d. Here y_m are suitable "truncation parameters". By the inclusion-exclusion principle it follows that the conditions (5.21) are satisfied no matter how the y_m are chosen.

Brun's construction can be motivated by Buchstab's iterations. The latter start from the following obvious identity:

$$S(\mathcal{A}, z) = |\mathcal{A}| - \sum_{p_1} S(\mathcal{A}_{p_1}, p_1) \tag{6.25}$$

where p_1 runs over the divisors of $P(z)$. Suppose we seek an upper-bound sieve. For large p_1, say $p_1 \geqslant y_1$, the subsequence $\mathcal{A}_{p_1}$ is short and the sifting range is relatively large so we may have nothing better to use than the trivial bound $S(\mathcal{A}_{p_1}, p_1) \geqslant 0$. Therefore, we drop these terms, hoping that not much is lost. We get an upper bound

$$|\mathcal{A}| - \sum_{p_1 < y_1} S(\mathcal{A}_{p_1}, p_1).$$

Next we apply the formula (6.25) for each remaining term $S(\mathcal{A}_{p_1}, p_1)$ getting

$$|\mathcal{A}| - \sum_{p_1 < y_1} |\mathcal{A}_{p_1}| + \sum\sum_{p_2 < p_1 < y_1} S(\mathcal{A}_{p_1 p_2}, p_2).$$

Now the sifting functions $S(\mathcal{A}_{p_1 p_2}, p_2)$ appear with positive sign so we cannot drop any of these. At this point we could use instead an upper bound but nothing good is available so we must apply the Buchstab formula to each and every $S(\mathcal{A}_{p_1p_2}, p_2)$, getting

$$|\mathcal{A}| - \sum_{p_1<y_1} |\mathcal{A}_{p_1}| + \sum\sum_{p_2<p_1<y_1} |\mathcal{A}_{p_1p_2}| - \sum\sum\sum_{p_3<p_2<p_1<y_1} S(\mathcal{A}_{p_1p_2p_3}, p_3).$$

Now again the sifting functions appear with negative sign so we can drop the terms $S(\mathcal{A}_{p_1p_2p_3}, p_3)$. We do so for large p_3, say $p_3 \geqslant y_3$. Continuing this procedure we end up with

$$S(\mathcal{A}, z) \leqslant \sum_{\substack{d|P(z) \\ d\in\mathcal{D}^+}} \mu(d)|\mathcal{A}_d| = S^+(\mathcal{A}, z), \tag{6.26}$$

say. Similarly, we arrive at the lower bound

$$S(\mathcal{A}, z) \geqslant \sum_{\substack{d|P(z) \\ d\in\mathcal{D}^-}} \mu(d)|\mathcal{A}_d| = S^-(\mathcal{A}, z). \tag{6.27}$$

Tracking the iteration steps we can recover the missing terms in the above inequalities. To this end we group divisors of $P(z)$ not in $\mathcal{D}^+$ or $\mathcal{D}^-$ respectively, in accordance with the first failure of the condition $p_m < y_m$, getting sums of type

$$S_n(\mathcal{A}, z) = \sum_{\substack{y_n\leqslant p_n<\cdots<p_1 \\ p_m<y_m, m<n, m\equiv n(2)}} \cdots \sum S(\mathcal{A}_{p_1\cdots p_n}, p_n). \tag{6.28}$$

Putting these back into (6.26) or (6.27), according to the parity of n, we replace these inequalities by two different identities for the same $S(\mathcal{A}, z)$:

$$S(\mathcal{A}, z) = S^+(\mathcal{A}, z) - \sum_{n \text{ odd}} S_n(\mathcal{A}, z), \tag{6.29}$$

$$S(\mathcal{A}, z) = S^-(\mathcal{A}, z) + \sum_{n \text{ even}} S_n(\mathcal{A}, z). \tag{6.30}$$

The same procedure can be applied to the expected main term $A(x)V(z)$ resulting in the formulas

$$V(z) = V^+(D, z) - \sum_{n \text{ odd}} V_n(z), \tag{6.31}$$

$$V(z) = V^-(D, z) + \sum_{n \text{ even}} V_n(z), \tag{6.32}$$

where

$$V^\pm(D, z) = \sum_{\substack{d|P(z) \\ d\in\mathcal{D}^\pm}} \mu(d)g(d), \tag{6.33}$$

$$V_n(z) = \sum_{\substack{y_n\leqslant p_n<\cdots<p_1 \\ p_m<y_m, m<n, m\equiv n(2)}} \cdots \sum g(p_1 \dots p_n)V(p_n). \tag{6.34}$$

By the construction of the sets $\mathcal{D}^{\pm}$ (the exclusion-inclusion process), it is easy to see that the sifted sums $S^{\pm}(\mathcal{A}, z) = S^{\pm}(\mathcal{A}, D, z)$ defined by (6.26), (6.27), can be truncated analogously to (6.3) as follows:

$$S^{\pm}(\mathcal{A}, D, z) = \sum_{\substack{d|P(z) \\ d\in\mathcal{D}^{\pm},\, \omega(d)<r}} \mu(d)|\mathcal{A}_d| \mp \sum_{\substack{d|P(z) \\ d\in\mathcal{D}^{\pm},\, \omega(d)=r}} S^{\pm}(\mathcal{A}_d, D/p(d), p(d)) \;, \tag{6.35}$$

where r is any positive integer, odd or even according to S^+ or S^- respectively.

One can arrange the above identities in a still more general form by applying an arbitrary partition of unity at each step of the iteration in place of the sharp cut at the assigned parameters. This leads us to the following identity.

PROPOSITION 6.4. *Let (ρ_d) be any sequence of real numbers with $\rho_1 = 1$. Put $\lambda_1 = 1, \sigma_1 = 0$ and for $d = p_1 \dots p_\ell$ with $p_1 > \dots > p_\ell$ put*

$$\lambda_d = \mu(d) \prod_{1\leqslant k\leqslant \ell} \rho_{p_1\dots p_k} \;,$$

$$\sigma_d = \mu(d)\,(1 - \rho_{p_1\dots p_\ell}) \prod_{1\leqslant k<\ell} \rho_{p_1\dots p_k} .$$

Then, we have

$$S(\mathcal{A}, z) = \sum_{d|P(z)} \lambda_d|\mathcal{A}_d| + \sum_{d|P(z)} \sigma_d S\left(\mathcal{A}_d, p(d)\right) \;,$$

$$V(z) = \sum_{d|P(z)} \lambda_d g(d) + \sum_{d|P(z)} \sigma_d g(d) V\left(p(d)\right).$$

There is a prospect of making good use of the above identities with diverse coefficients $0 \leqslant \rho_d \leqslant 1$ but we shall restrict ourselves to the two values $\rho_d = 0, 1$ since such a sieve is already capable of a lot. Even with arbitrary ρ_d the sieve weights λ_d derived from Proposition 6.4 cannot be completely general. It is convenient to think of the construction of these weights as taking place over time. They are special in this respect: at any given moment of construction one takes into account the past history but not the future. This temporal aspect of the construction will be exploited in Chapter 11 in our analysis of the relevant systems of differential-difference equations which arise in the search for optimal solutions.

In Chapter 7 we shall construct an upper-bound sieve due to Atle Selberg which is timeless and is based on principles of global optimization. This does not mean that Selberg's sieve produces optimal bounds in every case because it begins within a restricted class of weights. There are numerous advantages to the combinatorial sieves. One of these is that the weights λ_d depend only on the dimension κ and the level D. On the other hand, Selberg's construction does not depend explicitly on κ, but instead, it involves the density function $g(d)$ from the approximation (1.2) so it is somewhat linked to the particular sequence $\mathcal{A}$ being sifted. This feature is sometimes an advantage and sometimes a disadvantage.

6.3. Compositions using Brun's Sieve

Before proceeding in the next section to specific constructions, we pause to note some transformations of a general nature which show sieve techniques in a larger perspective.

Let θ_n^0 denote the "one-detector" symbol. It is at the basis of sieve methods to use this symbol in the convolution form

$$\theta_n^0 = \sum_{d|n} \mu(d) = \begin{cases} 1 & \text{if } n = 1\ , \\ 0 & \text{if } n > 1\ . \end{cases} \tag{6.36}$$

The key starting point is the replacement of the one-detector symbol by another expression of convolution type:

$$\theta_n = \sum_{d|n} \lambda_d\ , \tag{6.37}$$

where λ_d looks like the Möbius function but truncated to $d < D$, with D chosen at will. We shall write θ_n with superscripts $+$, $-$ respectively, to denote the property that $\theta_n^+ \geqslant \theta_n^0$ for every n in case of the upper-bound sieve, or $\theta_n^- \leqslant \theta_n^0$ for every n in case of the lower-bound sieve. In Sections 5.7 and 5.8 we discussed a number of features of the sieve weights $\Lambda = (\lambda_d)$ of level D in abstract terms. We now elaborate on this further in some important specific cases.

Recall that by Möbius inversion

$$\lambda_d = \sum_{ab=d} \mu(a)\theta_b\ . \tag{6.38}$$

If $g(d)$ is a sifting density we defined its relative density function $h(b)$ by

$$h(p) = g(p)\,(1 - g(p))^{-1} \tag{6.39}$$

and we derived the formula

$$V(D, z) = \sum_{d|P(z)} \lambda_d\, g(d) = V(z) \sum_{b|P(z)} h(b)\theta_b\ . \tag{6.40}$$

Now, consider the example of a combinatorial sieve $\Lambda = (\lambda_d)$ which is obtained by truncating the Möbius function as described in Section 6.2, that is, $\lambda_1 = 1$ and $\lambda_d = \mu(d)$ if

$$d = p_1 \dots p_n\ , \quad p_1 > \dots > p_n\ ,$$
$$p_m \leqslant y_m \text{ for all } m \text{ of fixed parity.}$$

Recall that y_m are the truncation parameters which can be quite arbitrary. Then we find, as in (6.31)–(6.34), that for $b > 1$, $b \mid P(z)$,

$$\theta_b = \pm \sum_{n \equiv \nu\ (2)} \ \sideset{}{^\star}\sum_{\substack{ap_1 \dots p_n = b \\ (a, P(p_n)) = 1}} 1\ , \tag{6.41}$$

where $\nu = \frac{1}{2}(1 \pm 1)$ restricts n to a fixed parity (odd for the upper-bound sieve, even for the lower-bound sieve), and $\sum^*$ restricts $p_1 \dots p_n$ by the conditions

$$p_1 > \dots > p_n \geqslant y_n\ , \qquad p_m < y_m \quad \text{if} \quad m < n,\ m \equiv n\ (\operatorname{mod} 2)\ . \tag{6.42}$$

Inserting (6.41) in (6.40) we derive

$$V(D, z) = V(z) \pm \sum_{n \equiv \nu\ (2)} V_n(z)\ , \tag{6.43}$$

where

$$V_n(z) = V(z) \sideset{}{^\star}\sum_{\substack{ap_1\dots p_n|P(z)\\(a,P(p_n))=1}} h(ap_1\dots p_n)$$

$$= V(z) \sum_{p_1\dots p_n|P(z)} h(p_1\dots p_n) \prod_{\substack{p>p_n\\ pp_1\dots p_n|P(z)}} (1+h(p))$$

Recalling (6.39) we find

$$V_n(z) = \sideset{}{^\star}\sum_{p_1\dots p_n|P(z)} g(p_1\dots p_n)V(p_n) \ , \tag{6.44}$$

which agrees with (6.34).

Let (λ_d^+), (λ_d^-) be the upper-bound and lower-bound sieves of level D with truncation parameters y_m and assume that the these parameters do not depend on d. One such possibility, due to Brun, is to take $y_m = D^{\alpha\gamma^m}$ (see (6.53) below) but other choices would be fine as well. One can see from (6.41) that for $n = pm$, $m > 1$, p larger than any prime in m, $pm \mid P(z)$, $z \leqslant D^\alpha$ we have

$$\theta_n^+ = \theta_m^+ - \theta_m^- \ . \tag{6.45}$$

Actually, this also holds for $m = 1$ because $\theta_p^+ = 1 - 1 = 0$. We shall make use of the observation (6.45) in Section 24.7.

Next we proceed to the composition of Brun's sieve. We begin with a general sieve $\Lambda = (\lambda_d)$, either an upper-bound or a lower-bound sieve, say of level D. Suppose $\Lambda^+ = (\lambda_d^+)$ is any upper-bound sieve, say of level D^+. Then one obtains a new sieve $\Lambda' = (\lambda_d')$ by the formula

$$\lambda_d' = \sum_{[d_1,d_2]=d} \lambda_{d_1}\lambda_{d_2}^+ \ , \tag{6.46}$$

which translates into

$$\theta_b' = \theta_b\,\theta_b^+ \ . \tag{6.47}$$

Hence it is natural to write $\Lambda' = \Lambda\Lambda^+$. It is clear that θ_b' has the same sign as θ_b for all b. Therefore, Λ' is a lower-bound sieve or an upper-bound sieve according to Λ being such a sieve. For the new sieve $\Lambda' = (\lambda_d')$, which has level DD^+, the corresponding main term equals

$$V'(DD^+,z) = \sum_{d|P(z)} \lambda_d' g(d) = V(z) \sum_{b|P(z)} h(b)\theta_b\,\theta_b^+ \ . \tag{6.48}$$

More extended combinations of sieves have also proven to be of value and an interesting example which is useful for multi-dimensional problems has been given by J. Brüdern and E. Fouvry in [**15**].

A very basic combination already emerges from the simple lower-bound sieve $\Lambda = (\lambda_d)$ with $\lambda_1 = 1$, $\lambda_p = -1$ and $\lambda_d = 0$ otherwise (which is just the Brun pure sieve for the single step $r = 1$). Note that

$$\theta_b = 1 - \sum_{p|b} 1 = 1 - \nu(b) \ .$$

This primitive sieve, when composed with a strong upper-bound sieve $\Lambda^+ = (\lambda_d^+)$ of level D, can yield a remarkably strong lower-bound sieve $\Lambda' = \Lambda\Lambda^+ = (\lambda'_d)$. We have

$$\lambda'_d = \sum_{[d_1,d_2]=d} \lambda_{d_1}\lambda^+_{d_2} = \lambda^+_d - \sum_{[p,d_2]=d} \lambda^+_{d_2}\ ,$$

so

$$\lambda'_d = \lambda^+_d - \sum_{p|d}\left(\lambda^+_d + \lambda^+_{d/p}\right)\ . \tag{6.49}$$

In this case the sieve has level zD and the main term (6.48) becomes

$$V'(zD,z) = V(z)\sum_{b|P(z)} h(b)\Big(1 - \sum_{p|b} 1\Big)\theta^+_b\ . \tag{6.50}$$

This construction of the lower-bound sieve $\Lambda' = \Lambda\Lambda^+$ will be considered in great detail in connection with the Selberg upper-bound sieve $\Lambda^+ = \Lambda^2$ (see Sections 7.5, 7.6). This is also interesting in connection with a combinatorial upper-bound sieve $\Lambda^+ = (\lambda^+_d)$ of level D. In that case, by (6.41), the main term (6.50) can be arranged as follows:

$$V'(zD,z) = V(z) + V(z)\sum_{n\,\text{odd}}\ \sideset{}{^\star}\sum_{\substack{ap_1\dots p_n|P(z)\\(a,P(p_n))=1}} h(ap_1\dots p_n)\Big(1 - \sum_{p|ap_1\dots p_n} 1\Big)\ .$$

Hence, by the arguments which gave us (6.44), we now get

$$V'(zD,z) = V(z) + \sum_{n\ \text{odd}} V'_n(z) \tag{6.51}$$

with

$$V'_n(z) = \sideset{}{^\star}\sum_{p_1\dots p_n|P(z)} g(p_1\dots p_n)V(p_n)\Bigg\{1 - n - \sum_{\substack{p>p_n\\pp_1\dots p_n|P(z)}} g(p)\Bigg\}\ , \tag{6.52}$$

where, as before, $\sum^*$ restricts $p_1\dots p_n$ by the conditions (6.42).

6.4. Choosing the Truncation Parameters

From now on our combinatorial sieve will depend alone on the truncation parameters y_m in (6.23). Brun tried various parameters; his best choice was

$$y_m = D^{\alpha\gamma^m} \tag{6.53}$$

for suitable constants $0 < \alpha, \gamma < 1$. Note that the parameters (6.53) do not depend on the primes $p_1, \dots, p_m$ but only on the number of these; that is, they are fixed in any given step of the exclusion-inclusion procedure. This feature is sometimes very convenient and we take advantage of it later, in Section 24.7. We shall not pursue the further development of Brun's choices. Excellent presentations of these are given in the books of Gelfond and Linnik [**62**] and of Halberstam and Richert [**79**].

Rather, we give a treatment of a Brun type sieve that follows from the choice

$$y_m = (D/p_1\dots p_m)^{\frac{1}{\beta}} \tag{6.54}$$

where $\beta \geqslant 1$ is a fixed number. Thus, the sieve weights are given by the Möbius function, restricted respectively to the sets

$$(6.55)\quad \begin{aligned} \mathcal{D}^+ &= \{d = p_1 \dots p_n : p_1 > \dots > p_n, p_1 \dots p_m p_m^\beta < D \quad \text{for all } m \text{ odd}\}, \\ \mathcal{D}^- &= \{d = p_1 \dots p_n : p_1 > \dots > p_n, p_1 \dots p_m p_m^\beta < D \quad \text{for all } m \text{ even}\}, \end{aligned}$$

where we make the convention that both sets contain $d = 1$.

We have already given in Section 6.2 a motivation for the introduction of the truncation parameters y_m in the combinatorial sieve in general. It is possible to elaborate on this and give a natural motivation for our particular choice (6.54) based upon the concepts of level of distribution and sifting limit. Recall how in Section 6.2 we were led to the inequalities

$$(6.56)\quad S^-(\mathcal{A}, z) \leqslant S(\mathcal{A}, z) \leqslant S^+(\mathcal{A}, z)$$

by deleting some of the $S(\mathcal{A}_{p_1 \dots p_m}, p_m)$ in the iteration procedure for $S(\mathcal{A}, z)$. Now we ask how to decide which terms to delete? To this end let us assume there is a number $\beta \geqslant 1$ (which depends only on the dimension κ) with the property that a good sieve $\Lambda^-(D)$ of level D is capable of showing that

$$(6.57)\quad S(\mathcal{A}, z) \gg XV(z)$$

for any $\mathcal{A}$ wherein the remainder of level D is under control (see (6.10)) provided the sifting range $\mathcal{P}$ is not too large, namely the sifting variable satisfies $s = \log D / \log z > \beta$. We shall refer to $\beta = \beta(\kappa)$ as the sifting limit for dimension κ. We do not attempt here a precise definition of sifting limit; its intuitive meaning is clear enough to explain our choice (6.54). If the sifting range falls below this threshold, that is, $s < \beta$, then we expect (6.57) to fail at least for some $\mathcal{A}$ and $\mathcal{P}$ of dimension κ.

Now we apply the concept of $\beta(\kappa)$ to the subsequence $\mathcal{A}_{p_1 \dots p_m}$ pretending it is a genuine sifting sequence of the appropriate level $D/p_1 \dots p_m$. Accordingly, we expect no loss by dropping any term $S(\mathcal{A}_{p_1 \dots p_m}, p_m)$ such that

$$(6.58)\quad s_{p_1 \dots p_m} = \frac{\log(D/p_1 \dots p_m)}{\log p_m} < \beta.$$

On the contrary, if the reverse of (6.58) holds we should retain this term since it yields a positive contribution, namely we have

$$S\left(\mathcal{A}_{p_1 \dots p_m}, p_m\right) \gg g(p_1 \dots p_m) V(p_m) X$$

by (6.57) (note that $g(p_1 \dots p_m)X$ is the expected approximation to $|\mathcal{A}_{p_1 \dots p_m}|$). The above reasoning indicates that the truncation parameter y_m which resolves (6.58) is critical. This suggests our choice (6.54).

We shall refer to the combinatorial sieve induced by the parameters (6.54) as the beta-sieve of level D. At first glance one tends to believe that our choice (6.54) must lead to the best possible results by taking the optimal $\beta = \beta(\kappa)$. This turns out to be true in some important cases, however, false for sieve problems of any dimension $\kappa > 1$. Hence the question: what goes wrong with the described heuristic when $\kappa > 1$? One of many possible explanations is that we were allowed to drop sifting functions only at every second step. Another potential weakness, already alluded to in Section 6.2, is the lack of global optimization. The decision made at a given step ignores the possibility of looking ahead to what will be decided in future steps. Whatever the reason, the optimal constructions for sieve problems of larger dimension remain a mystery.

6.5. Fundamental Lemma

Before we proceed, in several subsequent chapters but especially Chapter 11, to a detailed analysis of the beta-sieve, we give in this section a relatively simple treatment. This is already sufficient to produce results which are extremely strong when the sifting variable

$$s = \frac{\log D}{\log z} \tag{6.59}$$

is large. Such a result is called a "Fundamental Lemma" because it is useful in many parts of sieve theory, very often as a preliminary sieve to be followed by more specialized arguments.

Since $\beta \geqslant 1$ the sequences $\Lambda^+ = \{\mu(d) : d \in \mathcal{D}^+\}$ and $\Lambda^- = \{\mu(d) : d \in \mathcal{D}^-\}$ have level of support D except for $\mu(p)$ in λ^-. We control the latter by restricting the range $\mathcal{P}$ to primes $p < z < D$. By (6.26) and (6.27), we obtain

$$\begin{aligned} S(\mathcal{A}, \mathcal{P}, z) &\leqslant S^+(\mathcal{A}, \mathcal{P}, z) = XV^+(D, z) + R^+(\mathcal{A}, \mathcal{P}; D) \ , \\ S(\mathcal{A}, \mathcal{P}, z) &\geqslant S^-(\mathcal{A}, \mathcal{P}, z) = XV^-(D, z) + R^-(\mathcal{A}, \mathcal{P}; D) \ , \end{aligned} \tag{6.60}$$

where $R^+(\mathcal{A}, \mathcal{P}; D)$, $R^-(\mathcal{A}, \mathcal{P}; D)$ are the remainders of level D (see (5.26)) and where $V^+(D, z)$, $V^-(D, z)$ are defined in (5.25) and expressed in terms of $V(z)$ and $V_n(z)$ in (6.31) and (6.32) respectively. We seek an upper bound for $V^+(D, z)$ and a lower bound for $V^-(D, z)$; therefore, in both cases we need only upper bounds for $V_n(z)$.

Recall that

$$V_n(z) = \sum_{\substack{y_n \leqslant p_n < \cdots < p_1 < z \\ p_m < y_m, m < n, m \equiv n (2)}} \cdots \sum g(p_1 \dots p_n) V(p_n). \tag{6.61}$$

Since the terms of $V_n(z)$ are non-negative some of the summation conditions can be ignored. Before we do so, first we derive some inequalities for the graded divisors of the summation variables.

LEMMA 6.5. *Let $d = p_1 \dots p_n$ with $p_1 > \cdots > p_n$ be in the range of $V_n(z)$. Then, for any $1 \leqslant \ell \leqslant n-1$, $\ell \equiv n-1 \pmod 2$, we have*

$$p_1 \dots p_\ell < D z^{-(s-1)\left(\frac{\beta-1}{\beta+1}\right)^{[\ell/2]}} . \tag{6.62}$$

PROOF. By induction on ℓ. For $\ell = 1$ we have $p_1 < z$, a bound which just agrees with (6.62). Let $2 \leqslant \ell \leqslant n-1$, $\ell \equiv n-1 \pmod 2$. The summation condition $p_m < y_m$ in (6.61) for $m = \ell - 1$ gives $p_1 \dots p_{\ell-2} p_{\ell-1}^{\beta+1} < D$. Hence

$$p_1 \dots p_\ell < p_1 \dots p_{\ell-2} p_{\ell-1}^2 < p_1 \dots p_{\ell-2} \big(D / p_1 \dots p_{\ell-2}\big)^{2/(\beta+1)} .$$

Here, using (6.62) for $\ell - 2$, we obtain exactly the bound (6.62) for ℓ. □

Next, applying (6.62) for $\ell = n-1$ to the condition $p_n \geqslant y_n$ in (6.61), one gets

COROLLARY 6.6. *Let $d = p_1 \dots p_n$ with $p_1 > \dots > p_n$ be in the range of $V_n(z)$. Then*

$$p_n \geqslant z^{\delta_n} \quad \text{with} \quad \delta_n = \frac{s-1}{\beta-1}\Big(\frac{\beta-1}{\beta+1}\Big)^{[(n+1)/2]} . \tag{6.63}$$

Assume that $s \geqslant \beta$ if n is even and $s \geqslant \beta + 1$ if n is odd. Then,

$$\delta_n \geqslant \Big(\frac{\beta-1}{\beta+1}\Big)^{n/2} ,$$

regardless of the parity of n. Put

$$z_n = z^{\left(\frac{\beta-1}{\beta+1}\right)^{n/2}} . \tag{6.64}$$

Having the estimate $p_n \geqslant z_n$, we drop the original conditions in (6.61) and proceed to estimate $V_n(z)$ as follows:

$$\begin{aligned} V_n(z) &\leqslant \sum \dots \sum_{z_n \leqslant p_n < \dots < p_1 < z} g(p_1 \dots p_n) V(p_n) \\ &\leqslant \frac{V(z_n)}{n!}\Big(\sum_{z_n \leqslant p < z} g(p)\Big)^n \leqslant \frac{V(z_n)}{n!}\Big(\log \frac{V(z_n)}{V(z)}\Big)^n . \end{aligned}$$

By (5.38) we have

$$\frac{V(z_n)}{V(z)} \leqslant K\Big(\frac{\log z}{\log z_n}\Big)^{\kappa} = Ke^{\alpha n} ,$$

where

$$\alpha = \frac{\kappa}{2} \log \frac{\beta+1}{\beta-1} . \tag{6.65}$$

Hence

$$\frac{V_n(z)}{V(z)} \leqslant \frac{K}{n!}\left(e^{\alpha} \log Ke^{\alpha n}\right)^n .$$

Using the inequality $1 + x \leqslant e^x$ we get

$$\frac{V_n(z)}{V(z)} \leqslant \frac{1}{n!}\left(n\alpha e^{\alpha}\right)^n K^{1+\alpha^{-1}} . \tag{6.66}$$

We need to sum the bound (6.66) over n. Therefore, we must choose β sufficiently large that

$$a = \alpha e^{1+\alpha} < 1 , \tag{6.67}$$

the condition required for convergence. Let $c = 3.591\dots$ be the solution of (6.11), that is,

$$e^{1+c^{-1}} = c . \tag{6.68}$$

Then the condition $a < 1$ means $\alpha < c^{-1}$, or equivalently,

$$\kappa < \frac{2}{c}\Big(\log \frac{\beta+1}{\beta-1}\Big)^{-1} , \tag{6.69}$$

$$\beta > 1 + 2\left(e^{2/c\kappa} - 1\right)^{-1} . \tag{6.70}$$

Now we are ready to sum the bound (6.66) over n. But first notice that $V_n(z)$ vanishes for $n \leqslant s-\beta$ because the condition $y_n \leqslant p_n$ in (6.61) implies that we have $D \leqslant p_1^{\beta+n} < z^{\beta+n}$. For this reason we introduce two incomplete power series:

$$\psi^+(x,s) = \sum_{\substack{n>s \\ n \text{ odd}}} \frac{1}{n!}\Big(\frac{nx}{e}\Big)^n, \tag{6.71}$$

$$\psi^-(x,s) = \sum_{\substack{n>s \\ n \text{ even}}} \frac{1}{n!}\Big(\frac{nx}{e}\Big)^n . \tag{6.72}$$

By (6.66) we conclude the following (recall that $s = \log D/\log z$ and the sums $V^{\pm}(D,z)$ were introduced in (5.25)):

PROPOSITION 6.7. *Let Λ^+, Λ^- be the upper-bound, lower-bound beta-sieve of level D. Then, for any multiplicative function $g(d)$ satisfying* (5.38) *with κ bounded by* (6.69), *we have*

$$V^+(D,z) \leqslant \Big\{1 + \psi^+(a, s-\beta)K^{1+\alpha^{-1}}\Big\} V(z)\ , \quad \text{if } s \geqslant \beta+1\ , \tag{6.73}$$

$$V^-(D,z) \geqslant \Big\{1 - \psi^-(a, s-\beta)K^{1+\alpha^{-1}}\Big\} V(z)\ , \quad \text{if } s \geqslant \beta\ , \tag{6.74}$$

where a, α are given by (6.67) *and* (6.65) .

It remains to estimate the series $\psi^+(x,s)$, $\psi^-(x,s)$. First we do it quickly using the bound $n! \geqslant \frac{1}{2}e^2(n/e)^n$ if $n \geqslant 2$. We obtain

$$\psi^{\pm}(a, s-\beta) < 2e^{-2}\sum_{n>s-\beta}^{\pm} a^n < 2e^{-2}(1-a^2)^{-1}a^{s-\beta} \tag{6.75}$$

where $\sum^+, \sum^-$ mean sums restricted to odd, even numbers. Choosing $\alpha = 1/9$ we find that $a = \alpha e^{1+\alpha} < e^{-1}$ so (6.75) yields

$$\psi^{\pm}(a, s-\beta) < e^{\beta-1-s}\ ,$$

and $\beta < 9\kappa + 1$ by (6.65). Hence Proposition 6.7 implies:

LEMMA 6.8 (Fundamental Lemma). *Let $\Lambda^+(D), \Lambda^-(D)$ be the beta-sieve of level D for $\beta = 9\kappa+1$. Then for any multiplicative function $g(d)$ satisfying* (5.38) *we have*

$$V^+(D,z) \leqslant \big(1 + e^{9\kappa-s}K^{10}\big)\, V(z)\ , \tag{6.76}$$

$$V^-(D,z) \geqslant \big(1 - e^{9\kappa-s}K^{10}\big)\, V(z)\ , \tag{6.77}$$

for all $s \geqslant 9\kappa+1$ (recall that $s = \log D/\log z$).

Note that the Fundamental Lemma 6.8 is a statement about the sequence Λ of sieve weights and the density function g and is independent of the notion of any arithmetic sequence $\mathcal{A}$ to which it is intended to apply. We remark that a result of essentially the same strength can be derived along the above lines using Brun's weights (6.53).

By Lemma 6.8, for a sieve of dimension κ and level D, we obtain the following estimates for the sifting function $S(\mathcal{A}, z)$.

THEOREM 6.9. *Let $\kappa \geqslant 0, z \geqslant 2$ and $D \geqslant z^{9\kappa+1}$. Suppose* (5.38) *holds for all w with $2 \leqslant w < z$ and some $K > 1$. Then*

$$S(\mathcal{A}, z) \leqslant XV(z)\{1 + e^{9\kappa-s}K^{10}\} + R^{+}(\mathcal{A}, D) \ , \tag{6.78}$$

$$S(\mathcal{A}, z) \geqslant XV(z)\{1 - e^{9\kappa-s}K^{10}\} + R^{-}(\mathcal{A}, D) \ , \tag{6.79}$$

where $s = \log D/\log z$ and $R^{\pm}(\mathcal{A}, D)$ are the remainders

$$R^{\pm}(\mathcal{A}, D) = \sum_{\substack{d|P(z)\\ d<D}} \lambda_d^{\pm} r_d(\mathcal{A}) \ .$$

The condition $D \geqslant z^{9\kappa+1}$ in Theorem 6.9 can be replaced by $D \geqslant z$ if we weaken the statements a bit. Precisely, using the non-negativity of $S(\mathcal{A}, z)$ and its monotonicity in z, we derive by (6.78), (6.79) and (5.38):

COROLLARY 6.10. *Let $D \geqslant z \geqslant 2$. Suppose* (5.38) *holds for all w with $2 \leqslant w < z$ and some $\kappa \geqslant 0$, $K > 1$. Then*

$$S(\mathcal{A}, z) = XV(z)\{1 + 4\theta(9\kappa + 1)^{\kappa} e^{9\kappa-s}K^{11}\} + \theta R(\mathcal{A}, D) \tag{6.80}$$

where $s = \log D/\log z$, $|\theta| \leqslant 1$ and

$$R(\mathcal{A}, D) = \sum_{\substack{d|P(z)\\ d<D}} |r_d(\mathcal{A})|.$$

REMARK. By increasing κ to $\kappa+1$ one may, by (5.38), have $K = 1+L(\log z)^{-1}$ which is close to 1 for large z.

From the estimates (6.76), (6.77), one can see that $V^{+}(D, z), V^{-}(D, z)$ tend to $V(z)$ very rapidly as the sifting variable s tends to infinity and the dimension κ is kept fixed. There are situations in practice which require even stronger estimates than (6.76), (6.77) with respect to s. Therefore, we are going to improve these estimates for any beta-sieve which is restricted only by our convergence condition $a < 1$; see (6.67) with α given by (6.65). To this end we modify our treatment of (6.61) as follows.

First, we can assume that $n > s - \beta$, because otherwise the sum $V_n(z)$ is void. Moreover, assume that $s \geqslant \beta + 1$ so $n \geqslant 2$. Put $n = \ell + r$ with $1 \leqslant r \leqslant [s - \beta]$. We have

$$D \leqslant p_1 \dots p_n p_n^{\beta} \leqslant z^{r-1} p_r^{\ell+\beta+1} \leqslant D z^{-\beta} p_r^{\beta+n} \ ,$$

whence

$$p_r \geqslant z^{\beta/(\beta+n)} > z^{1/(\beta+n)} = w_n \ , \quad \text{say} \ .$$

This condition was not detected in our previous way of handling $V_n(z)$, but now we are going to exploit it. Retaining the old condition $p_n \geqslant z_n$ and the new one $p_r \geqslant w_n$ in (6.61) and dropping everything else, we proceed similarly. We have

$$\begin{aligned} V_n(z) &\leqslant \frac{1}{\ell! r!} \Big(\sum_{z_n \leqslant p < z} g(p) \Big)^{\ell} \Big(\sum_{w_n \leqslant p < z} g(p) \Big)^{r} V(z_n) \\ &\leqslant \frac{1}{\ell! r!} \big(\alpha n + \log K\big)^{\ell} \big(\kappa \log(\beta + n) + \log K\big)^{r} K e^{\alpha n} V(z) \ . \end{aligned}$$

Here we estimate $\log(\beta+n)$ crudely as follows:

$$\log(\beta+n) = \log(\beta+r+\ell) \leqslant 2\log(\beta+r+\sqrt{\ell})$$
$$\leqslant 2\log(\beta+r) + \frac{2\sqrt{\ell}}{\beta+r} \leqslant 2\log(\beta+r)e^{\sqrt{\ell}/r} \leqslant 2(\log s)e^{\sqrt{\ell}/r} ,$$

getting

$$V_n(z) \leqslant \frac{e^{\sqrt{\ell}}}{\ell! r!}\left(n\alpha e^{\alpha}\right)^{\ell}\left(2e^{\alpha}\log s^{\kappa}K\right)^{r} K^{\eta} V(z) ,$$

where $\eta = 1+\alpha^{-1}$. Next we estimate the factorials as follows:

$$\ell! r! \geqslant n^{-r} n! r! \geqslant n^{-r}\left(n/e\right)^{n}\left(r/e\right)^{r} = n^{\ell} r^{r} e^{-\ell-2r} \geqslant n^{\ell}(1+\beta+r)^{r} e^{-1-\beta-\ell-2r} ,$$

getting

$$V_n(z) \leqslant e^{1+\beta+\sqrt{\ell}} a^{\ell}\left(2e^{\alpha+2}(1+\beta+r)^{-1}\log s^{\kappa}K\right)^{r} K^{\eta} V(z) .$$

Note that $\alpha < c^{-1} < 2/7$, so $2e^{\alpha+2} < 20$. We choose $r = [s-\beta]$ and obtain

$$V_n(z) \leqslant e^{1+\beta+\sqrt{\ell}} a^{\ell}\left(20 s^{-1}\log s^{\kappa}K\right)^{s-\beta} K^{\eta} V(z) .$$

Finally, we sum over all $\ell \geqslant 0$. The series

$$A = e^{1+\beta}\sum_{\ell\geqslant 0} a^{\ell} e^{\sqrt{\ell}} \tag{6.81}$$

converges because $a < 1$. Hence, we conclude the following explicit estimate:

$$V^{+}(D,z) - V^{-}(D,z) \leqslant V(z)\left(20 s^{-1}\log s^{\kappa}K\right)^{s-\beta} A K^{\eta} , \tag{6.82}$$

for all $s \geqslant \beta+1$. Here η, A are positive constants which depend only on κ and β, specifically they are given by (6.65) and (6.81).

If one does not care about the explicit dependence on κ and β, then the bound (6.82) can be simplified.

LEMMA 6.11 (Fundamental Lemma). *Let $\Lambda^{+}(D), \Lambda^{-}(D)$ be the beta-sieve with β satisfying* (6.70) *and $\kappa > 0$. Then, for any multiplicative function $g(d)$ satisfying* (5.38) *we have*

$$V^{\pm}(D,z) = \left\{1 \pm K^{\eta}\exp\left(-s\log s + s\log\log sK + O(s)\right)\right\} V(z) , \tag{6.83}$$

for $s = \log D/\log z \geqslant \beta+1$, where η and the implied constant depend only on β and κ.

This lemma yields (note that the right-hand side of (6.70) is bounded by $c\kappa+1$):

THEOREM 6.12. *Let $\kappa > 0$. For $s = \log D/\log z \geqslant c\kappa + 2$ (recall that $c = 3.591\ldots$ is defined in* (6.58)*), we have*

$$\begin{aligned} S(\mathcal{A},z) = XV(z)\left\{1+\theta K^{\eta}\exp\left(-s\log s + s\log\log sK + O(s)\right)\right\} \\ + \theta R(\mathcal{A},D) , \end{aligned} \tag{6.84}$$

where $|\theta| \leqslant 1$, while η and the implied constants depend only on κ.

In Chapter 11 we shall develop more precise arguments (differential-difference equations) which yield somewhat sharper results. We could derive a bound almost as strong as (6.83) rather quickly from (6.73)–(6.75) by choosing a sufficiently small (that is, β sufficiently large) in terms of s. However, such a beta-sieve would depend on the variable s which is not a pleasant feature.

6.6. Improved Bounds for the Sifting Limits

In this section, which will be superseded by the results of Chapter 11, we investigate more precisely the series (6.71) and (6.72) to see what estimates are possible with this simple treatment of the combinatorial sieve.

First we modify the inequality (6.75) in two ways as follows:

$$\psi^+(a, s-\beta) < 2e^{-2}(1-a^2)^{-1}a^3 \quad \text{if } s \geqslant \beta + 1\,, \tag{6.85}$$

$$\psi^-(a, s-\beta) < 2e^{-2}(1-a^2)^{-1}a^2 \quad \text{if } s \geqslant \beta\,. \tag{6.86}$$

Now we apply Proposition 6.7 with $\alpha = \frac{1}{4}$, so $a = \frac{1}{4}e^{5/4} < \frac{7}{8}$ getting by (6.65),

$$\beta = \beta_\kappa = 1 + 2\left(e^{1/2\kappa} - 1\right)^{-1}\,. \tag{6.87}$$

By (6.86) we compute that $\psi^- < 2e(16\sqrt{e} - e^3)^{-1} = 0.8637\ldots < \frac{7}{8}$, then by (6.85) we get $\psi^+ < \frac{7}{8}$. Hence we obtain

COROLLARY 6.13. *For $\kappa > 0$ let β_κ be given by* (6.87). *Then we have*

$$V^+(D,z) \leqslant \left\{1 + \tfrac{7}{8}K^5\right\}V(z) \quad \text{if } s \geqslant \beta_\kappa + 1 \tag{6.88}$$

and

$$V^-(D,z) \geqslant \left\{1 - \tfrac{7}{8}K^5\right\}V(z) \quad \text{if } s \geqslant \beta_\kappa\,. \tag{6.89}$$

Note that (6.87) yields $\beta_1 = 4.082\ldots, \beta_2 = 8.041\ldots$ and for any $\kappa > 0$,

$$\beta_\kappa \leqslant 4\kappa + 1\,. \tag{6.90}$$

Actually, this is a strict inequality and it gives an easy to remember bound for the sifting limit, valid in any dimension.

COROLLARY 6.14. *Let $\kappa > 0$. Then, for any density function g satisfying* (5.38) *the sifting limit satisfies $\beta(\kappa) < 4\kappa + 1$, which means that, for z large,*

$$V^-(D,z) \gg V(z) \quad \text{if } D \geqslant z^{4\kappa+1}\,. \tag{6.91}$$

PROOF. This follows by (6.89) provided K is sufficiently close to one. Recall that K is the constant such that (5.38) holds and we can depress its size close to one by choosing a slightly larger value of κ; see (5.42). □

EXERCISE 6.1. Let $0 < \kappa \leqslant \frac{1}{4}$. Take $\alpha = \frac{1}{4}\sqrt{\kappa}$ and show that $\beta < 1 + 4\sqrt{\kappa}$, $a < \sqrt{\kappa}$. Then apply (6.73), (6.74), (6.85), (6.86) to derive the following bounds:

$$V^+(D,z) \leqslant \left(1 + \kappa K^{1+4/\sqrt{\kappa}}\right)V(z) \quad \text{if } s \geqslant 2 + 4\sqrt{\kappa} \tag{6.92}$$

and

$$V^-(D,z) \geqslant \left(1 - \kappa K^{1+4/\sqrt{\kappa}}\right)V(z) \quad \text{if } s \geqslant 1 + 4\sqrt{\kappa}\,. \tag{6.93}$$

Although it is not important for applications, it is still interesting to see the extent to which our bounds (6.87) and (6.90) for the sifting limit can be sharpened by a precise computation of the series (6.71) and (6.72). We want to determine β in terms of κ for which $\psi^-(a) = \psi^-(a,0) = 1$. Recall the relations

$$a = \alpha e^{1+\alpha} \qquad \text{and} \qquad \alpha = \frac{\kappa}{2} \log \frac{\beta+1}{\beta-1} \; ;$$

see (6.67) and (6.65). In other words,

$$\beta = \beta_\kappa = 1 + 2\left(e^{2\alpha/\kappa} - 1\right)^{-1} . \tag{6.94}$$

Our first quick choice of $\alpha = \frac{1}{4}$ gave $\psi^-\left(\alpha e^{1+\alpha}\right) < 1$, so now we must have $\alpha > \frac{1}{4}$. To compute the true value of α from the root of $\psi^-(x) = 1$ we are first going to establish an exact formula for $\psi^-(x)$. We begin with the series

$$\sum_{n=0}^{\infty} \frac{(nx)^n}{n!}$$

which converges absolutely for $|x| < e^{-1}$. Using the contour integral

$$\frac{n^n}{n!} = \frac{1}{2\pi i} \int_{|z|=1} e^{nz} z^{-n-1} dz$$

we find that

$$\sum_{n=0}^{\infty} \frac{(nx)^n}{n!} = \frac{1}{2\pi i} \int_{|z|=1} \left(z - x e^{z}\right)^{-1} dz .$$

Continue to assume $|x| < e^{-1}$. Contracting the circle to the origin we pass a simple pole at $z = u$ where u is the solution to

$$u e^{-u} = x$$

with residue $(1-u)^{-1}$, and we get

$$\sum_{n=0}^{\infty} \frac{(nx)^n}{n!} = \frac{1}{1-u} .$$

Applying this result for $-x$ we obtain

$$\sum_{n=0}^{\infty} \frac{(-nx)^n}{n!} = \frac{1}{1+u'}$$

where u' is the solution to

$$u' e^{u'} = x .$$

Now, adding these two series we arrive at

$$2 \sum_{\substack{n \geqslant 0 \\ n \text{ even}}} \frac{(nx)^n}{n!} = \frac{1}{1-u} + \frac{1}{1+u'} .$$

Finally, applying this for x/e we conclude

LEMMA 6.15. *Let $0 \leqslant x < 1$. Write x in two ways:*

$$x = \alpha e^{1+\alpha} = \alpha' e^{1-\alpha'} . \tag{6.95}$$

Then $0 \leqslant \alpha < \alpha' < 1$ *and*

$$\psi^-(x) = \sum_{\substack{n \geqslant 0 \\ n \text{ even}}} \frac{1}{n!}\Big(\frac{nx}{e}\Big)^n = \frac{1}{2}\Big(\frac{1}{1+\alpha} + \frac{1}{1-\alpha'}\Big) - 1 \, . \tag{6.96}$$

Now we are ready to compute the root of $\psi^-(x) = 1$ and the corresponding numbers α, α'. By (6.96) we get

$$\alpha' = \frac{2+3\alpha}{3+4\alpha} \, .$$

Introducing this to (6.95) we obtain a single equation

$$\alpha \, \frac{3+4\alpha}{2+3\alpha} \, e^{\alpha + (2+3\alpha)/(3+4\alpha)} = 1 \, .$$

Hence α is the zero of the function

$$\alpha + \frac{2+3\alpha}{3+4\alpha} + \log \alpha + \log \frac{3+4\alpha}{2+3\alpha} \, .$$

A numerical computation gives (use the Taylor expansion at $\frac{1}{4}$)

$$\alpha = 0.264904\ldots \, .$$

For this value of α the formula (6.94) gives the sifting limit β_κ somewhat smaller than (6.87) which is based on the quick choice $\alpha = \frac{1}{4}$. In particular, we obtain $\beta_1 = 3.8629\ldots$ and $\beta_2 = 7.5941\ldots$. For any $\kappa > 0$ we derive from (6.94)

$$\beta_\kappa < \alpha^{-1}\kappa + 1 < 3.775\,\kappa + 1 \, , \tag{6.97}$$

which sharpens the earlier bound (6.90).

These improved values for the sifting limit β_κ are deduced above from Proposition 6.7 by an exact computation; however, that proposition itself used simplifications which dropped some of the summation conditions in $V_n(z)$. Later on we shall see that saving these conditions gives a further improvement (the best possible), but for large κ it is a surprisingly small one.

6.7. Preliminary Sieving

One important use of the Fundamental Lemma is to perform a preliminary sieving of a sequence, preparing it for the subsequent application of a more powerful sieve. Because of the asymptotic nature of the Fundamental Lemma, not much is lost in this process. The need for the preliminary sieving comes about in the case of some inadequacy of the stronger sieve in the range of small primes, for example, caused by an erratic behaviour of the density function. For an illustration of this process see Section 14.8. Let $z_0 \leqslant z$ and suppose that the density function $g(d)$ satisfies

$$\prod_{w \leqslant p < z_0} \big(1 - g(p)\big)^{-1} \leqslant K_0 \Big(\frac{\log z_0}{\log w}\Big)^{\kappa_0} \tag{6.98}$$

for any $2 \leqslant w < z_0$ with $K_0 \geqslant 1$, $\kappa_0 \geqslant 0$.

PROPOSITION 6.16. *Let* $D_0 \geqslant z_0$ *and* $s_0 = \log D_0 / \log z_0$. *Let* $\Lambda^\pm = (\lambda_d^\pm)$ *be upper-bound and lower-bound sieves respectively, of level* D, *supported on divisors of* $P(z, z_0) = P(z)/P(z_0)$. *We have*

$$S(\mathcal{A},z) \leqslant XV(z_0)\Big\{\sum_{d|P(z,z_0)} \lambda_d^+ g(d) + O\big(e^{-s_0}V(z_0)/V(z)\big)\Big\}$$
$$+ \sum_{\substack{d_0|P(z_0)\\ d_0\leqslant D_0}} \sum_{\substack{d|P(z,z_0)\\ d\leqslant D}} |\lambda_d^+ r_{d_0 d}(\mathcal{A})| \tag{6.99}$$

and

$$S(\mathcal{A},z) \geqslant XV(z_0)\Big\{\sum_{d|P(z,z_0)} \lambda_d^- g(d) + O\big(e^{-s_0}V(z_0)/V(z)\big)\Big\}$$
$$- \sum_{\substack{d_0|P(z_0)\\ d_0\leqslant D_0}} \sum_{\substack{d|P(z,z_0)\\ d\leqslant D}} |\lambda_d^- r_{d_0 d}(\mathcal{A})| \ , \tag{6.100}$$

where the implied constant depends only on K_0 and κ_0.

Proof. Let $\mathcal{A}^*$ be the subsequence of $\mathcal{A}$ of terms a_n with $(n, P(z_0)) = 1$. Then we have

$$S(\mathcal{A},z) = S(\mathcal{A}^*,z) \leqslant \sum_{d|P(z,z_0)} \lambda_d^+ |\mathcal{A}_d^*| \ .$$

By the Fundamental Lemma, for some $|\theta| \leqslant 1$,

$$|\mathcal{A}_d^*| = \sum_{\substack{n\equiv 0\,(\mathrm{mod}\ d)\\ (n,P(z_0))=1}} a_n = g(d)XV(z_0)\{1+O(e^{-s_0})\}$$
$$+\,\theta \sum_{\substack{d_0|P(z_0)\\ d_0\leqslant D_0}} |r_{d_0 d}(\mathcal{A})| \ .$$

We also have

$$\sum_{d|P(z,z_0)} |\lambda_d^+| g(d) \leqslant \prod_{p|P(z,z_0)} \big(1+g(p)\big) \leqslant V(z_0)/V(z) \ .$$

Introducing these into the above inequality we complete the proof of (6.99). Similar arguments lead to (6.100). □

Whatever second sieve we may apply, possibly of different dimension κ and level D, we shall have bounds of the following type:

$$\sum_{d|P(z,z_0)} \lambda_d^+ g(d) \leqslant \frac{V(z)}{V(z_0)}\{F(s)+\eta(s,z)\}\ , \qquad s \geqslant \beta+1\ , \tag{6.101}$$

$$\sum_{d|P(z,z_0)} \lambda_d^- g(d) \geqslant \frac{V(z)}{V(z_0)}\{f(s)-\eta(s,z)\}\ , \qquad s \geqslant \beta\ , \tag{6.102}$$

where $F(s)$, $f(s)$ are functions of $s = \log D/\log z$ which are non-negative and bounded, while $\eta(s,z)$ is a relatively small error term. Introducing (6.101), (6.102) into (6.99), (6.100) respectively, and using the bound

$$\frac{V(z_0)}{V(z)} \ll \Big(\frac{\log z}{\log z_0}\Big)^{\kappa}\ ,$$

we obtain the following result.

THEOREM 6.17. *Let $z \geqslant z_0 \geqslant 2$. Suppose (6.98) holds in the range $2 \leqslant w < z_0$ with some $\kappa_0 \geqslant 0$, $K_0 \geqslant 1$ and (5.38) holds in the range $z_0 \leqslant w < z$ with some $\kappa \geqslant 0$, $K \geqslant 1$. Let $\Lambda^{\pm} = (\lambda_d^{\pm})$ be upper-bound and lower-bound sieves of level D such that (6.101) and (6.102) hold respectively. Then we have*

$$S(\mathcal{A}, z) \leqslant XV(z)\Big\{F(s) + \eta(s,z) + O\Big(e^{-s_0}\Big(\frac{\log z}{\log z_0}\Big)^{2\kappa}\Big)\Big\} + R(\mathcal{A}, DD_0)\ ,$$

if $s \geqslant \beta + 1$, and

$$S(\mathcal{A}, z) \geqslant XV(z)\Big\{f(s) - \eta(s,z) + O\Big(e^{-s_0}\Big(\frac{\log z}{\log z_0}\Big)^{2\kappa}\Big)\Big\} - R(\mathcal{A}, DD_0)\ ,$$

if $s \geqslant \beta$. Here $s = \log D / \log z$, $s_0 = \log D_0 / \log z_0$ and the implied constants depend on κ, K and κ_0, K_0.

6.8. A Cancellation in the Sieve Weights

The weights λ_d of the beta-sieve of level D satisfy, as already noted, the bound $|\lambda_d| \leqslant 1$ which is trivial, but is also sharp. We shall have need of a variety of bounds for sums of these weights which obtain savings as opposed to what would follow from a trivial application of the bound for the individual λ_d. By far the most important of these is that for the evaluation of the main term

$$\sum_{d \leqslant D} \lambda_d g(d) \asymp V \asymp (\log D)^{-\kappa}$$

which saves a crucial power of the logarithm as compared to the trivial bound V^{-1}.

Another useful bound for the linear sieve weights, which we shall obtain in Chapter 12, describes the support of these weights:

$$\big|\{d \in \mathcal{D}^{\pm};\ d|P(z)\}\big| \ll e^{-s} D (\log D)^{-2}\ ,$$

for $3 \leqslant z \leqslant \sqrt{D}$, which saves two powers of the logarithm as compared to the trivial bound D.

In this section we use arguments related to those developed earlier in this chapter to give a bound of somewhat similar nature which will find application in Section 6.10.

LEMMA 6.18. *Let $2 \leqslant z \leqslant D^{1/(\beta+1)}$, $\mathcal{P}$ a set of primes, and $P = P(z)$ the product of those primes $p \in \mathcal{P}$, $p < z$. Let λ_d denote either the upper-bound or lower-bound beta-sieve weights of level D (see Section 6.5) with $\beta \geqslant 8$. Then, we have*

$$\sum_{d|P} d\Big(\sum_{\substack{m|P \\ m \equiv 0(d)}} \frac{\lambda_m}{m}\Big)^2 \ll \prod_{p|P}\Big(1 - \frac{1}{p}\Big)\ , \tag{6.103}$$

the implied constant depending only on β.

REMARKS. Note that the trivial bound is $(\log D)^3$. Of course the improvement is due to the sign changes of the weights, yet the extent of this cancellation is a little surprising.

The inner sum in (6.103) was estimated in Corollaries 5.2 and 5.3 but only for d prime. Here, in the proof we evaluate it exactly for every $d \mid P$. However, even for d prime, our exact expression is not useful for the same purposes as there.

PROOF. Denote by W the sum on the left-hand side of (6.103) and by V the product on the right-hand side. Writing

$$\lambda_m = \sum_{ab=m} \mu(a)\theta_b \ ,$$

we compute the inner sum of W as follows (remember that $d|P$):

$$\begin{aligned}
\sum_{\substack{m|P\\ m\equiv 0(d)}} \frac{\lambda_m}{m} &= \sum_{\substack{ab|P\\ ab\equiv 0(d)}} \frac{\mu(a)\theta_b}{ab} = \sum_{k\ell=d} \sum_{\substack{ab|P\\ k|a,\, \ell|b}} \frac{\mu(a)\theta_b}{ab} \\
&= \sum_{k\ell=d} \sum_{\substack{b|P\\ \ell|b}} \frac{\theta_b}{b} \frac{\mu(k)}{k} \prod_{\substack{p|P\\ p\nmid bk}} \Big(1-\frac{1}{p}\Big) \\
&= V \sum_{b|P} \frac{\theta_b}{b} \sum_{\ell|(b,d)} \mu\Big(\frac{d}{\ell}\Big)\frac{\ell}{d} \prod_{p|bd}\Big(1-\frac{1}{p}\Big)^{-1} \\
&= V \sum_{b|P} \frac{\theta_b}{b} \frac{\mu(d)}{d} \mu((b,d))\varphi((b,d)) \prod_{p|bd}\Big(1-\frac{1}{p}\Big)^{-1} \\
&= V \frac{\mu(d)}{\varphi(d)} \sum_{b|P} \frac{\theta_b}{\varphi(b)} f((b,d)) \ ,
\end{aligned}$$

where $f(c) = \mu(c)\varphi^2(c)/c$. Hence, we get

$$W = \sum_{d|P} d\Big(\sum_{\substack{m|P\\ m\equiv 0(d)}} \frac{\lambda_m}{m}\Big)^2 = V^2 \sum_{b_1}\sum_{b_2} \frac{\theta_{b_1}\theta_{b_2}}{\varphi(b_1)\varphi(b_2)} \sum_d \frac{\mu(d)}{f(d)} f((b_1,d)) f((b_2,d)) \ .$$

If there is a prime which divides one of b_1, b_2 but not the other, then the corresponding local factor in the inner sum is $1-1=0$. Therefore, this sum vanishes unless $b_1 = b_2$, giving

$$W = V^2 \sum_b \Big(\frac{\theta_b}{\varphi(b)}\Big)^2 \sum_d \frac{\mu(d)}{f(d)} f^2((b,d)) \ .$$

Here the inner sum is equal to the product

$$\begin{aligned}
\prod_{p\nmid b}\Big(1-\frac{1}{f(p)}\Big) \prod_{p|b}(1-f(p)) &= \mu(b)f(b)\prod_p\Big(1-\frac{1}{f(p)}\Big) \\
&= \frac{\varphi^2(b)}{b}\prod_p\Big(1+\frac{p}{(p-1)^2}\Big) \ .
\end{aligned}$$

Finally, we get

$$W = \prod_{p|P}\Big(1-\frac{1}{p}+\frac{1}{p^2}\Big)\sum_{b|P}\frac{\theta_b^2}{b} \ .$$

It remains to estimate the last sum. To this end we use the trivial bound $|\theta_b| \leqslant \tau(b)$ in one of the two factors, which does not lose much since the other factor records that we are, in essence, supported on integers with few prime factors. We also use the fact that for $b \neq 1$, θ_b does not change sign. Therefore,

$$\sum_{b|P} \frac{\theta_b^2}{b} \leqslant 2 + \Big| \sum_{b|P} \frac{\tau(b)}{b} \theta_b \Big| .$$

Now we return from the θ's to the λ's getting

$$\sum_{b|P} \frac{\tau(b)}{b} \theta_b = \sum_{ac|P} \frac{\tau(ac)}{ac} \lambda_c = \sum_{c|P} \frac{\tau(c)}{c} \lambda_c \prod_{\substack{p|P \\ p\nmid c}} \Big(1 + \frac{2}{p}\Big)$$

$$= \prod_{p|P} \Big(1 + \frac{2}{p}\Big) \sum_{c|P} \lambda_c \, g(c) ,$$

where $g(c)$ is the multiplicative function with $g(p) = 2/(p+2)$. This density function satisifies the condition (5.38) of dimension $\kappa = 2$. Therefore, by the relevant bound in Proposition 6.7, we obtain

$$\sum_{c|P} \lambda_c \, g(c) \ll \prod_{p|P} (1 - g(p)) ,$$

because $\beta \geqslant 8$ satisfies (6.69) with $\kappa = 2$. (Note that, somewhat paradoxically, for this application we don't need to know that the lower bound in (6.74) is positive, that is $\psi^- < 1$, but only that ψ^- is bounded.)

This shows that

$$\sum_{b|P} \frac{\tau(b)}{b} \theta_b \ll 1$$

so that

$$\sum_{b|P} \frac{\theta_b^2}{b} \ll 1 ,$$

and finally that $W \ll V$, completing the proof of Lemma 6.18. □

REMARK. In Section 6.10 we shall apply this result to bound the number of primes in a short interval. This is a linear sieve problem and conceivably one would want to take $\kappa = 1$. Choosing $\beta = 8$ and $\kappa = 2$ is certainly allowable because it satisfies our one-sided dimension assumption, but it should lead to weaker results. Although the way we proved the lemma required this more restrictive condition (albeit less restrictive than that given in the earlier proof [**46**]), this was not entirely laziness on our part and it is intrinsic to the problem that there be some such limitation. In some sense, the sieve weight sum (6.103) turns the sieve problem into one of higher dimension. We shall make a closer examination of this phenomenon in Section 7.3.

6.9. Applications of Brun's Sieve

Perhaps the most frequent application of the sieve stems from the fact that, given even a modest level of distribution of $\mathcal{A}$, we get an upper bound for $S(\mathcal{A}, z)$ of the correct order of magnitude in a wide range of z. Furthermore, in a restricted range we get a lower bound of the correct order of magnitude as well. Assuming that

$$R(\mathcal{A}, \mathcal{P}, D) = o(XV(z)) \quad \text{for } D = X^{\alpha - \varepsilon}$$

we obtain

$$S(\mathcal{A}, z) \asymp XV(z) \quad \text{if } z \leqslant X^{\alpha/\beta - 2\varepsilon} \tag{6.104}$$

where $\alpha > 0$ is the exponent of distribution of $\mathcal{A}$ and $\beta = \beta(\kappa)$ is the sifting limit of dimension κ.

For comparison of the strength of the pure sieve with the beta-sieve just described we reconsider the problem of primes represented by polynomials. By the Fundamental Lemma, either Lemma 6.8 or Lemma 6.11, we derive

$$\pi_F(x) \ll x(\log x)^{-k}$$

in place of (6.21), where the implied constant depends on F. In particular, for the number of twin primes we get

$$\pi_2(x) \ll x(\log x)^{-2} \tag{6.105}$$

while it is conjectured that

$$\pi_2(x) \sim 2 \prod_{p>2} \left(1 - (p-1)^{-2}\right) x(\log x)^{-2}. \tag{6.106}$$

Moreover, we now know that (6.20) holds in a wider range, namely

$$\pi_F(x, z) \asymp x(\log z)^{-k} \quad \text{if } x \geqslant z^{4k+1}\ , \tag{6.107}$$

by (6.91). In particular, for $F(m) = m(m-2)$ with $1 \leqslant m \leqslant x$ this shows:

COROLLARY 6.19. *There are infinitely many m such that m and $m-2$ together have at most nine prime divisors.*

By the same method one derives

COROLLARY 6.20. *Every even number sufficiently large can be represented as the sum of two numbers which have together at most nine prime divisors.*

These applications are the original results of Viggo Brun [**17**], though our sieve weights are somewhat different. Had we used more accurate estimates, the number of prime divisors in both problems could be reduced easily from nine to eight. In 1939 V. A. Tartakovski [**152**] reduced it further down to four by applying his own refinement of Brun's sieve.

Another approach to the twin primes problem (introduced by A. Rényi [**137**]) applies the one-dimensional sieve to the numbers $n = m - 2$ where m takes prime values $\leqslant x$ (rather than the two-dimensional sieve as previously applied to the polynomial values $n = m(m-2)$). As a result he shows that there are infinitely many primes p such that $p - 2$ has a bounded number of prime divisors. Later, we shall provide stronger, quantitative versions of this statement.

A similar assertion is established for the Goldbach problem by considering the numbers $N - m$ with m prime, $2 < m < N$. Apart from a slight difference in the sifting range the twin primes and the Goldbach problems are indistinguishable by sieve methods until one reaches the most advanced techniques, after which the question of uniformity in N can be an additional problem.

We next apply the Fundamental Lemma to the sequence $\mathcal{A} = (a_n)$, the characteristic function of an arithmetic progression

$$n \equiv a (\operatorname{mod} q), \quad n \leqslant x\ , \quad (a, q) = 1\ ,$$

and the set $\mathcal{P}$ of primes $p \nmid q$. Suppose that $q \leqslant x \exp(-2\sqrt{\log x})$. Then, we can take $D = D(x,q) = xq^{-1}\exp(-\sqrt{\log x})$ as the level of distribution, the remainder term being bounded by $D(x,q)$. Hence, by (6.80) we obtain

COROLLARY 6.21. *Let $\pi(x,z;q,a)$ be the number of positive integers $n \leqslant x, n \equiv a \,(\mathrm{mod}\, q)$ which have no prime divisors $< z$. For $q \leqslant x\exp(-2\sqrt{\log x})$, $(a,q) = 1$, and $z \leqslant \exp\big((\log x)^{\frac{1}{3}}\big)$, we have*

$$\pi(x,z;q,a) = \frac{x}{q}\prod_{\substack{p<z \\ p\nmid q}}\Big(1-\frac{1}{p}\Big)\Big\{1+O\Big(\exp(-(\log x)^{\frac{1}{6}})\Big)\Big\}\ , \tag{6.108}$$

where the implied constant is absolute.

Next we are going to evaluate the number of solutions of the linear diophantine equation

$$am + h = bn \tag{6.109}$$

in positive integers m, n which have no prime divisors $p < z$ nor any $p|h$. Let $N_{abh}(x,z)$ denote the number of such integers with $am \leqslant x$. We shall assume that the coefficients of (6.109) satisfy

$$a,\ b,\ h > 0, \quad (a,b) = 1, \quad (ab,h) = 1, \quad 2|abh\ . \tag{6.110}$$

To this end we apply the sieve to the sequence $\mathcal{A} = (\alpha_\ell)$ where α_ℓ is the number of representations $mn = \ell$ with m, n satisfying (6.109) and $(m,h) = 1$, $am \leqslant x$. Note that $(\ell,h) = 1$ so the sifting range $\mathcal{P}$ consists of primes $p \nmid h$. Letting

$$P(z) = \prod_{\substack{p<z \\ p\nmid h}} p\ ,$$

we have $N_{abh}(x,z) = S(\mathcal{A},z)$. For every $d|P(z)$ we have (note that m and n are coprime)

$$A_d = \sum_{\substack{d_1d_2=d \\ (d_1,b)=1 \\ (d_2,a)=1}}\ \sum_{\substack{ad_1m+h=bd_2n \\ (m,h)=1 \\ ad_1m\leqslant x}} 1\ .$$

Here, the inner sum is just the number of positive integers $m \leqslant x/ad_1$ coprime with h in the arithmetic progression $m \equiv -h\overline{ad_1}\ (\mathrm{mod}\ bd_2)$, so it is equal to

$$\frac{\varphi(h)}{h}\frac{x}{abd} + O\big(\tau(h)\big)\ .$$

Hence

$$A_d = \frac{\varphi(h)}{h}\frac{x}{ab}\frac{\omega(d)}{d} + O\big(\tau(dh)\big)\ ,$$

where $\omega(d)$ is the number of representations $d_1d_2 = d$ with $(d_1,b) = 1$, $(d_2,a) = 1$. Thus ω is the multiplicative function given by

$$\omega(p) = \begin{cases} 2 & \text{if } p \nmid ab\ , \\ 1 & \text{if } p \mid ab\ . \end{cases} \tag{6.111}$$

By (6.80) we obtain

$$S(\mathcal{A},z) = XV(z)\{1+O(e^{-s})\} + O\big(\tau(h)D(\log z)^2\big) \tag{6.112}$$

with any $D \geqslant z$, where $s = \log D / \log z$, $X = x\varphi(h)/abh$ and

$$V(z) = \prod_{\substack{p<z \\ p\nmid h}} \Big(1 - \frac{\omega(p)}{p}\Big) = \prod_{\substack{p<z \\ p\nmid abh}} \Big(1 - \frac{2}{p}\Big) \prod_{\substack{p<z \\ p\mid ab}} \Big(1 - \frac{1}{p}\Big) . \tag{6.113}$$

Note that the dimension condition (5.38) holds with $\kappa = 2$ and with an absolute constant K, so the implied constants above are absolute. Choosing D somewhat smaller than x/ab, the formula (6.112) provides a pretty good approximation to $N_{abh}(x, z) = S(\mathcal{A}, z)$, particularly so if z is considerably smaller than x/ab in the logarithmic scale. Later on, in Chapter 15 we shall need only an upper bound for $N_{abh}(x, z)$, but one which holds uniformly in a, b, h, z in large segments with respect to x and for that purpose we can simplify (6.112) a bit. Note that

$$V(z) \leqslant \frac{ab}{\varphi(ab)} \Big(\frac{h}{\varphi(h)}\Big)^2 \prod_{2<p<z} \Big(1 - \frac{2}{p}\Big) .$$

Hence

$$XV(z) \leqslant \frac{hx}{\varphi(abh)} \prod_{2<p<z} \Big(1 - \frac{2}{p}\Big)$$

and (6.112) with $D = x/\tau(h)ab(\log z)^4$ yields:

PROPOSITION 6.22. *Let a, b, h be positive integers satisfying* (6.110). *Let $N_{abh}(x, z)$ denote the number of pairs of positive integers m, n with $am \leqslant x$, $(mn, h) = 1$, mn having no prime divisors $< z$, and satisfying $am + h = bn$. For $z \geqslant 2$ and $x \geqslant \tau(h)abz(\log z)^4$, we have*

$$N_{abh}(x, z) \ll \frac{hx}{\varphi(abh)} (\log z)^{-2} , \tag{6.114}$$

where the implied constant is absolute.

Another straightforward yet powerful application of Brun's upper-bound sieve shows that the set $\mathcal{S} = \mathcal{P} + \mathcal{P}$ of integers which are the sum of two primes has positive density. The proof goes as follows:

$$\frac{x^2}{(\log x)^2} \asymp \sum_{n\leqslant x} \Big(\sum_{p_1+p_2=n} 1 \Big) \leqslant \Big(\sum_{\substack{n\leqslant x \\ n\in\mathcal{S}}} 1 \Big)^{\frac{1}{2}} \Big(\sum\sum\sum\sum_{p_1+p_2=p_3+p_4} 1 \Big)^{\frac{1}{2}} .$$

By the sieve, the last sum is bounded by $x^3(\log x)^{-4}$, giving the result.

Combining this with his additive theory, Schnirelmann [**142**] deduced that the set of primes forms an additive basis and hence every sufficiently large number n is the sum of an absolutely fixed number of primes.

6.10. Extremely Short Intervals

The Fundamental Lemma applies of course to the sequence $\mathcal{A} = \big(a_n\big)$, the characteristic function of a short interval $(x - w, x]$. Indeed, the situation is rather similar to, but slightly simpler than, the case of an arithmetic progression, with w playing the role of x/q. We easily obtain a level of distribution $w(\log w)^{-2}$

(although, later in Chapter 13 we shall do somewhat better). If we consider in particular the case of upper bounds we deduce that

$$\pi(x) - \pi(x-w) \ll \frac{w}{\log w} \,. \tag{6.115}$$

The Fundamental Lemma does not give a very good implied constant here but in later chapters we shall do fairly well in this regard.

In the case that $w > x^\varepsilon$ the classical estimate (6.115) gives a result of the correct order of magnitude which still, nearly a century after Brun, seems remarkable, especially in light of our total ignorance of lower bounds as soon as $w < \sqrt{x}$. On the other hand, when w is smaller than a fixed power of x this upper bound also begins to lose its strength so that by the time we take w to be a fixed power of $\log x$ the bound (6.115) becomes $w/\log\log x$. Here, so long as $w \geqslant (\log x)^A$ with A not too small, we might have expected the bound $w/\log x$.

In this section we are going to show that this expected bound holds in a statistical sense, that is, for "most" intervals $\big(y-w, y\big]$ as y varies in a dyadic interval $(x, 2x]$, requiring only that $w/\log x \to \infty$, which are the shortest intervals for which one could reasonably conjecture this to happen.

Let $\mathcal{A} = \mathcal{A}(y)$ denote the set of integers in the interval $\big(y-w, y\big]$. We shall be interested in sifting (by a set of primes $p \in \mathcal{P}$, $p < z$) the intervals of this type as y varies between x and $2x$. For simplicity, we take $w = w(x)$ to depend only on x. In practice w will be near to $\log x$ so, if one preferred $w = w(y)$ to depend on y rather than being fixed by x, this would vary over the interval only by the additive constant $\log 2$ which is insignificant. The main term in our usual approximation formula is just w/d and we may write the remainder term, say $r_d(y)$, as

$$r_d(y) = \psi\left(\frac{y-w}{d}\right) - \psi\left(\frac{y}{d}\right)$$

where $\psi(t) = t - [\,t\,] - \frac{1}{2}$ is the sawtooth function.

We shall obtain an upper bound, on average over y, for the remainder

$$R^\Lambda(\mathcal{A}(y), D) = \sum_{d \leqslant D} \lambda_d r_d(y) \tag{6.116}$$

which allows, for most y, a non-trivial application of the Fundamental Lemma.

Note that, because we are going to take advantage of some cancellation in the remainder, in order to apply such a bound we need a sieve result which does not exhibit the remainder terms with absolute values; that is, we need Theorem 6.9 in the form of (6.78), or correspondingly (6.79), rather than (6.80). This is in contrast to the applications given in the previous section. Herein our cancellation in the remainder comes from the average over y; in other words we are dealing simultaneously with many sequences. In later chapters we shall obtain cancellation within a single sequence and that will require more subtle arguments.

We begin with a harmonic analysis of the remainder terms. For real t, denote as usual $e(t) = e^{2\pi i t}$ and by $\|t\|$, the distance from t to the nearest integer. We then have the following well-known approximation for $\psi(t)$; see for example Huxley [**99**].

LEMMA 6.23. *For $H > 0$ we have*

$$\psi(t) = -\frac{1}{2\pi i} \sum_{0<|h|\leqslant H} \frac{1}{h}\, e(ht) + O\left(\min\Big\{1, \frac{1}{H\|t\|}\Big\}\right) \,.$$

We shall also require a simple divisor sum bound.

LEMMA 6.24. *For τ the usual divisor function, we have*

$$\sum_{k\leqslant 2z}\frac{1}{k}\sum_{b\leqslant z}\tau(b)\tau(b+k)\ll z(\log z)^3 .$$

PROOF. An elementary counting of the solutions to the diophantine equation $x_1x_4-x_2x_3=k$ with $0<x_2x_3\leqslant z$ shows that the bound

$$\sum_{b\leqslant z}\tau(b)\tau(b+k)\ll\sigma_{-1}(k)z(\log z)^2 ,$$

holds uniformly for $k\leqslant 2z$ where $\sigma_{-1}(k)$ denotes the sum of the reciprocals of the divisors of k. After summation over k the result follows. □

Although all we need is a sufficiently good upper bound for the average of the sieve remainder (6.116), with only a little extra work we can give an asymptotic formula for the mean square of a more general sum (see [**47**]).

For $1\leqslant D\leqslant x$ let $\{\lambda_d,\ 1\leqslant d<D\}$ denote any set of reals with

$$\lambda_1=1,\qquad|\lambda_d|\leqslant 1 , \tag{6.117}$$

and let θ_n be the convolution function

$$\theta_n=\sum_{d|n}\lambda_d .$$

Our goal is to evaluate

$$\mathcal{R}=\int_x^{2x}\Big|\sum_{y-w<n\leqslant y}\theta_n-w\sum_{d<D}\frac{\lambda_d}{d}\Big|^2dy . \tag{6.118}$$

Note that in the case of sieve weights this integral is given, in the notation of (6.116), by

$$\mathcal{R}=\int_x^{2x}\big|R^\Lambda(\mathcal{A}(y),D)\big|^2dy .$$

PROPOSITION 6.25. *Let λ_d, $1\leqslant d<D$ satisfy* (6.117). *Then, for $x\geqslant 2$, $w\geqslant 2$, we have*

$$\mathcal{R}=2x\sum_d\gamma_d\Big(\sum_{\substack{m<D\\ m\equiv 0(d)}}\frac{\lambda_m}{m}\Big)^2+O\big(w^2D^2(\log x)^{A+8}+wx(\log x)^{-A}\big) , \tag{6.119}$$

where the coefficients are given by

$$\gamma_d=\sum_{\substack{h\geqslant 1\\ (h,d)=1}}\Big(\frac{d}{\pi h}\sin\frac{\pi hw}{d}\Big)^2 , \tag{6.120}$$

A is any positive number, and the implied constant depends only on A.

PROOF. We use Lemma 6.23 to approximate $\mathcal{R}$ by

$$\mathcal{M}=\frac{1}{\pi^2}\sum\sum_{d_1,d_2}\lambda_{d_1}\lambda_{d_2}\sum\sum_{\substack{0<|h_1|<H_1\\ 0<|h_2|<H_2}}\frac{1}{h_1h_2}\int_x^{2x}E(y)dy , \tag{6.121}$$

where $H_1 = d_1(\log x)^B$, $H_2 = d_2(\log x)^B$ for some $B > 0$, and

$$E(y) = \frac{1}{4}\Big\{e\Big(\frac{h_1}{d_1}y\Big) - e\Big(\frac{h_1}{d_1}(y-w)\Big)\Big\}\Big\{e\Big(-\frac{h_2}{d_2}y\Big) - e\Big(-\frac{h_2}{d_2}(y-w)\Big)\Big\}$$
$$= e\Big(\big(\frac{h_1}{d_1} - \frac{h_2}{d_2}\big)\big(y - \frac{w}{2}\big)\Big)\sin\Big(\frac{\pi h_1 w}{d_1}\Big)\sin\Big(\frac{\pi h_2 w}{d_2}\Big)\ . \tag{6.122}$$

We shall show that

$$\mathcal{R} = \mathcal{M} + O\big(wx(\log x)^{5-B}\big)\ . \tag{6.123}$$

For each of $j = 1, 2$, for given $d_j < D$ and $1 \leqslant m_j < H_j$ define $I(d_j, m_j)$ to be the set of y in $(x, 2x)$ such that either of the two conditions

$$\frac{m_j - 1}{H_j} \leqslant \Big\|\frac{y}{d_j}\Big\| < \frac{m_j}{H_j}\ , \qquad \frac{m_j - 1}{H_j} \leqslant \Big\|\frac{y-w}{d_j}\Big\| < \frac{m_j}{H_j}\ ,$$

hold and let $\mu(d_1, m_1; d_2, m_2)$ denote the measure of the intersection of $I(d_1, m_1)$ and $I(d_2, m_2)$. From Lemma 6.23 it follows that

$$\mathcal{R} = \mathcal{M} + O(T + U) \tag{6.124}$$

where

$$T = \sum_{d_1,d_2<D}\sum \ \sum_{\substack{m_1<H_1\\ m_2<H_2}}\sum \frac{1}{m_1 m_2}\,\mu(d_1, m_1; d_2, m_2)\ ,$$

and, choosing one of two similar sums,

$$U = \sum_{d_1,\, d_2<D}\sum \int_x^{2x} |r_y(d_1)|\Big(1 + H_2\Big\|\frac{y}{d_2}\Big\|\Big)^{-1} dy\ ,$$

with

$$r_y(d) = \sum_{\substack{y-w<n\leqslant y\\ n\equiv 0 \,(\mathrm{mod}\, d)}} \theta_n - w\,\frac{\lambda_d}{d}\ .$$

We first estimate T. If y is in the intersection of $I(d_1, m_1)$ and $I(d_2, m_2)$, then for $j = 1, 2$, there exist integers N_j such that

$$(\log x)^{-B}(m_j - 1) \leqslant |y - N_j d_j| \leqslant (\log x)^{-B} m_j$$

or similar inequalities where, in one or both, y is replaced by $y - w$. In all cases this forces $[y]$ into one of two residue classes modulo each d_j and hence one of at most four residue classes modulo the least common multiple of d_1 and d_2. Thus

$$T \ll x(\log x)^{2-B}\sum_{d_1,d_2<D}\sum \frac{(d_1, d_2)}{d_1 d_2}$$
$$= x(\log x)^{2-B}\sum_{\delta\leqslant D}\varphi(\delta)\Big(\sum_{\substack{d<D\\ d\equiv 0\,(\mathrm{mod}\,\delta)}}\frac{1}{d}\Big)^2 \ll x(\log x)^{5-B}\ . \tag{6.125}$$

To estimate U we define $J(d_1)$ to be the set of y in $(x, 2x)$ for which $\big\|\frac{y}{d_1}\big\| < \frac{w}{d_1}$. Note that the estimate $r_y(d_1) \ll 1$ holds for all y in $(x, 2x)$ and that the estimate $r_y(d_1) \ll w/d_1$ holds except on $J(d_1)$. It follows that $U \ll V + W$ with

$$V = w\,(\log x)\int_x^{2x}\sum_{d_2<D}\Big(1 + H_2\Big\|\frac{y}{d_2}\Big\|\Big)^{-1} dy$$

and

$$W = \sum\sum_{d_1, d_2 < D} \sum_{m_2 < H_2} \frac{1}{m_2} \mu(d_1; d_2, m_2) ,$$

where $\mu(d_1; d_2, m_2)$ denotes the measure of the intersection of $J(d_1)$ with $I(d_2, m_2)$.

Now,

$$V \ll w (\log x) \sum_{d<D} \sum_{m<H_2} \frac{1}{m} \frac{x}{d(\log x)^B} \ll wx(\log x)^{3-B} . \tag{6.126}$$

To estimate W we note that if y is in $J(d_1)$ then $[y]$ must be one of $\ll w$ residue classes modulo d_1 and if y is in $I(d_2, m_2)$ it is in one of two classes modulo d_2. Any y in this intersection thus has $[y]$ restricted to $\ll w$ classes modulo the least common multiple $[d_1, d_2]$, and so

$$W \ll wx(\log x)^{1-B} \sum\sum_{d_1, d_2 < D} \frac{(d_1, d_2)}{d_1 d_2} \ll wx(\log x)^{4-B} . \tag{6.127}$$

The combination of (6.124) to (6.127) completes the verification of (6.123).

We now turn to the evaluation of $\mathcal{M}$, defined in (6.121). We write $\mathcal{M} = \mathcal{M}_1 + \mathcal{M}_2$ where $\mathcal{M}_1$ denotes the sum over the diagonal terms, that is, those for which $h_1/d_1 = h_2/d_2$, and $\mathcal{M}_2$ is the sum over the remaining terms. For the latter sum we use the estimates $|\lambda_d| \leqslant 1$, $|\sin t| \ll t$ to deduce that

$$\mathcal{M}_2 \ll w^2 \sum\sum\sum\sum \frac{1}{h_1 d_2 - h_2 d_1}$$

where the sum runs over $d_1, d_2 < D$, $0 < |h_1| < H_1, 0 < |h_2| < H_2$ subject to $h_1 d_2 > h_2 d_1$. By Lemma 6.24 we conclude that

$$\mathcal{M}_2 \ll w^2 \sum_{k < 2D^2 (\log x)^B} \frac{1}{k} \sum_{b < D^2 (\log x)^B} \tau(b) \tau(b+k) \ll w^2 D^2 (\log x)^{3+B} . \tag{6.128}$$

The diagonal terms give the main contribution. For these we write $h_1/d_1 = h_2/d_2 = h/d$ where $(h, d) = 1$. Then

$$\mathcal{M}_1 = \frac{2x}{\pi^2} \sum_d d^2 \sum_{\substack{1 \leqslant h \leqslant d(\log x)^B \\ (h,d)=1}} \frac{1}{h^2} \sin^2\Big(\frac{\pi h w}{d}\Big) \Big(\sum_{\substack{m<D \\ m \equiv 0 \, (\mathrm{mod}\, d)}} \frac{\lambda_m}{m} \Big)^2 . \tag{6.129}$$

Here, we extend the summation to all $(h, d) = 1$ estimating the tail of the series by

$$\sum_d d^2 \sum_{h > d(\log x)^B} \frac{1}{h^2} \Big(\sum_{\substack{m<D \\ m \equiv 0 \, (\mathrm{mod}\, d)}} \frac{1}{m} \Big)^2 \ll (\log x)^{3-B} . \tag{6.130}$$

Proposition 6.25 follows from (6.123) and (6.128)–(6.130) with $B = A + 5$. □

Now we specialize the sequence $\Lambda = (\lambda_d)$ to be either an upper-bound or lower-bound beta-sieve with $\beta = 8$, of level D, and ranging over all primes $p < z$. By $\sin t \ll \min\{1, t\}$ we deduce that $\gamma_d \ll dw$ and hence Lemma 6.18 allows us to bound the main term in Proposition 6.25. Choosing $A = 1$ and $D = x^{1/2}/w^{1/2}(\log x)^5$ we obtain for these special sieve weights:

PROPOSITION 6.26. *Let* $2 \leqslant w, z \leqslant x^{1/20}$. *Then*

$$\int_x^{2x} \Big| \sum_{y-w<n\leqslant y} \theta_n - w \sum_d \frac{\lambda_d}{d} \Big|^2 dy \ll \frac{wx}{\log z}\,, \tag{6.131}$$

where the implied constant is absolute.

COROLLARY 6.27. *Let* $\mathcal{E}(x,w,z)$ *be the subset of* $y \in [x,2x]$ *for which*

$$\Big| \sum_{y-w<n\leqslant y} \theta_n - w \sum_d \frac{\lambda_d}{d} \Big| \geqslant \frac{\delta w}{\log z}\,,$$

where $0 < \delta \leqslant 1$. *Then*

$$|\mathcal{E}(x,w,z)| \ll \frac{x \log z}{\delta^2 w}\,,$$

where the implied constant is absolute.

In particular, for $w = w(x,z) = \eta(x)\log z$ *with* $\eta(x) \to \infty$ *as* $x \to \infty$ *and* $2 \leqslant z \leqslant x^{1/20}$,

$$\sum_{y-w<n\leqslant y} \theta_n = w \sum_d \frac{\lambda_d}{d} + O\Big(\frac{\delta w}{\log z}\Big)$$

for all y *with* $x < y \leqslant 2x$, *apart from a set of measure* $O\big(x/\delta^2\eta(x)\big)$.

Now we are ready to estimate the sifted sum

$$S(y,w,z) = \sum_{\substack{y-w<n\leqslant y\\ (n,P(z))=1}} 1\,. \tag{6.132}$$

By the Fundamental Lemma in the form of Corollaries 6.13 and 6.14 we find that $\sum_d \lambda_d/d \asymp (\log z)^{-1}$ since, in our situation, $\kappa = 1$ and $\beta \geqslant 5$. Moreover, we have

$$S(y,w,z) \leqslant \sum_{y-w<n\leqslant y} \theta_n\,,$$

or the opposite inequality, according to whether $\Lambda = (\lambda_d)$ is an upper-bound or lower-bound sieve. Hence we obtain

COROLLARY 6.28. *Let* $2 \leqslant z \leqslant x^{1/20}$ *and* $w = \eta(x)\log z$ *with* $\eta(x) \to \infty$ *as* $x \to \infty$. *Then*

$$S(y,w,z) \asymp \frac{w}{\log z} \tag{6.133}$$

for all y *with* $x < y < 2x$ *except for a set of measure* $O\big(x/\eta(x)\big)$, *thus for almost all* y.

Taking $z = x^{1/20}$ we infer that almost all intervals $(y-w, y]$, of length $w = \eta(x)\log x$ with $\eta(x) \to \infty$ as $x \to \infty$, contain integers having at most 19 prime factors. Moreover, we deduce that

$$\pi(y) - \pi(y-w) \ll \frac{w}{\log x}\,, \tag{6.134}$$

for almost all y with $x < y < 2x$.

Note that the intervals encompassed by this result are as short as one could reasonably expect to contain about the right order of primes. Without the condition $\eta(x) \to \infty$ one would certainly expect many such intervals to not have any primes whatsoever and even for the longer intervals of length up to about $(\log x)^2$ a well-known conjecture of Cramér predicts the existence of some intervals without any primes.

Also, in the case of the upper bound, the statistical estimate (6.134) should not be expected to hold for "all" intervals once the length becomes shorter than about $(\log x)^2$. Although these lengths are much too short for rigorous results such a bound would violate the probabilistic model by the following heuristic argument shown to us by A. Granville.

By this model one predicts that the probability that the interval $(y - \eta \log x, y]$ with $x < y \leqslant 2x$ contains k primes is about $e^{-\eta}\eta^k/k!$. If we reckon that there are, say, $x^{1+o(1)}$ such intervals and the extremal value for k should occur very rarely, then this leads to the approximation

$$e^{-\eta}\frac{\eta^k}{k!}\, x^{1+o(1)} = 1 \ .$$

Taking logarithms, we infer by Stirling's formula that $\log x \sim \eta + k\log(k/e\eta)$. If the interval length is small compared to $(\log x)^2$, more precisely $\eta = o(\log x)$, then this yields

$$k \sim (\log x)/\log\big(\frac{\log x}{\eta}\big) \ .$$

On the other hand, the Fundamental Lemma (6.115) yields the upper bound

$$k \ll (\eta \log x)/\log(\eta \log x)$$

which is essentially the same as the above asymptotic provided that $\eta(x) \to \infty$ very slowly. This means that, if these heuristics can be believed, the sieve upper bound, having lost the correct order of magnitude when the interval lengths become smaller than a fixed power of x, begins to approach the correct order of magnitude much later when the length $\eta(x)\log x$ steps below about $(\log x)^2$, and comes arbitrarily close to the right order as $\eta(x) \to \infty$ more and more slowly. Similar predictions follow from the assumption of the prime k-tuple conjecture with extremely wide uniformity which can be deduced from heuristic assumptions about the randomness of the Möbius function.

Corollary 6.28 illustrates an important sieve principle. The sieve problem considered is a linear problem, that is, $\kappa = 1$. As we shall see in Chapter 11, the optimal sieve bounds in this case are given by making the choice $\beta = 2$. However, for the problem considered here we require Lemma 6.18 which specifically postulates that $\beta \geqslant 8$. Hence, for this application it is very convenient that, given any sieve problem of dimension κ, a set of sieve weights more naturally associated to a sieve of dimension larger than κ will also serve. By this we mean that the upper bound (and, where positive, the lower bound) will still give results of the correct order of magnitude.

Later, in Chapter 11 we improve the linear sieve theory to the extent that we can claim Corollary 6.28 with $z = x^{(1-\varepsilon)/4}$, hence showing that almost all intervals $(y - w, y]$ of length $w = \eta(x)\log x$ with $\eta(x) \to \infty$ contain integers having at most four prime factors (and that the implied constant in the upper bound (6.134) can be taken to be $4 + o(1)$). One can go further by employing deep estimates for general bilinear forms in exponentials with Kloosterman fractions (see [**37**]). This allows us to choose D a little bit larger than $\sqrt{x/w}$ and consequently we can claim Corollary 6.28 with $z = x^{1/4+\delta}$ for some small $\delta > 0$. Hence we can conclude that almost all intervals $(y-w,y]$ contain, in correct order of magnitude, integers free from prime divisors $p<z$, and so having at most three prime divisors. It would be interesting to get integers with at most two prime divisors.

The fact that we can reach, in this statistical sense, intervals of the shortest possible size leads to another consequence. If one considers the set of primes p_n it seems reasonable to conjecture that, for fixed $\alpha \geqslant 0$,

$$\sum_{x<p_n<2x} (p_{n+1}-p_n)^\alpha \ll x(\log x)^{\alpha-1} ,$$

which would hold if the primes were roughly equally spaced and would be optimal. This bound does hold trivially by Hölder's inequality for $0 \leqslant \alpha \leqslant 1$ but is not known for any $\alpha>1$, even under assumption of the Riemann Hypothesis.

If we consider instead of primes the sequence of integers q_n with $(q_n, P(z))=1$ then, for say $z=x^{1/200}$, the same results (still optimal) can be reached for larger α. We first deduce from Proposition 6.26 the following bound.

PROPOSITION 6.29. *For $w \geqslant 1$ we have*

$$\sum_{\substack{x<q_n\leqslant 2x\\ q_{n+1}-q_n\geqslant w}} (q_{n+1}-q_n) \ll w^{-1}x\log x . \tag{6.135}$$

PROOF. For large w, say $w>x^{1/20}$, it follows from our Fundamental Lemma that the left-hand side vanishes, so we can assume to the contrary that $w \leqslant x^{1/20}$ as required by Proposition 6.26. Supposing that $x<q<q'\leqslant 2x$ are consecutive integers in our set such that $q'-q \geqslant w$ (we can ignore with an acceptable error the last gap for which we might have $q'>2x$), then for the lower bound sieve $\theta_m \leqslant 0$ for integers in between q and q' and hence

$$\int_q^{q'-\frac12 w}\Big|\sum_{y<m<y+\frac12 w}\theta_m - \frac12 w\sum_d \frac{\lambda_d}{d}\Big|^2 dy \gg \int_q^{q'-\frac12 w}\Big|\frac{w}{\log x}\Big|^2 dy \gg \frac{w^2}{(\log x)^2}(q'-q) .$$

Summing over those gaps $q'-q$ which exceed w, using the disjointness of the intervals of integration and applying Proposition 6.26, we find that

$$\frac{w^2}{(\log x)^2}\sum_{\substack{x<q_n\leqslant 2x\\ q_{n+1}-q_n\geqslant w}} (q_{n+1}-q_n) \ll \frac{wx}{\log x} ,$$

giving the result. □

COROLLARY 6.30. *Let q_n run through the integers $x<q_n<2x$ having no prime divisor less than $z=x^{1/200}$. Then, for $0<\alpha<2$ we have*

$$\sum_{x<q_n<2x} (q_{n+1}-q_n)^\alpha \ll x(\log x)^{\alpha-1} ,$$

where the implied constant depends on α.

PROOF. The cases $\alpha=0$ and $\alpha=1$ are trivial and the intermediate range follows from these by Hölder's inequality, so we can assume $\alpha>1$.

We treat trivially the contribution from the smaller than average gaps, namely,

$$\sum_{\substack{x<q_n\leqslant 2x\\ q_{n+1}-q_n\leqslant \log x}} (q_{n+1}-q_n)^\alpha \ll \frac{x}{\log x}(\log x)^\alpha \ll x(\log x)^{\alpha-1} .$$

For the larger gaps we write

$$\sum_{\substack{x<q_n\leqslant 2x\\ q_{n+1}-q_n\geqslant\log x}} (q_{n+1}-q_n)^\alpha = -\int_{\log x}^{x} w^{\alpha-1}dM(w)$$

where $M(w)$ is the sum occurring in (6.135) and the required bound follows by partial summation from Proposition 6.29.

□

Of course, as with our earlier results in this section, the exponent $1/200$ can be substantially improved.

CHAPTER 7

Selberg's Sieve – Kvadrater er Positive

7.1. General Upper-Bound Results

A powerful and elegant method for an upper-bound sieve originated with Atle Selberg (for his most detailed description of this see [**145**]). Selberg's method yields results of great generality; it is simpler than combinatorial sieves at the outset, although equally complex in its advanced forms.

Recall that an upper-bound sieve of level D is a sequence of real numbers λ_d for $d < D$ with $\lambda_1 = 1$ such that

$$\sum_{d|n} \lambda_d \geqslant 0 \qquad \text{for all } n \in \mathcal{N}. \tag{7.1}$$

(Hereafter, in this section, the superscript + is omitted for notational simplicity.) This positivity condition was quite difficult to get a hold of in the combinatorial sieve. By contrast, Selberg made it very easy by choosing λ_d such that

$$\sum_{d|n} \lambda_d = \Big(\sum_{d|n} \rho_d\Big)^2 \tag{7.2}$$

where (ρ_d) is another sequence of real numbers, satisfying

$$\rho_1 = 1. \tag{7.3}$$

Since the squares are non-negative such a choice guarantees (7.1) no matter what the ρ_d are! Sometimes we shall refer to these as the sieve weight constituents.

Although not apparent at first glance, Selberg's choice is of the form required for a sieve since it amounts to taking

$$\lambda_d = \sum_{[d_1,d_2]=d} \rho_{d_1}\rho_{d_2}. \tag{7.4}$$

In order to control the level we assume the coefficients ρ_d are supported on integers smaller than $\sqrt{D}$,

$$\rho_d = 0 \qquad \text{if } d \geqslant \sqrt{D}. \tag{7.5}$$

Hence, the resulting sieve (λ_d) has level of support D. Following Selberg, we call it the "Λ^2-sieve" of level D.

Applying the Λ^2-sieve to $\mathcal{A} = (a_n)$ in the sifting range $\{p \in \mathcal{P}, p < z\}$ we obtain

$$S(\mathcal{A}, z) = \sum_{(n,P)=1} a_n \leqslant S^+(\mathcal{A}, z)\ ,$$

where

$$S^+(\mathcal{A},z) = \sum_n a_n\Big(\sum_{d|(n,P)} \rho_d\Big)^2 \tag{7.6}$$
$$= \sum\sum_{d_1,d_2|P} \rho_{d_1}\rho_{d_2}|\mathcal{A}_{[d_1,d_2]}| = XG + R^+(\mathcal{A},\Lambda^2)\ ,$$

$P = P(z)$ is the product of the primes in the sifting range,

$$G = \sum\sum_{d_1,d_2|P} g([d_1,d_2])\rho_{d_1}\rho_{d_2} \tag{7.7}$$

and

$$R^+(\mathcal{A},\Lambda^2) = \sum\sum_{d_1,d_2|P} \rho_{d_1}\rho_{d_2} r_{[d_1,d_2]}(\mathcal{A}). \tag{7.8}$$

The task before us is to make this general inequality optimal. Forgetting for a moment about the remainder sum $R^+(\mathcal{A},\Lambda^2)$ we wish to minimize G with respect to the unknown numbers ρ_d subject to (7.3) and (7.5). A nice feature about the result is that the ensuing numbers will turn out to satisfy

$$|\rho_d| \leqslant 1. \tag{7.9}$$

Hence the remainder is automatically under control and in dealing with this sieve we shall usually use the simple bound $|R^+(\mathcal{A},\Lambda^2)| \leqslant R(\mathcal{A},\Lambda^2)$ where

$$R(\mathcal{A},\Lambda^2) = \sum\sum_{d_1,d_2|P} |r_{[d_1,d_2]}(\mathcal{A})|\ .$$

It is interesting that, although it is not known how to optimize the choice amongst all upper-bound weights satisfying (7.1), it is possible to do so amongst the more restricted class which also satisfy (7.2).

The expression (7.7) is a quadratic form in ρ_d. In order to find the minimum of G it helps to diagonalize. In the presentation below it goes without saying that ρ_d is supported on, and the relevant variables of summation run over, the divisors of P (thus over squarefree numbers). Sometimes we display this fact, but often not. Furthermore, we can assume that

$$0 < g(p) < 1 \qquad \text{if } p \mid P, \tag{7.10}$$
$$g(p) = 0 \qquad \text{if } p \nmid P. \tag{7.11}$$

Let $h(d)$ be the multiplicative function (relative density function) defined by

$$h(p) = \frac{g(p)}{1-g(p)} \quad \text{so} \quad g(p) = \frac{h(p)}{1+h(p)}\ . \tag{7.12}$$

We obtain

$$\begin{aligned} G &= \sum\sum\sum_{abc|P} g(abc)\rho_{ac}\rho_{bc} \\ &= \sum_c g(c) \sum\sum_{(a,b)=1} g(a)g(b)\rho_{ac}\rho_{bc} \\ &= \sum_c g(c) \sum_d \mu(d)g(d)^2 \Big(\sum_m g(m)\rho_{cdm}\Big)^2 \\ &= \sum_{d|P} h(d)^{-1} \Big(\sum_{m\equiv 0 \, (\mathrm{mod}\, d)} g(m)\rho_m \Big)^2 . \end{aligned}$$

Hence by the linear change of variables

$$y_d = \frac{\mu(d)}{h(d)} \sum_{m\equiv 0 \, (\mathrm{mod}\, d)} g(m)\rho_m \tag{7.13}$$

we obtain the diagonal form

$$G = \sum_{d|P} h(d) y_d^2. \tag{7.14}$$

We still have to reinterpret the condition (7.3) in terms of the new variables y_d. To this end we use Möbius inversion (see Section 2.2) to convert (7.13) into

$$\rho_\ell = \frac{\mu(\ell)}{g(\ell)} \sum_{d\equiv 0 (\mathrm{mod}\, \ell)} h(d) y_d. \tag{7.15}$$

In particular, for $\ell = 1$ this gives the linear equation

$$\sum_{d|P} h(d) y_d = 1. \tag{7.16}$$

Moreover, one observes by (7.13) and (7.15) that the support conditions for y_d are the same as those for ρ_d, that is, (7.5) is equivalent to

$$y_d = 0 \qquad \text{if } d \geqslant \sqrt{D}. \tag{7.17}$$

By the way, note that the convolution $1 * \rho$ in the new variables y_d becomes

$$\sum_{\ell|n} \rho_\ell = \sum_d \mu((d,n)) h\Big(\frac{d}{(d,n)}\Big) y_d \, . \tag{7.18}$$

Having made this change of variables our target now is to minimize (7.14) on the hyperplane (7.16). Applying Cauchy's inequality to (7.16) we derive

$$1 = \sum_{d|P} h(d) y_d \leqslant \Big(\sum_{d|P} h(d) y_d^2\Big)^{\frac{1}{2}} \Bigg(\sum_{\substack{d|P \\ d<\sqrt{D}}} h(d) \Bigg)^{\frac{1}{2}} = (GJ)^{\frac{1}{2}} \, ,$$

where we have put

$$J = J(D) = \sum_{\substack{d|P \\ d<\sqrt{D}}} h(d) \, . \tag{7.19}$$

Thus, we have shown that $GJ \geqslant 1$, so G cannot be smaller than J^{-1}. However, the equality

$$G = J^{-1} \tag{7.20}$$

holds for the choice

$$y_d = J^{-1} \qquad \text{if } d < \sqrt{D}\ , \tag{7.21}$$

so the minimal value of G is J^{-1}. Note that

$$J \leqslant \sum_{d|P} h(d) = \prod_{p|P}(1 + h(p)) = \prod_{p|P}(1 - g(p))^{-1} = V(z)^{-1} \tag{7.22}$$

so $G \geqslant V(z)$, as it has to be. From now on we assume that the y_d are the constant given by (7.21).

Next we compute ρ_ℓ by inserting (7.21) into (7.15) getting

$$\mu(\ell)g(\ell)\rho_\ell\, J = \sum_{\substack{m<\sqrt{D} \\ m\equiv 0 (\operatorname{mod} \ell)}} h(m)\ .$$

After pulling out the factor $h(\ell)$ we write this as the quotient

$$\rho_\ell = \mu(\ell)j(\ell)J_\ell(D)/J(D) \tag{7.23}$$

where

$$j(\ell) = \frac{h(\ell)}{g(\ell)} = \prod_{p|\ell}\big(1 - g(p)\big)^{-1}\ , \tag{7.24}$$

$$J_\ell(D) = \sum_{\substack{d<\sqrt{D}/\ell \\ (d,\ell)=1}} h(d)\ , \tag{7.25}$$

and $J(D) = J_1(D)$. For this choice we find that

$$\lambda_d = \mu(d)j(d) \sum_{abc=d} \mu(c)j(c)J_{ac}(D)J_{bc}(D)J(D)^{-2}\ .$$

Moreover, the convolution $1 * \rho$ becomes (see (7.18)),

$$\sum_{\ell|n} \rho_\ell = \frac{1}{J(D)} \sum_{q<\sqrt{D}} \mu((n,q))h\Big(\frac{q}{(n,q)}\Big)\ . \tag{7.26}$$

See Section 9.5 where further expressions, in terms of Ramanujan sums, are derived.

Remarks. The summation in $J_\ell(D)$ runs over divisors of P, so over squarefree numbers d, a condition which will occasionally be displayed. Sometimes we hide the support level D and write for brevity J_ℓ, J in place of $J_\ell(D)$, $J(D)$.

Now we show (7.9). To this end we group the terms in (7.19) according to the greatest common divisor of d and ℓ, getting

$$\begin{aligned} J(D) &= \sum_{k|\ell} \sum_{\substack{d<\sqrt{D} \\ (d,\ell)=k}} h(d) = \sum_{k|\ell} h(k) \sum_{\substack{m<\sqrt{D}/k \\ (m,\ell)=1}} h(m) \\ &\geqslant \Big(\sum_{k|\ell} h(k)\Big) \sum_{\substack{m<\sqrt{D}/\ell \\ (m,\ell)=1}} h(m) = j(\ell)J_\ell(D) \end{aligned}$$

and this proves (7.9) (this neat estimate is due to J.H. van Lint and H.-E. Richert [**156**]). From this one gets directly, by (7.4),

$$|\lambda_d| \leqslant \tau_3(d). \tag{7.27}$$

Actually, we have proved that $\mu(\ell)\rho_\ell = 1 - \delta_\ell$ with

$$\delta_\ell = \frac{1}{J(D)} \sum_{k|\ell} h(k) \sum_{\substack{\sqrt{D}/\ell \leqslant m < \sqrt{D}/k \\ (m,\ell)=1}} h(m) \leqslant 1, \tag{7.28}$$

so $0 \leqslant \mu(d)\rho_d \leqslant 1$.

Collecting the above results we conclude the following:

THEOREM 7.1. *Let $\mathcal{A} = (a_n)$ be a finite sequence of non-negative numbers and let P be a finite product of distinct primes. For every $d|P$ we write*

$$|\mathcal{A}_d| = \sum_{n \equiv 0 \,(\mathrm{mod}\, d)} a_n = g(d)X + r_d(\mathcal{A}) \ , \tag{7.29}$$

where $X > 0$ and $g(d)$ is a multiplicative function with $0 < g(p) < 1$ for $p|P$. Let $h(d)$ be the multiplicative function given by $h(p) = g(p)(1 - g(p))^{-1}$ and $J = J(D)$ given by (7.19) for some $D > 1$. Then we have

$$S(\mathcal{A}, z) = \sum_{(n,P)=1} a_n \leqslant XJ^{-1} + R^+(\mathcal{A}, \Lambda^2) \ , \tag{7.30}$$

where

$$R^+(\mathcal{A}, \Lambda^2) = \sum_{d|P} \lambda_d r_d(\mathcal{A}) \ , \tag{7.31}$$

with λ_d given by (7.4) and (7.23)–(7.25).

Actually, the right-hand side of (7.30) is exactly equal to the sifted sum $S^+(\mathcal{A}, z)$ for the optimal vector (7.21).

Using (7.27) one estimates the remainder crudely by

$$|R^+(\mathcal{A}, \Lambda^2)| \leqslant R(\mathcal{A}, \Lambda^2) \leqslant \sum_{\substack{d|P \\ d<D}} \tau_3(d)|r_d(\mathcal{A})|. \tag{7.32}$$

WARNING: Selberg, and also other writers, use λ_d to denote sometimes the sieve weights but also sometimes the Selberg sieve constituents, whereas we distinguish the latter with the notation ρ_d. As a result, care must be exercised when interpreting some results in the literature.

7.2. Comments on the Λ^2-Sieve

There is a great deal of substance to the Λ^2-sieve hidden beneath its surface and we find it worthwhile to excavate this, not necessarily for applications, but primarily for educational purposes. In this section, which might well be skipped on a first reading, we take the liberty to speak freely without feeling the need for complete proofs.

7.2.1. Fully Optimal Weights. The optimization of the quadratic form G in (7.7) can be accomplished in a number of different ways; for example, Selberg proposed (in conversations) the use of Lagrange multipliers.

One could also treat the optimization at an earlier stage, and even more generally for any finite sequence $\mathcal{A} = (a_n)$ of non-negative numbers, by considering the sum

$$S^+(\mathcal{A}) = \sum_n a_n \Big(\sum_{d|n} \rho_d\Big)^2 , \tag{7.33}$$

before the quadratic form (7.7) emerges from the approximation (7.29). To determine the optimal vector ρ_d, that is, the real numbers ρ_d with $d \in \mathcal{D}$ for which the quadratic form (7.33) assumes its minimum, subject to the normalization $\rho_1 = 1$, we consider the bilinear form

$$S(\rho, \xi) = \sum_n a_n \Big(\sum_{d|n} \rho_d\Big)\Big(\sum_{k|n} \xi_k\Big)$$

for any $(\rho_d), (\xi_k)$ with $d, k \in \mathcal{D}$ and $\rho_1 = \xi_1 = 1$. We have two formulae:

$$S(\xi, \xi) = S(\rho, \rho) + 2S(\xi - \rho, \rho) + S(\xi - \rho, \xi - \rho)$$

and

$$S(2\rho - \xi, 2\rho - \xi) = S(\rho, \rho) - 2S(\xi - \rho, \rho) + S(\xi - \rho, \xi - \rho) .$$

If (ρ_d) is the optimal vector then the two left-hand sides are $\geqslant S(\rho, \rho)$, hence

$$2|S(\xi - \rho, \rho)| \leqslant S(\xi - \rho, \xi - \rho) .$$

Since here the left side is linear and the right side is quadratic in the vector $\xi - \rho$, this implies that $S(\xi - \rho, \rho)$ is identically zero. In other words, the optimal vector (ρ_d) has the remarkable property

$$\sum_n a_n \Big(\sum_{d|n} \rho_d\Big)^2 = \sum_n a_n \Big(\sum_{d|n} \rho_d\Big)\Big(\sum_{k|n} \xi_k\Big) \tag{7.34}$$

for every (ξ_k) with $k \in \mathcal{D}$ and $\xi_1 = 1$. In particular, taking $\xi_k = 0$ for all $k \neq 1$ we get the somewhat surprising equation

$$\sum_n a_n \Big(\sum_{d|n} \rho_d\Big)^2 = \sum_n a_n \Big(\sum_{d|n} \rho_d\Big) . \tag{7.35}$$

Hence, the property (7.34) is equivalent to the system of equations

$$\sum_{n \equiv 0 \,(\mathrm{mod}\, k)} a_n \Big(\sum_{d|n} \rho_d\Big) = 0 \quad \text{for every } k \in \mathcal{D},\ k \neq 1 . \tag{7.36}$$

Denoting as usual,

$$A_m = \sum_{n \equiv 0 \,(\mathrm{mod}\, m)} a_n$$

we can write (7.36) as a system of linear equations in ρ_d:

$$\sum_d \rho_d A_{[d,k]} = 0, \quad \text{for every} \quad k \in \mathcal{D},\ k \neq 1 . \tag{7.37}$$

For general sequences $\mathcal{A}$, solving the system (7.37) is not an easy problem. One may try the method of successive approximations (starting with A_m replaced by $g(m)X$), but the results derived so far are not encouraging.

Let us now return to (7.35) which we write in the form

$$\sum_{d_1}\sum_{d_2}\rho_{d_1}\rho_{d_2}A_{[d_1,d_2]} = \sum_{d}\rho_d A_d\ . \tag{7.38}$$

Setting $A_d = g(d)X + r_d$, this becomes

$$\sum_{d_1}\sum_{d_2}\rho_{d_1}\rho_{d_2}\Big(g([d_1,d_2])X + r_{[d_1,d_2]}\Big) = \sum_{d}\rho_d\big(g(d)X + r_d\big)\ .$$

Ignoring the remainders r_d, we are led to the expectation that the equation

$$\sum_{d_1}\sum_{d_2}\rho_{d_1}\rho_{d_2}\, g\big([d_1,d_2]\big) = \sum_{d}\rho_d\, g(d) \tag{7.39}$$

is approximately true. Of course, it cannot be exactly true because the fully optimal weights $\rho_d = \rho_d(\mathcal{A})$ depend on the sequence $\mathcal{A}$. However, if we take $\rho_d = \rho_d(g)$ to be the genuine Selberg choice (7.23) which optimized the quadratic form (7.7) for the given density function g, then (7.39) does hold, as it must. This cute formula can also be verified directly.

7.2.2. Flexibility. The Λ^2-sieve is very general indeed. Its upper bound does not require any regularity in the distribution of the density function $g(p)$ over primes, therefore the dimension of a sifting problem does not explicitly play a role. If $g(p) = \omega(p)p^{-1}$, where $\omega(p)$ is the number of residue classes (mod p) which one wants to exclude, then $h(p) = \omega(p)(p-\omega(p))^{-1}$ is the ratio of the numbers of rejected to admitted classes, and J incorporates these ratios. Of course, the larger $\omega(p)$ is the smaller the upper bound (7.30). However, a few local deviations of $\omega(p)$ from its average have insignificant global effect. To the contrary our estimates derived by combinatorial sieves are quite sensitive in this respect because the hypothesis (5.38) is assumed to hold for all segments of primes $w \leqslant p < z$, no matter how short. Actually, one could weaken the sensitivity of hypothesis (5.38) considerably, but at the expense of massive complications in the statements and arguments. It seems excessive to work with such harsh conditions in this book. Instead, when needed, we clean things up in the combinatorial sieve by preliminary applications of the Fundamental Lemma.

The Λ^2-sieve yields fantastic results for sifting problems when the number of residue classes to be excluded is large; it can compete with the large sieve method of Linnik. We shall give a connection between the two methods in Section 9.5.

7.2.3. Smoothing the Möbius Function. Note that the Selberg weight constituents ρ_d depend on the multiplicative function $g(d)$ so, indirectly, on the sifting sequence $\mathcal{A}$. However, assuming some regularity of $g(d)$, as in the κ-dimensional sieve, one can establish a fairly good approximation (for small d at any rate) by a smoothed Möbius function:

$$\rho_d \sim \mu(d)\left(\frac{\log^+ \sqrt{D}/d}{\log\sqrt{D}}\right)^{\kappa} \tag{7.40}$$

(apply Theorem A.5 to the sum (7.19)). Hence, for small d we have $\rho_d \sim \mu(d)$ and

$$\lambda_d = \sum_{abc=d}\rho_{ac}\rho_{bc} \sim \sum_{abc=d}\mu(a)\mu(b) = \mu(d).$$

Putting

$$\rho_d^+ = \mu(d)\left(\frac{\log^+ \sqrt{D}/d}{\log\sqrt{D}}\right)^{\kappa} \qquad \text{and} \qquad \lambda_d^+ = \sum_{abc=d} \rho_{ac}^+\rho_{bc}^+$$

one gets an upper-bound sieve $\Lambda^+ = (\lambda_d^+)$ of level D. The smoothed sieve Λ^+ depends only on κ and D. It is not the optimal one, nevertheless, it produces asymptotically equivalent results. If $g(d)$ is a multiplicative function such that $g(p)p$ is κ on average over primes, then

$$G^+ = \sum_{d|P} \lambda_d^+ g(d) \sim \frac{1}{J}\ .$$

For $g(d) = d^{-1}$, in which case $\kappa = 1$, the smoothed weights satisfy (see [**4**] and [**67**])

$$\sum_{1\leqslant n\leqslant x}\Big(\sum_{d|n}\rho_d^+\Big)^2 \ll \frac{x}{\log\sqrt{D}}$$

if $x > \sqrt{D}$, whereas the optimal weights satisfy (see (9.35)),

$$\sum_{1\leqslant n\leqslant x}\Big(\sum_{d|n}\rho_d\Big)^2 < \frac{x+D}{\log\sqrt{D}}. \tag{7.41}$$

A smoothed truncation of the Möbius function is the essence of the Λ^2-sieve. With a sharp cut we would only achieve

$$\sum_{1\leqslant n\leqslant x}\Bigg(\sum_{\substack{d|n\\ d<\sqrt{D}}}\mu(d)\Bigg)^2 \asymp x\ .$$

Although this feature appears here only as the result of optimization of the quadratic form G, it needs to be said that the benefit of smoothing had been foreseen by Selberg in his earlier masterwork (for an easy to find reference see [**144**]) on the zeros of the Riemann zeta-function. One can find traces of Selberg's ideas in several of his other papers which anticipated the creation of the Λ^2-sieve. Selberg cultivated the smoothing technology (both in its analytic and arithmetic aspects) beyond what had been the state of the art. This is manifested clearly in his being able to show a positive density of zeros on the critical line by means of specially devised mollification factors. These mollifiers act as damping agents for the large values of the zeta-function, just as the sieve factors θ_n serve to dampen the highly composite numbers. On the critical line the smoothed cut-off is of critical importance, although not so much so when off the line.

Subsequently, these ideas have been adopted for the study of the non-vanishing of the special values of L-functions, in particular, for families of automorphic L-functions at the central point. A successful conclusion is drawn by comparing the first and the second power moments suitably mollified to arrive at estimates having equal order of magnitude (if treated simply without mollification the extreme values would contribute to a discrepancy in the logarithmic factors). For best results, as in the Λ^2-sieve, one is challenged with the optimization of quadratic forms. In certain situations two quadratic forms in the common unknown vector appear in a slanted combination and one must decide on which of the two to focus, because the optimization of the total expression is practically impossible. Sometimes one is lucky. For example, a situation occurs in [**112**] where a diagonalization is performed on one chosen quadratic form in an attempt to optimize its value while the other quadratic form, although not of diagonal shape in the new variables, turns out clearly to assume the best value at the same vectors. In this way an impressive fifty percent non-vanishing of the central values of the L-functions under consideration is achieved.

The treatment of L-functions described above is reminiscent of Selberg's method of producing almost-primes by measuring the ratio of two differently weighted sifted sums. An even more impressive exploitation of such ideas can be found in the recent works of Goldston, Pintz, and Yildirim [**64**] which are designed for the detection of small gaps between prime numbers. We are going to present their works in considerable detail in Sections 7.13 and 7.14.

By the trivial estimation $|\rho_d^+| \leqslant 1$ we immediately obtain the bound $|\lambda_d^+| \leqslant \tau_3(d)$. However, the true value of λ_d^+ is about $\mu(d)$. Indeed, we have

$$\lambda_d^+ = \mu(d) + O\Big(2^\kappa \tau_3(d)\Big(\frac{\log d}{\log D}\Big)^2\Big)\,, \tag{7.42}$$

where the implied constant is absolute.

For the proof of this approximation we can assume that $d < \sqrt{D}$. Then

$$\begin{aligned}
\lambda_d^+ &= \mu(d) \sum_{abc=d} \mu(c)\Big(\frac{\log \sqrt{D}/ac}{\log \sqrt{D}}\Big)^\kappa \Big(\frac{\log \sqrt{D}/bc}{\log \sqrt{D}}\Big)^\kappa \\
&= \mu(d) \sum_{abc=d} \mu(c)\bigg(\Big(1 - \frac{\log d}{\log \sqrt{D}}\Big)\Big(1 - \frac{\log c}{\log \sqrt{D}}\Big) + \frac{\log a}{\log \sqrt{D}}\frac{\log b}{\log \sqrt{D}}\bigg)^\kappa \\
&= \mu(d) \sum_{mc=d} \tau(m)\mu(c)\Big(1 - \kappa\frac{\log cd}{\log \sqrt{D}}\Big) + O\Big(2^\kappa \tau_3(d)\Big(\frac{\log d}{\log D}\Big)^2\Big)\,.
\end{aligned}$$

Since, by $\zeta^2\zeta^{-1} = \zeta$ and $\zeta^2(\zeta^{-1})' = -\,\zeta'$, we have

$$\sum_{mc=d} \tau(m)\mu(c) = 1\,, \qquad \sum_{mc=d} \tau(m)\mu(c)\log c = -\log d\,,$$

this completes the proof of (7.42).

EXERCISE 7.1. Show that, if $\kappa = 1$, then for $d < \sqrt{D}$,

$$\lambda_d^+ = \mu(d)\Big(1 - \sum_{p|d}\Big(\frac{\log p}{\log \sqrt{D}}\Big)^2\Big)\,. \tag{7.43}$$

Hint: Use the formula

$$\sum_{abc=d} \mu(c)(\log a)(\log b) = (\log d)^2 - \sum_{p|d}(\log p)^2$$

which follows from $\zeta^{-1}\zeta'\zeta' = \zeta'' - \zeta(\zeta'/\zeta)'$.

EXERCISE 7.2. Assuming sufficiently strong conditions on the density function $g(d)$, prove (7.42) for the genuine Selberg weights λ_d.

7.2.4. Sign and Size of the Weights. In some applications one may wish to have an upper-bound sieve $\Lambda = (\lambda_d)$ of level D with λ_m being exactly $\mu(m)$ for all $m < M = D^\gamma$ with some $0 < \gamma < 1$, rather than its approximation. To construct such weights one can initiate Selberg's sieve (7.4) with $\rho_d = \mu(d)$ for all $d < M$ and perform the optimization of the resulting quadratic form in the remaining range $M \leqslant d < \sqrt{D}$. Then, clearly we have

$$\lambda_m = \sum_{[d_1,d_2]=m} \mu(d_1)\mu(d_2) = \mu(m) \ ,$$

if $m < M$. Hence, by Selberg's sieve we may produce the desired effect with any $\gamma < \frac{1}{2}$. On the other hand, the upper-bound beta-sieve automatically has $\lambda_m = \mu(m)$ if $m < M = D^{1/(\beta+1)}$ because the set $\mathcal{D}^+$ contains all squarefree numbers $m < D^{1/(\beta+1)}$. Choosing $\beta = 1$ we accomplish the same goal with $\gamma = \frac{1}{2}$. This sieve has the effect of completely removing every number $1 < m < M$ because $(1 * \mu)(m) = 0$, creating a gap which is desirable in certain situations.

The bound (7.27) for the Selberg sieve weights λ_d is often sufficient in applications but it is certainly not the best possible one. We are going to improve it a bit. Put $\mu(\ell)\rho_\ell = \gamma_\ell + \frac{1}{2}$, so by the van Lint–Richert inequality $|\gamma_\ell| \leqslant \frac{1}{2}$. From (7.4) we find that

$$\lambda_d = \sum\sum_{abc=d} \rho_{ac}\rho_{bc} = \mu(d) \sum\sum_{abc=d} \mu(c)\big(\gamma_{ac} + \tfrac{1}{2}\big)\big(\gamma_{bc} + \tfrac{1}{2}\big) \ .$$

Hence,

$$\mu(d)\lambda_d = \tfrac{3}{4} + \sum\sum_{abc=d} \mu(c)\gamma_{ac}\gamma_{bc}$$

which gives

$$|\lambda_d| \leqslant \tfrac{1}{4}\big(3 + \tau_3(d)\big) \ . \tag{7.44}$$

We do know that the sieve weight constituents ρ_d always have the same sign as $\mu(d)$, but it is not clear if this is always true for the sieve weights λ_d. It certainly is true for small d and obviously for d prime we have $\lambda_p = \rho_p(2 + \rho_p)$ so that

$$-1 \leqslant \lambda_p \leqslant 0 \ . \tag{7.45}$$

A more complicated argument shows that $\lambda_{p_1p_2} \geqslant 0$. Of course, for the combinatorial sieve all of this is automatic.

7.3. Crossing Density Functions

The Selberg upper-bound sieve $\Lambda = (\lambda_d)$ given by (7.4) is intrinsically connected to the density function $g(d)$ of the sequence $\mathcal{A} = (a_n)$ in the process of optimization of the quadratic form (7.7). The result is very elegant, $G = J^{-1}$ with $J = J(D)$ given by (7.19). However, there are situations when one needs to apply this particular sieve to a sequence $\mathcal{A}' = (a'_n)$ which has a somewhat different density function, say $g'(d)$. The relevant quadratic form

$$G' = G'(D) = \sum_{d_1,d_2|P}\sum g'([d_1,d_2])\rho_{d_1}\rho_{d_2} \tag{7.46}$$

is no longer optimal and we are interested in evaluating the effect of this change.

PROPOSITION 7.2. *Let $g'(d)$ be a multiplicative function. Put*

$$f(p) = \big(g(p) - g'(p)\big)\big(1 - g(p)\big)^{-1}\ , \tag{7.47}$$

$$h'(p) = g'(p)\big(1 - g'(p)\big)\big(1 - g(p)\big)^{-2}\ . \tag{7.48}$$

Extend these to squarefree numbers by multiplicativity. We have

$$G'(D) = \frac{1}{J^2(D)} \sideset{}{^\flat}\sum_{d<\sqrt{D}} h'(d)\Big(\sideset{}{^\flat}\sum_{\substack{\ell<\sqrt{D}/d \\ (\ell,d)=1}} f(\ell)\Big)^2\ . \tag{7.49}$$

PROOF. For convenience we assume in the proof (as we may by continuity) that $g(p)$ is never zero and recall that we put $\xi_d = h(d)y_d$. Inserting (7.15) in (7.46), we have

$$G' = \sum_a \sum_b \sum_c \frac{\mu(abc)\mu(c)}{g(ac)g(bc)} g'(abc) \sum_{\substack{d_1\equiv 0\,(ac) \\ d_2\equiv 0\,(bc)}}\sum \xi_{d_1}\xi_{d_2} = \sum_{d_1}\sum_{d_2} \xi_{d_1}\xi_{d_2} E(d_1,d_2)\ ,$$

say, where

$$\begin{aligned} E(d_1,d_2) &= \sum_{a|d_1}\sum_{b|d_2} \mu(ab)\frac{g'(ab)}{g(ab)} \sum_{\substack{c|(d_1,d_2) \\ (c,ab)=1}} \mu^2(c)\frac{g'(c)}{g^2(c)} \\ &= \prod_{p|(d_1,d_2)} \Big(1 + \frac{g'(p)}{g^2(p)}\Big) \prod_{\substack{p|d_1d_2 \\ p\nmid(d_1,d_2)}} \Big(1 - \frac{g'(p)}{g(p)}\Big) \\ &\qquad \cdot \prod_{p|(d_1,d_2)} \Big(1 - 2\frac{g'(p)}{g(p)}\Big(1 + \frac{g'(p)}{g^2(p)}\Big)^{-1}\Big) \\ &= \prod_{\substack{p|d_1d_2 \\ p\nmid(d_1,d_2)}} \Big(1 - \frac{g'(p)}{g(p)}\Big) \prod_{p|(d_1,d_2)} \Big(1 - 2\frac{g'(p)}{g(p)} + \frac{g'(p)}{g^2(p)}\Big)\ . \end{aligned}$$

This yields

$$\begin{aligned} G' &= \sum_{\substack{a\ b\ c \\ (a,b)=1}}\sum\sum \xi_{ac}\xi_{bc} \prod_{p|ab}\Big(1 - \frac{g'(p)}{g(p)}\Big) \prod_{p|c}\Big(1 - 2\frac{g'(p)}{g(p)} + \frac{g'(p)}{g^2(p)}\Big) \\ &= \sum_d \sum_{cm=d} \mu(m) \prod_{p|m}\Big(1 - \frac{g'(p)}{g(p)}\Big)^2 \prod_{p|c}\Big(1 - 2\frac{g'(p)}{g(p)} + \frac{g'(p)}{g^2(p)}\Big) \\ &\qquad \cdot \Big(\sum_{(\ell,d)=1} \xi_{d\ell} \prod_{p|\ell}(1 - \frac{g'(p)}{g(p)})\Big)^2 \\ &= \sum_d \prod_{p|d}(1 - g'(p))\frac{g'(p)}{g^2(p)}\Big(\sum_{(\ell,d)=1} \xi_{d\ell} \prod_{p|\ell}(1 - \frac{g'(p)}{g(p)})\Big)^2\ . \end{aligned}$$

Hence, introducing (7.21) we obtain the result. □

When $g'(p) = g(p)$ for all $p \nmid q$ and $g'(p) = 0$ then (7.49) yields

$$\sum_{(d,q)=1} \lambda_d \, g(d) = \frac{1}{J(D)^2} \sideset{}{^\flat}\sum_{\substack{d<\sqrt{D} \\ (d,q)=1}} h(d) \Big(\sideset{}{^\flat}\sum_{\substack{\ell<\sqrt{D}/q \\ \ell|q}} h(\ell) \Big)^2 . \tag{7.50}$$

Note that for $q = 1$ (that is when $g' = g$) we recover (7.20). Subtracting (7.50) for $q = p$ prime from that for $q = 1$ we obtain

$$\sum_{d \equiv 0 (\mathrm{mod}\, p)} \lambda_d \, g(d) = -h(p) j(p) J_p(D) J(D)^{-2} \leqslant 0 \tag{7.51}$$

and hence

$$\sum_{d \equiv 0 (\mathrm{mod}\, p)} \lambda_d \, g(d) = \rho_p \, h(p) \sum_d \lambda_d \, g(d). \tag{7.52}$$

Next, we give various estimates for $G'(D)$ using the formula (7.49). We seek these mainly for educational purposes and not for application elsewhere in the book, so shall be sketchy in the arguments. Let us assume that both g and g' satisfy the conditions for density functions, of dimension κ and κ' respectively. The question is how $G'(D)$ differs from $G(D)$ in its order of magnitude when κ' differs from κ. First, it is easy to get a lower bound for $G'(D)$. Putting

$$J'(D) = \sum_{d<\sqrt{D}} h'(d) \tag{7.53}$$

we obtain by positivity

$$G'(D) \geqslant J^{-2}(D) \big(J'(D) - J'(D/4) \big) . \tag{7.54}$$

In practice we often have

$$J(D) \sim c (\log \sqrt{D})^{\kappa}, \qquad J'(D) \sim c' (\log \sqrt{D})^{\kappa'} . \tag{7.55}$$

Actually, asymptotic expansions hold with error terms lower than the leading terms by a factor $(\log D)^{-2}$, hence one gets

$$J'(D) - J'(D/4) \asymp (\log D)^{\kappa' - 1} .$$

One can also derive this as follows:

$$[J'(D) - J'(D/4)] \log \sqrt{D} \geqslant \sideset{}{^\flat}\sum_{\frac{1}{2}\sqrt{D}<d<\sqrt{D}} h'(d) \log d$$

$$= \sideset{}{^\flat}\sum \sum_{\substack{\frac{1}{2}\sqrt{D}<dp<\sqrt{D} \\ p \nmid d}} h'(d) h(p) \log p \gg \sideset{}{^\flat}\sum_{d<D^{1/4}} h'(d) \asymp (\log D)^{\kappa'} .$$

In this situation (7.54) yields

$$G'(D) \gg (\log D)^{\kappa' - 1 - 2\kappa} . \tag{7.56}$$

Interestingly enough, the lower bound (7.56) has the correct order of magnitude in many important cases. For example, let us consider the case

$$g'(p) = 2 g(p) . \tag{7.57}$$

Then we have (7.49) with

$$f(\ell) = \mu(\ell) h(\ell) \tag{7.58}$$

and

$$h'(d) = \tau(d)h(d)\prod_{p|d}\big(1 - h(p)\big)\ . \tag{7.59}$$

Furthermore, if $g(p)$ has dimension κ (hence, so does $h(p)$) a positive integer, then

$$\sum_{\substack{\ell\leqslant y\\(\ell,d)=1}} f(\ell) = \sum_{\substack{\ell\leqslant y\\(\ell,d)=1}} \mu(\ell)h(\ell) \ll \prod_{p|d}\big(1 + h(p)\big)(\log 2y)^{-1}\ ,$$

by arguments familiar from prime number theory. This leads to

$$G'(D) \ll J^{-2}(D)\sum_{d<\sqrt{D}}^{\flat} \tau(d)h(d)\prod_{p|d}\big(1 + h(p)\big)^2(\log 2\sqrt{D}/d)^{-2}\ .$$

Hence, by elementary arguments we derive the upper bound

$$G'(D) \ll (\log D)^{-1} \tag{7.60}$$

which is as sharp as the lower bound (7.56) because $\kappa' = 2\kappa$.

This particular example $g'(p) = 2g(p)$ reveals a somewhat surprising behaviour in the Selberg upper-bound sieve. To see this, consider simultaneously the two sums

$$S^+(\mathcal{A}) = \sum_{n\leqslant x} a_n\Big(\sum_{d|n}\rho_d\Big)^2\ , \tag{7.61}$$

$$S'(\mathcal{A}) = \sum_{n\leqslant x} a_n\tau(n)\Big(\sum_{d|n}\rho_d\Big)^2\ , \tag{7.62}$$

with the same sieve weights ρ_d in both sums. The first sum occurs in the sieve problem for the sequence $\mathcal{A} = (a_n)$ and, under our standard conditions, it is

$$S^+(\mathcal{A}) = XG + R\ . \tag{7.63}$$

The second sum occurs in the sieve problem for the sequence $\mathcal{A}' = (a'_n)$ with $a'_n = a_n\tau(n)$. Although this sequence may not satisfy the axiom (7.29), we can suppose it does so in the modified form

$$|\mathcal{A}'_d| = g'(d)X\log(\eta x/d) + r'_d(x) \tag{7.64}$$

where, naturally, $g'(d) = \tau(d)g(d)$. For example, let $a_n = 1$ and $a'_n = \tau(n)$ in which case $g(d) = d^{-1}$ and (7.64) holds with $g'(d) = \tau(d)d^{-1}$ (here $\log\eta = 2\gamma - 1$, γ the Euler constant). We get

$$S'(\mathcal{A}) = G'X\log\eta x - G''X + R'$$

where

$$G'' = \sum_d \lambda_d\, g'(d)\log d$$

with $\Lambda = (\lambda_d)$ being the upper-bound sieve given by (7.4). By (7.56) and (7.60) we obtain

$$G' \asymp (\log D)^{-1}\ .$$

The problem of estimating G'' reduces to that of G' due to the Monotonicity Principle (I), leading to

$$G'' \ll 1$$

by Corollary 5.5. Gathering the above estimates we deduce that

$$S'(\mathcal{A}) \asymp X \frac{\log x}{\log D} \asymp X \ . \tag{7.65}$$

On the other hand, (7.63) yields

$$S^+(\mathcal{A}) \asymp X(\log D)^{-\kappa} \ . \tag{7.66}$$

Concluding the above considerations we observe that twisting the sequence $\mathcal{A} = (a_n)$ by the divisor function $\tau(n)$ causes a full loss of the factor $(\log D)^{-\kappa}$ which had been saved by the sieve weights

$$\theta_n = \Big(\sum_{d|n} \rho_d\Big)^2 \ .$$

Hence, thinking that θ_n has the effect of capturing only numbers without small prime factors (almost-primes) is wrong in some respects. Yes, one knows (see Section 10.3) that the elements of the sequence $\mathcal{A} = (a_n)$ with n having small prime divisors contribute to $S^+(\mathcal{A})$ a negligible quantity, however, after twisting by the divisor function, this contribution has been amplified significantly.

We now come back to the general situation of g, g' being density functions of dimension κ, κ' respectively. Assuming that the dimension conditions are sufficiently strong we claim the following estimates:

$$G'(D) \asymp \big(\log D\big)^{-\kappa'} \qquad \text{if} \quad \kappa' < \kappa + \tfrac{1}{2} \ , \tag{7.67}$$

$$G'(D) \asymp \big(\log D\big)^{-\kappa'} \log\log D \qquad \text{if} \quad \kappa' = \kappa + \tfrac{1}{2} \ , \tag{7.68}$$

$$G'(D) \asymp \big(\log D\big)^{\kappa' - 2\kappa - 1} \qquad \text{if} \quad \kappa' > \kappa + \tfrac{1}{2} \ . \tag{7.69}$$

Here is the sketch of a proof. First we use the upper bound

$$\sum_{\ell < y}{}^{\flat} f(\ell) \ll \big(\log 2y\big)^{\kappa - \kappa'} \tag{7.70}$$

for $y \geqslant 1$. Hence we derive by (7.49)

$$\begin{aligned} G'(D) &\ll J(D)^{-2} \sum_{d < \sqrt{D}} g'(d)\big(\log 2\sqrt{D}/d\big)^{2(\kappa-\kappa')} \\ &\asymp \big(\log \sqrt{D}\big)^{-2\kappa} \int_1^{\sqrt{D}} \big(\log 2\sqrt{D}/x\big)^{2(\kappa-\kappa')} d(\log x)^{\kappa'} \\ &= \big(\log \sqrt{D}\big)^{-\kappa'} \int_0^1 \Big(1 - t + \frac{\log 4}{\log D}\Big)^{2(\kappa-\kappa')} dt^{\kappa'} \ . \end{aligned}$$

This yields the upper bounds in (7.67), (7.68), (7.69). For the lower bounds note that (7.69) is already established in (7.56) for any κ', κ. We need to improve (7.56) for $\kappa' \leqslant \kappa + \frac{1}{2}$. Under this condition, the upper bound (7.70) has the correct order of magnitude provided that y is sufficiently large. Therefore the estimates from above for $G'(D)$ represent the correct order of magnitude, giving (7.67) and (7.68). Incidentally, if $\kappa' < \kappa + \frac{1}{2}$, then the sum (7.70) is evaluated asymptotically in (A.25).

The appearance of this transition in the order of magnitude of $G'(D)$ at $\kappa' = \kappa + \frac{1}{2}$ is not obvious, to say the least. Retrospectively, we have noticed the relevant comment of Selberg [**145**] which he made in the context of the convergence issue of his sieve construction. We quote from page 154: "This is possible because the Λ^2-sieve which is optimal for density k_0 converges for k somewhat beyond k_0 (as is not hard to establish, it converges for $k < k_0 + \frac{1}{2}$)."

Something rather analogous happens in the case of the beta-sieve. There, because of the one-sided assumption (5.38), it is obvious that an upper-bound sieve which serves for a sequence with given dimension κ also serves for a sequence with dimension $\kappa' < \kappa$ and a glance at the statement of the Fundamental Lemma tells us that the upper bound (and, where positive, the lower bound) will still have the correct order of magnitude. What is less obvious is that the beta-sieve construction for dimension $\kappa > \frac{1}{2}$ also converges even if β is slightly below $\beta(\kappa)$ the largest zero of the adjoint function $q_\kappa(s-1)$ (see (11.55)). This may have potential value in connection with the upper bound. However, in the lower-bound sieve the resulting function $f_\kappa(s)$ would then be negative for $\beta < s < \beta(\kappa)$ which is obviously worse than the trivially attainable value $f_\kappa(s) = 0$. The reader is invited to further discussion on the distribution of the zeros of the adjoint functions in Section 3 of Appendix B.

The situation of (7.67) appears in the work of Goldston, Pintz, and Yildirim where the Selberg sieve weights of dimension $\kappa = k + \ell$ with $\ell \geqslant 0$ are used for two sieve problems of dimension $\kappa' = k$ and $\kappa' = k - 1$, respectively. One obtains two quadratic forms, each of the order of magnitude $(\log D)^{-\kappa'}$, as is shown in (7.67). This means that the excess by ℓ in the dimension of the sieve weights preserves the desired order of magnitude. A negative value of ℓ, that is a deficit in the dimension, would ruin that order, as shown in (7.69), which is a message worth remembering for future constructions. The benefit of introducing the excess parameter ℓ appears in the comparison of the two quadratic forms. Specifically, one gains a factor almost as large as 2; to see this compare (7.191) for $\ell = 0$ and with that for ℓ moderately large. Needless to say, were we optimizing the relevant expression with respect to only a single quadratic form, the best choice would have been $\kappa' = \kappa$, and this relative gain of the factor two would disappear.

7.4. Explicit Estimates for $J(D, z)$

To bring the Selberg upper bound (7.30) to a practical form we need a clear lower bound for the sum (7.19), which we now denote by

$$J(D, P) = \sum_{\substack{d|P \\ d<\sqrt{D}}} h(d). \tag{7.71}$$

The upper bound $J(D, P) \leqslant V(z)^{-1}$ holds in general (see (7.22)) but a good lower bound requires some restrictions on the density function g, so on h and the sifting range.

Let $I(D, P)$ denote the complementary sum so $J(D, P) + I(D, P) = V(z)^{-1}$. Therefore we need an upper bound for $I(D, P)$. Using Rankin's trick we estimate as follows:

$$I(D, P) = \sum_{d \geqslant \sqrt{D}, d|P} h(d) \leqslant D^{-\varepsilon} \sum_{d|P} h(d) d^{2\varepsilon} = D^{-\varepsilon} \prod_{p|P} (1 + h(p) p^{2\varepsilon}) \, ,$$

$$V(z) I(D, P) \leqslant D^{-\varepsilon} \prod_{p|P} (1 + g(p)(p^{2\varepsilon} - 1)) \leqslant D^{-\varepsilon} \exp\Big(\sum_{p|P} g(p)(p^{2\varepsilon} - 1)\Big).$$

Suppose the primes in $\mathcal{P}$ satisfy $p < z = D^{1/s}$ with $s \geqslant 1$ and $g(p)$ obeys the one-sided dimension bound (5.38). Let $\varepsilon = \alpha/2\log z$ with $\alpha > 0$ to be chosen. By Lemma 5.1 we have

$$\begin{aligned}\sum_{p|P} g(p)(p^{2\varepsilon}-1) &\leqslant \kappa\int_1^z \frac{w^{2\varepsilon}-1}{w\log w}dw + (z^{2\varepsilon}-1)\log K \\ &= \kappa c(\alpha) + (e^\alpha - 1)\log K\end{aligned}$$

where

$$c(\alpha) = \int_0^\alpha \frac{e^v - 1}{v} dv\ . \tag{7.72}$$

Hence we conclude that

$$V(z)I(D,P) \leqslant K^{e^\alpha - 1}\exp\Big(\kappa\, c(\alpha) - \frac{\alpha s}{2}\Big) = Q(\alpha)\ , \tag{7.73}$$

say. Note that

$$c(\alpha) = \sum_{n=1}^\infty \frac{\alpha^n}{n\ n!} \leqslant e^\alpha - 1 \tag{7.74}$$

so

$$Q(\alpha) \leqslant \exp\Big((e^\alpha - 1)k - \frac{\alpha s}{2}\Big)\ , \tag{7.75}$$

where we have put

$$k = \kappa + \log K\ . \tag{7.76}$$

The optimal choice of α in the upper bound (7.75) is

$$\alpha = \log\frac{s}{2k} \tag{7.77}$$

provided $s > 2k$, giving

$$V(z)I(D,P) \leqslant e^{-k}\Big(\frac{2ek}{s}\Big)^{s/2}\ . \tag{7.78}$$

Hence we conclude:

PROPOSITION 7.3. *For $s = \log D/\log z > 2k$ we have*

$$1 - e^{-k}\Big(\frac{2ek}{s}\Big)^{s/2} \leqslant V(z)J(D,P) \leqslant 1\ . \tag{7.79}$$

Note that the lower bound in (7.79) is positive for $s > 2k$. Inserting (7.79) into Theorem 7.1, one gets from Selberg's sieve an explicit upper bound for the sifting function $S(\mathcal{A}, z)$.

THEOREM 7.4. *Assume the conditions of Theorem 7.1 hold with a multiplicative function g which satisfies (5.38). Suppose the primes in $\mathcal{P}$ satisfy $p < z = D^{1/s}$ with $s > 2k$ where $k = \kappa + \log K$. Then*

$$S(\mathcal{A}, z) \leqslant XV(z)\bigg(1 - e^{-k}\Big(\frac{2ek}{s}\Big)^{s/2}\bigg)^{-1} + \sum_{\substack{d|P\\ d<D}} \tau_3(d)|r_d(\mathcal{A})|\ . \tag{7.80}$$

The upper bound in (7.73) is useful only if $Q(\alpha) < 1$, giving

$$(V(z)J(D,P))^{-1} \leqslant (1-Q)^{-1} \leqslant 1 + Q(\alpha)(1-Q)^{-1} \tag{7.81}$$

where Q is the minimum of $Q(\alpha)$. Clearly, if $s > 2k$ then any sufficiently small α gives $Q(\alpha) < 1$. The condition $s > 2k$ is necessary. If we assume that $k \geqslant 1$ and s is a little bit larger than $2k$, namely $s \geqslant 2k+2$, then by (7.75) we obtain

$$Q < e^{-k}\Big(\frac{2ek}{s}\Big)^{s/2} \leqslant e\Big(1+\frac{1}{k}\Big)^{-k-1} < 1 - \frac{1}{4k} \tag{7.82}$$

and (7.81) yields:

LEMMA 7.5. *If $k = \kappa + \log K \geqslant 1$ and $s \geqslant 2k+2$, then*

$$J(D,P)^{-1} < V(z)(1 + 4kQ(\alpha)) \tag{7.83}$$

for any $\alpha > 0$, where $Q(\alpha)$ is given by (7.73).

In particular, choosing $\alpha = \log 2$, we have

$$Q(\alpha) = Ke^{c\kappa}2^{-s/2}\ , \tag{7.84}$$

where

$$c = \int_0^1 \frac{2^v - 1}{v}\,dv\ . \tag{7.85}$$

Moreover, choosing $\alpha = \log(s/2k)$, we have

$$Q(\alpha) \leqslant e^{-k}\Big(\frac{2ek}{s}\Big)^{s/2}\ . \tag{7.86}$$

7.5. From the Upper-Bound to the Lower-Bound

Selberg's idea of the Λ^2-sieve leads naturally to an upper bound. However, there is no direct interpretation of this for a lower bound. This is in contrast to Brun's idea in which the two appear quite symmetrically.

From Buchstab's formula one can in general fashion go from a lower-bound sieve to an upper-bound sieve and vice-versa. In this section we use this technique to derive a lower bound from the Selberg upper bound. Although the Buchstab identity does not seem a natural tool in connection with the Selberg sieve, nevertheless, it produces quite strong results when the dimension is large, just as did the upper bound.

Recall the Buchstab identity

$$S(\mathcal{A},z) = |\mathcal{A}| - \sum_{p|P(z)} S(\mathcal{A}_p,p)\ . \tag{7.87}$$

Hence it is clear that if $\Lambda^+ = (\lambda_d^+(D))$ is an upper-bound sieve of level D, then $\Lambda^- = (\lambda_d^-(D))$ is a lower-bound sieve of level D where $\lambda_1^-(D) = 1$ and, for $d > 1$,

$$\lambda_d^-(D) = -\lambda_{d/p}^+(D/p), \tag{7.88}$$

where p is the largest prime factor of d. For this lower-bound sieve we obtain

$$S(\mathcal{A},z) \geqslant XG^-(z) + R^-(\mathcal{A},D) \tag{7.89}$$

where

$$G^-(z) = \sum_{d|P(z)} \lambda_d^-(D)g(d) \tag{7.90}$$

and

$$R^-(\mathcal{A}, D) = \sum_{d|P(z)} \lambda_d^-(D) r_d(\mathcal{A}) \ . \tag{7.91}$$

In the remainder we can apply (7.24) to deduce that

$$|\lambda_d(D)| \leqslant \tau_3(d) \tag{7.92}$$

and so

$$\left|R^-(\mathcal{A}, D)\right| \leqslant \sum_{\substack{d|P(z) \\ d<D}} \tau_3(d)|r_d(\mathcal{A})| \ , \tag{7.93}$$

just as was the case (see (7.32)) for the original upper-bound sieve.

Now we deal with the main term $G^-(z)$. By the construction (7.88) we obtain

$$G^-(z) = 1 - \sum_{p|P(z)} g(p) J_p^{-1} \tag{7.94}$$

where J_p is the sum corresponding to (7.20), that is,

$$J_p = \sum_{\substack{d|P(p) \\ d<\sqrt{D/p}}} h(d) \ . \tag{7.95}$$

We have already estimated J_p^{-1} in (7.84). That bound is applicable with

$$s_p = \frac{\log D/p}{\log p} = \frac{\log D}{\log p} - 1 \ .$$

Assuming $\kappa \geqslant 1$ and $s \geqslant 2k + 3$ we get by (7.84),

$$J_p^{-1} < V(p)\{1 + 4k\sqrt{2}Ke^{c\kappa}2^{-\log D/2\log p}\} \ . \tag{7.96}$$

Inserting this into (7.94), we obtain

$$\begin{aligned} G^-(z) &> 1 - \sum_{p|P(z)} g(p)V(p)\{1 + 4k\sqrt{2}Ke^{c\kappa}2^{-\log D/2\log p}\} \\ &= V(z) - 4k\sqrt{2}Ke^{c\kappa} \sum_{p|P(z)} g(p)V(p)2^{-\log D/2\log p} \ . \end{aligned}$$

Next, by Lemma 5.1 we have

$$\sum_{p|P(z)} g(p)V(p)2^{-\log D/2\log p} \leqslant KV(z) \int_1^\infty 2^{-st/2} dt^\kappa + (K-1)V(z)2^{-s/2} \ .$$

Note that $2^{-s/2}$ is bounded by the above integral (integration by parts). Hence

$$G^-(z)V(z)^{-1} > 1 - 4\kappa\sqrt{2}K^2(2K-1)e^{c\kappa} \int_1^\infty 2^{-st/2} dt^\kappa \ . \tag{7.97}$$

Here we also applied the inequality $k < \kappa K$, which holds because $\kappa, K \geqslant 1$. We think of K being close to one so that the above bound is near to

$$f_\kappa(s) = 1 - 4\kappa\sqrt{2}e^{c\kappa} \int_1^\infty 2^{-st/2} dt^\kappa \ . \tag{7.98}$$

Recall that c is the constant given by (7.85). In this notation (7.97) becomes

$$G^-(z)V(z)^{-1} > f_\kappa(s) - (K-1)(2K^2+K+1)\big(1 - f_\kappa(s)\big) \ . \tag{7.99}$$

This inequality is useless if $f_\kappa(s) < 0$ so we state our result only for s with $f_\kappa(s) > 0$.

THEOREM 7.6. *Let $\kappa \geqslant 1$ and $s = \log D/\log z$ be larger than the zero of $f_\kappa(s)$. Assume also that $s \geqslant 2k + 3$. Then the main term $G^-(z)$ in the lower-bound sieve* (7.89) *satisfies*

$$G^-(z) \geqslant V(z)\{f_\kappa(s) - 4K^2(K-1)\} . \tag{7.100}$$

Hence

$$S(\mathcal{A}, P) \geqslant XV(z)\{f_\kappa(s) - 4K^2(K-1)\} - \sum_{\substack{d|P(z)\\ d<D}} \tau_3(d)|r_d(\mathcal{A})| . \tag{7.101}$$

In the s-aspect we have $f_\kappa(s) = 1+O(2^{-s/2})$ as $s \to \infty$. This could be improved substantially if we had chosen the parameter $\alpha = \alpha(s)$ much larger than $\alpha = \log 2$, however, the results would have been less explicit.

An important consideration of course is for which values of s do we have $f_\kappa(s) > 0$ so that the lower bound (7.100) is positive (apart from the second term $4K^2(K-1)$ which is small since, in practice, K can be made close to one; see the comments in Section 5.2). Using Stirling's formula (see (8.327) of [**66**]) we can estimate the integral in (7.98) by

$$\int_0^\infty 2^{-st/2} dt^\kappa = \Big(\frac{2}{s\log 2}\Big)^\kappa \Gamma(\kappa+1) < \sqrt{2\pi\kappa}\Big(\frac{2\kappa}{es\log 2}\Big)^\kappa\Big(1+\frac{1}{6\kappa}\Big)$$

because $\kappa \geqslant 1$. Hence,

$$f_\kappa(s) \geqslant 1 - 9\kappa\sqrt{d\kappa}\Big(\frac{d\kappa}{s}\Big)^\kappa \tag{7.102}$$

where

$$d = \frac{2}{\log 2}e^{c-1} = \frac{2}{\log 2}\exp\Big(\int_0^1 \frac{2^\nu - 1}{\nu}d\nu - 1\Big) = 2.44\ldots . \tag{7.103}$$

This shows that $f_\kappa(s)$ is positive for

$$s > s_\kappa = d\kappa\big(9\kappa\sqrt{\pi\kappa}\big)^{1/\kappa} = d\kappa + O(\log 2\kappa) . \tag{7.104}$$

Interestingly enough, our constant d agrees with the one obtained by Ankeny and Onishi [**1**] who used the more advanced analysis of differential-difference equations.

7.6. Selberg's Lower-Bound Sieve Directly

Our derivation of a lower-bound sieve from the Selberg upper-bound sieve in the previous section can be refined a bit. One could start from the Buchstab identity

$$\mathcal{S}(\mathcal{A}, z) = \mathcal{S}(\mathcal{A}, w) - \sum_{w \leqslant p < z} \mathcal{S}(\mathcal{A}_p, p) \tag{7.105}$$

with w relatively small ($2 \leqslant w < z$, $w = 2$ being the previous case), and use a strong lower bound for $\mathcal{S}(\mathcal{A}, w)$ rather than the exact value $S(\mathcal{A}, w) = |\mathcal{A}|$ at $w = 2$. The result would be slightly better because for small p the Selberg upper bound which we applied to all $\mathcal{S}(\mathcal{A}_p, p)$ is weaker than what we already produced by the combinatorial sieve (this will be evident soon in Section 7.9 where we evaluate the main term of Selberg's sieve asymptotically by means of differential-difference equations). However, we did gain at many other places p for which $\mathcal{S}(\mathcal{A}_p, p)$ is better estimated by Selberg's sieve than by the combinatorial sieve (well, only if κ is large). For immediate comparison we recall that the sifting limits s_κ obtained by these methods satisfy as $\kappa \to \infty$,

$$s_\kappa \sim c\kappa\ , \qquad c = 3.59\dots\ ,$$
$$s_\kappa \sim d\kappa\ , \qquad d = 2.44\dots\ .$$

In this section we show a third way to create a lower-bound sieve from an upper-bound sieve $\Lambda^+ = (\lambda_d^+)$ and this one yields

$$s_\kappa \sim 2\kappa\ .$$

The idea of this sieve (which is due to Selberg [**145**]) can be explained by starting from the obvious inequality

$$\mathcal{S}(\mathcal{A}, z) \geqslant \sum_{(n,P(w))=1} a_n \Bigg\{1 - \sum_{\substack{p|n \\ p<z}} 1\Bigg\}\Bigg(\sum_{\substack{d|n \\ d|P(z)}} \lambda_d^+\Bigg)\ . \tag{7.106}$$

In other words, this inequality asserts that the sequence of numbers given by (6.49) is a lower-bound sieve. One may see a resemblance of this setting with the sieve derived from (7.105). However, there are some differences which are worth pointing out before we proceed further.

In the derivation of (7.101) we applied a lower-bound sieve for $\mathcal{S}(\mathcal{A}, w)$, say $(\lambda_d^-(D))$ of level D, and upper-bound sieves for every $\mathcal{S}(\mathcal{A}_p, p)$, say $(\lambda_d^+(D/p))$ of level D/p. On the other hand, in the case of (7.106) we apply a fixed upper-bound sieve $\Lambda^+ = (\lambda_d^+)$ at all places. At first glance it seems we lose at every place p because we cannot adjust the level of $\Lambda^+ = (\lambda_d^+)$ according to the value of p; it has to be fixed at D/z to leave room for every $p < z$. Hence, one may think that larger losses occur at smaller places. That may not be the case, because the sifting range is now also fixed to $z > p$, which compensates the loss and actually produces some gain in spite of the worse level. There is also another gain in the set-up of (7.106), namely we now apply the same upper-bound sieve to the positive term, rather than a smaller lower-bound sieve. When the sieve dimension κ is large all of these gains are substantial.

In this section we perform with $\Lambda^+ = \Lambda^2$ because it is the most powerful upper-bound sieve when κ is large and z is expected to be relatively large. Selberg has worked out this sieve quite effectively, so we follow his original arguments closely (except for adjustments toward our own notation and technical shortcuts). In the next section we perform, in somewhat different fashion, more general transformations which could prove very useful, but leave the choice of weights to the reader for her/his own experimentation.

Let $\Lambda^+ = (\lambda_d^+)$ be given by

$$\lambda_d^+ = \sum_{[d_1,d_2]=d} \rho_{d_1}\, \rho_{d_2}$$

where $\rho_1 = 1$ and $\rho_d = 0$ if $d > x$, so Λ^+ is an upper-bound sieve of level x^2. Let $\Lambda^- = (\lambda_d^-)$ be the sequence with $\lambda_1^- = 1$, $\lambda_p^- = -1$ if $p < z$ and $\lambda_d^- = 0$ otherwise. This is a lower-bound sieve of level z. We consider the sequence $\Lambda = \Lambda^-\Lambda^+ = (\lambda_d)$

where $I(X,z)$ denotes the partial sum

$$I(X,z) = \sum_{\substack{d|P(z)\\ d\geqslant X}} h(d) .$$

This has (with slightly different notation) been estimated in (7.78), giving

$$I(X,z)V(z) \leqslant e^{-k}\left(\frac{2ek}{t}\right)^{t/2} ,$$

provided $t = 2\log X/\log z > 2k$. We need this bound for $X = x/z$. Because the level of the sieve $\Lambda = \Lambda^-\Lambda^+$ is $D = x^2z = z^s$ (so $s = \log D/\log z$ as usual) we have $x = z^{(s-1)/2}$ and $t = s-3$. Hence,

$$J^2V(D,z)V(z) \geqslant 1 - \frac{s+3}{2e^k}\left(\frac{2ek}{s-3}\right)^{(s-3)/2} \tag{7.119}$$

provided $s > 2k+3$. To complete the job we should estimate the normalizing factor J which can be quickly done by inserting (7.114) into (7.112). Trivially, we find that

$$J = \sum_{\substack{d|P(z)\\ d\leqslant x}} h(d)\min\Big\{1, \frac{\log x/d}{\log z}\Big\} \leqslant V(z)^{-1} ,$$

so we conclude:

THEOREM 7.7. *Let $g(d)$ be a density function which satisfies (5.38). Associated with it let $\Lambda = \Lambda^-\Lambda^+ = \Lambda^-\Lambda^2$ be the lower-bound sieve of level D described above. Then we have, provided $s \geqslant 2k+3$,*

$$V(D,z) \geqslant \left\{1 - \frac{s+3}{2e^k}\left(\frac{2ek}{s-3}\right)^{(s-3)/2}\right\} V(z) \tag{7.120}$$

where $s = \log D/\log z$, and $k = \kappa + \log K$. Hence,

$$S(\mathcal{A},z) \geqslant XV(z)\left\{1 - \frac{s+3}{2e^k}\left(\frac{2ek}{s-3}\right)^{(s-3)/2}\right\} - 2R_4(\mathcal{A},D) , \tag{7.121}$$

where

$$R_4(\mathcal{A},D) = \sum_{\substack{d|P(z)\\ d<D}} \tau_4(d)|r_d(\mathcal{A})| .$$

We remark that the right-hand side of (7.120) is positive if $k \geqslant 1$ and

$$s > 2k + 2\sqrt{2k\log k} + \log k + 9 . \tag{7.122}$$

However, choosing s slightly larger we obtain, by combining Theorem 7.7 with Theorem 7.4, the following result.

COROLLARY 7.8. *For $s \geqslant 2k + (2+c)\sqrt{2k\log k}$ we have*

$$S(\mathcal{A},z) = V(z)\left\{1 + O\big(k^{-c}\big)\right\} + 2\theta R_4(\mathcal{A},D) , \tag{7.123}$$

where $|\theta| \leqslant 1$ and the implied constant is absolute.

This result is of Fundamental Lemma type and displays the strength of Selberg's sieve with respect to large dimension. We remark that S. Blight (thesis, Rutgers 2010) has sharpened this result using a Selberg-type combination but with a Brun lower-bound sieve supported on products up to three primes rather than one as in the right-hand side of (7.106).

7.7. Three Formulas for the Composite Sieve $\Lambda^-\Lambda^2$

In the previous section we used the composition $\Lambda = \Lambda^-\Lambda^2$ of a very simple Brun lower-bound sieve $\Lambda^- = (\lambda_q^-)$, vanishing on composite numbers, with a Selberg Λ^2-sieve. In this section we provide some transformations of the composition $\Lambda = \Lambda^-\Lambda^2$ of an arbitrary lower-bound sieve Λ^- with an arbitrary Λ^2-sieve. We shall apply the results of this construction no further in this book so the reader may skip this section but we expect that the three formulas obtained, see (7.124), (7.127) and (7.135), will be handy for future applications.

We begin with the inequality

$$S(\mathcal{A}, z) \geqslant S^-(\mathcal{A}, z) = \sum_n a_n \Big(\sum_{\substack{q|n \\ q|P(z)}} \lambda_q^- \Big) \Big(\sum_{\substack{\ell|n \\ \ell|P(z)}} \rho_\ell \Big)^2$$

where λ_q are supported on $q < \Delta$ and ρ_ℓ are supported on $\ell < \sqrt{D/\Delta}$ with $1 < \Delta < D$. We have

$$S^-(\mathcal{A}, z) = XV(D, z) + R^-(\mathcal{A}, D) \ ,$$

where $R^-(\mathcal{A}, D)$ is the remainder of level D and $V(D, z)$ is the main term sum

$$V(D, z) = \sum_{d|P(z)} \lambda_d g(d)$$

with

$$\lambda_d = \sum_{[q, \ell_1, \ell_2] = d} \lambda_q^- \rho_{\ell_1} \rho_{\ell_2} \ .$$

Our goal is to evaluate $V(D, z)$ in practical terms and we shall proceed in several directions. For notational simplicity we shall not always display that the variables run over divisors of $P(z)$, a property usually evident from the context.

We regard $V(D, z)$ as a quadratic form in the variables ρ_ℓ and we are going to express $V(D, z)$ in the variables y_d given by (7.13). As in the original Λ^2-sieve, which here represents the special case $\lambda_1^- = 1$, $\lambda_q^- = 0$ for $q \neq 1$, we shall use the alternative notation $V(D, z) = G$ to depict an analogy with the earlier arguments. We begin with the formula (5.81) for the sieves $\Lambda' = \Lambda^-$ and $\Lambda'' = \Lambda^2$ getting

$$G = \sum_q \lambda_q^- g(q) H(q), \quad \text{with} \quad H(q) = \sum_{m|q} G_m \ ,$$

where

$$G_m = \frac{1}{h(m)} \sum_{[\ell_1, \ell_2] \equiv 0 \, (\mathrm{mod}\, m)} \rho_{\ell_1} \rho_{\ell_2} g([\ell_1, \ell_2]) \ .$$

Note that G_m is a generalization of (7.7). Next, we are going to express G_m in terms of the vector (y_d) which diagonalizes the form G in the unadorned Λ^2-sieve, but of course we do not expect to get a diagonal form also for G_m. Inserting the choice (7.15) into G_m, we obtain

$$G_m = \frac{1}{h(m)} \sum_{d_1} \sum_{d_2} h(d_1)h(d_2) y_{d_1} y_{d_2} \sum_{\substack{\ell_1 | d_1 \\ [\ell_1,\ell_2]\equiv 0\,(m)}} \sum_{\ell_2 | d_2} \frac{\mu(\ell_1)\mu(\ell_2)}{g\big((\ell_1,\ell_2)\big)} .$$

Put $d = (d_1, d_2)$, $d_1 = dd'$, $d_2 = dd''$ with $dd'd''$ squarefree. The inner sum vanishes unless $d'd'' = m/(d,m)$ giving

$$G_m = \sum_d \frac{h^2(d)}{h\big((d,m)\big)} \sum\sum_{d'd''=m/(d,m)} y_{dd'} y_{dd''} \sum_{\ell_1} \sum_{\ell_2}$$

where the inner sum is

$$\sum_{\substack{\ell_1 | dd' \\ [\ell_1,\ell_2]\equiv 0\,(m)}} \sum_{\ell_2 | dd''} \frac{\mu(\ell_1)\mu(\ell_2)}{g\big((\ell_1,\ell_2)\big)} = \sum_{\substack{\ell\ell' | dd' \\ \ell\ell'\ell''\equiv 0\,(m)}} \sum_{\ell\ell'' | dd''} \frac{\mu(\ell'\ell'')}{g(\ell)}$$

$$= \sum_{\substack{k | [d,m] \\ k\equiv 0 (m)}} \mu(k) \sum_{\ell | (d,k)} \frac{\mu(\ell)}{g(\ell)} \tau\Big(\frac{(d,k)}{\ell}\Big) = \sum_{\substack{k | [d,m] \\ k\equiv 0 (m)}} \mu(k) \prod_{p | (d,k)} \Big(1 - \frac{1}{h(p)}\Big)$$

$$= \mu(m) \prod_{p | (d,m)} \Big(1 - \frac{1}{h(p)}\Big) \sum_{k | d/(d,m)} \mu(k) \prod_{p | k} \Big(1 - \frac{1}{h(p)}\Big)$$

$$= \mu(m) \frac{h((d,m))}{h(d)} \prod_{p | (d,m)} \Big(1 - \frac{1}{h(p)}\Big) .$$

Thus we obtain

$$G_m = \mu(m) \sum_d h(d) \prod_{p | (d,m)} \Big(1 - \frac{1}{h(p)}\Big) \sum\sum_{d'd''=m/(d,m)} y_{dd'} y_{dd''} .$$

This is a generalization of (7.14). Hence we compute the convolution

$$H(q) = \sum_{m|q} G_m = \sum_d h\Big(\frac{d}{(d,q)}\Big) \sum_{d'd''|q} \mu(d'd'') y_{dd'} y_{dd''}$$

$$= \sum_{(d,q)=1} h(d) \Big(\sum_{c|q} \mu(c) y_{cd}\Big)^2 .$$

This yields our first expression for the main term (see also Section 7.3.1 of [76]):

$$G = \sum_{qd|P} \lambda_q^- g(q) h(d) \Big(\sum_{c|q} \mu(c) y_{cd}\Big)^2 . \tag{7.124}$$

The formula (7.124) is a generalization of (7.109). Indeed, if $\lambda_1^- = 1$, $\lambda_p^- = -1$ and $\lambda_q^- = 0$ elsewhere in (7.124) then (7.124) can be easily arranged into the quadratic form (7.109).

Given the sieve (λ_q^-) of level Δ it would be interesting to determine the maximum of the quadratic form G in (7.124) with respect to the variables y_d for $d \mid P(z)$, $1 \leqslant d < \sqrt{D/\Delta}$, subject to the normalization

$$\sum_d h(d) y_d = 1 . \tag{7.125}$$

Next we present somewhat different transformations of the main term G which lead to nice formulas in terms of

$$\theta_b^+ = \sum_{q|b} \lambda_q^- ,$$

rather than λ_q^-. In this context we return to the more descriptive notation $V(D,z)$ for G. We start from the expression (5.80),

$$V(D,z) = \sum_{b|P(z)} h(b)\theta_b^- \theta_b^2$$

where by (7.18) we have

$$\theta_b = \sum_{\ell|b} \rho_\ell = \sum_{c|b} \mu(c) L_c(y) .$$

Here, $L_c(y)$ is the linear form in $y = (y_d)$ given by

$$L_c(y) = \sum_d h(d) y_{cd} .$$

Hence

$$V(D,z) = V(z) \sum_{b|P(z)} h(b)\theta_b^- \Big(\sum_{c|b} \mu(c) L_c(y)\Big)^2 . \tag{7.126}$$

Squaring out and changing the order of summation brings us to our second expression for the main term:

$$V(D,z) = \sum_c \mu(c) Q_c(y) Z(c) , \tag{7.127}$$

where $Q_c(y)$ is the quadratic form in $y = (y_d)$,

$$Q_c(y) = \sum_{[c_1,c_2]=c} \mu((c_1,c_2)) L_{c_1}(y) L_{c_2}(y) \tag{7.128}$$

and

$$Z(c) = V(z) \sum_{\substack{b|P(z)\\ b\equiv 0(c)}} h(b)\theta_b^- .$$

On opening the convolution $\theta^- = 1 * \lambda$ one could have immediately derived the alternative formula

$$Z(c) = \sum_q \lambda_q^- g([c,q]) , \tag{7.129}$$

but our intention is to keep this in terms of θ_b^- because of its pleasing properties.

For the remainder of this section we assume that $\Lambda^- = (\lambda_q^-)$ is the Brun sieve introduced in Section 6.2. Precisely, $\lambda_q^- = \mu(q)$ for $q \in \mathcal{D}^-$ (see (6.23)) and $\lambda_q^- = 0$ elsewhere. For $c = 1$ we have

$$Z(1) = V(z) \sum_{b|P(z)} h(b)\theta_b^-$$

which is also equal to the main term of the sieve Λ^-:

$$V^-(\Delta,z) = \sum_{q|P(z)} \lambda_q^- g(q) . \tag{7.130}$$

Moreover, we have by the normalization condition (7.125),

$$Q_1(y) = 1 \ . \tag{7.131}$$

Now let $c \neq 1$. Then $b \neq 1$ in $Z(c)$ so θ_b^- is given by the formula (6.41), that is

$$\theta_b = - \sum_{n \text{ even}} \sum_{\substack{ap_1 \dots p_n = b \\ (a, P(p_n))=1}}^{\star} 1 \ ,$$

where Σ^* restricts $p_1 \dots p_n$ by the conditions (6.42). Hence, writing $b = [c, p_1 \dots p_n] b'$ with $b' \mid P(z)$, $\big(b', cp_1 \dots p_n P(p_n)\big) = 1$, we find

$$Z(c) = -V(z) \sum_{n \, \text{even}} \sum_{\substack{p_1 \dots p_n | P(z) \\ (c, P(p_n))=1}}^{\star} h\big([c, p_1 \dots p_n]\big) \prod_{\substack{p | P(z), \, p \geqslant p_n \\ p \nmid cp_1 \dots p |_n}} (1 + h(p)) \ .$$

Note that $1 + h(p) = \big(1 - g(p)\big)^{-1}$. Completing the product, we arrange this as follows:

$$Z(c) = -g(c) \sum_{n \, \text{even}} V_n(z; c) \ , \tag{7.132}$$

where

$$V_n(z; c) = \sum_{\substack{y_n \leqslant p_n < \dots < p_1 \\ p_m < y_m, m < n, m \, \text{even} \\ (c, P(p_n)=1}} \cdots \sum g_c(p_1 \dots p_n) V(p_n) \tag{7.133}$$

and $g_c(d) = g\big(d/(c,d)\big)$. Note that $V_n(z;c)$ is a generalization of $V_n(z)$ as given in (6.34).

Finally, adding the above formulas we obtain

$$V(D, z) = V^-(\Delta, z) - \sum_{c \neq 1} \mu(c) g(c) Q_c(y) \sum_{n \, \text{even}} V_n(z; c) \tag{7.134}$$

where $Q_c(y)$ is given by (7.128) and $V_n(z;c)$ by (7.133). Here, for $c = 1$ we have $Q_1(y) = 1$ and $V_n(z; 1) = V_n(z)$ is given in (6.34). Therefore, adding and subtracting these terms in the right-hand side of (7.134), we conclude by (6.32) our third expression for the main term:

$$V(D, z) = V(z) - \sum_{c} \mu(c) g(c) Q_c(y) \sum_{n \, \text{even}} V_n(z; c) \ . \tag{7.135}$$

The formulas (7.124) and (7.135) for $G = V(D, z)$ are quite different; the first one is convenient when the Brun sieve $\Lambda^- = (\lambda_q^-)$ lives on almost-primes while the latter may serve better for the composition $\Lambda = \Lambda^- \Lambda^2$ with $\Lambda^- = (\lambda_q^-)$ being the full-fledged beta-sieve.

7.8. Clearing the Sifting Range and a Neat Bound for $J(D)$

Now we proceed to more precise estimates. It is easy to remove (for convenience) a few primes from the sifting range. Specifically, if $q|P$, then

$$J(D, P) \leqslant J(D, q) J(D, P/q) \ ,$$

whence

$$J(D, P/q) \geqslant V(q) J(D, P). \tag{7.136}$$

Let $P = P(z)$ be the product of all primes $p < z$. We begin by considering the sifting level $z \geqslant \sqrt{D}$, in which case the sifting range does not obstruct (7.71) so we have

$$J(D) = \sum_{d<\sqrt{D}}^{\flat} h(d).$$

We can estimate $J(D)$ strongly, elementarily, and neatly for the density functions $g(d)$ which satisfy the inequality $g(p)p \geqslant k$ for almost all p.

LEMMA 7.9. *Let g be a density function with*

$$g(p) \geqslant \frac{k}{p} \quad \text{if } p \nmid q\,, \tag{7.137}$$

for some positive integers k, q. Then

$$J(D) \geqslant \frac{(\log\sqrt{D})^k}{k!\,H_q}\Big(1 - \frac{k\ell(q)}{\log\sqrt{D}}\Big) \tag{7.138}$$

where

$$H_q = \prod_{p|q}(1 - g(p))\Big(1 - \frac{1}{p}\Big)^{-k} \tag{7.139}$$

and

$$\ell(q) = \sum_{p|q} g(p)\log p\,. \tag{7.140}$$

PROOF. We begin by factoring out the exceptional terms for which (7.137) may not hold. We have

$$J(D) = \sum_{\substack{ab<\sqrt{D}\\ a|q,\ (b,q)=1}}^{\flat}\sum^{\flat} h(a)h(b)\,.$$

For given $a < \sqrt{D}$ we need to estimate from below the sums

$$F(x) = \sum_{\substack{b<x\\ (b,q)=1}}^{\flat} h(b)\,.$$

Extending the definition of $g(d)$ from squarefree to all integers d by multiplicativity, we write

$$h(b) = \prod_{p|b}\big(g(p) + g(p^2) + \dots\big)\,.$$

Hence,

$$F(x) \geqslant \sum_{\substack{b<x\\ (b,q)=1}} g(b)\,.$$

Here, we have $g(b) \geqslant \tau_k(b)b^{-1}$ by (7.137), so we get

$$F(x) \geqslant \sum_{\substack{b<x\\ (b,q)=1}} \tau_k(b)b^{-1} \geqslant \Big(\frac{\varphi(q)}{q}\Big)^k \sum_{b<x}\tau_k(b)b^{-1}\,.$$

Since, for $x \geqslant 1$

$$\sum_{b<x} \tau_k(b)b^{-1} \geqslant \int \cdots \int_{\substack{x_1 \dots x_k < x \\ x_1,\dots,x_k \geqslant 1}} \frac{dx_1 \dots dx_k}{x_1 \dots x_k} = \frac{1}{k!}(\log x)^k$$

we obtain

$$F(x) \geqslant \frac{1}{k!}\Big(\frac{\varphi(q)}{q}\log x\Big)^k .$$

Hence

$$J(D) \geqslant \frac{1}{k!} \sideset{}{^\flat}\sum_{\substack{a<\sqrt{D} \\ a|q}} h(a)\Big(\frac{\varphi(q)}{q}\log\frac{\sqrt{D}}{a}\Big)^k .$$

Since $(1-y)^k \geqslant 1-ky$ for $0 \leqslant y \leqslant 1$, this gives

$$J(D) \geqslant \frac{1}{k!}\Big(\frac{\varphi(q)}{q}\log\sqrt{D}\Big)^k \sideset{}{^\flat}\sum_{a|q} h(a)\Big(1-\frac{k\log a}{\log\sqrt{D}}\Big) .$$

Now, the complete sums over the divisors of q are equal to

$$\sideset{}{^\flat}\sum_{a|q} h(a) = \prod_{p|q}(1+h(p)) = \prod_{p|q}(1-g(p))^{-1} = j(q)$$

say, and

$$\sideset{}{^\flat}\sum_{a|q} h(a)\log a = \sideset{}{^\flat}\sum_{a|q} h(a)\sum_{p|a}\log p = j(q)\ell(q) .$$

This completes the proof of (7.138). □

The lower bound (7.138) is very sharp; it agrees with the asymptotic we shall establish under less stringent conditions in the next section. Note that if $g(p)=0$ for $p \mid q$ then $\ell(q)=0$ so (7.138) becomes

$$J(D) \geqslant \frac{1}{k!}\Big(\frac{\varphi(q)}{q}\log\sqrt{D}\Big)^k . \tag{7.141}$$

Such a situation occurs when one is sifting the numbers in a segment of an arithmetic progression, in which case $k=1$, $g(d)=1/d$, and $h(d)=1/\varphi(d)$.

7.9. Asymptotic for $J(D,z)$

The argument in the previous section took advantage of the fact that we had an inequality for the density function g which held for all but finitely many primes. Now let $\kappa \geqslant 0$ and suppose we assume only that $g(p)p$ is κ, not exactly but on average, say we have

$$\sum_{p\leqslant x} g(p)\log p = \kappa\log x + O(1) \tag{7.142}$$

for all $x \geqslant 2$. Hence $g(p)\log p \ll 1$. Suppose also that

$$\sum_p g(p)^2\log p < \infty.$$

Since $h(p) = g(p) + O(g(p)^2)$ it follows that (7.142) holds for the relative density function h as well (with a different implied constant). Applying Theorem A.5 for $\mathcal{M}_h(\sqrt{D})$ we get

$$J(D) = c(\log \sqrt{D})^\kappa \left\{1 + O((\log D)^{-1})\right\}$$

where

$$c = \frac{1}{\Gamma(\kappa+1)} \prod_p \Big(1 - \frac{1}{p}\Big)^\kappa \big(1 - g(p)\big)^{-1}$$

and the implied constant depends on that in (7.142). If D is large in terms of this constant we can invert this approximation getting

$$J(D)^{-1} = 2^\kappa \Gamma(\kappa+1) H_g (\log D)^{-\kappa} \{1 + O((\log D)^{-1})\} \tag{7.143}$$

where

$$H_g = \prod_p \big(1 - g(p)\big)\Big(1 - \frac{1}{p}\Big)^{-\kappa}. \tag{7.144}$$

Now let $P = P(z)$ with $z \leqslant \sqrt{D}$. Suppose g satisfies (7.142). Then, by Theorem A.7 for $\mathcal{M}_h(x, z)$ with $x = \sqrt{D}$ we obtain

$$J(D, z)V(z) = \sigma(s) + O((\log D)^{-1}) \tag{7.145}$$

where $s = \log D / \log z \geqslant 2$ and $\sigma(s)$ is the continuous solution to the differential-difference problem

$$\begin{aligned} s^{-\kappa}\sigma(s) &= 2^{-\kappa} e^{-\gamma\kappa} \Gamma(\kappa+1)^{-1}, & &\text{if } 0 < s \leqslant 2, \\ s\sigma'(s) &= \kappa\sigma(s) - \kappa\sigma(s-2), & &\text{if } s > 2, \end{aligned} \tag{7.146}$$

that is, $\mathfrak{f}(s) = \sigma(2s)$ is the solution to (A.43). Note that the differential-difference equation (7.146) can be written as

$$\big(s^{-\kappa}\sigma(s)\big)' = \big(s^{-\kappa}\big)' \sigma(s-2) . \tag{7.147}$$

By (7.146), (7.32) and (7.30), we get a more highly developed estimate:

THEOREM 7.10. *Suppose g satisfies (7.142) and $D \geqslant z^2$. Then we have*

$$S(\mathcal{A}, z) < XV(z) \left\{ \frac{1}{\sigma(s)} + O\Big(\frac{1}{\log D}\Big) \right\} + R(\mathcal{A}, \Lambda^2) , \tag{7.148}$$

where $s = \log D / \log z$ and $\sigma(s)$ is given by (7.146). The implied constant depends only on g.

7.10. Explicit Estimates for the Main Term

The dependence on the sieve parameter s in the upper bound (7.148) is given by the function $s^\kappa/\sigma(s)$, $s \geqslant 2$. Since $\sigma(s)$ is not elementary we spend some time in this section giving more explicit estimates. Some bounds of this type follow already from Theorem 7.4 for $s > 2\kappa$ but here we are interested in s not so large.

Recall that for $0 < s \leqslant 2$ the function $s^\kappa/\sigma(s)$ is constant (see (7.146));

$$C(\kappa) = 2^\kappa e^{\gamma\kappa} \Gamma(\kappa+1) . \tag{7.149}$$

PROPOSITION 7.11. *For $\kappa \geqslant 0$ and $s \geqslant 2$ we have*

$$1 - \frac{s}{2}\Big(1 - \frac{2}{s}\Big)^\kappa \leqslant s^{-\kappa}\sigma(s)C(\kappa) \leqslant 1 .$$

PROOF. By (7.147) it follows that $s^{-\kappa}\sigma(s)$ is decreasing. Hence we get the upper bound. For the lower bound we integrate (7.147) obtaining

$$\begin{aligned} s^{-\kappa}\sigma(s) &= 2^{-\kappa}\sigma(2) + \int_2^s \sigma(t-2)dt^{-\kappa} \\ &\geqslant 2^{-\kappa}\sigma(2)\Big(1 + \int_2^s (t-2)^\kappa dt^{-\kappa}\Big) \\ &\geqslant 2^{-\kappa}\sigma(2)\Big(1 - \frac{s}{2}\Big(1-\frac{2}{s}\Big)^\kappa\Big) . \end{aligned}$$

To see the last inequality note that the function

$$\frac{s}{2}\Big(1-\frac{2}{s}\Big)^\kappa + \int_2^s (t-2)^\kappa dt^{-\kappa}$$

vanishes at 2 and is increasing since its derivative is

$$\frac{1}{2}\Big(1 + \frac{4\kappa}{s(s-2)}\Big)\Big(1-\frac{2}{s}\Big)^\kappa \geqslant 0 .$$

□

By the inequality $1 - x \leqslant e^{-x}$, Proposition 7.11 gives the simple bound

$$1 - \tfrac{1}{2}se^{-2\kappa/s} < s^{-\kappa}\sigma(s)C(\kappa) \leqslant 1 . \tag{7.150}$$

This lower bound is quite different from that in (7.79) and is of interest for smaller s relative to κ. For fixed s, (7.150) shows the exponential decay in the difference between upper and lower bounds as $\kappa \to \infty$. We can ask in which range of s are these asymptotically the same as $\kappa \to \infty$.

PROPOSITION 7.12. *For $\kappa \geqslant e$ and $2 \leqslant s \leqslant 2\kappa/\log\kappa$ we have*

$$1 - \frac{1}{\log\kappa} < s^{-\kappa}\sigma(s)C(\kappa) \leqslant 1 .$$

This result reveals that Selberg's upper bound sieve gives essentially the same outcome for $S(\mathcal{A}, z)$ in the whole range

$$D^{(\log\kappa)/4\kappa} < z \leqslant \sqrt{D} \tag{7.151}$$

as $\kappa \to \infty$. Hence, when κ is very large, sieving by primes above $D^{(\log\kappa)/4\kappa}$ has essentially no effect.

The same phenomenon occurs in more obvious fashion in the beta-sieve as we shall see in Chapter 11. There the function corresponding to $s^\kappa/\sigma(s)$ is $s^\kappa F(s)$ which is given by the constant $A(\kappa)$ in the larger range $s \leqslant \beta + 1$ with $\beta \sim c\kappa$ where $c = 3.591\ldots$ was defined in (6.11). Even though this range of stability is larger, the Selberg upper bound is much stronger when κ is large and s is relatively small because the constant $A(\kappa)$ exceeds $C(\kappa)$ by an exponential factor larger than e^κ. More precisely, by Stirling's formula and (11.198) we have

$$C(\kappa) \approx \big(2e^{\gamma-1}\kappa\big)^\kappa, \qquad 2e^{\gamma-1} = 1.310\ldots$$

and

$$A(\kappa) \approx (c\kappa)^\kappa, \qquad c = 3.591\ldots > 2e^\gamma .$$

On the other hand, when κ is fixed and $s \to \infty$, the Selberg upper bound is not as strong as that of the beta-sieve, even in the crude form of the Fundamental Lemma. Precisely, we have

PROPOSITION 7.13. *Let $\kappa \geqslant 0$. We have, for s sufficiently large,*

$$\sigma(s) = 1 - \exp(-\tfrac{1}{2}s\log s + O(s\log\log s)) \ ,$$

$$\frac{1}{\sigma(s)} = 1 + \exp\left(-\tfrac{1}{2}s\log s + O(s\log\log s)\right) \ ,$$

where the implied constant depends on κ.

PROOF. By (A.43) we have

$$s\mathfrak{f}'(s) = \kappa \int_{s-1}^{s} \mathfrak{f}'(t)dt \qquad \text{if } s > 1 \ ,$$

and $s\mathfrak{f}'(s) = s^{\kappa}e^{-\gamma\kappa}\Gamma(\kappa)^{-1}$ if $0 < s \leqslant 1$. Hence, it follows by Corollary 11.4, that

$$\mathfrak{f}'(s) = \exp(-s\log s + O(s\log\log s)) \ ,$$
$$\mathfrak{f}(s) = 1 - \exp(-s\log s + O(s\log\log s)) \ .$$

Hence

$$\sigma(s) = 1 - \exp(-\tfrac{1}{2}s\log s + O(s\log\log s)) \ ,$$

if s is sufficiently large. □

On the other hand, the corresponding function $F(s)$ in the upper bound of Theorem 11.13 satisfies (11.68) so it goes to 1 much more quickly (square the speed) than does $\sigma(s)^{-1}$.

7.11. Explicit Estimates for the Remainder

Once again we consider a class of sifting problems in which the individual remainder terms satisfy

$$|r_d(\mathcal{A})| \leqslant g(d)d. \tag{7.152}$$

Naturally, with this property one makes the condition

$$g(d)d \geqslant 1 \quad \text{if } d|P. \tag{7.153}$$

This implies $g([d_1,d_2])[d_1,d_2] \leqslant g(d_1)g(d_2)d_1d_2$. Therefore,

$$R(\mathcal{A},\Lambda^2) \leqslant \Big(\sum_{d<\sqrt{D}} |\rho_d| g(d)d\Big)^2 \leqslant \Big(\frac{1}{J}\sum_{m<\sqrt{D}} h(m)\sigma(m)\Big)^2$$

by (7.8), (7.15), and (7.21), where now $\sigma(m)$ denotes the sum of divisors of m.

Assuming that (see Section A.1 in Appendix A)

$$\sum_{y\leqslant p\leqslant x} g(p)\log p \ll \log(2x/y) \ , \tag{7.154}$$

we infer the same condition for $h(p)\sigma(p)p^{-1}$ and apply (A.8) obtaining

$$\sum_{m<\sqrt{D}} h(m)\sigma(m) \ll \frac{\sqrt{D}}{\log D}\sum_{m<\sqrt{D}} h(m)\sigma(m)m^{-1}.$$

Here we have

$$\sum_{m<\sqrt{D}} h(m)\sigma(m)m^{-1} \leqslant J\sum_{m} h(m)m^{-1} \ll J.$$

Hence we conclude that

$$R(\mathcal{A}, \Lambda^2) \ll D(\log D)^{-2}. \tag{7.155}$$

Combining with (7.30) we get

THEOREM 7.14. *Suppose the conditions of Theorem 7.1 hold. Moreover, assume (7.152), (7.153) and (7.154). Then we have*

$$S(\mathcal{A}, z) \leqslant \frac{X}{J} + O\left(\frac{D}{(\log D)^2}\right) \tag{7.156}$$

where $J = J(D)$ is given by (7.19), $D > 1$ is arbitrary, and the implied constant depends only on that in (7.154).

7.12. Selected Applications

Consider the sequence $\mathcal{A} = (a_n)$ which is the characteristic function of an arithmetic progression in a short interval

$$n \equiv a \pmod{q}, \qquad x < n \leqslant x + y \tag{7.157}$$

where $(a, q) = 1$ and $1 \leqslant q < y$. Let $\mathcal{P}$ consist of primes $p \leqslant \sqrt{y}$ with $p \nmid q$. Then

$$\pi(x+y; q, a) - \pi(x; q, a) \leqslant S(\mathcal{A}, \sqrt{y}) + \sqrt{y}\, q^{-1}. \tag{7.158}$$

On the other hand, the conditions of Theorem 7.14 are satisfied with $X = yq^{-1}$ and $g(d) = d^{-1}$ if $(d, q) = 1$. By (7.141) we have $J(D) > (\varphi(q)/2q) \log D$. Combining (7.156) with (7.158), we get

$$\pi(x+y; q, a) - \pi(x; q, a) < \frac{2y}{\varphi(q) \log D} + O\left(\frac{D}{(\log D)^2} + \frac{\sqrt{y}}{q}\right).$$

Choosing $D = yq^{-1}$ we conclude:

THEOREM 7.15. *For $(a, q) = 1$ and $1 \leqslant q < y$ we have*

$$\pi(x+y; q, a) - \pi(x; q, a) < \frac{2y}{\varphi(q) \log(y/q)} + O\left(\frac{y}{q(\log(y/q))^2}\right), \tag{7.159}$$

where the implied constant is absolute.

Using a quite different method, the large sieve inequality, H.L. Montgomery and R.C. Vaughan [**130**] have shown that the error term in (7.159) can be deleted.

Next we take $\mathcal{A} = (a_n)$ the characteristic function of the polynomial $n = (m - \alpha_1) \dots (m - \alpha_k)$ with $1 \leqslant m \leqslant x$, where all α_j are distinct. In this case $g(p) = \omega(p)p^{-1}$ where $\omega(p)$ is the number of roots modulo p. If p is sufficiently large then $\omega(p) = k$, so we have a k-dimensional sieve problem. By (7.146) and (7.148) we deduce

THEOREM 7.16. *Let $\mathfrak{a} = (\alpha_1, \dots, \alpha_k)$ be distinct integers which do not cover all residue classes to any prime modulus. Then, the number of integers $1 \leqslant m \leqslant x$ for which $m - \alpha_1, \dots, m - \alpha_k$ are all primes satisfies*

$$\pi(x; \mathfrak{a}) \leqslant 2^k k! H x (\log x)^{-k} \{1 + O(\log\log x / \log x)\}, \tag{7.160}$$

where

$$H = \prod_p \Big(1 - \frac{\omega(p)}{p}\Big)\Big(1 - \frac{1}{p}\Big)^{-k}. \tag{7.161}$$

REMARK. The upper bound (7.160) is larger by the factor $2^k\, k!$ than the conjectured asymptotic

$$\pi(x;\mathfrak{a}) \sim Hx(\log x)^{-k}.$$

To take the opposite extreme, if our polynomial is irreducible, say of degree r, then we have a sieve of dimension one and the upper bound provided by Theorem 7.10 exceeds the expected asymptotic by a factor of only $2r$.

7.13. Bounded Gaps Between Primes

One of the most exciting developments related to the Selberg sieve has come in the recent work of Goldston and Yildirim [**65**] and Goldston, Pintz and Yildirim [**64**] on small gaps between primes. In this section we illustrate their ideas, with emphasis on the sieve-theoretic aspects, to prove the following result of [**64**].

THEOREM 7.17. *Assume that the Bombieri–Vinogradov estimate* (9.83) *holds with* $Q = x^\theta$ *for some fixed exponent* $\theta > \frac{1}{2}$. *Then, there are infinitely many pairs of primes distanced by less than a constant depending only on* θ.

We begin by introducing a set $\mathcal{H} = \{h_1, \dots, h_k\}$ of integers, $1 \leqslant h_1 < \dots < h_k \leqslant M$. Let

$$Q(X) = \prod_{h\in\mathcal{H}} (X - h) \tag{7.162}$$

and $\omega(d)$ denote the number of solutions $\nu \pmod d$ of $Q(\nu) \equiv 0 \pmod d$, so $\omega(d)$ is multiplicative. We say that the set $\mathcal{H}$ is admissible if $\omega(p) < p$ for every prime p. Let Δ be a positive number which is divisible by every prime factor of

$$\det \mathcal{H} = \prod_{i<j} (h_i - h_j)^2 = \prod_{h\in\mathcal{H}} |Q'(h)|\ . \tag{7.163}$$

Since $\mathcal{H}$ is admissible, Δ is divisible by all primes $p \leqslant k$. Note that $\omega(p) = k$ if $p \nmid \Delta$. Associated with the set $\mathcal{H}$ we introduce two constants

$$\gamma(\mathcal{H}) = \prod_{p|\Delta} \Big(1 - \frac{\omega(p)}{p}\Big)\ , \qquad H(\mathcal{H}) = \prod_p \Big(1 - \frac{\omega(p)}{p}\Big)\Big(1 - \frac{1}{p}\Big)^{-k}. \tag{7.164}$$

For convenience we use the notation $\Lambda^\flat(n) = \mu^2(n)\Lambda(n)$, that is, $\log n$ if n is prime and zero otherwise. We consider the sequence $\mathcal{A} = (a_n)$ with

$$a_n = \sum_{h\in\mathcal{H}} \Lambda^\flat(n-h) - \log n\ .$$

Our goal is to prove that a_n is positive for infinitely many n. For such n it follows that at least two of the shifted integers $n - h$, $h \in \mathcal{H}$ are primes, therefore we get infinitely many solutions to the equation

$$p' - p = h' - h$$

with $p' > p$ primes and $h' > h$ in $\mathcal{H}$. Since the set $\mathcal{H}$ is contained in the segment $[1, M]$ we obtain infinitely often between pairs of primes, gaps which are bounded by M. We are not able to achieve this by summing the elements a_n themselves, but can do so by weighting the sequence in the style which is familiar from Selberg's lower-bound sieve (compare with (7.106)). Specifically, we consider the following weighted sum:

$$W(D,N) = \sum_{\substack{M<n\leqslant N \\ (Q(n),\Delta)=1}} \Big(\sum_{h\in\mathcal{H}} \Lambda^\flat(n-h) - \log N \Big) \Big(\sum_{d|Q(n)} \lambda_d \Big) , \tag{7.165}$$

where (λ_d) is an upper-bound sieve of level D. Note that we have replaced $\log n$ by $\log N$ which simplifies our presentation and goes in the right direction. We shall specify the sieve later and it will be a Λ^2–sieve. In anticipation of this we may assume (7.27). Since we restrict the outer summation to n with $\big(Q(n),\Delta\big)=1$ we shall assume throughout the section that (λ_d) is supported on squarefree numbers $d\leqslant D$, $(d,\Delta)=1$. Actually, the outer restriction $\big(Q(n),\Delta\big)=1$ is itself a sieve, which we separate from the Λ^2-sieve for technical reasons. It will of course be treated by Möbius inversion.

Our goal is to evaluate $W(D,N)$ asymptotically and choose the parameters so that it is positive. This implies that a_n is frequently positive since always

$$\sum_{d|Q(n)} \lambda_d \geqslant 0 ,$$

this being an upper-bound sieve. The role of this upper-bound sieve is to ensure that each of the shifted numbers $n-h$ is essentially an almost prime so we shall get

$$W(D,N) \asymp N(\log N)^{1-k} . \tag{7.166}$$

In fact, we shall establish Theorem 7.17 in a quantitative form.

THEOREM 7.18. *Assume that the Bombieri–Vinogradov bound* (9.83) *holds with* $Q=x^\theta$ *for some fixed* $\frac{1}{2}<\theta<1$. *Let* $\mathcal{H}=\{h_1,\dots,h_k\}$ *be an admissible set of* k *positive integers with* $k\geqslant 2(\theta-\frac{1}{2})^{-2}$. *Then, for* N *sufficiently large in terms of* θ *and* $\mathcal{H}$, *we have*

$$\sum\sum_{i<j} \sum_{n\leqslant N} \Lambda(n-h_i)\Lambda(n-h_j) \gg N(\log N)^{2-k} , \tag{7.167}$$

where the implied constant depends only on θ *and the set* $\mathcal{H}$.

PROOF. First, applying the general inequality $\sum_j c_j \leqslant 1+\sum\sum_{i<j} c_ic_j$, which holds for $0\leqslant c_j\leqslant 1$, we derive by (7.166) that

$$\sum_{n\leqslant N} \Big(\sum\sum_{i<j} \Lambda(n-h_i)\Lambda(n-h_j) \Big) \Big(\sum_{d|Q(n)} \lambda_d \Big) \gg N(\log N)^{1-k} .$$

Note that this lower bound has the right order of magnitude. Here, we can remove the sieve weights by Cauchy's inequality and by the standard sieve upper bound

$$\sum_{n\leqslant N} \Big(\sum_{d|Q(n)} \lambda_d \Big)^2 \ll N(\log N)^{-k} .$$

This yields (7.167). The correct order of magnitude in the lower bound (7.167) should be N, but we lost this when we removed the sieve weights.

For the evaluation of (7.165) we split the sum as

$$W(D,N)=\sum_{h\in\mathcal{H}}V_h(D,N)-U(D,N)\log N \tag{7.168}$$

where

$$V_h(D,N)=\sum_{\substack{M<n\leqslant N\\(Q(n),\Delta)=1}}\Lambda^\flat(n-h)\Big(\sum_{d|Q(n)}\lambda_d\Big) \tag{7.169}$$

and

$$U(D,N)=\sum_{\substack{M<n\leqslant N\\(Q(n),\Delta)=1}}\Big(\sum_{d|Q(n)}\lambda_d\Big)\,. \tag{7.170}$$

Both of these are sifted sums, however for different sequences, the first for $\Lambda^\flat$ and the second for the constant sequence. In the first case the linear factor $n-h$ in $Q(n)$ is already prime and the sieve is redundant for that factor so the sieve problem here is of dimension $k-1$. In the second sequence no factor is redundant so the sieve problem is of dimension k. In spite of this difference we are forced to use the same sieve weights for both in order to deduce the positivity of a_n for many n (the same idea as in (7.106)).

First we evaluate (7.170) because it can be done quickly. We split into arithmetic progressions $n\,(\text{mod }\Delta d)$, getting

$$U(D,N)=\sum_d\lambda_d\sum_{\substack{\alpha(\text{mod }\Delta)\\\big(Q(\alpha),\Delta\big)=1}}\;\sum_{\substack{\beta(\text{mod }d)\\Q(\beta)\equiv0\,(\text{mod }d)}}\left\{\frac{N-M}{\Delta d}+O(1)\right\}\,.$$

By Möbius inversion, the number of classes $\alpha\ (\text{mod }\Delta)$ with $(Q(\alpha),\Delta)=1$ is equal to $\gamma(\mathcal{H})\Delta$. Note that $\gamma(\mathcal{H})>0$ because $\mathcal{H}$ is an admissible set. For squarefree d with $(d,\Delta)=1$ we see that the number of classes $\beta\,(\text{mod }d)$ with $Q(\beta)\equiv0\,(\text{mod }d)$ is exactly $\tau_k(d)$. Hence, we conclude

$$U(D,N)=\gamma(\mathcal{H})NG(D)+O\left((\Delta D+M)(\log D)^{3k}\right)\,, \tag{7.171}$$

where

$$G(D)=\sum_{(d,\Delta)=1}\lambda_d\frac{\tau_k(d)}{d}$$

and the implied constant depends only on k.

Next we evaluate (7.169) in similar fashion. We have

$$V_h(D,N)=\sum_d\lambda_d\sum_{\substack{\alpha(\text{mod }\Delta)\\(Q(\alpha),\Delta)=1}}\;\sum_{\substack{\beta(\text{mod }d)\\Q(\beta)\equiv0(d)}}\;\sum_{\substack{M<n\leqslant N\\n\equiv\alpha(\Delta)\\n\equiv\beta(d)}}\Lambda^\flat(n-h)\,. \tag{7.172}$$

Here the condition $(\alpha-h,\Delta)=1$ in the inner sum is automatic since $\big(Q(\alpha),\Delta\big)=1$ while the condition $(\beta-h,d)=1$ is necessary because the $n-h$ are primes. This extra condition has the effect that the resulting sieve problem for V_h is of one dimension lower than the corresponding problem for U.

Note that when $(d,\Delta)=1$ with d squarefree, the number of classes $\beta\ (\mathrm{mod}\, d)$ such that $Q(\beta)\equiv 0\,(\mathrm{mod}\ d)$ and $(\beta-h,d)=1$ is equal to $\tau_{k-1}(d)$. Putting

$$E(N;q,a)=\sum_{\substack{n\leqslant N\\ n\equiv a(\mathrm{mod}\ q)}}\Lambda^\flat(n)-\frac{N}{\varphi(q)}$$

we get

$$V_h(D,N)=\gamma(\mathcal{H})\frac{\Delta N}{\varphi(\Delta)}\sum_d\lambda_d\frac{\tau_{k-1}(d)}{\varphi(d)}+R_h+O\left((\Delta D+M)(\log N)^{3k-2}\right),$$

where

$$R_h=\sum_d\lambda_d\sum_{\substack{\alpha(\Delta)\\ (Q(\alpha),\Delta)=1}}\sum_{\substack{\beta(d)\\ Q(\beta)\equiv 0(d)\\ (\beta-h,d)=1}}E(N;\Delta d,a)$$

and a is the residue class modulo Δd given by the congruences $a\equiv\alpha-h\ (\mathrm{mod}\,\Delta)$ and $a\equiv\beta-h\ (\mathrm{mod}\,d)$.

The above error term $O(\dots)$ takes care of a few terms where $n-h$ is a prime divisor of d and the correction for $\Lambda^\flat(m)$ with m between 1 and M. The implied constant depends only on k. To simplify this error term we now assume that $M\leqslant\sqrt{N}$.

The remainder sum R_h is estimated by Cauchy's inequality and the bound $E(N;q,a)\ll N/\varphi(q)$ as follows:

$$\begin{aligned}|R_h|&\leqslant\Delta\sum_{q\leqslant\Delta D}\tau_{k+3}(q)\max_{(a,q)=1}|E(N;q,a)|\\ &\ll\Delta\Big(\sum_{q\leqslant\Delta D}\tau_{k+3}^2(q)\varphi^{-1}(q)N\Big)^{\frac12}\Big(\sum_{q\leqslant\Delta D}\max_{(a,q)=1}|E(N;q,a)|\Big)^{\frac12}.\end{aligned}$$

Hence, using the Bombieri–Vinogradov type condition

$$\sum_{q\leqslant N^\theta}\max_{(a,q)=1}|E(N;q,a)|\ll N(\log N)^{-A},$$

where A is any positive number, we deduce that

$$R_h\ll\Delta N(\log N)^{-A}$$

provided $\Delta D\leqslant N^\theta$, where A is any positive number, the implied constant depending only on k and A. Hence,

$$V_h(D,N)=\gamma(\mathcal{H})NG'(D)+O\big(\Delta N(\log N)^{-A}\big),\tag{7.173}$$

where

$$G'(D)=\frac{\Delta}{\varphi(\Delta)}\sum_{(d,\Delta)=1}\lambda_d\frac{\tau_{k-1}(d)}{\varphi(d)}.$$

Combining the evaluations for U and V_h we conclude that

$$W(D,N)=\gamma(\mathcal{H})N\left\{kG'(D)-G(D)\log N\right\}+O\big(\Delta N(\log N)^{-A}\big),\tag{7.174}$$

provided that $\Delta D\leqslant N^\theta$ where θ is the assumed exponent of distribution for primes in arithmetic progressions.

It remains to choose the upper-bound sieve (λ_d) so that

$$kG'(D)-G(D)\log N\gg(\log N)^{1-k}\tag{7.175}$$

(see the asymptotic formula (7.195)). Here, the sieve problems for $G'(D)$, $G(D)$ are of dimension $k-1$ and k, respectively, with k rather large and experience shows that for large dimension the Λ^2-sieve is particularly strong. We model our strategy accordingly and choose

$$\lambda_d = \sum_{[d_1,d_2]=d} \rho_{d_1}\rho_{d_2} \ , \tag{7.176}$$

as in (7.4), where ρ_m are supported on squarefree numbers $m \leqslant \sqrt{D}$, $(m, \Delta) = 1$. In the following, even when not mentioned, all integer variables are squarefree and coprime to Δ.

By (7.1)–(7.14) we obtain

$$G(D) = \sum_d \frac{1}{h(d)} \Big(\sum_{m\equiv 0(d)} \rho_m \frac{\tau_k(m)}{m} \Big)^2 \tag{7.177}$$

where $h(d)$ is the multiplicative function such that $h(p) = k/(p-k)$. Similarly,

$$G'(D) = \frac{\Delta}{\varphi(\Delta)} \sum_d \frac{1}{h'(d)} \Big(\sum_{m\equiv 0(d)} \rho_m \frac{\tau_{k-1}(m)}{\varphi(m)} \Big)^2 \tag{7.178}$$

where $h'(d)$ is the multiplicative function such that $h'(p) = (k-1)/(p-k)$. We write

$$h(d) = \frac{\tau_k(d)}{f(d)} \ , \ h'(d) = \frac{\tau_{k-1}(d)}{f(d)} \ , \quad \text{where } f(d) = \prod_{p|d}(p-k) \ . \tag{7.179}$$

These expressions are convenient for diagonalization as in Section 7.1. Since the sieve weights ρ_m are the same in both expressions but the attached multiplicative functions are not, it is difficult to simultaneously diagonalize both quadratic forms $G(D)$, $G'(D)$. We make the decision to diagonalize $G(D)$ (because it has larger dimension) and then express the form $G'(D)$ in terms of the resulting new vectors.

As in (7.13) we put

$$y_d = \frac{\mu(d)}{h(d)} \sum_{m\equiv 0(d)} \rho_m \frac{\tau_k(m)}{m} \tag{7.180}$$

and we obtain the diagonal form

$$G(D) = \sum_{d\leqslant\sqrt{D}} \frac{\tau_k(d)}{f(d)} y_d^2 \ . \tag{7.181}$$

By Möbius inversion we express the original vector (ρ_m) in terms of the new vector (y_d) by

$$\rho_\ell = \mu(\ell)\frac{\ell}{\tau_k(\ell)} \sum_{b\equiv 0(\ell)} h(b) y_b \ , \tag{7.182}$$

see (7.15). In particular, the normalization condition $\rho_1 = 1$ becomes

$$\sum_d h(d) y_d = 1 \ . \tag{7.183}$$

Now, we are going to express the other form $G'(D)$ in these terms. We get

(7.184)
$$\sum_{\ell\equiv 0(d)} \rho_\ell \frac{\tau_{k-1}(\ell)}{\varphi(\ell)} = \sum_{\ell\equiv 0(d)} \mu(\ell)\frac{\ell}{\varphi(\ell)}\frac{\tau_{k-1}(\ell)}{\tau_k(\ell)}\Bigg(\sum_{m\equiv 0(\ell)} h(m)y_m\Bigg)$$
$$= \mu(d)\frac{d}{\varphi(d)}\frac{\tau_{k-1}(d)}{\tau_k(d)}\sum_{m\equiv 0(d)} h(m)y_m \sum_{cd|m}\mu(c)\frac{c}{\varphi(c)}\frac{\tau_{k-1}(c)}{\tau_k(c)},$$

on writing $\ell = cd$. The inner sum over c is equal to
$$\frac{f(m/d)}{\tau_k(m/d)\varphi(m/d)}.$$

Hence, (7.184) becomes

$$\sum_{\ell\equiv 0(d)} \rho_\ell \frac{\tau_{k-1}(\ell)}{\varphi(\ell)} = \mu(d)\frac{d}{f(d)}\tau_{k-1}(d)\sum_{m\equiv 0(d)}\frac{y_m}{\varphi(m)}, \tag{7.185}$$

giving

$$G'(D) = \frac{\Delta}{\varphi(\Delta)}\sum_{d\leqslant\sqrt{D}}\frac{d^2}{f(d)}\tau_{k-1}(d)\Bigg(\sum_{m\equiv 0(d)}\frac{y_m}{\varphi(m)}\Bigg)^2. \tag{7.186}$$

We need to remember the Δ coprimality conditions which, for notational simplicity, were omitted up to now. We choose y_d to be a continuous function of d, specifically,

$$y_d = \frac{1}{Y}\Big(\log\frac{\sqrt{D}}{d}\Big)^\ell \tag{7.187}$$

for $d \leqslant \sqrt{D}$, d squarefree, $(d,\Delta) = 1$ and $y_d = 0$ otherwise. Usually, in Selberg's sieve one chooses y_d to be constant, however, the above choice is essential for a successful conclusion; see the detailed discussion in the survey article [**151**] where the optimal choice of continuous function is presented. For the choice (7.187) the quadratic forms $G(D)$, $G'(D)$ become

$$Y^2G(D) = \sideset{}{^\flat}\sum_{\substack{d\leqslant\sqrt{D}\\(d,\Delta)=1}}\frac{\tau_k(d)}{f(d)}\Big(\log\frac{\sqrt{D}}{d}\Big)^{2\ell}, \tag{7.188}$$

(7.189)
$$Y^2G'(D) = \frac{\Delta}{\varphi(\Delta)}\sideset{}{^\flat}\sum_{\substack{d\leqslant\sqrt{D}\\(d,\Delta)=1}}\Big(\frac{d}{\varphi(d)}\Big)^2\frac{\tau_{k-1}(d)}{f(d)}\Bigg(\sideset{}{^\flat}\sum_{\substack{n\leqslant\sqrt{D}/d\\(n,\Delta d)=1}}\frac{1}{\varphi(n)}\Big(\log\frac{\sqrt{D}}{dn}\Big)^\ell\Bigg)^2,$$

and the normalization factor becomes (so that (7.183) holds):

$$Y = \sideset{}{^\flat}\sum_{\substack{d\leqslant\sqrt{D}\\(d,\Delta)=1}}\frac{\tau_k(d)}{f(d)}\Big(\log\frac{\sqrt{D}}{d}\Big)^\ell. \tag{7.190}$$

In order to achieve the positive lower bound (7.175) we need to compare $G'(D)$ and $G(D)$ within a quite precise factor (whereas the estimation of the normalization

Y does not need such precision). Therefore, we are going to evaluate asymptotically each of $G'(D)$, $G(D)$ to the extent that we can conclude

$$G'(D) = \frac{2\ell+1}{(\ell+1)(k+2\ell+1)} G(D)\big(\log D + O(1)\big) \ . \tag{7.191}$$

To this end we appeal to Corollary A.6 from Appendix A. From this corollary we derive first, for the inner sum in (7.189),

$$\sideset{}{^\flat}\sum_{\substack{n \leqslant \sqrt{D}/d \\ (n,\Delta d)=1}} \frac{1}{\varphi(n)}\Big(\log \frac{\sqrt{D}}{dn}\Big)^\ell = \frac{1}{\ell+1}\,\frac{\varphi(\Delta d)}{\Delta d}\Big(\log\frac{\sqrt{D}}{d}\Big)^\ell\Big\{\log\frac{\sqrt{D}}{d} + O(1)\Big\}\ .$$

Hence, $Y^2G'(D)$ is given by

$$\frac{1}{(\ell+1)^2}\,\frac{\varphi(\Delta)}{\Delta}\sum_{\substack{d\leqslant\sqrt{D}\\(d,\Delta)=1}}\frac{\tau_{k-1}(d)}{f(d)}\Big(\log\frac{\sqrt{D}}{d}\Big)^{2\ell}\Big\{\Big(\log\frac{\sqrt{D}}{d}\Big)^2 + O\Big(\log\frac{2\sqrt{D}}{d}\Big)\Big\}\ .$$

Applying Corollary A.6 again we derive

$$Y^2G'(D) = \frac{1}{(\ell+1)^2}\,\frac{(2\ell+2)!}{(2\ell+k+1)!}\,\frac{\gamma(\mathcal{H})}{H(\mathcal{H})}(\log\sqrt{D})^{k+2\ell+1}\Big\{1+O\Big(\frac{1}{\log D}\Big)\Big\}\ .$$

Again by the same corollary we derive

$$Y^2G(D) = \frac{(2\ell)!}{(2\ell+k)!}\,\frac{\gamma(\mathcal{H})}{H(\mathcal{H})}(\log\sqrt{D})^{k+2\ell}\Big\{1+O\Big(\frac{1}{\log D}\Big)\Big\}$$

and

$$Y = \frac{\ell!}{(\ell+k)!}\,\frac{\gamma(\mathcal{H})}{H(\mathcal{H})}(\log\sqrt{D})^{k+\ell}\Big\{1+O\Big(\frac{1}{\log D}\Big)\Big\}\ . \tag{7.192}$$

From these evaluations we conclude that

$$G(D) = \gamma(k,\ell)\frac{H(\mathcal{H})}{\gamma(\mathcal{H})}\big(\log\sqrt{D}\big)^{-k}\Big\{1+O\Big(\frac{1}{\log D}\Big)\Big\} \tag{7.193}$$

and

$$G'(D) = \frac{2(2\ell+1)}{(\ell+1)(k+2\ell+1)}\,\gamma(k,\ell)\,\frac{H(\mathcal{H})}{\gamma(\mathcal{H})}\big(\log\sqrt{D}\big)^{1-k}\Big\{1+O\Big(\frac{1}{\log D}\Big)\Big\}\ , \tag{7.194}$$

where

$$\gamma(k,\ell) = k!\binom{2\ell}{\ell}\binom{\ell+k}{\ell}\Big/\binom{2\ell+k}{\ell}\ .$$

Comparing (7.193) and (7.194) we obtain (7.191). Hence, by (7.168) we find that the weighted sum $W(D,N)$ defined in (7.165) satisfies

$$W(D,N) = \gamma(k,\ell)H(\mathcal{H})\{w(D,N)+O(1)\}N\big(\log\sqrt{D}\big)^{-k}\ , \tag{7.195}$$

where

$$w(D,N) = \frac{k}{k+2\ell+1}\,\frac{2\ell+1}{\ell+1}\log D - \log N\ . \tag{7.196}$$

This is the same expression obtained originally by Goldston, Pintz and Yildirim. Here, we can choose $\Delta D = N^\theta$ to be the level of distribution for primes in arithmetic progressions. Therefore, we can write

$$w(D,N) \asymp \log N \tag{7.197}$$

provided integers $1 \leqslant \ell \leqslant k$ can be found such that

$$\left(1+\frac{2\ell+1}{k}\right)\left(1+\frac{1}{2\ell+1}\right) < 2\theta \; . \tag{7.198}$$

Suppose the exponent of distribution for primes in arithmetic progressions is $\theta > \frac{1}{2}$. It is clear that, if ℓ is large and k is still considerably larger, then (7.198) holds and this proves (7.166), Theorem 7.18 and Theorem 7.17. □

As revealed by GPY, our choice of sieve weights has the effect of applying a $(k+\ell)$-dimensional sieve to a k-dimensional sieve problem.

Although we prefer to work with the diagonalizing vectors $\{y_m\}$ rather than with the original sieve weight constituents $\{\rho_m\}$ it is still worthwhile to see the latter explicitly. Letting $Y = Y_\Delta(\sqrt{D})$ be the sum (7.190) we find by (7.182), (7.180) and (7.187) that

$$\rho_m = \frac{\mu(m)m}{f(m)} Y_{\Delta m}(\sqrt{D}/m)/Y_\Delta(\sqrt{D}) \; .$$

Hence, by (7.192) we get the asymptotic formula

$$\rho_m = \mu(m)\left(1-\frac{\log m}{\log\sqrt{D}}\right)^{k+\ell}\left\{1+O\Big(\Big(\log\frac{\sqrt{D}}{m}\Big)^{-1}\Big)\right\} . \tag{7.199}$$

If one had chosen for ρ_m the exact values $\mu(m)(1-\log m/\log\sqrt{D})^{k+\ell}$ the final results would be asymptotically equivalent.

In the rest of this section we make a few observations with only a sketch of proofs.

Observation (A). One can check easily that, for given k, the best choice of ℓ is about $\frac{1}{2}\sqrt{k}$ and then for all θ larger than $\frac{1}{2}$ we have (7.198) if $k > 2\left(\theta-\frac{1}{2}\right)^{-2}$.

On the other hand, if $\theta < \frac{1}{2}$ then no such choice is possible, whereas, if $\theta = \frac{1}{2}$ we barely miss. Since $\theta = \frac{1}{2}$ is essentially available, by the Bombieri–Vinogradov Theorem 9.18, it was possible for Goldston, Pintz and Yildirim, using a courageous extension of the weighted sum $W(D,N)$, to prove unconditionally that

$$\liminf \frac{p_{n+1}-p_n}{\log p_n} = 0 \; . \tag{7.200}$$

We shall give a proof of this result in the next section. They later succeeded to prove unconditionally that

$$p_{n+1}-p_n \ll \sqrt{\log p_n}\,(\log\log p_n)^2 \tag{7.201}$$

infinitely often but we have no energy to include their sophisticated refinements in this book.

Observation (B). One can try to choose the Selberg parameters (7.187) as a continuous function other than a power of the logarithm. Indeed, Conrey and Soundararajan examined such a possibility, both finding and verifying the optimal choice, which interestingly is given in terms of Bessel functions. The results are sharper but the consequences are more or less the same. One can also try to attach to the parameters a more arithmetic factor, for example the multiplicative function $a^{\omega(d)}$ with $a>0$. This turns out to be pointless since analysis shows the best choice to be $a=1$.

OBSERVATION (C). The method of GPY can probably be generalized for many sequences in place of the primes, and sometimes unconditionally producing bounded gaps. A fascinating example is the theorem of Goldston, Graham, Pintz and Yildirim [**63**] that, infintely often, the gap between integers having precisely two prime factors is bounded by six.

With only a slight modification of the previous arguments we derive the following unconditional result.

THEOREM 7.19. *The gaps between numbers n of type*

$$n = p_1 p_2 \quad \text{with} \quad p_1 < p_2 < p_1^3 \tag{7.202}$$

are bounded by 660 *infinitely often.*

PROOF. Define $\Lambda''(n)$ to be the characteristic function of numbers of the type (7.202) multiplied by $\log n$. Replace $\Lambda^\flat(n-h)$ in the definition of $W(D,N)$ by $\Lambda''(n-h)$. As is the case for $\Lambda^\flat(n-h)$, the new sequence $\Lambda''(n-h)$ for $n \leqslant N$ has the level of distribution $D = N^{1/2-\varepsilon}$ by Theorem 9.17. However, the mean value of $\Lambda''(n)$ is a bit larger than that of $\Lambda^\flat(n)$, precisely we have

$$\sum_{n\leqslant N} \Lambda''(n) \sim \sum_{N^{1/4}<n<N^{1/2}} \frac{N\log N}{p\log(N/p)} \sim N\log 3\ .$$

The weights ρ_m of the Λ^2-sieve which are attached to $W(D,N)$ are supported on numbers $m < \sqrt{D} < N^{1/4}$, so they may be applied to $\Lambda''(n-h)$ in exactly the same way as was the case for $\Lambda^\flat(n-h)$. Therefore, the previous arguments yield (7.195) with $w(D,N)$ replaced by

$$\frac{k}{k+2\ell+1}\,\frac{2\ell+1}{\ell+1}(\log 3)\log D - \log N\ .$$

Now, for $D = N^{1/2-\varepsilon}$ this satisfies (7.197) provided

$$\Big(1+\frac{2\ell+1}{k}\Big)\Big(1+\frac{1}{2\ell+1}\Big) < \log 3.$$

Choosing $k = (2\ell+1)^2$ this condition becomes $(1+1/(2\ell+1))^2 < \log 3$, which holds for $\ell = 10$ and $k = 441$.

Now it remains to find an admissible set $\mathcal{H} = \{h_1,\dots,h_k\}$ with $1 \leqslant h_j \leqslant M$ for M as small as we can. One can easily see that any set of k distinct primes larger than k is admissible. Thus M can be determined by

$$\pi(M) = k + \pi(k)\ .$$

For $k = 441$ we find $\pi(M) = 441 + 85 = 526$, $M = p_{526} = 957$, $h_k = M = 957$, $h_1 = p_{k+1} = p_{442} = 297$. Hence, the set $\mathcal{H}$ is contained in an interval of length $h_k - h_1 = 660$. This proves Theorem 7.19. □

OBSERVATION (D). We briefly mentioned, at the construction of the weighted sum $W(D,N)$, that the role of the upper-bound sieve (λ_d) is to ensure that every number $n-h_1,\dots,n-h_k$ is essentially an almost prime (of course, not in the strict sense). To get more insight about how powerful this Λ^2-sieve is here, we try to determine how few prime divisors are left in $Q(n) = (n-h_1)\dots(n-h_k)$. Of course at least k, but interestingly enough, not many more. For this educational purpose we are going to evaluate the sum

$$U^*(D,N) = \sum_{\substack{M<n\leqslant N\\(Q(n),\Delta)=1}} \omega^*\big(Q(n)\big)\Big(\sum_{d|Q(n)}\lambda_d\Big), \tag{7.203}$$

where $\omega^*(Q)$ is the number of primes $p|Q$, $p\leqslant D$, which should be compared with the sum $U(D,N)$ given by (7.170).

Recall that for the latter we have the asymptotic formula

$$U(D,N)\sim\gamma(H)NG(D) \tag{7.204}$$

where

$$G(D)=\sum_d \lambda_d\frac{\tau_k(d)}{d}$$

(see (7.171)) and $G(D)$ is evaluated in (7.193).

Opening $\omega^*(Q)$, we derive in similar fashion that

$$U^*(D,N)\sim\gamma(H)NG^*(D) \tag{7.205}$$

where

$$G^*(D)=\sum_{\substack{p<D\\p\nmid\Delta}}\sum_d \lambda_d\frac{\tau_k([d,p])}{[d,p]}\ .$$

Here $[d,p]$ denotes the least common multiple of d and p. We also assume for simplicity that $D=N^\theta$ with $\theta<\frac{1}{2}$, so the remainder sums can be ignored. We have

$$G^*(D)=\sum_d\frac{\tau_k(d)}{d}\sum_{\substack{p<D\\p\nmid\Delta d}}\frac{k}{p}\left(\lambda_d+\lambda_{dp}\right).$$

For the λ's given by (7.176) we have (because $p\nmid d$)

$$\lambda_d+\lambda_{dp}=\sum_{[d_1,d_2]=d}\left(\rho_{d_1}\rho_{d_2}+2\rho_{pd_1}\rho_{d_2}+\rho_{pd_1}\rho_{pd_2}\right).$$

Hence we obtain, similarly to (7.177), the following expression:

$$G^*(D)=\sum_{\substack{p<D\\p\nmid\Delta}}\frac{k}{p}\sum_{(d,p)=1}\frac{1}{h(d)}\Bigg(\sum_{\substack{m\equiv0(d)\\(m,p)=1}}\left(\rho_m+\rho_{pm}\right)\frac{\tau_k(m)}{m}\Bigg)^2.$$

The condition $(m,p)=1$ is only needed for ρ_m because it is redundant for ρ_{pm}. One can remove this condition by the inclusion-exclusion principle and replace $\rho_m+\rho_{pm}$ by $\rho_m+\frac{p-k}{p}\rho_{pm}$. Then, by (7.180) we get

$$\sum_{m\equiv0(d)}\left(\rho_m+\rho_{pm}\right)\frac{\tau_k(m)}{m}=\mu(d)\,\frac{\tau_k(d)}{f(d)}\left(y_d-y_{pd}\right)$$

and so

$$G^*(D)=k\sum_d\frac{\tau_k(d)}{f(d)}\sum_{\substack{p<D\\p\nmid\Delta d}}p^{-1}\left(y_d-y_{pd}\right)^2.$$

One could try to minimize this quadratic form in the free variables y_d subject to the normalization condition (7.183), but we mustn't do so because we intend to compare $G^*(D)$ with $G(D)$. Therefore, we take the same values of y_d given by (7.187) getting

$$Y^2 G^*(D) = k \sum_{\substack{d<\sqrt{D} \\ (d,\Delta)=1}}^{\flat} \frac{\tau_k(d)}{f(d)}\Big(\log\frac{\sqrt{D}}{d}\Big)^{2\ell} \sum_{\substack{p<D \\ p\nmid \Delta d}} \frac{1}{p}\Big(1-\Big(\frac{\log^+\sqrt{D}/pd}{\log\sqrt{D}/d}\Big)^{\ell}\Big)^2 .$$

The contribution of primes $p<\sqrt{D}/d$ is asymptotically equal to

$$\eta(\ell)=\int_0^1 t^{-1}(1-(1-t)^\ell)^2 dt \tag{7.206}$$

and the remaining ones contribute

$$\sum_{\sqrt{D}/d\leqslant p<D} p^{-1} \sim \log\Big(\frac{\log D}{\log\sqrt{D}/d}\Big).$$

Next we execute the summation over $d<\sqrt{D}$ using (A.27) by partial summation (the first part is just a multiple of $G(D)$) getting

$$Y^2 G^*(D) \sim \frac{c}{\Gamma(k)}\int_1^{\sqrt{D}} \Big(\eta(\ell)+\log\Big(\frac{\log D}{\log\sqrt{D}/x}\Big)\Big)\Big(\log\frac{\sqrt{D}}{x}\Big)^{2\ell} d(\log x)^k .$$

This means

$$G^*(D)\sim \nu(k,\ell)k\, G(D) \tag{7.207}$$

where

$$\nu(k,\ell)=\eta(\ell)+\eta(k,\ell) \tag{7.208}$$

and

$$\eta(k,\ell)=\binom{2\ell+k}{k}\int_0^1 \log\Big(\frac{2}{1-t}\Big)(1-t)^{2\ell}\,dt^k. \tag{7.209}$$

It remains to compute the integrals $\eta(\ell)$ and $\eta(k,\ell)$. By (3.268.2) and (8.362.2) of [**66**] we get

$$\begin{aligned}\eta(\ell)&=2\psi(\ell+1)-\psi(2\ell+1)-\psi(\ell)\\ &=\sum_{0<n\leqslant\ell} n^{-1}-\sum_{\ell<n\leqslant 2\ell} n^{-1}=\log\frac{\ell}{2}+\gamma+O\Big(\frac{1}{\ell}\Big).\end{aligned}$$

Then, by (8.380.1) and (4.253.1) of [**66**] we have

$$\begin{aligned}\eta(k,\ell)&=\log 2+\psi(k+2\ell+1)-\psi(2\ell+1)\\ &=\log 2+\sum_{2\ell<n\leqslant k+2\ell} n^{-1}=\log\frac{k}{\ell}+O\Big(\frac{1}{\ell}+\frac{\ell}{k}\Big).\end{aligned}$$

Adding these integrals we find that

$$\nu(k,\ell)=\sum_{0<n\leqslant\ell} n^{-1}-\sum_{\ell<n\leqslant 2\ell} n^{-1}+\sum_{2\ell<n\leqslant k+2\ell} n^{-1}+\log 2=\log k+O(1)$$

if $1\leqslant\ell\leqslant k$, which is our case.

By (7.204), (7.205) and (7.207) we have

$$U^*(D,N) \sim \nu(k,\ell)kU(D,N). \tag{7.210}$$

This says that, after sieving, the average number of prime factors remaining in $Q(n) = (n-h_1)\dots(n-h_k)$ is about $k\log k$, so on average $\log k$ per factor. Actually, Heath-Brown [**89**] proved (in a similar context) that there are infinitely many n such that each number $n-h$ has about $O(\log k)$ prime factors.

7.14. Small Gaps Between Primes

In this section we prove the following unconditional result of Goldston, Pintz and Yildirim [**64**].

THEOREM 7.20. *The gaps between consecutive primes satisfy*

$$\liminf \frac{p_{n+1}-p_n}{\log p_n} = 0\ . \tag{7.211}$$

The proof involves an extension of the arguments of the previous section with interesting further innovations. We use the same notation as in the previous section. Let $\mathcal{M}$ be a set of positive integers $m \leqslant M$ containing the admissible set $\mathcal{H}$. For every $m \in \mathcal{M}\setminus\mathcal{H}$ we add to the weighted sum $W(D,N)$ defined in (7.165), the amount

$$V_m(D,N) = \sum_{\substack{M<n\leqslant N\\ (Q(n),\Delta)=1}} \Lambda^\flat(n-m)\Big(\sum_{d|Q(n)}\lambda_d\Big)\ ; \tag{7.212}$$

in other words, we consider the extended weighted sum

$$W_{\mathcal{M}}(D,N) = \sum_{\substack{M<n\leqslant N\\ (Q(n),\Delta)=1}} \Big(\sum_{m\in\mathcal{M}}\Lambda^\flat(n-m) - \log N\Big)\Big(\sum_{d|Q(n)}\lambda_d\Big)\ . \tag{7.213}$$

Once again, by the same reasoning, we infer that if $W_{\mathcal{M}}(D,N)$ is positive, then there are at least two primes $p<p'<N$ with $p'-p = m'-m < M$, $m', m\in\mathcal{M}$.

These new terms $V_m(D,N)$ with $m\in\mathcal{M}\setminus\mathcal{H}$ give individual contributions much smaller, by a factor of $\log N$, than those of $V_h(D,N)$ with $h\in\mathcal{H}$. The reason is that the primes $n-m$ are not redundant in the Selberg sieve weights. Their deficiency in size will be compensated by their abundance. Eventually, we shall choose $M=\delta(k)\log N$ with $\delta(k)\to 0$ as $k\to\infty$ but much more slowly than $N\to\infty$. This will yield an extra total contribution of size proportional to $\delta(k)U(D,N)$. Using the earlier argument but taking the exponent of distribution for primes in arithmetic progressions to be $\theta = \frac{1}{2}-\varepsilon$, as is provided by the Bombieri–Vinogradov theorem, our estimate for the weighted sum $W(D,N)$ comes up negative of size $-2k^{-1/2}U(D,N)$. This shortfall will be overcome by the extra terms as long as $\delta(k)$ is larger than $5k^{-1/2}$ which still leaves room for $\delta(k)\to 0$ as $k\to\infty$.

Recall that $f(d)$ is the multiplicative function with

$$f(p) = p-\omega(p)$$

where $\omega(p)$ is the number of roots of $Q(\alpha) \equiv 0 \,(\mathrm{mod}\; p)$ so $f(p)$ is the number of classes $\alpha\,(\mathrm{mod}\; p)$ with $(Q(\alpha), p) = 1$. For every $m \in \mathcal{M} \setminus \mathcal{H}$ we put

$$\gamma^{(m)}(\mathcal{H}) = \prod_{\substack{p|\Delta \\ p\nmid Q(m)}} \Big(1 - \frac{1}{f(p)}\Big) . \tag{7.214}$$

Recall that $\omega(p) = k < p$ if $p \nmid \Delta$ so $f(p) = p - k$. We denote

$$\gamma_k(\Delta) = \prod_{p\nmid\Delta}\Big(1 - \frac{1}{p-k}\Big)\Big(1 - \frac{1}{p}\Big)^{-1} . \tag{7.215}$$

The new ingredient in the proof of Theorem 7.20 is the following bound.

LEMMA 7.21. *For $m \notin \mathcal{H}$ we have*

$$V_m(D,N) \geqslant \gamma^{(m)}(\mathcal{H})\gamma_k(\Delta)\frac{\Delta N}{\varphi(\Delta)}U(D,N)\Big\{1 + O\big(\frac{1}{\log N}\big)\Big\} , \tag{7.216}$$

where the implied constant depends only on Δ and $\mathcal{H}$.

REMARKS. With extra work we could establish an asymptotic equivalence between $V_m(D,N)$ and $U(D,N)$ but this is not needed and we can take advantage of this fact to simplify the argument. Clearly, $\gamma_k(\Delta) < 1$ but it can be made close to one by choosing Δ to be divisible by every prime $p \leqslant K$ with K sufficiently large. Indeed, for $K > k$ we have

$$\gamma_k(\Delta) \geqslant \prod_{\ell > K}\Big(1 - \frac{1}{\ell - k}\Big)\Big(1 - \frac{1}{\ell}\Big)^{-1} = 1 - \frac{k}{K} . \tag{7.217}$$

Note that if $p = k+1$ is a prime which does not divide Δ, then $\gamma_k(\Delta) = 0$. This peculiar case occurs because we shall change $\varphi(d)$ to d in (7.220) to obtain (7.223); this could have been avoided with a little extra complication. Even if $K = k+1$ we can be sure that $\gamma_k(\Delta)$ is not dangerously small. We have $\gamma_k(\Delta) \geqslant \gamma_k$, where

$$\gamma_k = \prod_{p \geqslant k+2}\Big(1 - \frac{1}{p-k}\Big)\Big(1 - \frac{1}{p}\Big)^{-1} .$$

As an exercise, prove that

$$\sum_{K \leqslant k < 2K} \log\frac{1}{\gamma_k} \ll K$$

for $K \geqslant 1$, where the implied constant is absolute. Hence, γ_k is bounded below by a positive absolute constant for some k in any dyadic segment.

We begin the proof of Lemma 7.21 by splitting $V_m(D,N)$ into residue classes $\alpha\,(\mathrm{mod}\;\Delta)$ and $\beta\,(\mathrm{mod}\; d)$ in the same style as we did in (7.180) for m in $\mathcal{H}$. The condition $(Q(\alpha), \Delta) = 1$ is required for the same reason, but now for $m \notin \mathcal{H}$, the condition $(\alpha - m, \Delta) = 1$ is not automatic and it is necessary if $n - m$ is to be prime. The number of such classes is

$$\sum_{\substack{\alpha\,(\mathrm{mod}\;\Delta) \\ ((\alpha-m)Q(\alpha),\Delta)=1}} 1 = \Delta\sum_{\delta|\Delta}\frac{\mu(\delta)}{\delta}\Big(\sum_{\substack{\alpha\,(\mathrm{mod}\;\Delta) \\ (\alpha-m)Q(\alpha)\equiv 0(\delta)}} 1\Big)$$

$$= \Delta \prod_{\substack{p|\Delta \\ p\nmid Q(m)}}\Big(1 - \frac{\omega(p)+1}{p}\Big) \prod_{\substack{p|\Delta \\ p|Q(m)}}\Big(1 - \frac{\omega(p)}{p}\Big) = \Delta\,\gamma(\mathcal{H})\gamma^{(m)}(\mathcal{H}) .$$

We also need to revise the formula for the number of classes $\beta \,(\mathrm{mod}\, d)$ for d squarefree with $(d, \Delta) = 1$. Now, these are the classes which satisfy $Q(\beta) \equiv 0 \,(\mathrm{mod}\, d)$ with the new restriction $(\beta - m, d) = 1$. Therefore, the number of such classes is equal to

$$\tau_k^{(m)}(d) = \tau_k(a)\tau_{k-1}(b) \tag{7.218}$$

where $ab = d$ with $(a, Q(m)) = 1$ and $b \mid Q(m)$.

Now we apply the Bombieri–Vinogradov bound getting

$$V_m(D, N) = \gamma^{(m)}(\mathcal{H})\gamma(\mathcal{H})\frac{\Delta N}{\varphi(\Delta)}G^{(m)}(D) + O\big(\Delta N(\log N)^{-A}\big) \tag{7.219}$$

provided that ΔD does not exceed the level of distribution for primes in arithmetic progressions, the implied constant depending only on k and A. Here, the new sum over the sieve weights

$$G^{(m)}(D) = \sum_{(d,\Delta)=1} \lambda_d \frac{\tau_k^{(m)}(d)}{\varphi(d)} \tag{7.220}$$

closely resembles the sum $G(D)$ from (7.179) with d replaced by $\varphi(d)$ (counting all integers versus counting primes) and $\tau_k(d)$ replaced by $\tau_k^{(m)}(d)$. The sum V_m with $m \in \mathcal{M} \setminus \mathcal{H}$ is, despite the notation, more closely related to the sum U than it is to V_h with $h \in \mathcal{H}$, due to the sieve dimension, the factor $\tau_k(a)$ in (7.218) being more important than the factor $\tau_{k-1}(b)$. Consequently, in what follows we shall view V_m and $\tau_k^{(m)}$ as arithmetic perturbations of U and τ_k respectively. We shall try to compare $G^{(m)}(D)$ to

$$G(D) = \sum_{(d,\Delta)=1} \lambda_d \frac{\tau_k(d)}{d} \tag{7.221}$$

quite directly by exploiting properties of the sieve weights and without a complete asymptotic evaluation of either sum. This approach will ease our task significantly because the dependence on m will be clearly seen to factor out.

The diagonalization of the quadratic form in the Selberg sieve in the previous section was performed for $G(D)$ rather than $G'(D)$ and this motivates us to try here to reach a quadratic form for the perturbed sum $G^{(m)}(D)$ which is close to diagonal. First we take care of modifying the change from d to $\varphi(d)$. To accomplish this we use the Monotonicity Principle II; see (5.75). Fortunately, this applies in the desired direction and this is the place where we lose the possibility of an asymptotic evaluation.

We replace the function $\tau_k^{(m)}(d)/\varphi(d)$ by the smaller function

$$g^{(m)}(d) = \tau_k^{(m)}(d)/d\,. \tag{7.222}$$

Hence, by (5.75) we obtain

$$G^{(m)}(D) \geqslant \gamma_k(\Delta) \sum_{(d,\Delta)=1} \lambda_d g^{(m)}(d)\,. \tag{7.223}$$

Now we follow the process of diagonalization. Using (7.13), (7.14) we get

$$\sum_{(d,\Delta)=1} \lambda_d g^{(m)}(d) = \sum_{(d,\Delta)=1} \frac{1}{h^{(m)}(d)}\Big(\sum_{\ell \equiv 0 \,(\mathrm{mod}\, d)} g^{(m)}(\ell)\rho_\ell\Big)^2, \tag{7.224}$$

where $h^{(m)}$ is the companion to $g^{(m)}$, given by

$$h^{(m)}(p) = \frac{g^{(m)}(p)}{1-g^{(m)}(p)} = \begin{cases} k/(p-k) & \text{if } p \nmid Q(m), \\ (k-1)/(p-k+1) & \text{if } p \mid Q(m), \end{cases}$$

so that

$$h^{(m)}(d) \leqslant \tau_k^{(m)}(d)/f(d)\ . \tag{7.225}$$

Next we replace the vector ρ_ℓ by the new vector y_b as in (7.182). This gives

$$\begin{aligned}\sum_{\ell\equiv 0\,(\mathrm{mod}\,d)} g^{(m)}(\ell)\rho_\ell &= \sum_{\ell\equiv 0\,(\mathrm{mod}\,d)} \mu(\ell)\frac{\tau_k^{(m)}(\ell)}{\tau_k(\ell)} \sum_{b\equiv 0\,(\mathrm{mod}\,\ell)} h(b)y_b \\ &= \mu(d)\frac{\tau_k^{(m)}(d)}{\tau_k(d)} \sum_c h(cd)y_{cd} \sum_{q|c} \mu(q)\frac{\tau_k^{(m)}(q)}{\tau_k(q)}\ .\end{aligned}$$

Here, the last sum over $q \mid c$ vanishes unless $c \mid Q(m)$ in which case it is equal to $1/\tau_k(c)$ giving

$$\sum_{\ell\equiv 0\,(\mathrm{mod}\,d)} g^{(m)}(\ell)\rho_\ell = \mu(d)\frac{\tau_k^{(m)}(d)}{f(d)} \sum_{c|Q(m)} \frac{y_{cd}}{f(c)}\ .$$

Hence, by (7.224) and (7.225) we obtain

$$\sum_{(d,\Delta)=1} \lambda_d g^{(m)}(d) \geqslant \sum_d \frac{\tau_k^{(m)}(d)}{f(d)} \Big(\sum_{c|Q(m)} \frac{y_{cd}}{f(c)} \Big)^2\ .$$

Since y_d is non-negative and non-increasing we deduce that

$$\Big(\sum_{c|Q(m)} \frac{y_{cd}}{f(c)} \Big)^2 \geqslant y_d \sum_{c|Q(m)} \frac{y_{cd}}{f(c)} \geqslant \sum_{c|Q(m)} \frac{y_{cd}^2}{f(c)}\ .$$

Hence, we find that

$$\sum_{(d,\Delta)=1} \lambda_d g^{(m)}(d) \geqslant \sum_d \frac{\tau_k^{(m)}(d)}{f(d)} \sum_{c|Q(m)} \frac{y_{cd}^2}{f(c)} = \sum_{(\ell,\Delta)=1} \frac{\tau_k(\ell)}{f(\ell)} y_\ell^2 = G(D)\ ,$$

the last equality being (7.181) and the prior one following by

$$\sum_{\substack{cd=\ell \\ c|Q(m)}} \tau_k^{(m)}(d) = \tau_k(\ell)\ .$$

Applying this inequality to (7.223), we arrive at

$$G^{(m)}(D) \geqslant \gamma_k(\Delta)G(D)\ . \tag{7.226}$$

Finally, applying (7.226) to (7.219) and using (7.171), we complete the proof of Lemma 7.21.

Having estimated the contributions $V_m(D,N)$ for individual $m \notin \mathcal{H}$ we sum these. It is clear that we do best by making the set $\mathcal{M}$ as large as possible so we take it to consist of all integers $1 \leqslant m \leqslant M$. The only remaining dependence on

$m \in \mathcal{M}\setminus\mathcal{H}$ in the lower bound (7.216) appears in the arithmetic factor $\gamma^{(m)}(\mathcal{H})$ so we need to sum these. We find

$$\sum_{m\leqslant M}\gamma^{(m)}(\mathcal{H}) = \sum_{m\leqslant M}\prod_{\substack{p|\Delta\\ p\nmid Q(m)}}\Big(1-\frac{1}{f(p)}\Big) = \sum_{m\leqslant M}\sum_{\substack{\delta|\Delta\\ (\delta,Q(m))=1}}\frac{\mu(\delta)}{f(\delta)}$$
$$= \sum_{\delta|\Delta}\frac{\mu(\delta)}{f(\delta)}\sum_{\substack{\alpha(\mathrm{mod}\,\delta)\\ (Q(\alpha),\delta)=1}}\Big(\frac{M}{\delta}+O(1)\Big)$$
$$= \sum_{\delta|\Delta}\frac{\mu(\delta)}{f(\delta)}\Big(\frac{f(\delta)}{\delta}M+O(f(\delta))\Big) = \frac{\varphi(\Delta)}{\Delta}M+O(\tau(\Delta))\ ,$$

where the implied constant is absolute. Of course, for our application we need to exclude from the above sum those $m \in \mathcal{H}$, but their contribution is bounded by $|\mathcal{H}|$ which is relatively small and, in this situation, negligible.

We choose $M = \delta(k)\log N$. We input the result (7.195) from the previous section for $W_{\mathcal{H}}(D,N) = W(D,N)$ with the exponent of distribution $\theta = \frac{1}{2}$ which is unconditionally permitted by the Bombieri–Vinogradov theorem. We arrive at the following estimate:

$$W_{\mathcal{M}}(D,N) > \gamma(k,\ell)w(k,\ell)H(\mathcal{H})N(\log N)(\log\sqrt{D})^{-k}\ , \tag{7.227}$$

where

$$w(k,\ell) = \frac{k(2\ell+1)}{2(k+2\ell+1)(\ell+1)} + \delta(k)\gamma_k(\Delta) - 1 - 2\varepsilon\ , \tag{7.228}$$

provided that N is sufficiently large in terms of ε, Δ, and $\mathcal{H}$.

Recall that the factor $\gamma_k(\Delta)$ emerged after an application of the monotonicity property and that it can be made arbitrarily close to one. Taking $K = 2k$ in (7.217) we get $\gamma_k(\Delta) \geqslant \frac{1}{2}$. We choose $\ell = [\sqrt{k}/2]$ and $\delta(k) = 8/\sqrt{k}$ getting

$$w(k,\ell) > 1 - \frac{2}{\sqrt{k}} + \frac{\delta(k)}{2} - 1 - 2\varepsilon\ ,$$

and this is positive if $k > \varepsilon^{-2}$. This completes the proof of Theorem 7.20.

7.15. Gaps Between Primes and Powers

The set $\mathcal{M}$ in the GPY construction does not need to be the full segment of integers. There are interesting applications for other sets. Let us take the set of powers of a fixed integer $a \geqslant 2$, specifically

$$\mathcal{M} = \{a^m\ ;\ 1 \leqslant m \leqslant M\}$$

with $M = (\log N)/2\log a$, so $a^m \leqslant \sqrt{N}$. Let $\mathcal{H}$ be the subset of the first k elements:

$$\mathcal{H} = \{a^j\ ;\ 1 \leqslant j \leqslant k\}\ ,$$

for some $k \leqslant M$. In this situation we have the polynomial

$$Q(X) = \prod_{1\leqslant j\leqslant k}(X - a^j)\ .$$

If $p \mid a$ then $Q(X)$ has exactly one zero modulo p, that is $\omega(p) = 1$, and, if $p \nmid a$ then $\omega(p) < p$ because $Q(0) \not\equiv 0 \pmod p$. Therefore, the set $\mathcal{H}$ is admissible. Since

$$\det \mathcal{H} = \prod_{1 \leqslant i < j \leqslant k} \left(a^j - a^i\right)^2 = a^{k(k-1)} \prod_{1 \leqslant \nu < k} \left(a^\nu - 1\right)^{2(k-\nu)} ,$$

we can take Δ to be the product of all prime divisors of a and all primes p for which $a^\nu \equiv 1 \pmod p$ with some $1 \leqslant \nu \leqslant k$. Note that we have purposely included $\nu = k$ so, if $k+1$ is prime, it does divide Δ.

Now, for $a^m \in \mathcal{M}$ with $m > k$ we have

$$Q(a^m) = \prod_{1 \leqslant j \leqslant k} \left(a^m - a^j\right) = a^{k(k+1)/2} \prod_{m-k \leqslant \nu < m} \left(a^\nu - 1\right) .$$

Hence, it is clear that if $p \mid \Delta$ then $p \mid Q(a^m)$. Therefore, the product (7.214) is void, giving $\gamma^{(a^m)}(\mathcal{H}) = 1$ in this case, so

$$\sum_{m \leqslant M} \gamma^{(a^m)}(\mathcal{H}) = M + O(k) .$$

This is better than the amount $\varphi(\Delta)\Delta^{-1}M + O(\tau(\Delta))$ which we got in the previous case $\mathcal{M} = [1, M]$, so (7.227) holds with

$$w(k, \ell) = \frac{k(2\ell+1)}{2(k+2\ell+1)(\ell+1)} + \frac{\gamma_k}{2\log a} - 1 - 2\varepsilon ,$$

provided that N is sufficiently large in terms of ε, a, and k. Recall that γ_k is bounded below by a positive absolute constant for some k in any dyadic segment. Choosing such a k, sufficiently large in terms of ε and a, and as before $\ell = [\sqrt{k}/2]$, we find that $w(k, \ell)$ is positive. This leads to the following result.

THEOREM 7.22. *Let $a \geqslant 2$ be an integer. There are infinitely many solutions to*

$$p' - p = a^{m'} - a^m$$

in primes $p' > p$ and integers $m' > m$.

REMARK. We include this theorem for illustration only of the GPY ideas as there is another proof which is rather old. We can obviously reword the result as saying that infinitely many integers n have two different representations in the form $p + a^m$. Actually, the much stronger statement that for infinitely many n and some $c > 0$ there are $c \log\log n$ different such representations was proved already by P. Erdös [**40**]. The result is stated there only for $a = 2$ but the proof given carries over with natural changes.

We believe (although we did not check all details) that the method presented here can, when combined with the Fundamental Lemma, produce infinitely many almost-prime integers which have two different representations in the form $p + a^m$. The method of Erdös, although cute and simple, does not offer this possibility.

CHAPTER 8

Sifting Many Residue Classes

Molto Vivace

8.1. Sifting Arbitrary Classes

Selberg's method is particularly flexible for the treatment of more general sieve problems where the input data is given, not by simple divisibility, but rather by more involved congruence criteria. For example, it could be that we are interested in removing several, or even increasingly many, residue classes for each prime in the sifting range. Of course, the combinatorial sieve is also applicable by appropriate adjustments, however the results are not impressive.

Suppose that, for each prime p, we are given a set Ω_p of classes to be removed. Of course, we assume

$$\omega(p) = |\Omega_p| < p \tag{8.1}$$

so that we don't remove everything at one step. The problem becomes that of estimating the sifting function

$$S(\mathcal{A}, \Omega) = \sum_{\substack{n \,(\mathrm{mod}\, p) \notin \Omega_p \\ \text{for every } p \in \mathcal{P}}} a_n \, . \tag{8.2}$$

Now, Selberg's idea takes the form $S(\mathcal{A}, \Omega) \leqslant S^+(\mathcal{A}, \Omega)$ where

$$S^+(\mathcal{A}, \Omega) = \sum_n a_n \Big(\sum_{\substack{d \\ n \,(\mathrm{mod}\, p) \in \Omega_p \\ \text{for every } p \mid d}} \rho_d \Big)^2 , \tag{8.3}$$

for any real numbers ρ_d supported on $d \mid P$ with $\rho_1 = 1$.

Opening the square, we rearrange this as follows:

$$S^+(\mathcal{A}, \Omega) = \sum_{d_1} \sum_{d_2} \rho_{d_1} \rho_{d_2} A_{[d_1, d_2]}(\Omega) \tag{8.4}$$

where

$$A_d(\Omega) = \sum_{\substack{n \,(\mathrm{mod}\, p) \in \Omega_p \\ \text{for every } p \mid d}}^{\Omega} a_n \, .$$

This is the natural generalization of the congruence sum A_d which is the case of $\Omega_p = \{0\}$ for every p.

In practice, $A_d(\Omega)$ is not highly dependent on the classes in Ω but only on their number. Therefore it is natural to postulate that

$$A_d(\Omega) = g(d)X + r_d(\Omega) \, , \tag{8.5}$$

where g is a multiplicative density function and $r_d(\Omega)$ is a remainder term, presumably small. From this point on, the argument proceeds as before.

The new challenge occurs in the estimation of $r_d(\Omega)$, especially when the number of classes is large. Naturally, one could split into progressions modulo d and estimate the corresponding remainder for each class separately. For a sieve problem with just a few residue classes this approach is not much of a compromise but when the number of residue classes is large it is worthwhile to seek a cancellation in the sum of such remainders, even for each individual d. We have

$$A_d(\Omega) = \sum_{\nu}^{\Omega} \sum_{n\equiv\nu \,(\mathrm{mod}\, d)} a_n \,, \tag{8.6}$$

where the superscript Ω restricts the summation to classes $\nu\,(\mathrm{mod}\, d)$ such that $\nu\,(\mathrm{mod}\, p) \in \Omega$ for every $p \mid d$.

We suppose now that the residue classes in Ω are reduced (which is not the case in the traditional setting of $\Omega_p = \{0\}$). Then, we can detect the congruence $n \equiv \nu\,(\mathrm{mod}\, d)$ by multiplicative characters $\chi\,(\mathrm{mod}\, d)$, getting

$$A_d(\Omega) = \frac{1}{\varphi(d)} \sum_{\chi\,(\mathrm{mod}\, d)} \Omega_d(\bar{\chi}) A(\chi) \,, \tag{8.7}$$

where

$$\Omega_d(\chi) = \sum_{\nu\,(\mathrm{mod}\, d)}^{\Omega} \chi(\nu) \tag{8.8}$$

and

$$A(\chi) = \sum_{n} a_n \chi(n) \,. \tag{8.9}$$

Recall that d is squarefree. Writing uniquely $\chi = \prod_{p|d} \chi_p$, where χ_p is a character modulo p, we have

$$\Omega_d(\chi) = \prod_{p|d} \Omega_p(\chi_p) \,.$$

For the principal character we get $\Omega_d(\chi_0) = \omega(d)$, where

$$\omega(d) = \prod_{p|d} \omega(p) \,. \tag{8.10}$$

It is convenient to write $A_d(\Omega)$ in terms of primitive characters:

$$A_d(\Omega) = \frac{1}{\varphi(d)} \sum_{qr=d} \omega(r) \sum_{\chi\,(\mathrm{mod}\, q)}^{\star} \Omega_q(\bar{\chi}) A_r(\chi) \,, \tag{8.11}$$

where

$$A_r(\chi) = \sum_{(n,r)=1} a_n \chi(n) \,.$$

Our duplicate notation for $A_d(\Omega)$ and $A_r(\chi)$ should be distinguished by the context; in the first one Ω is a set, in the second χ is a character.

The principal character (so $q = 1$) contributes $\omega(d)\varphi(d)^{-1} A_d(1)$. Here we write

$$A_d(1) = \sum_{(n,d)=1} a_n = \frac{\varphi(d)}{d} X + r_d \,, \tag{8.12}$$

where X does not depend on d and is a good approximation to

$$A = \sum_n a_n \ , \tag{8.13}$$

so one may think of r_d as a reasonably small remainder term. Thus, we obtain (8.5) with the density function

$$g(d) = \frac{\omega(d)}{d} \tag{8.14}$$

and the remainder term

$$r_d(\Omega) = \frac{\omega(d)}{\varphi(d)} r_d + r_d^*(\Omega) \ , \tag{8.15}$$

where

$$r_d^*(\Omega) = \frac{1}{\varphi(d)} \sum_{\substack{\chi \,(\mathrm{mod}\, d) \\ \chi \neq \chi_0}} \Omega_d(\bar{\chi}) A(\chi) \ . \tag{8.16}$$

In some applications $A(\chi)$ is quite small for all $\chi \neq \chi_0$ and so is $r_d^*(\Omega)$. By Cauchy's inequality and the orthogonality of characters we get

$$\left| r_d^*(\Omega) \right|^2 \leqslant \frac{\omega(d)}{\varphi(d)} \sum_{\substack{\chi \,(\mathrm{mod}\, d) \\ \chi \neq \chi_0}} \left| A(\chi) \right|^2 \ . \tag{8.17}$$

From here one can proceed either by estimating $A(\chi)$ for every $\chi \neq \chi_0$ or by applying the large sieve inequality (see Section 9.6 for sums of $|A(\chi)|^2$ over characters and moduli. The first approach can produce (by an adequate harmonic analysis) very good estimates for $A(\chi)$ with $\chi \,(\mathrm{mod}\, d)$ of small conductor. However, if the sequence $\mathcal{A} = (a_n)$ is very lacunary the large sieve is not an option because the resulting bound could be worse than trivial.

It is not always clear which characters $\chi \,(\mathrm{mod}\, d)$ produce the main term of $A_d(\Omega)$. In the above setting we assumed tacitly that only the principal character $\chi_0 \,(\mathrm{mod}\, d)$ gives the main term but there are many natural examples of sequences $\mathcal{A} = (a_n)$ for which other characters also yield significant contributions. We shall illustrate this situation in Section 8.3 wherein all the real characters play a role.

8.2. Sieving for Squares

We are interested in estimating the sum

$$S^{\square}(\mathcal{A}) = \sum_{n = \square} a_n \ , \tag{8.18}$$

where $\mathcal{A} = (a_n)$ is a finite sequence of non-negative real numbers which will be lacunary for the applications we have in mind. The thinness of the sequence gives us no chance to establish an asymptotic formula, or even a positive lower bound so we are aiming for a reasonably strong upper bound. Because of the lacunarity the large sieve inequality (9.32) is not an option; it produces a bound worse than the trivial estimate $S^{\square}(\mathcal{A}) \leqslant A$. One could try the "square sieve" of Heath-Brown [**86**], but we wish to show how Selberg's sieve copes in such a situation. We shall return to the square sieve in Chapter 15.

We begin by specializing the general setting described in the previous section. To this end we apply the sieve to remove from each p in a set $\mathcal{P}$ residue classes which are not squares modulo p. Since we are looking for an upper bound the sifting range $\mathcal{P}$ does not need to include every prime. Therefore, we choose $\mathcal{P}$ to be a large finite set which does not contain a few inconvenient primes, and we denote by P the product of all primes in $\mathcal{P}$. Since the sieve weights are supported on divisors of P we do not need to display the condition $d \mid P$ in various summations to follow.

In our case, the number of residue classes to be removed is given by

$$\omega(p) = \tfrac{1}{2}(p-1) \qquad \text{for every } p \mid P\ . \tag{8.19}$$

We have

$$S^{\square}(\mathcal{A}) \leqslant S^{+}(\mathcal{A}, \Omega)\ ,$$

where $S^{+}(\mathcal{A}, \Omega)$ is given by (8.3). Our goal is to find good weights ρ_d (these will be almost optimal) and to evaluate $S^{+}(\mathcal{A}, \Omega)$ asymptotically for these weights.

We find that, for a non-principal character $\chi \pmod{p}$,

$$\Omega_p(\chi) = \sum_{\nu \in \Omega_p} \chi(\nu) = -\omega(p) = -\tfrac{1}{2}(p-1)$$

if $\chi = \chi_p$ is real and $\Omega_p(\chi) = 0$ if χ is complex. Introducing this to (8.11) we obtain

$$A_d(\Omega) = \frac{1}{\tau(d)} \sum_{qr=d} \mu(q) A_d(\chi_q)\ , \tag{8.20}$$

where χ_q is the real character of conductor q. Opening $A_d(\chi_q)$ by (8.9) and changing the order of summation, we find that

$$A_d(\Omega) = \sum_{(n,d)=1} a_n \psi_d(n) \tag{8.21}$$

where

$$\psi_d(n) = \frac{1}{\tau(d)} \sum_{q \mid d} \mu(q) \chi_q(n) = \prod_{p \mid d} \tfrac{1}{2}(1 - \chi_p(n))\ . \tag{8.22}$$

Note that for $(n, d) = 1$ this symbol is the characteristic function of numbers n which are quadratic non-residues for every $p \mid d$. For convenience we extend $\psi_d(n)$ to all n by setting $\psi_d(n) = 0$ if $(n, d) \neq 1$. Hence, we get

$$S^{+}(\mathcal{A}, \Omega) = \sum_n a_n \Big(\sum_d \rho_d \psi_d(n)\Big)^2\ . \tag{8.23}$$

Of course, the inequality $S^{\square}(\mathcal{A}) \leqslant S^{+}(\mathcal{A}, \Omega)$ can now be checked directly for any real numbers ρ_d, $d \mid P$ with $\rho_1 = 1$; however, we have taken this roundabout path in order to illuminate the large sieve concepts hidden behind the Selberg sieve. Note the similarity of the weights

$$\Theta_n = \Big(\sum_d \rho_d \psi_d(n)\Big)^2$$

to the original weights of Selberg's sieve in (7.6):

$$\theta_n = \Big(\sum_{d \mid n} \rho_d\Big)^2\ .$$

There, the targeted numbers (those coprime with P) are majorized by the weights θ_n of convolution type sums, while here, the targeted numbers (squares) are majorized by the weights Θ_n of character twist type sums. In each case the inequality for the sifting function comes from the positivity of squares. In each case the weights can be unwound for analysis, there to produce the congruence sums, here to produce the analogous character sums $A_d(\chi)$. This flexibility of the Selberg idea has the potential for other variations.

Next, we are going to perform detailed computations in the case of sequences $\mathcal{A} = (a_n)$ for which the main term in (8.5) has the density function $g(d)$ given by (8.14). Now, for the system Ω of quadratic non-residues this becomes

$$g(d) = \frac{\varphi(d)}{\tau(d)d} . \tag{8.24}$$

The main term $g(d)X$ in (8.5) when introduced to (8.4) yields

$$S^+(\mathcal{A}, \Omega) = GX + \text{ remainder },$$

where G is the quadratic form (7.7) for this specific density function. The optimization of the form G has already been done in general, giving $G = J^{-1}$ (see (7.20)), with J given by (7.19), where $h(d)$ is the multiplicative function defined by

$$h(p) = \frac{p-1}{p+1} . \tag{8.25}$$

This minimum of G is attained at the weights (7.23).

Before evaluating J in the main term X/J of $S^+(\mathcal{A}, \Omega)$ we need to say something about the sifting range P. We assume

$$P = \prod_{\substack{p<\sqrt{D} \\ p\nmid\Delta}} p , \tag{8.26}$$

where Δ is a number which can be adjusted for applications in order to verify (8.5) for $d \mid P$. Using elementary arguments (similar to the proof of Lemma 8.4 we derive for $1 \leqslant y \leqslant \sqrt{D}$, $(\ell, \Delta) = 1$:

$$\sum_{\substack{d\leqslant y,\, d|P \\ (d,\ell)=1}} h(d) = CC_\Delta C_\ell y + O\big(\tau(\Delta\ell)\sqrt{y}\big) , \tag{8.27}$$

where C_Δ, C_ℓ are the multiplicative functions with

$$C_p = \Big(1 + \frac{p-1}{p(p+1)}\Big)^{-1} \tag{8.28}$$

and C is a positive absolute constant given by

$$C = \prod_p \Big(1 + \frac{p-1}{p(p+1)}\Big)\Big(1 - \frac{1}{p}\Big) .$$

For $\ell = 1$ and $y = \sqrt{D}$ the sum (8.27) is just

$$J = CC_\Delta\sqrt{D} + O\big(\tau(\Delta)D^{\frac{1}{4}}\big) . \tag{8.29}$$

Hence, the main term of the sifted sum $S^+(\mathcal{A}, \Omega)$ is $X\,/J \sim X/\,CC_\Delta\sqrt{D}$. Moreover, the sieve weight constituents ρ_ℓ given by (7.23) simplify to

$$\rho_\ell = \mu(\ell)\frac{\tau(\ell)}{\ell}\prod_{p|\ell}\Big(1 + \frac{2}{p} - \frac{1}{p^2}\Big)^{-1} + O\Big(\frac{\tau(\Delta)\Delta}{\varphi(\Delta)}\,\frac{\tau^2(\ell)}{\ell^{\frac{1}{2}}D^{\frac{1}{4}}}\Big) . \tag{8.30}$$

These are distinctly smaller than in the sieve of fixed dimension, due to the larger number of classes being removed.

8.3. Ternary Quadratic Forms

Let $Q(r, s, t)$ be the quadratic form $ar^2 + bs^2 + ct^2$ with positive integral coefficients a, b, c. Our goal is to estimate the number $r(N, Q)$ of representations

$$ar^2 + bs^2 + ct^2 = N \tag{8.31}$$

in integers r, s, t. For simplicity we assume that

$$(abc, N) = 1, \qquad abc \quad \text{squarefree} . \tag{8.32}$$

If N is very large relative to a, b, c, this is a problem which is best treated by means of the spectral theory of metaplectic modular forms; cf [**38**] and [**105**]. W. Duke showed, in particular [**35**], subject to minor conditions, the asymptotic formula

$$r(N, Q) = \pi\mathfrak{S}(N, Q)\Big(\frac{N}{abc}\Big)^{\frac{1}{2}} + O\Big(N^{\frac{55}{111}}\Big) ; \tag{8.33}$$

see Theorem 20.15 of [**109**]. Here $\mathfrak{S}(N, Q)$ is given by a singular series in Gauss sums which satisfies the bound

$$\mathfrak{S}(N, Q) \ll \tau(\Delta) \log \Delta , \tag{8.34}$$

with

$$\Delta = 4abcN \tag{8.35}$$

and an absolute implied constant. Unfortunately, the implied constant in the error term of (8.33) does depend on a, b, c. Actually, Duke's arguments give (8.33) uniformly for $N \gg (abc)^{337}$ (see also [**6**] for a wider range).

We are going to examine the equation (8.31) with N not so large relative to a, b, c, so the free variables r, s, t have not much room: $|r| < R$, $|s| < S$, $|t| < T$ with

$$R = \sqrt{N/a}, \quad S = \sqrt{N/b}, \quad T = \sqrt{N/c} . \tag{8.36}$$

To get an asymptotic formula for the representation number is out of the question and even an upper bound of the correct order of magnitude seems to be far away by any means currently available. The most optimistic goal would be the bound

$$r(N, Q) \ll \Big(\frac{N}{abc}\Big)^{\frac{1}{2}} \tau(\Delta) \log N , \tag{8.37}$$

uniformly for $abc \leqslant N^{1-\varepsilon}$ with an implied constant depending only on ε.

On the other hand, given $|t| < T$ we have at most $O(N^\varepsilon)$ solutions in r, s, because the number of integral ideals in $\mathbb{Q}(\sqrt{-ab})$ of norm $N - ct^2$ is $O(N^\varepsilon)$. Hence we get the trivial bound

$$r(N, Q) \ll \min(R, S, T) N^\varepsilon \ll (abc)^{-\frac{1}{6}} N^{\frac{1}{2}+\varepsilon} , \tag{8.38}$$

which is, by a factor $(abc)^{\frac{1}{3}}$, weaker than the hoped-for bound (8.37).

In this section we shall do better than (8.38) by using Selberg's sieve for squares. As our primary objective is to demonstrate this sieve in action rather than seeking the strongest possible result we shall compromise some arguments and generality for convenience.

We are going to detect t^2 in the equation (8.31) by the sieve while treating r, s as free variables satisfying the congruence

$$ar^2 + bs^2 \equiv N \pmod{c} . \tag{8.39}$$

After the sieving has been completed we shall interchange r, s, t to give the bound a symmetric shape in terms of abc.

To simplify the forthcoming Fourier analysis we begin by applying the Selberg square sieve to the sequence of numbers

$$n = (N - ar^2 - bs^2)/c \tag{8.40}$$

with r, s satisfying (8.39) and weighted by a non-negative smooth function $F(r, s)$ supported on $|r| \leqslant 2R$, $|s| \leqslant 2S$, such that $F(r, s) = 1$ for $|r| \leqslant R$, $|s| \leqslant S$. We can also assume that the Fourier transform of this test function satisfies

$$\hat{F}(u, v) \ll RS(1 + |u|R)^{-2}(1 + |v|S)^{-2} . \tag{8.41}$$

Then, we have

$$r(N, Q) \leqslant S^{\square}(\mathcal{A}) \leqslant S^+(\mathcal{A}, \Omega) \tag{8.42}$$

where Ω is the system of quadratic non-residues for every prime $p \mid P$ and the sifting range P is at our disposal. We let P be the product (8.26) with $\Delta = 4abcN$, so every $d \mid P$ is coprime with $4abcN$. The value of D, to which we refer as the level of distribution, will be chosen at the end of our arguments to optimize the result by matching the main term and the remainder.

We need to establish the approximation (8.5) for every $A_d(\Omega)$ with an appropriate density function $g(d)$ and a small remainder term $r_d(\Omega)$. By (8.20) this reduces to the asymptotic evaluation of the character sums $A_d(\chi)$ for real characters $\chi \,(\mathrm{mod}\, d)$. In our case this is

$$A_d(\chi) = \chi(c) \sum\sum_{ar^2+bs^2 \equiv N \,(\mathrm{mod}\, c)} F(r, s)\chi(N - ar^2 - bs^2) . \tag{8.43}$$

We split the summation over r, s into residue classes modulo cd getting

$$A_d(\chi) = \frac{1}{c^2d^2} \sum_k \sum_\ell \hat{F}\Big(\frac{k}{cd}, \frac{\ell}{cd}\Big) V_{k\ell}(d/c) W^{\chi}_{k\ell}(c/d) \tag{8.44}$$

where

$$V_{k\ell}(d/c) = \sum\sum_{a\alpha^2+b\beta^2 \equiv N \,(\mathrm{mod}\, c)} e\Big(\frac{\bar{d}}{c}(\alpha k + \beta \ell)\Big) \tag{8.45}$$

and

$$W^{\chi}_{k\ell}(c/d) = \chi(c) \sum_\gamma \sum_\delta \chi(N - a\gamma^2 - b\delta^2) e\Big(\frac{\bar{c}}{d}(\gamma k + \delta \ell)\Big) . \tag{8.46}$$

These are complete character sums in α, β modulo c and in γ, δ modulo d. We shall evaluate them precisely.

LEMMA 8.1. *For any k, ℓ we have*

$$|V_{k\ell}(d/c)| \leqslant \tau(c) c^{\frac{1}{2}} (c, k, \ell)^{\frac{1}{2}} . \tag{8.47}$$

If $c \mid (k, \ell)$ then $V_{k\ell}(d/c)$ becomes

$$V(c) = c \prod_{p \mid c} (1 - \chi_p(-ab)p^{-1}) . \tag{8.48}$$

PROOF. By the multiplicativity property

$$V_{k\ell}(d/c) = V_{k\ell}(dc_1/c_2)V_{k\ell}(dc_2/c_1)$$

for $c = c_1c_2$, the proof reduces to the case of prime modulus.

Let c be prime, $c \nmid 2abdN$. Detecting the congruence $a\alpha^2 + b\beta^2 \equiv N \pmod c$ by additive characters we obtain

$$V_{k\ell}(d/c) = \frac{1}{c} \sum\sum\sum_{\alpha,\,\beta,\,\eta,\,(\mathrm{mod}\,c)} e_c\big(\eta(a\alpha^2 + b\beta^2 - N) + \bar{d}(\alpha k + \beta\ell)\big) .$$

For $\eta \equiv 0 \pmod c$ we get a contribution c if $c \mid (k, \ell)$ and nothing otherwise. Next, let $\eta \not\equiv 0 \pmod c$. Now, the sums over α, β are Gauss sums. Hence, the contribution from the terms $\eta \not\equiv 0 \pmod c$ is just

$$\begin{aligned} &\frac{\big(\varepsilon_c\sqrt{c}\big)^2}{c}\Big(\frac{ab}{c}\Big) \sum_{\eta\,(\mathrm{mod}\,c)} e_c\big(-\eta N - \overline{4\eta a d^2}k^2 - \overline{4\eta b d^2}\ell^2\big) \\ &\quad = \Big(\frac{-ab}{c}\Big) S\big(N, \overline{4abd^2}k^2(a\ell^2 + bk^2); c\big) , \end{aligned}$$

where $S(*, *; c)$ denotes the Kloosterman sum of modulus c. Since the Kloosterman–Ramanujan sum $S(N, 0; c) = \mu(c) = -1$ we conclude that

$$V_{k\ell}(d/c) = c - \chi_c(-ab) \quad \text{if } c \mid (k, \ell) \tag{8.49}$$

and

$$|V_{k\ell}(d/c)| \leqslant 2c^{\frac{1}{2}} \quad \text{if } c \nmid (k, \ell) , \tag{8.50}$$

by Weil's bound for the Kloosterman sum. These formulas yield (8.47) and (8.48). □

LEMMA 8.2. *For any k, ℓ and any real character $\chi \pmod d$, we have*

$$|W^{\chi}_{k\ell}(c/d)| \leqslant \tau(d)d(d, k, \ell) . \tag{8.51}$$

If $d \mid (k, \ell)$ then $W^{\chi}_{k\ell}(c/d)$ becomes

$$W^{\chi}(d) = \chi(\Delta)\, d \prod_{p|r} \big(p - 1 + \chi_p(-ab)p^{-1}\big) \tag{8.52}$$

where $d = qr$ and q is the conductor of χ.

PROOF. By the multiplicativity property

$$W^{\chi}_{k\ell}(c/d) = W^{\chi_1}_{k\ell}(cd_2/d_1)W^{\chi_2}_{k\ell}(cd_1/d_2)$$

for $d = d_1d_2$ and $\chi = \chi_1\chi_2$ with $\chi_1 \pmod{d_1}$, $\chi_2 \pmod{d_2}$, the proof reduces to the case of prime modulus.

Let d be prime with $d \nmid \Delta$. If χ is the principal character modulo d, then

$$W^{\chi}_{k\ell}(c/d) = \sum_{\gamma}\sum_{\delta} e_d\big(\bar{c}(\gamma k + \delta\ell)\big) - \sum\sum_{a\gamma^2 + b\delta^2 \equiv N\,(\mathrm{mod}\,d)} e_d\big(\bar{c}(\gamma k + \delta\ell)\big) .$$

The first sum vanishes unless $d \mid (k, \ell)$ in which case it is d^2. The second sum is just $V_{k\ell}(c/d)$, so that (8.49) yields $d - \chi_d(-ab)$ if $d \mid (k, \ell)$ while (8.50) yields the bound $2\sqrt{d}$ if $d \nmid (k, \ell)$.

Combining these results we conclude that, for the principal character $\chi \pmod{d}$, we have

$$W_{k\ell}^{\chi}(c/d) = d(d-1) + \chi_d(-ab) \quad \text{if } d \mid (k,\ell) \tag{8.53}$$

and

$$|W_{k\ell}^{\chi}(c/d)| \leqslant 2d^{\frac{1}{2}} \quad \text{if } d \nmid (k,\ell)\ . \tag{8.54}$$

Recall that χ_d above for d prime is just the Legendre symbol.

Next we evaluate $W_{k\ell}^{\chi}(c/d)$ for the character $\chi = \chi_d$. We expand χ into additive characters using Gauss sums

$$\chi(m) = \frac{1}{\tau_\chi} \sum_\eta \chi(\eta) e_d(\eta m)\ .$$

Hence

$$W_{k\ell}^{\chi}(c/d) = \frac{\chi(c)}{\tau_\chi} \sum_\eta \sum_\gamma \sum_\delta \chi(-\eta) e_d\Big(\eta(a\gamma^2 + b\delta^2 - N) + \bar{c}(\gamma k + \delta\ell)\Big)\ .$$

The sum over γ, δ was just computed in the context of $V_{k\ell}(d/c)$. Interchanging c and d we obtain

$$W_{k\ell}^{\chi}(c/d) = \chi(abc)\frac{d}{\tau_\chi} S_\chi\big(N, \overline{4abc^2}(a\ell^2 + bk^2); d\big)\ ,$$

where $S_\chi(*,*;d)$ denotes the Salié sum of modulus d. Hence, by Salié's estimate for this sum we conclude

$$|W_{k\ell}^{\chi}(c/d)| \leqslant 2d\ . \tag{8.55}$$

The case $d|(k,\ell)$ is special because the Salié sum then reduces to the Gauss sum $S_\chi(N,0;d) = \chi(N)\tau_\chi$, so

$$W_{k\ell}^{\chi}(c/d) = \chi(\Delta)d \quad \text{if} \quad d \mid (k,\ell)\ . \tag{8.56}$$

Combining (8.53)–(8.56), we conclude the proof of (8.51) and (8.52) by the multiplicativity property. □

Now, we are ready to evaluate the "congruence sums" $A_d(\Omega)$.

LEMMA 8.3. *For every d squarefree with $(d,\Delta) = 1$ we have*

$$A_d(\Omega) = g(d)X + r_d(\Omega) \tag{8.57}$$

with

$$g(d) = \frac{1}{\tau(d)} \prod_{p|d} \Big(1 - \frac{1+\chi_p(\Delta)}{p} + \frac{\chi_p(-ab)}{p^2}\Big)\ , \tag{8.58}$$

$$X = \hat{F}(0,0)c^{-1} \prod_{p|c} \Big(1 - \frac{\chi_p(-ab)}{p}\Big)\ , \tag{8.59}$$

and the remainder term satisfying

$$r_d(\Omega) \ll \frac{\tau^2(cd)}{\sqrt{c}}(cd + R + S)\ , \tag{8.60}$$

where R, S are given by (8.36) and $\hat{F}(0,0) \asymp RS$.

PROOF. First we evaluate the character sum $A_d(\chi)$ for any real character $\chi\,(\operatorname{mod} d)$ by using the Fourier expansion (8.44). The main term comes from the zero frequencies $k=\ell=0$ giving

$$\hat{F}(0,0)\frac{\chi(\Delta)}{cd}\prod_{p|c}\Big(1-\frac{\chi_p(-ab)}{p}\Big)\prod_{p|r}\Big(p-1+\frac{\chi_p(-ab)}{p}\Big)\,, \tag{8.61}$$

by (8.48) and (8.52). For the other frequencies we use (8.47) and (8.51), showing that they contribute at most

$$\frac{\tau(cd)}{c^{\frac{3}{2}}d}\sum_{(k,\ell)\neq(0,0)}\sum(cd,k,\ell)\Big|\hat{F}\Big(\frac{k}{cd},\frac{\ell}{cd}\Big)\Big|\,.$$

Estimating the Fourier transform by (8.41) we obtain the bound

$$\frac{\tau(cd)}{c^{\frac{3}{2}}d}RS\sum_{(k,\ell)\neq(0,0)}\sum(cd,k,\ell)\Big(1+\frac{|k|R}{cd}\Big)^{-2}\Big(1+\frac{|\ell|S}{cd}\Big)^{-2}\,,$$

which in turn is bounded by $\tau^2(cd)c^{-1/2}(cd+R+S)$. Adding this estimate to (8.61) we obtain an asymptotic formula for $A_d(\chi)$ for any real character $\chi\,(\operatorname{mod} d)$. Inserting this formula in (8.20) (note that $A_d(\chi)=A_d(\chi_q)$ and $\chi(\Delta)=\chi_q(\Delta)$) we arrive at (8.57). □

Before we introduce (8.57) into (8.4) we need to know the weights ρ_ℓ which are given by (7.23)–(7.25) and the sum J given by (7.19).

LEMMA 8.4. *Let g be the density function which is supported on squarefree numbers coprime with Δ given by (8.58) and let $h(d)$ be the multiplicative function supported on squarefree numbers, defined by $h(p)=g(p)(1-g(p))^{-1}$. For $(\ell,\Delta)=1$ we have*

$$\sum_{\substack{d<y\\(d,\ell)=1}}h(d)=\eta(\ell)c(\Delta)y+O\big(\tau(\ell\Delta)\sqrt{y}\big) \tag{8.62}$$

where

$$\eta(\ell)=\prod_{p|\ell}\Big(1+\frac{h(p)}{p}\Big)^{-1}\asymp\frac{\varphi(\ell)}{\ell}\,, \tag{8.63}$$

and

$$c(\Delta)=\prod_{p}\Big(1-\frac{1}{p}\Big)\Big(1+\frac{h(p)}{p}\Big)\asymp\frac{\varphi(\Delta)}{\Delta}\,. \tag{8.64}$$

PROOF. Note that $h(p)=1+O(p^{-1})$. We write $h=1*f$ so $f=\mu*h$ is the multiplicative function with $f(p)=h(p)-1$, $f(p^2)=-h(p)$ and $f(p^\alpha)=0$ for $\alpha\geqslant 3$. Hence, our sum is equal to

$$\begin{aligned}\sum_{\substack{ab<y\\(ab,\ell\Delta)=1}}\sum f(b)&=\sum_{\substack{b<y\\(b,\ell\Delta)=1}}f(b)\Big(\frac{\varphi(\ell\Delta)y}{\ell\Delta b}+O\big(\tau(\ell\Delta)\big)\Big)\\&=y\frac{\varphi(\ell\Delta)}{\ell\Delta}\sum_{(b,\ell\Delta)=1}\frac{f(b)}{b}+O\big(\tau(\ell\Delta)\sqrt{y}\big)\,.\end{aligned}$$

Here, the infinite series over $(b, \ell\Delta) = 1$ is equal to the product

$$\prod_{p\nmid \ell\Delta}\Big(1-\frac{1}{p}\Big)\Big(1+\frac{h(p)}{p}\Big)\,. \tag{8.65}$$

Hence, the result follows. □

In particular, taking $y = \sqrt{D}$ in (8.62) we get

$$J = \sum_{d<\sqrt{D}} h(d) = c(\Delta)\sqrt{D} + O\big(\tau(\Delta)D^{\frac14}\big)\,. \tag{8.66}$$

Moreover, by (8.62) with $y = \sqrt{D}/\ell$ for $1 \leqslant \ell < \sqrt{D}$ we derive from (7.24)

$$\rho_\ell = \mu(\ell)\frac{h(\ell)}{g(\ell)}\frac{\eta(\ell)}{\ell} + O\big(\tau^2(\ell\Delta)\ell^{-\frac12}D^{-\frac14}\big) \ll \frac{\tau^2(\ell\Delta)}{\ell}\,. \tag{8.67}$$

Hence

$$\lambda_d = \sum_{[d_1,d_2]=d} \rho_{d_1}\rho_{d_2} \ll \frac{\tau^2(d\Delta)}{d}\,, \tag{8.68}$$

and by (8.60) we find that the absolute remainder of level D satisfies

$$R(\mathcal{A},\Omega) = \sum_{d<D} |\lambda_d r_d(\Omega)| \ll \frac{\tau^2(c\Delta)}{\sqrt{c}}(cD+R+S)(\log 2D)^4\,. \tag{8.69}$$

By the above estimates we conclude the following formula:

$$S^+(\mathcal{A},\Omega) = \frac{X}{J} + R(\mathcal{A},\Omega)\,, \tag{8.70}$$

where X is given by (8.59), J is given by (8.66), and $R(\mathcal{A},\Omega)$ satisfies (8.69). Recall that $S^+(\mathcal{A},\Omega)$ is the sifting function for the sequence $\mathcal{A}$ of numbers (8.40) (smoothly weighted) with respect to the system Ω of quadratic non-residues. Hence we deduce

$$S^+(\mathcal{A},\Omega) \ll \frac{N^\varepsilon}{\sqrt{c}}\Big(\frac{RS}{\sqrt{cD}} + cD + R + S\Big)\,.$$

We choose $D = c^{-1}(RS)^{\frac23} + 1$ and conclude that

$$S^+(\mathcal{A},\Omega) \ll \bigg[\frac{N^{\frac23}}{(ab)^{\frac13}c^{\frac12}} + c^{\frac12} + \Big(\frac{N}{ac}\Big)^{\frac12} + \Big(\frac{N}{bc}\Big)^{\frac12}\bigg]N^\varepsilon\,. \tag{8.71}$$

Recall that $r(N,Q) \leqslant S^+(\mathcal{A},\Omega)$ by (8.42). For the estimation of $r(N,Q)$ we can combine (8.71) with the trivial bound $N^\varepsilon(N/c)^{1/2}$ so we can replace the term $c^{1/2}$ by $\min(c^{1/2},(N/c)^{1/2}) \leqslant N^{1/4}$. Having done so, we choose c to be the largest of a, b, c getting:

THEOREM 8.5. *Let abc be squarefree and coprime with N. Then, the number of solutions to* (8.31) *satisfies*

$$r(N,Q) \ll \big((abc)^{-\frac{7}{18}}N^{\frac23} + N^{\frac14}\big)N^\varepsilon\,, \tag{8.72}$$

where the implied constant depends only on ε.

REMARK. The bound (8.72) is better than (8.38) if $N^{3/4} < abc < N^{3/2}$.

CHAPTER 9

The Large Sieve

Intermezzo

9.1. The Basic Inequality

The large sieve was invented in a short paper of 1941 by Ju.V. Linnik [**124**]. While it is based on quite different principles from those of conventional sieves the method applies to the common sifting problem; it is a tool for examining sets of the form

$$(\mathcal{M};\mathcal{P},\Omega) = \{m \in \mathcal{M};\ m(\mathrm{mod}\ p) \notin \Omega_p \quad \text{for any } p \in \mathcal{P}\}.$$

The large sieve methods are much more powerful than the combinatorial sieve when the set $\mathcal{M}$ is an interval and the number of residue classes to be excluded, $\omega(p) = |\Omega_p|$, is relatively large in comparison to the modulus. Briefly speaking, the power of large sieve methods is supplied by harmonic analysis on the circle $\mathbb{R}/\mathbb{Z}$ and this offers a more precise counting of the relevant integers by an employment of characters.

After intensive studies (notably by Rényi, Roth, Barban, Bombieri, Davenport, Halberstam, Montgomery, Vaughan, Selberg, Gallagher) the large sieve developed into a collection of abstract inequalities for sets of well-spaced points, from which the origin of the subject is today barely recognizable. The analogous large sieve for Fourier coefficients of cusp forms [**31**] is powerful and quite useful as a supporting actor on the stage of modern sieve productions, but is not conceptually relevant to the topics considered here.

One of the many variants of the large sieve inequalities gives a bound for the trigonometric polynomial

$$S(\alpha) = \sum_n a_n e(\alpha n) \tag{9.1}$$

where $\mathcal{A} = (a_n)$ are arbitrary complex numbers in a segment

$$M < n \leqslant M + N. \tag{9.2}$$

By Cauchy's inequality we have

$$|S(\alpha)|^2 \leqslant N \sum_n |a_n|^2 \tag{9.3}$$

and this is, of course, best possible in general. However, by varying α one expects to be able to establish non-trivial estimates, even for general $\mathcal{A} = (a_n)$, because the additive characters $\psi_r(n) = e(\alpha_r n)$ are independent and almost orthogonal provided the points α_r are well-spaced modulo one. Precisely, suppose that

$$\|\alpha_r - \alpha_s\| \geqslant \delta \qquad \text{if } r \neq s \tag{9.4}$$

where $\|\alpha\|$ denotes the distance of α to the nearest integer. Note that the number of δ-spaced points cannot exceed $1+\delta^{-1}$. The large sieve inequality asserts that

$$\sum_r |S(\alpha_r)|^2 \leqslant D(\delta,N)\sum_n |a_n|^2 \,, \tag{9.5}$$

where $D(\delta,N)$ depends only on δ and N. In other words, the best possible value for $D(\delta,N)$ for arbitrary complex numbers a_n is just the norm of a linear operator acting between finite-dimensional vector spaces.

We shall reduce various sifting problems to the inequality (9.5) in the following sections. Before giving any examples we provide a comprehensive treatment of this basic inequality for general points $\alpha_r \in \mathbb{R}/\mathbb{Z}$.

It is clear by (9.3) that (9.5) cannot hold with $D(\delta,N)$ smaller than N. Moreover, $D(\delta,N)$ must exceed the number of points α_r which can be as large as δ^{-1}. It turns out that $D(\delta,N)$ need be no larger than the sum of these cardinalities (which are the dimensions of the relevant vector spaces).

THEOREM 9.1. *For any set of δ-spaced points $\alpha_r \in \mathbb{R}/\mathbb{Z}$ and any complex numbers a_n with $M < m \leqslant M+N$, where $0 < \delta \leqslant \frac{1}{2}$ and N is a positive integer, we have*

$$\sum_r |S(\alpha_r)|^2 \leqslant (\delta^{-1}+N-1)\sum_n |a_n|^2 \,. \tag{9.6}$$

This neat inequality, which is best possible, was proved by A. Selberg [**145**] and independently by H. Montgomery and R.C. Vaughan [**131**]. There are several ways of proving the large sieve inequality (9.5) which produce $D(\delta,N)$ in slightly weaker form than

$$D(\delta,N) = \delta^{-1}+N-1 \,, \tag{9.7}$$

but sufficient for the most essential applications. The first result of this kind, with

$$D(\delta,N) = \tfrac{11}{5}\max(\delta^{-1},N) \,,$$

was established by H. Davenport and H. Halberstam [**29**]. A very simple and elegant derivation of (9.6) with $D(\delta,N) = \delta^{-1}+\pi N$ was subsequently given by P. X. Gallagher [**59**]. In this chapter we present the proof of (9.6) following the method of Selberg. The proof given by Montgomery and Vaughan is quite different; cf. [**131**] or [**109**].

PROOF. By duality (the fact that the norm of a linear operator is equal to the norm of its dual) we need to show that the inequality

$$\sum_{M<n\leqslant M+N} \Big|\sum_r c_r\, e(\alpha_r n)\Big|^2 \leqslant (\delta^{-1}+N-1)\sum_r |c_r|^2$$

holds for any complex numbers c_r. Changing the coefficients c_r to $e(\alpha_r(M+1))c_r$ we reduce this to

$$\sum_{0\leqslant n\leqslant N-1} \Big|\sum_r c_r\, e(\alpha_r n)\Big|^2 \leqslant (\delta^{-1}+N-1)\sum_r |c_r|^2 \,.$$

The left-hand side is bounded by

$$\sum_n b(n)\Big|\sum_r c_r\, e(\alpha_r n)\Big|^2 \,,$$

where n runs over all integers and $b(x)$ is any function which majorizes the characteristic function of the interval $[0, N-1]$. Assuming that $b(x)$ is continuous and rapidly decreasing to zero as $|x| \to \infty$ we can write the latter sum as

$$\begin{aligned}\sum_r \sum_s c_r \bar{c}_s \sum_n b(n) e\big((\alpha_r - \alpha_s)n\big) \\ = \sum_r \sum_s c_r \bar{c}_s \sum_k \hat{b}_k(k - \alpha_r + \alpha_s)\end{aligned}$$

by Poisson's formula, where $\hat{b}(y)$ is the Fourier transform

$$\hat{b}(y) = \int_{-\infty}^{\infty} b(x) e(-xy) dx \ .$$

Put

$$B(\alpha) = \sum_k \hat{b}(k - \alpha) \ .$$

Note that $B(\alpha)$ is periodic in α of period one. Suppose that $\hat{b}(y) = 0$ for $|y| \geqslant \delta$. Since the points $\alpha_r \pmod 1$ are δ-spaced we get only the terms on the diagonal and the sum is equal to

$$B(0) \sum_r |c_r|^2 \ .$$

Thus, our task is to minimize

$$B(0) = \sum_k \hat{b}(k) = \hat{b}(0) = \int_{-\infty}^{\infty} b(x) dx$$

amongst functions $b(x)$ having the above properties.

LEMMA 9.2 (Beurling–Selberg). *Let $F(z)$ be the function of the complex variable z defined by*

$$F(z) = \Big(\frac{\sin \pi z}{\pi}\Big)^2 \Big\{ \sum_{n=0}^{\infty} \frac{1}{(z-n)^2} - \sum_{n=1}^{\infty} \frac{1}{(z+n)^2} + \frac{2}{z} \Big\} \ . \tag{9.8}$$

Then

$$F(x) \geqslant \operatorname{sgn} x \quad \text{if } x \text{ is real,} \tag{9.9}$$

$$\int_{-\infty}^{\infty} \big(F(x) - \operatorname{sgn} x\big) dx = 1 \ . \tag{9.10}$$

Moreover, $F(z)$ is entire and $F(z) \ll e^{2\pi |\operatorname{Im} z|}$.

PROOF OF LEMMA. The last two assertions are clear. By the formula

$$\Big(\frac{\sin \pi z}{\pi}\Big)^2 \Big(\sum_{n=-\infty}^{\infty} (z+n)^{-2} \Big) = 1 \ ,$$

we get two expressions:

$$\begin{aligned} F(x) &= 1 + 2\Big(\frac{\sin \pi x}{\pi}\Big)^2 \Big\{ \frac{1}{x} - \sum_{n=1}^{\infty} \frac{1}{(x+n)^2} \Big\} \ , \\ -F(-x) &= 1 + 2\Big(\frac{\sin \pi x}{\pi}\Big)^2 \Big\{ \frac{1}{x} - \sum_{n=0}^{\infty} \frac{1}{(x+n)^2} \Big\} \ . \end{aligned}$$

Hence (9.9) follows because, for $x > 0$ we have

$$\sum_{n=1}^{\infty} \frac{1}{(x+n)^2} < \frac{1}{x} < \sum_{n=0}^{\infty} \frac{1}{(x+n)^2} .$$

Finally, we show (9.10) as follows:

$$\int_{-\infty}^{\infty} (F(x) - \operatorname{sgn} x)dx = \int_0^{\infty} (F(x) + F(-x))dx = 2 \int_0^{\infty} \Big(\frac{\sin \pi x}{\pi x}\Big)^2 dx = 1 .$$

□

COMPLETION OF THE PROOF OF THEOREM 9.1. Put

$$G(x) = \frac{1}{2}F(x) + \frac{1}{2}F(d-x) \tag{9.11}$$

where $d > 0$ will be chosen. Obviously, $G(x)$ majorizes the characteristic function $\chi(x)$ of the segment $[0, d]$. More precisely, we have

$$G(x) = \chi(x) + \frac{1}{2}\big(F(x) - \operatorname{sgn} x\big) + \frac{1}{2}\big(F(d-x) - \operatorname{sgn}(d-x)\big) \geqslant \chi(x) \tag{9.12}$$

by (9.9). Moreover, we have $\hat{G}(t) = 0$ if $|t| \geqslant 1$ by the Paley–Wiener theorem. Next,

$$\hat{G}(0) = \int_{-\infty}^{\infty} G(x)dx = d + 1$$

by (9.10) and (9.12).

Now we put $d = \delta(N-1)$ and $b(x) = G(\delta x)$. By the above properties we see that $b(x)$ majorizes the characteristic function of the segment $[0, N-1]$, that $\hat{b}(t) = 0$ if $|t| \geqslant \delta$, and

$$\hat{b}(0) = \frac{1}{\delta}\hat{G}(0) = \frac{d+1}{\delta} = N - 1 + \frac{1}{\delta} .$$

□

REMARK. Selberg's choice of the majorizing function $b(x)$ is somewhat elaborate, but of course simpler ones also yield good results, not optimal though.

9.2. The Large Sieve Inequality for Additive Characters

In applications of the large sieve inequality (9.6) to number theory we often take the points α_r to be rationals a/q with $1 \leqslant q \leqslant Q$ and $(a, q) = 1$. These points are spaced by $\delta = Q^{-2}$; indeed, if $a/q \neq a'/q'$ then

$$\Big\|\frac{a}{q} - \frac{a'}{q'}\Big\| = \Big\|\frac{aq' - a'q}{qq'}\Big\| \geqslant \frac{1}{qq'} \geqslant \frac{1}{Q^2}.$$

Therefore, Theorem 9.1 yields:

THEOREM 9.3. *For any complex numbers a_n with $M < n \leqslant M + N$ where N is a positive integer we have*

$$\sum_{q \leqslant Q} \sideset{}{^\star}\sum_{a \,(\mathrm{mod}\, q)} \Big|S\Big(\frac{a}{q}\Big)\Big|^2 \leqslant (Q^2 + N - 1) \sum_n |a_n|^2. \tag{9.13}$$

Notice that if $\mathcal{A} = (a_n)$ is supported on an arithmetic progression $n \equiv \ell (\mathrm{mod}\ k)$ and $(k, q) = 1$, we can change the variables to derive

COROLLARY 9.4. *For any complex numbers a_n with $M < n \leqslant M+N$ we have*

$$\sum_{\substack{q\leqslant Q\\(q,k)=1}} \sum_{a\,(\mathrm{mod}\,q)}^{\star} \Big| \sum_{n\equiv \ell\,(\mathrm{mod}\,k)} a_n e\Big(\frac{an}{q}\Big)\Big|^2 \leqslant (Q^2 + k^{-1}N) \sum_{n\equiv\ell\,(\mathrm{mod}\,k)} |a_n|^2. \tag{9.14}$$

9.3. Equidistribution over Residue Classes

As an application of Theorem 9.3 we show that a general set of distinct integers $\mathcal{M} \subset (M, M+N]$ represents almost all residue classes for almost all prime moduli $p \leqslant \sqrt{N}$ provided only that $\mathcal{M}$ is numerous. More exactly we establish the equidistribution of a general sequence of real numbers $\mathcal{A} = (a_n)$ with $M < n \leqslant M+N$ over distinct residue classes $\nu(\mathrm{mod}\ q)$. Put

$$X = \sum_n a_n,$$
$$X(q,\nu) = \sum_{n\equiv\nu(\mathrm{mod}\ q)} a_n.$$

We show that

$$\Delta(q,\nu) = X(q,\nu) - \frac{X}{q} \tag{9.15}$$

is small for almost all $\nu(\mathrm{mod}\ q)$. Indeed, using additive characters we have

$$\Delta(q,\nu) = \frac{1}{q} \sum_{\substack{a(\mathrm{mod}\ q)\\ a\not\equiv 0(\mathrm{mod}\ q)}} e\Big(-\frac{\nu a}{q}\Big) S\Big(\frac{a}{q}\Big).$$

Hence, by orthogonality (Plancherel's theorem),

$$q \sum_{\nu(\mathrm{mod}\ q)} |\Delta(q,\nu)|^2 = \sum_{\substack{a(\mathrm{mod}\ q)\\ a\not\equiv 0(\mathrm{mod}\ q)}} \Big|S\Big(\frac{a}{q}\Big)\Big|^2. \tag{9.16}$$

Summing this over prime moduli we infer by (9.13) that

$$\sum_{p\leqslant Q} p \sum_{\nu(\mathrm{mod}\ p)} |\Delta(p,\nu)|^2 \leqslant (Q^2+N) \sum_n |a_n|^2. \tag{9.17}$$

In particular, if $\mathcal{A} = (a_n)$ is the characteristic function of a set $\mathcal{M}$, this yields

$$\sum_{p\leqslant\sqrt{N}} p \sum_{\nu(\mathrm{mod}\ p)} |X(p,\nu) - p^{-1}X|^2 \leqslant 2NX\ , \tag{9.18}$$

where now $X(p,\nu)$ is the number of $m \in \mathcal{M}$ with $m \equiv \nu(\mathrm{mod}\ p)$ and $X = |\mathcal{M}|$ is the number of all elements in $\mathcal{M}$. In the language of probability theory (as spoken by A. Rényi), this estimate resembles Tchebyshev's inequality for the variance.

Given $0 < \eta \leqslant 1$ we call a prime p exceptional if the number of residue classes $\nu(\mathrm{mod}\ p)$ covered by $\mathcal{M}$ does not exceed $(1-\eta)p$. Let $E_\eta(Q)$ denote the number of exceptional $p \leqslant Q$. By (9.18) we derive

$$E_\eta(\sqrt{N}) \leqslant \frac{2N}{\eta X}. \tag{9.19}$$

This bound is impressive if $\mathcal{M}$ has positive natural density, say $X \geqslant \delta N$ with $0 < \delta \leqslant 1$, in which case we conclude that the number of exceptional primes $p \leqslant \sqrt{N}$ is bounded, namely $E_\eta(\sqrt{N}) \leqslant 2/\delta\eta$. This was (apart from the constant) the original conclusion of Linnik. An estimation for the variance (somewhat weaker than (9.18)) was first established by Rényi using Linnik's large sieve method.

Linnik gave a striking application of his method to estimate the least quadratic non-residue (mod p), i.e., the smallest positive integer $q(p)$ such that $(q(p)/p) = -1$. Note that $q(p)$ is prime. It is conjectured that

$$q(p) \ll p^\varepsilon \tag{9.20}$$

for any $\varepsilon > 0$ with the implied constant depending on ε, whereas the best known estimate is that (9.20) holds for any $\varepsilon > 1/4\sqrt{e} = 0.1516\dots$. From the Riemann Hypothesis for the corresponding L-function one derives (due to Ankeny) that

$$q(p) \ll (\log p)^2.$$

THEOREM 9.5 (Linnik). *The number of primes $p \leqslant N$ such that $q(p) > N^\varepsilon$ is bounded by a constant depending on ε.*

PROOF. Consider the sifting problem $(\mathcal{M}, \mathcal{P}, \Omega)$ with

$$\begin{aligned} \mathcal{M} &= \{1, 2, \dots, N\}, \\ \mathcal{P} &= \{p \leqslant \sqrt{N} : (n/p) = 1 \quad \text{for all } n \leqslant N^\varepsilon\}, \\ \Omega_p &= \{\nu (\text{mod } p) : (\nu/p) = -1\}. \end{aligned}$$

Thus $\omega(p) = |\Omega_p| = \frac{1}{2}(p-1)$ and by (9.18),

$$\sum_{p \in \mathcal{P}} \Big(1 - \frac{1}{p}\Big) \leqslant 4NX^{-1} \tag{9.21}$$

where X is the number of elements in

$$(\mathcal{M}, \mathcal{P}, \Omega) = \{1 \leqslant m \leqslant N : (m/p) = 1 \quad \text{for any } p \in \mathcal{P}\}.$$

Note that this set contains all numbers $m \leqslant N$ free of prime divisors $> N^\varepsilon$, thus also all the numbers of type $m = np_1 \dots p_k \leqslant N$ with $N^{\varepsilon - \varepsilon^2} < p_j < N^\varepsilon$ for $1 \leqslant j \leqslant k = \varepsilon^{-1}$. Therefore,

$$X \geqslant \sum_{p_1 \dots p_k} \Big[\frac{N}{p_1 \dots p_k}\Big] \gg N.$$

Inserting this bound into (9.21) we obtain $|\mathcal{P}| \ll 1$ as claimed. □

Rather than sieving by many residue classes per modulus one may want to sieve by most classes. In this situation the large sieve becomes the "larger sieve". The name is due to P. X. Gallagher who discovered [**61**] an interesting method to handle such extreme sieving. We shall develop this in Section 9.7.

The estimate for the variance (9.18) can be extended to all moduli, see [**128**]. To this end we use the Ramanujan sum

$$c_q(n) = \sideset{}{^\star}\sum_{a (\text{mod } q)} e\Big(\frac{an}{q}\Big). \tag{9.22}$$

We have

$$c_q(n) = \sum_{d|(n,q)} d\mu(q/d) \ . \tag{9.23}$$

Hence, we derive

$$\sum_{d|q} d\mu(q/d)X(d,\nu) = \sum_{a(\text{mod } q)}^{\star} S\Big(\frac{a}{q}\Big)e\Big(\frac{\nu a}{q}\Big) \ .$$

Here, we can replace $X(d,\nu)$ by $\Delta(d,\nu)$ and d by q/d, getting

$$q\sum_{d|q} \frac{\mu(d)}{d}\Delta\Big(\frac{q}{d},\nu\Big) = \sum_{a(\text{mod } q)}^{\star} S\Big(\frac{a}{q}\Big)e\Big(\frac{\nu a}{q}\Big) \ .$$

This holds for very $\nu(\text{mod}\, q)$. Hence, by orthogonality,

$$q\Big|\sum_{d|q} \frac{\mu(d)}{d}\Delta\Big(\frac{q}{d},\nu\Big)\Big|^2 = \sum_{a\,(\text{mod}\, q)}^{\star} \Big|S^*\Big(\frac{a}{q}\Big)\Big|^2 \ . \tag{9.24}$$

Summing over q, we infer by (9.13) the following extension of (9.17).

THEOREM 9.6. *For any complex numbers a_n with $M < n \leqslant M+N$ we have*

$$\sum_{q\leqslant Q} q \sum_{\nu\,(\text{mod}\, q)} \Big|\sum_{d|q} \frac{\mu(d)}{d}\Delta\Big(\frac{q}{d},\nu\Big)\Big|^2 \leqslant (Q^2+N)\sum_n |a_n|^2. \tag{9.25}$$

9.4. Arithmetic Large Sieve

The large sieve inequality (9.13) can be used to derive an upper bound for the sifting function

$$Z = \sum_{n\in(\mathcal{N};\mathcal{P},\Omega)} a_n \tag{9.26}$$

with the number of residue classes $\omega(p) = |\Omega_p|$ not necessarily very large, and it is capable of producing results equivalent to these derived by the Λ^2-sieve. We may assume that

$$a_n = 0 \qquad \text{unless } n \in (\mathcal{N};\mathcal{P},\Omega) \ , \tag{9.27}$$

so $Z = X$. Let h be the multiplicative function supported on squarefree numbers with

$$h(p) = \frac{\omega(p)}{p-\omega(p)} \ . \tag{9.28}$$

LEMMA 9.7. *Let $S(\alpha)$ be the trigonometric series (9.1) with coefficients supported on $(\mathcal{N};\mathcal{P},\Omega)$. We have for any q,*

$$h(q)|S(0)|^2 \leqslant \sum_{a\,(\text{mod}\, q)}^{\star} \Big|S\Big(\frac{a}{q}\Big)\Big|^2. \tag{9.29}$$

PROOF. If $q = p$ is prime then $X(p,\nu) = 0$ for all $\nu \in \Omega_p$; therefore, by Cauchy's inequality we get

$$\begin{aligned}|S(0)|^2 &= \Big|\sum_{\nu (\mathrm{mod}\ p)} X(p,\nu)\Big|^2 \leqslant (p-\omega(p)) \sum_{\nu (\mathrm{mod}\ p)} |X(p,\nu)|^2 \\ &= \Big(1-\frac{\omega(p)}{p}\Big) \sum_{a (\mathrm{mod}\ p)} \Big|S(\frac{a}{p})\Big|^2 .\end{aligned}$$

Transferring the contribution of $a \equiv 0 \,(\mathrm{mod}\ p)$ from the left to the right side we get (9.29). In general, if $q = q_1 q_2$ with $(q_1, q_2) = 1$ we have

$$\sum_{a (\mathrm{mod}\ q)}^{\star} \Big|S(\frac{a}{q})\Big|^2 = \sum_{a_1 (\mathrm{mod}\ q_1)}^{\star} \sum_{a_2 (\mathrm{mod}\ q_2)}^{\star} \Big|S(\frac{a_1}{q_1}+\frac{a_2}{q_2})\Big|^2 .$$

Assuming (9.29) holds for q_1 and q_2 the above factorization yields (change a_n into $a_n e(na_1/q_1)$) the bound

$$\geqslant h(q_2) \sum_{a_1 (\mathrm{mod}\ q_1)}^{\star} \Big|S(\frac{a_1}{q_1})\Big|^2 \geqslant h(q_2)h(q_1)|S(0)|^2 .$$

This completes the proof of (9.29) by induction on the number of prime factors of q. □

Summing (9.29) over the moduli $q \leqslant Q$, we derive by (9.13) the following:

THEOREM 9.8. *For any complex numbers a_n with $n \in (\mathcal{N};\mathcal{P},\Omega)$ and $M < n \leqslant M+N$ we have*

$$J\Big|\sum a_n\Big|^2 \leqslant (N+Q^2)\sum |a_n|^2 \tag{9.30}$$

where

$$J = \sum_{q \leqslant Q} h(q). \tag{9.31}$$

Letting $\mathcal{A} = (a_n)$ be the characteristic function of numbers $n \in (\mathcal{N},\mathcal{P},\Omega)$ in an interval $M < n \leqslant M+N$ we get by (9.30):

COROLLARY 9.9. *The number of $n \in (\mathcal{N};\mathcal{P},\Omega)$ with $M < n \leqslant M+N$ satisfies*

$$Z \leqslant \frac{N+Q^2}{J}. \tag{9.32}$$

Note that, by Cauchy's inequality, the estimate (9.30) actually follows from its special case (9.32). This estimate is due to H.L. Montgomery [**129**]; it is neat and only slightly weaker than the estimate of Theorem 7.1 and Theorem 7.14 derived by Selberg's sieve.

Many further interesting arithmetical aspects of the large sieve inequality are developed in the recent book of O. Ramaré [**136**].

9.5. Where Linnik meets Selberg: A Duet

Actually, there is a connection between the Selberg upper-bound sieve and the large sieve inequality for additive characters. We begin by recalling the formula (7.26) for the convolution $1 * \rho$:

$$\sum_{\ell | n} \rho_\ell = \frac{1}{J} \sum_{q < \sqrt{D}} \mu((n,q)) h(q/(n,q)). \tag{9.33}$$

We proceed further in the particular case $g(d) = d^{-1}$ if $(d,k) = 1$ and $g(d) = 0$ otherwise. This gives $h(d) = \varphi(d)^{-1}$ if $(d,k) = 1$ and $h(d) = 0$ otherwise, so (9.33) becomes

$$\sum_{\ell | n} \rho_\ell = \frac{1}{J} \sum_{\substack{q < \sqrt{D} \\ (q,k)=1}} \varphi(q)^{-1} \mu((n,q)) \varphi((n,q)).$$

Here one can see the Ramanujan sum (for q squarefree)

$$c_q(n) = \mu(q)\mu((n,q))\varphi((n,q)).$$

Therefore,

$$\sum_{\ell | n} \rho_\ell = \frac{1}{J} \sum_{\substack{q < \sqrt{D} \\ (q,k)=1}} \frac{\mu(q)}{\varphi(q)} c_q(n)$$

and

$$\sum_n a_n \Big(\sum_{\ell | n} \rho_\ell \Big)^2 = \frac{1}{J^2} \sum_n a_n \Big(\sum_{\substack{q < \sqrt{D} \\ (q,k)=1}} \frac{\mu(q)}{\varphi(q)} c_q(n) \Big)^2. \tag{9.34}$$

If $\mathcal{A} = (a_n)$ is the characteristic function of numbers $n \equiv \nu (\text{mod } k)$ in the interval $M < n \leqslant M + N$, then the expression (9.34) is the dual form of

$$\sum_{\substack{q < \sqrt{D} \\ (q,k)=1}} \sideset{}{^\star}\sum_{a (\text{mod } q)} \Big| S\big(\frac{a}{q}\big) \Big|^2.$$

This connection was observed by I. Kobayashi [**118**].

In this particular case under consideration we have by (7.141),

$$J = J(D) = \sideset{}{^\flat}\sum_{\substack{d < \sqrt{D} \\ (d,k)=1}} \frac{1}{\varphi(d)} \geqslant \frac{\varphi(k)}{k} \log \sqrt{D}\ .$$

Applying the large sieve inequality (9.14) we conclude by the duality principle,

$$\sum_{\substack{M < n \leqslant M+N \\ n \equiv \nu\, (\text{mod } k)}} \Big(\sum_{\ell | n} \rho_\ell \Big)^2 \leqslant \frac{N + kD}{\varphi(k) \log \sqrt{D}}\ . \tag{9.35}$$

Note that the bound (9.35) is exactly the same as (9.32) with $Q = \sqrt{D}$ and N replaced by N/k.

The large sieve inequality for additive characters has strongest effect for the fullest sequences $A = (a_n)$ such as the set of all integers in an interval, in which case it produces a very precise bound as shown in (9.35). This can be attributed to the key feature of the particular situation at hand; the density function $g(d) = d^{-1}$, for which the Selberg's sieve weights ρ_d are designed, perfectly matches the density function for the integers in an interval. Now, for educational purposes, we are going to investigate how the Selberg sieve weights ρ_d, designed for a density function $g(d)$ of dimension κ, co-operate with the large sieve via duality. Something will get fouled up; the discrepancy will be of the same calibre as previously observed in Section 7.3.

Again we begin with (see (9.33)):

$$\sum_{\ell|n} \rho_\ell = \frac{1}{J} \sum_{q<\sqrt{D}} h(q)\psi_q(n) \ , \tag{9.36}$$

where

$$\psi_q(n) = \frac{\mu((n,q))}{h((n,q))}. \tag{9.37}$$

Here the function $\psi_q(n)$ is no longer the Ramanujan sum, nevertheless, being periodic in n modulo q, it has the Fourier expansion

$$\psi_q(n) = \sum_{a(\mathrm{mod}\, q)} \hat{\psi}_q(a) e\Big(\frac{an}{q}\Big)$$

with coefficients given by the inverse Fourier transform

$$\hat{\psi}_q(a) = \frac{1}{q} \sum_{b\,(\mathrm{mod}\, q)} \psi_q(b) e\Big(-\frac{ab}{q}\Big) = \frac{1}{q} \sum_{b\,(\mathrm{mod}\, q)} \frac{\mu((b,q))}{h((b,q))} e\Big(-\frac{ab}{q}\Big) \ .$$

For q squarefree this is multiplicative in q and for prime moduli we find

$$\hat{\psi}_p(a) = 1 - 1/g(p)p \quad \text{if} \quad p|a \ ,$$
$$\hat{\psi}_p(a) = -1/g(p)p \qquad \text{if} \quad p\nmid a \ .$$

Hence,

$$\hat{\psi}_q(a) = \frac{\mu(q)}{g(q)q} \sum_{t|(a,q)} \mu(t)g(t)t$$

and

$$\begin{aligned} \psi_q(n) &= \frac{\mu(q)}{g(q)q} \sum_{a(\mathrm{mod}\, q)} \Big(\sum_{t|(a,q)} \mu(t)g(t)t \Big) e\Big(\frac{an}{q}\Big) \\ &= \frac{\mu(q)}{g(q)q} \sum_{rs=q} c_r(n) \prod_{p|s} (1 - g(p)p) \ , \end{aligned}$$

where

$$c_r(n) = \sum_{a(\mathrm{mod}\, r)}^{\star} e\Big(\frac{an}{r}\Big)$$

is the Ramanujan sum. Introducing this to (9.36) we get

$$\sum_{\ell|n} \rho_\ell = \frac{1}{J} \sum_{rs<\sqrt{D}} c_r(n) \frac{\mu(rs)h(rs)}{g(rs)rs} \prod_{p|s} (1 - g(p)p) \ .$$

In other words,

$$\sum_{\ell|n}\rho_\ell=\frac{1}{J}\sum_{r<\sqrt{D}}\alpha_r c_r(n) \tag{9.38}$$

where

$$\alpha_r=\frac{\mu(r)h(r)}{g(r)r}\sideset{}{^\flat}\sum_{\substack{s<\sqrt{D}/r\\(s,r)=1}} f(s) \tag{9.39}$$

and

$$f(s)=\frac{h(s)}{g(s)}\prod_{p|s}\Big(g(p)-\frac{1}{p}\Big)\,. \tag{9.40}$$

The function f is multiplicative with

$$f(p)=\Big(g(p)-\frac{1}{p}\Big)\big(1-g(p)\big)^{-1}\,,$$

so $f(s)$, if regarded as a sieve density, has dimension $\kappa-1$.

Now we are ready to estimate the sifted sum

$$S^+(\mathcal{A})=\sum_{M<n\leqslant M+N}\Big(\sum_{\ell|n}\rho_\ell\Big)^2 \tag{9.41}$$

by applying the large sieve inequality (9.14). We get

$$S^+(\mathcal{A})\leqslant\frac{N+D}{J^2}\parallel\alpha\parallel^2, \tag{9.42}$$

where

$$\parallel\alpha\parallel^2=\sum_{r<\sqrt{D}}\varphi(r)|\alpha_r|^2=\sideset{}{^\flat}\sum_{r<\sqrt{D}}\varphi(r)\Big(\frac{h(r)}{g(r)r}\Big)^2\Big(\sideset{}{^\flat}\sum_{\substack{s<\sqrt{D}/r\\(s,r)=1}} f(s)\Big)^2.$$

For the rest of this section we shall be somewhat sketchy. We have

$$\sum_s f(s)\ll\Big(\log\frac{2\sqrt{D}}{r}\Big)^{\kappa-1}$$

and

$$\parallel\alpha\parallel^2\ll\sideset{}{^\flat}\sum_{r<\sqrt{D}}\frac{1}{r}\Big(\log\frac{2\sqrt{D}}{r}\Big)^{2\kappa-2}\asymp L(D)(\log D)^{2\kappa}$$

where

$$L(D)=\begin{cases}(\log D)^{-1} & \text{if } \kappa>1/2\,,\\ (\log D)^{-1}\log\log D & \text{if } \kappa=1/2\,,\\ (\log D)^{-2\kappa} & \text{if } \kappa<1/2\,.\end{cases} \tag{9.43}$$

Finally, introducing the bound $J=J(D)\asymp(\log D)^\kappa$ to (9.42) we arrive at

$$S^+(A)\ll(N+D)L(D). \tag{9.44}$$

These estimates agree with (7.67), (7.68), (7.69) for $\kappa'=1$.

Although we have skipped some details (not a lot), the above presentation sheds clear light on the nuances of the large sieve estimates as concerns the true order of magnitude. The large sieve is on a par with the Selberg sieve for segments of arithmetic progressions but it loses the competition when it comes to more general sequences.

These last remarks encourage the question as to what should be attainable by the large sieve as an estimate for the sum

$$\sum_{n\leqslant N}\Big|\sum_{q\leqslant Q}\ \sum_{a(q)}^{\star} c(q,a)e\Big(\frac{a}{q}\,Q(n)\Big)\Big|^2$$

where $Q(X)$ is a polynomial? In particular, for $Q(X)$ given by (7.162), would one be able to recover the Selberg sieve bound

$$\sum_{n\leqslant N}\Big(\sum_{\ell|Q(n)} \rho_\ell\Big)^2 \ll N(\log N)^{-\kappa}$$

which has the true order of magnitude?

9.6. The Large Sieve Inequality for Multiplicative Characters

The multiplicative characters $\chi(\text{mod } q)$ can be expanded into additive characters by means of Gauss sums:

$$\tau(\chi) = \sum_{a(\text{mod } q)} \chi(a)\, e\Big(\frac{a}{q}\Big). \tag{9.45}$$

Let s be the conductor of χ so $s|q$. If $s = q$ then χ is said to be primitive. For a primitive character $\chi(\text{mod } q)$ we have

$$\tau(\chi)\bar{\chi}(n) = \sum_{a(\text{mod } q)} \chi(a)\, e\Big(\frac{an}{q}\Big) \tag{9.46}$$

for all n. Hence by orthogonality it follows that

$$|\tau(\chi)|^2 = \sqrt{q}. \tag{9.47}$$

If $\chi(\text{mod } q)$ is not primitive then (9.46) requires the condition $(n, q) = 1$. Let χ_s be the primitive character of conductor s which is induced by χ. Suppose that $q = rs$ with $(r, s) = 1$. Then we have

$$\begin{aligned}\sum_{a(\text{mod } q)} \chi(a)\, e\Big(\frac{an}{q}\Big) &= \sum_{b(\text{mod } r)}^{\star}\ \sum_{c(\text{mod } s)}^{\star} \chi(bs+cr)\, e\Big(\frac{bn}{r}+\frac{cn}{s}\Big)\\ &= \bar{\chi}(n)\,\chi_s(r)\,c_r(n)\,\tau(\chi_s)\end{aligned} \tag{9.48}$$

where $c_r(n)$ is the Ramanujan sum. In particular, for $n = 1$ this gives

$$\tau(\chi) = \mu(r)\,\chi_s(r)\,\tau(\chi_s). \tag{9.49}$$

Hence

$$|\tau(\chi)|^2 = \mu^2(r)s \qquad \text{if } (r,s) = 1. \tag{9.50}$$

Given a finite sequence $\mathcal{A} = (a_n)$ of complex numbers we denote, for any arithmetic function $f : \mathcal{N} \to \mathbb{C}$,

$$T(f) = \sum_n a_n f(n).$$

In particular, if $f(n) = e(an/q)$ then $T(f) = S(a/q)$. By (9.48) we get

$$T(\bar{\chi}c_r) = \bar{\chi}_s(r)\tau(\chi_s)^{-1} \sum_{a (\mathrm{mod}\ q)} \chi(a) S\Big(\frac{a}{q}\Big).$$

Hence by the orthogonality of characters

$$\sum_{\substack{rs\leqslant Q\\(r,s)=1}}\sum \frac{s}{\varphi(rs)} \sideset{}{^\star}\sum_{\chi\,(\mathrm{mod}\ s)} |T(\chi c_r)|^2 \leqslant \sum_{q\leqslant Q} \frac{1}{\varphi(q)} \sum_{\chi (\mathrm{mod}\ q)} \Big| \sum_{a (\mathrm{mod}\ q)} \chi(a) S\Big(\frac{a}{q}\Big)\Big|^2$$

$$= \sum_{q\leqslant Q} \sideset{}{^\star}\sum_{a(\mathrm{mod}\ q)} \Big|S\Big(\frac{a}{q}\Big)\Big|^2.$$

Applying Theorem 9.3 we derive

THEOREM 9.10. *For any complex numbers a_n with $M < n \leqslant M + N$ where N is a positive integer we have*

$$\sum_{\substack{rs\leqslant Q\\(r,s)=1}}\sum \frac{s}{\varphi(rs)} \sideset{}{^\star}\sum_{\chi\,(\mathrm{mod}\ s)} |T(\chi c_r)|^2 \leqslant (Q^2 + N - 1) \sum_n |a_n|^2. \tag{9.51}$$

This result (with a slightly larger bound) is due to E. Bombieri and Davenport [**10**]. Taking $r = 1$ one gets

$$\sum_{q\leqslant Q} \frac{q}{\varphi(q)} \sideset{}{^\star}\sum_{\chi(\mathrm{mod}\ q)} \Big| \sum_{M<n\leqslant M+N} a_n \chi(n)\Big|^2 \leqslant (Q^2 + N) \sum_n |a_n|^2. \tag{9.52}$$

Subsequently, Bombieri and Davenport made good use of the summation over r. If $(n, r) = 1$ then $c_r(n) = \mu(r)$. Thus, assuming the condition

$$a_n = 0 \quad \text{if } n \text{ has a prime divisor} \leqslant Q, \tag{9.53}$$

we have

$$T(\chi c_r) = \mu(r) T(\chi). \tag{9.54}$$

Furthermore, we have by (7.141)

$$\sum_{\substack{r\leqslant Q/s\\(r,s)=1}} \frac{\mu(r)^2}{\varphi(r)} \geqslant \frac{\varphi(s)}{s} \log \frac{Q}{s}. \tag{9.55}$$

Combining (9.51), (9.54) and (9.55), we derive

THEOREM 9.11. *If $\mathcal{A} = (a_n)$ satisfies (9.53), then*

$$\sum_{s\leqslant Q} \Big(\log \frac{Q}{s}\Big) \sideset{}{^\star}\sum_{\chi(\mathrm{mod}\ s)} |T(\chi)|^2 \leqslant (Q^2 + N - 1) \sum_n |a_n|^2. \tag{9.56}$$

COROLLARY 9.12. *If $M > Q$ we have*

$$\sum_{s\leqslant Q} \Big(\log \frac{Q}{s}\Big) \sideset{}{^\star}\sum_{\chi(\mathrm{mod}\ s)} |\pi(M+N, \chi) - \pi(M, \chi)|^2 \leqslant (Q^2 + N)(\pi(M+N) - \pi(M)) \tag{9.57}$$

where

$$\pi(x, \chi) = \sum_{p\leqslant x} \chi(p). \tag{9.58}$$

Taking only the one term $s = 1$ on the left side of (9.57) we deduce the bound

$$\pi(M+N) - \pi(M) \leqslant \frac{N+Q^2}{\log Q} \tag{9.59}$$

for any $Q \leqslant M$. Choosing $Q^2 = N/\log N$ the resulting bound (9.59) is almost as good as (7.159) for $q = 1$ a fact which is remarkable because the contribution in (9.57) of everything except that of the principal character has been ignored.

9.7. The Larger Sieve of Gallagher

As we have seen, the large sieve inequality gives particularly strong upper bounds when the number of residue classes is relatively large, say a positive proportion of the total. For example, if we want to bound the number Z of squares in an interval of length N we remove half of the classes for each prime modulus and then (9.32) gives $Z \ll (N+Q^2)/Q \ll \sqrt{N}$, on choosing $Q = \sqrt{N}$. Of course this obvious result can be done simply without using the large sieve inequality but more serious examples do exist.

When the number of residue classes removed becomes even larger, so that relatively few of them remain behind, the large sieve loses much of its power. In this section we present the "larger sieve" of Gallagher [**61**]. Not only are the results stronger in this situation but the proof is elegant and much simpler. We follow Gallagher's argument but with a slight generalization.

Let Ω_q be a set of residue classes modulo q of cardinality $\omega(q) < q$. Let $\mathcal{A} = (a_n)$ be a sequence of real non-negative numbers supported on an interval of length N. Our goal is to estimate the sum

$$Z = \sum_n{}^{\Omega} a_n\ , \tag{9.60}$$

where the superscript Ω means the summation is restricted to n with

$$n(\operatorname{mod} q) \notin \Omega_q, \quad \text{for every } q\ . \tag{9.61}$$

Put

$$Z(q,\nu) = \sum_{n \equiv \nu (\operatorname{mod} q)} a_n\ .$$

Since

$$Z = \sum_{\nu \notin \Omega_q} Z(q,\nu)\ ,$$

we obtain by Cauchy's inequality,

$$\begin{aligned} Z^2 &\leqslant (q-\omega(q)) \sum_{\nu\notin\Omega_q} |Z(q,\nu)|^2 \\ &\leqslant (q-\omega(q)) \sum_{\nu (\operatorname{mod} q)} |Z(q,\nu)|^2 \\ &= (q-\omega(q)) \sum\sum_{m\equiv n (\operatorname{mod} q)} a_m a_n\ . \end{aligned}$$

Let $t(q)$ be a non-negative function of compact support. We write the above inequality as

$$\frac{t(q)}{q-\omega(q)} Z^2 \leqslant t(q) \sum\sum_{m\equiv n(\operatorname{mod} q)} a_m a_n$$

and sum over q. On the left-hand side we obtain

$$T_\omega = \sum_q \frac{t(q)}{q - \omega(q)} . \tag{9.62}$$

On the right-hand side the diagonal terms $m = n$ produce the factor

$$T = \sum_q t(q) . \tag{9.63}$$

The terms $m \neq n$ produce the factor

$$\sum_{q|(m-n)} t(q) . \tag{9.64}$$

Now, assume that we have a bound

$$t * 1 \leqslant u \tag{9.65}$$

where u is a non-decreasing function so that (9.64) is bounded by $u(|m-n|) \leqslant u(N)$. Gathering the above estimates we arrive at

$$T_\omega\Big(\sum_n a_n\Big)^2 \leqslant T \sum_n a_n^2 + u(N) \sum\sum_{m \neq n} a_m a_n ,$$

which can be written as

$$(T_\omega - u(N))\Big(\sum_n a_n\Big)^2 \leqslant (T - u(N)) \sum_n a_n^2 . \tag{9.66}$$

Now we specialize $a_n = 1$ for the integers in an interval of length N and satisfying (9.61) and $a_n = 0$ elsewhere. In this case (9.66) yields the following result.

PROPOSITION 9.13. *If $u(N) < T_\omega$ then*

$$Z \leqslant \frac{T - u(N)}{T_\omega - u(N)} . \tag{9.67}$$

Gallagher considered the case $t(q) = \Lambda(q)$ for q in some finite set $\mathcal{S}$, in which case we can take $u(N) = \log N$. Another choice would be $t(q) = \Lambda_k(q)$, in which case we can take $u(N) = (\log N)^k$.

To compare Gallagher's larger sieve bound (9.67) to the large sieve bound (9.32) we consider the case

$$\omega(q) = \alpha q + O(1)$$

for $t(q) = \Lambda(q)$ and $q \leqslant Q$, where $\frac{1}{2} < \alpha < 1$. In this case we have

$$T_\omega = \sum_{q \leqslant Q} \frac{\Lambda(q)}{q - \omega(q)} = \frac{\log Q}{1 - \alpha} + O(1)$$

and

$$T = \sum_{q \leqslant Q} \Lambda(q) \sim Q ,$$

so Gallagher's result gives

$$Z \ll Q\Big(\log QN^{\alpha - 1}\Big)^{-1}$$

provided $QN^{\alpha-1}$ is larger than some constant depending on α. On the other hand, in the large sieve inequality (9.32),

$$J \asymp Q(\log Q)^{(2\alpha-1)/(1-\alpha)} ,$$

so that

$$Z \ll \frac{N+Q^2}{Q}(\log Q)^{(1-2\alpha)/(1-\alpha)} ,$$

which is a much weaker bound once Q is somewhat smaller than $\sqrt{N}$. In the very extreme case that $\omega(q) = q + O(1)$ for $q \leqslant Q$ with $Q \gg \log N$ the larger sieve yields the bound $Z \ll 1$. Nothing better could one ask for!

We remark that the apparently more general problem wherein the interval of length N is replaced by an arithmetic progression of length N to modulus k reduces to the previous case by changing the residue classes provided that all the moduli q are relatively prime to k.

9.8. Equidistribution to Large Moduli

One of the most important applications of the large sieve has been to the distribution of primes in arithmetic progressions. The basic problem with regard to the latter is to derive the expected equidistribution when the modulus is large. Here, the known results (Siegel–Walfisz theorem) fall far short of what is expected and even the assumption of the GRH leads only half way to the goal. A great virtue of the theory of the large sieve is that it also leads us half way to the goal but unconditionally, without GRH. In this section we present two basic results of this type.

We begin by considering a general sequence (β_n) of complex numbers. We assume almost nothing about it other than its uniform distribution in residue classes to small moduli. Specifically, we assume

CONDITION S-W. For any $k \geqslant 1$ and $(\ell, k) = 1$ we have

$$\sum_{\substack{n\leqslant N\\ n\equiv \ell (\operatorname{mod} k)}} \beta_n - \frac{1}{\varphi(k)} \sum_{\substack{n\leqslant N\\ (n,k)=1}} \beta_n \ll \frac{(\beta(N)N)^{\frac{1}{2}}}{(\log N)^A} \tag{9.68}$$

with some $\beta(N)$ satisfying

$$\beta(N) \geqslant \sum_{n\leqslant N} |\beta_n|^2 ,$$

and any $A > 0$, the implied constant depending only on A.

Very often we can verify (9.68) with

$$\beta(N) = \sum_{n\leqslant N} |\beta_n|^2 .$$

However, such a stronger assumption may not be practical for verification; in particular, it could be difficult in the case of lacunary sequences. On the other hand, the results which we are going to draw from (9.68) do not say much in the case of very lacunary sequences.

Note that (9.68) is essentially trivial unless $k \leqslant (\log N)^A$. Therefore, we are not demanding much. In the case of $\beta_n = \Lambda(n)$ the Condition S-W is the Siegel–Walfisz theorem. Our first general result is

THEOREM 9.14. *Assume that* (β_n) *satisfies* (9.68). *Then, we have*

$$\sum_{q\leqslant Q}\ \sideset{}{^\star}\sum_{a\,(\mathrm{mod}\,q)}\Big|\sum_{\substack{n\leqslant N\\ n\equiv a\,(\mathrm{mod}\,q)}}\beta_n-\frac{1}{\varphi(q)}\sum_{\substack{n\leqslant N\\ (n,q)=1}}\beta_n\Big|^2\ll\frac{\beta(N)N}{(\log N)^A} \tag{9.69}$$

where $Q=N(\log N)^{-B}$. *Here* A *is any positive number while* B *and the implied constant depend on* A.

PROOF. Using characters we have

$$\sum_{\substack{n\leqslant N\\ n\equiv a(q)}}\beta_n-\frac{1}{\varphi(q)}\sum_{\substack{n\leqslant N\\ (n,q)=1}}\beta_n=\frac{1}{\varphi(q)}\sum_{\substack{\chi\,(\mathrm{mod}\,q)\\ \chi\neq\chi_0}}\overline{\chi}(a)\sum_{n\leqslant N}\chi(n)\beta_n\,.$$

Hence, by orthogonality of characters we get

$$\sideset{}{^\star}\sum_{a\,(\mathrm{mod}\,q)}\Big|\sum_{\substack{n\leqslant N\\ n\equiv a(q)}}\beta_n-\frac{1}{\varphi(q)}\sum_{\substack{n\leqslant N\\ (n,q)=1}}\beta_n\Big|^2=\frac{1}{\varphi(q)}\sum_{\substack{\chi\,(\mathrm{mod}\,q)\\ \chi\neq\chi_0}}\Big|\sum_{n\leqslant N}\chi(n)\beta_n\Big|^2\,.$$

Here the right–hand side is in a form ready for the large sieve inequality (9.52) except that the sum is not restricted to primitive characters. Every $\chi(\mathrm{mod}\,q)$ in the above sum is induced by a unique primitive character ψ mod f where $q=ef$, and $f>1$. Moreover, $\chi(n)=\psi(n)$ if $(n,e)=1$ and otherwise $\chi(n)=0$. Therefore the sum (9.69) is bounded by

$$\sum_{\substack{ef\leqslant Q\\ f>1}}\sum\frac{1}{\varphi(e)\varphi(f)}\sideset{}{^\star}\sum_{\psi(f)}\Big|\sum_{\substack{n\leqslant N\\ (n,e)=1}}\psi(n)\beta_n\Big|^2\,. \tag{9.70}$$

For each e with $1\leqslant e<Q$ we provide an individual estimate. Given e we choose F with $1<F\leqslant Q/e$. The sum over $F<f\leqslant Q/e$ is split into dyadic segments $F_1<f\leqslant 2F_1$. For each of these, (9.52) is applied, leading to the bounds

$$\Big(F_1+\frac{N}{F_1}\Big)\sum_{n\leqslant N}|\beta_n|^2 \tag{9.71}$$

and then, summing over F_1 with $F\leqslant F_1<Q/e$, we get

$$\Big(\frac{Q}{e}+\frac{N}{F}\Big)\Big(\sum_{n\leqslant N}|\beta_n|^2\Big)\,. \tag{9.72}$$

It remains to estimate the contribution of the primitive characters $\psi\,(\mathrm{mod}\,f)$ with $1<f\leqslant F$. For each of these we appeal to the hypothesis (9.68).

First we detect the coprimality condition $(e,n)=1$ by the formula

$$\sum_{\substack{c|e\\ c|n}}\mu(c)=\sum_{\substack{ab\leqslant C\\ ab|e\\ (b,n)=1}}\mu(a)+\sum_{\substack{c>C\\ c|e\\ c|n}}\mu(c).$$

Hence, the sum of $\psi(n)\beta_n$ over n coprime with e splits accordingly. In the first part, where $c=ab\leqslant C$, we split further into reduced residue classes modulo bf

getting the bound $C\tau(e)f\sqrt{\beta(N)N}(\log N)^{-6A}$ by (9.68). The contribution of the second part is estimated trivially by Cauchy's inequality as follows:

$$\sum_{\substack{c>C\\ c|e}}\sum_{\substack{n\leqslant N\\ n\equiv 0(c)}}|\beta_n|\leqslant\sum_{\substack{c>C\\ c|e}}\left(\frac{\|\beta\|N}{c}\right)^{\frac{1}{2}}\leqslant\tau(e)\left(\frac{\|\beta\|N}{C}\right)^{\frac{1}{2}}.$$

Adding these two bounds and then choosing $C=(\log N)^{-4A}$ we obtain

$$\sum_{\substack{n\leqslant N\\ (n,e)=1}}\psi(n)\beta_n\ll\tau(e)f\frac{(\beta(N)N)^{\frac{1}{2}}}{(\log N)^{2A}}. \tag{9.73}$$

Hence,

$$\sum_{1<f\leqslant F}\frac{1}{\varphi(f)}\sum_{\psi(f)}{}^{\star}\Big|\sum_{\substack{n\leqslant N\\ (n,e)=1}}\psi(n)\beta(n)\Big|^2\ll\tau^2(e)F^3\frac{\beta(N)N}{(\log N)^{4A}}. \tag{9.74}$$

Summing (9.72) and (9.74) over e we infer the following bound for our original sum in (9.69):

$$\{Q+F^{-1}N\log N+F^3N(\log N)^{4-4A}\}\,\beta(N).$$

Choosing $F=(\log N)^A$ we complete the proof of the theorem. □

Theorem 9.14 is a general form of the Barban–Davenport–Halberstam theorem about primes in arithmetic progressions which we state as:

COROLLARY 9.15. *We have for any* $A>0$,

$$\sum_{q\leqslant Q}\sum_{a\,(\mathrm{mod}\,q)}{}^{\star}\Big|\sum_{\substack{n\leqslant x\\ n\equiv a\,(\mathrm{mod}\,q)}}\Lambda(n)-\frac{x}{\varphi(q)}\Big|^2\ll x^2(\log x)^{-A} \tag{9.75}$$

where $Q=x(\log x)^{-B}$ *for some* $B=B(A)>0$, *and the implied constant depends only on* A.

The last result is remarkable in that it provides information about primes in progressions to extremely large moduli, virtually as large as one can hope. Unfortunately, because it requires one to average over the residue classes as well as the moduli its applicability is somewhat limited. Our next result, the Bombieri–Vinogradov theorem, gives an average which involves only a single class per modulus. Although the moduli are no longer as large the applications are far more numerous.

As before we begin with a more general form of the theorem.

THEOREM 9.16. *Let* $\alpha=(\alpha_m)$, $\beta=(\beta_n)$ *be sequences of complex numbers with* β *satisfying the Siegel–Walfisz condition* (9.68). *Then, for any* $A>0$ *we have*

$$\begin{aligned}&\sum_{q\leqslant Q}\max_{(a,q)=1}\Big|\sum\sum_{\substack{m\leqslant M,\,n\leqslant N\\ mn\equiv a\,(\mathrm{mod}\,q)}}\alpha_m\beta_n-\frac{1}{\varphi(q)}\sum\sum_{\substack{m\leqslant M,\,n\leqslant N\\ (mn,q)=1}}\alpha_m\beta_n\Big|\\ &\ll\left\{Q+\sqrt{M+N}\left(\log QMN\right)^2+\sqrt{MN}\left(\log MN\right)^{-A}\right\}\alpha(M)^{\frac{1}{2}}\beta(N)^{\frac{1}{2}},\end{aligned} \tag{9.76}$$

where

$$\alpha(M)=\sum_{m\leqslant M}|\alpha_m|^2$$

and the implied constant depends only on A and on the one in (9.68).

Note that this bound is $\ll (MN)^{\frac{1}{2}}(\log MN)^{-A}\alpha(M)^{\frac{1}{2}}\beta(N)^{\frac{1}{2}}$ provided that the moduli are limited by $Q=\sqrt{MN}\,(\log MN)^{-A}$ and that $M,N\gg(\log MN)^{2A+4}$. The last condition means we are dealing with a genuine bilinear form.

PROOF. Using characters we see that the left-hand side is bounded by

$$(9.77)\qquad \begin{aligned}&\sum_{q\leqslant Q}\frac{1}{\varphi(q)}\sum_{\substack{\chi\,(\mathrm{mod}\,q)\\ \chi\neq\chi_0}}\Big|\sum_{m\leqslant M}\alpha_m\chi(m)\Big|\Big|\sum_{n\leqslant N}\beta_n\chi(n)\Big|\\ &\leqslant\sum_{\substack{ef\leqslant Q\\ f>1}}\sum\frac{1}{\varphi(e)\,\varphi(f)}\sum_{\psi\,(\mathrm{mod}\,f)}^{\star}\Big|\sum_{\substack{m\leqslant M\\(m,e)=1}}\alpha_m\psi(m)\Big|\Big|\sum_{\substack{n\leqslant N\\(n,e)=1}}\beta_n\psi(n)\Big|\end{aligned}$$

where, as in the previous proof, ψ is the character of conductor f which induces χ. Fix e and split the sum over f into dyadic segments $F_1<f\leqslant 2F_1$. Then apply Cauchy's inequality getting

$$\leqslant\Big(\sum_{F_1<f\leqslant 2F_1}\frac{1}{\varphi(f)}\sum_{\psi}^{\star}\Big|\sum_m\Big|^2\Big)^{\frac{1}{2}}\Big(\sum_{F_1<f\leqslant 2F_1}\frac{1}{\varphi(f)}\sum_{\psi}^{\star}\Big|\sum_n\Big|^2\Big)^{\frac{1}{2}}.$$

By (9.52) we have

$$\begin{aligned}&\Big(F_1+\frac{M}{F_1}\Big)^{\frac{1}{2}}\Big(F_1+\frac{N}{F_1}\Big)^{\frac{1}{2}}\Big(\sum_m|\alpha_m|^2\Big)^{\frac{1}{2}}\Big(\sum_n|\beta_n|^2\Big)^{\frac{1}{2}}\\ &\ll\Big(F_1+\sqrt{M+N}+\frac{\sqrt{MN}}{F_1}\Big)\Big(\sum_m|\alpha_m|^2\Big)^{\frac{1}{2}}\Big(\sum_n|\beta_n|^2\Big)^{\frac{1}{2}}.\end{aligned}$$

Summing this over $F\leqslant F_1\leqslant Q/e$ we get

$$(9.78)\qquad \Big(\frac{Q}{e}+\sqrt{M+N}\,\log Q+\frac{\sqrt{MN}}{F}\Big)\Big(\sum_m|\alpha_m|^2\Big)^{\frac{1}{2}}\Big(\sum_n|\beta_n|^2\Big)^{\frac{1}{2}}.$$

For $1<f\leqslant F$ in (9.77) we apply (9.73) to the second factor and the trivial bound to the first factor obtaining

$$(9.79)\qquad \sum_{1<f\leqslant F}\frac{1}{\varphi(f)}\sum_{\psi(f)}^{\star}\Big|\sum_m\Big|\Big|\sum_n\Big|\ll\|\,\alpha\,\|\,F^2\big(\tau(e)\beta(N)MN\big)^{\frac{1}{2}}(\log N)^{-A}.$$

Adding (9.78) to (9.79), choosing $F=(\log N)^{A/3}$, and then summing over e we complete the proof of the theorem. □

We next derive a form of the previous theorem having special technical restrictions which make it more practical for applications.

THEOREM 9.17. *Let $h\geqslant 1$. Suppose $|\alpha_m|\leqslant\tau_h(m)$, $|\beta_n|\leqslant\tau_h(n)$ and (β_n) satisfies the Siegel–Walfisz condition*

$$\sum_{\substack{n\leqslant N\\ n\equiv\ell(\mathrm{mod}\,k)}}\beta_n-\frac{1}{\varphi(k)}\sum_{\substack{n\leqslant N\\(n,k)=1}}\beta_n\ll\frac{N}{(\log N)^c},$$

for any $k \geqslant 1$, $(\ell, k) = 1$, $N \geqslant 2$ and $c > 0$, the implied constant depending only on c and h. Let $A > 0$ and $\Delta = (\log x)^{-B}$, where B is sufficiently large in terms of A and h. Then

$$\sum_{q \leqslant \Delta\sqrt{x}} \max_{(a,q)=1} \Bigg| \sum\sum_{\substack{mn \leqslant x \\ m,n \leqslant \Delta x \\ mn \equiv a(q)}} \alpha_m \beta_n - \frac{1}{\varphi(q)} \sum\sum_{\substack{mn \leqslant x \\ m,n \leqslant \Delta x \\ (mn,q)=1}} \alpha_m \beta_n \Bigg| \ll \frac{x}{(\log x)^A}, \tag{9.80}$$

where the implied constant depends only on A and h.

Proof. We split the summation into short intervals $(1 - \delta)M < m \leqslant M$, $(1-\delta)N < n \leqslant N$ with $(1-\delta)^2 MN \leqslant x$ and $M, N \leqslant \Delta x$ where $0 < \delta \leqslant \Delta$ will be chosen. The number of such small boxes is $\ll (\log x)^2$. The condition $mn \leqslant x$ will be redundant unless $(1-\delta)^2 x < mn < (1-\delta)x$. For the latter boundary terms we estimate the contribution trivially by

$$\sum_{q \leqslant \Delta\sqrt{x}} \max_{(a,q)=1} \Bigg\{ \sum_{\substack{(1-\delta)x < \ell < (1+\delta)x \\ \ell \equiv a(q)}} \tau_h(\ell)\tau(\ell) + \frac{1}{\varphi(q)} \sum_{(1-\delta)x < \ell < (1+\delta)x} \tau_h(\ell)\tau(\ell) \Bigg\}.$$

This is bounded by $\delta x (\log x)^{C(h)}$, with some constant $C(h)$. For every remaining box we can drop the condition $mn \leqslant x$ so Theorem 9.16 applies giving a bound $\sqrt{\Delta}\, x (\log x)^{C(h)}$. Multiplying this bound by the number of boxes $O\big(\delta^{-2}(\log x)^2\big)$ and then choosing $\delta = \Delta^{1/6}$ we obtain the bound $\Delta^{1/6} x (\log x)^{C(h)+2} \ll x(\log x)^{-A}$ as claimed. □

We now apply Theorem 9.17 to prove the Bombieri–Vinogradov theorem.

Theorem 9.18. *We have, for any $A > 0$,*

$$\sum_{q \leqslant Q} \max_{(a,q)=1} \left| \pi(x; q, a) - \frac{\pi(x)}{\varphi(q)} \right| \ll x (\log x)^{-A} \tag{9.81}$$

where $Q = \sqrt{x}(\log x)^{-B}$ with some $B(A) > 0$ and the implied constant depends only on A.

Proof. We apply the Ramaré identity (see Section 17.3) for $n \equiv a (\operatorname{mod} q)$. We have

$$\sum_{\substack{n \leqslant x \\ n \equiv a (\operatorname{mod} q) \\ (n, P(z))=1}} 1 = \pi(x; q, a) + \sum\sum_{\substack{pm \leqslant x,\, pm \equiv a(q) \\ z \leqslant p \leqslant \sqrt{x} \\ (m, P(z))=1}} \gamma_m(x) + O\Big(\frac{x}{qz}\Big). \tag{9.82}$$

Here, we choose $z = \exp(\sqrt{\log x})$. By the Fundamental Lemma (see Section 6.5) we see that the sum on the left-hand side is equidistributed in reduced residue classes $a (\operatorname{mod} q)$ to level $q \leqslant \sqrt{x}$. On the other hand, by Theorem 9.17 the double sum on the right side is also equidistributed over reduced residue classes $\operatorname{mod} q$, but this time only on average over q and only to the level $q \leqslant Q$. Note that the Siegel–Walfisz condition in this double sum is satisfied by the variable p. This implies the same level Q of equidistribution for $\pi(x; q, a)$. □

Of course, from this follows, in the same notation as Theorem 9.18,

$$\sum_{q \leqslant Q} \max_{(a,q)=1} \left| \psi(x; q, a) - \frac{x}{\varphi(q)} \right| \ll x (\log x)^{-A}. \tag{9.83}$$

We remark that there have been a great many combinatorial identities for sums over primes, which have been used in the proof of Theorem 9.18. The main idea is to produce a bilinear form to which the large sieve inequality can be applied. In fact, very little is required in the shape of the bilinear form and we have chosen the Ramaré identity because it illustrates this point very clearly.

To gain some historical perspective we should point out that the early paths to Bombieri–Vinogradov type theorems went through the density theorems for zeros of L-functions. The proofs of these "zero-density theorems" are based on combinatorial decompositions of the L-functions which are reminiscent of those for $\Lambda(n)$. It was P. X. Gallagher [**59**], [**60**], who first by-passed the zeros and went directly to bilinear forms in characters. This short-cut which Gallagher created led to a better understanding of the roots of the subject and his ideas stimulated a search for new more flexible identities, from which we have a great stock to choose today (see Chapter 17).

9.9. Limitations to Equidistribution

The great virtue of the Bombieri–Vinogradov theorem is that it gives estimates for the error term in the asymptotic formula for primes in arithmetic progressions having moduli as large as those covered by the Riemann Hypothesis. Although it estimates these error terms only on average over the modulus, this is just as good for most applications of the sieve. Moreover, useful estimates for this average can be expected to hold for still larger values of the modulus, hence going beyond the Riemann Hypothesis. In case one restricts the sum to progressions having a fixed non-zero value of a then one can accomplish this and such results have been given, for example in [**44**] and in [**11**]. It is expected (the Elliott–Halberstam conjecture) that (9.83) holds even for $Q = x^{1-\varepsilon}$ for every $\varepsilon > 0$, and for some time it was conjectured to hold even up to $Q = x(\log x)^{-A}$.

We show that the latter is too much to hope for. There is more than one way to do this. Perhaps the simplest argument (which uses ideas from Theorem A2 of [**49**]) proceeds as follows. We are going to evaluate asymptotically the sum

$$S(x; Q, a) = \sum_{\substack{Q<q\leqslant(1+\varepsilon)Q\\(q,a)=1}} \psi(x; q, a) \ .$$

where $1 \leqslant a \leqslant x^{\frac{2}{3}} \leqslant Q \leqslant x(\log x)^{-3}$ and $\varepsilon = \varepsilon(x)$ satisfies $(\log x)^{-2} \leqslant \varepsilon \leqslant (\log x)^{-1}$.

By the Dirichlet involution we have

$$\begin{aligned} S(x; Q, a) &= \sum_{r\leqslant(x-a)/Q} \sum_{\substack{n\leqslant x,\, n\equiv a(r)\\((n-a)/r,a)=1\\Q<(n-a)/r\leqslant(1+\varepsilon)Q}} \Lambda(n) \\ &= \sum_{\substack{r\leqslant(x-a)/Q\\(r,a)=1}} \sum_{\substack{n\leqslant x,\, n\equiv a(r)\\Q<(n-a)/r\leqslant(1+\varepsilon)Q}} \Lambda(n) + \ O\left(\sqrt{x}(\log x)^2\right) \ . \end{aligned}$$

The last equation is an identity except for special terms in n. These exceptional numbers n are higher powers of primes, which trivially contribute $O(\sqrt{x}(\log x)^2)$, and prime numbers dividing a, which trivially contribute $O(Q^{-1}x\log x)$. Hence,

$$\begin{aligned} S(x;Q,a) &= \sum_{\substack{r\leqslant(x-a)/Q\\(r,a)=1}} \frac{1}{\varphi(r)}\min(x-rQ-a,\varepsilon rQ)+O(x(\log x)^{-A})\\ &= \sum_{\substack{r\leqslant R\\(r,a)=1}} \frac{1}{\varphi(r)}\min(x-rQ,\varepsilon rQ)+O(x(\log x)^{-A})\ , \end{aligned}$$

where $R = x/Q$ and A is any positive number. The above minimum is εrQ except for r in the short interval $R(1+\varepsilon)^{-1} < r \leqslant R$. We estimate the contribution of the latter trivially by

$$\varepsilon Q \sum_{R(1+\varepsilon)^{-1}<r\leqslant R} \frac{r}{\varphi(r)} \ll \varepsilon^2 x\ .$$

This proves the following result.

PROPOSITION 9.19. *Let* $1 \leqslant a \leqslant x^{\frac{2}{3}} \leqslant Q \leqslant x(\log x)^{-3}$ *and* $(\log x)^{-2} \leqslant \varepsilon(x) \leqslant (\log x)^{-1}$. *Then*

$$S(x;Q,a) = \varepsilon(x)Q \sum_{\substack{r\leqslant R\\(r,a)=1}} \frac{r}{\varphi(r)} + O(\varepsilon(x)^2 x)\ ,$$

where $R = x/Q$.

Now, we specialize. Let $a = P(z)$ be the product of all primes less than z, with $z = R^{1/s}$, $s > 1$. Note that, since $a \leqslant x^{2/3}$, therefore $z < 2\log x$. Also, since $(r, P(z)) = 1$ and $r \leqslant R$, we have

$$\frac{r}{\varphi(r)} = 1 + O\Big(\frac{s}{z}\Big)\ .$$

Hence, by Buchstab's asymptotic formula (see Lemma 12.1) for the frequency of integers without small prime factors, it follows that

$$S(x;Q,a) = \varepsilon(x)\omega(s)x\{1+O(s/z)\}\ ,$$

where $\omega(s)$ is the Buchstab function (see (12.19)).

On the other hand, if the Bombieri–Vinogradov theorem were to hold in the range $Q < q \leqslant (1+\varepsilon)Q$ we would get

$$\begin{aligned} S(x;Q,a) &= \varepsilon(x)x\frac{\varphi(a)}{a} + O\Big(\frac{x}{(\log x)^2}\Big)\\ &= \varepsilon(x)e^{-\gamma}x\{1+O(1/z)\} \end{aligned}$$

by Mertens' formula.

Note that, as we shall see in (12.23), the function $\omega(s) - e^{-\gamma}$ changes sign in every interval of length one. Therefore, the conclusion of the Bombieri–Vinogradov theorem cannot hold with $Q = x(\log x)^{-B}$ for any fixed B, no matter how large.

CHAPTER 10

Molecular Structure of Sieve Weights

Sleeping Beauty

10.1. Introduction

Our first priority has all along been in the distribution of arithmetically interesting sequences, especially the primes. In order to count their frequency we have introduced sieve weights which provide us with upper and lower bounds. Although they have been viewed as instruments and not our primary concern, these weights possess additional properties that are fascinating on their own. In this chapter we pause to take a look at some of them. A few others of similar nature have been applied elsewhere in the book and so are placed closer to those locations (see the links given in Section 10.5). The ones described here should prove useful in the future but, as we do not apply them now, this chapter is somewhat looser in style. We hope however, it provides a better understanding of the sieve mechanism.

We shall concentrate on the Selberg sieve and on the beta-sieve. This does not really restrict the generality in the sense that, to be honest, there is little else. We also, to some extent, emphasize the upper bounds since this is where the two are most naturally compared.

Parts of the discussion here appeal to facts about the beta-sieve to be developed later in Chapters 11 and 12 and some readers may prefer to glance at those first.

10.2. What the Upper Bound Leaves Behind

We began with our sifting function

$$S(\mathcal{A}, z) = \sum_{\substack{n \leqslant x \\ (n, P(z))=1}} a_n$$

whose estimation was our goal, and bounded it from above by the sifted sum

$$S^+(\mathcal{A}, z) = \sum_{n \leqslant x} a_n \Big(\sum_{\substack{d|n \\ d|P(z)}} \lambda_d^+ \Big) , \tag{10.1}$$

where (λ_d^+) is an upper-bound sieve of level $D \leqslant x$. If $z = \sqrt{D}$ and D is close to x then clearly, $S(\mathcal{A}, z)$ counts primes and essentially nothing else. The sifted sum $S^+(\mathcal{A}, z)$ counts these and something more. Let's examine just what that is.

We begin with the beta-sieve where the surplus is quickly visible. We have

$$S^+(\mathcal{A}, z) = \sum_{n \leqslant x} a_n \Big(\sum_{\substack{d|n \\ d \in \mathcal{D}^+}} \mu(d) \Big) ,$$

and, by (6.29) this gives

$$S^+(\mathcal{A}, z) = S(\mathcal{A}, z) + \sum_{n\,\mathrm{odd}} S_n(\mathcal{A}, z)\ ,$$

with $S_n(\mathcal{A}, z)$ given by (6.28). Taking $z = \sqrt{D}$ we obtain

$$S_1(\mathcal{A}, \sqrt{D}) = \sum_{D^{\frac{1}{3}} \leqslant p < D^{\frac{1}{2}}} \Big(\sum_{\substack{pq \leqslant x \\ (q, P(p)) = 1}} a_{pq} \Big)\ . \tag{10.2}$$

On the other hand, in the Λ^2-sieve (10.1) becomes

$$S^+(\mathcal{A}, z) = \sum_{n \leqslant x} a_n \Big(\sum_{\substack{d \mid n \\ d \mid P(z)}} \rho_d^+ \Big)^2\ , \tag{10.3}$$

and from here the surplus is not so readily observed. To compare it with (10.2) we take $\kappa = 1$ and extract from (10.3) the contribution of the subsequence over $n = pq$ with $p < \sqrt{D}$, $(q, P(\sqrt{D})) = 1$. This contribution is

$$S_1(\mathcal{A}, \sqrt{D}) = \sum_{p < D^{\frac{1}{2}}} \big(1 + \rho(p)\big)^2 \Big(\sum_{\substack{pq \leqslant x \\ (q, P(\sqrt{D})) = 1}} a_{pq} \Big)\ . \tag{10.4}$$

In (10.2) the surplus runs over a smaller range of primes. On the other hand, in (10.4) the primes are weighted by the damping factor $\big(1 + \rho(p)\big)^2 \leqslant 1$. Indeed, from the asymptotic (7.40) we find that

$$1 + \rho(p) \sim \frac{\log p}{\log \sqrt{D}}\ . \tag{10.5}$$

It is noteworthy that, although the surplus in the beta-sieve and that in the Λ^2-sieve come from different subsequences, the values of $S^+(\mathcal{A}, \sqrt{D})$ are asymptotically the same. Specifically, for $\kappa = 1$ we have

$$S^+(\mathcal{A}, \sqrt{D}) \sim \frac{2X}{\log D}\ . \tag{10.6}$$

Recall that the asymptotic sieve in Chapter 3 leads to

$$\sum_{p \leqslant x} \Big(\frac{\log p}{\log x} \Big)^2 a_p + \sum\sum_{p_1 p_2 \leqslant x} \frac{\log p_1}{\log x} \frac{\log p_2}{\log x} a_{p_1 p_2} \sim \frac{2X}{\log x}\ .$$

It is remarkable that these three very different ways of counting yield the same asymptotic value.

That none of them do better is somewhat understandable in the context of the parity barrier and shouldn't be disappointing in view of the enormity of the consequences which would ensue. We have in mind the question of existence or non-existence of exceptional characters. Were the factor 2 replaced by a smaller constant, that ghost would disappear.

10.3. Contribution of Small Prime Factors

One may operate as if the sieve weights

$$\theta_n = \sum_{d|n} \lambda_d$$

eliminate numbers n which have small prime divisors, or even more conveniently, think that the only survivors are almost-primes. Indeed, this is a good guideline when testing sieve arguments, but the truth is that a few numbers n having small prime factors are also left behind. Fortunately, their contribution is relatively small, so we may practice with this simplified image quite safely. In this section we give a quantitative measure of this feature. We restrict ourselves to two types of upper-bound sieve, namely the Λ^2-sieve and the beta-sieve (the latter of dimension $\kappa \geqslant 1$).

We consider the weighted sifted sum

$$S^+(\mathcal{A}, z; w) = \sum_n a_n \omega(n; w) \Big(\sum_{\substack{d|n \\ d|P(z)}} \lambda_d \Big)$$

where $\omega(n; w)$ denotes the number of prime divisors of n dividing $P(w)$. Our goal is to show that

$$S^+(\mathcal{A}, z; w) \ll S^+(\mathcal{A}, z) \frac{\log w}{\log D} , \tag{10.7}$$

which proves our point if w is relatively smaller than D in the logarithmic scale. We write

$$\begin{aligned} S^+(\mathcal{A}, z; w) &= \sum_n a_n \sum_{\substack{p|n \\ p|P(w)}} \Big(\sum_{\substack{d|n \\ d|P(z)}} \lambda_d \Big) = \sum_{p|P(w)} \sum_{d|P(z)} \lambda_d A_{[p,d]} \\ &= XG(D, z; w) + \text{ remainder } , \end{aligned}$$

where

$$G(D, z; w) = \sum_{p|P(w)} G_p(D, z) \tag{10.8}$$

with

$$G_p(D, z) = \sum_{d|P(z)} \lambda_d \, g\big([p, d]\big) .$$

Our task is to show that

$$G(D, z; w) \ll G(D, z) \frac{\log w}{\log D} . \tag{10.9}$$

Recall that

$$G(D, z) = \sum_{d|P(z)} \lambda_d \, g(d) . \tag{10.10}$$

We also have

$$G_p(D, z) = g(p) G(D, z) + \big(1 - g(p)\big) \sum_{d \equiv 0(p)} \lambda_d \, g(d) . \tag{10.11}$$

First, we treat the Selberg sieve

$$\lambda_d = \sum_{[d_1,d_2]=d} \rho_{d_1}\rho_{d_2} ,$$

because most of the work in this case has already been done in Section 7.6. We obtain

$$G_p(D,z) = \sum_{d_1|P(z)} \sum_{d_2|P(z)} \rho_{d_1}\rho_{d_2} g\big([p,d_1,d_2]\big) .$$

Following the diagonalization procedure we arrive at (see (7.109))

$$G_p(D,z) = g(p) \sum_{\substack{d|P(z)\\(d,p)=1}} h(d)\big(y_d - y_{pd}\big)^2$$

where y_d is given by (7.13). Now we make the original choice (7.21) of Selberg (as we must for comparison of the results), that is, we take $y_d = J^{-1}$ for $d < \sqrt{D}$ and $y_d = 0$ for $d \geqslant \sqrt{D}$, where $J = J(D)$ was defined in (7.19). This simple choice makes the above sum equal to

$$\sum_{\substack{\sqrt{D}/p \leqslant d < \sqrt{D}\\(d.p)=1}} \frac{h(d)}{J(D)^2} \leqslant \frac{J(D) - J(D/p^2)}{J(D)^2} \ll J(D)^{-1}\frac{\log p}{\log D} ,$$

by arguments similar to those used in the derivation of the lower bound (7.56). Recall that, in the Selberg sieve, $G(D,z) = J(D)^{-1}$, so we get the upper bound

$$G_p(D,z) \ll G(D,z)\, g(p)\, \frac{\log p}{\log D} . \tag{10.12}$$

Alternatively, (10.12) can be deduced from the exact formula

$$G_p(D,z) = g(p)(1+\rho_p)G(D,z) ,$$

which follows from (7.52) since $1+\rho_p \ll \log p/\log D$. Applying (10.12) to (10.8) and summing over p we deduce (10.9).

Next we consider the combinatorial upper-bound sieve which is $\lambda_d = \mu(d)$ for $d \in \mathcal{D}^+$ and $\lambda_d = 0$ otherwise, where $\mathcal{D}^+$ is the set defined at the beginning of Section 6.2. We have by (10.11)

$$G(D,z;w) = \sum_{p|P(w)} \Big(\sum_d \lambda_d g(d) - \big(1-g(p)\big) \sum_{(d,p)=1} \lambda_d g(d)\Big) .$$

Given p, we arrange the two inner sums in accordance with the identity (6.31), showing that $G(D,z;w)$ equals

$$\sum_{p|P(w)} \Big[V(z) - \big(1-g(p)\big)V^{(p)}(z) + \sum_{n\,\mathrm{odd}} \Big(V_n(z) - \big(1-g(p)\big)V_n^{(p)}(z)\Big)\Big] .$$

Here, the superscript (p) indicates that the prime p is to be removed from the sifting range. Assuming $w \leqslant z$ we have

$$V(z) = \big(1-g(p)\big)V^{(p)}(z) .$$

Comparing the parts $V_n(z)$ from (6.34) with those of $\big(1-g(p)\big)V_n^{(p)}(z)$ above for $d = p_1 \dots p_n$ with $p_1 > \dots > p_n > p$ we see that we are left with

$$G(D,z;w) = \sum_{n \text{ odd}} \sum_{\substack{y_n \leqslant p_n < \dots < p_1 < z \\ p_m < y_m, m<n, m \equiv n(2)}} \cdots \sum g(p_1 \dots p_n) V(p_n)$$

$$\Big\{ \sum_{p_n \leqslant p < w} 1 - \sum_{\substack{p_n \leqslant p < w \\ p \neq p_j}} \big(1-g(p)\big) \Big\} .$$

Here, the two sums over p are void unless $w > p_n$ in which case the difference is

$$\sum_{p=p_j<w} \big(1-g(p)\big) + \sum_{p_n \leqslant p < w} g(p) \leqslant n + \sum_{p_n \leqslant p < w} g(p) \ll n \frac{\log w}{\log p_n} .$$

Hence,

$$G(D,z;w) \ll (\log w) \sum_{n \text{ odd}} n\, H_n(D,z)$$

with

$$H_n(D,z) = \sum_{\substack{y_n \leqslant p_n < \dots < p_1 < z \\ p_m < y_m, m<n, m \equiv n(2)}} \cdots \sum g(p_1 \dots p_n) V(p_n) / \log p_n .$$

Suppose $\kappa \geqslant 1$. One can show that

$$\sum_{n \text{ odd}} n\, H_n(D,z) \ll V(z)\, (\log D)^{-1} . \tag{10.13}$$

Hence

$$G(D,z;w) \ll V(z)\, \frac{\log w}{\log D} , \tag{10.14}$$

which implies (10.9). The proof of (10.13) can be performed along the lines of the proof of the beta-sieve results in Chapter 11. To include this would require a substantial restructuring of those arguments and, especially since we do not apply (10.13) in this book, we felt it not worth this amount of extra effort.

We return again to the Λ^2-sieve

$$\theta_n = \sum_{d|n} \lambda_d = \Big(\sum_{d|n} \rho_d\Big)^2$$

to expand on the comments in Subsection 7.2.3. We shall make our point in the simplest case of the linear sieve problem

$$S^+(\mathcal{A}) = \sum_{n \leqslant x} \Big(\sum_{d|n} \rho_d\Big)^2 .$$

If the sieve weight constituents are given by the smoothly truncated Möbius function

$$\rho_d = \mu(d) \Big(1 - \frac{\log d}{\log \sqrt{D}}\Big) , \qquad 1 \leqslant d \leqslant \sqrt{D} ,$$

we know that

$$S^+(\mathcal{A}) \sim \frac{2x}{\log D} , \qquad \text{if} \quad x^\varepsilon < D < x^{1-\varepsilon} .$$

Since, from the start, ρ_d approximates $\mu(d)$ very well it is tempting to think that the weights θ_n eliminate numbers having small prime factors and that this is why $S^+(\mathcal{A})$

has order of magnitude $x/\log x$. However, this is not the only reason because, if the smoothing factor were omitted and replaced by the sharp cut instead, then $S^+(\mathcal{A})$ would have been of order x. Therefore, the saving factor $\log D$ is not exclusively due to the Möbius function but the smoothing device is equally responsible for this effect.

To look further along these lines we are going to consider a smooth truncation of $\mu(d)$, not just at the end point $\sqrt{D}$, but also at some middle point, say $Y = D^{1/6}$. Specifically, we take

$$\rho_d = \mu(d) F(\log d/\log Y) \tag{10.15}$$

with

$$F(y) = 1 - |2 - y| \,, \qquad \text{if} \quad 1 \leqslant y \leqslant 3 \tag{10.16}$$

and $F(y) = 0$ otherwise. Hence ρ_d is supported in the segment $Y < d < Y^3$ and is given there by

$$\rho_d = \frac{\mu(d)}{\log Y} \log \min\Big(\frac{d}{Y}\,,\frac{Y^3}{d}\Big)\,.$$

We shall prove the asymptotic formula

$$S^+(\mathcal{A}) = \frac{2x}{\log Y}\Big(1 + O\big(\frac{1}{\log Y}\big)\Big)\,. \tag{10.17}$$

This formula shows that $S^+(\mathcal{A})$ has the same order of magnitude as $S(\mathcal{A}, \sqrt{D})$ (the number of positive integers $n \leqslant x$ free from prime divisors $p < \sqrt{D}$), but the sets of integers registered in $S^+(\mathcal{A})$ and $S(\mathcal{A}, \sqrt{D})$ are disjoint. Clearly, $S^+(\mathcal{A})$ registers some integers n having many small prime factors. The result (10.17) implies that

$$\sum_{\substack{d|n\\ Y<d<Y^3}} \mu(d)\Big(1 - \Big|2 - \frac{\log d}{\log Y}\Big|\Big) \ll (\log Y)^{-\frac{1}{2}}$$

for almost all $n \leqslant x$ where $x^\varepsilon < Y < x^{1/6-\varepsilon}$. In other words, such integers n have many divisors in the range $Y < d < Y^3$ and these tend to be evenly distributed between the two parity classes.

Similar effects can be seen in the Λ^2-sieve of any dimension provided that the truncation of the Möbius function is made sufficiently smooth.

Proof. We now prove (10.17). We have

$$S^+(\mathcal{A}) = xG + O\big(D(\log D)^3\big) \tag{10.18}$$

with

$$\begin{aligned} G &= \sum_{d_1}\sum_{d_2} \frac{\mu(d_1)\mu(d_2)}{[d_1, d_2]} F\Big(\frac{\log d_1}{\log Y}\Big) F\Big(\frac{\log d_2}{\log Y}\Big) \\ &= \sum_{d}{}^{\flat}\, \frac{\varphi(d)}{d^2}\Big(\sum_{(a,d)=1} \frac{\mu(a)}{a} F\Big(\frac{\log ad}{\log Y}\Big)\Big)^2\,. \end{aligned}$$

We shall evaluate the inner sum, say $I(d)$, by means of the zeta-function. To this end we use the integral representation

$$F\Big(\frac{\log y}{\log Y}\Big) = -\frac{1}{2\pi i}\int_{(1)} \Big(\frac{Y}{y}\Big)^s \Big(\frac{Y^s - 1}{s}\Big)^2 \frac{ds}{\log Y}\,.$$

Hence

$$I(d)\log Y = \frac{1}{2\pi i}\int_{(1)}\Big[2\Big(\frac{Y^2}{d}\Big)^s - \Big(\frac{Y}{d}\Big)^s - \Big(\frac{Y^3}{d}\Big)^s\Big]\frac{ds}{s^2\zeta_d(s+1)}$$

where $\zeta_d(s)$ is the Riemann zeta-function with the local factors at primes dividing d being omitted.

If d exceeds Y^2, Y, Y^3 in the relevant term of the integrand, then we move that part of the integral to the right to $\mathrm{Re}\, s = \infty$ getting zero. Otherwise, we move it slightly into the left half-plane (staying within the zero-free region of the zeta-function), passing a simple pole at $s = 0$ with residue $d/\varphi(d)$ multiplied by 2, -1, -1, respectively. Hence,

$$I(d)\log Y = \frac{\varepsilon(d)d}{\varphi(d)} + r_d(Y)$$

where the contribution $r_d(Y)$ of the contour integrals is at most

$$r_d(Y) \ll \sigma_{-1/2}(d)/\log(3Y^\alpha/d) \qquad \text{if} \quad Y^{\alpha-1} \leqslant d < Y^\alpha \ , \quad \alpha = 1, 2, 3 \ ,$$

and where $\varepsilon(d) = 1$, -1 in the ranges $Y < d < Y^2$, $Y^2 < d < Y^3$ respectively, and $\varepsilon(d) = 0$ elsewhere. Finally, we deduce that

$$G(\log Y)^2 = \sideset{}{^\flat}\sum_{Y<d<Y^3} \frac{1}{\varphi(d)} + O(1)$$

and, inserting this into (10.18), we complete the proof of (10.17). □

10.4. Partial Contributions

We begin with the Λ^2-sieve. Before reading this section, we recommend a quick glance back at Sections 7.1 and 7.2. Seeing (7.33), (7.35) and (7.36) one might imagine that the computation of $S^+(\mathcal{A})$ requires the knowledge of the congruence sums A_d only to moduli $d < \sqrt{D}$ rather than up to D. This would be a very helpful feature, but it is not the case. Remember that for the equation (7.36) to hold, the choice of the sieve weights ρ_d is unique and depends on the sequence $\mathcal{A}$. Specifically, the fully optimal weights ρ_d are solutions to the linear system of equations (7.37) with coefficients $A_{[d,k]}$ where $[d,k]$ can be as large as $dk < D$.

Here is another related observation, more promising in practice, which concerns the choice of the genuine Selberg's sieve weights given by (7.23). Opening the square in (7.2) we write

$$\theta_n = \Big(\sum_{d|n}\rho_d\Big)^2 = \sum_{c|n}\theta_n^*(c) \tag{10.19}$$

say, where

$$\theta_n^*(c) = \sum\sum_{\substack{(a,b)=1\\ abc|n}} \rho_{ac}\rho_{bc}. \tag{10.20}$$

The sifted sum (7.33) splits accordingly,

$$S^+(\mathcal{A}) = \sum_c S^*(\mathcal{A}_c) \tag{10.21}$$

say, where

$$S^*(\mathcal{A}_c) = \sum_{n\equiv 0 \,(\mathrm{mod}\, c)} a_n \theta_n^*(c). \tag{10.22}$$

Note that the computation of $S^*(\mathcal{A}_c)$ for the subsequence $\mathcal{A}_c$ requires the knowledge of the level of distribution of $\mathcal{A}$ only up to Dc^{-1} because $abc < Dc^{-1}$. Therefore, unless c is small, one gains a bit in the requirement for the level of distribution.

What does make the above observation attractive is the fact that the small moduli c contribute relatively little to the main term

$$G = \sum_{\substack{a\,b\,c \\ (a,b)=1}}\sum\sum \rho_{ac}\rho_{bc} g(abc) \ .$$

Let us estimate these contributions explicitly. Put

$$G_c^* = \sum_{(a,b)=1}\sum \rho_{ac}\rho_{bc} g(abc) \ , \tag{10.23}$$

so G splits into

$$G = \sum_c G_c^* \ . \tag{10.24}$$

Proceeding with the diagonalization argument, as we did for the complete quadratic form G between the lines (7.1)–(7.14), we find that the partial quadratic form G_c^* is also a diagonal form,

$$G_c^* = g(c)^{-1} \sum_m \mu(m) h(cm) y_{cm}^2 \ .$$

For the vector y_d given by (7.21) this becomes

$$G_c^* = \frac{h(c)j(c)}{J(D)^2} \sum_{\substack{m<\sqrt{D}/c \\ (m,c)=1}} \mu(m) h^2(m) \ , \tag{10.25}$$

where we recall that

$$j(c) = \frac{h(c)}{g(c)} = \prod_{p|c} (1-g(p))^{-1}. \tag{10.26}$$

Let us put

$$K(D) = \sum_{m<\sqrt{D}}^{\flat} h^2(m) \ . \tag{10.27}$$

Estimating trivially, we derive by (10.25),

$$|G_c^*| \leqslant h(c)j(c)K(D)J(D)^{-2} \ . \tag{10.28}$$

Choose Δ with $1 < \Delta \leqslant D$ and denote

$$G^* = \sum_{c<\sqrt{\Delta}} G_c^* \ . \tag{10.29}$$

Of course, G^* depends on Δ; likewise G depends on D. For greater clarity we sometimes display these facts by writing $G^* = G^*(\Delta)$ and $G = G(D)$. Now, by (10.28) we get

$$|G_c^*(\Delta)| \leqslant \frac{K(D)}{J(D)^2} \sum_{c<\sqrt{\Delta}} h(c)j(c) \ . \tag{10.30}$$

Note that $hj = h + h^2$ at primes, whence $hj = h * h^2$ at squarefree numbers. Hence, (10.30) implies the following inequality:

$$|G^*(\Delta)| \leqslant J(D)^{-2}K(D)K(\Delta)J(\Delta) \ .$$

For comparison we write this in the form

$$|G^*(\Delta)| \leqslant G(D)\frac{J(\Delta)}{J(D)}K(\Delta)K(D) \ . \tag{10.31}$$

Let us examine the contributions to this inequality from particular factors. The sum (10.27) converges increasingly to the product

$$K = \prod_p (1 + h^2(p)) \ , \tag{10.32}$$

so $K(\Delta)K(D) \leqslant K^2$ in (10.31). This is certainly an acceptable bound, even when attention to the uniformity in h is needed. For example, if $g(p) = kp^{-1}$ for $p \geqslant k^2$ (as in Section 7.13), then K is absolutely bounded. More important is the ratio $J(\Delta)/J(D)$. Assuming $g(d)$ is a density function of dimension κ we get

$$\frac{J(\Delta)}{J(D)} \sim \left(\frac{\log \Delta}{\log D}\right)^{\kappa} = \delta^{\kappa} \tag{10.33}$$

for $\Delta = D^{\delta}$ with $0 < \delta < 1$, as $D \to \infty$. This is quite a small factor when δ is small for any $\kappa > 0$, while if κ is large it is still small even when δ is close to one. Inserting (10.32) and (10.33) into (10.31) we obtain

$$G^*(\Delta) \ll \delta^{\kappa} K^2 G(D) \ .$$

In other words, the contribution to the full sum G coming from the small values of c in (10.29) is insignificant.

One still needs to address the problem of having to estimate the partial sum $S^*(\mathcal{A}_c)$ for small c. Since we expect to get a small contribution one can afford to lose a large portion of the true value so as to make the job easier. For instance, one may contemplate replacing the elements a_n of $\mathcal{A}$ by somewhat larger ones b_n, forming a majorizing sequence $\mathcal{B}$ which admits a better level of distribution than does the primary sequence $\mathcal{A}$. Such a substitution can be applied to $S^+(\mathcal{A})$ but not directly to its components $S^*(\mathcal{A}_c)$, because the factors $\theta_n^*(c)$ are no longer positive. Probably, one needs to exploit some cancellation in the terms $a_n\theta_n^*(c)$ due to the sign changes of $\theta_n^*(c)$.

To see more clearly what is at stake we perform the following computation. Putting $\xi_d = h(d)y_d$ we derive by (10.20), (7.15), for $n \equiv 0(\bmod c)$ that

$$g(c)^2\theta_n^*(c) = \sum_u\sum_v \xi_{cu}\xi_{cv} \sum\sum_{\substack{ab|n\\ a|u,\, b|v}} \frac{\mu(ab)}{g(ab)} = \sum_u\sum_v \xi_{cu}\xi_{cv} \sum_{\ell|(n,uv)} \tau(\ell,u,v)\frac{\mu(\ell)}{g(\ell)}$$

$$= \sum_u\sum_v \xi_{cu}\xi_{cv} \prod_{p|(n,u)} (1-\frac{1}{g(p)}) \prod_{p|(n,v)} (1-\frac{1}{g(p)}) \prod_{p|(n,u,v)} (1-\frac{2}{g(p)})(1-\frac{1}{g(p)})^{-2}.$$

Since$(1-2/g)(1-1/g)^{-2} = 1-(1-g)^{-2}$ the last product is equal to the sum

$$\sum_{w|(n,u,v)} \mu(w)j(w)^2.$$

Hence,

$$\theta_n^*(c) = \frac{1}{g^2(c)} \sum_{w|n} \frac{\mu(w)}{g^2(w)} \Big(\sum_u \xi_{cuw}\frac{\mu((n,u))}{h((n,u))}\Big)^2. \tag{10.34}$$

For the vector $\xi_d = h(d)y_d$ given by (7.21) this becomes

$$\theta_n^*(c) = \Big(\frac{j(c)}{J(D)}\Big)^2 \sum_{\substack{w|n\\ (w,c)=1}} \mu(w)j^2(w)\Big(\sum_{\substack{(u,cw)=1\\ cuw<\sqrt{D}}} \mu((n,u))h\Big(\frac{u}{(n,u)}\Big)\Big)^2. \tag{10.35}$$

The innermost sum can be also written as

$$\sum_{\substack{d|n\\ (d,cw)=1}} \mu(d)j(d) \sum^{\flat}_{\substack{(u,cdw)=1\\ u<\sqrt{D}/cdw}} h(u) = \sum_{\substack{d|n\\ (d,cw)=1}} \mu(d)j(d)J_{cdw}(D)$$

(recall that $J_\ell(D)$ is defined in (7.25)). Hence,

$$\begin{aligned}\theta_n^*(c) &= \Big(\frac{j(c)}{J(D)}\Big)^2 \sum_{\substack{w|n\\ (w,c)=1}} \mu(w)j^2(w)\Big(\sum_{\substack{d|n\\ (d,cw)=1}} \mu(d)j(d)J_{cdw}(D)\Big)^2\\ &= \sum_{\substack{w|n\\ (w,c)=1}} \mu(w)\Big(\sum_{\substack{d|n\\ d\equiv 0\,(\bmod cw)}} \mu(d)j(d)J_d(D)/J(D)\Big)^2.\end{aligned} \tag{10.36}$$

One can also derive (10.36) directly from (10.20) by relaxing the condition $(a,b)=1$ with Möbius inversion and introducing (7.23) for ρ_ℓ, but proceeding as we have done gives us also the expression (10.34) which is valid for the general vector (ξ_d).

Next, introducing (10.36) to (10.22), we have

(10.37)
$$S^*(\mathcal{A}_c) = \Big(\frac{j(c)}{J(D)}\Big)^2 \sum_{n\equiv 0(c)} a_n \sum_{\substack{w|n\\ (w,c)=1}} \mu(w)j^2(w)\Big(\sum_{\substack{d|n\\ (d,cw)=1}} \mu(d)j(d)J_{cdw}(D)\Big)^2.$$

Obviously, summing (10.37) over all c we recover the formula (7.33) with ρ_d given by (7.23), that is,

$$S^+(\mathcal{A}) = \sum_n a_n \Big(\sum_{d|n} \mu(d) j(d) J_d(D)/J(D) \Big)^2 . \tag{10.38}$$

We do believe that the partial sums (10.22) satisfy

$$S^*(\mathcal{A}_c) \ll h(c) j(c) J(D)^{-2} X \ , \quad 1 \leqslant c \leqslant \sqrt{\Delta} \ , \tag{10.39}$$

for many interesting sequences $\mathcal{A}$, even if our knowledge of the level of distribution of $\mathcal{A}$ is limited to $D/\sqrt{\Delta}$. In particular, for $c = 1$ we expect that

$$S^*(\mathcal{A}) = \sum_n a_n \theta_n^* \ll J(D)^{-2} X \ . \tag{10.40}$$

Of course, if $\mathcal{A}$ admits a level of distribution D then the expected bound (10.39) is true as shown by (10.28), but proving (10.39) without improving the level of distribution remains a task for the future.

There is a similar phenomenon for the beta-sieve weights. We illustrate this briefly in its simplest aspects. We consider the sifted sum

$$S^+(\mathcal{A}, \sqrt{D}) = \sum_{\substack{d | P(\sqrt{D}) \\ d \in \mathcal{D}^+}} \mu(d) A_d(x)$$

in the case of the linear sieve, that is $\kappa = 1$ and $\beta = 2$. Of course, the moduli $d \in \mathcal{D}^+$ are smaller than D but unless d has many prime factors they are actually considerably smaller, namely

$$d < D^{1 - 3^{-(r+1)/2}} \quad \text{if } \omega(d) = r \ .$$

This can be shown by the arguments used for Lemma 6.5.

One can see from our induction arguments that the main contribution comes from d with $\omega(d)$ relatively small. Yet, when restricted to these d, the required level of distribution is somewhat less than D. Although the influence to the main term from large $\omega(d)$ is minor, the problem of actually bounding the contribution to S^+ from these without improving the level of distribution, remains open.

10.5. Related Instances

Finally, as we have mentioned at the beginning of this chapter, there are other places in the book where molecular characteristics of the sieve weights have played a role.

In Section 6.8 we developed a delicate bound for a sum over sieve weights which was then applied in Section 6.10 to the problem of counting primes in "most" very short intervals. In Section 7.3 we studied the interaction of sieves of competing dimensions and this appeared in more spectacular form in Sections 7.13 and 7.14 on the existence of small gaps between primes. In Section 12.7 we use a detailed description of the sets $\mathcal{D}^\pm$ to develop a bilinear form for the remainder in the linear sieve. This can be applied to numerous problems. In Section 13.3 we exploit isolated bits of the beta-sieve decomposition, strategically chosen to harmonize with the requirements of a Karatsuba bound for extremely short Kloosterman-type sums. This yields improved upper bounds for $\pi(x; q, a)$ in the full range of the modulus.

Any sieve weights (λ_d) (or their constituents) are in one way or another the Möbius function $\mu(d)$, stripped for accommodation below the available level of distribution. There are other compositions which act like a sieve but in tonality are really quite different. We are going to encounter the convolution $\theta = 1*\chi$ where χ is a real primitive character; this will be one of the main instruments in Section 24.2. Well, this one behaves like a sieve only in case χ is "exceptional". The point is that, under such a hypothetical assumption, false in the real world, $\theta_p = 1+\chi(p)$ vanishes for a full density of primes, so any desired additional sifting will be of dimension zero. Furthermore, the character values $\chi(d)$ are easier to work with than those of the Möbius function. After all, χ is periodic so one can apply a simple yet powerful harmonic analysis, whereas, at least until the Riemann Hypothesis is proven and maybe beyond, an adequate harmonic-analytic treatment of μ remains a wishful future project. The Dirichlet hyperbola method also works for χ, hence in order to evaluate the sequence $\{a_n\theta_n;\ n \leqslant x,\ \chi(n) = 1\}$, one needs the congruence sums $A_d(x)$ only to moduli $d \leqslant \sqrt{x}$. Unfortunately, these nice features come at a price. Unlike for λ_d, one cannot perform any truncation of $\chi(d)$ without losing the sieve effects and consequently the level of distribution cannot be adjusted at all.

CHAPTER 11

The Beta-Sieve

Concerto Grosso

11.1. Introduction: Buchstab Iterations

Sieve methods produce general inequalities of the following type:

$$S(\mathcal{A}, z) \geqslant XV(z)\{f(s) + o(1)\} - R(\mathcal{A}, D) \quad \text{if } s > 1 , \tag{11.1}$$

$$S(\mathcal{A}, z) \leqslant XV(z)\{F(s) + o(1)\} + R(\mathcal{A}, D) \quad \text{if } s > 0 , \tag{11.2}$$

where f, F are continuous functions in $s = \log D/\log z$ such that

$$0 \leqslant f(s) < 1 < F(s), \tag{11.3}$$

$f(s)$ is increasing and $F(s)$ is decreasing to 1 exponentially. For example, the Fundamental Lemma (Theorem 6.9) yields the above results with

$$f(s) = 1 + O(e^{-s}), \quad F(s) = 1 + O(e^{-s}). \tag{11.4}$$

One of the main objectives of sieve theory is to find the best pair (f, F). In this section we give motivation for our construction of a pair (f, F) which will turn out to be optimal in some of the most important cases.

In 1938 A. A. Buchstab [**19**] discovered, by applying earlier results from Brun's sieve to each $S(\mathcal{A}_p, p)$ in the recurrence formula

$$S(\mathcal{A}, z) = |\mathcal{A}| - \sum_{p<z} S(\mathcal{A}_p, p) , \tag{11.5}$$

that the new estimates resulting for $S(\mathcal{A}, z)$ were improved in wide ranges, always so for s sufficiently large.

More generally, given a pair of functions (f, F) for which (11.1) and (11.2) hold, one can produce a new pair $(\hat{f}, \hat{F})$ by passing through the recurrence formula (11.5). To accelerate the process we first subtract from $S(\mathcal{A}, z)$, $S(\mathcal{A}_p, p)$ the expected values $XV(z)$, $g(p)XV(p)$ respectively, and we work with the recurrence formula

$$S(\mathcal{A}, z) - XV(z) = |\mathcal{A}| - X - \sum_{p<z} (S(\mathcal{A}_p, p) - g(p)XV(p)) , \tag{11.6}$$

rather than (11.5). Suppose the lower bound (11.1) is true for any $\mathcal{A}$ (which satisfies the sieve axioms of dimension κ) in the range $s > 1$, in particular, the inequality

$$S(\mathcal{A}_p, p) \geqslant g(p)XV(p)\{f(s_p) + o(1)\} - R(\mathcal{A}_p, D_p)$$

is true for all $p < z$ provided $s > 2$, where $D_p = D/p$ and $s_p = \log D/\log p$. Inserting this into (11.6) we obtain

$$S(\mathcal{A}, z) - XV(z) \leqslant X \sum_p g(p)V(p)\{1 - f(s_p) + o(1)\} + R(\mathcal{A}, D).$$

By Lemma 5.1 the above sum over $p < z$ is bounded by $V(z)\mathcal{I}$ where

$$\begin{aligned}\mathcal{I} &= K\int_0^z (1-f(s_t))\,d\left(\frac{\log z}{\log t}\right)^\kappa + (K-1)(1-f(s-1)) + o(1)\\ &= -1 + K\hat{f}(s) - (K-1)f(s-1) + o(1) = -1 + \hat{f}(s) + O(K-1) + o(1)\ .\end{aligned}$$

Here $\hat{f}(s)$ is defined to be the following integral transfom of f:

$$\hat{f}(s) = 1 + s^{-\kappa}\int_s^\infty (1-f(t-1))dt^\kappa \quad \text{if } s > 2\ . \tag{11.7}$$

Hence, we get the upper bound

$$S(\mathcal{A}, z) \leqslant XV(z)\{\hat{f}(s) + O(K-1) + o(1)\} + R(\mathcal{A}, D) \tag{11.8}$$

in the range $s > 2$. Extend this by setting

$$s^\kappa \hat{f}(s) = 2^\kappa \hat{f}(2) \quad \text{if } 0 < s \leqslant 2\ . \tag{11.9}$$

Thus, (11.8) holds for $0 < s \leqslant 2$ as well, because $S(\mathcal{A}, z)$ is non-increasing in z. Similarly, if the upper bound (11.2) is true for any $\mathcal{A}$ of dimension κ in the range $s > 0$, then we derive the lower bound

$$S(\mathcal{A}, z) \geqslant XV(z)\left\{\hat{F}(s) + O(K-1) + o(1)\right\} - R(\mathcal{A}, D) \tag{11.10}$$

in the range $s > 1$, where $\hat{F}(s)$ is defined by the same formula (11.7) but for all $s > 1$. Keep in mind that K is close to 1 so the new upper and lower bounds (11.8), (11.10) have main terms $XV(z)\hat{f}(s)$, $XV(z)\hat{F}(s)$ up to a small correction.

We shall call the map $(f, F) \mapsto (\hat{F}, \hat{f})$ the Buchstab transform. One can refine the new pair by taking $(\tilde{f}, \tilde{F})$ with

$$\tilde{f} = \max(f, \hat{F})\ , \qquad \tilde{F} = \min(F, \hat{f})\ .$$

Usually this refinement makes a difference only for small s since the transforms $\hat{F}, \hat{f}$ improve on f, F if s is sufficiently large.

Continuing this process we produce a sequence of admissible pairs $\{f_j, F_j\}$ each one at least as good as the predecessor, i.e., $0 \leqslant f_0 \leqslant f_1 \leqslant \cdots < 1 < \cdots F_1 \leqslant F_0$. The functions $f_j(s)$ are defined for $s > 1$ and $F_j(s)$ for $s > 0$.

The question is what one gets after an infinite number of iterations. If the process converges one obtains in the limit a pair (f, F) which is the relevant fixed point of the Buchstab transform, that is f, F solve the system of integral equations

$$\hat{f}(s) = F(s)\ , \tag{11.11}$$

$$\hat{F}(s) = f(s)\ , \tag{11.12}$$

for sufficiently large s at any rate. This is equivalent to the system of differential-difference equations

$$(s^\kappa f(s))' = \kappa s^{\kappa-1}F(s-1)\ , \tag{11.13}$$

$$(s^\kappa F(s))' = \kappa s^{\kappa-1}f(s-1)\ , \tag{11.14}$$

subject to the decay conditions (11.4) at infinity. With no initial conditions at hand there are many solutions to this system, the obvious one being $f(s) = F(s) = 1$. In practice these constant functions are never achieved because the initial conditions are never realized. Usually one is given lower and upper bounds for $S(\mathcal{A}, z)$ which are quite sharp for s large (think of the Fundamental Lemma, Theorem 6.9). The iteration process improves these bounds step-by-step up to some point, say, in the range $s \geqslant \beta$ for the lower bound and $s \geqslant \beta + 1$ for the upper bound, while always producing worse results for smaller s. Therefore, for the new upper bound in the range $s < \beta + 1$ one should better choose the result at $s = \beta + 1$. This is possible because $S(\mathcal{A}, z)$ is increasing in $s = \log D / \log z$. In other words, we perform the iterations by applying only the trivial lower bound $S(\mathcal{A}, z) \geqslant 0$ in the range $s \leqslant \beta$. Actually, when analyzing this process, it is convenient not to claim any lower bound for $S(\mathcal{A}, z)$ in the range $s < \beta$.

Clearly, one cannot perform any sifting by primes p beyond the level of distribution D because we assume no knowledge of the congruence sums $|\mathcal{A}_d|$ for moduli $d > D$. Thus one cannot expect improvements below the point $\beta = 1$.

According to the above observations, when Brun's sieve is used to ignite the iteration process one is naturally led to postulate initial conditions of the form

$$s^{\kappa} F(s) = A \quad \text{if} \quad \beta - 1 \leqslant s \leqslant \beta + 1 , \tag{11.15}$$

$$s^{\kappa} f(s) = B \quad \text{at} \quad s = \beta , \tag{11.16}$$

where β, A, B are constants to be determined in terms of κ. The initial condition (11.15) rules out the possibility $F(s) = 1$. The parameter β is best chosen as the smallest number $\geqslant 1$ for which the system (11.13), (11.14) with the initial conditions (11.15), (11.16) has a solution for $A > 0$, $B \geqslant 0$ with the property (11.4). Given β, A, B one can find $f(s), F(s)$ in step-by-step integrations of (11.13), (11.14) starting from (11.15), (11.16). The constants A, B are determined by the conditions that f and F approach 1 at infinity. Therefore, there is exactly one solution to this initial value problem for a given β. There is really no obligation to choose β as the smallest value which yields $A > 0, B \geqslant 0$, but we believe that our choice does this.

In order to establish results rigorously by iterating Brun's sieve we must take care of the error terms and that would be a painstaking enterprise. It is far better, both for the elegance of the argument and the quality of the remainders, to directly apply the beta-sieve. In this chapter we proceed in this fashion. The main terms in our results are precisely the same as those which could be achieved by an infinite number of iterations.

One can ignite Buchstab's iterations with any pair of lower and upper bounds, in which case the initial conditions could differ from (11.15), (11.16). Ankeny and Onishi [**1**] performed one step of the iteration employing Selberg's upper bound and the trivial lower bound. Subsequently, Diamond, Halberstam and Richert [**32**] reached the limit of iterations employing Selberg's sieve for the upper bounds and the beta-sieve for the lower bounds. The latest results on this procedure receive a detailed treatment in the very recent book [**33**] of Diamond and Halberstam. See also an earlier analysis in [**113**]. These results are interesting for $\kappa > 1$; however, if $\kappa \leqslant 1$ there is no advantage to starting the iterations with extra sharp bounds in the initial segments.

In this chapter we consider the general case of the combinatorial sieve of beta type. In trying to draw the line between elegance and generality we are forced, both here and elsewhere, to make difficult choices as to what to include. We give full proofs of the beta-sieve for a restricted range of the dimension, including the most important cases $\kappa \leqslant 1$, where we are able to give a very clean version of the proof.

For the sieve of higher dimension the proof using the same type of arguments is quite complicated (although full details are given in [**104**]). Hence, for these we have chosen, in Section 11.16, a different method which involves composing two different types of beta-sieve weights. This simplifies the proof and the final theorems are of equal strength. The only disadvantage to this approach is that the resulting sieve weights, not being the pure beta-weights, are less aesthetic and do not enjoy all of the same properties discussed in Chapter 10.

Recall that the beta-sieve of level D is the combinatorial sieve whose truncation parameters (see (6.23)) are

$$y_m(p_1,\dots,p_m) = (D/p_1\dots p_m)^{1/\beta} \tag{11.17}$$

where $\beta \geqslant 1$. In other words, the sieve weights are given by the Möbius function, restricted respectively, to the sets

$$\begin{aligned} \mathcal{D}^+ &= \{d = p_1\dots p_n : p_1 > \dots > p_n,\ p_1\dots p_m p_m^\beta < D \quad \text{for all } m \text{ odd}\}, \\ \mathcal{D}^- &= \{d = p_1\dots p_n : p_1 > \dots > p_n,\ p_1\dots p_m p_m^\beta < D \quad \text{for all } m \text{ even}\}\ . \end{aligned} \tag{11.18}$$

Recall that, by convention, both sets contain $d = 1$.

We have already given a quick and simple treatment of this sieve in Section 6.5. In this chapter we establish the optimal results (as far as the main terms are concerned); this amounts to determining the best $\beta = \beta(\kappa)$. An account of the beta-sieve, rather different but also based on the choice (11.17), was given independently by A. Selberg [**145**] (Selberg attributes his introduction to these ideas as originating with B. Rosser).

The case of $\beta = 1$ is somewhat different in technical details, therefore, we first consider the beta-sieve with $\beta > 1$. This turns out to be the appropriate condition for the sieve of any dimension $\kappa > \frac{1}{2}$ while $\beta = 1$ is the best choice for $0 \leqslant \kappa \leqslant \frac{1}{2}$. The semi-linear sieve ($\kappa = \frac{1}{2}$) is very interesting and we just miss it in Theorem 11.18, although one can get the results for this case indirectly by increasing the dimension slightly. Instead, in Sections 11.12–11.14 we complete the development of the beta-sieve for any dimension $0 \leqslant \kappa \leqslant \frac{1}{2}$ by a different argument.

11.2. Cast of Characters

Before we can state the main results (for which see Section 11.8) we need to define some series of functions which play a role. We begin with the sums

$$V_n(D,z) = \sum_{\substack{p_n < \dots < p_1 < z \\ p_1\dots p_m p_m^\beta < D,\, m<n,\, m\equiv n (2) \\ p_1\dots p_n p_n^\beta \geqslant D}} \cdots \sum g(p_1\dots p_n) V(p_n)\ , \tag{11.19}$$

which were introduced more generally in Section 6.2 and which form the building blocks for the main term in our sieve bounds. Our crude estimates from Section 6.5 were derived by discarding a lot of the summation conditions. In spite of this extravagant treatment the results were quite good, particularly if the sifting variable

$$s = \log D/\log z \tag{11.20}$$

is large. Here, we shall no longer discard these conditions. Compatible with the sum (11.19) is the integral (for $s > 0$)

$$\kappa^{-n} s^{\kappa} f_n(s) = \int \cdots \int_{\substack{0<t_n<\cdots<t_1<1/s \\ t_1+\cdots+t_m+\beta t_m<1, m<n, m\equiv n(2) \\ t_1+\cdots+t_n+\beta t_n>1}} t_n^{-\kappa} \frac{dt_1 \dots dt_n}{t_1 \dots t_n}. \tag{11.21}$$

Naturally, one expects that $f_n(s)V(z)$ approximates to $V_n(D, z)$. Note that the range of summation in (11.19) and the range of integration in (11.21) are void if $s \geqslant \beta + n$, so

$$V_n(D, z) = 0 \qquad \text{if } s \geqslant \beta + n, \tag{11.22}$$

$$f_n(s) = 0 \qquad \text{if } s \geqslant \beta + n. \tag{11.23}$$

We shall examine $V_n(D, z)$ and $f_n(s)$ in the range

$$s \geqslant \beta_n = \beta - \varepsilon_n \quad \text{with } \varepsilon_n = \begin{cases} 1 & \text{if } n \text{ is odd,} \\ 0 & \text{if } n \text{ is even.} \end{cases} \tag{11.24}$$

For $n = 1$ and $\beta - 1 \leqslant s < \beta + 1$ we have

$$V_1(D, z) = V\left(D^{1/(\beta+1)}\right) - V(z) \tag{11.25}$$

and

$$s^{\kappa} f_1(s) = (\beta + 1)^{\kappa} - s^{\kappa}. \tag{11.26}$$

For $n \geqslant 2$ and $\beta_n \leqslant s \leqslant \beta + n$ we have the following recurrence formulas:

$$V_n(D, z) = \sum_{z_n \leqslant p < x_n} g(p) V_{n-1}\left(D/p, p\right) \tag{11.27}$$

where $z_n = D^{1/(\beta+n)}$, $x_n = \min\left(z, D^{1/(\beta+\varepsilon_n)}\right) = D^{1/\max(s,\beta+\varepsilon_n)}$ and

$$s^{\kappa} f_n(s) = \int_{\max(s,\beta+\varepsilon_n)}^{\beta+n} f_{n-1}(t-1) dt^{\kappa}. \tag{11.28}$$

Hence, if $n > 1$, n odd, we have in the range $\beta - 1 \leqslant s \leqslant \beta + 1$:

$$V_n(D, z) = V_n\left(D, D^{1/(\beta+1)}\right), \tag{11.29}$$

$$s^{\kappa} f_n(s) = (\beta + 1)^{\kappa} f_n(\beta + 1). \tag{11.30}$$

Notice that $s^{\kappa} f_n(s)$ is continuous and non-decreasing for $s \geqslant \beta_n$.

We need upper bounds for two sums of the $V_n(D, z)$, over n odd and over n even (see (6.31) and (6.32) respectively). To this end we shall first examine the corresponding partial sums of the continuous functions $f_n(s)$:

$$T_N(s) = \sum_{\substack{n \leqslant N \\ n \equiv N(2)}} f_n(s) \quad \text{for } s \geqslant \beta_N, \tag{11.31}$$

that is, for $s \geqslant \beta - 1$ if N is odd and $s \geqslant \beta$ if N is even.

11.3. The Functions $F(s), f(s)$

Let κ be any real number, $\kappa \geqslant 0$. Our first task is to make a specific choice of $\beta = \beta(\kappa) \geqslant 1$ for which the partial sums (11.31) converge. We have already seen in our treatment of Brun's sieve that any $\beta \geqslant 4\kappa + 1$ does the job but for better results we want a smaller value. For the choice we shall make, we are able to give the infinite series relatively simple integral representations which will be very useful in applications. In this section we begin this analysis.

Ignoring for the time being the question of convergence, we form two series:

$$F(s) = 1 + \sum_{n \text{ odd}} f_n(s) \qquad \text{if } s \geqslant \beta - 1, \tag{11.32}$$

$$f(s) = 1 - \sum_{n \text{ even}} f_n(s) \qquad \text{if } s \geqslant \beta. \tag{11.33}$$

Postulating the condition (11.4) at infinity, we find by the recurrence formula (11.28) that the pair F, f reverses itself under the Buchstab integral transform (11.7) or, equivalently, F, f satisfy the system of differential-difference equations

$$\begin{aligned} (s^\kappa F(s))' &= \kappa s^{\kappa-1} f(s-1) \quad \text{if } s > \beta + 1, \\ (s^\kappa f(s))' &= \kappa s^{\kappa-1} F(s-1) \quad \text{if } s > \beta. \end{aligned} \tag{11.34}$$

Moreover, $s^\kappa F(s)$ is constant in the segment $\beta - 1 \leqslant s \leqslant \beta + 1$. Therefore, we are led to the initial conditions

$$\begin{aligned} s^\kappa F(s) &= A \qquad \text{if } \beta - 1 \leqslant s \leqslant \beta + 1, \\ s^\kappa f(s) &= B \qquad \text{at } s = \beta. \end{aligned} \tag{11.35}$$

We shall see that our problem is posed correctly, namely the constants A, B are determined by κ and β due to the condition (11.4). As a curiosity, we point out that the weaker condition

$$\begin{aligned} F(s) &= 1 + o(s^{-2\kappa}), \\ f(s) &= 1 + o(s^{-2\kappa}), \end{aligned} \tag{11.36}$$

would suffice; however, the eventual solution functions F, f, then gain the much stronger condition (11.68).

Letting for $s \geqslant \beta$,

$$\begin{aligned} P(s) &= F(s) + f(s), \\ Q(s) &= F(s) - f(s), \end{aligned} \tag{11.37}$$

we turn the system (11.34) into two independent equations:

$$\begin{aligned} sP'(s) &= -\kappa P(s) + \kappa P(s-1), \\ sQ'(s) &= -\kappa Q(s) - \kappa Q(s-1), \end{aligned} \tag{11.38}$$

for $s > \beta + 1$. The initial conditions (11.35) become

$$\begin{aligned} s^\kappa P(s) &= A + B + A \int_\beta^s (t-1)^{-\kappa} dt^\kappa, \\ s^\kappa Q(s) &= A - B - A \int_\beta^s (t-1)^{-\kappa} dt^\kappa, \end{aligned} \tag{11.39}$$

for $\beta \leqslant s \leqslant \beta+1$, and the conditions at infinity give

$$P(s) = 2 + o(s^{-2\kappa}), \quad Q(s) = o(s^{-2\kappa}). \tag{11.40}$$

Now we can treat each of the problems for $P(s)$ and $Q(s)$ separately. We solve these problems by the method of adjoint equation. A general account of this method is given in Appendix B. The equations adjoint to (11.38) are

$$\begin{aligned}(sp(s))' &= \kappa p(s) - \kappa p(s+1),\\ (sq(s))' &= \kappa q(s) + \kappa q(s+1).\end{aligned} \tag{11.41}$$

The first of these holds (see (B.10) with $a = -b = \kappa$) for

$$p(s) = \int_0^\infty \exp\left(-sz - \kappa\int_0^z \frac{1-e^{-u}}{u}du\right)dz. \tag{11.42}$$

A solution to the second equation is given in Lemma B.1 for $a = b = \kappa$ with $N = [2\kappa]$. In particular, this gives

$$q(s) = \frac{1}{\Gamma(1-2\kappa)}\int_0^\infty e^{-sz}\exp\left(\kappa\int_0^z \frac{1-e^{-u}}{u}du\right)z^{-2\kappa}dz \tag{11.43}$$

if $0 \leqslant \kappa < \frac{1}{2}$ by the formula (B.10), and

$$q(s) = s^{2\kappa-1} + \frac{1}{\Gamma(1-2\kappa)}\int_0^\infty e^{-sz}\left\{\exp(\kappa\int_0^z \frac{1-e^{-u}}{u}du) - 1\right\}z^{-2\kappa}dz \tag{11.44}$$

if $\frac{1}{2} < \kappa < 1$ by the formula (B.11). For $\kappa = \frac{1}{2}$ we have $q(s) = 1$ and for $\kappa = 1$ we have $q(s) = s - 1$. For any $\kappa \geqslant 0$ we have (see (B.12)),

$$p(s) \sim s^{-1}, \quad q(s) \sim s^{2\kappa-1} \quad \text{as } s \to \infty. \tag{11.45}$$

Moreover, we have as $s \to 0+$,

$$s^{1-\kappa}p(s) \sim e^{-\gamma\kappa}\Gamma(1-\kappa)\ , \qquad \text{if } \kappa < 1, \tag{11.46}$$

$$s^{1-\kappa}q(s) \sim e^{\gamma\kappa}\Gamma(1-\kappa)/\Gamma(1-2\kappa)\ , \qquad \text{if } \kappa \leqslant \tfrac{1}{2}, \tag{11.47}$$

by Lemma B.2 (in case $\kappa = \frac{1}{2}$ the last limit is zero).

Next we compute the inner products (see (B.5) and (B.6))

$$\langle P, p\rangle = sP(s)p(s) - \kappa\int_{s-1}^s P(x)p(x+1)dx \tag{11.48}$$

$$= (s-1)P(s-1)p(s-1) + \int_{s-1}^s x^{1-\kappa}p(x)dx^\kappa P(x) \tag{11.49}$$

and

$$\langle Q, q\rangle = sQ(s)q(s) + \kappa\int_{s-1}^s Q(x)q(x+1)dx \tag{11.50}$$

$$= (s-1)Q(s-1)q(s-1) + \int_{s-1}^s x^{1-\kappa}q(x)dx^\kappa Q(x). \tag{11.51}$$

First, by the behaviour at infinity (see (11.40) and (11.45)) we derive from the first expressions in (11.48) and (11.50) that

$$\langle P, p\rangle = 2, \qquad \langle Q, q\rangle = 0. \tag{11.52}$$

These equations hold for all $s \geqslant \beta+1$. Putting $s=\beta+1$ in the second expressions for the inner products we derive by inserting the initial conditions (11.39) that

$$\begin{aligned}\langle P,p\rangle &= (A+B)\beta^{1-\kappa}p(\beta)+A\int_{\beta}^{\beta+1} x^{1-\kappa}p(x)(x-1)^{-\kappa}dx^{\kappa}\\ &= (A+B)\beta^{1-\kappa}p(\beta)-A\int_{\beta-1}^{\beta} dx^{1-\kappa}p(x)\end{aligned}$$

by (11.41). Hence

$$\langle P,p\rangle = A(\beta-1)^{1-\kappa}p(\beta-1)+B\beta^{1-\kappa}p(\beta)=2. \tag{11.53}$$

Similarly, we derive

$$\langle Q,q\rangle = A(\beta-1)^{1-\kappa}q(\beta-1)-B\beta^{1-\kappa}q(\beta)=0. \tag{11.54}$$

First we consider the range $\kappa>\frac{1}{2}$. We know by Lemma B.4 that $q(s)$ has fewer than 2κ real zeros. In fact, $q(s)$ has a positive zero, and we choose $\beta-1$ to be the largest zero of $q(s)$:

$$q(\beta-1)=0, \qquad \beta=\beta(\kappa)>1 \quad \text{if} \quad \kappa>\tfrac{1}{2}\ . \tag{11.55}$$

This yields by (11.54),

$$B=0. \tag{11.56}$$

Then by (11.53) we get

$$A=2(\beta-1)^{\kappa-1}p(\beta-1)^{-1}. \tag{11.57}$$

Remarks. It turns out that our choice of β is the smallest possible one which works. Any value for $\beta-1$ between the last two zeros of $q(s)$ would yield negative B by (11.54); consequently $f(s)$ would be negative at $s=\beta$ (see (11.35)) which is worse than the value zero, trivially available for any lower-bound sieve. Furthermore, the second to largest zero of $q(s)$ (if it exists) is not a good choice, because it is too small for the convergence of the series (11.32) and (11.33) (the two largest zeros of $q(s)$ are distanced by at least 1, actually the distance increases to infinity with κ). We do not bother to prove all of these properties here; we are only required to show that the choice (11.55) does secure the convergence in question. Incidentally, if κ is half of an integer then β is an algebraic number since $q(s)$ is a monic polynomial of degree $2\kappa-1$ with rational coefficients.

We still have to consider the range $0\leqslant\kappa\leqslant\frac{1}{2}$. In this case we simply choose

$$\beta=\beta(\kappa)=1 \qquad \text{if} \quad 0\leqslant\kappa\leqslant\tfrac{1}{2}\ . \tag{11.58}$$

To compute the constants A, B in this case one must take the limit values (11.46), (11.47) in place of $(\beta-1)^{1-\kappa}p(\beta-1)$ and $(\beta-1)^{1-\kappa}q(\beta-1)$ in the linear equations (11.53) and (11.54), respectively. We also need values of $p(1)$ and $q(1)$. By (11.42) we get

$$p(1)=\int_0^{\infty}\exp\left(-z-\kappa\int_0^z\frac{1-e^{-u}}{u}\,du\right)dz \tag{11.59}$$

for all $\kappa\geqslant 0$ and by (11.43) we get

$$q(1)=\frac{1}{\Gamma(1-2\kappa)}\int_0^{\infty}\exp\left(-z+\kappa\int_0^z\frac{1-e^{-u}}{u}\,du\right)z^{-2\kappa}dz \tag{11.60}$$

if $0 \leqslant \kappa < \frac{1}{2}$. For $\kappa = \frac{1}{2}$ we have $q(s) = 1$, so $q(1) = 1$.

Therefore, for $0 \leqslant \kappa \leqslant \frac{1}{2}$, the system of linear equations (11.53), (11.54) for the unknown constants A, B becomes

$$Ae^{-\gamma\kappa}\Gamma(1-\kappa) + Bp(1) = 2\ ,$$

$$Ae^{\gamma\kappa}\frac{\Gamma(1-\kappa)}{\Gamma(1-2\kappa)} - Bq(1) = 0\ .$$

If $\kappa = \frac{1}{2}$ this gives

(11.61) $$A = 2\sqrt{e^{\gamma}/\pi}\ , \qquad B = 0\ .$$

If $0 \leqslant \kappa < \frac{1}{2}$ it gives

(11.62) $$A = \frac{2e^{\gamma\kappa}}{\Gamma(1-\kappa)}\frac{q}{p+q}\ , \qquad B = \frac{2e^{\gamma\kappa}}{p+q}\ ,$$

where

(11.63) $$\begin{aligned} p &= e^{\gamma\kappa}\int_0^\infty \exp\Big(-z-\kappa\int_0^z \frac{1-e^{-u}}{u}\,du\Big)dz,\\ q &= e^{-\gamma\kappa}\int_0^\infty \exp\Big(-z+\kappa\int_0^z \frac{1-e^{-u}}{u}\,du\Big)z^{-2\kappa}dz\ . \end{aligned}$$

REMARK. Note that the constant B, given by (11.62), is positive, so the sieve of dimension $\kappa < \frac{1}{2}$ gives a positive lower bound for the sifting function $S(\mathcal{A}, z)$ for z all the way up to the level of distribution.

Having chosen β, A, B as above one can improve the conditions at infinity (11.36) quite substantially. Indeed, by the inner product formulas we have two integral equations:

(11.64) $$sq(s)Q(s) = \kappa\int_{s-1}^{s} q(x+1)Q(x)dx,$$

(11.65) $$sp(s)P(s) + \kappa\int_{s-1}^{s} p(x+1)P(x)dx = 2.$$

The latter holds also for $P(s) = 2$ (because it is another solution to (11.38) and (11.40)) so by subtracting we get

(11.66) $$sp(s)(P(s)-2) = -\kappa\int_{s-1}^{s} p(x+1)(P(x)-2)dx.$$

Using (11.45) we infer

$$sQ(s) \ll \int_{s-1}^{s} |Q(x)|dx$$

and

$$s|P(s)-2| \ll \int_{s-1}^{s} |P(x)-2|dx\ .$$

From these integral inequalities it is easy to derive

(11.67) $$Q(s) = O(s^{-s}), \qquad P(s) = 2 + O(s^{-s})\ ,$$

whence

(11.68) $$F(s) = 1 + O(s^{-s}), \qquad f(s) = 1 + O(s^{-s}).$$

Indeed, we have more general results.

LEMMA 11.1. *Let $\kappa > 0$ and*

$$m(t) = \Big(\frac{e\kappa}{t\log t}\Big)^t, \quad \text{for } t > 1\ . \tag{11.69}$$

Then, for all s sufficiently large in terms of κ we have

$$\kappa\int_{s-1}^{s} m(t)dt = sm(s)\Big(1+\frac{\log\log s}{\log s}\Big)^{-1}\Big(1+O(\frac{1}{\log s})\Big)\ . \tag{11.70}$$

PROOF. We have

$$-\frac{m'(t)}{m(t)} = \log(t\log t) - \log\kappa + (\log t)^{-1}\ .$$

Hence

$$m(t) = -m'(t)\big(\log(t\log t)+O(1)\big)^{-1}\ ,$$

$$\int_{s-1}^{s} m(t)dt = m(s-1)\big(\log(s\log s)+O(1)\big)^{-1}\ .$$

By the mean-value theorem there exists ξ with $s-1\leqslant \xi\leqslant s$ such that

$$\frac{m(s-1)}{m(s)} = \exp\Big(-\frac{m'(\xi)}{m(\xi)}\Big) = \frac{\xi}{\kappa}\big(\log\xi + O(1)\big) = \frac{s}{\kappa}\big(\log s+O(1)\big)\ ,$$

which completes the proof. □

LEMMA 11.2. *Let $\kappa > 0$ and let $M(t)$ be a continuous non-negative function such that*

$$\big(s+O(1)\big)M(s) \leqslant \kappa\int_{s-1}^{s} M(t)dt \tag{11.71}$$

for all s sufficiently large. Then

$$M(s) \leqslant bm(s) \tag{11.72}$$

for all s sufficiently large, where $m(t)$ is given by (11.69) *and b is a large positive constant.*

PROOF. Suppose the assertion is not true. Then, because b is sufficiently large, there exists a large s for which the inequality (11.72) becomes an equation and the inequality holds for all smaller arguments. Hence (11.72) yields

$$\big(s+O(1)\big)M(s) \leqslant b\kappa\int_{s-1}^{s} m(t)dt = bsm(s)\Big(1+\frac{\log\log s}{\log s}\Big)^{-1}\Big(1+O(\frac{1}{\log s})\Big)$$

by Lemma 11.1. This shows that $M(s) < bm(s)$, which is the desired contradiction. □

LEMMA 11.3. *Let $\kappa > 0$ and let $M(t)$ be a continuous positive function such that*

$$\big(s+O(1)\big)M(s) \geqslant \kappa\int_{s-1}^{s} M(t)dt \tag{11.73}$$

for all s sufficiently large. Then

$$M(s) \geqslant am(s)\Big(1+2\frac{\log\log s}{\log s}\Big)^{-s} \tag{11.74}$$

for all s sufficiently large, where $m(s)$ is given by (11.69) *and a is a small positive constant.*

PROOF. Suppose the assertion is not true. Then, because a is sufficiently small, there exists a large s for which the inequality (11.74) becomes an equation and the inequality holds for all smaller arguments. Hence (11.74) yields

$$\big(s+O(1)\big)M(s) \geqslant a\kappa \int_{s-1}^{s} m(t)c^{-t}dt$$

where $c = 1 + 2\log\log s/\log(s-1)$. By Lemma 11.1 with κ replaced by κ/c we obtain

$$\big(s+O(1)\big)M(s) \geqslant asm(s)c^{1-s}\Big(1+\frac{\log\log s}{\log s}\Big)^{-1}\Big(1+O(\frac{1}{\log s})\Big)\,.$$

We have $c = \big(1+2(\log\log s)/\log s\big)\big(1+O(1/s\log s)\big)$. Hence

$$c^{1-s} = \Big(1+2\frac{\log\log s}{\log s}\Big)^{1-s}\Big(1+O\Big(\frac{1}{\log s}\Big)\Big)\,,$$

which shows that $M(s)$ satisfies the strong inequality (11.74). This contradiction completes the proof. □

Combining the upper and lower bounds from the previous two lemmas we derive:

COROLLARY 11.4. *Let $\kappa > 0$ and let $M(t)$ be a continuous positive function such that*

$$\big(s+O(1)\big)M(s) = \kappa \int_{s-1}^{s} M(t)dt \tag{11.75}$$

for all s sufficiently large. Then

$$M(s) = \exp\Big(-s\log s - s\log\log s + s\log e\kappa + O\Big(\frac{s\log\log s}{\log s}\Big)\Big)\,. \tag{11.76}$$

REMARK. Since $p(s)$ is positive for all $s > 0$ (see the formula (11.42)), it follows by (11.66) that $P(s) - 2$ changes sign in every interval of length one.

We are now done with the definition of the functions $f(s), F(s)$, except that we still need to show that the condition of the convergence of the series (11.32), (11.33) is justified in the case of our choice of the parameters β, A, B. Our method involves the explicit construction of a function which majorizes the partial sums (11.31).

11.4. The Functions $H(s), h(s)$ for $\kappa > \frac{1}{2}$

Let $\kappa > \frac{1}{2}$ and $\beta = \beta(\kappa) > 1$ be as chosen in the previous section. We consider the pair of functions (H, h) defined as the continuous solution to the system of differential-difference equations

$$\begin{aligned} (s^{\kappa+1}H(s))' &= -\kappa s^{\kappa}h(s-1) &\quad& \text{if } s > \beta+1,\\ (s^{\kappa+1}h(s))' &= -\kappa s^{\kappa}H(s-1) && \text{if } s > \beta, \end{aligned} \tag{11.77}$$

with the initial conditions

$$\begin{aligned} s^{\kappa+1}H(s) &= (\beta-1)^{\kappa} &\quad& \text{if } \beta-1 \leqslant s \leqslant \beta+1,\\ s^{\kappa+1}h(s) &= \beta^{\kappa} && \text{if } s = \beta. \end{aligned} \tag{11.78}$$

We shall prove that these functions majorize the partial sums (11.31) for n odd and even respectively, and this shows the convergence. The plan is simple, yet its realization requires some skill. There will be no room for waste in the induction inequalities. Somewhere we have to appeal to our choice of β as the largest root of (11.55), and we do it neatly by employing again the method of adjoint equations. This time, however, our targets are reversed, we construct the inner products to deduce conditions for H, h at infinity from the initial values (11.78). We should like to say that it is a gift of nature that the initial values $(\beta-1)^\kappa, \beta^\kappa$ in (11.78) seem simple. Actually, they are carefully chosen (uniquely up to a constant normalizing factor) so that the solutions $H(s), h(s)$ have exponential decay at infinity, or even better, as in Corollary 11.4.

As before with (11.34), we turn the system (11.77) into two independent equations by forming the linear combinations

$$\begin{aligned} U(s) &= H(s) + h(s), \\ V(s) &= H(s) - h(s), \end{aligned} \tag{11.79}$$

for any $s \geqslant \beta$. These satisfy

$$\begin{aligned} (sU(s))' &= -\kappa U(s) - \kappa U(s-1), \\ (sV(s))' &= -\kappa V(s) + \kappa V(s-1), \end{aligned} \tag{11.80}$$

for $s > \beta + 1$, and the initial conditions for $\beta \leqslant s \leqslant \beta + 1$ are

$$\begin{aligned} s^{\kappa+1}U(s) &= (\beta-1)^\kappa + \beta^\kappa + (\beta-1)^\kappa \int_\beta^s t^\kappa d(t-1)^{-\kappa}, \\ s^{\kappa+1}V(s) &= (\beta-1)^\kappa - \beta^\kappa - (\beta-1)^\kappa \int_\beta^s t^\kappa d(t-1)^{-\kappa}. \end{aligned} \tag{11.81}$$

The adjoint equations are

$$\begin{aligned} su'(s) &= \kappa u(s) + \kappa u(s+1), \\ sv'(s) &= \kappa v(s) - \kappa v(s+1). \end{aligned} \tag{11.82}$$

Equivalently,

$$\begin{aligned} (s^{-\kappa}u(s))' &= \kappa s^{-\kappa-1}u(s+1), \\ (s^{-\kappa}v(s))' &= -\kappa s^{-\kappa-1}v(s+1). \end{aligned} \tag{11.83}$$

Note that the system (11.82) is related to (11.41) by differentiation of the latter.

In what follows we use the standard solutions $u(s)$, $v(s)$ which are derived from (B.9) with $(a,b) = (\kappa+1,\kappa)$ and $(a,b) = (\kappa+1,-\kappa)$, respectively. These satisfy

$$\begin{aligned} u'(s) = 2\kappa q(s)\ , \qquad u(s) &\sim s^{2\kappa}, \\ v(s) &= 1\ . \end{aligned} \tag{11.84}$$

By (11.55) we have $u'(\beta-1) = 0$ whence, by (11.82), we have $u(\beta-1) = -u(\beta)$. Using this fact we shall show that the inner product (see (B.5))

$$\langle U, u\rangle = su(s)U(s) - \kappa \int_{s-1}^s u(x+1)U(x)dx$$

vanishes. Indeed, from the general theory, $\langle U, u\rangle$ is constant for all $s \geqslant \beta+1$. By the other expression (see (B.6)),

$$\langle U, u\rangle = (s-1)u(s-1)U(s-1) + \int_{s-1}^s x^{-\kappa}u(x)dx^{\kappa+1}U(x)$$

at $s = \beta + 1$ we infer by inserting the initial conditions (11.81)

$$\langle U, u\rangle = \beta u(\beta)U(\beta) + (\beta - 1)^\kappa \int_\beta^{\beta+1} u(x)d(x-1)^{-k}.$$

Using (11.83) we get

$$\langle U, u\rangle = \beta u(\beta)U(\beta) + (\beta - 1)^\kappa[(\beta-1)^{-\kappa}u(\beta-1) - \beta^{-\kappa}u(\beta)] = 0$$

as claimed. In other words, we have for $s \geqslant \beta + 1$,

$$su(s)U(s) = \kappa \int_{s-1}^{s} u(x+1)U(x)dx, \tag{11.85}$$

Similarly, one shows that $\langle V, v\rangle$ vanishes. In other words, we have have $s \geqslant \beta + 1$,

$$sV(s) = -\kappa \int_{s-1}^{s} V(x)dx. \tag{11.86}$$

Of course, this can also be checked directly using the initial conditions (11.81).

The next result, which is rather delicate, appeals to the fact that $\beta - 1$ is the largest root of the adjoint function $q(s)$. A crude approach using an asymptotic evaluation of β would not succeed.

LEMMA 11.5. *Let*

$$\gamma = \left(\frac{\beta}{\beta-1}\right)^\kappa + \int_\beta^{\beta+1} t^\kappa d(t-1)^{-\kappa}\ . \tag{11.87}$$

Then we have $0 < \gamma < 1$.

PROOF. The upper bound is obvious (replace t^κ by β^κ). For the proof of the lower bound we need to show that

$$-\int_\beta^{\beta+1} t^\kappa d(t-1)^{-\kappa} < \left(\frac{\beta}{\beta-1}\right)^\kappa\ . \tag{11.88}$$

To do this we make use of the properties of $u(s)$ already established. First note that $u(s)$ is increasing for $s > \beta - 1$ because $u'(s) = 2\kappa q(s) > 2\kappa q(\beta-1) = 0$. Hence, the function

$$\delta(t) = \frac{1}{2}\left(\frac{\beta}{t}\right)^\kappa \left(1 + \frac{u(t)}{u(\beta)}\right)$$

is increasing for $t \geqslant \beta$ because

$$\delta'(t) = \frac{\kappa}{2t}\left(\frac{\beta}{t}\right)^\kappa \left(\frac{u(t+1)}{u(\beta)} - 1\right) > 0.$$

Therefore, $\delta(t) \geqslant \delta(\beta) = 1$ for $t \geqslant \beta$ and so the left-hand side of (11.88) is strictly bounded by

$$-\int_\beta^{\beta+1} \delta(t)t^\kappa d(t-1)^{-\kappa} = -\frac{\beta^\kappa}{2}\int_{\beta-1}^{\beta}\left(1 + \frac{u(s+1)}{u(\beta)}\right) ds^{-\kappa}.$$

Integrating the first of the two equations (11.83) over $\beta - 1 < s < \beta$ one gets from the above:

$$\frac{1}{2}\left(\frac{\beta}{\beta-1}\right)^\kappa \left(1 - \frac{u(\beta-1)}{u(\beta)}\right) = \left(\frac{\beta}{\beta-1}\right)^\kappa,$$

which is exactly the right-hand side of (11.88). □

Note that for $\beta \leqslant s \leqslant \beta + 1$ we have

$$s^{\kappa+1}h(s) \geqslant (\beta+1)^{\kappa+1}h(\beta+1) = \gamma(\beta-1)^{\kappa} > 0,$$

so $h(s)$ is positive in the initial segment.

LEMMA 11.6. *There exists a constant* $0 < \eta < 1$, *which depends on* κ, *such that*

$$|V(s)| < \eta U(s) \qquad \text{if } s \geqslant \beta. \tag{11.89}$$

PROOF. By the initial conditions (11.78) we get

$$\frac{(\beta-1)^{\kappa} - \beta^{\kappa}}{(\beta-1)^{\kappa} + \beta^{\kappa}} \leqslant \frac{V(s)}{U(s)} \leqslant \frac{1-\gamma}{1+\gamma}$$

in $\beta \leqslant s \leqslant \beta + 1$ so (11.89) is true in the initial segment. We shall prove that (11.89) holds true for all $s \geqslant \beta + 1$ with the same constant η. If it failed there would exist $s > \beta + 1$ such that $|V(t)| < \eta U(t)$ for all $t < s$ and $|V(s)| = \eta V(s)$. By (11.86) and (11.85) we derive

$$\begin{aligned} s|V(s)| &\leqslant \kappa \int_{s-1}^{s} |V(t)|dt < \eta\kappa \int_{s-1}^{s} U(t)dt \\ &< \frac{\eta\kappa}{u(s)} \int_{s-1}^{s} U(t)u(t+1)dt = \eta s U(s), \end{aligned}$$

which is the desired contradiction. □

The immediate consequences of Lemma 11.6 are the following bounds:

$$\tfrac{1}{2}(1-\eta)U(s) < H(s), h(s) < \tfrac{1}{2}(1+\eta)U(s) \ , \tag{11.90}$$

which show that both functions $H(s)$, $h(s)$ are positive in the whole range $s \geqslant \beta$ and have the same order of magnitude as $U(s)$.

The function $U(s)$ decays to zero exponentially, as do $V(s)$, $H(s)$, $h(s)$. Indeed, by (11.82) we have

$$\kappa \int_{s-1}^{s} U(x)dx = \big(s + O(1)\big)U(s) \quad \text{if } s \geqslant \beta + 1 \tag{11.91}$$

because $u(s) > 0$, $u(s) = s^{2\kappa}\left\{1 + O(s^{-1})\right\}$ if $s \geqslant \beta$. By Lemma 11.2,

$$U(s) \ll s^{-s} \ . \tag{11.92}$$

By (11.77) it follows that $s^{\kappa+1}H(s)$, $s^{\kappa+1}h(s)$ are decreasing, as are $H(s), h(s)$ and $U(s) = H(s) + h(s)$. Hence (11.91) implies

$$sU(s) \ll U(s-1) \ . \tag{11.93}$$

We also see by (11.91) that $U\big(s - \frac{1}{2}\big) \ll sU(s)$ and combining this estimate with that for $s - \frac{1}{2}$ we get

$$U(s-1) \ll s^2 U(s) \ . \tag{11.94}$$

The estimates (11.92)–(11.94) are not perfect, but they are sufficient for most of our applications. One can prove by elementary arguments that

$$U(s-1) \asymp U(s) s \log s \ , \quad \text{if } s \geqslant \beta + 1 \ . \tag{11.95}$$

We shall show only the one-sided bound

$$U(s-1) \gg U(s) s \log s \ , \quad \text{if } s \geqslant \beta + 1 \ , \tag{11.96}$$

because this is all that we shall need in our forthcoming arguments. We begin by means of the following:

LEMMA 11.7. *There exists a constant $\nu > 0$, which depends on κ, such that the function $(\nu s)^s U(s)$ is decreasing for $s \geqslant \beta + 1$.*

PROOF. Let $r(s) = (\nu s)^s$ so $R(s) = r(s)U(s)$ is the function in question. Suppose $s \geqslant \beta + 1$ is the first point at which $R(s)$ is not decreasing. We can assume that s is arbitrarily large by taking ν sufficiently small. At this critical point s we have $R'(s) = 0$ which yields

$$\frac{r'}{r}(s) = -\frac{U'}{U}(s) = \frac{\kappa U(s-1)}{sU(s)}\Big(1 + O(1/s)\Big) \gg 1 \tag{11.97}$$

by (11.80) and (11.93). Hence,

$$sU(s)\frac{r'}{r}(s) = \kappa U(s-1)\Big(1 + O(1/s)\Big) . \tag{11.98}$$

Note that, for all x with $s - 1 \leqslant x \leqslant s$, we have

$$\frac{r'}{r}(s) = \log \nu e s = \Big(1 + O(1/s)\Big)\frac{r'}{r}(x) .$$

On the other hand, (11.91) implies

$$\begin{aligned} s\, U(s)\frac{r'}{r}(s) &= \kappa\Big(1 + O(1/s)\Big)\frac{r'}{r}(s)\int_{s-1}^{s}\frac{R(x)}{r(x)}\,dx \\ &\leqslant \kappa\Big(1 + O(1/s)\Big)\frac{r'}{r}(s)R(s-1)\int_{s-1}^{s}\frac{dx}{r(x)} \\ &= \kappa\Big(1 + O(1/s)\Big)R(s-1)\int_{s-1}^{s}\frac{r'(x)}{r(x)^2}\,dx \\ &= \kappa\Big(1 + O(1/s)\Big)U(s-1)\Big(1 - \frac{r(s-1)}{r(s)}\Big) \\ &= \kappa U(s-1)\Big(1 - \frac{1}{\nu e s}\Big)\Big(1 + O(1/s)\Big) . \end{aligned}$$

This contradicts (11.98), because ν is small and s is large. □

Now (11.96) follows by the expression

$$\frac{R'}{R}(s) = \log(\nu e s) - \frac{\kappa U(s-1)}{sU(s)}\Big(1 + O(1/s)\Big) ,$$

which is negative by Lemma 11.7.

Using (11.90), (11.94) and (11.95) we conclude that, for $s \geqslant \beta + 1$,

$$H(s)s\log s \ll h(s-1) \ll s^2 H(s) , \tag{11.99}$$

$$h(s)s\log s \ll H(s-1) \ll s^2 h(s) . \tag{11.100}$$

11.5. Connections with the Buchstab and Dickman Functions

We do not need the results of this section for the theory of the beta-sieve; nevertheless it is worthwhile to point out that, in the case of $\kappa = 1$, some of our functions are related to classical ones. Let $\kappa > \frac{1}{2}$. The functions $U(s), V(s)$ are derivatives of $Q(s), P(s)$ respectively, up to a normalization factor; precisely we have

$$Q'(s) = -CU(s), \tag{11.101}$$

$$P'(s) = -CV(s), \tag{11.102}$$

if $s > \beta$, where the constant C is given by

$$C = \kappa(\beta - 1)^{-\kappa} A = 2\kappa/(\beta - 1)p(\beta - 1)\ . \tag{11.103}$$

Indeed, $Q'(s)$ and $U(s)$ satisfy the same differential-difference equation (see (11.38) and (11.80)) for $s > \beta + 1$. In the initial segment $\beta \leqslant s \leqslant \beta + 1$ we have

$$s^\kappa Q(s) = A - A\kappa \int_\beta^s (t-1)^{-\kappa} t^{\kappa-1} dt,$$

$$s^{\kappa+1} U(s) = (\beta - 1)^\kappa + \beta^\kappa - (\beta - 1)^\kappa \int_\beta^s t^\kappa (t-1)^{-\kappa-1} dt$$

(see (11.39) and (11.81)). Hence it is easy to check that (11.101) holds. Similarly, one checks (11.102).

Now we assume $\kappa = 1$. In this case $\beta = 2$, $A = 2e^\gamma$ and (11.101) holds with $C = A$. We have

$$sQ(s) = A - A\log(s-1)\ , \quad \text{if } 2 \leqslant s \leqslant 3\ ,$$

$$sQ'(s) = -Q(s) - Q(s-1)\ , \quad \text{if } s > 3\ .$$

Hence one can verify that

$$Q(s) = A\rho'(s)\ , \quad \text{if } s \geqslant 2 \tag{11.104}$$

where $\rho(s)$ is the well-known Dickman function which is defined for $s > 0$ as the continuous solution to

$$\begin{aligned} \rho(s) &= 1\ , && \text{if } 0 < s \leqslant 1\ , \\ s\rho'(s) &= -\rho(s-1)\ , && \text{if } s > 1\ . \end{aligned} \tag{11.105}$$

Indeed, for $1 \leqslant s \leqslant 2$ we have $\rho(s) = 1 - \log s$, whence for $2 \leqslant s \leqslant 3$ we get $s\rho'(s) = -1 + \log(s-1)$ which agrees with (11.104). For $s > 3$ both functions $\rho'(s)$ and $Q(s)$ satisfy the same differential-difference equation; therefore (11.104) holds for all $s \geqslant 2$ by continuity.

The case of $P(s)$ is similar. We have

$$sP(s) = A + A\log(s-1)\ , \quad \text{if } 2 \leqslant s \leqslant 3\ ,$$

$$sP'(s) = -P(s) + P(s-1)\ , \quad \text{if } s > 2\ .$$

Hence one can verify that

$$P(s) = A\omega(s) \quad \text{if } s \geqslant 2 \tag{11.106}$$

where $\omega(s)$ is the well-known Buchstab function defined in (12.19). We shall return to both $\rho(s)$ and $\omega(s)$ in Section 12.2 when we discuss the optimality of the linear sieve.

The Dickman function was studied intensively, using the Laplace transform method, by N.G. de Bruijn [**30**]. Since $\rho(s) \ll s^{-s}$ (see Lemma 11.2) the Laplace transform

$$L(z) = \int_0^\infty e^{-sz}\rho(s)ds$$

is entire, and $zL(z) \to 1$ as $z \to \infty$. We have

$$\begin{aligned} zL'(z) &= \int_0^\infty e^{-sz}\big(s\rho(s)\big)'ds \\ &= -\int_0^\infty e^{-sz}\big(\rho(s) - \rho(s-1)\big)ds = \big(e^{-z} - 1\big)L(z) \end{aligned}$$

where we put $\rho(s) = 0$ if $s < 0$, so $L(z)$ satisfies the ordinary differential equation

$$\big(zL(z)\big)' = e^{-z}L(z)\ .$$

Writing this in the form $\big(\log zL(z)\big)' = e^{-z}z^{-1}$ we find

$$zL(z) = \exp\big(-\mathcal{E}(z)\big)$$

where

$$\mathcal{E}(z) = \int_z^\infty e^{-t}t^{-1}dt = \int_0^z (1 - e^{-t})t^{-1}dt - \log z - \gamma$$

(see (B.15)). Hence the Laplace transform of the Dickman function is given by

$$L(z) = \exp\Big(\gamma + \int_0^s (e^{-t} - 1)t^{-1}dt\Big)\ .$$

By Laplace inversion we conclude that

$$e^{-\gamma}\rho(s) = \frac{1}{2\pi i}\int_{(\lambda)} \exp\Big(-sz + \int_0^z (e^t - 1)t^{-1}dt\Big)dz \tag{11.107}$$

where the integration is over the vertical line $\operatorname{Re} z = \lambda$ and λ is any real number. We choose the line of steepest descent, which is given by the root of the equation

$$\frac{e^\lambda - 1}{\lambda} = s\ .$$

Writing $z = \lambda + iy$ this gives

$$\rho(s) = R(s)\exp\Big(\gamma - \lambda s + \int_0^\lambda (e^t - 1)t^{-1}dt\Big)$$

where

$$R(s) = \frac{1}{2\pi}\int_{-\infty}^\infty \exp\left(-isy + \int_0^y \frac{e^{\lambda+ix} - 1}{\lambda + ix}dx\right)dy\ .$$

One can show by lengthy estimations that

$$R(s) \sim \big(2\pi s\big)^{-\frac12}\ , \qquad \text{as } s \to \infty\ .$$

Hint: The main contribution to $R(s)$ comes from integration over a small segment centred at $y = 0$ in which range the Taylor expansion simplifies the matter.

PROPOSITION 11.8 (de Bruijn). *As $s \to \infty$ we have*

$$\rho(s) \sim (2\pi s)^{-\frac{1}{2}} \exp\Big(\gamma - \lambda s + \int_0^\lambda (e^t - 1)t^{-1}dt\Big) \tag{11.108}$$

where $\lambda = \lambda(s) \sim \log s$ is the function given by $e^{\lambda(s)} = 1 + s\lambda(s)$.

The asymptotic formula (11.108) easily implies

$$\rho(s-1) \sim \rho(s) s \log s \ . \tag{11.109}$$

Then, the differential-difference equation

$$s\rho^{(n+1)}(s) = -n\rho^{(n)}(s) - \rho^{(n)}(s-1) \ ,$$

which is obtained by differentiating (11.105) n times, implies by induction on n that

$$\rho^{(n)}(s) \sim \rho(s)\big(-\log s\big)^n \ . \tag{11.110}$$

Applying (11.110) to (11.104) we get

$$Q^{(n)}(s) \sim Q(s)\big(-\log s\big)^n \ . \tag{11.111}$$

Next, applying (11.111) to (11.101) we get

$$U^{(n)}(s) \sim U(s)\big(-\log s\big)^n \ . \tag{11.112}$$

Applying (11.111) for $n = 1$ to the differential-difference equation (11.38) with $\kappa = 1$ we get

$$Q(s-1) \sim Q(s) s \log s \ . \tag{11.113}$$

Applying (11.112) for $n = 1$ to the differential-difference equation (11.80) with $\kappa = 1$ we get

$$U(s-1) \sim U(s) s \log s \ . \tag{11.114}$$

This asymptotic, derived here for $\kappa = 1$ by the Laplace transform technique, is more precise than our lower bound (11.96) which we derived by elementary arguments for any $\kappa > \frac{1}{2}$.

11.6. The Convergence Problem for $\kappa > \frac{1}{2}$

As planned, we now proceed to the estimation of the partial sums of $F(s)$, $f(s)$, given by (11.31).

LEMMA 11.9. *There exists a constant $\mu > 0$ depending on κ, such that*

$$T_N(s) \leqslant \mu H(s) \quad \text{if } N \text{ is odd, } s \geqslant \beta - 1 \ , \tag{11.115}$$

$$T_N(s) \leqslant \mu h(s) \quad \text{if } N \text{ is even, } s \geqslant \beta \ . \tag{11.116}$$

PROOF. By induction in N. If $N = 1$ we have for $\beta - 1 \leqslant s \leqslant \beta + 1$,

$$s^\kappa T_1(s) = (\beta+1)^\kappa - s^\kappa < (\beta+1)^\kappa \leqslant (\beta+1)^{\kappa+1}(\beta-1)^{-\kappa} s^\kappa H(s).$$

Now suppose $N \geqslant 2$ and that the result holds for $N - 1$. If N is even we get by the recurrence formula (11.28) and (11.115),

$$s^\kappa T_N(s) = \int_s^\infty T_{N-1}(t-1)dt^\kappa \leqslant \frac{\mu}{s}\int_s^\infty H(t-1)\kappa t^\kappa dt = \mu s^\kappa h(s)$$

by (11.77). The same argument works for N odd if $s \geqslant \beta + 1$, just interchange H and h. If $\beta - 1 \leqslant s \leqslant \beta + 1$ we get along similar lines,

$$\begin{aligned} s^\kappa T_N(s) &= (\beta+1)^\kappa T_N(\beta+1) + s^\kappa f_1(s) \leqslant \mu(\beta+1)^\kappa H(\beta+1) + (\beta+1)^\kappa - s^\kappa \\ &\leqslant \frac{\mu}{\beta+1}(\beta-1)^\kappa + (\beta+1)^\kappa - s^\kappa \leqslant \frac{\mu}{s}(\beta-1)^\kappa = \mu s^\kappa H(s) \end{aligned}$$

by (11.78), provided μ is sufficiently large. □

By (11.115), (11.116) and (11.99), we deduce that

$$T_N(s-1) \ll s^2 H(s)\,. \tag{11.117}$$

By Lemma 11.9 we conclude the proof that, in the case $\kappa > \frac{1}{2}$, the series (11.32) and (11.33) converge and that the resulting functions $F(s)$, $f(s)$ have all the properties claimed previously in Section 11.3.

REMARKS. It remains to treat the convergence problem also for $0 \leqslant \kappa \leqslant \frac{1}{2}$. For $\kappa = \frac{1}{2}$ the above arguments could be modified to yield the convergence by continuity as $\kappa \to \frac{1}{2}+$. If $0 \leqslant \kappa < \frac{1}{2}$ the convergence follows from that for $\kappa = \frac{1}{2}$ by monotonicity since $\beta = \beta(\kappa) = 1$ throughout that range (so that the beta-sieve does not depend on κ there). Nevertheless, we shall treat all $0 \leqslant \kappa \leqslant \frac{1}{2}$ from scratch in the next section, proving a few estimates which will be used in later chapters.

11.7. Estimation of $f_n(s)$ for $0 \leqslant \kappa \leqslant \frac{1}{2}$

Our next goal is the solution of the convergence problem in the remaining range $0 \leqslant \kappa \leqslant \frac{1}{2}$, that is, we need to prove the convergence of the series (11.32) and (11.33). We shall do this by majorizing the functions $f_n(s)$ by a geometric series. Now, cruder estimates are sufficient because, when κ is as small as it is here, the convergence is not as delicate an issue. We need the following technical lemma.

LEMMA 11.10. *Let $0 \leqslant \kappa \leqslant \frac{1}{2}$. There exists a function $E(s) > 0$ for $s \geqslant 0$, which satisfies*

$$E(s) \asymp e^{-s} \quad \text{for } s > 0, \tag{11.118}$$

$$E(2) \leqslant E(s) \quad \text{for } 0 \leqslant s \leqslant 2, \tag{11.119}$$

and

$$\int_s^\infty E(t-1)\Big(1-\frac{1}{t}\Big)^{-\frac{1}{6}} \frac{dt^\kappa}{(t-1)^\kappa} \leqslant b\,E(s), \quad \text{for } s \geqslant 1, \tag{11.120}$$

where b is an absolute constant, $0 < b < 1$.

PROOF. We consider the discontinuous but quite simple function defined by

$$E(s) = \begin{cases} e^{-s} & \text{if } s \geqslant 1, \\ e^{-2} & \text{if } 0 \leqslant s < 1. \end{cases} \tag{11.121}$$

Here, the conditions (11.118) and (11.119) are obvious, while (11.120) becomes

$$\kappa e^s \int_s^\infty E(t-1)\Big(1-\frac{1}{t}\Big)^{-\kappa-\frac{1}{6}} t^{-1} dt \leqslant b \tag{11.122}$$

for $s \geqslant 1$.

The left side is increasing in κ so it suffices to prove (11.122) for $\kappa = \frac{1}{2}$, that is,

$$I(s) = \frac{e^{s+1}}{2} \int_s^\infty E(t) t^{-\frac{2}{3}} (t+1)^{-\frac{1}{3}} dt \leqslant b$$

for $s \geqslant 0$. Since $I(s) \to 0$ as $s \to \infty$, it suffices to prove the strict inequality

$$I(s) < 1 \quad \text{for } s \geqslant 0. \tag{11.123}$$

First note that, for $s \geqslant 1$ we have the inequality

$$\begin{aligned} \int_s^\infty e^{-t} t^{-1} dt &= -e^{-s} \log s + \int_s^\infty e^{-t} \log t \, dt \\ &< -e^{-s} \log s + \int_s^\infty e^{-t} \Big(\log(t+1) - \frac{1}{t+1} \Big) dt = e^{-s} \log \Big(1 + \frac{1}{s} \Big) . \end{aligned}$$

Hence, for $s \geqslant 1$ we get

$$I(s) < \frac{e^{s+1}}{2} \int_s^\infty E(t) t^{-1} dt < \frac{e}{2} \log \frac{s+1}{s} \leqslant \frac{e}{2} \log 2 < 1 .$$

It remains to verify (11.123) for $0 \leqslant s \leqslant 1$. In this range we use the formula

$$\begin{aligned} I(s) &= \frac{e^{s-1}}{2} \int_s^1 t^{-\frac{2}{3}} (t+1)^{-\frac{1}{3}} dt + e^{s-1} I(1) \\ &< \frac{e^s}{2} \left[\frac{3}{e} \Big(1 - s^{\frac{1}{3}} \Big) + \log 2 \right] = J(s), \quad \text{say.} \end{aligned}$$

For $s = 0$ we have

$$J(0) = \frac{3}{2e} + \frac{\log 2}{2} < \frac{e}{3} .$$

Hence,

$$J(s) \leqslant J(0) e^s < \frac{e^{s+1}}{3} \leqslant 1 ,$$

if $s \leqslant \log \frac{3}{e}$. Now, suppose that $J(s)$ has its maximum attained at a point s with $\log \frac{3}{e} \leqslant s \leqslant 1$. Since

$$J'(s) = J(s) - \tfrac{1}{2} e^{s-1} s^{-\frac{2}{3}} = 0 ,$$

we find that

$$J(s) = \frac{1}{2} e^{s-1} s^{-\frac{2}{3}} \leqslant \frac{3}{2e^2} \Big(\log \frac{3}{e} \Big)^{-\frac{2}{3}} < 1 .$$

This completes the proof. □

We now proceed to bound the functions $f_n(s)$ in terms of the auxiliary function $H(s)$ just introduced.

LEMMA 11.11. *For $n \geqslant 1$ and $s \geqslant \frac{1}{2}(1 + (-1)^n)$ we have*

$$s^\kappa f_n(s) \leqslant c b^n E(s) , \tag{11.124}$$

where c is an absolute constant and b is the absolute constant appearing in (11.120).

PROOF. By induction of n. For $n = 1$ the assertion is obvious from (11.21) and (11.118) by choosing c sufficiently large.

Assuming (11.124) holds for $n-1$, we shall prove that it holds for n. First, for $s \geqslant \frac{1}{2}(3-(-1)^n)$ we use the recurrence formula (11.28), obtaining the bound

$$\begin{aligned} s^\kappa f_n(s) &\leqslant cb^{n-1}\int_s^\infty E(t-1)\frac{dt^\kappa}{(t-1)^\kappa} \\ &\leqslant cb^{n-1}\int_s^\infty E(t-1)\Big(1-\frac{1}{t}\Big)^{-\frac{1}{6}}\frac{dt^\kappa}{(t-1)^\kappa} \leqslant cb^n E(s) \end{aligned}$$

by (11.120), which completes the proof of the induction step. There still remains the case of n being odd and $0 < s < 2$, but for this we have

$$s^\kappa f_n(s) = 2^\kappa f_n(2) \leqslant cb^n E(2) \leqslant cb^n E(s),$$

completing the proof. □

This also completes the proof of the convergence of the series (11.32) and (11.33).

11.8. Statement of the Main Theorems of the Beta-Sieve

We have now introduced all definitions and concepts necessary to present the main statements of the beta-sieve.

Let $\kappa \geqslant 0$ and $\beta = \beta(\kappa) \geqslant 1$ as given in the previous sections. Let $\Lambda^+(D), \Lambda^-(D)$ be the beta-sieves of level $D \geqslant 2$ and

$$V^\pm(D,z) = \sum_{d|P(z)} \lambda_d^\pm g(d)\ . \tag{11.125}$$

We have (see(6.31), (6.32) of Chapter 6)

$$V^+(D,z) = V(z) + \sum_{n\ \mathrm{odd}} V_n(D,z), \tag{11.126}$$

$$V^-(D,z) = V(z) - \sum_{n\ \mathrm{even}} V_n(D,z), \tag{11.127}$$

where

$$V(z) = \prod_{p<z}(1-g(p)) \tag{11.128}$$

and $V_n(D,z)$ is given by (11.19). As we expect $V_n(D,z)$ are well-approximated by $f_n(s)V(z)$, it should not be surprising that the sums $V^+(D,z)$, $V^-(D,z)$ are well-approximated by $F(s)V(z)$, $f(s)V(z)$ respectively. Indeed, the main theorems of the beta-sieve confirm this, giving also a reasonable bound for the error term in this approximation.

Before stating these we recall that $g(d)$ is a multiplicative function such that

$$0 \leqslant g(p) < 1$$

and we assume throughout that

$$\prod_{\substack{w\leqslant p<z\\ p\in\mathcal{P}}}(1-g(p))^{-1} \leqslant \Big(\frac{\log z}{\log w}\Big)^\kappa\Big(1+\frac{L}{\log w}\Big) \tag{11.129}$$

for every $2 \leqslant w < z$, where $L \geqslant 1$ is constant.

THEOREM 11.12. *For every $\kappa \geqslant 0$ we have*

$$V^+(D,z) \leqslant \left\{F(s) + O((\log D)^{-\frac{1}{6}})\right\} V(z) \qquad \text{if } s \geqslant \beta - 1\ , \tag{11.130}$$

$$V^-(D,z) \geqslant \left\{f(s) + O((\log D)^{-\frac{1}{6}})\right\} V(z) \qquad \text{if } s \geqslant \beta\ , \tag{11.131}$$

where $s = \log D/\log z$ and $F(s), f(s)$ are the continuous solutions to the following system of differential-difference equations

$$\begin{cases} s^\kappa F(s) = A & \text{if } \beta - 1 \leqslant s \leqslant \beta + 1\ , \\ s^\kappa f(s) = B & \text{at } s = \beta\ , \end{cases} \tag{11.132}$$

$$\begin{cases} (s^\kappa F(s))' = \kappa s^{\kappa-1} f(s-1) & \text{if } s > \beta - 1\ , \\ (s^\kappa f(s))' = \kappa s^{\kappa-1} F(s-1) & \text{if } s > \beta\ , \end{cases} \tag{11.133}$$

and $\beta = \beta(\kappa)$, $A = A(\kappa)$ and $B = B(\kappa)$ are constants defined in (11.55)–(11.63). *The implied constant depends (continuously) on κ and on L.*

It should be mentioned that the upper and lower bounds for $V^+(D,z), V^-(D,z)$ in the theorem actually represent asymptotic values (as $D \to \infty$), but we are not able to prove it because our hypothesis (11.129) provides only a one-sided bound for the density function g.

On the other hand, such an inequality offers a great advantage because it does not require great regularity in the function $g(p)$; for instance, one can use the result even when $g(p)$ has lots of gaps and clusters. It also makes it possible to employ the result for a sieve problem where the natural dimension is less than κ and this has practical uses.

The functions $F(s)$, $f(s)$, in the upper and lower bounds merge to one as $s \to \infty$, at exponential speed, as also was the case for the Fundamental Lemma in Section 6.5. In fact, one can show the following quite precise formulas:

$$\begin{aligned} F(s) &= 1 + \exp\left\{-s\log s - s\log\log s + s\log e\kappa + O\left(\frac{s\log\log s}{\log s}\right)\right\}\ , \\ f(s) &= 1 - \exp\left\{-s\log s - s\log\log s + s\log e\kappa + O\left(\frac{s\log\log s}{\log s}\right)\right\}\ , \end{aligned} \tag{11.134}$$

for $s \geqslant \beta + 2$, where the implied constant depends on κ.

However, the greatest interest in Theorem 11.12 is for smaller s and now the values of the constants A and B matter.

If $\kappa \geqslant \frac{1}{2}$, we have $B = 0$ so that it is no surprise that $\beta(\kappa)$ is called the sieving limit since $f(s)$ is positive precisely for $s > \beta$. In this range the constant A is given by (11.57). For particular values we get

$$\begin{aligned} A &= 2e^\gamma && \text{when } \kappa = 1\ , \\ A &= 2\sqrt{e^\gamma/\pi} && \text{when } \kappa = \tfrac{1}{2}\ . \end{aligned}$$

For $\kappa < \frac{1}{2}$, B is positive so our lower bound (11.131) is positive for all $s \geqslant 1$, that is, for all $z \leqslant D$, provided that D is sufficiently large. For $\kappa = 0$ we have $F(s) = f(s) = 1$.

From Theorem 11.12 we deduce the corresponding estimates for the sifted sums and hence for the sifting function.

THEOREM 11.13. *Let $\mathcal{A} = (a_n)$ be a sequence of non-negative reals and let r_d be defined by*

$$|\mathcal{A}_d| = g(d)X + r_d\ ,$$

where $g(d)$ satisfies (11.129) with $\kappa \geqslant 0$ and $L \geqslant 1$. Put

$$R^{\pm}(D,z) = \sum_{d|P(z)} \lambda_d^{\pm} r_d\ .$$

Then, for $s \geqslant \beta - 1$ and $s \geqslant \beta$ we have

$$S(\mathcal{A},z) \leqslant S^{+}(\mathcal{A},z) \leqslant XV(z)\Big\{F(s) + O\big((\log D)^{-\frac{1}{6}}\big)\Big\} + R^{+}(D,z),$$

$$S(\mathcal{A},z) \geqslant S^{-}(\mathcal{A},z) \geqslant XV(z)\Big\{f(s) + O\big((\log D)^{-\frac{1}{6}}\big)\Big\} + R^{-}(D,z),$$

respectively, where the implied constant depends on κ and L.

Note that $R^{\pm}$ are bounded by

$$R(D,z) = \sum_{d|P(z)} |r_d|\ .$$

A. Granville and K. Soundararajan (Corollary 1.1 of [**71**]) showed that, as $s \to \infty$, the functions $F(s)$ and $f(s)$ in the above estimates for $S(\mathcal{A},z)$ are best possible up to the third term $s\log e\kappa$ in (11.134), interestingly, even if the sifting set $\mathcal{P}$ of primes is quite irregular.

REMARKS. After the arguments presented so far it remains only to give suitable upper bounds for the components $V_n(D,z)$. All proofs of this go by induction on n and are rather difficult.

Up to now our treatment of the beta-sieve applies for general dimension κ but, as indicated in Section 11.1, we now depart from this. We are going to give the last part of the proof for a restricted range of the dimension, which however, includes the cases required for the most important applications, and where we have found pleasing versions of the proof. In the next two sections we present two different arguments, one for small n and any $\kappa > \frac{1}{2}$, and one for large n but only when $\frac{1}{2} < \kappa \leqslant 1$. As a result, we complete the proofs of Theorem 11.12 and Theorem 11.13 in the range $\frac{1}{2} < \kappa \leqslant 1$. Our arguments for $0 \leqslant \kappa \leqslant \frac{1}{2}$ are of a different nature, so we postpone them to later sections of the chapter. In this way we complete the proofs of Theorem 11.12 and Theorem 11.13 for all $0 < \kappa \leqslant 1$.

For $\kappa > 1$ we have decided not to include complete proofs of Theorem 11.12 and Theorem 11.13. Complete proofs of these are presented in [**104**]. They are very complicated and we did not find a way to simplify them to our satisfaction. In place of those, we provide in Section 11.16 a proof of Theorem 11.22 which gives upper and lower bounds for the sifing function $S(\mathcal{A},z)$ of similar quality to those in Theorem 11.13 (but without giving bounds for the sifted sums). The advantage of the approach we use is its comparative simplicity; the disadvantage is the need to employ modifications of the beta-weights which render them less natural.

11.9. The Inductive Estimate for $V_n(D,z)$

Let $\kappa > \frac{1}{2}$. The argument we present here works for all $n \geqslant 1$, but the results fade in quality as n increases. Recall that we have defined two functions $H(s), h(s)$ in Section 11.4 which we are now going to employ in the inductive argument.

Although they were useful in the convergence argument, they are really designed for controlling an error term in the upper bounds for $V_n(D,z)$.

Recall that the density function g satisfies (11.129). Let $\beta = \beta(\kappa)$ be the largest zero of the adjoint function $q(s-1)$ and

$$\alpha_1 = \sqrt{\frac{\beta+1}{\beta}}\,, \quad \alpha_2 = \sqrt{\frac{\beta+2}{\beta+1}}\,, \quad \alpha = \alpha_1\alpha_2 = \sqrt{\frac{\beta+2}{\beta}}\,.$$

Because the parity of n matters in the inductive argument, we set $\alpha_n = \alpha_1$ if n is odd and $\alpha_n = \alpha_2$ if n is even. Moreover, we set $H_n(s) = H(s)$, $h(s)$ if n is odd, even, respectively. We have to keep careful track of the implied constant in the proof and to do so we require the following delicate but artificial-looking inequality.

PROPOSITION 11.14. *Let $\kappa > \frac{1}{2}$, $\beta = \beta(\kappa)$ and $D \geqslant z \geqslant 2$. Then, for $n \geqslant 1$ restricted by*

$$\alpha^n \leqslant \alpha_n \log Dz\,, \tag{11.135}$$

we have

$$V_n(D,z) \leqslant V(z)\left\{f_n(s) + C\frac{n}{n+1}H_n(s)\frac{(\log\log 3D)^5}{\log z}\right\} \tag{11.136}$$

for $s \geqslant \beta - 1$ if n is odd, $s \geqslant \beta$ if n is even. Here C is a large constant depending only on κ and L.

PROOF. First we show the result is obvious for small D by adjusting the constant C. To this end we choose a real number D_0 large in terms of L, as large as will be required in the induction argument. Irrespective of this number D_0 there exists a constant $C = C(L, D_0)$ such that (11.136) holds for all $D \leqslant D_0$ with this choice of C. Clearly, such a C exists, because the variables z, s, n are also then bounded in terms of D_0 while $H_n(s)$ is positive and continuous. In what follows we also assume, as we may, that $C = C(L, D_0)$ is large.

Now we are ready to start the induction. For $n = 1$ it follows by (11.25), (11.26) and (11.129) with $w = D^{1/(\beta+1)}$ that

$$V_1(D,z) \leqslant V(z)\left\{f_1(s) + (\beta+1)L/\log D\right\}, \tag{11.137}$$

which gives (11.136).

Let $n \geqslant 1$. We assume the theorem holds for n and we are going to prove it for $n+1$, that is,

$$V_{n+1}(D,z) \leqslant V(z)\left\{f_{n+1}(s) + C\frac{n+1}{n+2}H_{n+1}(s)\frac{(\log\log 3D)^5}{\log z}\right\} \tag{11.138}$$

for $s \geqslant \beta$ if n is odd, $s \geqslant \beta - 1$ if n is even, and

$$\alpha^{n+1} \leqslant \alpha_{n+1} \log Dz\,. \tag{11.139}$$

First we consider the range $s \geqslant \beta$ if n is odd, $s \geqslant \beta + 1$ if n is even. In this range we have

$$\log Dz = \frac{s+1}{s}\log D \leqslant \alpha_n^2 \log D\,.$$

Inserting this into (11.139) we obtain $\alpha^n \leqslant \alpha_n \log D$, which is the condition (11.135) with D, z replaced by D/p, p. Hence the recurrence formula (see (11.27))

$$V_{n+1}(D,z) = \sum_{D^{1/(\beta+1+n)} \leqslant p < D^{1/s}} g(p)V_n\left(\frac{D}{p}, p\right)$$

and the inductive hypothesis (11.136) apply, giving

$$V_{n+1}(D,z) \leqslant \sum_p g(p)V(p)\left\{f_n(s_p) + C\frac{n}{n+1}H_n(s_p)\frac{(\log\log 3D)^5}{\log p}\right\}$$

where p runs over primes in the segment $D^{1/(\beta+1+n)} \leqslant p < D^{1/s}$ and

$$s_p = \frac{\log D/p}{\log p} = \frac{\log D}{\log p} - 1\ . \tag{11.140}$$

Note that, as a function of p, the content of $\{\dots\}$ is increasing because $f_n(t-1)$ and $tH_n(t-1)$ are decreasing in t. Thus, we can apply Lemma 5.1 with

$$K = 1 + L\frac{\beta+1+n}{\log D}$$

showing that $V_{n+1}(D,z)/V(z)$ is bounded by

$$\left(1 + L\frac{\beta+1+n}{\log D}\right)\int_s^{\beta+1+n}\left\{f_n(t-1) + C\frac{n}{n+1}\,\frac{tH_n(t-1)}{s\log z}(\log\log 3D)^5\right\}d\left(\frac{t}{s}\right)^\kappa$$
$$+ L\frac{\beta+1+n}{\log D}\left\{f_n(s-1) + \frac{H_n(s-1)}{\log z}(\log\log 3D)^5\right\}\ .$$

Here we have

$$\int_s^{\beta+1+n} f_n(t-1)d\left(\frac{t}{s}\right)^\kappa = f_{n+1}(s)\ ,$$
$$\int_s^{\infty} tH_n(t-1)d\left(\frac{t}{s}\right)^\kappa = sH_{n+1}(s)\ ,$$

and $f_n(s-1) \ll s^2H_{n+1}(s)$, $H_n(s-1) \ll s^2H_{n+1}(s)$; see (11.117), (11.100). In fact, the above estimates hold with $s\log s$ in place of s^2, but we do not need such sharp results. Hence, by (11.136) we obtain

$$\frac{V_{n+1}(D,z)}{V(z)} \leqslant f_{n+1}(s) + C\frac{n}{n+1}\,\frac{H_{n+1}(s)}{\log z}\left\{(\log\log 3D)^5 + O(n^2)\right\}\ , \tag{11.141}$$

because $s \leqslant \beta+1+n \ll \log\log 3D$.

Recall that this estimate still needs to be established for n even and $\beta - 1 \leqslant s < \beta+1$. In this range we have by (11.141),

$$V_{n+1}(D,z) = V_{n+1}\left(D, D^{1(\beta+1)}\right) \leqslant V\left(D^{1(\beta+1)}\right)$$
$$\left[f_{n+1}(\beta+1) + C\frac{n}{n+1}H_{n+1}(\beta+1)\frac{\beta+1}{\log D}\left\{(\log\log 3D)^5 + O(n^2)\right\}\right]$$

and herein, we use the following:

$$V\left(D^{1/(\beta+1)}\right) \leqslant V(z)\left(\frac{\beta+1}{s}\right)^\kappa\left(1 + L\frac{\beta+1}{\log D}\right)\ ,$$
$$(\beta+1)^\kappa f_{n+1}(\beta+1) = s^\kappa f_{n+1}(s)\ ,$$
$$(\beta+1)^{\kappa+1}H_{n+1}(\beta+1) = s^{\kappa+1}H_{n+1}(s)\ .$$

From this we see that (11.141) holds in the extended range (possibly with a different implied constant). Since

$$\frac{n}{n+1} = \frac{n+1}{n+2}\left(1 - \frac{1}{(n+1)^2}\right)$$

and

$$\Big(1 - \frac{1}{(n+1)^2}\Big)\Big(1 + O\Big(\frac{n^2}{(\log\log 3D)^5}\Big)\Big) < 1 ,$$

this completes the proof of (11.138). □

Because of the sensitivity of the induction steps with respect to the implied constants we were obliged to carry somewhat artificial factors in the error terms. Now, the induction arguments being executed, we can formulate the result somewhat more simply.

COROLLARY 11.15. *Let $\kappa > \frac{1}{2}$, $\beta = \beta(\kappa)$ and $D \geqslant z \geqslant 2$. Then for*

$$1 \leqslant n \leqslant \frac{\log\log D}{\log \alpha} \tag{11.142}$$

we have

$$V_n(D,z) \leqslant V(z)\left\{f_n(s) + O\left(H(s)\frac{(\log\log 3D)^5}{\log z}\right)\right\} \tag{11.143}$$

for $s \geqslant \beta - 1$ if n is odd, $s \geqslant \beta$ if is n even, where the implied constant depends only on κ and L.

Recall that $s = \log D/\log z$ and $H(s)$ is defined in Section 11.4; it decays to zero exponentially as $s \to \infty$; specifically we have

$$H(s) \ll s^{-s-1} , \tag{11.144}$$

so the error term in (11.143) is $O\left(s^{-s}(\log\log 3D)^5(\log D)^{-1}\right)$. This is a pretty strong inequality, but it holds only for n restricted by (11.142). In the next section we establish a bound for $V_n(D,z)$ by direct arguments, which will be useful for large n.

11.10. Direct Estimation of $V_n(D,z)$

In this section we let $0 \leqslant \kappa \leqslant 1$. First we establish some general estimates for sums of the density function g. As before we assume the dimension condition (11.129). From this we derive for $z \geqslant w \geqslant 2$, by Lemma 5.1,

$$\begin{aligned}\sum_{w \leqslant p < z} g(p) \leqslant \log\left(\frac{V(w)}{V(z)}\right) &\leqslant \kappa \log\left(\frac{\log z}{\log w}\right) + \frac{L}{\log w}\\ &\leqslant \left(\kappa \log\frac{z}{w} + L\right)(\log w)^{-1} .\end{aligned} \tag{11.145}$$

Moreover, also by Lemma 5.1,

$$\sum_{p<z} g(p)V(p)\log p \leqslant 3LV(z)(\log z)\left(1 - \int_2^z \frac{\log w}{\log z} d\left(\frac{\log z}{\log w}\right)^{\kappa}\right) .$$

Here, we have

$$-\int_2^z \frac{\log w}{\log z} d\left(\frac{\log z}{\log w}\right)^{\kappa} = \frac{\kappa}{1-\kappa}\left[1 - \left(\frac{\log 2}{\log z}\right)^{1-\kappa}\right] \ll \kappa \log\log 2z .$$

Hence,

$$\sum_{p<z} g(p)V(p)\log p \ll LV(z)(\log z)(1 + \kappa\log\log 2z) . \tag{11.146}$$

Now we are ready to estimate the sum $V_n(D,z)$; see (11.19) for the definition. For $n \geqslant 3$ we have (note that $z \geqslant 5$, or else the sum is void)

$$V_n(D,z) \leqslant \sum_{\substack{m=p_2\cdots p_{n-1} \\ 3\leqslant p_j<z}} g(m) \sum_{p_n<z} g(p_n)V(p_n) \sum_{p_1} g(p_1) \tag{11.147}$$

where p_1 is restricted by $Dm^{-1}p_n^{-1-\beta} \leqslant p_1 < Dm^{-1}p_n^{1-\beta}$ and $p_1 > D^{1/(\beta+n)}$. Hence, by (11.145) we get

$$\sum_{p_1} g(p_1) \leqslant (2\kappa \log p_n + L)\frac{\beta+n}{\log D} \leqslant 2L(\beta+n)\frac{\log p_n}{\log D} . \tag{11.148}$$

Then, by (11.146) we have

$$\sum_{p_n} g(p_n)V(p_n)\log p_n \ll LV(z)(\log z)(1+\kappa\log\log z) ,$$

where the implied constant is absolute. Finally by (11.145) we obtain

$$\sum_m g(m) \leqslant \frac{1}{(n-2)!}\Big(\sum_{3\leqslant p<z} g(p)\Big)^{n-2} \leqslant \Big(\frac{e}{n-2}(\kappa\log\log z + L)\Big)^{n-2} .$$

Gathering the above estimates we arrive at

PROPOSITION 11.16. *Let $0 \leqslant \kappa \leqslant 1$, $\beta=\beta(\kappa)$ and $D \geqslant z \geqslant 5$. Then for $n \geqslant 3$ we have*

$$V_n(D,z) \ll L^2n^2V(z)\Big(\frac{e}{n}(\kappa\log\log z + L)\Big)^{n-1} \tag{11.149}$$

where the implied constant is absolute.

REMARK. In (11.147) we isolated the two primes p_1, p_n in order to gain in (11.148) the factor $\log p_n/\log D$ which leads to an improvement in the bound for $V_n(D,z)$. Without this improvement the range of κ for which our two methods overlap would be narrowed and we would miss covering the important case $\kappa=1$.

Note that (11.149) implies that the slightly larger bound

$$V_n(D,z) \ll n^3V(z)\Big(\frac{e}{n}(\kappa\log\log D + L)\Big)^{n} \tag{11.150}$$

holds for every $n \geqslant 1$ and $D \geqslant z \geqslant 2$, with an implied constant depending only on L.

COROLLARY 11.17. *Let $\frac{1}{2} \leqslant \kappa \leqslant 1$ and $D \geqslant z \geqslant 2$. Then, for $n \geqslant \nu\log\log D$ with $\nu \geqslant e\kappa$ we have*

$$V_n(D,z) \ll n^{-2}V(z)(\log D)^{-\delta}(\log\log 3D)^5 \tag{11.151}$$

where

$$\delta = \nu\log(\nu/e\kappa) \tag{11.152}$$

and the implied constant depends only on ν and L.

PROOF. We can assume that D is large in terms of κ and L. Putting

$$\ell(u) = u^5\Big(\frac{e}{u}(\kappa\log\log D + L)\Big)^u ,$$

we have

$$\frac{\ell'}{\ell}(u) = \frac{5}{u} + \log\frac{\kappa\log\log D + L}{u} .$$

Hence, for $u \geqslant u_0 = \nu\log\log D$ we have

$$\frac{\ell'}{\ell}(u) \leqslant \frac{5}{u_0} + \log\Big(1 + \frac{\nu L}{\kappa u_0}\Big) + \log\frac{\kappa}{\nu} < 0$$

showing that $\ell(u)$ is decreasing for $u \geqslant u_0$. Therefore,

$$\ell(u) \leqslant \ell(u_0) \leqslant e^{\nu L/\kappa}(\log D)^{-\delta}(\nu\log\log D)^5 ,$$

which proves (11.151). □

11.11. Completion of the Beta-Sieve for $\frac{1}{2} < \kappa \leqslant 1$

Summing the inductive estimates (11.143) over n in the range (11.142) and the direct estimates (11.151) over the remaining range we derive

THEOREM 11.18. *Let $\frac{1}{2} < \kappa \leqslant 1$, $\beta = \beta(\kappa)$ and $\Lambda^+(D), \Lambda^-(D)$ be the upper-bound and lower-bound beta-sieves of level $D \geqslant 2$. Suppose the density function g satisfies the condition (11.129) with some constant $L \geqslant 1$. Then we have*

$$V^+(D,z) \leqslant \Big\{F(s) + O\Big((\log D)^{-1/6}\Big)\Big\}V(z), \quad \text{if } s \geqslant \beta - 1 , \tag{11.153}$$

$$V^-(D,z) \geqslant \Big\{f(s) + O\Big((\log D)^{-1/6}\Big)\Big\}V(z), \quad \text{if } s \geqslant \beta , \tag{11.154}$$

where $s = \log D/\log z$ and $F(s), f(s)$ are as in Theorem 11.12. The implied constant depends only on κ and L.

PROOF. Recall that $\alpha = \sqrt{(\beta+2)/\beta}$. In Corollary 11.17 choose $\nu = (\log\alpha)^{-1}$ getting

$$\delta = -\log(e\kappa\log\alpha)/\log\alpha . \tag{11.155}$$

In the range $\frac{1}{2} < \kappa \leqslant 1$, this is smallest at $\kappa = 1$, in which case $\beta = 2$, $\alpha = \sqrt{2}$ and

$$\delta = \frac{2}{\log 2}\log(2/e\log 2) = 0.1721426\cdots > \frac{1}{6} .$$

Hence, the result follows. □

REMARKS. With some technical refinements this method would have been capable of extending the range of validity slightly to the right of $\kappa = 1$. The argument reveals that the choice of $\beta = \beta(\kappa)$ is not dictated by issues of convergence because we do have some room to spare in the above estimations. Rather, our choice of β as the largest zero of the adjoint function $q(s-1)$ is made in order to attain the optimal leading functions $F(s)$, $f(s)$.

11.12. Specializing to $\kappa \leqslant \frac{1}{2}$

In the next three sections of this chapter we are going to prove the main Theorems 11.12 and 11.13 in the range $0 \leqslant \kappa \leqslant \frac{1}{2}$. In this case the sieving limit is always $\beta(\kappa) = 1$ throughout the full range and hence we have the same beta-sieve weights (11.17) for every κ. For ease of reference we reformulate here some previous expressions (combinatorial partitions, recurrence identities, goals) in this simpler context.

Let $\mathcal{D}^+, \mathcal{D}^-$ be the sets of support for the beta-sieves with $\beta = 1$ and level D, that is,

$$\begin{aligned} \mathcal{D}^+ &= \{d = p_1 \dots p_n : p_1 > \dots > p_n, p_1 \dots p_m p_m < D \quad \text{for } m \text{ odd}\} , \\ \mathcal{D}^- &= \{d = p_1 \dots p_n : p_1 > \dots > p_n, p_1 \dots p_m p_m < D \quad \text{for } m \text{ even}\} . \end{aligned} \tag{11.156}$$

Note that, if $d \in \mathcal{D}^+$ then $d \leqslant D$. If both $d \in \mathcal{D}^-$ and $d|P(D)$ then $d \leqslant D$.

The sifted sums which arise are

$$S^+(\mathcal{A}, z) = \sum_{\substack{d|P(z)\\ d\in\mathcal{D}^+}} \mu(d)|\mathcal{A}_d| ,$$

$$S^-(\mathcal{A}, z) = \sum_{\substack{d|P(z)\\ d\in\mathcal{D}^-}} \mu(d)|\mathcal{A}_d|$$

with $P = P(z)$ being the product of primes $p < z, p \in \mathcal{P}$, so the expected main terms are $XV^+(D, z)$ and $XV^-(D, z)$ respectively, where

$$V^+(D, z) = \sum_{\substack{d|P(z)\\ d\in\mathcal{D}^+}} \mu(d)g(d) ,$$

$$V^-(D, z) = \sum_{\substack{d|P(z)\\ d\in\mathcal{D}^-}} \mu(d)g(d) .$$

Precisely, we have

$$S^+(\mathcal{A}, z) = XV^+(D, z) + R^+(\mathcal{A}, D) ,$$

$$S^-(\mathcal{A}, z) = XV^-(D, z) + R^-(\mathcal{A}, D) ,$$

where the remainders are

$$R^+(\mathcal{A}, D) = \sum_{\substack{d|P(z)\\ d\in\mathcal{D}^+}} \mu(d)r_d(\mathcal{A}) ,$$

$$R^-(\mathcal{A}, D) = \sum_{\substack{d|P(z)\\ d\in\mathcal{D}^-}} \mu(d)r_d(\mathcal{A}) .$$

Unlike the choice of β, the corresponding functions $F(s), f(s)$ do depend on κ; they are the continuous solutions to the system

$$\begin{cases} s^\kappa F(s) = A & \text{if } 0 < s \leqslant 2 , \\ s^\kappa f(s) = B & \text{at } s = 1 , \end{cases} \tag{11.157}$$

$$\begin{cases} (s^\kappa F(s))' = \kappa s^{\kappa-1} f(s-1) & \text{if } s > 2 , \\ (s^\kappa f(s))' = \kappa s^{\kappa-1} F(s-1) & \text{if } s > 1 , \end{cases} \tag{11.158}$$

where A and B are defined in (11.61), (11.62). In this case, which we have not yet established, Theorem 11.12 asserts that

$$V^+(D,z) < \Big\{F(s) + O\big((\log D)^{-1/6}\big)\Big\} V(z) \qquad \text{if } s > 0 , \tag{11.159}$$

$$V^-(D,z) > \Big\{f(s) + O\big((\log D)^{-1/6}\big)\Big\} V(z) \qquad \text{if } s \geqslant 1 , \tag{11.160}$$

where $s = \log D/\log z$. Applying these sieves to the sequence $\mathcal{A}$ we obtain

$$S^+(\mathcal{A},z) < XV^+(D,z) + R^+(\mathcal{A},D) , \tag{11.161}$$

$$S^-(\mathcal{A},z) > XV^-(D,z) + R^-(\mathcal{A},D) . \tag{11.162}$$

These estimates hold also for $S(\mathcal{A},z)$ because $S^- \leqslant S \leqslant S^+$, by suspending terms $S_n(\mathcal{A},z)$ in (6.29) and (6.30). Some of the suspended terms can be shown to be positive by taking advantage of recent advances so we reclaim them for future improvements. Recall that these were the identities

$$S(\mathcal{A},z) = S^+(\mathcal{A},z) - \sum_{n \text{ odd}} S_n(\mathcal{A},z) , \tag{11.163}$$

$$S(\mathcal{A},z) = S^-(\mathcal{A},z) + \sum_{n \text{ even}} S_n(\mathcal{A},z) , \tag{11.164}$$

where

$$S_n(\mathcal{A},z) = \sum_{\substack{p_n < \cdots < p_1 < z \\ p_1 \cdots p_m p_m < D, m < n, m \equiv n(2) \\ p_1 \cdots p_n p_n \geqslant D}} \cdots \sum S(\mathcal{A}_{p_1 \ldots p_n}, p_n) . \tag{11.165}$$

We want to evaluate $V^\pm(D,z)$. We decompose these as

$$V^+(D,z) = V(z) + \sum_{n \text{ odd}} V_n(D,z) ,$$

$$V^-(D,z) = V(z) - \sum_{n \text{ even}} V_n(D,z) ,$$

where, for $n \geqslant 1$,

$$V_n(D,z) = \sum_{\substack{p_n < \cdots < p_1 < z \\ p_1 \cdots p_m p_m < D, m < n, m \equiv n(2) \\ p_1 \cdots p_n p_n \geqslant D}} \cdots \sum g(p_1 \ldots p_n) V(p_n). \tag{11.166}$$

These are defined for $s \geqslant 0$ if n is odd and for $s \geqslant 1$ if n is even.

For $n = 1$ we have

$$V_1(D,z) = \sum_{D^{\frac12} \leqslant p_1 < z} g(p)V(p) = V\big(D^{\frac12}\big) - V(z)$$

if $0 < s \leqslant 2$, and $V_1(D,z) = 0$ if $s > 2$. For every $n > 1$ and $s \geqslant \frac12(1 + (-1)^n)$ we have the following recurrence formulas.

$$V_n(D,z) = \sum_{p < D^{1/s_n}} g(p) V_{n-1}\Big(\frac{D}{p}, p\Big) , \tag{11.167}$$

where $s_n = \max\big(s, \frac{1}{2}(3 - (-1)^n)\big)$.

Parallel to $V_n(D, z)$ we have defined the sequence of functions $f_n(s)$ which summed to give the series (11.32), (11.33). In our case the functions $f_n(s)$ are defined by

$$\kappa^{-n} s^\kappa f_n(s) = \underset{\substack{0<t_n<\cdots<t_1<1/s \\ t_1+\cdots+t_m+t_m<1, m<n, m\equiv n(2) \\ t_1+\cdots+t_n+t_n>1}}{\int \cdots \int} t_n^{-\kappa} \frac{dt_1 \ldots dt_n}{t_1 \ldots t_n}$$

for $s \geqslant \frac{1}{2}(1 + (-1)^n)$. Note that $f_n(s) = 0$ if $s \geqslant n + 1$ because the domain of integration is empty in this range.

For convenience we recall that the functions $F(s)$ and $f(s)$ were introduced by the convergent series

$$F(s) = 1 + \sum_{n \text{ odd}} f_n(s) \qquad \text{if } s > 0 \ , \tag{11.168}$$

$$f(s) = 1 - \sum_{n \text{ even}} f_n(s) \qquad \text{if } s \geqslant 1 \ . \tag{11.169}$$

For $n = 1$ we have

$$s^\kappa f_1(s) = \begin{cases} 2^\kappa - s^\kappa & \text{if } 0 < s \leqslant 2 \ , \\ 0 & \text{if } s > 2 \ . \end{cases}$$

Parallel to (11.167), for every $n > 1$ and $s \geqslant \frac{1}{2}(1 + (-1)^n)$, we have the recurrence formula

$$s f_n(s) = \int_{s_n}^{\infty} f_{n-1}(t - 1) dt^\kappa \ .$$

In estimating $V^\pm(D, z)$ we shall work with the partial sums

$$T_N(D, z) = \sum_{\substack{n \leqslant N \\ n \equiv N (\mathrm{mod}\, 2)}} V_n(D, z) \quad \text{for } s \geqslant \tfrac{1}{2}(1 + (-1)^N) \ ,$$

$$G_N(s) = \sum_{\substack{n \leqslant N \\ n \equiv N (\mathrm{mod}\, 2)}} f_n(s) \qquad \text{for } s \geqslant \tfrac{1}{2}(1 + (-1)^N) \ .$$

Note that $T_N(D, z) = 0$ if $s \geqslant N + 1$ and $G_N(s) = 0$ if $s \geqslant N + 1$. Our goal is to prove by induction on N the following inequalities:

$$T_N(D, z) \leqslant V(z)\big\{G_N(s) + AE(s)(\log z)^{-1/6}\big\} \tag{11.170}$$

for $s \geqslant \frac{1}{2}(1 + (-1)^N)$, where $A \geqslant 1$ is a constant depending on L and $E(s) > 0$ is a function decreasing exponentially to zero as s tends to infinity. Specifically, we choose the function defined by (11.121). Because the proof goes by induction the function $E(s)$ has been chosen carefully to absorb the error terms occurring in each induction step. In other words, the relevant integral transformation of $E(s)$ has to be contracting rather than invariant.

11.13. Estimation of $V_n(D,z)$

We assume that $0 \leqslant \kappa \leqslant \frac{1}{2}$. Define $E_0(s) = s^{-\kappa}E(s)$ where $E(s)$ is the specific function defined in (11.121). Recall that we are assuming the bound (11.129) for the density function and that we showed in Lemma 11.10 the existence of an absolute constant $0 < b < 1$ such that (11.120) holds. We shall prove by induction on n the following result.

PROPOSITION 11.19. *Let $D \geqslant 2$, $z \geqslant 2$ and $s = \log D/\log z$. Then, for some absolute constant a with $b < a < 1$, and some constant A depending only on L we have*

$$V_n(D,z) \leqslant V(z)\{f_n(s) + E_0(s)a^n A(\log D)^{-1/6}\} \tag{11.171}$$

for all $n \geqslant 1$ and $s \geqslant \frac{1}{2}(1+(-1)^n)$.

PROOF. Note that $V_n(D,z)$ vanishes for $s \geqslant n+1$, so we may assume throughout that $s \leqslant n+1$. For $n = 1$ we have (11.25) from which we get

$$V_1(D,z) \leqslant V(z)\left\{\left(\frac{2}{s}\right)^{\kappa} - 1 + O\left(\frac{L}{\log D}\right)\right\}, \tag{11.172}$$

by (11.129) and by (11.26), giving the result.

Let $n \geqslant 2$. Assuming the result for $n-1$, we prove it for n. First, let $s \geqslant \frac{1}{2}(3-(-1)^n)$ so $s_n = s$ in (11.167). Note that this is always the case when n is even and when n is odd, the only missing range will be $0 < s < 2$. By the recurrence formula (11.167) and the induction hypothesis we obtain

$$\begin{aligned} V_n(D,z) \leqslant &\sum_{\substack{D^{1/(n+1)} \leqslant p < D^{1/s} \\ p < D/2}} g(p)V(p)\left\{f_{n-1}\left(\frac{\log \frac{D}{p}}{\log p}\right) + E_0\left(\frac{\log \frac{D}{p}}{\log p}\right)\frac{a^{n-1}A}{(\log \frac{D}{p})^{\frac{1}{6}}}\right\} \\ &+ \sum_{D/2 \leqslant p < D^{1/s}} g(p)V_{n-1}\left(\frac{D}{p},\, p\right). \end{aligned}$$

Here we have split off the range $D/2 \leqslant p < D^{1/s}$ which is empty except when s is close to one; specifically, using $\max(2, D/2) \geqslant \sqrt{D}$ we get

$$1 < s < 1 + \frac{\log 4}{\log D} = s_0,$$

say. We do so because we cannot apply the induction hypothesis for p very close to D. Since, in this extreme range, $D/p \leqslant 2$ the sum $V_{n-1}(D/p,p)$ is empty unless $n = 2$ in which case we have $V_{n-1}(D/p,p) = V_1(1,p) = 1 - V(p) \leqslant 1$, so that

$$\sum_{D/2 \leqslant p < D^{1/s}} g(p)V_{n-1}\left(\frac{D}{p}, p\right) \leqslant \sum_{D/2 \leqslant p < D} g(p) \ll L(\log D)^{-1}.$$

By Lemma 5.1 with $K = 1 + O\big(Ln/\log D\big)$ we have

$$V_n(D,z) \leqslant KV(z)\int_s^\infty \Big\{f_{n-1}(t-1) + E_0(t-1)\frac{a^{n-1}A}{\big(\log D^{(1-1/t)}\big)^{1/6}}\Big\}d\big(t/s\big)^\kappa$$
$$+ (K-1)V(z)\Big\{f_{n-1}(s'-1) + E_0(s'-1)\frac{a^{n-1}A}{\big(\log D^{(1-1/s)}\big)^{1/6}}\Big\} + O\big(\frac{L}{\log D}\big)$$

where $s' = \max\{s, s_0\}$ and the last term can be dropped except from $n = 2$. Writing $K = 1 + (K-1)$ in the first term and combining the latter with the second term we get

$$V_n(D,z) \leqslant V(z)\int_s^\infty \{f_{n-1}(t-1) + E_0(t-1)(1-\frac{1}{t})^{-\frac{1}{6}}\frac{a^{n-1}A}{(\log D)^{1/6}}\}d\big(t/s\big)^\kappa$$
$$+ O\Big(V(z)nL\frac{Aa^n}{\sqrt{\log D}}\,E_0(s)\Big).$$

Note, in reaching this step we have used the bounds

$$f_n(s'-1),\ E_0(s'-1) \ll E_0(s)\big(\frac{s}{s'-1}\big)^\kappa \ll E_0(s)(\log D)^{\frac{1}{2}}\ ,$$

and the error term peculiar to $n = 2$ can now be incorporated into the other one.

Using the recurrence formula (11.28) and the inequality (11.120) we obtain

$$V_n(D,z) \leqslant V(z)\{f_n(s) + E_0(s)\frac{ba^{n-1}A}{(\log D)^{1/6}} + O\big(\frac{nLAa^nE_0(s)}{\sqrt{\log D}}\big)\}\ . \tag{11.173}$$

This also holds if n is odd and $0 < s < 2$. Indeed, by the case $s = 2$,

$$V_n(D,z) = V_n(D,\sqrt{D}) \leqslant V(\sqrt{D})\Big\{f_n(2) + E_0(2)\frac{ba^{n-1}A}{(\log D)^{1/6}} + O\big(\frac{nLAa^nE_0(2)}{\sqrt{\log D}}\big)\Big\}\ .$$

Here, we have

$$V(\sqrt{D}) \leqslant V(z)(2/s)^\kappa(1 + O(L/\log D))\ ,$$
$$(2/s)^\kappa f_n(2) = f_n(s), \text{ and } (2/s)^\kappa E_0(2) \ll E_0(s)\ ,$$

so that (11.173) holds in all cases.

Now we are ready to complete the proof of (11.171). If the inequality $\log D > (CnL)^3$ holds, where C is a sufficiently large absolute constant, we can ignore the error term in (11.173) by increasing b to a getting (11.171). In case this does not hold we use the direct estimation (11.150) getting

$$V_n(D,z) \ll V(z)n^3\big(\frac{3e}{2n}(\log CnL + L)\big)^n \ll V(z)e^{-2n}\ ,$$

where the implied constant depends only on L. Having in mind that $s \leqslant n+1$ and

$$E_0(s) \gg e^{-s}s^{-\kappa} \gg e^{-n}n^{-\frac{1}{2}}$$

this direct evaluation easily gives (11.171) (without the main term). This completes the proof of Proposition 11.19. □

11.14. Completion of the Beta-Sieve for $0 \leqslant \kappa \leqslant \frac{1}{2}$

Summing the inductive estimates over n, we derive

THEOREM 11.20. *Let $0 \leqslant \kappa \leqslant \frac{1}{2}$, and $\Lambda^+(D), \Lambda^-(D)$ be the upper-bound and lower-bound beta-sieves with $\beta = 1$ of level $D \geqslant 2$. Suppose the density function g satisfies the condition* (11.129) *with some constant $L \geqslant 1$. Then we have*

$$V^+(D,z) \leqslant \left\{F(s) + O\left(e^{-s}(\log D)^{-1/6}\right)\right\} V(z), \quad \text{if } s \geqslant 0\ , \tag{11.174}$$

$$V^-(D,z) \geqslant \left\{f(s) + O\left(e^{-s}(\log D)^{-1/6}\right)\right\} V(z), \quad \text{if } s \geqslant 1\ , \tag{11.175}$$

where $s = \log D/\log z$ and $F(s), f(s)$ are as in Theorem 11.12. The implied constant depends only on L.

REMARK. The factor e^{-s} in the error term is not very important but costs us nothing to include in the case $0 \leqslant \kappa \leqslant \frac{1}{2}$. It was not present in our statement for the range $\frac{1}{2} < \kappa \leqslant 1$ because the ad hoc argument we employed for large n cost us the opportunity to do so.

This completes the proofs of the main Theorems 11.12 and 11.13 of the beta-sieve for all $\kappa \leqslant 1$.

11.15. An Extension of the Sifting Range for $\kappa < \frac{1}{2}$

Before leaving the beta-sieve of small dimension we draw attention to a special feature which is particularly useful when κ is strictly less than one half. In this case $f(1) = B$ is positive (see (11.62)), so the lower bound (11.131) remains positive in the full range $z \leqslant D$ and therefore

$$S(\mathcal{A}, D) \geqslant S^-(\mathcal{A}, D) \gg XV(D) \tag{11.176}$$

provided that

$$R(\mathcal{A}, D) = \sum_{\substack{d|P(D)\\ d<D}} |r_d(x)| \ll X(\log x)^{-1/2}\ . \tag{11.177}$$

Unfortunately, in applications of these low-dimensional sieves one really needs a bound for $S(\mathcal{A}, x)$; that is, one needs to sieve out by the bad primes all the way up to x (see for example Section 11.20). Since the choice of D as large as x makes it impossible to handle the sieve remainder we need an extra ingredient. In case D is quite close to x we can succeed by sacrificing some positive portion of $f(1)$. When $\kappa \geqslant \frac{1}{2}$ there is nothing available for sacrifice.

THEOREM 11.21. *Let $\mathcal{P}$ be a set of primes. Let $\mathcal{A} = (a_n)$ be a sequence of non-negative reals with density function $g(d)$ satisfying* (11.129) *with $0 \leqslant \kappa < \frac{1}{2}$. Moreover, assume that*

$$\prod_{w \leqslant p < z} (1 - g(p)) \ll \frac{\log w}{\log z} \tag{11.178}$$

for all $z > w \geqslant 2$ (here, the product running over all primes, not restricted to those in $\mathcal{P}$), and

$$\prod_{p \notin \mathcal{P}} (1 + g(p^2) + g(p^3) + \dots) < \infty\ . \tag{11.179}$$

Suppose the remainder $R(\mathcal{A}, D)$ satisfies (11.177) for some D with $x^{1-\vartheta} \leqslant D < x$ where $\vartheta > 0$ is sufficiently small in terms of κ. Suppose also that

$$\sum_{d<x^{1/3}} \tau(d)|r_d(x)| \ll X(\log x)^{-1/2} . \tag{11.180}$$

Then, for all x sufficiently large in terms of κ we have

$$S(\mathcal{A}, x) \asymp X \prod_{p|P(x)} (1 - g(p)) . \tag{11.181}$$

PROOF. Suppose $x^{5/6} \leqslant D < x$. We have

$$S(\mathcal{A}, D) - S(\mathcal{A}, x) = \sum_{\substack{m \leqslant M \\ (m,P(M))=1}} \sum_{mp \leqslant x} a_{mp} ,$$

where $M = xD^{-1} \leqslant x^{1/6}$. Next we relax the condition of p being prime by applying an upper-bound sieve (λ_δ) of level $\Delta = x^{1/6}$. We obtain the upper bound

$$\sum_{\substack{m \leqslant M \\ (m,P(M))=1}} \sum_{m\ell \leqslant x} a_{m\ell} \Big(\sum_{\substack{\delta | \ell \\ (\delta,m)=1}} \lambda_\delta \Big) .$$

Here δ runs over all squarefree numbers restricted by $\delta < \Delta$ and $(\delta, m) = 1$ (the latter condition is retained for technical convenience). Changing the order of summation we arrive at the sum

$$\sum_{\substack{m \leqslant M \\ (m,\, \delta P(M))=1}} \sum_{\delta < \Delta} \lambda_\delta A_{m\delta}(x) .$$

Since $m\delta \leqslant M\Delta \leqslant x^{1/3}$ we can use the assumption (11.180) getting

$$X \sum_{\substack{m \leqslant M \\ (m,P(M))=1}} g(m) \sum_{\substack{\delta < \Delta \\ (\delta,m)=1}} \lambda_\delta g(\delta) + O\big(X(\log x)^{-\frac{1}{2}}\big) .$$

Then we apply a crude bound from Brun's sieve:

$$\sum_{\substack{\delta < \Delta \\ (\delta,m)=1}} \lambda_\delta g(\delta) \ll \prod_{\substack{p < \Delta \\ p \nmid m}} (1 - g(p)) = \prod_{p|m} (1 - g(p))^{-1} \prod_{p<\Delta} (1 - g(p))$$

followed by the rough bound

$$\begin{aligned} \sum_{\substack{m \leqslant M \\ (m,P(M))=1}} g(m) \prod_{p|m} (1 - g(p))^{-1} &\leqslant \prod_{\substack{p \leqslant M \\ p \notin \mathcal{P}}} \big(1 + (g(p) + g(p^2) + \dots)(1 - g(p))^{-1}\big) \\ &\leqslant \eta \prod_{\substack{p \leqslant M \\ p \notin \mathcal{P}}} (1 - g(p))^{-1} , \end{aligned}$$

where η is the constant given by the infinite product (11.179). Next we rearrange the resulting products as

$$\prod_{p<\Delta} (1 - g(p)) \prod_{\substack{p \leqslant M \\ p \notin \mathcal{P}}} (1 - g(p))^{-1} = U(M, \Delta)V(\Delta)$$

where

$$U(M,\Delta)=\prod_{\substack{M<p<\Delta\\ p\notin\mathcal{P}}}\big(1-g(p)\big)\ .$$

We recall that

$$V(\Delta)=\prod_{\substack{p<\Delta\\ p\in\mathcal{P}}}\big(1-g(p)\big)\asymp V(x)\gg(\log x)^{-\kappa}$$

by (11.129). Hence, we conclude that

$$S(\mathcal{A},x)=S(\mathcal{A},D)+O\big(XV(x)U(M,\Delta)+X(\log x)^{-1/2}\big)\ .$$

By (11.129) and (11.178) with $w=M,z=\Delta$ we find that

$$U(M,\Delta)\ll\Big(\frac{\log M}{\log\Delta}\Big)^{1-\kappa}\ll(1-\vartheta)^{\frac{1}{2}}\ ,$$

which yields a contribution sufficiently small so as to be absorbed by the lower bound (11.176). This gives the lower bound in (11.181). The corresponding upper bound follows from the upper-bound sieve of level D without all this extra work. □

REMARKS. It is important to recognize that this extension of the sieve range holds for the lower bound for $S(\mathcal{A},z)$ but it does not apply to $S^-(\mathcal{A},z)$. The additional requirement (11.180) refers to d not necessarily squarefree and not necessarily composed of primes in the sifting set, yet in practice this is easily verifiable.

11.16. Higher Dimension: Composing Beta-Sieves

So far our treatment of the beta-sieve has provided full coverage only for dimension $\kappa\leqslant 1$. That part of the treatment dealing with the continuous functions $f_n(s),F(s),f(s),H(s),h(s)$ in Sections 11.3, 11.4, 11.6, 11.7, was however, accomplished for all $\kappa\geqslant 0$. The arithmetic counterparts $V_n(D,z)$ involve sums over primes and provide greater difficulty. So far we have succeeded for every κ but only for limited n. With the larger values of κ the induction broke down (although still more delicate arguments would have completed it, see [**104**]) and we introduced ad hoc methods to finish the job, in Section 11.10 for $\frac{1}{2}<\kappa\leqslant 1$ and in Section 11.13 for $0\leqslant\kappa\leqslant\frac{1}{2}$. The ranges of n where the precise induction and the rough bound overlap, as they need to for success, include κ up to a little greater than one but fail thereafter (barely so if sufficient effort is applied). There is another option which easily covers all κ which we are going to present next. This involves combining the optimal beta-sieve in the sifting range of large primes with the Fundamental Lemma in the range of small primes. The result of this combination is described in general terms in Theorem 6.17. Our Fundamental Lemma is itself a beta-sieve but with a larger, non-optimal choice of β, so indeed the outcome is not a genuine beta-sieve but a composition of two of them. The result also can be made to cover $\kappa\leqslant 1$ but for this most important range we wanted a pure beta-sieve.

THEOREM 11.22. *Let $\kappa\geqslant 1$ and let $F(s),f(s)$ be the functions given by* (11.132) *and* (11.133). *Let $\mathcal{A}=(a_n)$ be a sequence of non-negative reals with density function $g(d)$ satisfying* (11.129). *Then, for $D\geqslant 3,z\geqslant 2$, we have*

$$S(\mathcal{A},z)\leqslant XV(z)\big\{F(s)+O\big(e^{-\sqrt{\log\log D}}\big)\big\}+R(\mathcal{A},DD_0)\quad\textit{if } s\geqslant\beta-1,\tag{11.182}$$

$$S(\mathcal{A},z)\geqslant XV(z)\big\{f(s)+O\big(e^{-\sqrt{\log\log D}}\big)\big\}-R(\mathcal{A},DD_0)\quad\textit{if } s\geqslant\beta,\tag{11.183}$$

where $s = \log D/\log z$, $D_0 = D^{1/\sqrt{\log\log D}}$, *and the remainder is*

$$R(\mathcal{A}, DD_0) = \sum_{\substack{d|P(z)\\ d<DD_0}} |r_d| \ .$$

The implied constant depends only on κ *and* L.

PROOF. We apply Theorem 6.17 with $D_0 = D^\varepsilon$ and $z_0 = D^{\varepsilon^2/2}$, where $\varepsilon = (\log\log D)^{-1/2}$. The conditions (6.101) and (6.102) are satisfied by Corollary 11.15 because

$$V_n(D,z) \leqslant \sum_{\substack{z_0\leqslant p_n<\cdots<p_1<z\\ p_n\cdots p_1<D}} g(p_1\dots p_n)$$

which vanishes for $n > 2\varepsilon^{-2}$. Note that, in (11.143), have

$$\log\alpha = \tfrac{1}{2}\log\bigl(1+\tfrac{2}{\beta}\bigr) \leqslant \tfrac{1}{2}\log 2 < \frac{1}{2} \ .$$

Hence, (11.182) and (11.183) hold with the error term majorized as

$$\ll sH(s)\frac{(\log\log D)^5}{\varepsilon^2\log D} + e^{-2/\varepsilon}(\varepsilon s)^{-2\kappa} \ll e^{-\sqrt{\log\log D}} \ ,$$

which completes the proof. □

11.17. Estimation of $\mathcal{L}^+(D,z)$

Let $0 \leqslant \kappa \leqslant \frac{1}{2}$. For use in Chapter 14, in connection with the asymptotic semi-linear sieve, we shall require certain technical estimates. Although these were not needed for the principal theorems of this chapter, the proofs bear a strong resemblance to the arguments presented above. Hence we include them in this section and the next.

For any $n > 0$, n even, $z \geqslant 2, D \geqslant 1$, we define

$$\mathcal{L}_n(D,z) = \sum_{\substack{p_n<\cdots<p_1<z\\ p_1\cdots p_m p_m<D,\ \text{all } m \text{ odd}\\ p_1\cdots p_n p_n^2\geqslant D}} \cdots \sum g(p_1\dots p_n)(\log 4D/p_1\dots p_n)^{-2/3} \ . \tag{11.184}$$

Note that the least prime p_n in the last condition $p_1\dots p_n p_n^2 \geqslant D$ appears with multiplicity three, not two as in the preceding conditions. The last but one condition $p_1\dots p_{n-1}p_{n-1} < D$ implies $d = p_1\dots p_n \leqslant D$, so (11.184) makes sense. We also define

$$\mathcal{L}_0(D,z) = (\log 4D)^{-2/3} \quad \text{if } z \geqslant D^{1/2} \tag{11.185}$$

and $\mathcal{L}_0(D,z) = 0$ otherwise. Our goal is to estimate the sum

$$\mathcal{L}^+(D,z) = \sum_{\substack{n\geqslant 0\\ n \text{ even}}} \mathcal{L}_n(D,z) \ . \tag{11.186}$$

PROPOSITION 11.23. *For* $z \geqslant 2$, $D \geqslant 1$ *we have*

$$\mathcal{L}^+(D,z) \ll e^{-s}\bigl(\log 4D\bigr)^{-2/3} \tag{11.187}$$

where $s = \log D/\log z$ *and the implied constant depends only on* L.

When proving the upper bounds (11.171) for $V_n(D,z)$ by induction on n we have introduced specific error terms whose task was to absorb the excess in estimations of the main terms. Now the same induction arguments which control these error terms work perfectly for the estimation of $\mathcal{L}_n(D,z)$.

In order to carry out the induction we also need $\mathcal{L}_n(D,z)$ for n odd. If $n \geqslant 1$, n odd, $z \geqslant 2$, $D \geqslant z$, we define

$$\mathcal{L}_n(D,z) = \sum_{\substack{p_n<\dots<p_1<z \\ p_1\dots p_m p_m<D,\ \text{all } m \text{ even} \\ p_1\dots p_n p_n^2 \geqslant D}} \cdots \sum g(p_1\dots p_n)\big(\log 4D/p_1\dots p_n\big)^{-2/3}. \tag{11.188}$$

This is well defined because $d = p_1 \dots p_n < D$. We are going to estimate $\mathcal{L}_n(D,z)$ for every $n \geqslant 0$. To state the result we recall that $E(s)$ is the function defined by (11.121) and $b<1$ is the constant satisfying (11.120).

LEMMA 11.24. *Let $z \geqslant 2$, $D \geqslant 1$, $s = \log D/\log z$. We have*

$$\mathcal{L}_n(D,z) \leqslant a^n A E(s)\big(\log 4D\big)^{-2/3} \tag{11.189}$$

if $\frac{1}{2}\big(1-(-1)^n\big) \leqslant s \leqslant n+2$, where a is an absolute constant, $b<a<1$ and A is a constant depending only on L.

PROOF. For $n=0$ we have (11.189) with $A=e^2$ by (11.185). Let $n \geqslant 1$. Assuming (11.189) for $n-1$, we prove it for n. First, let $s \geqslant \frac{1}{2}\big(3+(-1)^n\big)$. Note that this is always the case when n is odd and, when n is even, the only missing range will be $0 \leqslant s < 2$. By the recurrence formula

$$\mathcal{L}_n(D,z) = \sum_{D^{1/(n+2)}<p<D^{1/s}} g(p)\mathcal{L}_{n-1}(D/p,p)$$

and by the induction hypothesis (11.189) we get

$$\mathcal{L}_n(D,z) \leqslant a^{n-1}A \sum_{D^{1/(n+2)}<p<D^{1/s}} g(p)E\Big(\frac{\log D/p}{\log p}\Big)\Big(\log 4D/p\Big)^{-\frac{2}{3}}.$$

Note that the sum is void unless $s<n+2$. Moreover, we have

$$p \geqslant \max\big(2, D^{1/(n+2)}\big) \geqslant \big(4D\big)^{1/(n+4)}.$$

By (5.47) with $\kappa = 1/2$ and $K = 1+(n+4)L/\log 4D$ we obtain

$$\begin{aligned}\mathcal{L}_n(D,z) \leqslant a^{n-1}\frac{A}{2}\int_s^{n+2} & E(t-1)\big(\log 4D^{1-1/t}\big)^{-2/3}t^{-1}dt \\ &+ O\big(Lna^{n-1}Ae^{-s}(\log 4D)^{-1}\big).\end{aligned}$$

Using the estimate

$$\frac{1}{2}\int_s^\infty E(t-1)\Big(1-1/t\Big)^{-2/3}t^{-1}dt \leqslant bE(s)$$

which follows by (11.120), we arrive at

$$\mathcal{L}_n(D,z) \leqslant ba^{n-1}AE(s)\big(\log 4D\big)^{-2/3}\Big\{1+O\big(nL(\log 4D)^{-1/3}\big)\Big\}. \tag{11.190}$$

In the remaining case $0 \leqslant s < 2$, with $n \geqslant 2$ and even, we have $\mathcal{L}_n(D,z) = \mathcal{L}_n(D,\sqrt{D})$. Therefore (11.190) holds in all cases, because $E(2) \leqslant E(s)$ for $0 \leqslant s \leqslant 2$ (see (11.119)).

If $\log 4D > (CnL)^3$ where C is a sufficiently large absolute constant, the error term in (11.190) can be ignored by increasing b to a. If that inequality does not hold we use the direct estimation

$$\mathcal{L}_n(D,z) \leqslant \frac{1}{n!}\Big(\sum_{p<D} g(p)\Big)^n < \Big(\frac{e}{n}\log\big(2L\log 4D\big)\Big)^n$$
$$\leqslant \Big(\frac{4e}{n}\log CnL\Big)^n \ll e^{-2n}$$

where the implied constant depends only on L. Having in mind that $s \leqslant n+2$, we have $E(s) \gg e^{-s} \gg e^{-n}$ and $e^{-n} \ll a^n n^{-2} \ll a^n(\log 4D)^{-2/3}$. This direct estimation easily gives (11.189) with A sufficiently large, depending only on L. □

Summing (11.189) over even numbers we obtain (11.187).

11.18. Asymptotic Evaluation of $V_n(D,z)$

Let $0 \leqslant \kappa \leqslant \frac{1}{2}$. Our upper bounds for the sums $V_n(D,z)$ actually represent the true state of affairs in natural cases. However, to prove this it is necessary to make a stronger assumption about the density function $g(d)$, namely that the upper bound condition (11.129) also represents the true state of affairs. Precisely, we assume

$$\text{(11.191)}\qquad \Big(1+\frac{L}{\log w}\Big)^{-1} \leqslant \Big(\frac{\log z}{\log w}\Big)^{\kappa} \prod_{w\leqslant p<z}(1-g(p)) \leqslant 1+\frac{L}{\log w},$$

for all $2 \leqslant w < z$ where $L \geqslant 1$ is a constant.

Proposition 11.25. *Let $0 \leqslant \kappa \leqslant \frac{1}{2}$. Suppose the density function satisfies (11.191). Let $D > 1$, $z = D^{1/s}$ with $s > \frac{1}{2}(1+(-1)^n)$. Then we have*

$$\text{(11.192)}\qquad |V_n(D,z) - V(z)f_n(s)| \leqslant V(z)E_0(s)a^n A(\log 3D)^{-1/6}$$

for some absolute constant $0 < a < 1$ and some $A = A(L) \geqslant 1$ which depends only on L.

Here $E_0(s)$ is the function we have been working with in Section 11.13, that is $E_0(s) = s^{-\kappa}E(s)$ with $E(s)$ given by (11.121).

Proof. The proof of the lower bound for $V_n(D,z)$ in (11.192) goes along the same lines as that of the upper bound (11.171) by the induction just presented. The required modifications are quite obvious, so we only mention a few of them.

One postulates in the induction that

$$\text{(11.193)}\qquad V_n(D,z) \geqslant V(z)\Big\{f_n(s) - E_0(s)\frac{a^n A}{(\log D)^{1/6}}\Big\}$$

in place of (11.171). For $n=1$ one gets (11.172) with $\leqslant$ replaced by $\geqslant$ using the lower-bound condition (11.191). Then, by the recurrence formula and the induction hypothesis we find

$$V_n(D,z) \geqslant \sum_{\substack{D^{1/(n+1)}\leqslant p<D^{1/s}\\ p<D/2}} g(p)V(p)\Big\{f_{n-1}\Big(\frac{\log\frac{D}{p}}{\log p}\Big) - E_0\Big(\frac{\log\frac{D}{p}}{\log p}\Big)\frac{a^{n-1}A}{(\log\frac{D}{p})^{\frac{1}{6}}}\Big\}.$$

Note that, unlike in the upper bound following (11.172), we have dropped the sum over $D/2 \leqslant p < D^{1/s}$ by positivity (there is no need to estimate this part). Here we need a lower bound rather than an upper bound for the terms with f_{n-1} while we need an upper bound for the terms with h as before. For the lower bound we use the inequality (5.45) with $\leqslant$ replaced by $\geqslant$ and K replaced by K^{-1}. Since $K^{-1} \geqslant 1-(K-1)$ the rest of the proof of (11.193) is identical to that of (11.171). □

By Proposition 11.25, summing over n we obtain

PROPOSITION 11.26. *Let $0 \leqslant \kappa \leqslant \frac{1}{2}$. Suppose the density function satisfies* (11.191) *for all $2 \leqslant w \leqslant z$, where $L \geqslant 1$ is a constant. Let $D \geqslant 2$ and $z \geqslant 2$. We have*

$$V^+(D,z) = V(z)\left\{F(s) + O\big(e^{-s}(\log D)^{-1/6}\big)\right\}, \tag{11.194}$$
$$V^-(D,z) = V(z)\left\{f(s) + O\big(e^{-s}(\log D)^{-1/6}\big)\right\}$$

for $s > 0$ and $s \geqslant 1$ respectively, where $s = \log D/\log z$. The implied constants depend only on L.

11.19. Estimates and Numerical Tables for the Sieve Constants

Let $\kappa > \frac{1}{2}$. Recall that the adjoint function $q(s)$ has a real positive zero and we have chosen $\beta - 1$ to be the largest such. In this section we collect a few insightful inequalities and a table of numerical values for $\beta = \beta(\kappa)$ and $A = A(\kappa)$. More general results are presented, with proofs, in Appendix B.

One can show the following asymptotics:

$$\beta - 1 \sim \pi e^{\gamma}(2\kappa - 1)^2 \qquad \text{as } \kappa \to \tfrac{1}{2}^{+} \tag{11.195}$$

(so that $\beta(\kappa) < 2\kappa$ for κ close to $\frac{1}{2}$),

$$\beta \sim c\kappa \qquad \text{as } \kappa \to \infty, \tag{11.196}$$

where $\gamma = .577\ldots$ is the Euler constant and $c = 3.591\ldots$ is the number which solves the equation $(c/e)^e = e$, that is, (6.68). We also have the following neat inequalities (see (B.16) and take $a = b = \kappa$)

$$\begin{aligned} 0 &< \beta - 1 < \kappa && \text{if } \tfrac{1}{2} < \kappa < 1,\\ \kappa &< \beta - 1 < \kappa + \sqrt{\kappa(\kappa-1)} && \text{if } 1 < \kappa < \tfrac{3}{2},\\ \kappa + \sqrt{\kappa(\kappa-1)} &< \beta - 1 && \text{if } \kappa > \tfrac{3}{2}. \end{aligned} \tag{11.197}$$

For the constant $A = A(\kappa)$ in the initial conditions for $F(s)$ we have

$$1 + \frac{\kappa}{\beta} < \frac{A}{2(\beta-1)^{\kappa}} < 1 + \frac{\kappa}{\beta - 1}. \tag{11.198}$$

For $\kappa = 1$ (the linear sieve) we have $q(s) = s - 1$ so $\beta(1) = 2$, and $A = 2p(1)^{-1}$ by (11.57) where $p(s)$ is given by (11.42). To compute $p(1)$ we write

$$p(s) = e^{-\gamma}\int_0^\infty \exp\big(-sz - \mathcal{E}(z)\big)z^{-1}dz$$

where $\mathcal{E}(z)$ is given by (B.15). Hence, and by the differential-difference equation (11.41), we get

$$p(s+1) = -sp'(s) = e^{-\gamma}\int_0^\infty \exp\big(-z - \mathcal{E}(z/s)\big)dz \to e^{-\gamma}$$

as $s \to 0$. Therefore, $p(1) = e^{-\gamma}$ and $A = 2e^{\gamma}$. The initial conditions (11.35) become

$$F(s) = 2e^{\gamma}s^{-1} \qquad \text{if } 1 \leqslant s \leqslant 3\,,$$
$$f(2) = 0\,.$$

The following numerical values were computed by S. Blight at Rutgers in 2009.

κ	β	A
0.5	1.0	$2(e^{\gamma}/\pi)^{\frac{1}{2}}$
0.55	1.0340771100	1.6066049515
0.6	1.1042161305	1.7254774009
0.65	1.1922077070	1.8631458759
2/3	1.2242102011	1.9134761067
0.7	1.2912892849	2.0212453075
0.75	1.3981115251	2.2020138267
0.8	1.5107489225	2.4082113624
5/6	1.5884497546	2.5614047810
0.85	1.6279798714	2.6431268159
0.9	1.7489723058	2.9106243480
0.95	1.8731283112	3.2152143347
1.0	2.0	$2e^{\gamma}$
1.05	2.1292406269	3.9575133160
1.1	2.2605745188	4.4084002595
1.15	2.3937776845	4.9230275019
1.2	2.5286648100	5.5109450686
1.25	2.6650802364	6.1832512370
1.3	2.8028915201	6.9528515564
1.35	2.9419847168	7.8347636745
1.4	3.0822608556	8.8464761755
1.45	3.2236332483	10.0083711532
1.5	$\frac{1}{2}(5+\sqrt{3})$	11.3442221190
2	4.8339865967	21.7484437308

Note that it appears from the tables that $\beta(\kappa) < 2\kappa$ for all $\frac{1}{2} < \kappa < 1$. This is interesting because of the conjecture (cf. [**145**]) that the sifting limit of any lower-bound sieve is larger than 2κ if the dimension κ is large. For the beta-sieve we have the neat estimate $\beta < c\kappa - \frac{c-1}{c+1}$ if κ is sufficiently large, due to H.G. Diamond, H. Halberstam and H.-E. Richert.

We are now finished with the general theory of the beta-sieve. In the last three sections of the chapter we show applications for "fractional" dimensions $0 < \kappa < 1$ which require the full strength of the results. The beta-sieves of dimensions $\kappa = 1$ and $\kappa = \frac{1}{2}$ are distinguished by their special features and by many important applications. We shall develop these sieves considerably further in the following chapters.

11.20. p-adic Zeros of Ternary Quadratic Forms

Our first application illustrates the use of the beta-sieve of dimension $\kappa \leqslant \frac{1}{2}$ to solve a basic arithmetic question. In Chapter 14 we shall address many other applications with $\kappa = \frac{1}{2}$.

We consider a problem proposed by J-P. Serre [**147**] which concerns the quadratic form

$$\varphi_{ab}(X, Y, Z) = aX^2 + bY^2 - Z^2 \ .$$

The original question of Serre asks for how many positive a and b does the form φ_{ab} have a non-trivial rational zero. For simplicity we consider a, b odd, squarefree and relatively prime. By the Minkowski local-global principle (see page 42 of [**146**]) the answer is yes if and only if it does p-adically for every p. The place $p = \infty$ is automatic because our form is indefinite. The existence of a p-adic solution is described by the requirement that the Hilbert symbol satisfy

$$\left(\frac{a, b}{p}\right) = 1 \ . \tag{11.199}$$

We are going to take $b \equiv 1 \pmod 4$ so that the condition (11.199) is automatically satisfied at $p = 2$. It is also satisfied if $p \nmid ab$. For the remaining primes, that is $p|ab$, our local requirement becomes

$$\left(\frac{a}{p}\right) = 1 \quad \text{if } p|b \ , \tag{11.200}$$

$$\left(\frac{b}{p}\right) = 1 \quad \text{if } p|a \ . \tag{11.201}$$

Let $\mathcal{P}$ be a set of odd primes (in Serre's case it was the full set). Given b, we are interested in estimating the number of $a \leqslant A$ satisfying these local conditions for every $p \in \mathcal{P}$. This is a typical sieve problem because the requirement is that a have no prime factors in the set

$$\mathcal{P}_b = \left\{p \in \mathcal{P};\ \left(\frac{b}{p}\right) = -1\right\} \tag{11.202}$$

and that a satisfy (11.200) for every $p \in \mathcal{P}$. To this end we are going to estimate the sifted sum

$$S(\mathcal{A}, z) = \sideset{}{^\flat}\sum_{\substack{a \leqslant A \\ (a, 2bP(z)) = 1}} w(a) \tag{11.203}$$

where $P(z)$ is the product of all primes $p \in \mathcal{P}$, $p < z$, and

$$w(a) = \prod_{\substack{p|b \\ p \in \mathcal{P}}} \frac{1}{2}\left(1 + \left(\frac{a}{p}\right)\right) . \tag{11.204}$$

Note that $w(a)$ is the characteristic function of the condition (11.200) for primes $p \in \mathcal{P}$. Therefore, for $z = A$, the sifted sum $S(\mathcal{A}, A)$ counts exactly those squarefree integers $a \leqslant A$, coprime with $2b$, for which φ_{ab} has p-adic zeros for every $p \in \mathcal{P}$.

One may treat $S(\mathcal{A}, A)$ as the mean value of a multiplicative function for which the general estimate (A.8) applies, giving

$$S(\mathcal{A}, z) \ll \frac{A}{\log A} \prod_{\substack{p<z \\ p\notin\mathcal{P}_b}} \Big(1+\frac{1}{p}\Big) \ll A \prod_{\substack{p<A \\ p\in\mathcal{P}_b}} \Big(1+\frac{1}{p}\Big)^{-1} . \tag{11.205}$$

COROLLARY 11.27. *Suppose the set $\mathcal{P}_b$ is sufficiently large that*

$$\sum_{p\in\mathcal{P}_b} \frac{1}{p} = \infty . \tag{11.206}$$

Then, for almost all a which are squarefree and coprime with $2b$, the quadratic form $\varphi_{ab}(X, Y, Z)$ fails to have a p-adic zero for at least one $p \in \mathcal{P}$.

It is not obvious how to get the corresponding lower bound along similar lines from the theory of multiplicative functions, especially when the set $\mathcal{P}_b$ of primes is not regularly distributed. In such erratic situations sieve methods can show to advantage over analytic arguments.

Note that $\mathcal{P}_b$ is included in the set of all primes p with $(b/p) = -1$ and that the latter has density exactly one half. Therefore, our sieve problem is one of dimension no more than one half.

We shall soon see that the sequence $\mathcal{A} = \big(w(a)\big)$ with a squarefree, $(a, 2b) = 1$ has exponent of distribution one. Hence the beta-sieve produces upper and lower bounds for $S(\mathcal{A}, z)$ in the range $z \leqslant A^{1-\varepsilon}$. If the dimension is $\kappa < 1/2$ then one can cover the full range $z \leqslant A$ by using the extension established in Theorem 11.21. Moreover, if we request some regularity in the distribution of the set $\mathcal{P}_b$ of primes then there is a possibility to deduce an asymptotic formula for $S(\mathcal{A}, A)$ by the Eratosthenes sieve (see Theorem 4.2).

The case $\kappa = 1/2$ is somewhat different. In particular, when $\mathcal{P}$ is the full set of odd primes, the most important case, our sequence $\mathcal{A} = \big(w(a)\big)$ has special features. First of all, $w(a) \neq 0$ implies $(a/b) = 1$. Hence a can have only an even number of prime divisors in the set $\mathcal{P}_b$. Therefore, the conditions of Theorem 14.5 are satisfied giving an asymptotic formula for $S(\mathcal{A}, A)$.

THEOREM 11.28. *Let $\mathcal{P}$ be a set of odd primes and let $b > 1$, $b \equiv 1 \pmod 4$ be squarefree. Suppose that (5.41) holds with $\kappa < 1/2$ for $g(p) = (p+1)^{-1}$ over primes $p \in \mathcal{P}_b$, that is,*

$$\prod_{\substack{w\leqslant p<z \\ p\in\mathcal{P}_b}} \Big(1+\frac{1}{p}\Big) \ll \Big(\frac{\log z}{\log w}\Big)^{\kappa}\Big(1+\frac{L}{\log w}\Big) \tag{11.207}$$

for all $z > w \geqslant 2$. Then, for any A sufficiently large in terms of κ, b and L, we have

$$\sideset{}{^\flat}\sum_{\substack{a\leqslant A \\ (a,2bP_b)=1}} w(a) \asymp A \prod_{\substack{p\leqslant A \\ p\in\mathcal{P}_b}} \Big(1+\frac{1}{p}\Big)^{-1} . \tag{11.208}$$

THEOREM 11.29. *Let $\mathcal{P}$ be the set of all odd primes and $b > 1$, $b \equiv 1 \pmod 4$ be squarefree. Then, as $A \to \infty$, we have*

$$\sideset{}{^\flat}\sum_{\substack{a\leqslant A \\ (a,2bP_b)=1}} w(a) \sim c(\chi)\frac{A}{\sqrt{\log A}} \tag{11.209}$$

where $\chi = \chi_b$ *is the Jacobi symbol of modulus* b *and*

$$c(\chi) = \frac{1}{4}\Big(5 - \chi(2)\Big)\left(\frac{L(1,\chi)}{\pi\zeta(2)}\prod_{p|b}\Big(1+\frac{1}{p}\Big)^{-1}\prod_{\chi(b)=1}\Big(1-\frac{1}{p^2}\Big)\right)^{\frac{1}{2}} . \tag{11.210}$$

For the proofs of these theorems we need to establish good approximations to the congruence sums

$$A_d = \sum_{\substack{a\leqslant A\\ (a,2b)=1\\ a\equiv 0\,(d)}}^{\flat} w(a)$$

for d squarefree, $(d, 2b) = 1$. Since $w(a)$ is periodic with period b we have

$$A_d = \sum_{\beta\,(\mathrm{mod}\ 2b)}^{\star} w(\beta) \sum_{\substack{a\leqslant A\\ a\equiv\beta\,(2b)\\ a\equiv 0\,(d)}}^{\flat} 1 .$$

The inner sum is equal to

$$\begin{aligned}
\sum^{\flat} = \sum_{\substack{a\leqslant A/d\\ a\equiv\beta\bar{d}\,(2b)\\ (a,2bd)=1}}^{\flat} 1 &= \sum_{(\alpha,2bd)=1}\mu(\alpha)\sum_{\delta|d}\mu(\delta)\sum_{\substack{a\leqslant A/d\alpha^2\delta\\ \alpha\equiv\beta'\,(\,2b)}} 1\\
&= \sum_{\substack{\alpha<\sqrt{A/d}\\ (\alpha,2bd)=1}}\mu(\alpha)\sum_{\delta|d}\mu(\delta)\Big(\frac{A}{2bd\alpha^2\delta} + O(1)\Big)\\
&= \frac{A}{2bd}\frac{\varphi(d)}{d}\prod_{p\nmid 2bd}\Big(1-\frac{1}{p^2}\Big) + O\big(\tau(d)\sqrt{A/d}\big)\\
&= \frac{2A}{3\,\zeta(2)bd}\prod_{p|b}\Big(1-\frac{1}{p^2}\Big)\prod_{p|d}\Big(1+\frac{1}{p}\Big)^{-1} + O\big(\tau(d)\sqrt{A/d}\big) .
\end{aligned}$$

We have

$$\sum_{\beta\,(\mathrm{mod}\ 2b)}^{\star} w(\beta) = \frac{1}{\tau_{\mathcal{P}}(b)}\sum_{\beta\,(\mathrm{mod}\ 2b)}^{\star}\sum_{\substack{m|b\\ m|P}}\Big(\frac{\beta}{m}\Big) = \frac{\varphi(b)}{\tau_{\mathcal{P}}(b)}$$

where $\tau_{\mathcal{P}}(b)$ denotes the number of divisors of b composed of primes in $\mathcal{P}$. Hence, we conclude that

$$A_d = \frac{2bg(bd)A}{3\,\zeta(2)\tau_{\mathcal{P}}(b)} + O\big(\tau(d)\sqrt{A/d}\big) , \tag{11.211}$$

where g is the multiplicative function given by

$$g(p) = (p+1)^{-1} , \tag{11.212}$$

and the implied constant depends on b. The approximation (11.211) also holds trivially for d not squarefree in which case we set $g(d) = 0$ because $A_d = 0$ and hence so does the remainder term. In other words, the sequence $\mathcal{A}$ satisfies (1.2) with

$$X = \frac{2bg(b)A}{3\,\zeta(2)\tau_{\mathcal{P}}(b)}, \quad r_d(\mathcal{A}) \ll \tau(d)\sqrt{A/d} . \tag{11.213}$$

Hence, $\mathcal{A}$ has level of distribution

$$D = A(\log A)^{-4} \ . \tag{11.214}$$

From Theorem 11.21 we complete the proof of Theorem 11.28. Using Theorem 14.5 in place of Theorem 11.21 we also complete the proof of Theorem 11.29 apart from the computation of the main term. We get $S(\mathcal{A}, A) \sim XV(A)F(1)$ where $F(1) = 2\sqrt{e^\gamma/\pi}$ and

$$\begin{aligned} V(A) = \prod_{\substack{p<A \\ p\in\mathcal{P}_b}} (1 - g(p)) = \prod_{\substack{p<A \\ p\in\mathcal{P}_b}} \Big(1 + \frac{1}{p}\Big)^{-1} &= \frac{1}{4}(5 - \chi(2)) \prod_{\substack{p<A \\ \chi(p)=-1}} \Big(1 + \frac{1}{p}\Big)^{-1} \\ = \frac{1}{4}(5 - \chi(2)) \prod_{\substack{p<A \\ p|b}} \Big(1 + \frac{1}{p}\Big)^{\frac{1}{2}} \prod_{\substack{p<A \\ \chi(p)=1}} \Big(1 - \frac{1}{p^2}\Big)^{\frac{1}{2}} \\ \prod_{p<A} \Big(1 - \frac{1}{p^2}\Big)^{-\frac{1}{2}} \Big(1 - \frac{\chi(p)}{p}\Big)^{-\frac{1}{2}} \Big(1 - \frac{1}{p}\Big)^{\frac{1}{2}} \ . \end{aligned}$$

Applying Mertens' formula (2.21) and inserting X given by (11.213) we arrive at (11.209), concluding the proof.

REMARK. In both Theorems 11.28 and 11.29 it is possible to give a considerable range of uniformity in the parameter, wider than those available by analytic means. Further results and history on Serre's problem are in [**58**]. A different treatment of the question is given in Section 4.5 of the recent book [**24**] of A. Cojocaru and R. Murty.

11.21. Representations by Norm Forms

The sieve of fractional dimension $\frac{1}{2} < \kappa < 1$ is attractive for some arithmetical questions, a natural case being the problem of representability of integers in a given set by norms of ideals in the ring of integers in a number field.

Let $K/\mathbb{Q}$ be a Galois extension of degree $k \geqslant 2$. The set of rational primes which are norms of ideals has density $1/k$. Hence, if we want to select from a sequence $\mathcal{A} = (a_n)$ the elements a_n with

$$n = N\mathfrak{a}, \quad \mathfrak{a} \text{ an integral ideal,} \tag{11.215}$$

it suffices to sieve out those elements which have a prime divisor in the set $\mathcal{P}$ which is complementary to

$$\mathcal{P}_K = \{p\,;\ p = N\mathfrak{p}, \quad \mathfrak{p} \text{ a prime ideal}\}.$$

Therefore, the set of forbidden primes in the sifting range has density $\kappa = 1 - \frac{1}{k}$.

We have

$$\sum_{\substack{n\leqslant x \\ n=N\mathfrak{a}}} a_n \geqslant S(\mathcal{A}, \mathcal{P}, x) \ .$$

Note that we do not count the integers n in (11.215) with the multiplicity of their representations.

As we have already experienced, no sieve can produce a positive lower bound for $S(\mathcal{A}, \mathcal{P}, z)$ with z as large as $z = x$. Fortunately, under some circumstances we can significantly reduce the size of the sifting range z by exploiting arithmetical features of the field K and the sequence $\mathcal{A}$.

Suppose $K/\mathbb{Q}$ is abelian of conductor Δ. By class field theory, there exists a subgroup H of $(\mathbb{Z}/\Delta\mathbb{Z})^*$ of index k such that the rational prime p splits completely in K if and only if $p \,(\operatorname{mod} \Delta)$ belongs to H. Such primes are norms, that is, are in $\mathcal{P}_K$. Suppose also that the sequence $\mathcal{A} = (a_n)$ is supported on the residue classes $n \,(\operatorname{mod} \Delta) \in H$. Then, if n has one of the forbidden prime factors it must have at least one more. In these circumstances

$$S(\mathcal{A}, \mathcal{P}, x) = S(\mathcal{A}, \mathcal{P}, \sqrt{x}) \ . \tag{11.216}$$

Now, the sieve has a much better chance to succeed.

Suppose our sequence $\mathcal{A} = (a_n)$ has level of absolute distribution $D = x^{\alpha-\varepsilon}$ with some $0 < \alpha \leqslant 1$. By the beta-sieve of dimension $\kappa = 1 - 1/k$ (see Theorem 11.13), we have

$$S(\mathcal{A}, \mathcal{P}, z) \gg XV(z)$$

if $D \geqslant z^{\beta+\varepsilon}$, where β is the sieving limit. This gives us a lower bound for (11.216) of the correct order of magnitude if $D \geqslant x^{\beta/2+\varepsilon}$. This gives the following result.

Proposition 11.30. *Suppose that $\mathcal{A} = (a_n)$ is supported on the residue classes $n \,(\operatorname{mod} \Delta) \in H$ and that it has a density function $g(d)$ which satisfies* (11.129) *over primes $p \notin \mathcal{P}_K$ with $\kappa = 1 - 1/k$. If the exponent of absolute distribution is $\alpha > \beta/2$, then*

$$\sum_{\substack{n \leqslant x \\ n = N\mathfrak{a}}} a_n \asymp X \prod_{\substack{p < x \\ p \notin \mathcal{P}_K}} (1 - g(p)) \ ,$$

provided that x is sufficiently large in terms of K.

We give a quick example for norms of normal cubic fields. In this case

$$k = 3, \quad \kappa = \tfrac{2}{3}, \quad \beta = 1.2242\ldots \ , \tag{11.217}$$

from the table in Section 11.19.

Hence, we require the exponent of distribution to be $\alpha > 0.6121\ldots$. We shall prove in Section 22.6 that the sequence $\mathcal{A} = (a_n)$ with a_n the number of representations

$$n = \ell^6 + m^2, \quad \Delta | m, \quad (\ell, m) = 1 \ ,$$

has exponent of distribution $\alpha = 2/3$. Hence, we obtain

Theorem 11.31. *Let $K/\mathbb{Q}$ be a normal cubic extension. Then*

$$|\{(\ell, m);\ \ell^6 + m^2 = N\mathfrak{a} \leqslant x\}| \asymp \Big(\frac{x}{\log x}\Big)^{\frac{2}{3}} \ . \tag{11.218}$$

For the proof of the lower bound we use only the subsequence with $(\ell, m) = 1$ and $\Delta | m$. Note that $\ell^6 \,(\operatorname{mod} \Delta)$ belongs to the group H. The upper bound of the same order of magnitude follows easily along the same lines.

Exercises. Assume the Elliott–Halberstam conjecture. Let $K/\mathbb{Q}$ be an abelian extension of degree k. Then, for $c \neq 0$ we have

$$\sum_{\substack{n \leqslant x \\ n = c + N\mathfrak{a}}} \Lambda(n) \asymp x (\log x)^{\frac{1}{k} - 1} \quad \text{and} \quad \sum\sum_{\substack{\mathfrak{a} \subset K,\, n \leqslant x \\ n = c + N\mathfrak{a}}} \Lambda(n) \sim \mathfrak{S}(c, K) x \ .$$

11.22. Sums of Two Cubic Norms

Here we show another application of the sieve of dimension $\kappa = \frac{2}{3}$. Our problem is to estimate the number of solutions to the equation

$$a + b = N\,, \quad \text{with } (ab, N) = 1 \tag{11.219}$$

for a given large number N, where a, b are norms of integral ideals in a cubic field K,

$$a = N\mathfrak{a}\,, \quad b = N\mathfrak{b}\,. \tag{11.220}$$

Let $T(N)$ denote the number of such solutions. Note that we do not count a, b with the multiplicity of representation as a norm.

As in the previous section we assume $K/\mathbb{Q}$ is normal and we use the same notation. Recall that Δ denotes the conductor of $K/\mathbb{Q}$ and $H \subset (\mathbb{Z}/\Delta\mathbb{Z})^*$ is the subgroup of index 3 of the admissible residue classes for primes which split completely in K. These primes are norms of prime ideals of degree one; they form the set

$$\mathcal{P}_K = \{p \nmid \Delta;\ p = N\mathfrak{p}\} = \{p;\ p(\operatorname{mod}\Delta) \in H\}$$

which has density $\frac{1}{3}$.

Actually, we count only the solutions a, b of (11.219) which have all their prime factors in $\mathcal{P}_K$. Our goal is to estimate

$$T^*(N) = \sum\sum_{\substack{a+b=N\\(ab,N)=1}} C_K(a)C_K(b)\,, \tag{11.221}$$

where $C_K(a)$ denotes the characteristic function of numbers which are products of primes in $\mathcal{P}_K$. By the above characterisation we have $T(N) \geqslant T^*(N)$.

Clearly, N must satisfy the compatibility congruence condition

$$N \equiv \alpha + \beta \,(\operatorname{mod}\Delta) \tag{11.222}$$

for some α, β in H. If N is odd there is one more condition:

$$2(\operatorname{mod}\Delta) \in H, \tag{11.223}$$

because one of a, b is even.

Our problem translates into the estimation of the cardinality $S_{\alpha\beta}(N)$ of the set

$$\mathcal{S}_{\alpha\beta}(N) = \{(a,b);\ 0 < a,\ b < N,\ a + b = N,\ a \equiv \alpha(\Delta),\ b \equiv \beta(\Delta) \quad (ab, N) = 1,\ p|ab \Rightarrow p \notin \mathcal{P}\}\,, \tag{11.224}$$

where

$$\mathcal{P} = \{p \nmid 2\Delta N;\ p(\operatorname{mod}\Delta) \notin H\}\,. \tag{11.225}$$

By the above considerations we have

$$T^*(N) = \sum_{\substack{\alpha,\beta\in H\\ \alpha+\beta\equiv N(\Delta)}} S_{\alpha\beta}(N)\,. \tag{11.226}$$

Therefore the problem falls into a sieve question as we are asking for solutions $a + b = N$ free of prime divisors in a set of density $\kappa = \frac{2}{3}$.

THEOREM 11.32. *Let $K/\mathbb{Q}$ be a normal cubic extension. Suppose N satisfies the compatibility conditions (11.222), (11.223). Then, the number $T^*(N)$ of representations $N = a + b$ with ab having only prime divisors which split completely in K and $(ab, N) = 1$ satisfies*

$$T^*(N) \asymp \mathfrak{S}(N)N(\log N)^{-\frac{4}{3}} \tag{11.227}$$

provided N is sufficiently large in terms of Δ and where the implied constants depend only on Δ. Here

$$\mathfrak{S}(N) = \prod_{\substack{p|N \\ p\nmid\Delta}} \left(1 - \frac{\varepsilon_p}{p}\right) , \tag{11.228}$$

with $\varepsilon_p = \pm 1$ according to whether $p \,(\mathrm{mod}\, \Delta)$ is or is not in H.

Actually, we shall only prove the lower bound in (11.227), as the upper bound is routine. We begin the proof by applying the sieve to the sequence

$$\mathcal{A} = \left\{N - a;\ 1 \leqslant a < N,\ (a, N) = 1,\ a \equiv \alpha(\Delta),\ p|a \Rightarrow p \in \mathcal{P}_K\right\} .$$

For $(d, 2\Delta N) = 1$ we write

$$A_d = g(d)|\mathcal{A}| + r_d(\mathcal{A}) \tag{11.229}$$

with $g(d) = \varphi(d)^{-1}$, and by Theorem 9.17 it follows that

$$\sum_{\substack{d\leqslant D \\ (d,2\Delta N)=1}} \left|r_d(\mathcal{A})\right| \ll N(\log N)^{-C} , \tag{11.230}$$

where

$$D = \sqrt{N}\,(\log N)^{-C'} \tag{11.231}$$

for any $C \geqslant 0$ with C' depending on C. In other words, D given by (11.231) is the level of absolute distribution for the sequence $\mathcal{A}$ (this is an analogue of the Bombieri–Vinogradov theorem for primes in arithmetic progressions). Hence by Theorem 11.13 with $\kappa = \frac{2}{3}$ we get

$$S(\mathcal{A}, \mathcal{P}, z) \geqslant |\mathcal{A}|V(z)\left\{f(s) + o(1)\right\} \tag{11.232}$$

where $z = N^{1/2s}$ and $f(s)$ is the function given by

$$s^{\frac{2}{3}} f(s) = AI(s) \tag{11.233}$$

in the segment $\beta \leqslant s \leqslant \beta + 2$, with

$$I(s) = \int_\beta^s (t - 1)^{-\frac{2}{3}} dt^{\frac{2}{3}} .$$

Recall that A is the constant given by (11.57) and

$$\beta = 1.2242... \tag{11.234}$$

is the sieving limit for the beta-sieve of dimension $\kappa = \frac{2}{3}$. Here, the numerical value of $A = 1.9134\ldots$ does not matter, however, β being sufficiently close to 1 does matter. Moreover, we have

$$V(z) = \prod_{\substack{p<z, \\ p\in\mathcal{P}}} \left(1 - \frac{1}{p-1}\right) \asymp \mathfrak{S}^+(N)(\log N)^{-\frac{2}{3}} , \tag{11.235}$$

where

$$\mathfrak{S}^{+}(N)=\prod_{\substack{p\mid N\\ p\notin\mathcal{P}_K}}\Big(1+\frac{1}{p}\Big)\,.$$

We want a positive lower bound of the true order of magnitude for $S_{\alpha\beta}(N)=S(\mathcal{A},\mathcal{P},\sqrt{N})$, but of course the direct application of the sieve yields (11.232) only for $s>\beta$. Therefore we apply the Buchstab formula

$$S(\mathcal{A},\mathcal{P},\sqrt{N})=S(\mathcal{A},\mathcal{P},z)-\sum_{z\leqslant p<\sqrt{N}}S(\mathcal{A}_p,\mathcal{P},p)\,.$$

We shall succeed with $z=N^{\frac{1}{3}}$, that is, with $s=\frac{3}{2}$. Applying the "switching trick", as we shall see in Section 14.6, we obtain

$$\sum_{z\leqslant p<\sqrt{N}}S(\mathcal{A}_p,\mathcal{P},p)=S(\mathcal{B},\mathcal{P},\sqrt{N})\leqslant S(\mathcal{B},\mathcal{P},z)$$

where $\mathcal{B}$ is the sequence

$$\begin{aligned}\mathcal{B}=\big\{N-b;\ 1\leqslant b<N,\ b=cp_1p_2\equiv\beta(\Delta),\ z\leqslant p_1<p_2,\ p_1,p_2\in\mathcal{P},\\ (c,N)=1,\ p|c\Rightarrow p\in\mathcal{P}_K\big\}\,,\end{aligned}$$

which we may say is "reversed" to $\mathcal{A}$. This sequence has the same density function $g(d)=\varphi(d)^{-1}$ and the same level of distribution (11.231) by Theorem 9.17. Now applying Theorem 11.13 with $\kappa=\frac{2}{3}$ again we get

$$S(\mathcal{B},\mathcal{P},z)\leqslant|\mathcal{B}|V(z)\big\{F(s)+o(1)\big\}\tag{11.236}$$

where $F(s)$ is the function given by

$$s^{\frac{2}{3}}F(s)=A\tag{11.237}$$

in the segment $\beta-1\leqslant s\leqslant\beta+1$.

We want the lower bound (11.232) to be larger than the upper bound (11.236). To see that this is the case we need to evaluate the cardinalities of the sets $\mathcal{A}$ and $\mathcal{B}$. For this purpose we use the following formula which can be established by a standard contour integration in the zero-free region for Dirichlet L-functions.

LEMMA 11.33. *For any $\alpha\in H$ we have*

$$\sum_{\substack{n\leqslant x\\ n\equiv\alpha(\Delta),(n,N)=1}}C_K(n)\sim\gamma_K\mathfrak{S}^{-}(N)x(\log x)^{-2/3}\tag{11.238}$$

as $x\to\infty$ uniformly for $N\leqslant x$, where γ_K is a positive constant depending on the field K and which is the same for every $\alpha\in H$, and

$$\mathfrak{S}^{-}(N)=\prod_{\substack{p\mid N\\ p\in\mathcal{P}_K}}\Big(1-\frac{1}{p}\Big)\,.$$

By Lemma 11.33 we get immediately

$$|\mathcal{A}|\sim\gamma_K\mathfrak{S}^{-}(N)N(\log N)^{-2/3}\,.\tag{11.239}$$

Next, by Lemma 11.33 we derive

$$|\mathcal{B}| = \sum_{\substack{z\leqslant p_1<p_2\\ p_1,p_2\in\mathcal{P}\\ p_1p_2(\mathrm{mod}\,\Delta)\in H}} \sum_{\substack{c<N/p_1p_2\\ c\equiv\beta/p_1p_2(\mathrm{mod}\,\Delta)}} C_K(c)$$

$$\sim \gamma_K\mathfrak{S}^-(N)N\sum_{p_1}\sum_{p_2}(p_1p_2)^{-1}\big(\log N/p_1p_2\big)^{-2/3}.$$

Hence,

$$|\mathcal{B}| \sim J(s)\gamma_K\mathfrak{S}^-(N)N(\log N)^{-2/3}\,, \tag{11.240}$$

where $J(s)$ is given by the integral

$$J(s) = \frac{1}{9}\iint\limits_{\substack{u+v<1\\ u,v>1/2s}} \frac{du\,dv}{uv(1-u-v)^{2/3}}\,. \tag{11.241}$$

Now it remains to show that, for some s with $\beta < s \leqslant \frac{3}{2}$,

$$I(s) > J(s)\,. \tag{11.242}$$

One can show by elementary estimates that for $s = \frac{3}{2}$ we have $I\big(\frac{3}{2}\big) > 0.32$ and $J\big(\frac{3}{2}\big) < 0.3$. Note that $\mathfrak{S}^-(N)\mathfrak{S}^+(N) = \mathfrak{S}(N)$ and so this completes the proof of Theorem 11.32.

Remark. Using far more sophisticated inputs, the sieve is capable (see [**111**]) of also handling additive problems with quartic norms (still in abelian extensions), for example, the equation $a = b + 1$ where a, b are norms of integral ideals. In this case the sieve dimension is $\kappa = 3/4$. It remains an open problem to treat extensions of higher degree k, which would give examples of results for quite natural sequences with sieve dimension $\kappa = 1 - 1/k$.

CHAPTER 12

The Linear Sieve

Encore

The sieve of dimension $\kappa = 1$ is undoubtedly the most important case and is required in central areas of analytic number theory such as the distribution of prime numbers. In this chapter we take a closer look at some of these linear sieve problems. In this case the beta-sieve rather generally developed in the previous chapter gives an optimal main term which makes it worthwhile to expend more effort on this central case and sharpen the result so that the error term is also optimal.

12.1. A Summary of Previous Results

We begin by recalling some results of the previous chapter in the context of $\kappa = 1$. In this case the sifting limit is $\beta(1) = 2$ and the corresponding functions $F(s), f(s)$ are the continuous solutions to the system

$$\begin{cases} sF(s) = 2e^{\gamma} & \text{if } 1 \leqslant s \leqslant 3 \ , \\ sf(s) = 0 & \text{at } s = 2 \ , \end{cases} \tag{12.1}$$

$$\begin{cases} (sF(s))' = f(s-1) & \text{if } s > 3 \ , \\ (sf(s))' = F(s-1) & \text{if } s > 2 \ . \end{cases} \tag{12.2}$$

Suppose that

$$\prod_{w \leqslant p < z} \big(1 - g(p)\big)^{-1} \leqslant \frac{\log z}{\log w}\Big(1 + \frac{L}{\log w}\Big). \tag{12.3}$$

Then, by Theorem 11.12,

$$V^{+}(D,z) < \Big\{F(s) + O\big((\log D)^{-1/6}\big)\Big\} V(z) \qquad \text{if } s \geqslant 1 \ , \tag{12.4}$$

and

$$V^{-}(D,z) > \Big\{f(s) + O\big((\log D)^{-1/6}\big)\Big\} V(z) \qquad \text{if } s \geqslant 2 \ , \tag{12.5}$$

where $s = \log D / \log z$ and the implied constant depends only on L. It is the exponent of the logarithm in (12.4) and (12.5) that we seek to improve. Applying these sieves to the sequence $\mathcal{A}$ we obtain

$$S(\mathcal{A},z) \leqslant S^{+}(\mathcal{A},z) = XV^{+}(D,z) + R^{+}(\mathcal{A},D) \ , \tag{12.6}$$

$$S(\mathcal{A},z) \geqslant S^{-}(\mathcal{A},z) = XV^{-}(D,z) + R^{-}(\mathcal{A},D) \ , \tag{12.7}$$

where

$$R^{\pm}(\mathcal{A}, D) = \sum_{d|P} \lambda_d^{\pm} r_d(\mathcal{A}) \tag{12.8}$$

and $P = P(z)$ is the product of primes $p \in \mathcal{P}$, $p < z$. These estimates hold also for $S(\mathcal{A}, z)$ because $S^- \leqslant S \leqslant S^+$, by suspending the terms S_n in the identities

$$S(\mathcal{A}, z) = S^+(\mathcal{A}, z) - \sum_{n \text{ odd}} S_n(\mathcal{A}, z) \ , \tag{12.9}$$

$$S(\mathcal{A}, z) = S^-(\mathcal{A}, z) + \sum_{n \text{ even}} S_n(\mathcal{A}, z) \ , \tag{12.10}$$

where

$$S_n(\mathcal{A}, z) = \sum_{\substack{p_n < \dots < p_1 < z \\ p_1 \dots p_m p_m^2 < D, m < n, m \equiv n (2) \\ p_1 \dots p_n p_n^2 \geqslant D}} \cdots \sum S(\mathcal{A}_{p_1 \dots p_n}, p_n) \ . \tag{12.11}$$

Some parts of the terms S_n can be shown to be positive by a use of modern technology so we reclaim them.

From (12.6), (12.7), (12.9) and (12.10) we obtain

$$S(\mathcal{A}, z) < \big(F(s) + O((\log D)^{-1/6})\big) X V(z) + R(\mathcal{A}, D) \ , \tag{12.12}$$

$$S(\mathcal{A}, z) > \big(f(s) + O((\log D)^{-1/6})\big) X V(z) - R(\mathcal{A}, D) \ , \tag{12.13}$$

where $s = \log D / \log z$. The remainder is

$$R(\mathcal{A}, D) = \sum_{\substack{d|P(z) \\ d<D}} \big| r_d(\mathcal{A}) \big| \ . \tag{12.14}$$

General estimates of the type (12.12), (12.13) with the same functions (12.1) and (12.2) in the leading terms were first established in 1965 by W.B. Jurkat and H.-E. Richert [**116**]. Actually, they got a slightly weaker estimate for the exponent of $\log D$ and a slightly worse expression for $R(\mathcal{A}, D)$, however, these differences are not essential for most applications. What needs to be emphasized is that Jurkat and Richert took a somewhat reverse approach to the one we have taken for the derivation of the beta-sieve in general. They started Buchstab's iterations from the interval $1 \leqslant s \leqslant 2$ in which the initial upper bound for $S(\mathcal{A}, z)$ is given by the Λ^2-sieve of Selberg. It is important that in this range Selberg's upper bound has the same function $\sigma(s)^{-1} = F(s) = 2e^{\gamma} s^{-1}$ in the leading term (see (7.146)), which itself is an interesting coincidence, well, only for $\kappa = 1$ which is the case in this chapter. However, for $s > 2$ the Selberg sieve yields a weaker bound because $\sigma(s) F(s) < 1$, and it requires infinitely many Buchstab iterations to reach the functions $F(s), f(s)$ in the whole range $s \geqslant 2$. One may say that we have also passed infinitely many iterations but starting from large s for which the Brun sieve provided very good estimates. Of course, in reality, the iterations are replaced by explicit constructions in either approach.

Recall that the functions $F(s)$ and $f(s)$ were introduced by the series (11.32) and (11.33) respectively, and we have proved in Section 11.6, for every $\kappa > \frac{1}{2}$, that these series converge absolutely. In particular, for $\kappa = 1$ we have

$$F(s) = 1 + \sum_{n \text{ odd}} f_n(s) \qquad \text{if } s \geqslant 1\,, \tag{12.15}$$

$$f(s) = 1 - \sum_{n \text{ even}} f_n(s) \qquad \text{if } s \geqslant 2\,, \tag{12.16}$$

where, for $s \geqslant \frac{1}{2}(3 + (-1)^n)$,

$$sf_n(s) = \underset{\substack{0<t_n<\cdots<t_1<1/s \\ t_1+\cdots+t_m+2t_m<1, m<n, m\equiv n(2) \\ t_1+\cdots+t_n+2t_n>1}}{\int \cdots \int} (t_1 \dots t_n)^{-1} t_n^{-1} dt_1 \dots dt_n\,.$$

We remark that G. Greaves [**74**] developed the following series representations:

$$F(s) = 2e^{\gamma} \sum_{n \text{ odd}} I_n(s) \qquad \text{if } s \geqslant 1\,,$$

$$f(s) = 2e^{\gamma} \sum_{n \text{ even}} I_n(s) \qquad \text{if } s \geqslant 2\,,$$

where

$$sI_n(s) = \underset{\substack{u_1>\cdots>u_n>1/s \\ u_1+\ldots+u_n=1}}{\int \cdots \int} d\mu_n$$

and $d\mu_n = (u_1 \dots u_n)^{-1} du_1 \dots du_{n-1}$ is the measure on the set

$$T_n = \{(u_1, \dots, u_n) : u_1 > \cdots > u_n, \quad u_1 + \cdots + u_n = 1\}$$

if $n \geqslant 2$ and it is the point measure if $n = 1$. Note that $I_n(s) = 0$ if $n \geqslant s$, $sI_1(s) = 1$ if $s \geqslant 1$, $sI_2(s) = \log(s-1)$ if $s \geqslant 2$ and

$$(sI_n(s))' = I_{n-1}(s-1) \quad \text{if} \quad s > \tfrac{1}{2}(3 + (-1)^n),\ n \geqslant 2.$$

Therefore, the Greaves series converge absolutely to functions which satisfy (12.1) and (12.2); thus, by uniqueness, they yield $F(s)$ and $f(s)$.

12.2. The True Asymptotics for Special Sifting Functions

It is interesting to compare the linear sieve estimates with the actual asymptotic for the sum

$$\Phi(x, z) = |\{n \leqslant x : (n, P(z)) = 1\}|. \tag{12.17}$$

LEMMA 12.1 (Buchstab). *For $s > 1$ we have*

$$\Phi(x, x^{1/s}) \sim s\omega(s) x (\log x)^{-1} \tag{12.18}$$

as $x \to \infty$ where $\omega(s)$, the Buchstab function, is the continuous solution to

$$\begin{cases} s\omega(s) = 1 & \text{if } 1 \leqslant s \leqslant 2\,, \\ (s\omega(s))' = \omega(s-1) & \text{if } s > 2\,. \end{cases} \tag{12.19}$$

PROOF. For $x > z \geqslant \sqrt{x}$ we have $\Phi(x,z) = \pi(x) - \pi(z) + 1$ and hence the asymptotic formula (12.18) holds for $1 < s \leqslant 2$ by the Prime Number Theorem. For $z < \sqrt{x}$ we have the recurrence formula

$$\Phi(x,z) = \Phi(x,\sqrt{x}) + \sum_{z \leqslant p < \sqrt{x}} \Phi\Big(\frac{x}{p}, p\Big).$$

Hence, by the induction hypothesis (assume the result for $s \leqslant N$ to deduce it for $s \leqslant N+1$), we derive

$$\begin{aligned}\Phi(x,z) &\sim \frac{x}{\log x} + x \sum_{z \leqslant p < \sqrt{x}} \omega\Big(\frac{\log x}{\log p} - 1\Big)\frac{1}{p \log p} \\ &\sim \Big(1 + \int_2^s \omega(t-1)dt\Big)x(\log x)^{-1} = s\,\omega(s)x(\log x)^{-1}.\end{aligned}$$

□

REMARKS. By the Mertens formula

$$V(z) = \prod_{p<z}\Big(1 - \frac{1}{p}\Big) \sim e^{-\gamma}(\log z)^{-1}$$

we can write (12.18) with $s = \log x/\log z$ as

$$\Phi(x,z) \sim e^{\gamma}\omega(s)xV(z). \tag{12.20}$$

Note that $e^{\gamma}\omega(s)$ is half of $F(s)$ in the range $1 < s \leqslant 2$. For $s \geqslant 2$ we have

$$\omega(s) = e^{-\gamma} + O(s^{-s}). \tag{12.21}$$

This follows from the general theory of the adjoint equation developed in Appendix B. The equation adjoint to (12.19) is

$$sq'(s) = -q(s+1)$$

whose standard solution is given by

$$q(s) = \int_0^\infty \exp\Big(-sz - \int_0^z (1-e^{-u})u^{-1}du\Big)dz; \tag{12.22}$$

see (B.10) with $(a,b) = (1,-1)$. Note that $q(s) > 0$ for $s > 0$ and $sq(s) \sim 1$ as $s \to \infty$ by (B.12). Hence (B.5) yields the inner product $\langle \omega, q\rangle = \omega(\infty) = e^{-\gamma}$. The constant function $\omega(\infty)$ has the same properties, therefore subtracting (B.5) for $\omega(\infty)$ from that for $\omega(s)$ we get the following integral formula:

$$sq(s)(\omega(s) - \omega(\infty)) = -\int_{s-1}^s (\omega(x) - \omega(\infty))q(x+1)dx \tag{12.23}$$

for $s \geqslant 2$. This shows that $\omega(s) - \omega(\infty)$ changes sign in each interval of length 1. Accordingly, the true asymptotic value for $\Phi(x,z)$ fluctuates about the expected value $xV(z)$ as often as $s = \log x/\log z$ moves by one. We have already exploited this behaviour of $\Phi(x,z)$ in Section 9.9 to show irregularities in the distribution of primes over residue classes to large moduli.

Let $\lambda(n)$ be the totally multiplicative function such that $\lambda(p) = -1$ (the Liouville function). Put

$$\Phi^*(x,z) = -\sum_{\substack{1 \leqslant n \leqslant x \\ (n,P(z))=1}} \lambda(n). \tag{12.24}$$

This sum coincides with $\Phi(x, z)$ if $x > z \geqslant \sqrt{x}$ and for $z < \sqrt{x}$ it satisfies the recurrence formula

$$\Phi^*(x, z) = \Phi^*(x, \sqrt{x}) - \sum_{z \leqslant p < \sqrt{x}} \Phi^*\left(\frac{x}{p}, p\right).$$

Hence we derive by the same argument as for the Buchstab sum:

LEMMA 12.2. *For $s > 1$ we have*

$$\Phi^*(x, x^{1/s}) \sim -s\rho'(s)x(\log x)^{-1} \tag{12.25}$$

as $x \to \infty$ where $\rho(s)$ is the Dickman function, defined in (11.105).

Note that that $\rho'(s)$ is the continuous solution to

$$\begin{cases} s\rho'(s) = -1 & \text{if } 1 \leqslant s \leqslant 2, \\ (s\rho'(s))' = -\rho'(s-1) & \text{if } s > 2. \end{cases} \tag{12.26}$$

By Mertens' formula we can write (12.25) as

$$\Phi^*(x, z) \sim -e^{\gamma}\rho'(s)xV(z). \tag{12.27}$$

Combining (12.20) and (12.27) we find that for $s \geqslant 1$,

$$\begin{cases} \omega(s) - \rho'(s) = e^{-\gamma}F(s), \\ \omega(s) + \rho'(s) = e^{-\gamma}f(s). \end{cases} \tag{12.28}$$

The Dickman function $\rho(s)$ appeared originally in the asymptotic formula

$$\Psi(x, y) = \sum_{\substack{n \leqslant x \\ p \mid n \Rightarrow p \leqslant y}} 1 \sim \rho(s)x$$

where $s = \log x / \log y$, which one can prove by the same inductive argument. There is an extensive literature on $\Psi(x, y)$; see [**95, 70**]. The problem becomes especially challenging for y relatively small and where one requires both uniformity and great precision. The strongest results were first accomplished by a brilliant insight of Hildebrand [**94**], who modified Buchstab's recurrence formula along Tchebyshev type lines.

REMARKS. Throughout the book the Euler constant γ appears and then disappears in various places. It comes from two different sources, one the constant occurring in the Mertens formula (2.21). We see it more frequently through its appearance in (B.14) in the analysis of differential-difference equations. We have already used without comment the fact that these two coincide. The results of this section give us the opportunity to see that this is indeed so. The γ occurring in (12.27) comes from the differential-difference equation. Therefore, they must coincide because

$$\Phi(x, z) \sim xV(z) \quad \text{as } s \to \infty$$

by combining the Fundamental Lemma with the Mertens formula.

12.3. The Optimality of the Linear Sieve

We shall show that the upper and the the lower bounds for $S(\mathcal{A}, z)$ obtained by the beta-sieve for $\kappa = 1$ are essentially best possible. We already have some clue of this from the equations (12.28). Consider the sequences $\mathcal{A}^+ = (a_n^+)$ and $\mathcal{A}^- = (a_n^-)$ with

$$a_n^+ = \tfrac{1}{2}\big(1 - \lambda(n)\big), \qquad a_n^- = \tfrac{1}{2}\big(1 + \lambda(n)\big) \tag{12.29}$$

for $1 \leqslant n \leqslant x$ and $a_n^+ = a_n^- = 0$ if $n > x$. Therefore, $\mathcal{A}^+$ is supported on integers $n \leqslant x$ having an odd number of prime divisors and $\mathcal{A}^-$ on those having an even number of prime divisors. We have

$$S(\mathcal{A}^+, z) + S(\mathcal{A}^-, z) = \Phi(x, z),$$
$$S(\mathcal{A}^+, z) - S(\mathcal{A}^-, z) = \Phi^*(x, z).$$

Hence, by (12.20), (12.27) and (12.28) we obtain

$$\begin{aligned} S(\mathcal{A}^+, z) &\sim \tfrac{1}{2} x F(s) V(z), \\ S(\mathcal{A}^-, z) &\sim \tfrac{1}{2} x f(s) V(z). \end{aligned} \tag{12.30}$$

On the other hand, we show that both sequences $\mathcal{A}^+, \mathcal{A}^-$ satisfy the hypotheses of the linear sieve. Indeed, we have

$$\sum_{m \leqslant y} \lambda(m) \ll y (\log y)^{-3} \tag{12.31}$$

by the Prime Number Theorem, whence for $d < x$ we derive

$$|\mathcal{A}_d^{\pm}| = \sum_{m \leqslant x/d} \tfrac{1}{2}\big(1 \pm \lambda(d)\lambda(m)\big) = \frac{x}{2d} + O\Big(\frac{x}{d}\Big(\log \frac{x}{d}\Big)^{-3}\Big).$$

Therefore, the approximation (1.2) holds with $g(d) = 1/d$, $X = x/2$ and the remainder is bounded by

$$R(\mathcal{A}, D) \ll \sum_{d < D} \frac{x}{d} \Big(\log \frac{x}{d}\Big)^{-3} \ll x (\log x)^{-2}$$

for $D = x^{1-\varepsilon}$. Hence, the upper bound for $S(\mathcal{A}^+, z)$ and the lower bound for $S(\mathcal{A}^-, z)$ given in (12.12) and (12.13) both coincide with the asymptotics (12.30). This shows that for $\kappa = 1$ Theorem 11.12 is essentially the best possible; that is, $F(s)$ cannot be reduced nor $f(s)$ increased for any $s \geqslant 2$.

The above examples of optimal sequences $\mathcal{A}^+$ and $\mathcal{A}^-$ were found by A. Selberg [**145**].

12.4. A Refinement of Estimates for the Error Terms

Some small, yet interesting, improvements of Theorem 11.13 are still possible, both in the secondary main terms and the remainder. Let $\mathcal{D}^+, \mathcal{D}^-$ be the sets of support for the linear sieves $\Lambda^+ = (\lambda_d^+)$ and $\Lambda^- = (\lambda_d^-)$ of level D, that is,

$$\begin{aligned} \mathcal{D}^+ &= \{d = p_1 \dots p_n : p_1 > \dots > p_n,\ p_1 \dots p_m p_m^2 < D \quad \text{for all } m \text{ odd}\}\ , \\ \mathcal{D}^- &= \{d = p_1 \dots p_n : p_1 > \dots > p_n,\ p_1 \dots p_m p_m^2 < D \quad \text{for all } m \text{ even}\}\ . \end{aligned} \tag{12.32}$$

Recall that, by convention, both sets contain $d = 1$.

We shall show the following:

LEMMA 12.3. *If* $2 \leqslant z \leqslant \sqrt{D}$ *then*

$$|\{d \in \mathcal{D}^+ \cup \mathcal{D}^-;\ d|P(z)\}| \ll D(\log D)^{-2}. \tag{12.33}$$

Hence, if the individual remainder terms $r_d(\mathcal{A})$ *are bounded, then the total remainder satisfies*

$$\sum_{d|P(z)} \lambda_d^{\pm} r_d(\mathcal{A}) \ll D(\log D)^{-2}. \tag{12.34}$$

Furthermore, we shall show that

LEMMA 12.4. *If* $2 \leqslant z \leqslant \sqrt{D}$ *then*

$$\begin{aligned} V^+(D,z) &< \{F(s) + O((\log D)^{-1})\}V(z),\\ V^-(D,z) &> \{f(s) + O((\log D)^{-1})\}V(z), \end{aligned} \tag{12.35}$$

where $s = \log D/\log z$.

Applying (12.34) and (12.35) with $D = X$ one obtains

THEOREM 12.5. *Suppose* (1.2) *holds and* $|r_d| \leqslant 1$. *Then, for* $2 \leqslant z \leqslant \sqrt{X}$ *we have*

$$\begin{aligned} S(\mathcal{A},z) &\leqslant \{F(s) + O((\log X)^{-1})\}XV(z),\\ S(\mathcal{A},z) &\geqslant \{f(s) + O((\log X)^{-1})\}XV(z), \end{aligned} \tag{12.36}$$

where $s = \log X/\log z$ *and* $F(s), f(s)$ *are the continuous solutions to* (12.1)–(12.2). *The implied constant depends only on* L.

Here, our objective is to prove Theorem 12.5, actually in a form which is stronger in the s-aspect. Although not apparently much stronger than the corresponding Theorems 11.12, 11.13 on the general beta-sieve, it has the best-possible error term which is especially worthwhile here since, in the case of the linear sieve, the main term is also best possible. Here, the argument (based on [**101**]) is much more delicate. The proof is by induction and in order to carry out the inductive step we are required to prove a somewhat stronger but more technical estimate.

We begin by recalling some of the material from the previous chapters which was established more generally for the beta-sieve. Hence

$$V^+(D,z) = \sum_{\substack{d|P(z)\\ d\in\mathcal{D}^+}} \mu(d)g(d)\ ,$$

$$V^-(D,z) = \sum_{\substack{d|P(z)\\ d\in\mathcal{D}^-}} \mu(d)g(d)\ ,$$

and the remainders are

$$R^+(\mathcal{A},D) = \sum_{\substack{d|P(z)\\ d\in\mathcal{D}^+}} \mu(d)r_d(\mathcal{A})\ ,$$

$$R^-(\mathcal{A},D) = \sum_{\substack{d|P(z)\\ d\in\mathcal{D}^-}} \mu(d)r_d(\mathcal{A})\ .$$

The sifted sums become

$$S^+(\mathcal{A}, z) = \sum_{\substack{d|P(z)\\ d\in\mathcal{D}^+}} \mu(d)|\mathcal{A}_d| \ ,$$

$$S^-(\mathcal{A}, z) = \sum_{\substack{d|P(z)\\ d\in\mathcal{D}^-}} \mu(d)|\mathcal{A}_d| \ .$$

Recall that for $n \geqslant 1$ we defined $S_n(\mathcal{A}, z)$ by (6.28) and so had the two identities (6.29), (6.30), and

$$S^+(\mathcal{A}, z) = XV^+(D, z) + R^+(\mathcal{A}, D) \ ,$$

$$S^-(\mathcal{A}, z) = XV^-(D, z) + R^-(\mathcal{A}, D) \ .$$

We want to evaluate $V^{\pm}(D, z)$. We decompose these as

$$V^+(D, z) = V(z) + \sum_{n \text{ odd}} V_n(D, z),$$

$$V^-(D, z) = V(z) - \sum_{n \text{ even}} V_n(D, z),$$

where, for $n \geqslant 1$,

$$V_n(D, z) = \sum_{\substack{p_n<\cdots<p_1<z\\ p_1\dots p_m p_m^2<D,\ m<n\ , m\equiv n (2)\\ p_1\dots p_n p_n^2\geqslant D}} g(p_1 \dots p_n)V(p_n) \ . \tag{12.37}$$

These are defined for

$$s \geqslant \tfrac{1}{2}(3 + (-1)^n) = \begin{cases} 1 & \text{if } n \text{ odd,} \\ 2 & \text{if } n \text{ even.} \end{cases}$$

For $n = 1$ we have

$$V_1(D, z) = \sum_{D^{1/3}\leqslant p_1<z} g(p)V(p) = V\big(D^{\frac{1}{3}}\big) - V(z)$$

if $1 \leqslant s \leqslant 3$ and $V_1(D, z) = 0$ if $s > 3$. For every $n > 1$ and $s \geqslant \frac{1}{2}(3 + (-1)^n)$ we have the following recurrence formula:

$$V_n(D, z) = \sum_{p<D^{1/s_n}} g(p)V_{n-1}\Big(\frac{D}{p}, p\Big)$$

where $s_n = \max\big(s, \frac{1}{2}(5 - (-1)^n)\big)$.

Parallel to $V_n(D, z)$ we have defined the sequence of functions $f_n(s)$ which summed to give the series (11.32), (11.33). Note that $f_n(s) = 0$ if $s \geqslant n + 2$, because the domain of integration is empty in this range.

For $n = 1$ we have

$$sf_1(s) = \begin{cases} 3 - s & \text{if } 1 \leqslant s \leqslant 3, \\ 0 & \text{if } s > 3. \end{cases}$$

For every $n > 1$ and $s \geqslant \frac{1}{2}(3+(-1)^n)$ we have the recurrence formula

$$sf_n(s) = \int_{s_n}^{\infty} f_{n-1}(t-1)dt .$$

Next, for every $N \geqslant 1$ and $s \geqslant \frac{1}{2}(3+(-1)^N)$ we put

$$T_N(D,z) = \sum_{\substack{n\leqslant N \\ n\equiv N (\mathrm{mod}\, 2)}} V_n(D,z)$$

and

$$G_N(s) = \sum_{\substack{n\leqslant N \\ n\equiv N\ (\mathrm{mod}\, 2)}} f_n(s) .$$

Our goal is to prove by induction on N the following inequalities:

$$T_N(D,z) \leqslant V(z)\Big\{G_N(s) + C\,\delta_N(s,z)\,\frac{H_N(s)}{\log z}\Big\}, \quad \text{for } s \geqslant \tfrac{1}{2}(3+(-1)^N),$$

where C is a constant depending only on L and $\delta_N(s,z)$ is a compensation factor depending on the parity of N, a factor which will be determined in the induction step.

The functions $H_N(s)$ which also depend only on the parity of N are quite special. These are the continuous solutions of the equations

$$\begin{aligned} s^2H_1(s) &= 1, \quad \text{if } 1 \leqslant s \leqslant 3,\\ s^2H_2(s) &= 2, \quad \text{at } s=2\ ,\\ \big(s^2H_1(s)\big)' &= -sH_2(s-1), \qquad \text{if } s>3,\\ \big(s^2H_2(s)\big)' &= -sH_1(s-1), \qquad \text{if } s>2\ . \end{aligned}$$

Denoting

(12.38) $$H(s) = H_1(s) + H_2(s), \qquad \text{if} \qquad s \geqslant 2\ ,$$

we see that $H(s)$ is the continuous solution of the differential-difference equation

(12.39) $$\big(s^2H(s)\big)' = -sH(s-1)\ , \quad s>3,$$

with the initial condition

(12.40) $$s^2H(s) = 2 + \frac{1}{s-1} - \log(s-1)\ , \quad 2 \leqslant s \leqslant 3\ .$$

By (11.85) we find that

(12.41) $$\Big[(s-1)^2 - \tfrac{1}{2}\Big]sH(s) = \int_{s-1}^{s} \Big(t^2 - \tfrac{1}{2}\Big)H(t)dt\ , \quad \text{if } s \geqslant 3\ .$$

Hence we see by Lemma 11.2 that

(12.42) $$0 < H(s) \ll s^{-s}\ .$$

Actually, it yields the following formula (see Corollary 11.4):

(12.43) $$H(s) = \exp\big(-s\log s - s\operatorname{loglog} s + O(s)\big)\ .$$

We also have the local property $H(s-1) \asymp H(s)s\log s$, from which we only require the lower bound

(12.44) $$H(s-1) \gg H(s)s\log s\ ,$$

established in (11.96); see also (11.90). By (11.90)

$$H(s) \ll H_1(s), H_2(s) \ll H(s) .$$

By (11.115), (11.116),

$$G_N(s) \ll H_1(s) \quad \text{if } s \geqslant 1,\ N \text{ odd} , \tag{12.45}$$

$$G_N(s) \ll H_2(s) \quad \text{if } s \geqslant 2,\ N \text{ even} . \tag{12.46}$$

Before we begin the induction procedure we establish a direct estimate for $V_n(D,z)$ which will be useful for extremely large n. Recall that we are assuming (12.3).

PROPOSITION 12.6. *For $s \geqslant e \operatorname{loglog} z + L + 3$ we have*

$$T_N(D,z) \ll s^4 \Big(\frac{e}{s}(\operatorname{loglog} z + L + 1)\Big)^s \tag{12.47}$$

where the implied constant is absolute.

PROOF. We have by (11.145) and (12.37)

$$\begin{aligned} V_n(D,z) &= \sum_{\substack{p_n < \cdots < p_1 < z \\ \cdots\cdots}} g(p_1 \dots p_n) V(p_n) \\ &\leqslant \frac{1}{n!}\Big(\sum_{p<z} g(p)\Big)^n \leqslant \Big(\frac{e}{n}\,(\operatorname{loglog} z + L + 1)\Big)^n \\ &= e^2\Big(\frac{(\operatorname{loglog} z + L + 1)}{n}\Big)^2 \Big(\frac{e}{n}\,(\operatorname{loglog} z + L + 1)\Big)^{n-2} . \end{aligned}$$

If we assume $s \geqslant e \operatorname{loglog} z + L + 3$ then the latter factor is decreasing in $n \geqslant s-2$. However, if $n \leqslant s-2$ then $V_n(D,z) = 0$. Hence

$$\begin{aligned} V_n(D,z) &\ll \Big(\frac{\operatorname{loglog} z + L + 1}{n}\Big)^2 \Big(\frac{e}{s}\,(\operatorname{loglog} z + L + 1)\Big)^{s-4} \\ &\ll \frac{s^4}{n^2}\Big(\frac{e}{s}\,(\operatorname{loglog} z + L + 1)\Big)^s . \end{aligned}$$

Summing over n, we obtain (12.47). □

Although the bound (12.47) is rather weak in the z-aspect, not even capturing the correct order of magnitude, it does give an estimate which is quite strong in the s-aspect, as strong as the leading term $-s\log s$ for $H(s)$ in (12.43). When we apply this crude bound only the second term $-s \operatorname{loglog} s$ will remain and this will ensure the decay needed to complete the induction argument.

For the induction argument we need a variant of Lemma 5.1 which capitalizes on a stronger dimension assumption than that given in (5.38). We introduce the following assumption:

For any $z > w \geqslant 2$ we have

$$\prod_{w \leqslant p < z} \big(1 - g(p)\big)^{-1} \leqslant \frac{\log z}{\log w}\big(1 + L\exp(-\sqrt{\log w})\big) , \tag{12.48}$$

where L is a constant.

Note that this is essentially as strong as the error term in the classical Prime Number Theorem. This hypothesis could be weakened considerably and we accept it (with some hesitation) because it helps tremendously with the proof.

LEMMA 12.7. *Let g be a multiplicative function with $0 \leqslant g(p) < 1$ such that (12.48) holds for all w with $y \leqslant w \leqslant z$. Let h be a continuous non-negative and non-decreasing function on the interval $[y, z]$. Then we have*

$$\sum_{y \leqslant p < z} g(p)h(p)V(p) \leqslant -V(z)\big(1 + L\exp(-\sqrt{\log y})\big) \int_y^z h(w)d\Big(\frac{\log z}{\log w}\Big) + L\exp(-\sqrt{\log y})h(z)V(z) \ . \tag{12.49}$$

PROOF. Apply Lemma 5.1 with $K = 1 + L\exp(-\sqrt{\log y})$. □

The following proposition, although rather technical both in statement and in proof, forms the core of the argument needed to prove our main results which are stated in Section 12.6.

PROPOSITION 12.8. *For any $D \geqslant 2$, $N \geqslant 1$, we have*

$$T_N(D, z) \leqslant V(z)\Big\{G_N(s) + \frac{CN}{N+1}\delta(s, z)\frac{H_N(s)}{\log z}\Big\} \tag{12.50}$$

for $s \geqslant 1$ if N is odd and $s \geqslant 2$ if N is even, where $s = \log D/\log z$, and

$$\delta(s, z) = \Big(1 + \frac{(\log s)^5}{(\log z)^2}\Big)^{5(s+1)} \ . \tag{12.51}$$

Here C is a large constant which depends only on L.

We shall prove the proposition by induction on N. Certainly the factor $\delta(s, z)$ looks strange but something like this is needed for the inductive step. After the proposition is proved we shall hurry to remove it from sight. This can be viewed as a compensation factor; it is invisible until s becomes extremely large and then it overcomes the decay in $H_N(s)$ (more precisely the second term $s \log\log s$ in (12.43)). At this point we appeal to Proposition 12.6 to complete the proof.

Although the factor $\delta(s, z)$ is needed for the proof, its presence complicates matters in some respects. Before proceeding to the induction we establish some technical estimates made necessary by these complications.

LEMMA 12.9. *Let $X \geqslant 3$ and $Y \geqslant BX^2 \log^4 X$ where B is a large absolute constant. Then, for $3 \leqslant s \leqslant X$ we have*

$$\int_s^X \Big(1 + \frac{t^2 \log^5 t}{Y}\Big)^{5t} H_N(t-1)t\,dt \leqslant \Big(1 + B\frac{s^2 \log^4 s}{Y}\Big)\Big(1 + \frac{s^2 \log^5 s}{Y}\Big)^{5s} H_{N+1}(s)s^2 \ . \tag{12.52}$$

PROOF. Let $R(s)$ be the function on the right-hand side of (12.52). We have

$$-\frac{R'}{R}(s) = \frac{-(H_{N+1}(s)s^2)'}{H_{N+1}(s)s^2} + O\Big(Y^{-1}(B + s\log s)s(\log s)^4\Big) \ .$$

For the first term we have the lower bound

$$\frac{-(H_{N+1}(s)s^2)'}{H_{N+1}(s)s^2} = \frac{H_N(s-1)s}{H_{N+1}(s)s^2} \gg \log s$$

by (11.99) and (11.100). Hence

$$-\frac{R'}{R}(s) \geqslant \frac{H_N(s-1)s}{H_{N+1}(s)s^2}(1 - \delta(s))$$

where

$$\delta(s) \ll Y^{-1}(B + s\log s)s(\log s)^3 \ .$$

Now, multiplying through by $R(s)$, we obtain

$$-R'(s) \geqslant \Big(1 + \frac{s^2 \log^5 s}{Y}\Big)^{5s} H_N(s-1)s \tag{12.53}$$

because

$$\big(1 + BY^{-1}s^2(\log s)^4\big)\big(1 - \delta(s)\big) \geqslant 1,$$

provided s is sufficiently large. Integrating (12.53) we obtain (12.52).

For small s the result follows from that for larger s_1 by increasing the constant B and using the formula

$$\int_s^{s_1} H_N(t-1)dt = s^2 H_{N+1}(s) - s_1^2 H_{N+1}(s_1) \ .$$

□

PROOF OF PROPOSITION 12.8. Note that $T_N(D,z) = 0$ and $G_N(s) = 0$ if $s \geqslant N+2$. Therefore we need to prove the result only for $s \leqslant N+2$.

We consider a real number D_0 large in terms of L, as large as will be needed in the argument. Irrespective of the size of D_0 there exists a constant $C = C(L, D_0)$ such that (12.50) holds for all $D \leqslant D_0$ with this choice of C. Clearly such a choice is possible because the variables z, s and N are also then bounded while $H_N(s)$ is positive and continuous. We also assume, as we may, that C is large.

For $N = 1$, we have $1 \leqslant s \leqslant 3$ or else $T_1(D,z)$ vanishes. In this range we have $G_1(s) = \frac{3}{s} - 1$ and

$$\begin{aligned} T_1(D,z) &= \sum_{D^{1/3} \leqslant p < D^{1/s}} g(p)V(p) = V\big(D^{1/3}\big) - V\big(D^{1/s}\big) \\ &\leqslant \Big[G_1(s) + 3\,L\exp\Big(-\tfrac{1}{3}\sqrt{\log D}\Big)\Big]\, V(z) \ , \end{aligned} \tag{12.54}$$

giving (12.50) in stronger form. Given $N \geqslant 1$ we assume (12.50) holds for N and we are going to prove it for $N+1$ in all of the appropriate ranges.

If $D \leqslant D_0$ we are done. From now on let $D > D_0$. Put

$$X = (\log D)(\operatorname{loglog} D)^{-\frac{11}{5}} \ . \tag{12.55}$$

If $s > X$ we easily have $s > e \operatorname{loglog} z + L + 3$ so Proposition 12.6 can be applied, yielding

$$T_{N+1}(D,z) \ll s^4 \Big(\frac{e}{s}(\operatorname{loglog} z + L + 1)\Big)^s \ . \tag{12.56}$$

Hence, by (12.43) we get

$$T_{N+1}(D,z) \ll \delta(s,z)\frac{H(s)}{\log z} V(z) \ . \tag{12.57}$$

Here we used the fact that our choice of the exponents $\frac{11}{5}$ in (12.55) and 5 in (12.51) ensure that the right-hand side of (12.57) far exceeds the right-hand side of (12.56) in the range $s > X$, essentially by a factor of size $\exp(2s \operatorname{loglog} s)$.

Now let $s \leqslant X$ and $s \geqslant 2$ if N odd, $s \geqslant 3$ if N even. We appeal to the recurrence formula

$$T_{N+1}(D,z) = \sum_{D^{1/X} \leqslant p < D^{1/s}} g(p) T_N\Big(\frac{D}{p}, p\Big) + T_{N+1}\big(D, D^{1/X}\big) . \tag{12.58}$$

The last term was estimated in (12.56) so

$$T_{N+1}\big(D, D^{1/X}\big) \ll X^4 \Big(\frac{e}{X}\big(\log\frac{\log D}{X} + L + 1\big)\Big)^X . \tag{12.59}$$

For the terms in the sum in (12.58) we apply the induction hypothesis (12.50) obtaining the bound

$$\sum_{D^{1/X} \leqslant p < D^{1/s}} g(p) V(p) h(p) \tag{12.60}$$

where

$$h(w) = G_N(t_w - 1) + \frac{CN}{N+1} \frac{H_N(t_w - 1)}{\log w} \Big(1 + \frac{(\log t_w)^5}{(\log w)^2}\Big)^{5t_w} \tag{12.61}$$

with $t_w = (\log D)(\log w)^{-1} - 1$. More explicitly we have

(12.62)

$$\begin{aligned} T_{N+1}(D,z) \leqslant & \sum_{D^{1/X} \leqslant p < D^{1/s}} g(p) V(p) G_N(t_p - 1) + \Big(\frac{6}{X} \log\log D\Big)^X \\ & + \frac{CN}{N+1} \sum_{D^{1/X} \leqslant p < D^{1/s}} g(p) \frac{V(p)}{\log p} H_N(t_p - 1) \Big(1 + \frac{(\log t_p)^5}{(\log p)^2}\Big)^{5t_p} . \end{aligned}$$

Making the change of variable $w = D^{1/t}$ we see that

$$h(w) = G_N(t-1) + \frac{CN}{N+1} \Big(1 + \frac{t^2 \log^5 t}{\log^2 D}\Big)^{5t} \frac{H_N(t-1)t}{\log D} . \tag{12.63}$$

We are going to show that $h(w)$ is decreasing in t and hence increasing in w. This is obvious for $G_N(t-1)$ but less so for the other term. We verify it as follows. Let

$$Q(t) = \Big(1 + \frac{t^2 \log^5 t}{\log^2 D}\Big)^{5t} H_N(t-1)t.$$

We need to show that $Q(t)$ is decreasing. We check that

$$\begin{aligned} -\frac{Q'(t)}{Q(t)} &= -\frac{\big[H_N(t-1)(t-1)^2\big]'}{H_N(t-1)(t-1)^2} - \frac{1}{t} + \frac{2}{t-1} + O\Big(\frac{t^2 \log^5 t}{\log^2 D}\Big) \\ &= \frac{H_{N+1}(t-2)}{(t-1)H_N(t-1)} + \frac{t+1}{t(t-1)} + O\Big(\frac{t^2 \log^5 t}{\log^2 D}\Big) \\ &= \frac{H_{N+1}(t-2)}{(t-1)H_N(t-1)} \Big\{1 + O\Big(\frac{t^2 \log^4 t}{\log^2 D}\Big)\Big\} + \frac{t+1}{t(t-1)} > 0 , \end{aligned}$$

so $Q'(t) < 0$ as required. This makes Lemma 12.7 applicable giving

(12.64)

$$\sum_{D^{1/X} \leqslant p < D^{1/s}} g(p)V(p)h(p) \leqslant V(z)\Big\{1 + L\exp\Big(-\sqrt{\frac{\log D}{X}}\Big)\Big\}$$

$$\cdot \int_s^X \Big\{G_N(t-1) + \frac{CN}{N+1}\Big(1+\frac{t^2\log^5 t}{\log^2 D}\Big)^{5t}\frac{H_N(t-1)t}{s\log z}\Big\}\frac{dt}{s}$$

$$+ V(z)\,L\exp\Big(-\sqrt{\frac{\log D}{X}}\Big)\Big\{G_N(s-1) + \frac{CN}{N+1}\Big(1+\frac{s^2\log^5 s}{\log^2 D}\Big)^{5s}\frac{H_N(s-1)}{\log z}\Big\}$$

and in the right-hand side we use the bounds:

$$L\exp\Big(-\sqrt{\frac{\log D}{X}}\Big) = L\exp(-(\operatorname{loglog} D)^{11/10}) \ll (\log D)^{-2005}\ ,$$

$$G_N(s-1) \ll s^2 H_{N+1}(s) \ll (\log D)^2 H_{N+1}(s)\ ,$$

$$H_N(s-1) \ll s^2 H_{N+1}(s) \ll (\log D)^2 H_{N+1}(s)\ ,$$

$$\int_s^X G_N(t-1)dt \leqslant \int_s^\infty G_N(t-1)dt = sG_{N+1}(s)\ .$$

Moreover, by (12.52) with $Y = (\log D)^2$, we have

$$\int_s^X \Big(1+\frac{t^2\log^5 t}{\log^2 D}\Big)^{5t} H_N(t-1)t\,dt \leqslant \Big(1+O\Big(\frac{s^2\log^4 s}{\log^2 D}\Big)\Big)\Big(1+\frac{s^2\log^5 s}{\log^2 D}\Big)^{5s} H_{N+1}(s)s^2$$

by (12.52).

Using these we find that the sum (12.64) augmented by (12.59) is

$$T_{N+1}(D,z) \leqslant G_{N+1}(s)$$

(12.65)

$$+ \Big[1 + O\Big(\frac{s^2\log^4 s}{\log^2 D}\Big)\Big]\frac{CN}{N+1}\Big(1+\frac{s^2\log^5 s}{\log^2 D}\Big)^{5s}\frac{H_{N+1}(s)}{\log z}\ .$$

Note that we have

$$\Big[1 + O\Big(\frac{s^2\log^4 s}{\log^2 D}\Big)\Big]\Big(1 - \frac{1}{(N+1)^2}\Big) \leqslant 1 + \frac{s^2\log^5 s}{\log^2 D}$$

for $2 \leqslant s \leqslant X$ and $N \ll \log D/\operatorname{loglog} D$, $D \geqslant D_0$ (considering large and small s separately). We have established the following bound:

(12.66) $$T_{N+1}(D,z) \leqslant V(z)\Big\{G_{N+1}(s) + C\frac{N+1}{N+2}\Big(1+\frac{s^2\log^5 s}{\log^2 D}\Big)^{5s+1}\frac{H_{N+1}(s)}{\log z}\Big\}$$

for $s \geqslant 2$ if N is odd and for $s \geqslant 3$ if N is even.

This completes the proof of (12.50) for $T_{N+1}(D,z)$ except when N is even and the sifting variable satisfies $1 \leqslant s \leqslant 3$. In this remaining situation we write

$$T_{N+1}(D,z) = T_1(D,z) + T_{N+1}(D,D^{1/3})\ .$$

Using (12.54) for the first term and (12.66) for the second we incorporate the error term in the second part inside the factor

$$\Big(1+\frac{s^2\log^5 s}{\log^2 D}\Big)^4$$

for which the first term has room to spare. This completes the proof of Proposition 12.8. □

12.5. Bounds for the Remainder

In this section we are going to estimate

$$R^+(D,z) = \sum_{d|P(z)} |\lambda_d^+| g(d)d \quad \text{for } s \geqslant 1$$

and

$$R^-(D,z) = \sum_{d|P(z)} |\lambda_d^-| g(d)d \quad \text{for } s \geqslant 2\ .$$

To this end we introduce the partial sums

$$R_N^\pm(D,z) = \sum_{\substack{d|P(z),\, d\in\mathcal{D}^\pm \\ \nu(d)\leqslant N}} g(d)d\ .$$

Note that in this section our notation $R^\pm$, R for the remainders departs from our usual practice.

We begin with a direct estimation of

$$R(D,z) = R^+(D,z) + R^-(D,z)\ , \tag{12.67}$$

which will be useful for z relatively small. Let $\delta \geqslant 0$. By Rankin's trick we get

$$\begin{aligned} R(D,z) &\leqslant \sum_{\substack{d|P(z)\\ d<D}} g(d)d \leqslant D^{1-\delta} \sum_{d|P(z)} g(d)d^\delta \\ &= D^{1-\delta}\prod_{p<z}\big(1+g(p)p^\delta\big) \leqslant D^{1-\delta}\exp\Big(\sum_{p<z} g(p)p^\delta\Big)\ . \end{aligned}$$

By (5.47),

$$\begin{aligned} \sum_{p<z} g(p)p^\delta &\leqslant \int_{\log 2}^{\log z} e^{\delta t}t^{-1}dt + z^\delta \log K \\ &\leqslant \int_0^{\delta\log z} (e^t-1)t^{-1}dt + \log\frac{\log z}{\log 2} + z^\delta \log K\ . \end{aligned}$$

For any $x \geqslant 0$ we have

$$\int_0^x (e^t-1)t^{-1}dt \leqslant e^x - 1\ .$$

Hence

$$\sum_{3\leqslant p<z} g(p)p^\delta \leqslant z^\delta \log eK + \log\log z\ .$$

Taking $\delta = \log s/\log z$ we conclude

LEMMA 12.10. *For $D \geqslant z \geqslant 2$ we have*

$$R(D,z) \leqslant \Big(\frac{eK}{s}\Big)^s D\log z\ . \tag{12.68}$$

Now we are ready to prove by induction on N the following estimates.

PROPOSITION 12.11. *For any* $D \geqslant z \geqslant 2$ *and* $N \geqslant 1$ *we have*

$$R_N^+(D,z) \leqslant \frac{CDN}{N+1}\Big(1+\frac{(\log s)^5}{(\log z)^2}\Big)^{5(s+1)} \frac{H_1(s)}{\log z} V(z)\ , \quad \text{if } s \geqslant 1 \tag{12.69}$$

and

$$R_N^-(D,z) \leqslant \frac{CDN}{N+1}\Big(1+\frac{(\log s)^5}{(\log z)^2}\Big)^{5(s+1)} \frac{H_2(s)}{\log z} V(z)\ , \quad \text{if } s \geqslant 2\ , \tag{12.70}$$

where C *is a large constant depending only on* L.

PROOF. The arguments are similar to those in the proof of Proposition 12.8, actually a bit shorter since here there are no main terms. We consider a constant D_0 large in terms of L, as large as will be needed in the induction argument. Irrespective of the size of D_0, there exists a constant $C = C(L, D_0)$, such that (12.69) and (12.70) hold for all $D \leqslant D_0$ with this choice of C. Such a choice is clearly possible because the variables z, s, N are also bounded while $H_1(s) \asymp H_2(s) \asymp H(s)$ are positive and continuous. We also assume, as we may, that C is large.

From now on, let $D \geqslant D_0$. For $N = 1$ we have

$$R_1^{\pm}(D,z) \leqslant 1 + \sum_{p<\sqrt{D}} g(p)p \leqslant 1 + \sqrt{D}\,\big|\log V(\sqrt{D})\big| \ll \sqrt{D}\log D$$

because $V(z) \leqslant 3L\log z$ by (12.48). On the other hand, using

$$H_1(s) \asymp H_2(s) \asymp H(s) \asymp \exp\big(-s\log s - s\log\log 2s + O(s)\big)\ ,$$

(see (12.43)), one can verify that the right-hand sides of (12.69) and (12.70) are bounded below by $D^{2/3}$. This completes the proof for $N = 1$.

Given $N \geqslant 1$ we assume (12.69) and (12.70) hold for N and we are going to prove these estimates for $N+1$ in all relevant ranges. As we have already said, we can assume $D \geqslant D_0$. Let X be given by (12.55). If $s > X$ we have

$$\log D \ll s\,(\log s)^{11/5}$$

so the right-hand sides of (12.69) and (12.70) are bounded below by

$$\Big(\frac{\alpha(\log s)^2}{s}\Big)^s D(\log z)^{-2}$$

for some constant $\alpha > 0$. Hence the estimates (12.69), (12.70) follow by (12.68) with some room to spare.

Now, let $s \leqslant X$. We appeal to the recurrence formulas

$$R_{N+1}^{\pm}(D,z) = \sum_{D^{1/X}\leqslant p<D^{1/s^{\pm}}} g(p)\,p\,R_N^{\mp}(D/p,p) + R_{N+1}^{\pm}(D,D^{1/X})$$

where $s^- = s \geqslant 2$ and $s^+ = \max(s,3)$. Note the similarity to (12.58). Using the induction hypothesis for $R_N^{\mp}(D/p,p)$ and the estimate (12.68) for $R_{N+1}^{\pm}(D,D^{1/X})$, we get

$$R_{N+1}^{\pm}(D,z) < \frac{CDN}{N+1}\sum_{D^{1/X}\leqslant p<D^{1/s^{\pm}}} g(p)\,\frac{V(p)}{\log p}\,H_{1,2}(t_p-1)\Big(1+\frac{(\log t_p)^5}{(\log p)^2}\Big)^{5t_p}$$
$$+\Big(\frac{eK}{X}\Big)^X \frac{D}{X}\log D\ .$$

Note the similarity with (12.62) except that here we have no main term.

From here one can complete the induction argument in identical fashion to that given in the proof of Proposition 12.8. □

12.6. Linear Sieve with Sharp Error Term

In this section we draw some consequences to Propositions 12.8 and 12.11. In order to simplify the statements we sacrifice a little of the strength of these. First, as a consequence of Proposition 12.8 we state the following.

THEOREM 12.12. *Let $\Lambda^+ = \{\lambda_d^+\}$, $\Lambda^- = \{\lambda_d^-\}$ be the linear sieve of level D, that is the Möbius function restricted to the sets (12.32). Let the density function g satisfy (12.48). Let*

$$V^{\pm}(D,z) = \sum_{d|P(z)} \lambda_d^{\pm} g(d) .$$

Then, for $z = D^{1/s}$ with $\frac{1}{2}(3 \mp 1) \leqslant s \leqslant \sqrt{\log D}$, we have

$$V^+(D,z) \leqslant \{F(s) + O(H(s)(\log z)^{-1})\} V(z) , \tag{12.71}$$

$$V^-(D,z) \geqslant \{f(s) + O(H(s)(\log z)^{-1})\} V(z) , \tag{12.72}$$

where the implied constant depends only on L.

The functions $F(s), f(s)$ are defined in (12.1), (12.2) and the function $H(s)$ was defined in (12.39), (12.40), and it satisfies the bound (12.43).

As a consequence of Proposition 12.11 we have the following.

THEOREM 12.13. *Let g satisfy (12.48) and $\frac{1}{2}(3 \mp 1) \leqslant s \leqslant \sqrt{\log D}$. Then the remainder satisfies the bound*

$$\sum_{\substack{d\in\mathcal{D}^{\pm} \\ d|P(z)}} g(d)d \ll D\frac{H(s)}{\log z} V(z) , \tag{12.73}$$

where the implied constant depends only on L.

In particular, for $g(d) = 1/d$ we obtain the bound (12.33) of Lemma 12.3 for the cardinalities of the sets $\mathcal{D}^{\pm}$:

$$\left|\{d \in \mathcal{D}^{\pm} : d|P(\sqrt{D})\}\right| \ll D(\log D)^{-2} .$$

This is best possible because these sets contain the numbers $d = p_1p_2$ with $p_2 < p_1 < \sqrt{D}$.

By combining the last two theorems we deduce estimates for the sifting function.

THEOREM 12.14. *Let $\mathcal{A} = (a_n)$ be a sequence of non-negative numbers, and $g(d)$ the associated density function. Suppose that remainder terms satisfy the bound*

$$\left|r_d(\mathcal{A})\right| \leqslant h(d)d , \tag{12.74}$$

where h might not be same function as g but satisfies the same sieve axioms; in particular, they both satisfy (12.48). Then we have

$$S(\mathcal{A},z) \leqslant \{F(s) + O(H(s)(\log z)^{-1})\} A(x)\, V(z) , \tag{12.75}$$

$$S(\mathcal{A},z) \geqslant \{f(s) + O(H(s)(\log z)^{-1})\} A(x)\, V(z) , \tag{12.76}$$

for $\frac{1}{2}(3 \mp 1) \leqslant s \leqslant \sqrt{\log D}$, respectively.

All this effort to get a very strong error term pays dividends in the lower bound. Since $f(s) \gg s-2$ in the interval $2 < s < 4$ the lower bound (12.76) is positive if $s-2 \gg (\log D)^{-1}$ giving the bound

$$S(\mathcal{A}, \eta\sqrt{D}) \gg A(x)(\log D)^{-2}$$

for a sufficiently small constant η. For example, this implies:

COROLLARY 12.15. *Let c be a sufficiently large positive constant. For $z \geqslant 2$, any interval of length $x = cz^2$ contains more than $(z/\log z)^2$ integers having no prime factor less than z.*

12.7. The Remainder in a Well-Factorable Form

In this section we modify the sums $S^+(A,z)$, $S^-(A,z)$ in the combinatorial identities (12.9), (12.10), so that the upper and lower bounds (12.12) and (12.13) will hold with essentially the same main terms but with a more flexible shape to the remainder.

The idea is driven by the following decomposition property of the linear sieve weights. Recall that these were

$$\begin{aligned} \lambda_d^+ &= \mu(d) \text{ if } d \in \mathcal{D}^+, \\ \lambda_d^- &= \mu(d) \text{ if } d \in \mathcal{D}^-, \end{aligned} \tag{12.77}$$

and $\lambda_d^\pm = 0$ otherwise. Here $\mathcal{D}^\pm$ are the sets of squarefree numbers $d = p_1 \dots p_r$ with $p_1 > p_2 > \dots > p_r$ which were defined in (12.32).

LEMMA 12.16. *Let $d = p_1 \dots p_r, p_r < \dots < p_1$, and*

$$p_1 \dots p_m p_m^2 \leqslant D\ , \tag{12.78}$$

for every $m \leqslant r$. Then, for any $D_1 \geqslant 1, D_2 \geqslant 1$ with $D_1 D_2 = D$, there are integers $d_1 \leqslant D_1, d_2 \leqslant D_2$ such that $d_1 d_2 = d$.

PROOF. By induction on r. For $r = 0$ and $r = 1$ the statement is clear. Assuming the statement for r we prove it for $r+1$. We write $d/p_{r+1} = d_1 d_2$ with $d_1 \leqslant D_1, d_2 \leqslant D_2$ by the induction hypothesis. Now, either $p_{r+1} d_1 \leqslant D_1$ or $p_{r+1} d_2 \leqslant D_2$ or else we would contradict (12.78) for $m = r+1$. □

COROLLARY 12.17. *The factorization property of Lemma 12.16 holds for every $d \in \mathcal{D}^+$ and, provided $p_1 < \sqrt{D}$, it holds for every $d \in \mathcal{D}^-$.*

This decomposition property is inherited by the sieve remainder

$$R^\pm = \sum_{d \in \mathcal{D}^\pm} \mu(d) r_d(x) \tag{12.79}$$

but is not very useful because the variables $p_1, \dots, p_r$ are badly intertwined. Our goal will be to make the summations over these variables run independently. To do so we split the range of summation into many small boxes. In doing so we lose a little precision in the main term but, with the aid of the fundamental lemma, the losses can be kept manageable.

For simplicity, we present the details of proof only for R^-. The case of R^+ is very similar (see [**103**]). From formula (12.10) we see that for any $N \geqslant 1$ we have

$$S(\mathcal{A}, z) \geqslant S^-(\mathcal{A}, z) + \sum_{\substack{n \leqslant N \\ n \text{ even}}} S_n(\mathcal{A}, z). \tag{12.80}$$

We apply this inequality, not for $\mathcal{A} = (a_n)$ itself but rather for the subsequence $\tilde{\mathcal{A}} = (\tilde{a}_n)$ where $\tilde{a}_n = a_n$ if $(n, P(u)) = 1$ and zero elsewhere. Here we take

$$u = D^{\varepsilon^2}$$

and later we shall return to $\mathcal{A}$ by means of the fundamental lemma. Let $z = D^{1/s}$ with $2 \leqslant s \leqslant \varepsilon^{-1}$. Note that when we replace $\mathcal{A}$ by $\tilde{\mathcal{A}}$, the only thing in (12.80) which changes is the term $S^-(\mathcal{A}, z)$, provided that N is not too large in terms of ε. Specifically, we require the lower bound (6.63) for p_n, which for the linear sieve ($\beta = 2$) yields

$$p_n \geqslant D^{1/2 \cdot 3^{N/2}} \tag{12.81}$$

and we require this to be larger than $u = D^{\varepsilon^2}$, which certainly holds provided

$$N \leqslant \frac{1}{2} \log \frac{1}{\varepsilon} . \tag{12.82}$$

Now it remains to evaluate $S^-(\tilde{\mathcal{A}}, z)$ which is

$$S^-(\tilde{\mathcal{A}}, z) = |\tilde{\mathcal{A}}| + \sum_{1 \leqslant r \leqslant \varepsilon^{-2}} (-1)^r \sum_{\substack{u \leqslant p_r < \cdots < p_1 < z \\ p_1 \ldots p_m p_m^2 < D \\ m \leqslant r, m \equiv r(2)}} |\tilde{\mathcal{A}}_{p_1 \ldots p_r}| . \tag{12.83}$$

For each r, we break the range in the inner sum into boxes. We let $D_1, \ldots, D_r$ run over numbers of type

$$D^{\varepsilon^2 (1+\eta)^j}, \ j = 0, 1, 2, \ldots$$

with $\eta = \varepsilon^9$. We denote by $\mathcal{D}_r$ the set of r-tuples

$$\begin{aligned} \mathcal{D}_r = \{(D_1, \ldots, D_r); D_r \leqslant \cdots \leqslant D_1 \leqslant \sqrt{D}, \\ D_1 \ldots D_m D_m^2 < D^{1/(1+\eta)}, m \leqslant r, m \equiv r(2)\} \end{aligned} \tag{12.84}$$

or else

$$\begin{aligned} \mathcal{D}_r = \{(D_1, \ldots, D_r); D_r \leqslant \cdots \leqslant D_1 \leqslant \sqrt{D}, \\ D_1 \ldots D_m D_m^2 \leqslant D, m \leqslant r, m \equiv r(2)\} , \end{aligned} \tag{12.85}$$

according to whether r is even or odd, respectively. Hence, by positivity,

$$S^-(\tilde{\mathcal{A}}, z) \geqslant \sum_{0 \leqslant r \leqslant \varepsilon^{-2}} (-1)^r \sum_{(D_1, \ldots, D_r) \in \mathcal{D}_r} \gamma(D_1, \ldots, D_r)^{-1} S_{(D_1, \ldots, D_r)} \tag{12.86}$$

where

$$S_{(D_1, \ldots, D_r)} = \sum_{\substack{D_j \leqslant p_j < D_j^{1+\eta} \\ p_1 \ldots p_r | P(z)}} |\tilde{\mathcal{A}}_{p_1 \ldots p_r}| . \tag{12.87}$$

Here $\gamma(D_1, \dots, D_r) = k_1! \dots k_\ell!$ where the k_j are the various multiplicities of the components. For $r = 0$ in (12.86) we intend the term $|\tilde{\mathcal{A}}|$ which corresponds to the single value $d = p_1 \dots p_r = 1$.

Now, by the Fundamental Lemma, we have an upper (and a lower) bound

$$\begin{aligned}|\tilde{\mathcal{A}}_{p_1 \dots p_r}| &= S(\mathcal{A}_{p_1 \dots p_r}, u) \\ &\leqslant g(p_1 \dots p_r) X V(u)\{1 + O(e^{-1/\varepsilon})\} + \sum_{b \leqslant D^\varepsilon} \lambda_b^{(r)} r_{bp_1 \dots p_r}(x)\end{aligned}$$

(with $\leqslant$ replaced by $\geqslant$ for the lower bound), where $\lambda_b^{(r)}$ are the upper (lower) bound beta-sieves of level D^ε according to whether r is even or odd. Inserting this into (12.86) we get

$$S^-(\tilde{\mathcal{A}}, z) \geqslant \sum_{0 \leqslant r \leqslant \varepsilon^{-2}} (-1)^r \sum_{(D_1, \dots, D_r) \in \mathcal{D}_r} \gamma(D_1, \dots, D_r)^{-1}$$

$$\cdot \sum_{\substack{D_j \leqslant p_j < D_j^{1+\eta} \\ p_1 \dots p_r | P(z)}} \left\{ g(p_1 \dots p_r) X V(u)\{1 + O(e^{-1/\varepsilon})\} + \sum_{\substack{b \leqslant D^\varepsilon \\ b | P(u)}} \lambda_b^{(r)} r_{bp_1 \dots p_r}(x) \right\}.$$

Here, the main term may look somewhat complicated but we can compare it to what we had before we split it into boxes:

$$V^-(D, z) = \sum_{d | P(z)/P(u)} \lambda_d^- g(d).$$

The difference between the two is accounted for by numbers d with two close prime factors, within a ratio D^η, and those d near the boundary. The contribution from the first class is bounded by

$$\sum_{\substack{d | P(z)/P(u) \\ pp' | d \\ p \leqslant p' < pD^\eta}} g(d) \leqslant \Big(\sum_{\substack{u \leqslant p < z \\ p \leqslant p' < pD^\eta}} g(pp') \Big) \prod_{u \leqslant p < z} (1 + g(p)) . \tag{12.88}$$

By (12.3) the product is $O(\varepsilon^{-2})$. For given p, the sum over p' is $O(\eta \frac{\log D}{\log p})$ and then the resulting sum satisfies

$$\sum_{u \leqslant p < z} \frac{g(p)}{\log p} \ll \frac{1}{\log u} ,$$

both of these following from Lemma 5.1. Hence, the sum in (12.88) is bounded by $\eta \varepsilon^{-4} = \varepsilon^5$. The second part of the discrepancy in the main terms, the boundary, is majorized by

$$\sum_r \sum_{\substack{u < p_r < \dots < p_1 < z \\ D^{1/(1+\eta)} < p_1 \dots p_m p_m^2 < D}} g(p_1 \dots p_r) , \tag{12.89}$$

where m is the first index ($m \leqslant r$) for which this occurs. For given m, we estimate the contribution of the sum over p_m by

$$\sum_{\max(A, u) \leqslant p_m < B} g(p_m) \ll \log \frac{\log B}{\log(\max(A, u))} + \frac{1}{\log u} ,$$

where $B = (D/p_1 \dots p_{m-1})^{1/3}$ and $A = BD^{-\eta/3}$, so this is bounded by

$$\sum_{\max(A,u) \leqslant p_m < B} g(p_m) \ll \frac{\log(B/A)}{\log u} = \frac{\eta}{3\varepsilon^2}.$$

Now, summing over the other p_j and then r we see that the total sum in (12.89) is bounded by

$$\eta\varepsilon^{-2} \prod_{u \leqslant p < z} (1 + g(p)) \ll \eta\varepsilon^{-4} = \varepsilon^5.$$

Finally, using formula (11.127) for $V^-(D, z)$ in the case of the linear sieve, we conclude the following result.

PROPOSITION 12.18. *Let $D \geqslant 1$, and $\varepsilon > 0$ be sufficiently small. Let $\mathcal{D}_r$ be defined by* (12.84), (12.85). *Let $\lambda_b^{(r)}$ be the lower bound linear beta-sieve weights of level D^ε. Then,*

$$S(\mathcal{A}, z) \geqslant XV(z)\{f(s) + O(\varepsilon^5)\} + \sum_{\substack{1 \leqslant n \leqslant N \\ n \text{ even}}} S_n(\mathcal{A}, z)$$
$$+ \sum_{0 \leqslant r \leqslant \varepsilon^{-2}} (-1)^r \sum_{(D_1, \dots, D_r) \in \mathcal{D}_r} \gamma(D_1, \dots, D_r)^{-1} \sum_{\substack{D_j \leqslant p_j < D_j^{1+\eta} \\ p_1 \dots p_r | P(z)}} \sum_{\substack{b \leqslant D^\varepsilon \\ b | P(u)}} \lambda_b^{(r)} r_{bp_1 \dots p_r}(x),$$

where $u = D^{\varepsilon^2}$, $\eta = \varepsilon^9$, $N = \frac{1}{2} \log 1/\varepsilon$.

We are next going to simplify Proposition 12.18, but we stated it first in the above form because the simplified version will lose some of the content. We begin by introducing the concept of well-factorable function.

DEFINITION 12.1. An arithmetic function $\lambda(d)$ is said to be well-factorable of level $D \geqslant 1$ if for any $M, N \geqslant 1$ with $MN = D$ there are functions $\alpha(m)$, $\beta(n)$ with $|\alpha(m)| \leqslant 1$, $|\beta(n)| \leqslant 1$ supported on $m \leqslant M$, $n \leqslant N$, such that

$$\lambda(d) = \sum_{mn=d} \alpha(m)\beta(n). \tag{12.90}$$

The same argument as in Lemma 12.16 shows that for any $M, N \geqslant 1$ with $MN = D$, every set $(D_1, \dots, D_r)$ in $\mathcal{D}_r$ can be partitioned into two subsets within which the two subproducts are bounded by M, N respectively.

By this property it is clear that, for every vector $(D_1, \dots, D_r)$ in $\mathcal{D}_r$, the remainder in Proposition 12.18 can be written as the sum of $\gamma(D_1, \dots, D_r)$ sums of the type

$$\sum_{\substack{b \leqslant D^\varepsilon \\ b | P(u)}} \sum_{d | P(z)/P(u)} \lambda_b(d) r_{bd}(x) = \sum_{d | P(z)} \lambda(d) r_d(x), \tag{12.91}$$

where $\lambda_b(d)$ is well-factorable of level $D^{1+\eta}$ and $\lambda(d)$ is well-factorable of level $D^{1+\eta+\varepsilon}$. Hence we conclude:

THEOREM 12.19. *Let $D \geqslant 1$, $s \geqslant 2$, $z = D^{1/s}$, and $f(s)$ the function defined by the system (12.1), (12.2). Let $\mathcal{A} = (a_n)$ be a sequence of non-negative numbers satisfying (5.38) with the density function $g(d)$ satisfying (5.38) with $\kappa = 1$. Let $V(z)$ be given by (5.12) and $S_n(\mathcal{A}, z)$ by (12.11). Let $\varepsilon > 0$ be sufficiently small, $N(\varepsilon) = \frac{1}{2}\log\frac{1}{\varepsilon}$, and $J(\varepsilon) = e^{-1/\varepsilon^3}$. Then, for D sufficiently large, in terms of ε, we have*

$$S(\mathcal{A}, z) \geqslant XV(z)\{f(s) + O(\varepsilon)\} + \sum_{\substack{1\leqslant n\leqslant N(\varepsilon)\\ n \text{ even}}} S_n(\mathcal{A}, z) + \sum_{1\leqslant j\leqslant J(\varepsilon)} \sum_{d|P(z)} \lambda^{(j)}(d) r_d(x) , \tag{12.92}$$

where $\lambda^{(j)}(d)$ is a well-factorable function of level D, and the implied constant depends only on K from (5.38).

PROOF. If $s \geqslant 2(1+\eta+\varepsilon)$, this follows immediately from Proposition 12.18 by changing D into $D^{1/(1+\eta+\varepsilon)}$. Of course this changes s into $s/(1+\eta+\varepsilon)$ but, by continuity, this change in $f(s)$ is within $O(\varepsilon)$. In the remaining range $2 \leqslant s \leqslant 2(1+\eta+\varepsilon)$ we cannot appeal to the previous theorem, so we apply Buchstab's formula

$$S(\mathcal{A}, z) = S(\mathcal{A}, \sqrt{D_0}) - \sum_{\substack{\sqrt{D_0}\leqslant p<\sqrt{D}\\ p|P(z)}} S(\mathcal{A}_p, p) ,$$

where $D_0 = D^{1/(1+\eta+\varepsilon)}$. For the first term we already have the result. For every $z_0 \leqslant p < z$ we estimate $S(\mathcal{A}_p, p)$ by a straightforward upper-bound sieve of absolute level $D^{1/4}$ getting

$$S(\mathcal{A}_p, p) \leqslant O(g(p)XV(p)) + \sum_{d\leqslant D^{1/4}} \xi_d r_{dp}(x) .$$

Here the first term sums to

$$X \sum_{\sqrt{D_0}\leqslant p<\sqrt{D}} g(p)V(p) \ll \varepsilon XV(z) .$$

It is clear that the resulting remainder

$$\sum_{\sqrt{D_0}\leqslant p<\sqrt{D}} \sum_{\substack{d\leqslant D^{1/4}\\ dp|P(z)}} \xi_d r_{dp}(x)$$

is of well-factorable form. Indeed, if $MN = D$, $M \geqslant N \geqslant 1$, then $M \geqslant \sqrt{D}$ and we consider d, p as separate variables if $N \geqslant D^{1/4}$. Otherwise, $dp < M$ and we consider it as a single variable. □

We state the corresponding upper bound which is proved in the same way (see [**103**] for a slightly different proof).

THEOREM 12.20. *Under the same conditions as in Theorem 12.19 we have*

$$S(\mathcal{A}, z) \leqslant XV(z)\{F(s) + O(\varepsilon)\} - \sum_{\substack{1 \leqslant n \leqslant N(\varepsilon) \\ n \text{ odd}}} S_n(\mathcal{A}, z) + \sum_{1 \leqslant j \leqslant J(\varepsilon)} \sum_{d | P(z)} \lambda^{(j)}(d) r_d(x), \tag{12.93}$$

where $\lambda^{(j)}(d)$ is a well-factorable function of level D, and the implied constant depends only on K from (5.38). *Here, $F(s)$ is defined by the same system* (12.1), (12.2) *as was $f(s)$.*

Let $D = MN$ with $M, N \geqslant 1$. Note that the remainder in (12.91) can also be written as

$$\sum_{d|P(z)} \lambda(d) r_d(x) = \sum_{\substack{m \leqslant M \\ mn|P(z)}} \sum_{\substack{n \leqslant N \\ (n,P(u))=1}} \alpha(m)\beta(n) r_{mn}(x) \ .$$

Here, the condition $mn|P(z)$ implies $(m,n) = 1$ and this constraint is sometimes helpful, but sometimes harmful. We can remove the condition $(m,n) = 1$ by Möbius inversion so that (12.91) becomes

$$\sum_{\substack{b \leqslant D^\varepsilon \\ b|P(u)}} \sum_{\ell | P(z)/P(u)} \mu(\ell) \sum_{\substack{\ell m \leqslant M \\ \ell m | P(z)/P(u)}} \sum_{\substack{\ell n \leqslant N \\ \ell n | P(z)/P(u)}} \alpha(\ell m) \beta(\ell n) r_{b\ell^2 mn}(x) \ .$$

The number of ℓ's is bounded by $2^{1/2\varepsilon^2}$. For each ℓ we consider $m' = b\ell m$ as one variable and $n' = \ell n$ as the other variable, obtaining the bilinear form

$$\sum_{\substack{m' \leqslant MD^\varepsilon \\ m'|P(z)}} \sum_{\substack{n' \leqslant N \\ n'|P(z)/P(u)}} \alpha_\ell(m') \beta_\ell(n') r_{m'n'}(x).$$

Hence we obtain by the same arguments alternative (recognizably bilinear) versions of Theorem 12.19 and Theorem 12.20.

THEOREM 12.21. *Let $\varepsilon > 0$, $D = MN$ with $M, N \geqslant 1$, $u = D^{\varepsilon^2}$. There are functions $\alpha_j(m), \beta_j(n)$, for $1 \leqslant j \leqslant J(\varepsilon)$, which depend on ε, M, N, satisfying $|\alpha_j(m)| \leqslant 1$, $|\beta_j(n)| \leqslant 1$ and such that the statements of Theorem 12.19 and Theorem 12.20 hold with the remainders replaced by sums*

$$\sum_{1 \leqslant j \leqslant J(\varepsilon)} \sum_{\substack{m \leqslant M \\ m|P(z)}} \sum_{\substack{n \leqslant N \\ (n,P(u))=1}} \alpha_j(m) \beta_j(n) r_{mn}(x) \ . \tag{12.94}$$

Moreover, the remainder (12.94) *can be amended to include the additional constraint $(m,n) = 1$.*

Note that we retain the condition $(n, P(u)) = 1$, that is, n has no small prime divisors (and at most ε^{-2} of them), because experience shows that this is often helpful in applications.

CHAPTER 13

Applications to Linear Sequences

Harmonic Variations I

13.1. Introduction

We have already seen how to apply the basic versions of the linear sieve to arithmetic progressions and to short intervals, reaching the limit of sieve methods in this format. In this chapter we investigate how the results for these linear sequences can be sharpened when one takes advantage of cancellation in the remainder terms achieving a level of distribution beyond the absolute level. To do this one requires two things: a flexibility in the shape of the sieve weights and the applicability of the harmonic analysis. Because of the latter, the scope of the method is more limited to the most basic sequences.

First we consider the sifting sequence

$$\mathcal{A} = \{n \leqslant x;\ n \equiv a \,(\mathrm{mod}\ q)\}\ , \tag{13.1}$$

for given $q \geqslant 1$ and $(a, q) = 1$. We know that this has the level of (absolute) distribution

$$D = D(x, q) = q^{-1}x^{1-\varepsilon}\ , \tag{13.2}$$

so a direct application of the linear sieve (either combinatorial, or Selberg's) yields

$$\pi(x; q, a) \leqslant \frac{(2+\varepsilon)x}{\varphi(q)\log(x/q)}\ , \tag{13.3}$$

for $1 \leqslant q < x^{1-\varepsilon}$, provided x is sufficiently large in terms of ε (see Theorem 7.15). H. L. Montgomery and R. C. Vaughan [**130**] obtained this bound with $\varepsilon = 0$; their arguments do not directly rely on the sieve and are quite delicate.

Earlier applications of the sieve had produced the same estimates with the factor $2+\varepsilon$ replaced by somewhat larger absolute constants. With any constant, this is known as the Brun–Titchmarsh theorem. The result has proved to be important for numerous applications due to the striking uniformity in the parameter q, huge by comparison to the range $q \leqslant (\log x)^A$ given by the Siegel–Walfisz theorem

$$\pi(x; q, a) = \frac{\pi(x)}{\varphi(q)} + O(x(\log x)^{-A})\ , \tag{13.4}$$

and even compared to the range $q \leqslant \sqrt{x}(\log x)^{-2}$ which follows from the assumption of the Grand Riemann Hypothesis

$$\pi(x; q, a) = \frac{\pi(x)}{\varphi(q)} + O(\sqrt{x}\log x). \tag{13.5}$$

The factor $c = 2+\varepsilon$ in (13.3) is the limit of the method in several respects. An improvement in the constant $c = 2$ for small q would have striking consequences for the problem of exceptional zeros of L-functions, hence for class numbers, etc.

In view of this limitation it was a great achievement when Y. Motohashi [**133**] was able to improve the constant $c = 2$ for certain larger ranges of q. His work has had a significant impact on subsequent developments in general sieve theory; in particular, it inspired the development of the bilinear form of the remainder which we derived in Section 12.7. In Section 13.2 we shall estimate the relevant bilinear form and apply the result to derive the following

THEOREM 13.1. *For* $1 \leqslant q \leqslant x^{\frac{2}{5}}$ *we have*

$$\pi(x;q,a) \leqslant \frac{(2+\varepsilon)x}{\varphi(q)\log(x/\sqrt{q})} \tag{13.6}$$

and for $x^{\frac{2}{5}} < q \leqslant x^{\frac{2}{3}}$ *we have*

$$\pi(x;q,a) \leqslant \frac{(2+\varepsilon)x}{\varphi(q)\log\left(x^{\frac{3}{2}}/q^{\frac{7}{4}}\right)}\ , \tag{13.7}$$

provided that x *is sufficiently large in terms of* ε.

Note that (13.6) does not improve (13.3) for small q. Our main ingredient in the analysis of the remainder is the Weil bound for Kloosterman sums. The first estimate (13.6) coincides with the original result of Y. Motohashi [**133**] who applied L-function estimates to the Selberg sieve rather than Kloosterman sum estimates to the combinatorial sieve.

The restriction $q \leqslant x^{\frac{2}{3}}$ in Theorem 13.1 is set by the range of applicability of Weil's bound for incomplete Kloosterman sums. To cover the remaining moduli $x^{\frac{2}{3}} < q < x$ we need bounds for extremely short Kloosterman sums and nobody knows how to do this. However, it is possible to proceed differently (see [**52**]). We shall take advantage of some of the extra terms in the combinatorial identity for the sifting function, terms which traditionally had been ignored by positivity because there were no known tools for handling them. These extra terms also require estimates for very short Kloosterman type sums, however, it suffices to provide them only in special ranges and the flexibility offered by the sieve formulas allows us to choose these ranges at will. This is important because the technique we shall use to estimate these short sums, based on ideas of Karatsuba [**117**], works only in specific isolated ranges. These ideas lead us in Section 13.3 to the following result.

THEOREM 13.2. *For* $x^{\frac{6}{11}+\varepsilon} \leqslant q \leqslant x^{1-\varepsilon}$ *we have*

$$\pi(x;q,a) \leqslant \frac{(2-\vartheta)x}{\varphi(q)\log(x/q)}\ , \tag{13.8}$$

with

$$\vartheta = \left(\frac{\log(x/q)}{4\log x}\right)^6\ , \tag{13.9}$$

for any x *sufficiently large in terms of* ε.

13.2. Large Moduli

For the sequence (13.1) we define the remainder term

$$r_d = A_d(x) - \frac{x}{d} \tag{13.10}$$

for any $(d, q) = 1$. Our goal is to estimate the bilinear remainder

$$R(M, N) = \sum_{\substack{m \leqslant M \\ (mn,q)=1}} \sum_{n \leqslant N} \alpha_m \beta_n r_{mn}(x) \ , \tag{13.11}$$

for any complex numbers α_m, β_n, with $|\alpha_m| \leqslant 1$, $|\beta_n| \leqslant 1$.

PROPOSITION 13.3. *Let* $1 \leqslant q \leqslant x^{\frac{2}{3}}$ *and*

$$M \leqslant q^{-1} x^{1-\varepsilon} \ , \tag{13.12}$$

$$\sqrt{M} N \leqslant q^{-\frac{5}{4}} x^{1-\varepsilon} \ , \tag{13.13}$$

$$MN \leqslant q^{-\frac{1}{2}} x^{1-\varepsilon} \ , \tag{13.14}$$

for $\varepsilon > 0$*, sufficiently small. Then*

$$R(M, N) \ll q^{-1} x^{1-\delta} \tag{13.15}$$

for some $\delta = \delta(\varepsilon) > 0$*, the implied constant depending only on* ε *(indeed,* $\delta(\varepsilon) = \varepsilon/3$ *is fine).*

For the proof we may assume that α_m, β_n are supported on the dyadic segments $\frac{1}{2}M < m \leqslant M$, $\frac{1}{2}N < n \leqslant N$, respectively. We begin by introducing a smooth function $f(u)$ such that

$$\begin{aligned} \operatorname{supp} f = [y,\, x + y], \quad & 0 \leqslant f(u) \leqslant 1 \ , \\ f(u) = 1, \quad & \text{if } \ 2y \leqslant u \leqslant x \ , \\ f^{(j)}(u) \ll y^{-j}, \quad & \text{for } \ j = 0, 1, 2, \dots \ , \end{aligned}$$

where $y = x^{1-2\varepsilon}$. Then, we have

$$R(M, N) = \sum_{(mn,q)=1} \sum \alpha_m \beta_n \Big(\sum_{\ell mn \equiv a(q)} f(\ell mn) - \frac{\hat{f}(0)}{qmn} \Big) + O\big(q^{-1} y x^{\varepsilon}\big) \ , \tag{13.16}$$

where $\hat{f}(v)$ is the Fourier transform of $f(u)$, so that

$$\hat{f}(0) = \int f(u) du = x + O(y) \ . \tag{13.17}$$

By Poisson's formula we have the Fourier series expansion

$$\sum_{\ell mn \equiv a(q)} f(\ell mn) - \frac{\hat{f}(0)}{qmn} = \frac{1}{qmn} \sum_{h \neq 0} \hat{f}\Big(\frac{h}{qmn}\Big) e\Big(\frac{a h \overline{mn}}{q}\Big) \ . \tag{13.18}$$

Integrating by parts the Fourier integral

$$\hat{f}(v) = \int f(u)e(-uv)du \ , \tag{13.19}$$

we find that

$$\hat{f}(v) \ll x(1+|v|y)^{-j}$$

for any $j \geqslant 0$. Hence, the series on the right side of (13.18) can be truncated to

$$0 < |h| \leqslant H = qMNy^{-1}x^{\varepsilon} \ , \tag{13.20}$$

with negligible error term. Changing the variable $u = mt$ in (13.19) we arrive at

$$R(M,N) \ll \int \sum_{m\sim M} f(mt) \Big| \sum_{0<|h|\leqslant H} \sum_{n} \frac{\beta_n}{qn} e\Big(\frac{ah\overline{mn}}{q} - \frac{ht}{qn}\Big)\Big| dt \ + q^{-1}yx^{\varepsilon}.$$

Applying the Cauchy-Schwarz inequality we get

$$R(M,N) \ll \frac{x}{qN\sqrt{M}} \Big(\sum_m \Big|\sum_h \sum_n \gamma(h,n)\, e\Big(\frac{ah\overline{mn}}{q}\Big)\Big|^2\Big)^{\frac{1}{2}} + q^{-1}yx^{\varepsilon} \ , \tag{13.21}$$

where $|\gamma(h,n)| \asymp |\beta_n|$ for $0 < |h| \leqslant H$, $\frac{1}{2}N < n \leqslant N$, $\frac{1}{2}M < m \leqslant M$. Squaring out and changing the order of summation we find

$$\begin{aligned} R(M,N) \ll \frac{x}{qN\sqrt{M}} \Big(\sum\sum_{n_1h_2\neq n_2h_1} \Big|\sum_m e\Big(a(h_1n_2 - h_2n_1)\frac{\overline{mn_1n_2}}{q}\Big)\Big|\Big)^{\frac{1}{2}} \\ + q^{-1}x^{1-\varepsilon} + (q^{-1}Mx)^{\frac{1}{2}}x^{\varepsilon}, \end{aligned}$$

where the last term accounts for the contribution of the diagonal terms $n_1h_2 = n_2h_1$.

The inner sum over m is an incomplete Kloosterman sum which we estimate by an appeal to the following proposition (a basic consequence of Weil's bound for complete Kloosterman sums and Ramanujan sums).

PROPOSITION 13.4. *For any b, q, $M \geqslant 1$ we have*

$$\sum_{\substack{1\leqslant m\leqslant M \\ (m,q)=1}} e\Big(\frac{b\overline{m}}{q}\Big) \ll (b,q)^{\frac{1}{2}}\tau(q)q^{\frac{1}{2}}\log q + (b,q)q^{-1}M \ , \tag{13.22}$$

where the implied constant is absolute.

Applying (13.22) with $b = a(h_1n_2 - h_2n_1)\,\overline{n_1n_2}$ and using the elementary estimate

$$\sum\sum\sum\sum_{n_1h_2\neq n_2h_1} (h_1n_2 - h_2n_1, q) \leqslant q^{\varepsilon}(HN)^{2+\varepsilon} \ , \tag{13.23}$$

we get

$$R(M,N) \leqslant \frac{Hx^{1+\varepsilon}}{q\sqrt{M}}\Big(q^{\frac{1}{2}} + \frac{M}{q}\Big)^{\frac{1}{2}} + \frac{x^{1-\varepsilon}}{q} + \Big(\frac{Mx}{q}\Big)^{\frac{1}{2}}x^{\varepsilon}.$$

Now, inserting the value of H from (13.20) we verify (13.15) subject to (13.12), (13.13), (13.14) (where ε is not the same as above). This completes the proof of Proposition 13.3.

Now we can prove Theorem 13.1. By Proposition 13.3, with $M = q^{-1}x^{1-\varepsilon}$ and

$$N = \min\left(q^{\frac{1}{2}}, q^{-\frac{3}{4}}x^{\frac{1}{2}}\right)\,, \tag{13.24}$$

we show that the level of bilinear distribution of the sequence (13.1) is

$$D(x,q) = \min\left(q^{\frac{1}{2}}, q^{-\frac{3}{4}}x^{\frac{1}{2}}\right)q^{-1}x^{1-\varepsilon}. \tag{13.25}$$

Therefore, Theorem 13.1 follows from Theorem 12.20.

13.3. Larger Moduli

We begin by considering a bilinear form

$$S = \sum\sum_{(mn,q)=1} \alpha_m\, \beta_n\, e_q\left(a\overline{mn}\right) \quad \text{with } (a,q)=1, \tag{13.26}$$

where α_m, β_n are arbitrary complex numbers satisfying $|\alpha_m| \leqslant 1$, $|\beta_n| \leqslant 1$ and supported on the dyadic intervals $M < m \leqslant 2M$, $N < n \leqslant 2N$. As usual

$$e(t) = e^{2\pi i t}, \quad e_q(t) = e(t/q), \quad d\overline{d} \equiv 1 \,(\mathrm{mod}\, q).$$

We also assume that α_m, β_n are supported on primes. This gives cleaner, stronger estimates and in any case is what occurs in our application.

PROPOSITION 13.5. *Let k, ℓ be positive integers with $k < N$, $\ell < M$. Suppose*

$$k(2N)^{2k-1} < q, \qquad \ell(2M)^{2\ell-1} < q\,. \tag{13.27}$$

Then we have

$$|S| \leqslant k\ell MN\left(qM^{-\ell}N^{-k}\right)^{\frac{1}{2k\ell}}\,. \tag{13.28}$$

The proof of this result is completely elementary, no use of anything of the depth of Weil's bound. The argument is simpler than, but reminiscent of, Vinogradov's treatment of exponential sums.

PROOF. For notational simplicity we do not record (but keep in mind) the relevant ranges (dyadic intervals) for the variables m and n. By Hölder's inequality we obtain

$$\begin{aligned} |S|^k &\leqslant M^{k-1}\sum_m\Big|\sum_n \beta_n\, e_q(a\overline{mn})\Big|^k \\ &\leqslant M^{k-1}\sum_{n_1}\cdots\sum_{n_k}\Big|\sum_m \varepsilon_m\, e_q\big(a\overline{m}(\overline{n}_1+\cdots+\overline{n}_k)\big)\Big| \\ &= M^{k-1}\sum_{b\,(\mathrm{mod}\,q)}\nu(b)\Big|\sum_m \varepsilon_m\, e_q(ab\overline{m})\Big| \end{aligned}$$

with some ε_m having $|\varepsilon_m| = 1$ and where $\nu(b)$ denotes the number of k-tuples $(n_1,\ldots,n_k)$ such that

$$\overline{n}_1 + \cdots + \overline{n}_k \equiv b\,(\mathrm{mod}\, q)\,. \tag{13.29}$$

Again by Hölder's inequality,

$$|S|^{2k\ell} \leqslant M^{2\ell(k-1)}\Big(\sum_b \nu(b)\Big)^{2\ell-2}\Big(\sum_b \nu^2(b)\Big)\sum_b\Big|\sum_m \varepsilon_m\, e_q(ab\overline{m})\Big|^{2\ell}\,.$$

Here, the first sum is $\leqslant N^k$ and the second sum is $\leqslant \mu_k(N)$, the number of $2k$-tuples $(n_1, \dots, n_{2k})$ such that

$$\overline{n}_1 + \cdots + \overline{n}_k \equiv \overline{n}_{k+1} + \cdots + \overline{n}_{2k} \pmod q \ . \tag{13.30}$$

Also, for the third sum we have

$$\sum_{b \,(\mathrm{mod}\, q)} \Big| \sum_m \varepsilon_m \, e_q(ab\overline{m}) \Big|^{2\ell} \leqslant \mu_\ell(M) q \ .$$

From these estimates we obtain

$$|S|^{2k\ell} \leqslant \mu_\ell(M)\mu_k(N) M^{2\ell(k-1)} N^{2k(\ell-1)} q \ . \tag{13.31}$$

We require a bound for $\mu_k(N)$. To this end we note that from the assumption (13.27) it follows that every solution of the congruence (13.30) is also a solution of the equation

$$\frac{1}{n_1} + \cdots + \frac{1}{n_k} = \frac{1}{n_{k+1}} + \cdots + \frac{1}{n_{2k}} \tag{13.32}$$

and hence it suffices to estimate the number of solutions of (13.32). We shall also make the assumption, which is really pretty weak, that $k < N$. Under this condition we can say that $n_1^{-1} + \cdots + n_k^{-1}$ may be written as a sum of fractions an^{-1} having distinct n, where $1 \leqslant a \leqslant k$, and hence $(a, n) = 1$. Recall that the n_j are primes. Thus,

$$\frac{1}{n_1} + \cdots + \frac{1}{n_k} = \frac{b}{[n_1, \dots, n_k]}$$

with $(b, n_1 \dots n_k) = 1$ where $[n_1, \dots, n_k]$ denotes the least common multiple, and a similar expression holds for the right-hand side of (13.32). Hence, for any solution of (13.32) we have $n_j = n_1$ for some $k < j \leqslant 2k$. By induction, the solutions of (13.32) are diagonal up to permutation and so

$$\mu_k(N) \leqslant k! \, N^k \ , \tag{13.33}$$

with a similar bound for $\mu_\ell(M)$. Inserting these into the bound (13.31) we conclude that

$$|S|^{2k\ell} \leqslant k! \, \ell! \, M^{2k\ell - \ell} N^{2k\ell - k} q \ . \tag{13.34}$$

This yields (13.28). □

Now, we proceed to the sieve problem. Our goal is to prove Theorem 13.2. Given $(a, q) = 1$ we consider the sifting sequence

$$\mathcal{A} = \{n \leqslant x : n \equiv a \,(\mathrm{mod}\, q)\}$$

and the sifting function

$$S(\mathcal{A}, z) = |\{n \in \mathcal{A} : (n, P(z)) = 1\}| \ ,$$

where $P(z)$ denotes the product of all primes $p < z$, $p \nmid q$.

Let $w < y < z$. Then, by an iteration of Buchstab's identity,

$$\begin{aligned} S(\mathcal{A}, z) = S(\mathcal{A}, w) &- \sum_{y \leqslant p < z} S(\mathcal{A}_p, p) - \sum_{w \leqslant p < y} S(\mathcal{A}_p, w) \\ &+ \sum\sum_{w \leqslant p_2 < p_1 < y} S(\mathcal{A}_{p_1 p_2}, p_2) \ , \end{aligned} \tag{13.35}$$

where $\mathcal{A}_d = \{n \in \mathcal{A} : n \equiv 0 \pmod{d}\}$. We choose $z = (x/q)^{\frac{1}{3}}$. If we were to apply classical estimates of the linear sieve, that is (12.12) and (12.13), for all of the sifting functions on the right, we would get the same bound as we would by applying such estimates to the left-hand side because it is in the nature of these bounds that they are, with the above choice of z, stable under transformations via the Buchstab identity. Applying these to either side of (13.35) we obtain

$$S(\mathcal{A}, z) < \frac{2x}{\varphi(q)\log D} \quad \text{with} \quad D = \frac{x^{1-\varepsilon}}{q} \tag{13.36}$$

for any $\varepsilon > 0$, provided x is sufficiently large in terms of ε. Thus, to improve (13.36) it suffices to improve any part on the right-hand side, and Proposition 13.5 allows us to do this for the double sum, if y and w are chosen suitably.

The classical bound for each sifting function in this double sum on the right side of (13.35) is, for $w > D^{\frac{1}{5}}$,

$$S(\mathcal{A}_{p_1 p_2}, p_2) < \frac{2x}{\varphi(q)p_1\, p_2 \log D_{12}} \quad \text{with} \quad D_{12} = \frac{D}{p_1\, p_2} \tag{13.37}$$

because the sequence $\mathcal{A}_{p_1p_2}$ has level of distribution D_{12} and $p_2^3 > D_{12}$; the latter follows from the stronger condition $p_2^3 > D_{12}'$ below. Later in this section we shall show that, on average over p_1, p_2, this sequence has a larger level of distribution.

Let $q = x^\theta$ with $\frac{6}{11} < \theta < 1$ and let $\ell \geqslant 3$ be an integer satisfying

$$1 + \frac{3\theta}{1-\theta} < 2\ell < \frac{5\theta}{1-\theta}\ ; \tag{13.38}$$

such an ℓ exists. We choose

$$w = q^{\frac{1}{2\ell}+\varepsilon} \quad \text{and} \quad y = q^{\frac{1-\delta}{2\ell-1}-\varepsilon}\ , \tag{13.39}$$

where $\delta = 1/2\ell(2\ell^2 - \ell + 1)$. Now, having fixed ℓ, w, y we can tell that for primes p_1, p_2 with $w < p_2 < p_1 < y$ the improved level of distribution for $\mathcal{A}_{p_1p_2}$ is

$$D_{12}' = \left(p_1\, p_2\, q^{-1/\ell}\right)^{\frac{1}{2\ell}} D_{12} \tag{13.40}$$

(actually, slightly larger; see (13.42)). Therefore, on average, we have

$$S(\mathcal{A}_{p_1\, p_2}, p_2) < \frac{2x}{\varphi(q)p_1\, p_2 \log D_{12}'}\ , \tag{13.41}$$

provided that $p_2^3 > D_{12}'$. It suffices to verify this last condition in the worst scenario $p_1 = p_2 = w$, in which case it is just the upper bound in (13.38). The lower bound in (13.38) suffices to imply the requirement $y < z$ for our choice of y and z.

The new bound (13.41) is smaller than the classical one (13.37) by the amount

$$\frac{2x}{\varphi(q)p_1\, p_2}\left(\frac{1}{\log D_{12}} - \frac{1}{\log D_{12}'}\right) > \frac{2x}{\varphi(q)p_1\, p_2}\, \frac{\log\left(p_1\, p_2\, q^{-1/\ell}\right)}{2\ell \log^2\left(xq^{-1-1/\ell}\right)}$$

because $D_{12}' < xq^{-1-1/\ell}$. Summing over p_1, p_2 we get

$$\sum\sum_{w \leqslant p_2 < p_1 < y} (p_1\, p_2)^{-1} \log\left(p_1\, p_2\, q^{-1/\ell}\right) \sim c(\ell) \log q$$

where

$$c(\ell) = \iint\limits_{\frac{1}{2\ell}<\alpha<\beta<\frac{1-\delta}{2\ell-1}} \Big(\alpha+\beta-\frac{1}{\ell}\Big)\, d\alpha\, d\beta$$

$$= \frac{1}{2}\Big(\frac{1-2\delta\ell}{2\ell(2\ell-1)}\Big)^3 = \frac{1}{16(2\ell^2-\ell+1)^3}.$$

Therefore, our improvement is

$$\frac{c(\ell)x\log q}{\ell\varphi(q)\log^2(xq^{-1-1/\ell})} = \frac{c(\theta,\ell)x}{\varphi(q)\log x/q}$$

where $c(\theta,\ell) = \ell c(\ell)\theta(1-\theta)\big((1-\theta)\ell-\theta\big)^{-2}$.

Choose the smallest integer ℓ from the interval (13.38). This is less than $3/2(1-\theta)$, giving

$$c(\theta,\ell) > c\Big(\theta, \frac{3}{2(1-\theta)}\Big) > \Big(\frac{1-\theta}{4}\Big)^6,$$

which proves Theorem 13.2, subject to the verification of the level of distribution (13.40).

To prove this latter statement it suffices (rounding up D'_{12}) to prove that the subsequence $\mathcal{A}_{p_1\, p_2}$ has level of distribution

(13.42) $$D''_{12} = \Big(\frac{P_1\, P_2}{q^{1/\ell}}\Big)^{1/2\ell}\frac{x^{1-3\varepsilon}}{q\,P_1\,P_2}$$

on average over p_1, p_2 in the dyadic box $P_1 < p_1 \leqslant 2P_1$, $P_2 < p_2 \leqslant 2P_2$, for any P_1, P_2 with $w \leqslant P_1$, $P_2 < y$. To this end we estimate the remainder sum

(13.43) $$R_d = \sum_{p_1}\sum_{p_2}\Big(|\,\mathcal{A}_{d\,p_1\,p_2}\,| - \frac{x}{qdp_1\,p_2}\Big)$$

over p_1, p_2 in the above box for every individual d with $(d,q)=1$. We shall prove that

(13.44) $$R_d \ll \frac{x^{1-\varepsilon}}{qd} \qquad \text{if} \qquad d < D''_{12}\,.$$

We write

$$|\mathcal{A}_{dp_1\,p_2}| = \sum_{\substack{n\equiv 0\,(\mathrm{mod}\,dp_1\,p_2)\\ n\equiv a\,(\mathrm{mod}\,q)}} f(n) + O\Big(\sum_{\substack{n\equiv 0\,(\mathrm{mod}\,dp_1\,p_2)\\ n\equiv a\,(\mathrm{mod}\,q)}}{}' \; 1\Big)\,,$$

where $f(t)$ is a smooth non-negative function supported on $x^{1-\varepsilon} < t < x + x^{1-\varepsilon}$ such that $t^j f^{(j)}(t) \ll x^\varepsilon$ for $j \geqslant 0$ and $\hat{f}(0) = x$, while $\sum'$ restricts the summation to n in either of the two short intervals $x < n < x + x^{1-\varepsilon}$ or $0 < n < x^{1-\varepsilon}$. By Poisson summation

$$\sum_n f(n) = \frac{1}{qdp_1\,p_2}\sum_h \hat{f}\Big(\frac{h}{qdp_1\,p_2}\Big)e_q\big(-ah\overline{dp_1p_2}\big)\,.$$

The zero frequency $h=0$ gives the main term x/qdp_1p_2, whereas the contribution from $|h| \geqslant H$, with

$$H = qdP_1P_2x^{2\varepsilon-1},$$

is negligible, as is seen by partial integration. Therefore, we have

$$R_d = \sum_{0<|h|<H} \sum_{p_1}\sum_{p_2} (qdp_1p_2)^{-1}\hat{f}\Big(\frac{h}{qdp_1p_2}\Big)e_q(-ah\overline{dp_1p_2}) + O\Big(\frac{x^{1-\varepsilon}}{qd}\Big).$$

To separate the variables p_1, p_2 we write the Fourier transform as

$$\hat{f}\Big(\frac{h}{qdp_1p_2}\Big) = p_2 \int f(tp_2)e\Big(\frac{ht}{qdp_1}\Big)dt\ .$$

Hence,

$$R_d \ll \frac{x}{qdP_1P_2}\sum_{0<|h|<H}\Big|\sum_{p_1}\sum_{p_2}\alpha_{p_1}\beta_{p_2}e_q(ah\overline{dp_1p_2})\Big| + \frac{x^{1-\varepsilon}}{qd}$$

for some $|\alpha_{p_1}| \leqslant 1$ and $|\beta_{p_2}| \leqslant 1$. To the double sum over p_1, p_2 we apply Proposition 13.5 with $k = \ell$ and the modulus $q/(q,h) \geqslant q/H$. The condition (13.27) requires

$$q > \ell(2P)^{2\ell-1}H \tag{13.45}$$

for both $P = P_1$ and $P = P_2$. Employing $d < D''_{12}$ with D''_{12} given by (13.42) and $P_1, P_2 < y$ with y given by (13.39), one verifies (13.45). Here the presence of δ in the choice of y corresponds to the presence of H in the condition (13.45) and the latter emerges because it is necessary to put the fraction h/q into lowest terms. Now, since (13.45) has been verified, we can derive by Proposition 13.5 that

$$R_d \ll \frac{xH}{qd}\left(\frac{q^{1/\ell}}{P_1P_2}\right)^{\frac{1}{2\ell}} + \frac{x^{1-\varepsilon}}{qd}\ ,$$

which gives (13.44).

Note that

$$S(\mathcal{A}_{p_1p_2}, p_2) \leqslant S(\mathcal{A}_{p_1p_2}, P_2)\ .$$

Applying the linear sieve to the larger sifting function, with the level of distribution D''_{12} we derive by (13.44) the estimate (13.41), on average over p_1, p_2 in every relevant dyadic box. This completes the proof of Theorem 13.2.

13.4. Short Intervals

The second basic example of a linear sequence where the sieve shows to advantage, in comparison to zeta-function techniques, is that of a short interval:

$$\mathcal{A} = \{n;\ x - y < n \leqslant x\}\ , \tag{13.46}$$

where $y = x^\theta$, $0 < \theta < 1$. By a direct application of the linear sieve one obtains

$$\pi(x) - \pi(x-y) \leqslant \frac{(2+\varepsilon)y}{\log y}, \tag{13.47}$$

for any $\varepsilon > 0$, provided that x is sufficiently large in terms of θ and ε.

The technique of bilinear forms allows us to improve the constant 2 in the full range $0 < \theta < 1$ (although, for $\theta \geqslant \frac{7}{12}$ the result is less interesting due to the current state of the Hoheisel prime number theorem). Our arguments are rather similar to those for arithmetic progressions given above, except that, rather than Kloosterman sums, we encounter an exponential sum of van der Corput type. The problem is made easier here because this sum, unlike the Kloosterman sum, has been successfully estimated by van der Corput even over extremely short ranges. In our case we need estimates of the following type:

$$\sum_{M<m\leqslant 2M} e\Big(\frac{X}{m}\Big) \ll X^{\alpha}M^{1-\beta}+M^2X^{-1} \tag{13.48}$$

for any $M,\ X\geqslant 1$, where $\alpha,\ \beta$ are positive constants.

LEMMA 13.6 (van der Corput). *The estimates* (13.48) *hold with*

$$(\alpha,\beta)=\Big(\frac{1}{2^k-2},\ \frac{k+1}{2^k-2}\Big)\ , \tag{13.49}$$

for every integer $k\geqslant 2$, the implied constant depending only on k.

This is proved, for example, as Theorem 2.9 of Graham and Kolesnik [**69**].

For the sequence (13.46) we define the remainder term:

$$r_d(x)=A_d(x)-\frac{y}{d}\ .$$

Our goal is to estimate the bilinear form

$$R(M,N)=\sum_{m\leqslant M}\sum_{n\leqslant N}\alpha_m\beta_n r_{mn}(x) \tag{13.50}$$

for any complex numbers $\alpha_m,\ \beta_n$ with $|\alpha_m|,\ |\beta_n|\leqslant 1$.

PROPOSITION 13.7. *Let $0<\alpha\leqslant\frac12$ and $0<\beta\leqslant 2$ be any fixed exponents such that* (13.48) *holds. Suppose*

$$M\leqslant yx^{-\varepsilon}\ , \tag{13.51}$$

with some $0<\varepsilon\leqslant\theta$,

$$MN\leqslant\sqrt{xy}\ , \tag{13.52}$$

and

$$N\leqslant y^{\beta/2}x^{-\alpha/2}\ . \tag{13.53}$$

Then we have

$$R(M,N)\ll yx^{-\delta} \tag{13.54}$$

for some $\delta=\delta(\alpha,\beta,\varepsilon)>0$, the implied constant depending on α, β and ε.

For the proof we may assume throughout that $\alpha_m,\ \beta_n$ are supported on the dyadic segment $\frac12M<m\leqslant M,\ \frac12N<n\leqslant N$, respectively. We begin by introducing a smooth majorizing function $f(u)$ such that

$$\begin{aligned}\operatorname{supp} f=[x-y-z,\ x+z],&\quad 0\leqslant f(u)\leqslant 1\ ,\\ f(u)=1,&\quad\text{if }\ x-y\leqslant u\leqslant x\ ,\\ f^{(j)}(u)\ll z^{-j},&\quad\text{for }\ j=0,1,2,\dots\ ,\end{aligned}$$

where $z=yx^{-2\varepsilon}$. Then, clearly we have

$$R(M,N)=\sum_m\sum_n\alpha_m\beta_n\Big(\sum_\ell f(\ell mn)-\frac{\hat f(0)}{mn}\Big)+O(zx^{\varepsilon})$$

where $\hat{f}(v)$ is the Fourier transform of $f(u)$, so

$$\hat{f}(0) = \int f(u)du = y + O(z) \ .$$

By Poisson's formula we have the Fourier series expansion

$$\sum_{\ell} f(\ell mn) - \frac{\hat{f}(0)}{mn} = \frac{1}{mn} \sum_{h \neq 0} \hat{f}\Big(\frac{h}{mn}\Big) \ .$$

Integrating by parts the Fourier integral (13.19), we find that

$$\hat{f}(v) \ll y(1 + |v|z)^{-j}$$

for any $j \geqslant 0$ and so the above Fourier series can be truncated to

$$0 < |h| \leqslant H = MNz^{-1}x^{\varepsilon} \tag{13.55}$$

with negligible error. Doing so, we obtain

$$R(M,N) \ll \frac{1}{M} \int f(u) \sum_{m} \Big| \sum_{0<|h|\leqslant H} \sum_{n} \frac{\beta_n}{n} e\Big(\frac{hu}{mn}\Big)\Big| du + zx^{\varepsilon} \ . \tag{13.56}$$

Applying the Cauchy–Schwarz inequality we get

$$R(M,N) \ll \frac{y}{N\sqrt{M}} \Big(\sum_{m} \Big| \sum_{h} \sum_{n} \gamma(h,n) e\Big(\frac{hu}{mn}\Big)\Big|^2\Big)^{\frac{1}{2}} + zx^{\varepsilon} \ , \tag{13.57}$$

where $|\gamma(h,n)| \asymp |\beta_n|$ for $0 < |h| \leqslant H$, $\frac{1}{2}N < n \leqslant N$, $\frac{1}{2}M < m \leqslant M$, for some u with $x - y - z < u < x + z$. Squaring out and changing the order of summation we arrive at

$$R(M,N) \ll \frac{y}{N\sqrt{M}} \Big(\sum\sum\sum\sum_{n_1h_2 \neq n_2h_1} \Big| \sum_{m} e\Big(\Big(\frac{h_1}{n_1} - \frac{h_2}{n_2}\Big)\frac{u}{m}\Big)\Big|\Big)^{\frac{1}{2}}$$
$$+ zx^{\varepsilon} + \sqrt{yM}x^{2\varepsilon},$$

where the last term accounts for the diagonal quadruples, i. e. with $h_1n_2 = h_2n_1$.

By Lemma 13.6 the sum over m satisfies

$$\sum_{m} \ll \Big(\frac{|h_1n_2 - h_2n_1|}{N^2}x\Big)^{\alpha} M^{1-\beta} + \frac{M^2N^2}{|h_1n_2 - h_2n_1|x} \ .$$

Summing over $h_1n_2 \neq h_2n_1$, we get

$$\sum_{n_1}\sum_{n_2}\sum_{h_1}\sum_{h_2} |\sum_{m}| \leqslant (HN)^2 \Big(\frac{Hx}{N}\Big)^{\alpha} M^{1-\beta} + M^2N^2HNx^{\varepsilon-1}.$$

Finally, bearing in mind the choice of H, we obtain

$$R(M,N) \ll MN\big[\big(Mxy^{-1}\big)^{\alpha}M^{-\beta} + yx^{-1}\big]^{\frac{1}{2}}x^{5\varepsilon} + yx^{-\varepsilon} + (yM)^{\frac{1}{2}}x^{2\varepsilon}.$$

This gives us (13.54) subject to (13.51), (13.52) and (13.53), which completes the proof of Proposition 13.7.

REMARK. Examining the sequence of admissible pairs (α, β) in (13.49) we note that $\alpha/\beta \to 0$ as $k \to \infty$. Therefore, for any $y = x^{\theta}$ with $0 < \theta < 1$, we can choose α, β such that (13.53) becomes $N \leqslant x^{\gamma}$ with

$$\gamma = \tfrac{1}{2}(\theta\beta - \alpha) > 0.$$

By Proposition 13.7 with $M = yx^{-\varepsilon}$ and $N = \min\big((x/y)^{\frac{1}{2}},\ (y^\beta/x^\alpha)^{1/2}\big)$ we show that the level of bilinear distribution of the sequence (13.46) is

$$D(x,y) = \min\big((x/y)^{\frac{1}{2}},\ (y^\beta/x^\alpha)^{\frac{1}{2}}\big)yx^{-\varepsilon}\ , \tag{13.58}$$

provided that (13.48) holds. Hence, from Theorem 12.20 we infer the following result.

THEOREM 13.8. *Let $0<\theta<1$. Then, for $x^\theta \leqslant y \leqslant x$, we have*

$$\pi(x)-\pi(x-y) \leqslant (2-\delta)\frac{y}{\log y}$$

for some $\delta=\delta(\theta)>0$, provided that x is sufficiently large in terms of θ.

REMARK. We have actually shown the result with $\delta = 2\gamma/(\gamma+\theta)$. For small θ one can do better using Vinogradov's bounds in place of van der Corput's.

For sums of the remainder terms with special coefficients we can go beyond the ranges of Proposition 13.7. We begin by considering the sums

$$R(N) = \sum_{n\leqslant N} r_n(x)\ , \tag{13.59}$$

which is the special case of $R(M,N)$ in (13.50) with $M=1$ and $\beta_n = 1$. As before we may restrict the summation to the dyadic segment $N/2 < n \leqslant N$. Now (13.56) becomes

$$R(N) \ll \int f(u)\Big|\sum_{0<|h|\leqslant H}\ \sum_{\frac{1}{2}N<n\leqslant N}\frac{1}{n}e\Big(\frac{hu}{n}\Big)\Big|\,du + yx^{-\varepsilon}\ ,$$

where $H = Ny^{-1}x^{3\varepsilon}$. Applying (13.48) we infer that

$$\begin{aligned}R(N) &\ll y\sum_h\big[(hx)^\alpha N^{-\beta} + N^2(hx)^{-1}\big] + yx^{-\varepsilon}\\ &\ll N^{1-\beta}(Nx/y)^\alpha x^{5\varepsilon} + N^2yx^{\varepsilon-1} + yx^{-\varepsilon}\ .\end{aligned}$$

Hence, we conclude

PROPOSITION 13.9. *Let $0<\alpha\leqslant\frac{1}{2}$ and $0<\beta\leqslant 2$ be any fixed exponents such that* (13.48) *holds. Let*

$$N^{1+\alpha-\beta} \ll y^{1+\alpha}x^{-\alpha-6\varepsilon},\qquad N \ll x^{\frac{1}{2}-\varepsilon}\ . \tag{13.60}$$

Then,

$$R(N) \ll yx^{-\varepsilon}\ , \tag{13.61}$$

the implied constant depending only on α, β, and ε.

The result can be easily generalized to the sums

$$R_k(N) = \sum_{\substack{n\leqslant N\\ n\equiv 0 \,(\mathrm{mod}\, k)}} r_n(x)\ .$$

Applying Proposition 13.9 with x, y, N replaced by x/k, y/k, N/k we obtain

$$R_k(N) \ll k^{-1}yx^{-\varepsilon}\ , \tag{13.62}$$

uniformly over $k \leqslant K$, subject to the following conditions:

$$K^{\beta-\alpha}N^{1+\alpha-\beta} \ll yx^{-\alpha-6\varepsilon}, \qquad N \ll x^{\frac{1}{2}-\varepsilon} . \tag{13.63}$$

Summing over k we derive

COROLLARY 13.10. *Let λ_k be real numbers with $|\lambda_k| \leqslant 1$ for $1 \leqslant k \leqslant K$. Then*

$$\sum_{n\leqslant N} \Big(\sum_{k|n} \lambda_k\Big)\, r_n(x) \ll yx^{-\varepsilon} \tag{13.64}$$

provided that K and N satisfy (13.63), *the implied constant depending on α, β, ε.*

13.5. Large Primary Factors

In Section 2.2 we introduced the Tchebyshev method for detecting integers having large primary factors. Recall that primary integers are powers of primes.

Let, as usual, $\mathcal{A} = (a_n)$ be a sequence of non-negative numbers having the exponent of absolute distribution $0 < \theta < 1$. Using the formula (2.24), which we rewrite here:

$$S(x,y) = \sum_{n\leqslant x} a_n\Big(\sum_{\substack{d|n\\ d>y}} \Lambda(d)\Big) = \sum_{n\leqslant x} a_n \log n - \sum_{d\leqslant y} \Lambda(d) A_d(x) , \tag{13.65}$$

we succeeded to directly derive the asymptotic formula

$$S(x,y) = A(x)\Big\{\log\frac{x}{y} + O(1)\Big\} ,$$

but only with $y \leqslant D = x^{\theta}$. Now, being equipped with sieve results, we can expand the method beyond this natural limit. To this end we take advantage of the fact that the sum over $d \leqslant y$ occurs with a negative sign. Therefore, in the excess range $D < d \leqslant y$, it suffices to have an upper bound. Even a crude bound will work if we take the range to be rather short.

It is worthwhile to work out the arguments in a general context which will allow us to illustrate the concept of special bilinear level of distribution. Then we shall apply the results for integers in very short intervals.

Assume that, for every p and any $\varepsilon > 0$,

$$\mathcal{A}_{p^2}(x) \ll p^{-2}A(x) + x^{\varepsilon} , \tag{13.66}$$

the implied constant depending only on ε. This is a rather crude estimate which can be verified easily for most popular sequences.

We know that $A_d(x)$ is well approximated by $g(d)A(x)$ for $d \leqslant x^{\theta}$. If, for some $\vartheta > \theta$, we could use this approximation for $d \leqslant x^{\vartheta}$, then

$$\sum_{x^{\theta}<d\leqslant x^{\vartheta}} A_d(x)\Lambda(d) = (\vartheta-\theta)A(x)\log x + O\big(A(x)\big) . \tag{13.67}$$

Hence we deduce

$$\sum_{n\leqslant x} a_n\Big(\sum_{\substack{d|n\\ d>x^{\vartheta}}} \Lambda(d)\Big) = (1-\vartheta)A(x)\log x + O\big(A(x)\big) . \tag{13.68}$$

This is positive for any $\vartheta < 1$.

Approximating $A_d(x)$ by $g(d)A(x)$ in the range $x^{\vartheta} < d \leqslant x^{\theta}$ is of course too much to hope for, especially when $\mathcal{A} = (a_n)$ is a sparse sequence. Here, the sieve method comes in handy because it offers an upper bound for $A_p(x)$ of the true order of magnitude, perhaps not for every p, but on average over p in ranges beyond the level of absolute distribution. When applying the sieve, one loses some factor, say $c > 1$, so that the upper bound

$$\sum_{x^\theta < p \leqslant x^\vartheta} A_p(x) \log p \leqslant c(\vartheta - \theta)A(x)\log x + O\big(A(x)\big) \tag{13.69}$$

holds in place of the asymptotic (13.67). Note that the terms $A_{p^\alpha}(x)\log p$ with $\alpha \geqslant 2$, $x^\theta < p^\alpha \leqslant x^\vartheta$ contribute at most $A(x)x^{-\theta/2} + x^{\varepsilon+\vartheta/2}$ by (13.66). Assuming $\vartheta < 2\theta$ this contribution is negligible so we get

$$\sum_{n \leqslant x} a_n\Big(\sum_{\substack{d|n\\ d > x^\vartheta}} \Lambda(d)\Big) > \Big(1 - \theta - c(\theta - \vartheta)\Big)A(x)\log x + O\big(A(x)\big)\ , \tag{13.70}$$

which is still positive if

$$\theta < \vartheta \leqslant \theta + c^{-1}(1-\theta)\ . \tag{13.71}$$

Now we are going to illustrate how the sieve method can produce (13.69). To simplify the arguments we first estimate $\log p$ by $\vartheta \log x$ and then apply an upper-bound sieve $\Lambda = (\lambda_k)$ to the log-free sum

$$T(\theta, \vartheta; x) = \sum_{x^\theta < p \leqslant x^\vartheta} A_p(x)\ , \tag{13.72}$$

getting

$$T(\theta, \vartheta; x) \leqslant \sum_{x^\theta < n \leqslant x^\vartheta} A_n(x)\Big(\sum_{k|n} \lambda_k\Big)\ . \tag{13.73}$$

On introduction of the approximations (1.2) this becomes

$$T(\theta, \vartheta; x) \leqslant A(x) \sum_{x^\theta < n \leqslant x^\vartheta} g(n)\Big(\sum_{k|n} \lambda_k\Big) + R\ , \tag{13.74}$$

where R is the corresponding remainder

$$R = \sum_{x^\theta < n \leqslant x^\vartheta} r_n(x)\Big(\sum_{k|n} \lambda_k\Big) = \sum_{x^\theta < k\ell \leqslant x^\vartheta} r_{k\ell}(x)\lambda_k\ . \tag{13.75}$$

Of course, we have not escaped from the forbidden range of moduli for the remainder terms $r_n(x)$ which run over the segment $x^\theta < n \leqslant x^\vartheta$ just above the level of absolute distribution x^θ. However, what we have accomplished is to relax the sum by the more friendly sieve factors $\lambda * 1$. If the level of (λ_k) is relatively small, say (λ_k) are supported on $k \leqslant K$, then $n = k\ell$ with ℓ quite large. Because ℓ runs over all integers with virtually no restriction, there is a chance to get some cancellation of the terms in the sum (13.75), even for each fixed k.

Let us make the following hypothesis:

$$\sum_{k \leqslant K}\Big|\sum_{\substack{n \leqslant N\\ n \equiv 0 \,(\mathrm{mod}\, k)}} r_n(x)\Big| \ll A(x)(\log x)^{-1}\ . \tag{13.76}$$

Recall from Section 5.4 that the couple $\{K, N\}$ are referred to as the "level of special bilinear distribution". We have seen, and will again later, many interesting sequences for which one can verify (13.76) with N somewhat larger than the level of absolute distribution. This typically requires an application of a Fourier expansion for $r_n(x)$ followed by an estimation for the resulting exponential sum over n which amounts to a second application of Fourier analysis.

Assuming that (13.76) holds with

$$K = x^\eta, \quad N = x^\vartheta \ , \tag{13.77}$$

we see that the remainder R in (13.74) contributes a negligible amount $A(x)/\log x$. Suppose, just for simplicity, that

$$g(n) = \frac{1}{n} \ . \tag{13.78}$$

Hence, the main factor in (13.74) attached to $A(x)$ is

$$\begin{aligned}\sum_k \lambda_k g(k) \sum_{x^\theta < k\ell \leqslant x^\vartheta} g(\ell) &= \Big(\sum_k \lambda_k g(k)\Big)\Big((\vartheta-\theta)\log x + O\big((\log x)^{-2}\big)\Big)\\ &\leqslant 2(\vartheta-\theta)\frac{\log x}{\log K} + O\big((\log x)^{-1}\big) \ ,\end{aligned}$$

the last line following from the linear sieve bound, say (12.4), with $s = 2$. Hence,

$$T(\theta,\vartheta;x) \leqslant 2\eta^{-1}(\vartheta-\theta)A(x) + O\big(A(x)(\log x)^{-1}\big) \ ,$$

so (13.69) holds with $c = 2\vartheta\eta^{-1}$. We have proved the following result.

THEOREM 13.11. *Let $\mathcal{A} = (a_n)$ be a sequence of non-negative numbers which has exponents of special bilinear distribution η, ϑ and exponent of absolute distribution θ where these exponents satisfy*

$$0 < \eta < \theta < \vartheta < 2\theta \ , \qquad \vartheta < \theta + \eta(1-\theta)/2\vartheta \ . \tag{13.79}$$

Suppose also that (13.66) *holds. Then*

$$\sum_{n\leqslant x} a_n\Big(\sum_{\substack{d\mid n\\ d>x^\vartheta}} \Lambda(d)\Big) \gg A(x)\log x \ , \tag{13.80}$$

provided that x is sufficiently large.

As an application of Theorem 13.11 we take the sequence (13.46) of all integers in a short interval (to be consistent with the current notation, change θ into $\theta - \varepsilon$). In Corollary 13.10 we showed that $\mathcal{A}$ has exponents of special bilinear distribution η, ϑ determined by

$$0 < \eta < \theta < \vartheta < \tfrac{1}{2} \ , \tag{13.81}$$

$$\eta(\beta-\alpha) + \vartheta(1+\alpha-\beta) < \theta - \alpha \ , \tag{13.82}$$

where α, β are the exponents in (13.48). Observe that, for any $0 < \theta < 1$, one can find α, β such that (13.81) and (13.82) hold. Indeed, we can take β and α/β very small; see the van der Corput pair (13.49). Hence we conclude:

THEOREM 13.12. *Let $0 < \theta < 1/2$. Any interval $[x - x^\theta, x]$ with x sufficiently large contains integers having a primary divisor greater than x^ϑ, where $\vartheta > \theta$ depends only on θ. More precisely, we have*

$$\sum_{x-x^\theta<n\leqslant x} \Big(\sum_{\substack{d\mid n\\ d>x^\vartheta}} \Lambda(d)\Big) \gg x^\theta \log x \ . \tag{13.83}$$

CHAPTER 14

The Semi-Linear Sieve

Mister Tambourine

14.1. Introduction

For smaller dimensions the beta-sieve has a number of interesting features and is the strongest sieve known in many such situations. We have already covered this in Chapter 11; however, as with the linear sieve, we feel that the semi-linear sieve $\kappa = \frac{1}{2}$ deserves special consideration. For one thing there are very interesting applications, especially for problems involving binary quadratic forms, such as for example the sum of two squares. Of theoretical as well as practical interest is also the possibility of obtaining asymptotic formulae. In this chapter we concentrate on these aspects.

14.2. A Summary of Previous Results

We begin by recalling some results of Chapter 11 in the context of dimension $\kappa = \frac{1}{2}$. In this case the sifting limit is $\beta = 1$ and the sieve weights λ_d^+, λ_d^- are the restrictions of the Möbius function to the sets

$$\begin{aligned} \mathcal{D}^+ &= \{d = p_1 \dots p_n : p_1 > \dots > p_n,\, p_1 \dots p_m p_m < D \quad \text{for all } m \text{ odd}\}\ , \\ \mathcal{D}^- &= \{d = p_1 \dots p_n : p_1 > \dots > p_n,\, p_1 \dots p_m p_m < D \quad \text{for all } m \text{ even}\}\ . \end{aligned} \tag{14.1}$$

Recall that, by convention, both sets contain $d = 1$. Note that if $d \in \mathcal{D}^\pm$ and $d\,|\,P(D)$ then $d \leqslant D$.

The corresponding functions $F(s), f(s)$ are the continuous solutions to the system which begins

$$\begin{cases} \sqrt{s}F(s) = A & \text{if } 0 < s \leqslant 2\ , \\ f(s) = 0 & \text{at } s = 1\ , \end{cases} \tag{14.2}$$

where $A = 2\sqrt{e^\gamma/\pi}$, and continues with

$$\begin{cases} 2\sqrt{s}(\sqrt{s}F(s))' = f(s-1) & \text{if } s > 2\ , \\ 2\sqrt{s}(\sqrt{s}f(s))' = F(s-1) & \text{if } s > 1\ . \end{cases} \tag{14.3}$$

Here, for $1 \leqslant s \leqslant 3$ we get

$$2\sqrt{s}f(s) = A\int_1^s \frac{dt}{\sqrt{t(t-1)}} = A\log\left(1 + 2(s-1) + 2\sqrt{s(s-1)}\right)\ .$$

Our usual supposition about the dimension now reads as

$$\prod_{w\leqslant p<z}\big(1-g(p)\big)^{-1}\leqslant\Big(\frac{\log z}{\log w}\Big)^{\frac{1}{2}}\Big(1+\frac{L}{\log w}\Big)\ . \tag{14.4}$$

In this case, the main Theorem 11.12 of the beta-sieve asserts that

$$V^{+}(D,z)<\Big\{F(s)+O\big((\log D)^{-1/6}\big)\Big\}V(z)\qquad\text{if } s>0 \tag{14.5}$$

and

$$V^{-}(D,z)>\Big\{f(s)+O\big((\log D)^{-1/6}\big)\Big\}V(z)\qquad\text{if } s\geqslant 1\ , \tag{14.6}$$

where $s=\log D/\log z$ and the implied constant depends only on L. Applying these sieves to the sequence $\mathcal{A}$ we obtain

$$S^{+}(\mathcal{A},z)<XV^{+}(D,z)+R(\mathcal{A},D,z) \tag{14.7}$$

and

$$S^{-}(\mathcal{A},z)>XV^{-}(D,z)-R(\mathcal{A},D,z)\ , \tag{14.8}$$

where

$$R(\mathcal{A},D,z)=\sum_{\substack{d|P(z)\\ d<D}}|r_d(\mathcal{A})| \tag{14.9}$$

and $P(z)$ is the product of primes $p<z$, $p\in\mathcal{P}$. Note that we have used the trivial bound $|\lambda_d|\leqslant 1$ in the remainder because, in this chapter, we do not take advantage of any cancellation. The estimates (14.7) and (14.8) give bounds also for $S(\mathcal{A},z)$ because $S^{-}\leqslant S\leqslant S^{+}$. In fact, we have the identities

$$S(\mathcal{A},z)=S^{+}(\mathcal{A},z)-T^{+}(\mathcal{A},z)\ , \tag{14.10}$$

$$S(\mathcal{A},z)=S^{-}(\mathcal{A},z)+T^{-}(\mathcal{A},z)\ , \tag{14.11}$$

where

$$T^{+}(\mathcal{A},z)=\sum_{n\text{ odd}}S_n(\mathcal{A},z) \tag{14.12}$$

and

$$T^{-}(\mathcal{A},z)=\sum_{n\text{ even}}S_n(\mathcal{A},z) \tag{14.13}$$

with

$$S_n(\mathcal{A},z)=\sum_{\substack{p_n<\cdots<p_1<z\\ p_1\cdots p_m p_m<D;\ m<n,m\equiv n(2)\\ p_1\cdots p_n p_n\geqslant D}}\cdots\sum S(\mathcal{A}_{p_1\cdots p_n},p_n)\ . \tag{14.14}$$

14.3. First Application

A very natural problem for the semi-linear sieve arises from the sequence of integers which can be written as the sum of two squares. We present an application which, although immediate, indicates the power of the semi-linear sieve.

We are going to estimate the number of odd integers in a short interval which are properly represented as the sum of two squares. Let $b(n)$ denote the characteristic function of such integers and

$$B(x) = \sum_{n \leqslant x} b(n) \,, \tag{14.15}$$

so we are interested in

$$B(x) - B(x-y) = \sum_{x-y<n\leqslant x} b(n) \,, \tag{14.16}$$

for $2 \leqslant y \leqslant x$. Note that $b(n) = 1$ if and only if all of its prime divisors are $\equiv 1 \,(\mathrm{mod}\ 4)$. Therefore, this is a sieve problem for the sequence

$$\mathcal{A} = \{x - y < n \leqslant x;\ n \equiv 1 \,(\mathrm{mod}\ 4)\} \tag{14.17}$$

and the set of primes

$$\mathcal{P} = \{p \equiv 3 \,(\mathrm{mod}\ 4)\} \,. \tag{14.18}$$

For d odd we have

$$|\mathcal{A}_d| = \frac{y}{4d} + O(1) \,, \tag{14.19}$$

so the density function is $g(d) = 1/d$ and the sequence $\mathcal{A}$ has level of absolute distribution $D = y/\log y$. Hence, from the upper bound (14.7) with $z = D$ we deduce

$$B(x) - B(x-y) \leqslant \tfrac{1}{4}\, A\, y\, V(D) + O\big(y(\log y)^{-\frac{2}{3}}\big), \tag{14.20}$$

where $A = 2\sqrt{e^\gamma/\pi}$ and

$$V(D) = \prod_{\substack{p<D \\ p\in\mathcal{P}}} \left(1 - \frac{1}{p}\right) . \tag{14.21}$$

We compute the product as follows:

$$V(D) = \prod_{2<p<D} \left(1 - \frac{1}{p}\right)^{\frac{1}{2}} \prod_{p<D} \left(1 - \frac{\chi(p)}{p}\right)^{-\frac{1}{2}} \prod_{\substack{p<D \\ p\equiv 3\,(\mathrm{mod}\ 4)}} \left(1 - \frac{1}{p^2}\right)^{-\frac{1}{2}} , \tag{14.22}$$

where $\chi = \chi_4$ is the non-principal character modulo four. The first product is

$$\left(\frac{2e^{-\gamma}}{\log D}\right)^{\frac{1}{2}} \left(1 + O\left(\frac{1}{\log D}\right)\right)$$

by the Mertens formula, the second product is

$$\frac{\sqrt{\pi}}{2} \left(1 + O\left(\frac{1}{\log D}\right)\right)$$

by Leibniz' formula $L(1,\chi) = \pi/4$, and the third product is $c + O(D^{-1})$, where

$$c = \prod_{p\equiv 3 \,(\mathrm{mod}\ 4)} \left(1 - \frac{1}{p^2}\right)^{-\frac{1}{2}} . \tag{14.23}$$

Hence, we find

$$V(D) = c\left(\frac{\pi}{2e^\gamma}\right)^{\frac{1}{2}} \frac{1}{\sqrt{\log D}} + O\left(\frac{1}{\log D}\right) . \tag{14.24}$$

Inserting this in (14.20) we obtain

THEOREM 14.1. *For $2 \leqslant y \leqslant x$ we have*

$$B(x) - B(x-y) \leqslant c\sqrt{2}\frac{y}{\sqrt{\log y}} + O\left(\frac{y}{(\log y)^{\frac{2}{3}}}\right) . \tag{14.25}$$

In particular, (14.25) gives

$$B(y) \leqslant c\sqrt{2}\frac{y}{\sqrt{\log y}} + O\left(\frac{y}{(\log y)^{\frac{2}{3}}}\right) .$$

Later in this chapter we shall show, using the asymptotic semi-linear sieve, that this holds with equality, namely

THEOREM 14.2. *For $y \geqslant 3$ we have*

$$B(y) = c\sqrt{2}\frac{y}{\sqrt{\log y}} + O\left(\frac{y(\log\log y)^{\frac{1}{6}}}{(\log y)^{\frac{2}{3}}}\right) , \tag{14.26}$$

where the constant c is given by the product (14.23).

Alternatively, since $b(n)$ is multiplicative, Theorem 14.2 also follows from the asymptotic formula for multiplicative functions, such as that given in Theorem A.7 with $k = \frac{1}{2}$. However, Theorem 14.1 has no such alternative proof. The key feature is that the result holds for very short intervals, reinforcing once again an important aptitude of sieve methods. From Theorem 14.2 it follows that the inequality (14.25) of Theorem 14.1 can be stated in a nice form:

COROLLARY 14.3. *For $2 \leqslant y \leqslant x$, we have*

$$B(x) - B(x-y) \leqslant B(y) + O\left(\frac{y(\log\log y)^{\frac{1}{6}}}{(\log y)^{\frac{2}{3}}}\right) . \tag{14.27}$$

In view of the asymptotic formula (14.26), the upper bound (14.25) obtained from the semi-linear sieve is essentially best possible. In the next section we show that this feature holds in much greater generality.

14.4. Asymptotic Semi-Linear Sieve

Examining the relationship between $S^+(\mathcal{A},z)$ and $S(\mathcal{A},z)$ given in (14.10), we see that the claim that the upper bound for $S(\mathcal{A},z)$ is essentially best possible amounts to showing that the sum $T^+(\mathcal{A},z)$ defined in (14.12) is negligible. This would not be possible in complete generality. We are going to make an assumption (for this section only), which turns out to be very valuable in applications:

PARITY CONDITION. *The sequence $\mathcal{A} = (a_n)$ is supported on integers having an even number of prime divisors in the set $\mathcal{P}$.*

The particular example considered in the previous section was a sequence $\mathcal{A} = (a_n)$ supported on integers $n \equiv 1 \pmod 4$ and the sifting set $\mathcal{P}$ had only primes $p \equiv 3 \pmod 4$. Hence, the parity condition holds for this example.

Let $D \geqslant z \geqslant 2$. We take elements of $\mathcal{A} = (a_\ell)$ with $\ell \leqslant x$ where $x \geqslant 4D$. For $d > 1$, let $p(d)$ denote the least prime factor of d and, for notational convenience, set $p(1) = z$. Then, for n odd, the congruence sums in (14.14) are to moduli $p_1 \dots p_n = dp$, say, where $d = p_1 \dots p_{n-1}$ satisfies

$$\nu(d) = n-1,\ d \in \mathcal{D}^+,\ d|P(z),\ p(d) \geqslant \sqrt{D/d} \tag{14.28}$$

and $p = p_n \geqslant \sqrt{D/d}$, $p \nmid d$. Note that, since $D \geqslant z$, every d occurring in (14.28) satisfies $d \leqslant D$. From the parity condition it follows that

$$\sqrt{D/d} \leqslant p < \sqrt{x/d}\ . \tag{14.29}$$

We have in mind that $D = D(x)$ is close to x so p runs over a relatively short interval. This is the reason we expect the sums $S_n(\mathcal{A}, z)$ to be small. By the above observations we obtain

$$\mathcal{T}^+(\mathcal{A}, z) \leqslant \sum_{\substack{d \in \mathcal{D}^+,\, d|P(z) \\ \mu(d)=1,\, p(d) \geqslant \sqrt{D/d}}} \ \sum_{\substack{\sqrt{D/d} \leqslant p < \sqrt{x/d} \\ p \nmid d,\, p \in \mathcal{P}}} S(\mathcal{A}_{dp}, p)\ .$$

To each $S(\mathcal{A}_{dp}, p)$ we apply the semi-linear upper-bound sieve of level $p^{1/3}$, getting

$$S(\mathcal{A}_{dp}, p) \leqslant g(dp)V(p)\, X\Big\{F\big(\tfrac{1}{3}\big) + O\Big((\log p)^{-\frac{1}{6}}\Big)\Big\} + \sum_{\delta < p^{\frac{1}{3}},\, \delta | P(p)} |r_{\delta dp}(x)|\ .$$

Note that the remainder terms are to moduli δdp satisfying

$$\delta dp \leqslant dp^{\frac{4}{3}} < d\left(\frac{x}{d}\right)^{\frac{2}{3}} \leqslant D^{\frac{1}{3}} x^{\frac{2}{3}}\ .$$

Hence, we obtain

$$\mathcal{T}^+(\mathcal{A}, z) \leqslant CX \sideset{}{^+}\sum_d g(d) \sum_p g(p)V(p) + R\Big(\mathcal{A}, D^{\frac{1}{3}} x^{\frac{2}{3}}\Big)\ ,$$

where $\sum^+$ runs over the numbers satisfying (14.28), p runs over the primes of $\mathcal{P}$ in the interval (14.29), and C is a constant depending only on L in (14.4).

First, we are going to estimate the inner sum over p. It is helpful to keep in mind the redundant condition $2 \leqslant p < z$, together with (14.29). Hence $w \leqslant p < z_1$, where

$$w = \max(2, \sqrt{D/d}), \qquad z_1 = \min(z, \sqrt{x/d})\ .$$

We have

$$\sum_{w \leqslant p < z_1} g(p)V(p) = V(w) - V(z_1) = V(z)\frac{V(z_1)}{V(z)}\left[\frac{V(w)}{V(z_1)} - 1\right]\ .$$

By (14.4), we get

$$\frac{V(z_1)}{V(z)} \leqslant \left(\frac{2\log z}{\log x/d}\right)^{\frac{1}{2}}\left(1 + \frac{2L}{\log x/d}\right) \leqslant 4L\left(\frac{2\log z}{\log x/d}\right)^{\frac{1}{2}}$$

and

$$\frac{V(w)}{V(z_1)} \leqslant \left(\frac{\log z_1}{\log w}\right)^{\frac{1}{2}}\left(1 + \frac{L}{\log w}\right)\ .$$

Here,

$$\frac{\log z_1}{\log w} = 1 + \frac{\log z_1/w}{\log w} \leqslant 1 + \frac{1}{2}\frac{\log x/D}{\log w} \leqslant 1 + 2\,\frac{\log x/D}{\log 4D/d}$$

because $w \geqslant (4D/d)^{\frac{1}{4}}$. Using the inequality $(1+v)^{\frac{1}{2}} \leqslant 1 + v^{\frac{1}{2}}$ we find

$$\frac{V(w)}{V(z_1)} - 1 \leqslant 4L\left(\frac{\log x/D}{\log 4D/d}\right)^{\frac{1}{2}} .$$

Hence,

$$\begin{aligned}\sum_p g(p)V(p) &\leqslant cV(z)\left(\frac{\log z}{\log x/d}\right)^{\frac{1}{2}}\left(\frac{\log x/D}{\log 4D/d}\right)^{\frac{1}{2}}\\ &\leqslant c\,V(z)\left(\frac{\log z}{\log 4D/d}\right)^{\frac{1}{2}}\left(\frac{\log x/D}{\log x/d}\right)^{\frac{1}{6}}\\ &\leqslant cV(z)(\log z)^{\frac{1}{2}}(\log x/D)^{\frac{1}{6}}(\log 4D/d)^{-\frac{2}{3}} ,\end{aligned}$$

where c is a constant depending only on L; $c = 16L^2$ is fine. This gives

$$T^+(\mathcal{A}, z) \leqslant c\,CXV(z)\,(\log z)^{\frac{1}{2}}(\log x/D)^{\frac{1}{6}}\mathcal{L}^+(D, z) \;+\; R(\mathcal{A}, D^{\frac{1}{3}}x^{\frac{2}{3}}) , \tag{14.30}$$

where

$$\mathcal{L}^+(D, z) = \sum_d{}^{+} g(d)(\log 4D/d)^{-\frac{2}{3}} , \tag{14.31}$$

and the sum $\sum^+$ runs over numbers satisfying (14.28).

The sum $\mathcal{L}^+(D, z)$ was estimated in Proposition 11.23 by the bound

$$\mathcal{L}^+(D, z) \ll e^{-s}(\log D)^{-\frac{2}{3}} ,$$

with an implied constant depending only on L. Inserting this bound in (14.30), we conclude the following result.

Proposition 14.4. *Let $\mathcal{A} = (a_n)$ be a sequence of non-negative real numbers for $n \leqslant x$ satisfying the Parity Condition and the conditions of the upper bound semi-linear sieve; see* (14.4). *Let $2 \leqslant z \leqslant D \leqslant x/4$. We have*

$$S^+(\mathcal{A}, z) - S(\mathcal{A}, z) \leqslant ce^{-s}XV(z)\left(\frac{\log x/D}{\log D}\right)^{\frac{1}{6}} + R(\mathcal{A}, D^{\frac{1}{3}}x^{\frac{2}{3}}) , \tag{14.32}$$

where $s = \log D \,/\, \log z$ and c is a positive constant which depends only on L.

Recall that we also have $S^+(\mathcal{A}, z) - S(\mathcal{A}, z) \geqslant 0$. Moreover, Proposition 11.26 gives a formula for $S^+(\mathcal{A}, z)$ which is asymptotic up to the remainder $R(\mathcal{A}, D)$. This requires a two-sided dimension condition such as (11.191), which, in case $\kappa = \frac{1}{2}$, is

$$\left(1 + \frac{L}{\log w}\right)^{-1} \leqslant \left(\frac{\log z}{\log w}\right)^{\frac{1}{2}} \prod_{w \leqslant p < z}(1 - g(p)) \leqslant 1 + \frac{L}{\log w} , \tag{14.33}$$

for all $2 \leqslant w \leqslant z$, where $L \geqslant 1$ is a constant.

Thus, by combining Proposition 14.4 with Proposition 11.26 we establish the asymptotic semi-linear sieve.

THEOREM 14.5. *Let $\mathcal{A} = (a_n)$ be a sequence of non-negative real numbers for $n \leqslant x$, satisfying the Parity Condition with a density function obeying the two-sided sieve dimension condition* (14.33). *Let $2 \leqslant z \leqslant D \leqslant x/4$. Then*

$$S(\mathcal{A}, z) = XV(z)\left\{F(s) + O\left(e^{-s}\left(\frac{\log x/D}{\log D}\right)^{\frac{1}{6}}\right)\right\} + 2\theta R(\mathcal{A}, D^{\frac{1}{3}}x^{\frac{2}{3}}) \ , \tag{14.34}$$

where $F(s)$ is given by (14.2) *and* (14.3), $s = \log D / \log z$ *and* $|\theta| \leqslant 1$.

Note that Theorem 14.2 follows immediately from Theorem 14.5 (take $D = y(\log y)^{-2}$), and so the proof of Corollary 14.3 is complete.

14.5. Applications of the Asymptotic Semi-Linear Sieve

We give two quick applications of Theorem 14.5 for the function $b(n)$ introduced in Section 14.3. The first one is concerned with its distribution in short intervals and replaces the upper bound in Corollary 14.3 by an asymptotic.

THEOREM 14.6. *For $y = x^{1-\varepsilon(x)}$ where $\varepsilon(x) \to 0$ as $x \to \infty$, we have*

$$B(x) - B(x-y) \sim B(y) \ . \tag{14.35}$$

Our second application is to $b(n)$ over an arithmetic progression. Let

$$B(x; q, a) = \sum_{\substack{n \leqslant x \\ n \equiv a \,(\mathrm{mod}\ q)}} b(n) \ , \tag{14.36}$$

where

$$q \geqslant 2, \quad (a, q) = 1, \quad a \equiv 1\,(\mathrm{mod}(4, q)) \ . \tag{14.37}$$

THEOREM 14.7. *For q and a satisfying* (14.37), *we have for $x \geqslant q$,*

$$B(x; q, a) = \frac{c_q}{q}\frac{x}{\sqrt{\log x}}\left\{1 + O\left(\left(\frac{\log q}{\log x}\right)^{\frac{1}{7}}\right)\right\} \ , \tag{14.38}$$

where c_q is a positive constant and the implied constant is absolute.

For simplicity we give the proof only for $q \equiv 0\,(\mathrm{mod}\,4)$. In this case we have

$$c_q = \sqrt{2} \prod_{p \equiv 3\,(\mathrm{mod}\,4)} \left(1 - \frac{1}{p^2}\right)^{-\frac{1}{2}} \prod_{\substack{p \mid q \\ p \equiv 3\,(\mathrm{mod}\,4)}} \left(1 - \frac{1}{p}\right)^{-1} . \tag{14.39}$$

PROOF. We can assume that x is large and $q < \sqrt{x}/\log x$ since otherwise (14.38) is trivial.

We apply Theorem 14.5 for the sequence

$$\mathcal{A} = \{n \leqslant x;\ n \equiv a\,(\mathrm{mod}\,q)\} \ .$$

For every $(d, q) = 1$ we have

$$|\mathcal{A}_d| = \frac{x}{dq} + O(1)$$

so our sieve problem has density function $g(d) = 1/d$. We can take the level of absolute distribution to be

$$D = D(x, q) = x/q\log x > \sqrt{x} \ .$$

We obtain

$$B(x;q,a) = S(\mathcal{A},\mathcal{P},D) = XV_q(D)\left\{F(1) + O\left(\left(\frac{\log(q\log x)}{\log x}\right)^{\frac{1}{6}}\right)\right\},$$

with $X = x/q$, $F(1) = 2\sqrt{e^\gamma/\pi}$ and

$$V_q(D) = \prod_{\substack{p<D,\, p\nmid q \\ p\equiv 3\,(\bmod 4)}}\left(1-\frac{1}{p}\right) = V_1(D)\prod_{\substack{p\mid q \\ p\equiv 3\,(\bmod 4)}}\left(1-\frac{1}{p}\right)^{-1}.$$

Since $V_1(D)$ was computed in (14.24) this completes the proof of Theorem 14.7. □

We remark that the asymptotic formula (14.38) of Theorem 14.7 is meaningful for $q \leqslant x^{\varepsilon(x)}$ with $\varepsilon(x)$ tending to zero arbitrarily slowly. No analytic method to date is capable of producing such a wide range of uniformity. On the other hand, the formula (14.35) for short intervals can be derived by analytic arguments à la Hoheisel–Landau.

14.6. Shifted Primes as a Sum of Two Squares

In this section and the next we are going to give further applications of the semi-linear sieve. As before we restrict our attention to sieving by the set of primes $p \equiv 3 \,(\bmod 4)$. The odd numbers which survive this sieve are just those, all of whose prime factors are congruent to one modulo four, and so, just those which have a primitive representation as the sum of two squares. This example has a natural generalization to the primitive representation of integers by a binary quadratic form.

We recall that $b(n)$ denotes the characteristic function on positive integers all of whose prime factors are congruent to one modulo four. The semi-linear sieve is capable of giving good upper and lower bounds for the sums

$$\sum_{n\leqslant x} a_n\, b(n)$$

for many interesting sequences $\mathcal{A} = (a_n)$. If the sequence $\mathcal{A}$ has level of absolute distribution somewhat larger than $x^{\frac{1}{2}}$ this is often straightforward. We shall give examples of the more interesting situation where this level is just missed and an extra sieve ingredient of a different sort needs to be included.

THEOREM 14.8. *Let* $\alpha \geqslant 1$, $\beta \neq 0$, $(\alpha,\beta) = 1$, $\alpha+\beta \equiv 1 \,(\bmod 4)$. *Then, for* $x \geqslant 2|\beta|\alpha^{-1}$, $x \geqslant \alpha^{68}$, *we have*

$$\sum_{p\leqslant x} b(\alpha p+\beta) \asymp \frac{x}{(\log x)^{\frac{3}{2}}}\prod_{\substack{p\mid\alpha\beta \\ p\equiv 3(\bmod 4)}}\left(1+\frac{1}{p}\right).$$

PROOF. We begin the proof by applying the semi-linear sieve to the sequence $\mathcal{A}$ of numbers $\alpha p+\beta$ with $p\leqslant x$ and $p\equiv 1 \bmod(4/(\alpha,2))$ and sifting set of primes $\mathcal{P} = \{p\equiv 3(\bmod 4)\}$. Note that, in our set, $\alpha p+\beta \equiv 1(\bmod 4)$ and $\alpha p+\beta \leqslant 2\alpha x$, so we obtain

$$\sum_{p\leqslant x} b(\alpha p+\beta) = S(\mathcal{A},\mathcal{P},\sqrt{2\alpha x})\,. \tag{14.40}$$

For every $(d, 2\alpha\beta) = 1$, we have

$$A_d(x) = \pi\big(x; 4d/(2,\alpha), \gamma\big) ,$$

for some reduced class γ. Therefore, by the Bombieri-Vinogradov theorem our sequence $\mathcal{A}$ has level of distribution $x^{\frac{1}{2}}(\log x)^{-A}$ for some A. Hence, the lower bound semi-linear sieve (Theorem 11.13) yields

$$S(\mathcal{A}, \mathcal{P}, z) \geqslant \frac{\pi(x)}{\varphi(4/(2,\alpha))} V(z)\Big\{f(s) + O\big(\frac{1}{(\log x)^{1/6}}\big)\Big\} , \tag{14.41}$$

where

$$V(z) = \prod_{\substack{p<z \\ p\equiv 3(4) \\ (p,\alpha\beta)=1}} \Big(1 - \frac{1}{p-1}\Big)$$

and $s = \log x / 2 \log z$ with $1 \leqslant s \leqslant 2$. This lower bound is positive for any $s \geqslant 1+\varepsilon$, that is for $z \leqslant x^{(1-\varepsilon)/2}$. Thus, we just miss the target $z = \sqrt{2\alpha x}$ required to give a positive lower bound for (14.40) .

By Buchstab's identity we have

$$S(\mathcal{A}, \mathcal{P}, z) - S(\mathcal{A}, \mathcal{P}, \sqrt{2\alpha x}) = \sum_{\substack{z<p_1<\sqrt{2\alpha x} \\ p_1\equiv 3 \,(\mathrm{mod}\ 4)}} S(\mathcal{A}_{p_1}, \mathcal{P}, p_1) = T \tag{14.42}$$

say, and we need an upper bound for this. The standard approach of applying the semi-linear upper-bound sieve would again just fail because of the nature of the sieve weights. Instead, we reinterpret the question (the switching trick we saw in Section 11.22 and shall see again; it is especially useful in connection with the semilinear sieve). Assuming that $z > (2\alpha x)^{\frac{1}{4}}$ it follows that any number counted in the sum T must have exactly one prime factor $p_2 \equiv 3 \,(\mathrm{mod}\ 4)$ in addition to p_1, since the numbers in our sequence are congruent to one modulo four (so there are an even number of such factors) and there is no room for four of them. Hence, the numbers counted in T are of the form

$$\alpha p + \beta = a p_1 p_2 \tag{14.43}$$

where

$$p \leqslant x, \quad p \equiv 1 \,(\mathrm{mod}\ 4/(2,\alpha)), \quad p'|a \Rightarrow p' \equiv 1 \,(\mathrm{mod}\ 4)$$

and

$$z \leqslant p_1 < \sqrt{2\alpha x}, \quad p_1 < p_2, \quad p_1 \equiv p_2 \equiv 3 \,(\mathrm{mod}\ 4) .$$

These conditions imply that

$$a \leqslant 2\alpha x z^{-2}, \quad (a, \alpha\beta) = 1 \tag{14.44}$$

and

$$z \leqslant p_1 < \sqrt{2\alpha x/a} \leqslant \sqrt{2\alpha x}. \tag{14.45}$$

We choose (λ_ν), (λ_{ν_2}) to be upper-bound sieves of level $D < z$ and with λ_ν supported on numbers coprime to β. We shall choose D later. In equation (14.43), for given a, p_1, we attach factors $\sum_{\nu|p} \lambda_\nu$, $\sum_{\nu_2|p_2} \lambda_{\nu_2}$, which are redundant. Now, we drop the requirement that p, p_2 are prime, majorizing the result. Hence,

$$
\begin{aligned}
T \leqslant \sum_{\substack{a\leqslant 2\alpha x z^{-2}\\ (a,\alpha\beta)=1}} b(a) \sum_{\substack{z\leqslant p_1\leqslant\sqrt{2\alpha x/a}\\ p_1\equiv 3(4)}} \sum_{\nu}\sum_{\nu_2} \lambda_\nu\lambda_{\nu_2} \sum_{\substack{n\leqslant x\\ n\equiv 1(4/(2,\alpha))\\ \alpha n+\beta\equiv 0(ap_1\nu_2)\\ n\equiv 0(\nu),\,(n,\beta)=1}} 1 \\
= \sum_a b(a) \sum_{p_1} \sum_{\substack{(\nu,\nu_2)=1\\ (\nu,\alpha\beta)=1\\ (\nu_2,\alpha\beta)=1}}\sum \lambda_\nu\lambda_{\nu_2}\Big\{\frac{\varphi(\beta)x}{2\beta a p_1\nu\nu_2} + O(\tau(\beta))\Big\}\,.
\end{aligned}
\tag{14.46}
$$

Choosing $D = \sqrt{z/\alpha}x^{-\varepsilon}$ and linear sieve weights, say of beta type, so that $|\lambda_d| \leqslant 1$ for all d, we see that the error term gives a contribution to the sum which is $\ll x^{1-\varepsilon}$.

Evaluating explicitly the constants in Theorem 11.12 and Theorem 5.9 and noting that the constant C in (5.85) satisfies $C < 3$, we have

$$
\sum_{\nu}\sum_{\nu_2} \frac{\lambda_\nu\lambda_{\nu_2}}{\nu\nu_2} \leqslant \big(12+o(1)\big)\frac{\alpha a}{\varphi(\alpha a)}\Big(\frac{\beta}{\varphi(\beta)}\Big)^2 (\log D)^{-2}\,.
$$

Introducing this into (14.46), we get

$$
T \leqslant \frac{(6+o(1))\alpha\beta x}{\varphi(\alpha\beta)(\log D)^2}\sum_a \frac{b(a)}{\varphi(a)}\sum_{p_1}\frac{1}{p_1} + O\big(x(\log x)^{-2}\big)\,.
$$

Here, we have

$$
\sum_{p_1}\frac{1}{p_1} \sim \frac{1}{2}\log\Big(\frac{\log\alpha z}{2\log z}\Big) < \frac{\log\alpha x z^{-2}}{4\log z}\,.
$$

Next we sum over a as follows:

$$
\sum_{\substack{a\leqslant 2\alpha x z^{-2}\\ (a,\alpha\beta)=1}} \frac{b(a)}{\varphi(a)} \leqslant W(2\alpha x z^{-2}) \sim W(z)\Big(\frac{\log\alpha x z^{-2}}{\log z}\Big)^{\frac{1}{2}}
$$

where $W(z)$ is the corresponding product

$$
W(z) = \prod_{\substack{p<z\\ p\equiv 1\ (\mathrm{mod}\ 4)\\ p\nmid\alpha\beta}} \Big(1-\frac{1}{p}\Big)^{-1}\Big(1+\frac{1}{p(p-1)}\Big)\,.
\tag{14.47}
$$

Hence, we have proved that

$$
T \leqslant \Big(\frac{3}{2}+o(1)\Big)\frac{\alpha\beta}{\varphi(\alpha\beta)}\,xW(z)\Big(\frac{\log\alpha x z^{-2}}{\log z}\Big)\Big(\log\frac{z}{\alpha}\Big)^{-2}\,.
\tag{14.48}
$$

Recall that our target sum (14.40) is expressed in (14.42) as $S(\mathcal{A},\mathcal{P},z)-T$ and $S(\mathcal{A},\mathcal{P},z)$ is given a lower bound in (14.41). In this bound we have the function $f(s)$ which satisfies (14.2), (14.3); in particular, for $s=(\log x)/2(\log z)$ this gives

$$
f(s) > 2\Big(\frac{2e^\gamma}{\pi}\Big)^{\frac{1}{2}}(\log z)^{\frac{1}{2}}\Big(\log\frac{x}{z^2}\Big)^{\frac{1}{2}}(\log x)^{-1}\,.
\tag{14.49}
$$

Putting these estimates together we obtain

$$(14.50)\quad \begin{aligned} S(\mathcal{A},\mathcal{P},\sqrt{2\alpha x}) > \Big(\frac{2e^{\gamma}}{\pi}\Big)^{\frac12} xV(z)(\log z)^{\frac12}\Big(\log\frac{x}{z^2}\Big)^{\frac12}(\log x)^{-2}\\ -\Big(\frac32+o(1)\Big)\frac{\alpha\beta}{\varphi(\alpha\beta)}\frac{xW(z)}{(\log z/\alpha)^2}\Big(\frac{\log \alpha x z^{-2}}{\log z}\Big)^{\frac32}. \end{aligned}$$

We shall see that the local factors at the prime factors of $\alpha\beta$ match nicely, as they should. Indeed, we have

$$\begin{aligned} \frac{\alpha\beta W(z)}{\varphi(\alpha\beta)V(z)} &= \prod_{p\mid\alpha\beta}\Big(1-\frac1p\Big)^{-1}\prod_{\substack{p<z\\ p\nmid\alpha\beta}}\Big(1-\frac1p\Big)^{-1}\prod_{\substack{p<z\\ p\equiv1\ (\mathrm{mod}\ 4)\\ p\nmid\alpha\beta}}\Big(1+\frac{1}{p(p-1)}\Big)\\ &< c\prod_{p<z}\Big(1-\frac1p\Big)^{-1}\sim ce^{\gamma}\log z \end{aligned}$$

by the Mertens formula (2.21), where c is an absolute constant, given by

$$c=\prod_{p>2}\Big(1+\frac{1}{p(p-2)}\Big).$$

The fact that the local factors at the other primes do not match perfectly, although they are pretty close, should be expected because we did not bother to evaluate the sum $\sum b(a)/\varphi(a)$ asymptotically. For this reason neither can we match the transcendental factor e^{γ}/π. We estimate the constant c by "ad hoc" arguments as follows:

$$\begin{aligned} c &< \frac43\prod_{p\geqslant5}\Big(1+\frac{1}{p(p-2)}\Big) < \frac43\prod_{p\geqslant3}\Big(1+\frac{1}{p(p+2)}\Big)\\ &< \frac43\prod_{p\geqslant3}\Big(1-\frac{1}{p^2}\Big)^{-1} = \zeta(2)=\frac{\pi^2}{6}<\frac53. \end{aligned}$$

We also use the bound $\sqrt{\pi e^{\gamma}}<12/5$. We simplify further by requiring $z\leqslant\sqrt{x/\alpha}$, that is, $\alpha\leqslant xz^{-2}$. Now (14.50) implies

(14.51)

$$S(\mathcal{A},\mathcal{P},\sqrt{2\alpha x}) > \Big(\frac{2e^{\gamma}}{\pi}\Big)^{\frac12} xV(z)(\log z)^{\frac12}\Big(\log\frac{x}{z^2}\Big)^{\frac12}(\log x)^{-2}\{1-\delta+o(1)\},$$

where

$$\delta = 12\Big(\frac{\log x}{\log z^3x^{-1}}\Big)^2\frac{\log xz^{-2}}{\log z} = 12(s-1)\Big(\frac3s-2\Big)^{-2}$$

and, as before, $s=(\log x)/2(\log z)$. For $s=20/19$ we have $\delta<1$ and the estimate (14.51) completes the proof of Theorem 14.8. □

REMARKS. The above estimates for the absolute constants do not matter. At a quick look, it is not apparent from (14.50) that the lower bound (14.41) will be larger than the upper bound (14.48) yielding any result whatsoever, so one can be excused for thinking that the fact that we get a theorem is the result of a fortunate numerical accident (which sometimes happens in other sifting situations). However, in terms of the sieve parameter $s=\log(x/z^2)$ which is slightly greater than one, the lower bound (14.41) behaves like $\sqrt{s-1}$ while the upper bound (14.48) behaves like $(s-1)^{3/2}$. Hence it is inevitable that once $s>1$ is sufficiently close to one we shall get a positive bound. The explicit calculations are only needed to give the explicit range of uniformity in the statement of Theorem 14.8.

In passing, we mention without proof the following interesting theorem due to G. Greaves [**72**], which is similar in spirit to Theorem 14.8.

THEOREM 14.9. *For* $N \equiv 2, 3, 4, 6, 7 \pmod 8$ *we have*

$$\sum_{p_1^2+p_2^2\leqslant N}\sum b\big(N-p_1^2-p_2^2\big) \asymp N(\log N)^{-\frac{5}{2}} .$$

A highlight of Greaves' arguments is his proof that the sequence $N-p_1^2-p_2^2$ has level of distribution $\sqrt{N}(\log N)^{-C}$ which he deduces from the Barban–Davenport–Halberstam theorem (Corollary 9.15).

14.7. Hyperbolic Prime Number Theorem

As we know from Lagrange, every integer, and so a fortiori every prime, is a sum of four squares. In this section we are going to use the semi-linear sieve to study the number of representations

$$x_1^2+x_2^2+x_3^2+x_4^2 = p \leqslant X ,$$

subject also to the determinant equation $x_1x_4 - x_2x_3 = 1$.

We shall make use of an unproved assumption about primes in arithmetic progressions. Let

$$\psi(x;q,a) = \sum_{\substack{n\leqslant x\\ n\equiv a (\operatorname{mod} q)}} \Lambda(n)$$

as usual and, for a and q coprime, define the "remainder term"

$$E(x;q,a) = \psi(x;q,a) - \frac{x}{\varphi(q)} , \tag{14.52}$$

and the "remainder of level Q"

$$E(x,Q) = \sum_{q\leqslant Q} \max_{(a,q)=1}\max_{y\leqslant x} |E(y;q,a)| . \tag{14.53}$$

We introduce, for $0<\theta\leqslant 1$, the following:

ASSUMPTION A(θ). *The bound*

$$E(x,Q) \ll x(\log x)^{-A}$$

holds for $Q = x^{\theta-\varepsilon}$, *for every* $\varepsilon>0$ *and every* $A>0$, *with an implied constant that depends only on* ε *and* A.

Thus, as discussed in Section 9.8, $A(\frac{1}{2})$ is the Bombieri–Vinogradov theorem, and the expectation that $A(1)$ holds is the conjecture of Elliott and Halberstam. The extra condition mentioned here that the bound holds for all $y \leqslant x$ follows by an elementary splitting argument.

THEOREM 14.10. *If* $A(\theta)$ *holds for some* θ *sufficiently close to one, then*

$$\sum_{\substack{x_1^2+x_2^2+x_3^2+x_4^2\leqslant X\\ x_1x_4-x_2x_3=1}} \Lambda\big(x_1^2+x_2^2+x_3^2+x_4^2\big) \asymp X . \tag{14.54}$$

Note, that we do not need the assumption $A(\theta)$ for $\theta = 1$. Actually, we need this with some particular θ less than one and only for the lower bound; the proof of the upper bound is unconditional (and much easier).

One can give (see [**57**]) an interpretation of the above formula as a twin prime problem in the hyperbolic disc. We shall study it by giving another, more elementary reformulation.

We can reinterpret the system

$$\begin{cases} x_1^2 + x_2^2 + x_3^2 + x_4^2 = n \,, \\ x_1x_4 - x_2x_3 = 1 \,, \end{cases} \tag{14.55}$$

as the system

$$\begin{cases} y_1^2 + y_2^2 = n + 2 \,, \\ (y_1^2 + y_2^2) - (y_3^2 + y_4^2) = 4 \,, \end{cases} \tag{14.56}$$

by making the linear transformation $(x_1, x_2, x_3, x_4) \to (y_1, y_2, y_3, y_4)$ where

$$\begin{cases} y_1 = x_1 + x_4 & y_2 = x_2 + x_3 \,, \\ y_3 = x_1 - x_4 & y_4 = x_2 - x_3 \,. \end{cases}$$

The inverse is given by

$$\begin{cases} x_1 = \frac{1}{2}(y_1 + y_3) & x_2 = \frac{1}{2}(y_2 + y_4) \,, \\ x_3 = \frac{1}{2}(y_2 - y_4) & x_4 = \frac{1}{2}(y_1 - y_3) \,. \end{cases}$$

These provide a one-to-one correspondence as real numbers. Assume now that n is an odd integer. If x_1, x_2, x_3, x_4 are integers then so are y_1, y_2, y_3, y_4. Conversely, given y_1, y_2, y_3, y_4, we learn from the first equation in (14.56) that y_1 and y_2 have opposite parity, and then we learn from the second equation that so do y_3 and y_4. Hence, given a pair of integer solutions y_1, y_2, y_3, y_4 and y_1, y_2, y_4, y_3, precisely one of them gives rise to an integer solution in the x's. Therefore, the number of integer solutions in the x's is equal to one half of the number of integer solutions in the y's. (On the other hand, if n is even then automatically y_1 and y_2 have the same parity as do y_3 and y_4, so the correspondence between x's and y's is one-to-one.)

Denote as usual by $r(m)$, the number of representations of m as the sum of two squares. From the above discussion it follows that the number of solutions to the system (14.55), for n odd, is given by

$$s_n = \tfrac{1}{2} r(n-2) r(n+2) \,.$$

Hence, the statement of our main theorem now becomes:

THEOREM 14.11. *If $A(\theta)$ holds for some θ sufficiently close to one, then*

$$S(x) = \sum_{n \leqslant x} r(n-2) r(n+2) \Lambda(n) \asymp x \,. \tag{14.57}$$

It is in this form that we shall give the proof. Here too, we can note the similarity of the formula (14.57) to the twin prime conjecture.

To begin, we need results about the distribution of the function $r(m)$ over arithmetic progressions in a wide range of moduli. If one uses the formula

$$r(m) = 4 \sum_{d \mid m} \chi(d) \,, \tag{14.58}$$

where as before $\chi = \chi_4$, then not surprisingly one controls the error term by means of Kloosterman sums. The Weil bound for these produces results more than sufficient for our needs. We refer to the result of Smith [**149, 150**]; see also [**109**].

LEMMA 14.12. *Let* $(a,q)=1$. *Then, for* $q \not\equiv 0 \pmod 4$ *we have*

$$\sum_{\substack{m\leqslant x\\ m\equiv a \,(\mathrm{mod}\, q)}} r(m) = \frac{\pi x}{q}\prod_{p|q}\Big(1-\frac{\chi(p)}{p}\Big) + O\left(q^{-\frac12}x^{\frac23+\varepsilon}\right) \tag{14.59}$$

and for $q\equiv 0 \pmod 4$ *this holds with the main term multiplied by* $\big(1+\chi(a)\big)$.

Note that this gives the asymptotic within the range $q < x^{\frac23-\varepsilon}$. We also need a bound for the shifted convolution of r against itself.

LEMMA 14.13. *We have, uniformly for d odd and* $x \geqslant 3$,

$$\sum_{\substack{n\leqslant x\\ n\equiv 3\,(\mathrm{mod}\,4)\\ n\equiv 0\,(\mathrm{mod}\,d)}} r(n-2)r(n+2) = g(d)8x + O\Big(x^{\frac{11}{12}+\varepsilon}\Big) \tag{14.60}$$

where

$$g(d) = \frac1d\prod_{p|d}\frac{p-\chi(p)}{p+\chi(p)}\,. \tag{14.61}$$

Note that (14.60) accounts for all odd n since, if $n\equiv 1 \pmod 4$, both of $r(n\pm 2)$ vanish.

PROOF. Since, for $m\equiv 1 \pmod 4$,

$$r(m) = 8\sum_{\substack{k\ell=m\\ k<\sqrt m}}\chi(k) + 4\chi(\sqrt m)\,, \tag{14.62}$$

the sum (14.60) is equal to $8\sum_k\sum_n + O\left(x^{\frac12+\varepsilon}\right)$ where

$$\begin{aligned}\sum_k\sum_n &= \sum_{\substack{k<\sqrt x\\ (k,d)=1}}\chi(k)\sum_{\substack{k^2<n\leqslant x\\ n\equiv 3\,(\mathrm{mod}\,4)\\ n\equiv 0\,(\mathrm{mod}\,d)\\ n\equiv 2\,(\mathrm{mod}\,k)}} r(n+2)\\
&= 2\pi\sum_{\substack{k<\sqrt x\\ (k,d)=1}}\chi(k)\left\{\frac{x-k^2}{4dk}\prod_{p|dk}\Big(1-\frac{\chi(p)}{p}\Big) + O\left((dk)^{-\frac12}x^{\frac23+\varepsilon}\right)\right\}\\
&= \frac{\pi x}{2d}\prod_{p|d}\Big(1-\frac{\chi(p)}{p}\Big)\sum_{(k,d)=1}\frac{\chi(k)}{k}\prod_{p|k}\Big(1-\frac{\chi(p)}{p}\Big) + O\left(x^{\frac{11}{12}+\varepsilon}\right)\\
&= \frac{x}{d}\prod_{p|d}\Big(1-\frac{\chi(p)}{p}\Big)\Big(1+\frac{\chi(p)}{p}\Big)^{-1} + O\left(x^{\frac{11}{12}+\varepsilon}\right)\,,\end{aligned}$$

giving Lemma 14.13. □

We now have enough to prove the upper bound of Theorem 14.11.

PROOF. This is a linear sieve problem for the sequence $\mathcal{A} = \big(a_n\big)$ with $a_n = r(n-2)r(n+2)$. We have seen in Lemma 14.13 that this sequence has a level of distribution $D = D(x) = x^{1/12-2\varepsilon}$; therefore any upper-bound sieve, such as Brun's, yields the result. □

For our lower bound we shall also need an asymptotic formula for a shifted convolution of r with the von Mangoldt function. In this case we leave the remainder term untreated for the time being.

LEMMA 14.14. *Let $(a,q)=1$, $q \equiv 0 \,(\mathrm{mod}\, 4)$ and $a \equiv 3 \,(\mathrm{mod}\, 4)$. Then we have*

$$\sum_{\substack{n\leqslant x\\ n\equiv a \,(\mathrm{mod}\, q)}} r(n+2)\Lambda(n) = \frac{Hx}{\varphi(q)} \prod_{p\mid q}\Big(1+\frac{\chi(p)}{p(p-1)}\Big)^{-1} + O\Big(\sum_{k\leqslant\sqrt{x}} E_{\max}(x,qk) + \sqrt{x}\log x\Big)\,,$$

where

$$H = 2\pi\prod_p\Big(1+\frac{\chi(p)}{p(p-1)}\Big)$$

and where

$$E_{\max}(x,q) = \max_{(a,q)=1}\,\max_{y\leqslant x}\big|E(y;q,a)\big|\,,$$

the latter quantity being as defined in (14.52).

PROOF. Using (14.62) we obtain

$$\sum_{\substack{n\leqslant x\\ n\equiv a (\mathrm{mod}\, q)}} r(n+2)\Lambda(n) = 8\sum_{k\leqslant\sqrt{x}}\chi(k)\sum_{\substack{k^2<n\leqslant x\\ n\equiv a(q)\\ n\equiv -2(k)}}\Lambda(n) + O(\sqrt{x}\,\log x)$$

$$= 8\sum_{k\leqslant\sqrt{x}}\chi(k)\Big\{\frac{x-k^2}{\varphi(kq)} + O\big(E_{\max}(x,qk)\big)\Big\} + O(\sqrt{x}\,\log x)$$

$$= \frac{8x}{\varphi(q)}\sum_{k\geqslant 1}\frac{\chi(k)}{k}\prod_{\substack{p\mid k\\ p\nmid q}}\Big(1-\frac{1}{p}\Big)^{-1} + O\Big(\sum_{k\leqslant\sqrt{x}} E_{\max}(x,qk) + \sqrt{x}\,\log x\Big)\,,$$

giving the lemma. □

We are now ready to begin the proof of the lower bound, so our goal is to show that $S(x) \gg x$. We are going to use the semi-linear sieve and hence it will be convenient to replace the function r in one factor of (14.57) by the characteristic function of the support of r. Define $b(n)=1$ if all prime factors of n are congruent to $1 \,(\mathrm{mod}\, 4)$ and zero otherwise (that is, it records odd numbers which admit primitive representations as the sum of two squares). Note that b is totally multiplicative.

In case $b(n)\neq 0$ then $\chi(d)=1$ for every $d \mid n$. Thus, for every positive integer m we have by (14.58),

$$r(m) \geqslant 4b(m)\tau(m)\,. \tag{14.63}$$

Hence

$$\begin{aligned} S(x) &\geqslant 4\sum_{n\leqslant x} b(n-2)\tau(n-2)r(n+2)\Lambda(n) \\ &\geqslant 4\sum_{\delta\leqslant\Delta} b(\delta) \sum_{\substack{n\leqslant x\\ n\equiv 2\,(\mathrm{mod}\,\delta)}} b(n-2)r(n+2)\Lambda(n) \end{aligned} \tag{14.64}$$

for any $\Delta=\Delta(x)$, which will be chosen to be relatively small, $\Delta=x^{\vartheta}$.

We detect the factor $b(n-2)$ by the semi-linear sieve with respect to the set $\mathcal{P}=\{p\equiv 3\,(\mathrm{mod}\,4)\}$. To this end we consider the sifting sequence

$$\mathcal{A}^{(\delta)}=\big(a_n\big)\ ,\quad a_n=r(n+4)\Lambda(n+2)$$

for $n\leqslant x-2$, $n\equiv 1\,(\mathrm{mod}\,4)$, $n\equiv 0\,(\mathrm{mod}\,\delta)$, and so

$$\sum_{\substack{n\leqslant x\\ n\equiv 2\,(\mathrm{mod}\,\delta)}} b(n-2)r(n+2)\Lambda(n)=S\big(\mathcal{A}^{(\delta)},\mathcal{P},\sqrt{x}\big)\ , \tag{14.65}$$

where, in the usual sieve notation, this is the sum of the a_n in $\mathcal{A}^{(\delta)}$ over those n having no prime factor in $\mathcal{P}$ up to $\sqrt{x}$. Here $\sqrt{x}$ suffices because, if $n\equiv 1\,(\mathrm{mod}\,4)$ has one prime factor from $\mathcal{P}$, it must have a second one.

We are going to apply the lower bound semi-linear sieve to the sequence $\mathcal{A}^{(\delta)}$ for every $\delta\leqslant\Delta$. To this end we need to verify approximations of the type

$$\mathcal{A}_d^{(\delta)}(x)=g(d)X^{(\delta)}+r_d(x)\ , \tag{14.66}$$

for d squarefree and having all prime factors congruent to $3\,(\mathrm{mod}\,4)$. Note that this implies $(d,2\delta)=1$. Lemma 14.14, with $q=4\delta d$, provides such an approximation where

$$X^{(\delta)}=\frac{Hx}{2\varphi(\delta)}\prod_{p|\delta}\Big(1+\frac{1}{p(p-1)}\Big)^{-1} \tag{14.67}$$

and

$$g(d)=\frac{1}{\varphi(d)}\prod_{p|d}\Big(1-\frac{1}{p(p-1)}\Big)^{-1}\ . \tag{14.68}$$

The remainder term in (14.66) satisfies, by Lemma 14.14,

$$r_d(x)\ll\sum_{k\leqslant\sqrt{x}} E_{\max}(x,4\delta dk)+x^{\frac12}\log x\ . \tag{14.69}$$

The function $g(d)$ in (14.68) is a linear sieve density but we shall be sifting by only half of the primes, so this becomes a semi-linear sieve problem. Since the sieving limit for the semi-linear sieve is $\beta=1$ and our level of distribution is $D=D(x)<x^{\frac12}$, we cannot successfully apply the lower bound sieve to $S\left(\mathcal{A}^{(\delta)},\mathcal{P},\sqrt{x}\right)$, but we can come close to this.

Let $z=D^{1/s}$ with $1<s<3$. Then we have, by (14.2)–(14.6),

$$\begin{aligned} S\big(\mathcal{A}^{(\delta)},\mathcal{P},z\big) &\geqslant X^{(\delta)}V(z)\{f(s)+o(1)\}-R_\delta(x) \\ &\geqslant c\frac{x}{\delta}\big(\log x\big)^{-\frac12}\left\{\sqrt{s-1}+o(1)\right\}-R_\delta(x) \end{aligned} \tag{14.70}$$

where c is a positive absolute constant and

$$R_\delta(x) = \sum_{k\leqslant\sqrt{x}} \sum_{d\leqslant D} E_{\max}(x, 4\delta dk) . \tag{14.71}$$

We shall choose

$$D = D(x) = x^{\theta-\frac{1}{2}-\vartheta-\varepsilon} , \tag{14.72}$$

so that the modulus $4\delta dk$ in (14.71) does not exceed $4x^{\theta-\varepsilon}$.

To finish the job we are going to reinterpret the sieve problem. In order to turn our lower bound (14.70) into a lower bound for the sum in (14.65), it remains to estimate the difference

$$T\big(\mathcal{A}^{(\delta)}, \mathcal{P}, z\big) = S\big(\mathcal{A}^{(\delta)}, \mathcal{P}, z\big) - S\big(\mathcal{A}^{(\delta)}, \mathcal{P}, \sqrt{x}\big) ,$$

and, for the purpose of utilizing (14.64), we are only required to do this on average over δ.

We behave as if $\theta < 1$ is quite close to 1 and ϑ is very small, so the level of distribution $D = D(x)$ of $\mathcal{A}^{(\delta)}$ is only slightly smaller than $x^{\frac{1}{2}}$ and is, in particular, greater than $x^{\frac{1}{3}}$. Assuming $1 < s \leqslant \frac{4}{3}$ we have $z = D^{1/s} > x^{1/4}$; therefore $T\big(\mathcal{A}^{(\delta)}, \mathcal{P}, z\big)$ counts only integers having exactly two prime factors in $\mathcal{P}$. Precisely, we have

$$T\big(\mathcal{A}^{(\delta)}, \mathcal{P}, z\big) = \sum_{\substack{z\leqslant p_1<p_2 \\ p_1,p_2\in\mathcal{P}}} \sideset{}{'}\sum_{\substack{n\leqslant x+2 \\ n\equiv 1\,(\mathrm{mod}\,4) \\ n\equiv 4\,(\mathrm{mod}\,\delta p_1p_2)}} r(n)\Lambda(n-2) ,$$

where $\sum'$ indicates that the summation over n is also restricted by the condition $\big((n-4)/\delta p_1 p_2,\, P(z)\big) = 1$. Since we are looking for an upper bound we can ignore some constraints. We reinterpret the congruence $n \equiv 4\,(\mathrm{mod}\,\delta p_1 p_2)$ as the equation

$$n = 4 + \delta a p_1 p_2$$

so that a runs through integers having only prime factors $\equiv 1 (\mathrm{mod}\, 4)$ and

$$a \leqslant \frac{x}{\delta z^2} \leqslant \frac{x}{z^2} .$$

We record the fact that $z < p_1 \leqslant \sqrt{x}$ and, having done that, we weaken the condition for the prime p_2 by ignoring its range and giving to it a new name: $q_2 = (n-4)/\delta a p_1$. We attach to q_2 the following upper-bound linear sieve factor:

$$\sum_{\nu_2\mid q_2} \lambda_{\nu_2} ,$$

and forget that q_2 was prime. Similarly, we relax $\Lambda(n-2)$, majorizing it as follows:

$$\Lambda(n-2) \leqslant (\log x) \sum_{\nu\mid(n-2)} \lambda_\nu .$$

In both these applications (λ_ν) stands for a sieve supported on odd numbers and of relatively small level, say $x^{1/60}$, so that we need to remember that $2 \nmid \nu\nu_2$ and $\nu,\, \nu_2 \leqslant x^{1/60}$. As a result, we find that $T\left(\mathcal{A}^{(\delta)}, \mathcal{P}, z\right)$ is bounded by

$$(\log x) \sum_{a\leqslant xz^{-2}} b(a) \sum_{\substack{z<p_1\leqslant\sqrt{x} \\ p_1\equiv 3\,(\mathrm{mod}\,4)}} \sum\sum_{(\nu,\delta a\nu_2)=1} \lambda_\nu\lambda_{\nu_2} \sum_{\substack{n\leqslant x+2 \\ n\equiv 1\,(\mathrm{mod}\,4) \\ n\equiv 2\,(\mathrm{mod}\,\nu) \\ n\equiv 4\,(\mathrm{mod}\,\delta a p_1 \nu_2)}} r(n) .$$

At this point we cannot apply crude estimates because we cannot afford to lose the sign changes in $\lambda_\nu, \lambda_{\nu_2}$. Therefore, we apply Lemma 14.12 to the inner sum above, getting

$$\sum_n r(n) = \frac{\pi x}{4\nu\delta a p_1 \nu_2} \prod_{p\mid\delta a\nu\nu_2}\left(1 - \frac{\chi(p)}{p}\right) + O\left(z^{-\frac{1}{2}} x^{\frac{2}{3}+\varepsilon}\right) .$$

Hence,

$$T(\mathcal{A}^{(\delta)}, \mathcal{P}, z) \ll \frac{x\log x}{\delta}\prod_{p\mid\delta}\left(1-\frac{1}{p}\right)\Bigl(\sum_{z<p_1<\sqrt{x}}\frac{1}{p_1}\Bigr)\sum_{a\leqslant xz^{-2}}\frac{b(a)}{a}$$
$$\cdot\left|\sum\sum_{(\nu,\delta a\nu_2)=1}\frac{\lambda_\nu\lambda_{\nu_2}}{\nu\nu_2}\prod_{\substack{p\mid\nu\nu_2\\ p\nmid\delta a}}\left(1-\frac{\chi(p)}{p}\right)\right| + O\left(z^{-\frac{5}{2}}x^{\frac{11}{5}+\varepsilon}\right) .$$

Now, for the sum over ν, ν_2 we apply Theorem 5.9, getting

$$\sum_\nu\sum_{\nu_2} \ll \frac{\delta a}{\varphi(\delta a)}(\log x)^{-2} .$$

Hence,

$$T(\mathcal{A}^{(\delta)}, \mathcal{P}, z) \ll \frac{x}{\delta\log x}\log\left(\frac{\log\sqrt{x}}{\log z}\right)\left(\log\frac{x}{z^2}\right)^{\frac{1}{2}} + z^{-\frac{5}{2}}x^{\frac{11}{5}+\varepsilon} .$$

For $z = D^{1/s} = x^{(\theta-\frac{1}{2}-\vartheta-\varepsilon)/s}$ the above estimation simplifies to

$$T(\mathcal{A}^{(\delta)}, \mathcal{P}, z) \leqslant \frac{Cx}{\delta\sqrt{\log x}}\left(1-\frac{2}{s}(\theta-\frac{1}{2}-\vartheta)+\varepsilon\right)^{\frac{3}{2}} , \tag{14.73}$$

for some large absolute constant C and all x sufficiently large in terms of ε.

We want the upper bound (14.73) to be small when compared to the lower bound (14.70). This is possible because, when θ is less than one but near to one and ϑ is even closer to zero, then the above factor behaves like $(s-1)^{\frac{3}{2}}$ which tends to zero faster than $(s-1)^{\frac{1}{2}}$ (in the same fashion as already remarked in connection with Theorem 14.8). Specifically, we choose $\vartheta = \frac{1}{2}(1-\theta)$ and $s = 2-\theta$ so that

$$1-\frac{2}{s}(\theta-\frac{1}{2}-\vartheta) = 4\,\frac{1-\theta}{2-\theta} = 4\,\frac{s-1}{s} .$$

Subtracting (14.73) from (14.70) we get

$$S(\mathcal{A}^{(\delta)}, \mathcal{P}, \sqrt{x}) \geqslant \frac{cx\sqrt{1-\theta}}{2\delta\sqrt{\log x}} - R_\delta(x) .$$

Now, summing over δ we conclude from (14.64) that $S(x) \gg x$ subject to the assumption of $A(\theta)$ with $\theta < 1$, but sufficiently close to one.

Note that we actually require $A(\theta)$ for the sum weighted by a divisor function, but this extra factor can be handled by an application of Cauchy's inequality and elementary estimates for sums of divisor functions.

14.8. Prime Points on the Sphere – Secret Challenge

A celebrated result of W. Duke [**34, 35**] asserts that the integer points on the sphere

$$a^2 + b^2 + c^2 = N$$

are equidistributed as $N \to \infty$ over some fixed residue classes modulo eight. One of the main ingredients in the proof is a sub-convexity bound for automorphic L-functions, or a corresponding bound for Fourier coefficients of metaplectic cusp forms. It would be extremely interesting if one could make the same statement when the co-ordinates a, b, c are restricted to prime numbers (of course, only for $N \equiv 9 (\mathrm{mod}\, 24)$). Duke's argument is based on the harmonic analysis of the sphere (a Riemann surface of positive curvature) and it cannot be adapted to the selection of co-ordinates having such stringent side conditions. In this case the sieve gives a chance; for example it may produce points with almost-prime co-ordinates (see [**7**]). The problem of making all of the co-ordinates prime is too ambitious for a complete solution, so we consider the equation

$$a^2 + b^2 + p^2 = N \tag{14.74}$$

where we ask only that one co-ordinate be prime. Even so, we cannot solve this problem unconditionally, but will be able to do so subject to the Riemann Hypothesis for $L(s, \chi_N)L(s, \chi_4\chi_N)$ and the Elliott–Halberstam conjecture (that (9.83) holds with $Q = x^{1-\varepsilon}$).

First, let us try to count the solutions to (14.74) in all integers a, b and primes p by heuristic arguments. The number of these is given by the sum

$$\sum_{p \leqslant \sqrt{N}} r(N - p^2)$$

where $r(n)$ is the number of representations of n as the sum of two squares. For $n \equiv 1 \,(\mathrm{mod}\ 4)$ we have

$$r(n) = 4 \sum_{d|n} \chi_4(d) = 8 \sum_{\substack{d|n \\ d<\sqrt{n}}} \chi_4(d) + 4\chi_4(\sqrt{n}) \ .$$

Assume $N \equiv 2 \,(\mathrm{mod}\ 4)$. We have

$$\begin{aligned}\sum_{p \leqslant \sqrt{N}} r(N - p^2) &= 8 \sum_{d<\sqrt{N}} \chi_4(d) \sum_{\substack{p<\sqrt{N-d^2} \\ p^2 \equiv N \,(\mathrm{mod}\ d)}} 1 + O\big(r(N) + r(N-4)\big) \\ &= 8 \sum_{\substack{d<\sqrt{N} \\ (d,N)=1}} \chi_4(d) \sum_{\nu^2 \equiv N \,(\mathrm{mod}\ d)} \pi\big(\sqrt{N-d^2}; d, \nu\big) + O\big(r(N(N-4))\big) \ .\end{aligned}$$

For the contribution from $d < D = N^{\frac{1}{2}-\varepsilon}$ we can use the Prime Number Theorem with admissible error term since we assume the Elliott–Halberstam conjecture. Even with this help, we cannot do so in the range $D \leqslant d < \sqrt{N}$, but it is reasonable to ignore this contribution due to the expected cancellation coming from the sign changes of $\chi_4(d)$. Of course, proving this fact rigorously is impossible with current technology. Nevertheless, if we take this fact for granted we obtain

$$\sum_{p\leqslant\sqrt{N}} r(N-p^2) = 8 \sum_{\substack{d<\sqrt{D}\\(d,N)=1}} \chi_4(d)\frac{\rho(d)}{\varphi(d)}\,\mathrm{li}\big(\sqrt{N-d^2}\big) + O\Big(N^{\frac{1}{2}}(\log N)^{-A}\Big)\,,$$

where $\rho(d)$ denotes the number of solutions to the congruence

$$\nu^2 \equiv N\,(\mathrm{mod}\ d)\ . \tag{14.75}$$

Because the series

$$C(N) = \sum_{(d,N)=1} \chi_4(d)\frac{\rho(d)}{\varphi(d)}$$

converges, we arrive at the following:

CONJECTURE. *For $N \equiv 2\,(\mathrm{mod}\ 4)$ we have*

$$\sum_{p\leqslant\sqrt{N}} r(N-p^2) \sim 8C(N)\sqrt{N}(\log N)^{-1}\ .$$

By the way, the series $C(N)$ has the following product representation:

$$\frac{\pi}{4}L(1,\chi_4\chi_N)\prod_{p|N}\Big(1-\frac{\chi_4(p)}{p}\Big)\prod_{p\nmid N}\Big\{1+\frac{\chi_4(p)(1+\chi_N(p))-1}{p(p-1)}-\frac{\chi_N(p)}{p^2(p-1)}\Big\}\,,$$

where $\chi_N(p) = (N/p)$. (Here $N \neq 4$.) Hence,

$$C(N) \asymp L(1,\chi_4\chi_N)\prod_{p|N}\Big(1-\frac{\chi_4(p)}{p}\Big)\ .$$

Our assumption $N \equiv 2\,(\mathrm{mod}\ 4)$ was only introduced for illustration of the ideas. The conjecture should hold for all $N \not\equiv 0\,(\mathrm{mod}\ 4)$, but with a slightly different factor $C(N)$. On the other hand, if $N \equiv 0\,(\mathrm{mod}\ 4)$ then $n = N - p^2 \equiv 3\,(\mathrm{mod}\ 4)$ and no such number is a sum of two squares.

The equation (14.74) is engraved on the tombstone of Ju. V. Linnik (however, in a different notation, $n = x^2 + y^2 + p^2$). This is quite a mysterious message since Linnik actually worked with the equation $n = x^2 + y^2 + p$, using the ideas of his dispersion method [**126**] and with the equation $n = x^2 + y^2 + z^2$, using the ideas of his ergodic method.

Because we want to be able to use the letter b as an integer variable, we are going to alter our earlier notation for the function $b(n)$. In this section and again in Section 15.4, denote by $B(n)$ the characteristic function of positive integers having prime factors only in the set

$$\mathcal{Q} = \{q \text{ prime};\ q \equiv 1(\mathrm{mod}\,4)\}. \tag{14.76}$$

Therefore $B(n) = 1$ if $n = a^2 + b^2 \equiv 1(\mathrm{mod}\,4)$ with $(a,b) = 1$ and $B(n) = 0$ otherwise.

THEOREM 14.15. *Let $N \equiv 2\,(\mathrm{mod}\,4)$, $N \not\equiv 1\,(\mathrm{mod}\,3)$. Assume the Riemann Hypothesis for $L(s,\chi_N)L(s,\chi_4\chi_N)$ and the Elliott–Halberstam conjecture. Then*

$$\sum_{p<\sqrt{N}} B(N-p^2) \asymp \mathfrak{S}(N)N(\log N)^{-\frac{3}{2}} \tag{14.77}$$

where

$$\mathfrak{S}(N) = \left(\frac{L(1,\chi_4\chi_N)}{L(1,\chi_N)}\right)^{\frac{1}{2}} \prod_{\substack{p|N \\ p\equiv 3(4)}} \left(1+\frac{1}{p}\right). \tag{14.78}$$

REMARKS. Because we count only the primitive representations of $N-p^2$ as a sum of two squares, that implies $N \not\equiv p^2 (\text{mod}\, 3)$. On the other hand, $p^2 \equiv 1 (\text{mod}\, 3)$ for $p \neq 3$; hence the condition $N \not\equiv 1 (\text{mod}\, 3)$ is necessary. The proof of (14.77) uses the sieve and we have no chance to derive an asymptotic formula. Note that χ_N and $\chi_4\chi_N$ are non-principal characters.

First we prove the upper bound

$$\sum_{p<\sqrt{N}} B(N-p^2) \ll \ell(N) \prod_{\substack{p|N \\ p\equiv 3(4)}} \left(1+\frac{1}{p}\right) N(\log N)^{-\frac{3}{2}} \tag{14.79}$$

with

$$\ell(N) = \prod_{p\leqslant N^{\frac{1}{5}}} \left(1-\frac{\chi_N(p)}{p}\right)^{\frac{1}{2}} \left(1-\frac{\chi_4\chi_N(p)}{p}\right)^{-\frac{1}{2}} \tag{14.80}$$

because it is unconditional and much simpler. To this end we consider the sequence of numbers

$$n = N - p^2, \qquad 2 < p < \sqrt{N}.$$

Note that $n \equiv 1 (\text{mod}\, 4)$ and $(n, 3N) = 1$, so the sifting range consists of primes in

$$\mathcal{P} = \{p \equiv 3 (\text{mod}\, 4);\ p \nmid 3N\}. \tag{14.81}$$

For any $(d, N) = 1$ we have

$$\begin{aligned} A_d(N) &= \sum_{\substack{2<p<\sqrt{N} \\ p^2\equiv N (\text{mod}\, d)}} 1 = \sum_{\nu^2\equiv N (\text{mod}\, d)} \pi(\sqrt{N}; d, \nu) + O(1) \\ &= \frac{\rho(d)}{\varphi(d)} \operatorname{li}(\sqrt{N}) + r_d(N) \end{aligned}$$

where, by the Bombieri–Vinogradov theorem, the remainder terms $r_d(N)$ satisfy

$$\sum_{\substack{d<N^{\frac{1}{4}-\varepsilon} \\ (d,N)=1}} |r_d(N)| \ll \sqrt{N}(\log N)^{-A}$$

for any $\varepsilon > 0$, $A > 0$, the implied constant depending on ε and A. In other words, our sequence has level of distribution $N^{\frac{1}{4}-\varepsilon}$. Recall that $\rho(d)$ stands for the number of roots of (14.75); thus $\rho(d)$ is the multiplicative function with

$$\rho(p) = 1 + \chi_N(p), \quad \text{where } \chi_N(p) = \left(\frac{N}{p}\right). \tag{14.82}$$

There is a small snag with the density function

$$g(d) = \frac{\rho(d)}{\varphi(d)}. \tag{14.83}$$

Heuristically, one expects $\chi_N(p)$ to behave very randomly as ± 1 so that one may think that the sieve dimension condition (5.41) holds with $\kappa = \frac{1}{2}$. This would be true if we allowed L to depend on N, but this is not permitted. Luckily, we require only an upper bound axiom and since $|\chi_N(p)| = 1$ we have (5.41) with $\kappa = 1$ and L an absolute constant. Hence we can sacrifice the dimension and apply the linear sieve upper bound

$$S(\mathcal{A}, \mathcal{P}, N^{\frac{1}{5}}) \ll V(N^{\frac{1}{5}})\sqrt{N}(\log N)^{-1} ,$$

where

$$V(z) = \prod_{\substack{p<z \\ p\in\mathcal{P}}} (1 - g(p)) = \prod_{\substack{p<z \\ p\equiv 3(4) \\ p\nmid N}} \left(1 - \frac{1+\chi_N(p)}{p-1}\right)$$
$$\asymp \prod_{\substack{p|N \\ p\equiv 3(4)}} \left(1 + \frac{1}{p}\right) \prod_{p<z} \left[1 - \frac{(1+\chi_N(p))(1-\chi_4(p))}{2p}\right].$$

By Mertens' formula

$$V(z) \asymp \ell(N) \prod_{\substack{p|N \\ p\equiv 3(4)}} \left(1 + \frac{1}{p}\right) (\log z)^{-\frac{1}{2}}. \tag{14.84}$$

This gives (14.79). Now, if we assume the Riemann Hypothesis we can complete the finite Euler product (14.80), getting

$$\ell(N) \asymp L(1, \chi_4\chi_N)^{\frac{1}{2}} L(1, \chi_N)^{-\frac{1}{2}}. \tag{14.85}$$

This completes the proof of the upper bound of (14.77).

The lower bound requires a lot more analysis. For technical convenience we only take the solutions of (14.74) with p restricted by $\sqrt{3N/4} < p < \sqrt{N}$, so our sequence $\mathcal{A}$ consists of numbers $n = N - p^2$ with $n < N/4$. Therefore, the full sum (14.77) is bounded below by

$$\sum_{\sqrt{3N/4}<p<\sqrt{N}} B(N - p^2) = S(\mathcal{A}, \mathcal{P}, \sqrt{N}), \tag{14.86}$$

where $\mathcal{P}$, the sifting range is, as before, the set of primes (14.81). For any $(d, N) = 1$ the congruence sum $A_d(N)$ is well approximated by

$$A_d(N) = \frac{\rho(d)}{\varphi(d)} X + r_d(N),$$

with

$$X = \mathrm{li}(\sqrt{N}) - \mathrm{li}(\sqrt{3N/4}) \asymp \sqrt{N}(\log N)^{-1} .$$

Now we need a much larger level of distribution than is provided by the Bombieri–Vinogradov theorem. We assume the Elliott–Halberstam conjecture, which gives

$$\sum_{\substack{d<N^{\frac{1}{2}-\varepsilon} \\ (d,N)=1}} |r_d(N)| \ll \sqrt{N}(\log N)^{-A}. \tag{14.87}$$

Therefore our sequence has level of distribution

$$D = D(N) = N^{\frac{1}{2}-\varepsilon}. \tag{14.88}$$

Because the density function (14.83) at small primes is hard to control with respect to N we apply a preliminary sieve of level $D_0 = N^\varepsilon$ and then the lower-bound semi-linear sieve of level $N^{\frac{1}{2}-2\varepsilon}$. In the range of the preliminary sieve we use the trivial bound

$$g(p) \leqslant \frac{2}{p-1} \quad \text{if } 3 < p < z_0 = N^{1/\log\log N}$$

and the Fundamental Lemma with $\kappa = 1$. In the range $z_0 \leqslant p < z = N^{1/2s}$ for $1 < s < \frac{3}{2}$ we deduce from the Riemann Hypothesis that

$$\sum_{\substack{w \leqslant p < z \\ p \in \mathcal{P}}} g(p) = \frac{1}{2} \log \frac{\log z}{\log w} + O\left(\frac{1}{\log w}\right)$$

for $z_0 \leqslant w < z$ with the implied constant being absolute. Hence the dimension condition (14.4) holds with an absolute constant L. Note that

$$s_0 = \log D_0 / \log z_0 = \varepsilon \log\log N,$$

$$V(z_0)/V(z) \ll \log z / \log z_0 < \log\log N\ ,$$

and

$$e^{-s_0} V(z_0)/V(z) \ll (\log N)^{-\varepsilon} \log\log N.$$

Therefore by Theorem 6.17, together with (14.6) which verifies (6.102), we obtain

$$S(\mathcal{A}, \mathcal{P}, z) \geqslant XV(z)\{f(s) + O(\varepsilon)\}\ .$$

Here, we have by (14.84) and (14.85)

$$V(z) = \prod_{\substack{p < z \\ p \in \mathcal{P}}} (1 - g(p)) \asymp \mathfrak{S}(N)(\log N)^{-\frac{1}{2}}.$$

Moreover, we have $f(s) \gg \sqrt{s-1}$. Hence we conclude that

$$S(\mathcal{A}, \mathcal{P}, z) \gg \frac{\mathfrak{S}(N)\sqrt{N}}{(\log N)^{\frac{3}{2}}} \sqrt{s-1}. \tag{14.89}$$

We shall choose $s > 1$ very close to one, so $z = N^{1/2s}$ is close to our target value $\sqrt{N}$. To reach this target we apply Buchstab's formula

$$S(\mathcal{A}, \mathcal{P}, \sqrt{N}) = S(\mathcal{A}, \mathcal{P}, z) - T(\mathcal{A}, \mathcal{P}, z) \tag{14.90}$$

where

$$T(\mathcal{A}, \mathcal{P}, z) = \sum_{\substack{b < Nz^{-2} \\ (b, 3N) = 1}} B(b) U_b \tag{14.91}$$

and U_b denotes the number of solutions of

$$bp_1p_2 + p^2 = N \tag{14.92}$$

in primes p, p_1, p_2 with $p_2 \geqslant p_1 \geqslant z$, $p_1 \equiv p_2 \equiv 3 (\bmod 4)$ and $4bp_1p_2 < N$.

The equation (14.92) is not so different from the original one (14.74). Given b, which is relatively small, we are looking for prime points on the two-sheeted hyperboloid

$$bXY + Z^2 = N\ . \tag{14.93}$$

Hence, one could be forgiven for coming to the conclusion that we have simply transformed one difficult problem into one which is even more so, since now we ask for all of the co-ordinates to be prime rather than just one of them. Fortunately, sieve methods allow us to proceed further, as in the two previous sections. Since we need only an upper bound for U_b of the correct order of magnitude, we can afford to lose any large (but absolute) factor and hence can relax the condition that the co-ordinates be prime by attaching upper-bound sieve weights, even those of a quite small level. In this way the problem reduces to counting the integral points on another two-sheeted hyperboloid

$$cXY + d^2Z^2 = N \tag{14.94}$$

with coefficients c, d^2, relatively small. Having liberated the variables X, Y, Z from any stringent condition one could be excused for thinking the problem is now accessible; yes indeed, but it does not carry forward as easily as in the previous sections. We need an asymptotic formula for the number of solutions to (14.94) which holds uniformly in c, d, so that one can receive the desired cancellation when summing the main term over c, d, with sieve weights, while the remainder term needs some margin of saving to absorb the loss.

Let's have a look at the literature. The work of Duke and Schulze-Pillot [**38**] deals with positive definite forms, so it does not apply to (14.94) which is an indefinite ternary quadratic form. The indefinite case, $-4XY+Z^2$, has been treated by R. Chelluri in his Ph.D. thesis [**21**]. We believe that the extension to $cXY + d^2Z^2$, with coefficients c, d^2 relatively small, is possible with some modifications of Chelluri's work, and perhaps Dr. Raju Chelluri would gladly do this for us were it not for his premature death. We have already mentioned in Section 9.3 that, in the case of positive coefficients a, b, c, W. Duke established the asymptotic formula for the number of integral points on the surface $aX^2 + bY^2 + cZ^2 = N$, uniformly for $abc < N^{1/337}$. Perhaps his method could also be modified to handle (14.94) but we propose to follow a shorter route by taking advantage of some special features of the particular problem (14.92).

First, let us shed some light on the equation (14.92), and point out what is the key problem. Note that p_1, p_2 are located in a relatively short segment in the logarithmic scale. Indeed, we have $z < p_1, p_2 < Nz^{-1}$ which, for $z = N^{1/2s}$ with s close to one, is a narrow range. If we were to ignore that feature and consider $m = p_1p_2$ as a single variable with $(m, P(z)) = 1$, then by an upper-bound sieve one can easily show that

$$U_b \ll \mathcal{U}_b \tag{14.95}$$

with

$$\mathcal{U}_b = \frac{\rho(b)}{b} \prod_{p|b}\Big(1+\frac{3}{p}\Big) \frac{\sqrt{N}}{(\log N)^2} \prod_{p|N}\Big(1+\frac{1}{p}\Big) L(1,\chi_N)^{-1}. \tag{14.96}$$

Next we sum over b, getting

$$\sum_{\substack{b<Nz^{-2}\\(b,3N)=1}} B(b)\frac{\rho(b)}{b}\prod_{p|b}\Big(1+\frac{3}{p}\Big) \ll \prod_{\substack{p<Nz^{-2}\\p\nmid 3N}}\left(1+\frac{(1+\chi_4(p))(1+\chi_N(p))}{2p}\right)$$

$$\ll \prod_{\substack{p|N\\p\equiv 1(\mathrm{mod}\,4)}}\Big(1+\frac{1}{p}\Big)^{-1} L(1,\chi_N)^{\frac12} L(1,\chi_4\chi_N)^{\frac12}\Big(\log\frac{N}{z^2}\Big)^{\frac12}.$$

Hence, (14.91), (14.95) and (14.96) yield

$$T(\mathcal{A},\mathcal{P},z) \ll \frac{\mathfrak{S}(N)\sqrt{N}}{(\log N)^{3/2}}\sqrt{s-1}. \tag{14.97}$$

This upper bound matches the order of the lower bound (14.89), but with an unspecified (and presumably unsatisfactory) implied constant, so there is no way to conclude that the difference (14.90) is positive. To succeed, one needs to improve slightly on (14.95); in other words, we cannot afford to ignore the close proximity of the prime factors p_1, p_2 in (14.92). Our goal is to prove the following bound,

$$U_b \ll (s-1)\mathcal{U}_b. \tag{14.98}$$

Assuming (14.98) we can quickly finish the proof of Theorem 14.15. First, the upper bound (14.97) gets the extra factor $s-1$. Next we obtain by (14.89) and (14.90):

$$S(\mathcal{A},\mathcal{P},\sqrt{N}) \geqslant \frac{\mathfrak{S}(N)\sqrt{N}}{(\log N)^{3/2}}\{c_1\sqrt{s-1} - c_2(s-1)^{\frac{3}{2}}\}$$

where c_1, c_2 are absolute positive constants. Finally, choosing $s > 1$ close to one we conclude that

$$S(\mathcal{A},\mathcal{P},\sqrt{N}) \gg \mathfrak{S}(N)\sqrt{N}(\log N)^{-3/2}. \tag{14.99}$$

This completes the proof of the lower bound of (14.77).

It remains to prove (14.98). We can ease the job by counting more solutions of (14.92) than those accounted for by U_b. Clearly, we have $U_b \leqslant S_b$ for some

$$S_b = \sum_{bp_1p_2+p^2=N} F(p_1)G(p) \tag{14.100}$$

where F and G are two non-negative smooth functions with supp $F = [z/2, \sqrt{2N}]$, supp $G = [\sqrt{N/2}, \sqrt{2N}]$ and

$$x^j F^{(j)}(x) \ll 1, \quad y^j G^{(j)}(y) \ll 1.$$

Note that we let p_2 run free, but of course the equation (14.92) provides an adequate localization of p_2. We have also given up the congruence conditions $p_1 \equiv p_2 \equiv 3 \pmod 4$.

Having made the above preparations we are now ready to apply an upper-bound sieve to relax the requirement of p_1, p_2, p being primes. We obtain $S_b \leqslant S_b^+$, where S_b^+ is the sifted sum

$$S_b^+ = \sum_{\substack{bn_1n_2+n^2=N\\(n_1,bN)=1}} F(n_1)G(n)\Big(\sum_{d_1|n_1}\lambda_{d_1}\Big)\Bigg(\sum_{\substack{d_2|n_2\\(d_2,bn_1)=1}}\lambda_{d_2}\Bigg)\Bigg(\sum_{\substack{d|n\\(d,N)=1}}\lambda_d\Bigg).$$

Here $\Lambda = (\lambda_d)$ is an upper-bound sieve of relatively small level, say $d \leqslant \Delta = N^\delta$, where δ is a small positive absolute constant.

Note that the equation $bn_1n_2 + n^2 = N$ implies the following coprimality conditions:

$$d_1d_2d \text{ squarefree}, \quad (d_1d_2d, b) = 1, \quad (d_1d, N) = 1\ . \tag{14.101}$$

Before proceeding further one would like to see that replacing S_b by S_b^+ does not cause any great loss, that is to be sure that the solutions n_1, n_2, n which have relatively small prime divisors are effectively eliminated. Of course, this will be seen at the end of our rigorous estimates, but it is possible to make a prediction by analyzing the effects of the convolutions with sieve weights. We propose for the reader to go through such an analysis to confirm his understanding of sieve mechanisms. Certainly, it does not matter that we applied the same sieve $\Lambda = (\lambda_d)$ to each of the variables n_1, n_2, n.

Now, having convinced oneself that nothing substantial has been missed in the arrangement of the sifted sum S_b^+ we can confidently open the weight convolutions getting

$$S_b^+ = \sum_{d_1}\sum_{d_2}\sum_{d} \lambda_{d_1}\lambda_{d_2}\lambda_d S(d_1, d_2, d) \tag{14.102}$$

where

$$S(d_1, d_2, d) = \sum\sum\sum_{\substack{bd_1d_2m_1m_2+d^2m^2=N \\ (m_1, Nbdd_2)=1}} F(d_1m_1)G(dm) \tag{14.103}$$

for d_1, d_2, d satisfying (14.101). Note that m_2 is absolutely free so that the above equation can be interpreted as a congruence of modulus

$$q = bd_1d_2m_1, \quad \text{with} \quad (m_1, Nbdd_2) = 1 \ . \tag{14.104}$$

We have

$$S(d_1, d_2, d) = \sum_q F(d_1m_1) \sum_{d^2m^2\equiv N \pmod q} G(dm) \ . \tag{14.105}$$

Note that $(q, d) = 1$ but q may not be coprime with N, precisely $(q, N) = (d_2, N)$.

Next we split the sum over m into classes $m \equiv \nu \pmod q$ and for each class we apply Poisson's formula, getting

$$\sum_{d^2m^2\equiv N(q)} G(dm) = \frac{1}{dq}\sum_h \hat{G}\Big(\frac{h}{dq}\Big) W_{h\bar{d}}(N; q) \tag{14.106}$$

where $\hat{G}(u)$ is the Fourier transform of $G(x)$ and

$$W_h(N; q) = \sum_{\nu^2\equiv N \pmod q} e\Big(\frac{\nu h}{q}\Big) \ , \tag{14.107}$$

which is the Weyl sum of frequency h for the roots $\nu \pmod q$. We have

$$\hat{G}(u) \ll \sqrt{N}(1 + |u|\sqrt{N})^{-A}$$

for any $A \geqslant 0$, the implied constant depending on A, and

$$|W_h(N; q)| \leqslant \rho(q) \leqslant \tau(q).$$

Therefore, the Fourier expansion (14.106) converges rapidly. In fact only terms with $h \ll dqN^{\varepsilon-1/2}$ play a role; the contribution from the others can be trivially estimated. Note that

$$dq = bdd_1d_2m_1 \ll bdd_2\sqrt{N} \leqslant z^{-2}N^{\frac{3}{2}}\Delta^2 = N^{-\frac{1}{s}+\frac{3}{2}+2\delta} \ ,$$

so for s close to one we see that h is relatively small. Specifically, if $1 < s < (1-\delta)^{-1}$ then we are left with the frequencies

$$|h| \leqslant H = N^{4\delta} , \tag{14.108}$$

the remaining ones contributing $O(N^{-A})$ with any $A \geqslant 0$. Introducing (14.106) to (14.105) we deduce that

$$S(d_1,d_2,d) = \sum_q \frac{F(d_1m_1)}{dq} \sum_{|h|\leqslant H} \hat{G}\Big(\frac{h}{dq}\Big) W_{h\bar{d}}(N;q) + O(N^{-A}) , \tag{14.109}$$

where q runs over integers given by (14.104). Have in mind that q is a multiple of bd_1d_2, so actually the summation runs over the variable m_1 coprime with $Nbdd_2$.

The main term comes from $h = 0$ and is given by

$$S_0(d_1,d_2,d) = \hat{G}(0) \sum_q \frac{\rho(q)}{dq} F(d_1m_1) . \tag{14.110}$$

Recall that $\rho(q) = W_0(N;q)$ is the number of roots $\nu \,(\mathrm{mod}\, q)$ of $\nu^2 \equiv N (\mathrm{mod}\, q)$.

Estimating the remaining terms trivially we would miss the target by a relatively small factor $N^{5\delta}$, so we need some further improvement. Typically, after an application of Poisson's formula we need to save a bit more than the cardinality of the active frequencies, in our case a bit more than H. This extra saving is possible due to the cancellation in the sum of Weyl sums over the moduli q, but the arguments involved go far beyond the scope of this book. Here is the result whose proof we hope to present on a suitable future occasion.

PROPOSITION 14.16. *Let Ψ be a smooth function supported in $[Q,2Q]$ with $|\Psi^{(j)}| \leqslant Q^{-j}$ for $0 \leqslant j \leqslant 10$. Then for positive integers d,h,k,N we have*

$$\sum_{\substack{(q,d)=1\\ q\equiv 0 \,(\mathrm{mod}\, k)}} \Psi(q) W_{h\bar{d}}(N;q) \ll (dhk)^{10}(Q+\sqrt{N})Q^{-\eta} \tag{14.111}$$

where η is a positive absolute constant and the implied constant in $\ll$ is absolute.

Proposition 14.16 is reminiscent of a result we present later as Proposition 19.4. In fact, the proof of (14.111), like that of (19.8), uses the spectral theory of automorphic forms. To see how it proceeds we first note that the Weyl sums can be expressed by Salié sums. For example, if $(q,2hN) = 1$ then Lemma 12.4 of [**109**] yields

$$W_h(N;q) = \frac{\overline{\varepsilon_q}}{\sqrt{q}} \sum_{x (\mathrm{mod}\, q)} \Big(\frac{x}{q}\Big) e\Big(\frac{4Nx + \overline{h^2x}}{q}\Big).$$

In other cases, the expression is a bit different. Next, the sum of Salié sums over $q \equiv 0(k)$ can be expressed by a spectral sum in the Fourier coefficients of metaplectic forms. Finally, a subconvexity bound for the Fourier coefficients $\lambda_j(N)$ of these forms gives the vital saving $Q^{-\eta}$ in (14.111).

By Proposition 14.16 one can show that the terms of (14.109) with $0 < |h| \leqslant H$ contribute $O(N^{\frac{1}{2}(1-\eta)+80\delta})$, so we are left with

$$S(d_1,d_2,d) = S_0(d_1,d_2,d) + O\big(N^{\frac{1}{2}-\frac{\eta}{3}}\big) \tag{14.112}$$

provided that δ, the exponent of distribution of the sieve (λ_d), is a small fraction of η, which we can and do assume.

We now proceed to the evaluation of the main term

$$S_0(d_1,d_2,d)=\hat{G}(0)\frac{\rho(bd_2)}{d\,bd_2}\sum_{m_1}{}^{'}\frac{\rho(d_1m_1)}{d_1m_1}F(d_1m_1)\ ,$$

where $\sum'$ runs over positive integers m_1 with $(m_1,Nbdd_2)=1$; see (14.110) and (14.104). To evaluate $\sum'$ we use the corresponding zeta-function $Z(s)$ given by

$$\sum_{m_1}{}^{'}\rho(d_1m_1)(d_1m_1)^{-s}=\frac{\rho(d_1)}{d_1^s}\prod_{p|d_1}\Big(1-\frac{1}{p^s}\Big)^{-1}\prod_{p\nmid Nbdd_1d_2}\Big(1+\frac{\rho(p)}{p^s}\Big(1-\frac{1}{p^s}\Big)^{-1}\Big)\,.$$

Recall that $\rho(p)=1+\chi_N(p)$, hence

$$Z(s)=\zeta(s)\,\frac{\rho(d_1)}{d_1^s}\prod_{p|Nbdd_2}\Big(1-\frac{1}{p^s}\Big)\prod_{p\nmid Nbdd_1d_2}\Big(1+\frac{\chi_N(p)}{p^s}\Big)\,.$$

The last product can be compared to $L(s,\chi_N)$ so that $Z(s)$ can be rewritten as

$$\frac{\zeta(s)L(s,\chi_N)}{\zeta(2s)}\prod_{p|N}\Big(1+\frac{1}{p^s}\Big)^{-1}\frac{\rho(d_1)}{d_1^s}\prod_{p|bdd_1d_2}\Big(1+\frac{\chi_N(p)}{p^s}\Big)^{-1}\prod_{\substack{p|bdd_2\\ p\nmid N}}\Big(1-\frac{1}{p^s}\Big)\,.$$

Here, we can take advantage of the presence of factors $\rho(d_1)$ and $\rho(bd_2)$, so we can assume that $\chi_N(p)=1$ if $p\mid bd_1d_2,\ p\nmid(d_2,N)$. This observation leads to

$$\begin{aligned}Z(s)=\ &\frac{\zeta(s)L(s,\chi_N)}{\zeta(2s)}\prod_{p|N}\Big(1+\frac{1}{p^s}\Big)^{-1}\prod_{p|b}\Big(1-\frac{1}{p^s}\Big)\Big(1+\frac{1}{p^s}\Big)^{-1}\frac{\rho(d_1)}{d_1^s}\\ &\prod_{p|d_1}\Big(1+\frac{1}{p^s}\Big)^{-1}\prod_{\substack{p|d_2\\ p\nmid N}}\Big(1-\frac{1}{p^s}\Big)\Big(1+\frac{1}{p^s}\Big)^{-1}\prod_{p|d}\Big(1-\frac{1}{p^s}\Big)\Big(1+\frac{\chi_N(p)}{p^s}\Big)^{-1}.\end{aligned}$$

Hence, $Z(s)$ is analytic in Re $s\geqslant\frac{1}{2}$ with a simple pole at $s=1$ of residue

$$\begin{aligned}R(d_1,d_2,d)=\ &\frac{L(1,\chi_N)}{\zeta(2)}\prod_{p|N}\Big(1+\frac{1}{p}\Big)^{-1}\prod_{p|b}\Big(1-\frac{1}{p}\Big)\Big(1+\frac{1}{p}\Big)^{-1}\frac{\rho(d_1)}{d_1}\\ &\prod_{p|d_1}\Big(1+\frac{1}{p}\Big)^{-1}\prod_{\substack{p|d_2\\ p\nmid N}}\Big(1-\frac{1}{p}\Big)\Big(1+\frac{1}{p}\Big)^{-1}\prod_{p|d}\Big(1-\frac{1}{p}\Big)\Big(1+\frac{\chi_N(p)}{p}\Big)^{-1}.\end{aligned}$$

On the critical line Re $s=\frac{1}{2}$ we have the bound

$$Z(s)\ll|s|N^{\frac{3}{16}+\varepsilon}d_1^{-\frac{1}{2}}$$

by the Burgess estimate for $L(s,\chi_N)$, the implied constant depending only on ε.

Remarks. Since we assume the Riemann Hypothesis for $L(s,\chi_N)$ in other places we could use it here as well and remove the Burgess exponent 3/16, but it is not necessary although the convexity exponent 1/4 would not suffice. The best-known unconditional exponent is 1/6 due to [**25**].

By complex integration we derive

$$\sum{}' = \frac{1}{2\pi i}\int_{(1)} \tilde{F}(s)Z(s+1)\,ds = \tilde{F}(0)R(d_1,d_2,d) + O\Big((d_1z)^{-\frac{1}{2}}N^{\frac{3}{16}+\varepsilon}\Big)$$

by moving the contour to the line Re $s=-1/2$. Here $\tilde{F}(s)$ stands for the Mellin transform of $F(x)$; it satisfies

$$\tilde{F}(s) \ll z^{-\frac{1}{2}}|s|^{-3} \quad \text{if Re } s=-1/2$$

and

$$\tilde{F}(0) = \int F(x)x^{-1}dx \ll \log(Nz^{-2})\ .$$

Hence

$$S_0(d_1,d_2,d) = \tilde{F}(0)\hat{G}(0)\frac{\rho(bd_2)}{bdd_2}R(d_1,d_2,d) + E_0(d_1,d_2,d)$$

where

$$E_0(d_1,d_2,d) \ll (bdd_2)^{-1}(d_1z)^{-\frac{1}{2}}N^{\frac{11}{16}+\varepsilon}\ .$$

Note that this is smaller than the error term in (14.112). Introducing our formula for the residue $R(d_1,d_2,d)$, we arrange the main term in the following fashion:

$$S_0(d_1,d_2,d) = \tilde{F}(0)\hat{G}(0)\frac{L(1,\chi_N)}{\zeta(2)}\prod_{p|N}\Big(1+\frac{1}{p}\Big)^{-1}$$
$$\frac{\rho(b)}{b}\prod_{p|b}\Big(1-\frac{1}{p}\Big)\Big(1+\frac{1}{p}\Big)^{-1} g_1(d_1)g_2(d_2)g(d) + E_0(d_1,d_2,d)\ ,$$

where

$$g_1(d_1) = \frac{\rho(d_1)}{d_1}\prod_{p|d_1}\Big(1+\frac{1}{p}\Big)^{-1},$$

$$g_2(d_2) = \frac{\rho(d_2)}{d_2}\prod_{\substack{p|d_2\\ p\nmid N}}\Big(1-\frac{1}{p}\Big)\Big(1+\frac{1}{p}\Big)^{-1},$$

and

$$g(d) = \frac{1}{d}\prod_{p|d}\Big(1-\frac{1}{p}\Big)\Big(1+\frac{\chi_N(p)}{p}\Big)^{-1}.$$

Hence, by (14.112) and (14.102) we arrive at

$$S_b^+ = \tilde{F}\,(0)\hat{G}(0)\,\frac{L(1,\chi_N)}{\zeta(2)}\prod_{p|N}\Big(1+\frac{1}{p}\Big)^{-1}$$
$$\frac{\rho(b)}{b}\prod_{p|b}\Big(1-\frac{1}{p}\Big)\Big(1+\frac{1}{p}\Big)^{-1}(G_1*G_2*G) + O\Big(N^{\frac{1}{2}-\frac{\eta}{3}+3\delta}\Big)\ ,$$

where

$$G_1*G_2*G = \sum\sum\sum_{\substack{d_1d_2d\,\text{squarefree}\\ (d_1d_2d,b)=1\\ (d_1d,N)=1}} \lambda_{d_1}\lambda_{d_2}\lambda_d\, g_1(d_1)g_2(d_2)g(d).$$

In Section 5.9 we treated the reduced compositions of two sieves $G'*G''$. Similar arguments work for reduced compositions of any number of upper-bound sieves producing results which are analogous to Proposition 5.8. What this tells us is, as might have been expected but is not so quick to be proved, the mutual coprimality condition in the composition of sieves does not affect the bound apart from an insignificant factor. In our case, we obtain

$$G_1 * G_2 * G \ll V_1 V_2 V$$

where

$$V_1 = \prod_{\substack{p<\Delta \\ p\nmid bN}} (1 - g_1(p)), \quad V_2 = \prod_{\substack{p<\Delta \\ p\nmid b}} (1 - g_2(p)), \quad V = \prod_{\substack{p<\Delta \\ p\nmid bN}} (1 - g(p)),$$

and $\Delta = N^\delta$ is the level of our sieve (λ_d). For our particular density functions g_1, g_2, g we derive

$$\begin{aligned} V_1 &\asymp \prod_{\substack{p<\Delta \\ p\nmid bN}} \Big(1 - \frac{\rho(p)}{p}\Big) = \prod_{\substack{p<\Delta \\ p\nmid bN}} \Big(1 - \frac{1+\chi_N(p)}{p}\Big) \\ &\asymp \prod_{p<\Delta} \Big(1 - \frac{\chi_N(p)}{p}\Big) \prod_{p<\Delta} \Big(1 - \frac{1}{p}\Big) \prod_{p|b} \Big(1 - \frac{2}{p}\Big)^{-1} \prod_{p|N} \Big(1 - \frac{1}{p}\Big)^{-1}. \end{aligned}$$

We complete the first product to $L(1,\chi_N)^{-1}$ using the Riemann Hypothesis to get

$$V_1 \asymp (L(1,\chi_N)\log N)^{-1} \prod_{p|b} \Big(1 - \frac{2}{p}\Big)^{-1} \prod_{p|N} \Big(1 - \frac{1}{p}\Big)^{-1}.$$

Similarly, we compute that

$$V_2 \asymp (L(1,\chi_N)\log N)^{-1} \prod_{p|b} \Big(1 - \frac{2}{p}\Big)^{-1}$$

and

$$V \asymp (\log N)^{-1} \prod_{p|bN} \Big(1 - \frac{1}{p}\Big)^{-1}.$$

Hence,

$$S_b^+ \ll \tilde{F}(0)\,\hat{G}(0)\,\frac{\rho(b)}{b} \prod_{p|b} \Big(1 + \frac{3}{p}\Big) \prod_{p|N} \Big(1 + \frac{1}{p}\Big) L(1,\chi_N)^{-1} (\log N)^{-3}.$$

Recall that $\hat{G}(0) \ll \sqrt{N}$ and $\tilde{F}(0) \ll \log(Nz^{-2}) = (1 - 1/s)\log N$. This completes the proof of (14.98) and hence that of Theorem 14.15 (for fairness we repeat that a proof of the required estimate (14.111) for Salié sums is left for a future occasion).

REMARKS. A curious reader may observe that we have made recourse to very advanced harmonic analysis to derive a crude bound, just of the proper order of magnitude, for the number of solutions to the equation (14.92). This might provoke the question as to whether such a result lies within the scope of more direct, yet clever, arguments. Indeed that would be nicer but, for the time being, this provides an example of an apparently elementary problem which is difficult to treat without heavy industrial-strength material.

CHAPTER 15

Applications – Choice but not Prime

Throughout, our main interest when it comes to applications of sieve methods is on prime numbers, or failing that, on almost-primes. However, the sieve is a versatile tool which has many other uses. We have already seen a few of these in the last three sections of Chapter 11. In this chapter we shall highlight several more of them, ranging from a topic as old as squarefree numbers to one as new as quantum ergodicity.

15.1. Squarefree Values of Quadratic Polynomials

We have already seen in Chapter 1 that sieve ideas (Möbius inversion) are useful in estimating the number of squarefree integers in a lacunary sequence and, in particular, in Example 1.6 we studied, à la Estermann [**41**] and Ricci [**139**], the squarefree values of the polynomial m^2+1, obtaining an asymptotic of high precision. The problem of counting these becomes much more challenging when the quadratic polynomial has large coefficients and uniformity in these is wanted. In this section we consider the polynomial m^2+D and restrict the variable to integers $1 \leqslant m \leqslant M$, where M is relatively small compared to D.

It is easy to predict the asymptotic formula

$$\sum_{m\leqslant M} \mu^2\big(m^2+D\big) \sim cM\ , \tag{15.1}$$

where $c = c(D)$ is given by

$$c = \sum_d \mu(d)\rho\big(d^2\big)d^{-2} = \prod_p \Big(1-\rho(p^2)p^{-2}\Big)\ , \tag{15.2}$$

and $\rho(q)$ denotes the number of solutions to $\nu^2+D \equiv 0 (\operatorname{mod} q)$. Obviously, $\rho(p^2) \leqslant 2$ if $p^2 \nmid D$ and $\rho(p^2) \leqslant p$ if $p^2 \mid D$ so c is a positive constant. Using Möbius inversion we have

$$\sum_{m\leqslant M} \mu^2\big(m^2+D\big) = \sum_d \mu(d)\big|\mathcal{A}_{d^2}\big| \tag{15.3}$$

where

$$\big|\mathcal{A}_{d^2}\big| = \sum_{\substack{m\leqslant M\\ m^2+D\equiv 0(\operatorname{mod} d^2)}} 1 = \rho\big(d^2\big)d^{-2}M + O\Big(\rho\big(d^2\big)\Big)\ . \tag{15.4}$$

For relatively small d this simple formula does the trick. Indeed, we get

$$\sum_{d\leqslant Y} \rho\big(d^2\big) \ll \tau(D)Y\log Y\ , \tag{15.5}$$

which shows that (15.4) is valuable for $d \leqslant Y \ll MD^{-\varepsilon}$.

For larger d we do not need an asymptotic formula for $|\mathcal{A}_{d^2}|$ but only a rough upper bound because the squares are so sparse that these terms can only contribute to the error term. However, establishing sufficient bounds is not easy if M is small. As one possible approach we can consider the congruence $m^2 + D \equiv 0 (\operatorname{mod} d^2)$ as the norm equation

$$d^2 k - m^2 = D \tag{15.6}$$

in the real quadratic field $\mathbb{Q}(\sqrt{k})$ and sum over $k \leqslant (D + M^2)Y^{-2}$, obtaining the bound

$$\sum_{d>Y} |\mathcal{A}_{d^2}| \ll \tau(D)(D+M^2)Y^{-2} \log DM \ . \tag{15.7}$$

Here, $\tau(D)$ comes from the estimation of the number of ideals of norm D and $\log D$ from the estimation of the units. Hence, the total error term is

$$\Big[MY^{-1} + Y + (D+M^2)Y^{-2}\Big]\tau(D) \log DM \ , \tag{15.8}$$

where the first term comes from extending the sum in the main term to all $d > Y$. Taking $Y = (D+M^2)^{1/3}$ we conclude:

THEOREM 15.1. *For any $D, M \geqslant 1$ we have*

$$\sum_{m \leqslant M} \mu^2(m^2 + D) = cM + O\Big(\tau(D)(D+M^2)^{\frac{1}{3}} \log 2DM\Big) \ . \tag{15.9}$$

Note that (15.9) gives the asymptotic formula for $M \geqslant D^{\frac{1}{3}+\varepsilon}$.

As an alternative approach to (15.6) we can ask for the values of $d^2 k - D$ which are squares in the short interval $[1, M^2]$. This combines two different goals. The detection of numbers in a short interval lends itself to standard Fourier analysis (which in our case boils down to counting the lattice points (d, k) in a narrow region). However, detecting squares is not as routine. One method for doing so, the square sieve, has been devised by R. Heath-Brown [**86**]. What we adopt from his method is the following.

PROPOSITION 15.2. *For a_n with $n \geqslant 1$ non-negative reals and $P \geqslant 2$ we have*

$$\sum_{n=\square} a_n \leqslant 9P^{-2} \sum_n a_n \Big(\Big(\sum_{P<p\leqslant 2P} \Big(\frac{n}{p}\Big) \log p\Big)^2 + 2(\log n)^2\Big) \ .$$

Here, of course, (n/p) denotes the Legendre symbol.

PROOF. Since the weights are non-negative it suffices to check that these are $\geqslant 1$ on the square values $n = m^2$. Using the inequality $a^2 + 8b^2 \geqslant \frac{8}{9}(a+b)^2$ we find that

$$\begin{aligned}\Big(\sum_{P<p\leqslant 2P} \Big(\frac{n}{p}\Big) \log p\Big)^2 + 2(\log n)^2 &\geqslant \frac{8}{9}\Big(\sum_{\substack{P<p\leqslant 2P\\ p\nmid m}} \log p + \log m\Big)^2 \\ &\geqslant \frac{8}{9}\Big(\sum_{P<p\leqslant 2P} \log p\Big)^2 \geqslant \frac{1}{9} P^2 \ ,\end{aligned}$$

which completes the proof. □

Using this square sieve we can sharpen the bound (15.7) and improve (15.9) to give the following result.

THEOREM 15.3. *For* $1 \leqslant M \leqslant D$ *we have*

$$\sum_{m\leqslant M} \mu^2(m^2+D) = cM + O\Big(\tau(D)M^{\frac{3}{5}}D^{\frac{1}{10}}(\log 2D)^2\Big) . \tag{15.10}$$

Note that (15.10) gives an asymptotic formula valid in the wider region $M \geqslant D^{\frac{1}{4}+\varepsilon}$.

It suffices to establish the same result (15.10) for the sum over the dyadic segments $M < m \leqslant 2M$. We can also assume that

$$D^{\frac{1}{4}} \leqslant 4M \leqslant D^{\frac{1}{2}}$$

since the result is otherwise either trivial or follows from (15.9). We shall prove

$$\sum_{d>Y} |\mathcal{A}_{d^2}| \ll \tau(D)\Big(\frac{MD^{\frac{1}{6}}}{Y^{\frac{2}{3}}} + \frac{D^{\frac{1}{3}}}{Y^{\frac{1}{3}}} + \frac{D^{\frac{1}{2}}}{M} + \frac{M^2}{Y^2} + \frac{M^{\frac{4}{3}}}{D^{\frac{1}{3}}}\Big)(\log D)^2 , \tag{15.11}$$

which is somewhat better than (15.7) and which, when combined with (15.5), gives (15.10) on choosing $Y = M^{\frac{3}{5}}D^{\frac{1}{10}}$.

For the proof of (15.11) we shall divide the range of d into dyadic segments $\Delta < d \leqslant 2\Delta$ for some Δ satisfying

$$Y \leqslant \Delta \leqslant D^{\frac{1}{2}} . \tag{15.12}$$

Now, from the square sieve we infer that

$$\sum_{\Delta<d\leqslant 2\Delta} |\mathcal{A}_{d^2}| \leqslant 9P^{-2} \sum\sum_{\substack{\Delta<d\leqslant 2\Delta,\ k \\ M^2<|kd^2-D|\leqslant 4M^2}} [\ldots] \tag{15.13}$$

where the content of $[\ldots]$ is given by

$$\Big(\sum_{P<p\leqslant 2P} \big(\frac{kd^2-D}{p}\big)\log p\Big)^2 + 8\left(\log 2M\right)^2 .$$

Here, we can include the restriction $p \nmid dD$ at the cost of an extra term $O\big((\log D)^2\big)$. Next, before opening the square, we introduce a smooth product $\eta(d)\xi(kd^2-D)$ where $\eta(v)$ and $\xi(u)$ are functions supported on $v \asymp \Delta$ and $u \asymp M^2$, respectively, with $v^j\eta^{(j)}(v) \ll 1$ and $u^j\xi^{(j)}(u) \ll 1$ for $j = 0, 1, 2, 3, 4$. Then (15.13) yields

$$\sum_{\Delta<d\leqslant 2\Delta} |\mathcal{A}_{d^2}| \ll \Big(\frac{\log P}{P}\Big)^2 \sum\sum_{\substack{P<p_1\neq p_2\leqslant 2P \\ (p_1p_2,D)=1}} |S_{p_1p_2}| + \frac{(\log D)^2}{P} S_1 , \tag{15.14}$$

where

$$S_q = \sum_{(d,q)=1} \sum_k \eta(d)\xi(kd^2-D)\big(\frac{kd^2-D}{q}\big) .$$

Next we evaluate S_q using Poisson's formula

$$S_q = \frac{1}{q^2} \sum_h \sum_\ell F\big(\frac{h}{q}, \frac{\ell}{q}\big) G_q(h,\ell) , \tag{15.15}$$

where $F(x,y)$ is the Fourier integral

$$F(x,y) = \iint \eta(v)\xi(uv^2-D)e(-xu-yv)du\, dv \tag{15.16}$$

and $G_q(h,\ell)$ is the complete character sum

$$G_q(h,\ell)=\sum_{\alpha\,(q)}\sideset{}{^\star}\sum_{\beta\,(q)}\Big(\frac{\alpha\beta^2-D}{q}\Big)\,e\Big(\frac{\alpha h+\beta\ell}{q}\Big)\,. \tag{15.17}$$

To estimate $F(x,y)$ we make a change of variables and write

$$F(x,y)=\iint\eta(v)\xi(u)e\big(-x\frac{u+D}{v^2}-yv\big)\frac{du\,dv}{v^2}\,.$$

Integrating by parts in the v-variable (zero or four times) and then estimating trivially in each variable we derive

$$F(x,y)\ll\frac{M^2}{\Delta}\Big(1+\frac{|x|D}{\Delta^2}+|y|\Delta\Big)^{-4}\ll\frac{M^2}{\Delta}\Big(1+\frac{|x|D}{\Delta^2}\Big)^{-2}\Big(1+|y|\Delta\Big)^{-2}\,. \tag{15.18}$$

This estimate is valid for all x,y except in the region

$$|x|D\asymp|y|\Delta^3\,,\quad xy\neq0\,. \tag{15.19}$$

In this range the exponential function may have a stationary point at $2xDv^{-3}=y$ so that partial integration in the v-variable is not possible. But, we can integrate by parts with respect to the u-variable (zero or two times). Having done so, we apply the second derivative test (say Lemma 4.5 of [**154**] or Lemma 8.10 of [**109**]) and estimate trivially in the u-variable getting

$$F(x,y)\ll\frac{M^2}{\Delta^2}\Big(1+\frac{|x|M^2}{\Delta^2}\Big)^{-2}\Big(\frac{\Delta^4}{|x|D}\Big)^{\frac12}=\frac{M^2}{(|x|D)^{\frac12}}\Big(1+\frac{|x|M^2}{\Delta^2}\Big)^{-2}\,. \tag{15.20}$$

We shall apply (15.18) and (15.20) for $(x,y)=(h/q,\ell/q)$ and sum these estimates for $|F(h/q,\ell/q)|$ over all integers h,ℓ. However, before being able to execute such a summation we require a bound for $G_q(h,\ell)$ for each $q=p_1p_2$. As our estimates for the Fourier integrals do not depend specifically on q but only on its order of magnitude ($q=1$ or $q\asymp P^2$) we may first deal with the sum

$$E(h,\ell)=\sum\sum_{\substack{P<p_1\neq p_2\leqslant 2P\\(p_1p_2,D)=1}}|G_{p_1p_2}(h,\ell)|\,. \tag{15.21}$$

As a matter of fact, one could estimate every one of the individual sums $G_q(h,\ell)$ by an appeal to the Riemann Hypothesis for curves in the form of Weil's bound for suitable exponential sums. However, we can get the needed result in an elementary fashion by taking advantage of the averaging over $p_1\neq p_2$.

Using the twisted multiplicativity

$$G_{p_1p_2}(h,\ell)=G_{p_1}(\bar p_2h,\bar p_2\ell)G_{p_2}(\bar p_1h,\bar p_1\ell)\,,$$

we derive by Cauchy's inequality

$$E(h,\ell)\leqslant2\sum_{\substack{P<p\leqslant2P\\p\nmid D}}\ \sideset{}{^\star}\sum_{a(\mathrm{mod}\,p)}|G_p(ah,a\ell)|^2\,.$$

For the sum of prime modulus we have

$$G_p(h,\ell)=G_p(h)K_p(hD,\ell)\,,$$

where

$$G_p(h)=\sum_{\alpha(p)}\Big(\frac{\alpha}{p}\Big)\,e\Big(\frac{\alpha h}{p}\Big)=\varepsilon_p\sqrt{p}\,\Big(\frac{h}{p}\Big)$$

is the Gauss sum, and

$$K_p(h,\ell)=\sideset{}{^\star}\sum_{\beta(p)} e\Big(\frac{\beta^2h+\bar\beta\ell}{p}\Big)$$

is a hybrid Gauss-Kloosterman sum. We may assume that $p\nmid hD$ and compute as follows:

$$\sideset{}{^\star}\sum_{a(p)} |G_p(ah,a\ell)|^2=p\sideset{}{^\star}\sum_{a(p)} |K_p(ahD,a\ell)|^2\leqslant \nu(h,\ell)p^2$$

where $\nu(h,\ell)$ is the number of solutions of

$$(\beta_1^2-\beta_2^2)hD+(\bar\beta_1-\bar\beta_2)\ell\equiv 0\,(\operatorname{mod} p)$$

in $\beta_1,\beta_2\,(\operatorname{mod} p)$, $(\beta_1\beta_2,p)=1$. There are $p-1$ solutions if $\beta_1\equiv\beta_2$. For $\beta_1\not\equiv\beta_2$ we have

$$(\beta_1+\beta_2)\beta_1\beta_2\equiv\ell\,\overline{hD}\,(\operatorname{mod} p)\ .$$

After a change of variables we need to count the solutions of

$$\gamma(\gamma+1)\beta^2\equiv\ell\,\overline{hD}\,(\operatorname{mod} p)$$

in $(\beta\gamma,p)=1,\gamma\not\equiv 1$. If $p\mid\ell$ then $\gamma\equiv -1\,(\operatorname{mod} p)$ and we get exactly $p-1$ solutions. If $p\nmid\ell$ then the number of solutions is equal to

$$\sum_{\gamma\neq 0,\pm1}\Big(1+\Big(\frac{\ell hD\gamma(\gamma+1)}{p}\Big)\Big)=p-3+\Big(\frac{\ell hD}{p}\Big)\sum_{\gamma\neq 1}\Big(\frac{\gamma(\gamma+1)}{p}\Big)$$
$$=p-3-\Big(\frac{\ell hD}{p}\Big)\Big(1+\Big(\frac{2}{p}\Big)\Big)\leqslant p-1\ .$$

Adding up the above counts we find that $\nu(h,\ell)\leqslant 2(p-1)$. Hence we conclude:

LEMMA 15.4. *For all h,ℓ we have*

$$E(h,\ell)\leqslant 16P^4\ . \tag{15.22}$$

Since the bound (15.22) does not depend on h or ℓ, we can sum the estimates (15.18) and (15.20) for $|F(h/q,\ell/q)|$ for $q=1$ and $q\asymp P^2$ without any conditions on those frequencies h,ℓ. We obtain

$$\sum_h\sum_\ell |F(h/q,\ell/q)|\ll\frac{M^2}{\Delta}\sum_h\Big(1+\frac{|h|D}{q\Delta^2}\Big)^{-2}\sum_\ell\Big(1+\frac{|\ell|\Delta}{q}\Big)^{-2}$$
$$+M^2\sum\sum_{\substack{|h|D\asymp|\ell|\Delta^3\\ h\ell\neq 0}}\Big(1+\frac{|h|M^2}{q\Delta^2}\Big)^{-2}\Big(\frac{q}{|h|D}\Big)^{\frac12}\ .$$

On the right-hand side the first sum over h is $\ll(1+q\Delta^2D^{-1})$, the second sum over ℓ is $\ll(1+q\Delta^{-1})$, and the third sum over $h\ell\neq 0$ is bounded by

$$M^2(qD)^{\frac12}\Delta^{-3}\sum_h\Big(1+\frac{|h|M^2}{q\Delta^2}\Big)^{-2}|h|^{\frac12}\ll q^2D^{\frac12}M^{-1}\ .$$

Hence, we can conclude:

LEMMA 15.5. *We have*

$$\sum_h\sum_\ell |F(h/q,\ell/q)|\ll\frac{M^2}{\Delta}\Big(1+\frac{q\Delta^2}{D}\Big)\Big(1+\frac{q}{\Delta}\Big)+q^2\frac{D^{\frac12}}{M}\ . \tag{15.23}$$

Now, combining (15.22) and (15.23) we get by (15.14) and (15.15)

$$\sum_{\Delta<d\leqslant 2\Delta} |\mathcal{A}_{d^2}| \ll \frac{(\log D)^2}{P^2}\Big[\frac{M^2}{\Delta}\Big(1+\frac{P^2\Delta^2}{D}\Big)\Big(1+\frac{P^2}{\Delta}\Big)+P^4\frac{D^{\frac12}}{M}\Big]$$
$$+\frac{(\log D)^2}{P}\Big[\frac{M^2}{\Delta}\Big(1+\frac{\Delta^2}{D}\Big)\Big(1+\frac{1}{\Delta}\Big)+\frac{D^{\frac12}}{M}\Big]$$
$$\ll \Big(\frac{M^2}{\Delta P}+P^2\frac{D^{\frac12}}{M}+\frac{M^2}{\Delta^2}+\frac{M^2\Delta}{D}\Big)(\log D)^2 .$$

We choose $P = 2 + M(\Delta D^{\frac12})^{-\frac13}$ and the bound becomes

$$\ll \Big(MD^{\frac16}\Delta^{-\frac23}+D^{\frac13}\Delta^{-\frac13}+D^{\frac12}M^{-1}+M^2\Delta^{-2}+M^2\Delta D^{-1}\Big)(\log D)^2 .$$

Here, we improve the last term $M^2\Delta D^{-1}$ by combining it with $D\Delta^{-2}$ in (15.7) obtaining $\min\{M^2\Delta D^{-1}, D\Delta^{-2}\} \leqslant M^{4/3}D^{-1/3}$. Having done so we can see that the worst case for Δ is its smallest value $\Delta = Y$ giving (15.11). This completes the proof of Theorem 15.9.

15.2. Additive Convolution of Multiplicative Functions

Let f, g be multiplicative functions and h a fixed integer. In this section we are going to estimate the sum

$$S_{fg}(x;h) = \sum_{n\leqslant x} f(n)g(n+h). \tag{15.24}$$

If $h = 0$ the product $f(n)g(n)$ is a multiplicative function and there are many powerful methods to approach the problem, first of all by means of the corresponding zeta-function

$$Z_{fg}(s) = \sum_{1}^{\infty} f(n)g(n)n^{-s} ,$$

but also some elementary arguments are sufficiently good; see Appendix A.

If $h \neq 0$, this is an entirely different issue. Assume that $1 \leqslant h \leqslant x$. Of course, one can apply Cauchy's inequality

$$|S_{fg}(x;h)|^2 \leqslant \Big(\sum_{n\leqslant x}|f(n)|^2\Big)\Big(\sum_{n\leqslant 2x}|g(n)|^2\Big)$$

which reduces the problem to the case $h = 0$, but this quick step does not come without a price. For instance, if f, g change sign in a random fashion, then one loses the possibility of cancellation of the terms in the shifted convolution (15.24). Even if f, g are non-negative, after Cauchy's inequality one typically loses the true order of magnitude for $S_{fg}(x;h)$. For example, if $f = g = \tau$ is the divisor function we have (see [**100**]),

$$S_{\tau\tau}(x;h) = \sum_{n\leqslant x} \tau(n)\tau(n+h) \sim \frac{6}{\pi^2}\sigma_{-1}(h)x(\log x)^2$$

while by Cauchy's inequality we only get (see (2.8))

$$S_{\tau\tau}(x;0) = \sum_{n\leqslant x} \tau^2(n) \sim \pi^{-2}x(\log x)^3 .$$

Let us recall that the average value of $\tau(n)$ is $\log n$ (see (2.1)), so this is preserved in the mean value of the shifted product $\tau(n)\tau(n+h)$ but lost in the mean value of $\tau^2(n)$. The point is that $\tau(n)$ can take values quite a bit larger than $\log n$, but not very often; nevertheless, when we take $\tau^2(n)$ these larger values are amplified so much that their contribution is no longer absorbed by the more typical terms. The non-zero shift in $f(n)g(n+h)$ makes the multiplicative structure of $f(n)$ independent from that of $g(n+h)$ and the larger values of $f(n)$ do not conspire with the large values of $g(n+h)$.

In this section we show how the Fundamental Lemma of sieve theory can be used to estimate $S_{fg}(x;h)$ for quite general multiplicative functions f, g with almost true order of magnitude. This will be achieved by an appeal to Proposition 6.22.

For technical simplification, we assume throughout that f, g are real, non-negative, supported on squarefree numbers and bounded at primes, specifically

$$0 \leqslant f(p),\ g(p) \leqslant 2. \tag{15.25}$$

Hence Ingham's estimate for the shifted product of the divisor function implies

$$S_{fg}(x;h) \ll \frac{h}{\varphi(h)} x(\log x)^2$$

but, of course, we want to do better when $f(p)$ or $g(p)$ has an average value less than 2.

First, we assume for simplicity that f, g are supported on numbers coprime with h, and later we shall remove this condition. Let $P(z)$ denote the product of all primes $p < z$ where $4 \leqslant z \leqslant y = x^{\frac{1}{8}}$. We write

$$S_{fg}(x;h) = \sum\sum_{\substack{ad-bc=h \\ bc\leqslant x}}\sum f(bc)g(ad)$$

where here and hereafter we make the convention that a, b, c, d run over integers with $ab|P(z)$ and $(cd, P(z)) = 1$. This means that a and b have only small prime factors while c and d have only large ones.

Now we split this into four sums $S_1 + S_2 - S_{12} + S$, say, according to the following summation conditions

$$a > y; \quad b > y; \quad a, b > y; \quad a, b \leqslant y.$$

Note that S_{12} is bounded by S_1 and by S_2, but we can discard it by positivity.

First we are going to estimate S_2, the case of S_1 is similar by symmetry. We expect to get from S_2 a small contribution so we do not hesitate to use Cauchy's inequality giving

$$S_2^2 \leqslant \Big(\sum_{\substack{bc\leqslant x \\ b>y}}\sum f^2(bc)\Big)\Big(\sum_{n\leqslant 2x} g^2(n)\Big)$$

Next by Corollary A.2 for the multiplicative function f^2 we get

$$\sum_b\sum_c f^2(bc) \ll C^{-s}\frac{x}{\log x}\prod_{p<x}\Big(1+\frac{f^2(p)}{p}\Big)$$

where $s = \log x/\log z \geqslant 8$ and $C \geqslant 1$ is any fixed constant. Moreover, by (A.8) for the multiplicative function g^2 we get

$$\sum_n g^2(n) \ll \frac{x}{\log x} \prod_{p<x} \left(1 + \frac{g^2(p)}{p}\right).$$

Choosing $C = e^2$ (any larger C does not improve the results by much) we conclude that

$$S_2 \ll e^{-s} x \left(E_{f^2}(x) E_{g^2}(x)\right)^{\frac{1}{2}} , \tag{15.26}$$

where the implied constant is absolute. Here we use the notation

$$E_f(x) = \frac{1}{\log x} \prod_{p \leqslant x} \left(1 + \frac{f(p)}{p}\right) \tag{15.27}$$

for $x \geqslant 2$ and any non-negative function f. The same bound (15.26) holds for the first sum S_1.

Now we proceed to an estimation of the last sum

$$S = \sum\sum\sum_{\substack{ad-bc=h \\ bc\leqslant x,\ a,b\leqslant y}} f(bc) g(ad) .$$

Here we relax $f(c), g(d)$ by estimating them through divisor functions. There will be no great loss, because c, d are almost-primes, precisely they are free of prime divisors $p < z = x^{1/s}$. We could simply estimate $f(c) \leqslant \tau(c) \leqslant 2^s$ and $g(d) \leqslant \tau(d) \leqslant 2^s$, which would work, but one can do better by applying Proposition 22.15 getting

$$S \leqslant 2^{36} \sum\sum\sum\sum\sum_{\substack{akm-b\ell n=h \\ a,b\leqslant y,\ k,\ell\leqslant w,\ akm\leqslant 2x \\ ab|P(z),\ (m,h)=1 \\ (k\ell,hP(z))=1}} f(b)g(a)\tau(k)\tau(\ell)$$

where $y = x^{\frac{1}{8}}$ and $w = (2x)^{\frac{1}{4}}$. By Proposition 6.22 with the coefficients a, b replaced by $ak, b\ell$, respectively, we get

$$\begin{aligned} S &\ll \frac{h}{\varphi(h)} \frac{x}{(\log z)^2} \sum\sum_{\substack{ab|P(z) \\ a,b\leqslant y}} \frac{f(b)g(a)}{\varphi(b)\varphi(a)} \sum\sum_{\substack{(k\ell,P(z))=1 \\ k,\ell\leqslant w}} \frac{\tau(k)\tau(\ell)}{\varphi(k)\varphi(\ell)} \\ &\ll \frac{h}{\varphi(h)} \frac{x}{(\log z)^2} \prod_{p<z} \left(1 + \frac{f(p)}{p}\right)\left(1 + \frac{g(p)}{p}\right) \prod_{z\leqslant p<w} \left(1 + \frac{2}{p}\right)^2 . \end{aligned}$$

Hence, we conclude that

$$S \ll s^6 \frac{hx}{\varphi(h)} E_f(x) E_g(x), \tag{15.28}$$

where the implied constant is absolute.

Adding the above estimates for S_1, S_2 and S we arrive at

$$S_{fg}(x;h) \ll e^{-s} x \left(E_{f^2}(x) E_{g^2}(x)\right)^{\frac{1}{2}} + s^6 \frac{hx}{\varphi(h)} E_f(x) E_g(x) \tag{15.29}$$

with any $s \geqslant 8$. This restriction for s is not necessary, because, by Cauchy's inequality and by (A.8), we get

$$S_{fg}(x;h) \ll x\left(E_{f^2}(x)E_{g^2}(x)\right)^{\frac{1}{2}},$$

so our result is true for all $s \geqslant 0$. Before choosing s we remove the condition that f, g are supported on integers coprime with h. This amounts to removing the summation condition $(mn, h) = 1$. Dividing the equation $m = n + h$ by the common factor $d = (m, h) = (n, h)$ we find that the unrestricted sum is bounded by the sum of the restricted sums

$$\sum_{d|h} f(d)g(d)S_{f_h g_h}(x/d; h/d)$$

where f_h, g_h are just the restrictions of f, g to numbers coprime with h. Assume $x \geqslant 2h^2$. We have

$$E_{f_h^2}\left(\frac{x}{d}\right)E_{g_h^2}\left(\frac{x}{d}\right) \leqslant 4\, E_{f^2}(x)E_{g^2}(x)\prod_{p|h}\left(1+\frac{f^2(p)}{p}\right)^{-1}\left(1+\frac{g^2(p)}{p}\right)^{-1}.$$

Hence, after summing over $d|h$ the first term in (15.29) receives the extra factor

$$\prod_{p|h}\left(1+\frac{f(p)g(p)}{p}\right)\left(1+\frac{f^2(p)}{p}\right)^{-\frac{1}{2}}\left(1+\frac{g^2(p)}{p}\right)^{-\frac{1}{2}} \leqslant 1.$$

Similarly, we have

$$E_{f_h}\left(\frac{x}{d}\right)E_{g_h}\left(\frac{x}{d}\right) \leqslant 4E_f(x)E_g(x)\prod_{p|h}\left(1+\frac{f(p)}{p}\right)^{-1}\left(1+\frac{g(p)}{p}\right)^{-1},$$

so the second term in (15.29) gets the extra factor

$$\prod_{p|h}\left(1+\frac{f(p)g(p)}{p}\right)\left(1+\frac{f(p)}{p}\right)^{-1}\left(1+\frac{g(p)}{p}\right)^{-1} \leqslant 1\,.$$

Therefore, we have proved that (15.29) holds for f, g unrestricted. Now we choose $s = \frac{1}{2}(L_f(x) + L_g(x))$, where

(15.30) $$L_f(x) = \log\left(1 + E_{f^2}(x)/E_f^2(x)\right)$$

and conclude the following result.

THEOREM 15.6. *Let f, g be non-negative, multiplicative functions supported on squarefree numbers with $0 \leqslant f(p), g(p) \leqslant 2$. Then, for $h \geqslant 1$ and $x \geqslant 2h^2$ we have*

(15.31) $$\sum_{n\leqslant x} f(n)g(n+h) \ll \frac{h}{\varphi(h)}xE_f(x)E_g(x)\left(L_f(x)+L_g(x)\right)^6,$$

where the implied constant is absolute.

Note that $E_f(x) \gg (\log x)^{-1}$ and $E_{f^2}(x) \ll (\log x)^3$, so

(15.32) $$L_f(x) \ll \log\log 2x$$

and similarly for $L_g(x)$, so these factors are insignificant in practice. In fact, by a different, more advanced argument, M. Nair and G. Tenenbaum [**135**] showed the above bound without the log log factors, but we do not need that.

An interesting case is that of $f(n) = g(n) = |\lambda(n)|$, where $\lambda(n)$ are the Fourier coefficients of a cusp form normalized so that $|\lambda(n)| \leqslant \tau(n)$ by the Ramanujan conjecture (proved by P. Deligne for the classical holomorphic cusp forms). The Rankin–Selberg zeta-function

$$\sum_1^\infty \lambda^2(n) n^{-s}$$

has a simple pole at $s = 1$ and has the standard zero-free region giving

$$\sum_{p \leqslant x} \lambda^2(p) p^{-1} = \log\log x + O(1), \tag{15.33}$$

which shows that $\lambda^2(p)$ is one on average. Hence the average value of $|\lambda(p)|$ is at most one and

$$E_{\lambda^2}(x) \asymp 1, \quad E_{|\lambda|}(x) \ll 1.$$

Theorem 15.6 implies

$$\sum_{n \leqslant x} |\lambda(n)\lambda(n+h)| \ll \frac{hx}{\varphi(h)} . \tag{15.34}$$

But the truth is that $|\lambda(p)|$ has an average value strictly smaller than one unless the cusp form is of complex multiplication type. Indeed, by the Sato–Tate conjecture one predicts that

$$\sum_{p \leqslant x} |\lambda(p)| p^{-1} = c \log\log x + O(1) \tag{15.35}$$

with the constant c given by

$$c = \frac{4}{\pi} \int_0^\pi |\cos\theta| (\sin\theta)^2 d\theta = \frac{8}{3\pi} = 0.8488\ldots .$$

Hence,

$$E_{|\lambda|}(x) \asymp (\log x)^{\frac{8}{3\pi}-1}.$$

Corollary 15.7. *Let $\lambda(n)$ be the Fourier coefficients of a Hecke cusp form for which the Sato–Tate conjecture holds. Then, for $h \geqslant 1$ and $x \geqslant 2h^2$ we have*

$$\sum_{n \leqslant x} |\lambda(n)\lambda(n+h)| \ll \frac{h}{\varphi(h)} x (\log x)^{-2(1-8/3\pi)} (\log\log 2x)^6 \tag{15.36}$$

where the implied constant depends only on the cusp form.

The Sato–Tate conjecture has been recently established for all non-CM elliptic curves E/Q by R. Taylor, so in this case the estimate (15.36) is unconditional. In fact, one can improve (15.34) unconditionally for any non-CM cusp form. To this end one does not need the Sato–Tate conjecture in full force. It suffices to use analytic properties of the two symmetric power L-functions $L(\mathrm{sym}^2 f, s)$ and $L(\mathrm{sym}^4 f, s)$. It is known that the zeta-function

$$\sum_1^\infty \lambda^4(n) n^{-s}$$

has analytic continuation with a double pole at $s = 1$ and it has the standard zero-free region, giving

$$\sum_{p \leqslant x} \lambda^4(p) p^{-1} = 2 \log\log x + O(1). \tag{15.37}$$

This shows that $\lambda^4(p)$ is two on average. Now, using the inequality

$$|\lambda(p)| \leqslant \frac{1}{18}\left(8 + 11\lambda^2(p) - \lambda^4(p)\right),$$

which is valid for any number $-2 \leqslant \lambda(p) \leqslant 2$, we deduce by (15.33) and (15.37) that

$$\sum_{p \leqslant x} |\lambda(p)| p^{-1} < \frac{17}{18} \log\log x + O(1). \tag{15.38}$$

Hence, we get

$$E_{|\lambda|}(x) \ll (\log x)^{-1/18}$$

and conclude the following unconditional result.

COROLLARY 15.8. *Let $\lambda(n)$ be the Fourier coefficients of a classical Hecke cusp form which is not of CM type. Then, for $h \geqslant 1$ and $x \geqslant 2h^2$ we have*

$$\sum_{n \leqslant x} |\lambda(n)\lambda(n+h)| \ll \frac{hx}{\varphi(h)} (\log x)^{-1/10}, \tag{15.39}$$

where the implied constant depends only on the cusp form.

This section is based on the original work of R. Holowinsky [**97**] who established many important results, some of which are stronger than (15.39). Holowinsky showed that similar bounds for the sums

$$\sum_{n \leqslant x} |\lambda_j(n)\lambda_j(n+h)|,$$

where the λ_j are Hecke eigenvalues of cusp forms $u_j(z)$, hold with respectable uniformity in j and h. Subsequently, using these ideas in combination with some arguments due to Soundararajan they together made significant progress toward the Quantum Unique Ergodicity conjecture. It was already known that the Riemann hypothesis for the symmetric power L-functions would imply this conjecture. The work of [**98**] shows that the result also follows from the Ramanujan–Petersson conjecture. Using Deligne's theorem they thus complete the proof of the QUE conjecture unconditionally in the case of classical holomorphic cusp forms. Their work also constitutes important progress toward the case of Maass cusp forms.

15.3. Points on Elliptic Curves

The twin prime problem has stimulated much of the development in sieve theory. Although a problem of interest on its own for amateurs and professionals alike, it does not at first blush seem very connected to the main stream. However, mathematics does not work that way; problems which are interesting on their own turn out with surprising frequency to have relevance in unexpected places. Here we present problems of twin prime type which emerge from fundamental questions about elliptic curves.

Given an elliptic curve $E/\mathbb{Q}$ we are going to examine the reductions $E/\mathbb{F}_p$. The points on this reduction form a group and the most basic questions concern the order and the structure of this group. The torsion subgroup of $E/\mathbb{Q}$ by reduction mod p injects into $E/\mathbb{F}_p$. A conjecture of N. Koblitz [**119**] predicts that for any elliptic curve $E/\mathbb{Q}$ there are infinitely many p such that the order of $E/\mathbb{F}_p$, after the injection of torsion has been divided out, is a prime number. Here, we use the sieve to approach this conjecture. We consider the curve

$$E : y^2 = x^3 - x. \tag{15.40}$$

For $p \equiv 3 (\bmod 4)$ we have supersingular reduction in which case $|E/\mathbb{F}_p| = p+1$, so the question becomes "when do we have $p + 1 = 4p'$, where p and p' are both primes?", which is strongly reminiscent of the classical twin prime problem.

The remaining situation $p \equiv 1 (\bmod 4)$ is still more revealing. In this case

$$|E/\mathbb{F}_p| = p + 1 - 2a$$

where a is determined by

$$p = a^2 + b^2, \quad a + bi \equiv 1 \,(\bmod 2(1 + i))\,.$$

Hence,

$$|E/\mathbb{F}_p| = (a - 1)^2 + b^2 \equiv 0 \ (\bmod 8)$$

and 8 is the order of torsion injection. Therefore, the problem is to estimate

$$S(x) = |\{p \leqslant x;\, p \equiv 1 (\bmod 4),\, |E/\mathbb{F}_p| = 8p',\, p' \text{prime}\}|. \tag{15.41}$$

The connection to twin primes will be more transparent if we move to the Gaussian domain $\mathbb{Z}[i]$. Let $\pi = a + bi$. Then $|E/\mathbb{F}_p| = N(\pi - 1)$, so we are looking for solutions to the equation

$$\pi - 1 = 2(1 + i)\pi'$$

in Gaussian primes π, π'.

It is too much to expect sieve methods to reach this goal, but we shall succeed when π' is replaced by a Gaussian integer π'' having at most two primes factors. To this end we use an almost-prime sieve in a fashion similar to the argument of J.R. Chen [**22**] in his treatment of the classical twin prime problem (which is, however, treated somewhat differently in Chapter 24).

THEOREM 15.9. *For $x \geqslant 5$ we have*

$$|\{\pi \in \mathbb{Z}[i];\, N\pi \leqslant x,\, \pi - 1 = 2(1 + i)\pi''\}| \gg x\,(\log x)^{-2}\ , \tag{15.42}$$

where by π'' we mean a Gaussian integer having at most two prime factors.

Actually we shall succeed with π'' having no prime factors of norm $\leqslant x^{\frac{1}{8}}$, so the lower bound (15.42) for such restricted solutions represents the true order of magnitude. This result was established in [**107**]; here we follow their arguments. Going back to the elliptic curve we infer by Theorem 15.9:

COROLLARY 15.10. *For the elliptic curve* (15.40) *we have*

$$|\{p \leqslant x;\, p \equiv 1 (\bmod 4),\, |E/\mathbb{F}_p| = 8p'',\, \}| \gg x(\log x)^{-2}\ ,$$

where by p'' we mean a positive integer having at most two prime factors.

Earlier results for other curves, but with more prime factors, are given in [**23**].

In the proof of Theorem 15.9 we consider the sequence of $\mathcal{A} = (a_n)$ where a_n is the number of solutions of

$$|\pi - 1|^2 = 8n \tag{15.43}$$

in primes $\pi \equiv 1(\bmod 2(1+i))$ (such Gaussian primes are called primary). Our goal is to show that

$$W(x,z) = \sum_{\substack{n\leqslant x\\(n,P(z))=1\\\omega(n)=1,2}} a_n \gg \frac{x}{(\log x)^2}\ . \tag{15.44}$$

Here, as usual, $P(z)$ denotes the product of all primes $< z$ and $\omega(n)$ is the number of distinct prime divisors of n. We shall succeed with $z = x^{\frac{1}{8}}$. Clearly, (15.44) implies (15.42) (the contribution of n's having square prime divisors being trivially estimated by $O(x/z)$).

To establish (15.44) we consider the weighted sum à la Chen:

$$W(x,y,z) = \sum_{\substack{n\leqslant x\\(n,P(z))=1}} a_n\Big\{1 - \frac{1}{2}\sum_{\substack{p\mid n\\z\leqslant p<y}} 1 - \frac{1}{2}\sum_{\substack{p_1p_2p_3=n\\z\leqslant p_3<y\leqslant p_2<p_1}} 1\Big\}. \tag{15.45}$$

We take $y = x^{\frac{1}{3}}$ so it is easy to see that the weight attached to a_n is not positive unless n has either one or two distinct prime factors. By this observation we have $W(x,z) \geqslant W(x,y,z)$ and so it suffices to prove that

$$W(x,y,z) \gg x(\log x)^{-2}\ . \tag{15.46}$$

We write

$$W(x,y,z) = S(\mathcal{A},z) - \frac{1}{2}\sum_{z\leqslant p<y} S(\mathcal{A}_p,z) - \frac{1}{2}W_3(x,y,z) \tag{15.47}$$

where $S(\mathcal{A},z), S(\mathcal{A}_p,z)$ are the usual sifting functions and

$$W_3(x,y,z) = \sum\sum\sum_{z\leqslant p_3<y<p_2<p_1} a_{p_1p_2p_3}. \tag{15.48}$$

Therefore, we need a lower bound for $S(\mathcal{A},z)$ and upper bounds for $S(\mathcal{A}_p,z)$ (on average over p) and $W_3(x,y,z)$. These will be linear sieve problems.

First we have to establish approximations of type

$$A_d(x) \sum_{\substack{n\leqslant x\\n\equiv 0(d)}} a_n = g(d)X + r_d$$

for d squarefree, with some density function $g(d)$ and an appropriate X so that the remainder terms r_d are quite small. By Möbius inversion we write

$$A_d(x) = \sum_{c\mid d}\mu(c)\sum_{\substack{n\leqslant x\\(n,c)=1}} a_n.$$

For $n = |(\pi-1)/2(1+i)|^2$ (see (15.43)) the condition $(n,c)=1$ is equivalent to

$$\Big(\frac{\pi-1}{2(1+i)}, c\Big) = 1\ .$$

Detecting this again by Möbius inversion, but this time in the Gaussian domain $\mathbb{Z}[i]$, we obtain

$$A_d(x) = \sum_{c\mid d}\mu(c)\sum_{\mathfrak{b}\mid c}\mu(\mathfrak{b})\sum_{\substack{|\pi-1|^2\leqslant 8x\\\pi\equiv 1(2(1+i)\mathfrak{b})}} 1$$

where $\mu(\mathfrak{b})$ is the Möbius function on ideals of $\mathbb{Z}[i]$. Writing $d = km$ with k having all prime factors $\equiv 3(\bmod 4)$ and m having no such prime factors, we find that $\mathfrak{b}$ runs over the ideals $k\mathfrak{m}$ with $N\mathfrak{m} = m$ and $c = d$, giving

$$A_d(x) = \sum_{N\mathfrak{m}=m} \sum_{\substack{|\pi-1|^2 \leqslant 8x \\ \pi\equiv 1(2(1+i)k\mathfrak{m})}} .$$

Since $|\pi - 1|^2 = |\pi|^2 + O(|\pi|)$ the above expression becomes

$$A_d(x) = \sum_{N\mathfrak{m}=m} \sum_{\substack{|\pi|^2 \leqslant 8x \\ \pi\equiv 1(\bmod 2(1+i)k\mathfrak{m})}} 1 + O(\tau(m)\sqrt{x}). \tag{15.49}$$

For any ideal $\mathfrak{a} \subset \mathbb{Z}[i]$ and $\alpha \in \mathbb{Z}[i]$ with $(\alpha, \mathfrak{a}) = 1$ we denote by $\pi_G(x; \mathfrak{a}, \alpha)$ the number of Gaussian primes π with $|\pi|^2 \leqslant x$ and $\pi \equiv \alpha(\bmod \mathfrak{a})$. We also denote by $\pi_G(x)$ the number of Gaussian primes π with $|\pi|^2 \leqslant x$. Let $\Phi(\mathfrak{a}) = |(\mathbb{Z}[i]/\mathfrak{a})^*|$ be the number of classes $\alpha(\bmod \mathfrak{a})$ with $(\alpha, \mathfrak{a}) = 1$. Naturally, one expects that every such class contains the same proportion of primes, so

$$\Delta_G(x; \mathfrak{a}, \alpha) = \pi_G(x; \mathfrak{a}, \alpha) - \frac{1}{\Phi(\mathfrak{a})}\pi_G(x) \tag{15.50}$$

is quite small. The Riemann hypothesis for L-functions with characters on $(\mathbb{Z}[i]/\mathfrak{a})^*$ yields

$$\Delta_G(x; \mathfrak{a}, \alpha) \ll \sqrt{x}(\log xN\mathfrak{a})^2.$$

However, we have the following unconditional result à la Bombieri–Vinogradov theorem

$$\sideset{}{^\flat}\sum_{N\mathfrak{a}\leqslant Q} \max_{(\alpha,\mathfrak{a})=1} |\Delta_G(x; \mathfrak{a}, \alpha)| \ll x(\log x)^{-A}, \tag{15.51}$$

where A is any positive number and $Q = \sqrt{x}(\log x)^{-B}$ with B and the implied constant depending only on A. One can prove (15.51) by extending the arguments used in Theorem 9.18 to the Gaussian domain. More general results for arbitrary number fields were established by D. Johnson [**115**].

Before applying (15.51) to (15.49) we compute $\Phi(\mathfrak{a})$ for $\mathfrak{a} = 2(1+i)k\mathfrak{m}$. Note that $N\mathfrak{a} = 8k^2m$ (recall $d = km$). Suppose d is squarefree. Then $(d, 2) = (m, 2)$ and

$$\#\{\mathfrak{m}; N\mathfrak{m} = m\} = \tau(m/(m, 2)).$$

Moreover, by the multiplicativity of $\Phi(\mathfrak{a})$ we find for $\mathfrak{a} = 2(1+i)k\mathfrak{m}$,

$$\Phi(\mathfrak{a}) = 4(d, 2) \prod_{\substack{p|d \\ p\equiv 1(4)}} (p-1) \prod_{\substack{p|d \\ p\equiv 3(4)}} (p^2 - 1). \tag{15.52}$$

Hence,

$$\sum_{N\mathfrak{m}=m} \frac{1}{\Phi(\mathfrak{a})} = \frac{1}{4}g(d) \tag{15.53}$$

where $g(d)$ is the multiplicative function with $g(2) = 1/2$ and

$$g(p) = \frac{2}{p-1}, \quad \text{if } p \equiv 1(\bmod 4), \tag{15.54}$$

$$g(p) = \frac{1}{p^2-1}, \text{ if } p \equiv 3 (\mathrm{mod}\, 4) . \tag{15.55}$$

Now applying (15.51) to (15.49) we deduce that

$$\sideset{}{^\flat}\sum_{\substack{d=km \\ 8k^2m \leqslant Q}} |A_d(x) - g(d)X| \ll x(\log x)^{-A}$$

where

$$X = \frac{1}{4}\pi_G(8x) . \tag{15.56}$$

Since the contribution of the moduli $d = km$ with $k > (\log x)^A$ can be estimated trivially by $O(x(\log x)^{-A})$ we conclude that

$$\sideset{}{^\flat}\sum_{d \leqslant D} |A_d(x) - g(d)X| \ll x(\log x)^{-A} . \tag{15.57}$$

In other words,

$$D = D(x) = \sqrt{x}(\log x)^{-B} , \tag{15.58}$$

with $B = B(A) > 0$, is the level of distribution of our sequence $\mathcal{A} = (a_n)$. Note that

$$X = 2\pi(8x; 4, 1) + O(\sqrt{x}) \sim \frac{8x}{\log x} \tag{15.59}$$

because π_G counts Gaussian primes so the corresponding prime ideals are counted four times.

Now we are ready to apply the linear sieve estimates (12.6), (12.7). We get

$$S(\mathcal{A}, z) > V(z)X\{f(4) + o(1)\} \tag{15.60}$$

and

$$\sum_{z \leqslant p < y} S(\mathcal{A}_p, z) < V(z)X\Big\{ \sum_{z \leqslant p < y} g(p)F\left(\frac{\log \sqrt{x}/p}{\log z}\right) + o(1)\Big\} , \tag{15.61}$$

where

$$V(z) = \prod_{p<z}(1 - g(p)) = \frac{1}{2} \prod_{2<p<z} (1 - g(p)) . \tag{15.62}$$

In the lower bound (15.60) we have

$$f(4) = \frac{e^\gamma}{2} \log 3 . \tag{15.63}$$

In the upper bound (15.61) the sum over primes is asymptotically equal to the corresponding integral

$$I(F) = \int_{x^{1/8}}^{x^{1/3}} F\left(\frac{8 \log \sqrt{x}/t}{\log x}\right) d \log \log t = \int_1^{8/3} F(4-u)\frac{du}{u} .$$

Since $F(s) = 2e^\gamma/s$ in the segment $1 \leqslant s \leqslant 3$, we have

$$I(F) = \frac{e^\gamma}{2} \log 6 . \tag{15.64}$$

Next we proceed to the estimation of $W_3(x,y,z)$ which is given by (15.48). This translates to counting solutions of $|\pi-1|^2=8p_1p_2p_3$ with $\pi\equiv 1(\operatorname{mod} 2(1+i))$ and primes p_1,p_2,p_3 with

$$z\leqslant p_3<y\leqslant p_2<p_1,\ \ p_1p_2p_3\leqslant x. \tag{15.65}$$

We can restrict the solutions to primes $\equiv 1(\operatorname{mod} 4)$, because the contribution of the other solutions can be estimated trivially by $O(x/z)$. For such primes the equation $|\pi-1|^2=8p_1p_2p_3$ with $\pi\equiv 1(\operatorname{mod} 2(1+i))$ is equivalent to

$$\pi-1=2(1+i)\pi_1\pi_2\pi_3 \quad\text{with}\quad |\pi_j|^2=p_j. \tag{15.66}$$

Since $\pi_1,\bar\pi_1,\pi_2,\bar\pi_2,\pi_3,\bar\pi_3$ are pairwise coprime the products $\pi_1\pi_2\pi_3$ are all distinct up to a unit. Two of these factors can be changed freely by units while the third factor is then determined uniquely. Therefore,

$$W_3(x,y,z)=\frac{1}{16}B(x,y,z)+O(x/z) \tag{15.67}$$

where $B(x,y,z)$ denotes the number of solutions to (15.66) in primes π,π_1,π_2,π_3 restricted by (15.65).

Now we consider the sequence $\mathcal{B}=(b_\ell)$ with b_ℓ being the number of solutions to

$$|2(1+i)\pi_1\pi_2\pi_3+1|^2=\ell\ . \tag{15.68}$$

Since π is a primary prime if and only if $\ell=|\pi|^2$ is an odd prime we find that

$$B(x,y,z)=\sum_{\ell\text{ prime}} b_\ell. \tag{15.69}$$

Hence, we have a sieve problem for the sequence $\mathcal{B}=(b_\ell)$.

In order to use sieve methods we need approximations of type

$$\sum_{\ell\equiv 0(d)} b_\ell=g(d)Y+r_d$$

for d squarefree, odd. Arguing as in (15.49) we derive

$$\sum_{\ell\equiv 0(d)} b_\ell=\sum_{N\mathfrak{m}=m}\Bigg(\sum\sum\sum_{2(1+i)\pi_1\pi_2\pi_3\equiv -1(\operatorname{mod} k\mathfrak{m})} 1\Bigg) \tag{15.70}$$

where $d=km,k$ having prime divisors $\equiv 3(\operatorname{mod} 4)$ and m having prime divisors $\equiv 1(\operatorname{mod} 4)$. Here π_1,π_2,π_3 run over primes with $|\pi_j|^2=p_j$ satisfying (15.65). Let $\pi_G(x,y,z)$ denote the number of such triplets. These triplets π_1,π_2,π_3 satisfy the following analogue of the Bombieri–Vinogradov theorem:

$$\sum_{N\mathfrak{a}\leqslant Q}\max_{(\alpha,\mathfrak{a})=1}\Bigg|\sum\sum\sum_{\pi_1\pi_2\pi_3\equiv\alpha(\operatorname{mod}\mathfrak{a})} 1-\frac{1}{\Phi(\mathfrak{a})}\pi_G(x,y,z)\Bigg|\ll x(\log x)^{-A} \tag{15.71}$$

with $Q=\sqrt{x}(\log x)^{-B}$. The proof of (15.71) is similar to that of (15.51), in fact a little simpler, because one does not need any combinatorial identity for primes to decompose the composite numbers $\pi_1\pi_2\pi_3$.

From (15.70) and (15.71) we derive

$$\sum_{\substack{d\leqslant D\\ d \text{ odd}}}^{\flat}\Big|\sum_{\ell\equiv 0(d)} b_\ell - g(d)Y\Big| \ll x(\log x)^{-A} \tag{15.72}$$

where $g(d)$ is the same density function as in (15.57) and

$$Y = \pi_G(x,y,z). \tag{15.73}$$

In other words, $D = D(x)$ given by (15.58), is the level of distribution of our sequence $\mathcal{B} = (b_\ell)$.

By the linear upper-bound sieve we get

$$B(x,y,z) < 2V(D)Y\{F(1)+o(1)\}\ .$$

Here the constant 2 compensates for the local factor of $V(D)$ at $p = 2$ which is absent. We have $F(1) = 2e^\gamma$ and $V(D) \sim \frac{1}{4}V(z)$, so

$$B(x,y,z) < V(z)Y\{e^\gamma + o(1)\}. \tag{15.74}$$

We still need to relate Y to X. We compute Y asymptotically. By (15.73) we have

$$\begin{aligned} Y &= 4^3 2^3\,\#\{p_1\equiv p_2\equiv p_3\equiv 1 (\mathrm{mod}\, 4); (15.65) \text{ holds}\}\\ &\sim 4^3 \sum\sum_{z\leqslant p_3<y\leqslant p_2<\sqrt{x/p_3}} \pi(x/p_2p_3)\\ &\sim 4^3 \sum\sum_{z\leqslant p_3<y\leqslant p_2<\sqrt{x/p_3}} \frac{x}{p_2p_3}\Big(\log\frac{x}{p_2p_3}\Big)^{-1} \sim 4^3\frac{cx}{\log x}\ , \end{aligned}$$

where c is the corresponding double integral

$$c = \int_{1/8}^{1/3}\int_{1/3}^{(1-\alpha)/2} (\alpha\beta(1-\alpha-\beta))^{-1} d\alpha\, d\beta = 0.36308\ldots. \tag{15.75}$$

Comparing this with (15.59) we find that

$$Y \sim 8\,c\,X. \tag{15.76}$$

By (15.67), (15.74) and (15.76) we arrive at

$$W_3(x,y,z) < V(z)X\{\tfrac{1}{2}\,c\,e^\gamma + o(1)\}. \tag{15.77}$$

Finally, by (15.47), (15.60)–(15.64), and (15.77) we get

$$W(x,y,z) > V(z)\,X\,\frac{e^\gamma}{4}\Big\{\log\frac{3}{2} - c + o(1)\Big\} \asymp V(z)X \asymp x(\log x)^{-2}$$

because $c < \log\frac{3}{2}$. This completes the proof of (15.46) and Theorem 15.9.

Recall that the original problem concerns the sum $S(x)$ given by (15.41). We close this section with the asymptotic evaluation of $S(x)$ which can be derived quickly by the heuristical arguments from Section 3.1. We have by (3.7)

$$S(8x) \sim \frac{1}{2}\sum_{\substack{n\leqslant x\\ n\,\text{prime}}} a_n \sim \frac{HX}{2\log x}$$

where X satisfies (15.59) and H is the infinite product (3.6). Here the factor $\frac{1}{2}$ reduces our having counted the primary primes $\pi, \bar{\pi}$ with $\pi\bar{\pi} = p$ as two different solutions. Hence $S(x) \sim \frac{1}{2}Hx(\log x)^{-2}$. Using (15.53)–(15.55) we find

$$HL(1,\chi_4) = \prod_{p\equiv 1(4)} \left(1 - \frac{2}{p-1}\right)\left(1 - \frac{1}{p}\right)^{-2} \prod_{p\equiv 3(4)} \left(1 - \frac{1}{p^2-1}\right).$$

The above computations lead us to:

CONJECTURE. The number of primes $p \leqslant x$, $p \equiv 1(\mathrm{mod}\, 4)$ for which $\frac{1}{8}|E/\mathbb{F}_p|$ is prime satisfies the asymptotic formula

$$S(x) \sim Cx(\log x)^{-2} \tag{15.78}$$

with the constant

$$C = \frac{2}{\pi} \prod_{p\equiv 1(4)} \left(1 - \frac{3p-1}{(p-1)^3}\right) \prod_{p\equiv 3(4)} \left(1 - \frac{1}{p^2-1}\right). \tag{15.79}$$

15.4. Rational Points on a Cubic Surface

Let $F(x)$ be a cubic polynomial and $\varphi(u,v)$ a binary quadratic form, both with rational coefficients. We are interested in counting rational points on the surface

$$F(x) = \varphi(u,v). \tag{15.80}$$

Without loss of generality, one can assume that $\varphi(u,v) = u^2 - cv^2$ by completing the square. In this form the equation was considered by Châtelet [**20**], who addressed various natural aspects of the problem, such as the Hasse local-global principle. In a recent work [**110**] it was shown under some mild conditions that there are infinitely many rational points on Châtelet surfaces, and in fact some strong estimates were given for the number of such points with height below a given quantity. In this section we rely on their work, however, we handle some of the arguments differently.

Let $f(x,y) = y^3F(x/y)$ be the corresponding cubic form. Note that the integral points (x,y,u,v) on the surface

$$yf(x,y) = \varphi(u,v) \tag{15.81}$$

yield the rational points

$$(x/y, u/y^2, v/y^2) \tag{15.82}$$

on (15.80), so we study the problem of counting integral solutions to (15.81). Because our primary goal is to illustrate how sieve methods work in this situation we do not need to cover all such surfaces. We consider the equation (15.81) with $\varphi(u,v) = u^2 + v^2$ and we assume that $f(x,y)$ is an irreducible cubic form having integer coefficients. Because the Gaussian field has class number one, our task reduces to counting integers a, b with

$$b = \square + \square,$$
$$f(a,b) = \square + \square,$$

and hence it is not surprising that the semi-linear sieve offers an appropriate tool. Note that from now on we shall use the letters a, b as the variables of the cubic form $f(x,y)$, reserving the letters x, y for our more customary usage.

Recall from Section 14.8 that $B(b)$ denotes the characteristic function of positive integers b having prime factors only in the set

$$\mathcal{Q} = \{q \text{ prime};\ q \equiv 1 (\mathrm{mod}\, 4)\}. \tag{15.83}$$

One possibility is to estimate the sum

$$\sum_{a\sim x}\sum_{b\sim x} B(f(a,b))B(b) \tag{15.84}$$

while another would be instead the sum

$$\sum_{a\sim x}\sum_{b\sim x} r(f(a,b))r(b) \tag{15.85}$$

where $r(b)$ is the number of representations of b as a sum of two squares. Each of the functions $B(b), r(b)$ have some nice features and we shall take advantage of this by considering the product $r(f(a,b))B(b)$, that is we shall count values of $f(a,b)$ with multiplicity, while b without.

Because we are going to appeal to Theorem 22.22 we need some compatibility conditions. We assume that

$$f(1,0) = 1, \qquad f(a,1) \equiv 1 (\mathrm{mod}\, 4) \tag{15.86}$$

for some odd integer a. These conditions are not very essential, but they help us in the presentation of our arguments. In particular, we can ensure that the values of $f(a,b)$ are positive if the size of b is taken to be somewhat smaller than the size of a by a constant factor η.

THEOREM 15.11. *Let f be an irreducible binary cubic form with integer coefficients satisfying* (15.86). *Let $\eta > 0$ be a sufficiently small number depending on f. Then, for every x sufficiently large, we have*

$$\sum\sum_{\substack{x<a<2x\\ \eta x<b<2\eta x\\ (a,b)=1}} r(f(a,b))B(b) \gg x^2(\log x)^{-\frac{3}{2}}, \tag{15.87}$$

where the implied constant depends only on f.

By Theorem 15.11 we infer that the number of integers u, v such that the equation $bf(a,b) = u^2 + v^2$ has integral solutions with $(a,b) = 1$ and $a \asymp x, b \asymp x$ is bounded below by $x^2(\log x)^{-3/2}$. Then, by the transformation (15.82) these points yield distinct rational points on our Châtelet surface.

COROLLARY 15.12. *Let $F(X) = X^3 + \alpha X^2 + \beta X + \gamma$ with integral coefficients satisfying $\alpha + \beta + \gamma \equiv 0 \,(\mathrm{mod}\, 4)$. Then the cubic surface*

$$F(x_1) = x_2^2 + x_3^2$$

has infinitely many rational points (x_1, x_2, x_3). The number of such points having the maximum of their denominators at most y is $\gg y(\log y)^{-3/2}$.

REMARKS. One should expect that the sum (15.87) has order $x^2(\log x)^{-1/2}$. In our estimate we lost a factor $\log x$ for technical reasons. Then, in moving from the theorem to the corollary we sacrificed much more in the bound by counting only special rational points.

There is some similarity between our result and Heath-Brown's result about prime values of $f(a,b)=a^3+2b^3$. Rather than asking $f(a,b)$ to be prime we only ask it to be the sum of two squares, but simultaneously we also ask b to be the sum of two squares. This is reminiscent of the hyperbolic prime number theorem which we compared to the twin prime problem.

The factor $r(f)$ in the sum is very helpful in some parts of the arguments due to its decomposition

$$r(f)=4(1*\chi_4)(f), \tag{15.88}$$

where χ_4 is the non-principal character mod 4. However, in other places the multiplicity of representations in $r(f)$ hurts us. To mollify this we introduce the extra factor

$$\theta(f)=\sum_{k|f}\lambda_k, \tag{15.89}$$

where (λ_k) with $1\leqslant k\leqslant K$ is an upper-bound sieve of relatively small level K. This has the effect of making f behave as if it is an almost-prime, thus damping the multiplicity $r(f)$. The need to later remove these weights $\theta(f)$ leads to the loss of the logarithmic factor; because of the weights we are only counting some of the representations. Since f is counted with multiplicity $r(f)$ there is no need to sieve it by primes in the sifting set

$$\mathcal{P}=\{p \text{ prime};\ p\equiv 3(\operatorname{mod} 4)\}\ . \tag{15.90}$$

To simplify the notation in what follows we take the case $f(1,1)\equiv 1(4)$, the other case $f(-1,1)\equiv 1(4)$ being almost identical. We begin by introducing a smooth function $W(u,v)$ supported on the dyadic box $u\sim x, v\sim \eta x$ with derivatives satisfying

$$\frac{\partial^{i+j}W(u,v)}{\partial u^i\partial v^j}\ll x^{-i-j},\ 0\leqslant i,j\leqslant 2\ ,$$

and consider the sum

$$S(x)=\sum\sum_{\substack{(a,b)=1\\ a\equiv b\equiv 1(4)}} r(f(a,b))B(b)\theta(f(a,b))W(a,b)\ .$$

Since $f(1)=1$ we can choose the support of $W(a,b)$ so that a dominates (by a constant factor) over b and hence $f(a,b)>0$ throughout.

Since $\mathcal{P}$ is given by (15.90), therefore we take λ_k in (15.89) to be supported on

$$1\leqslant k\leqslant K,\quad k\mid Q(\sqrt{K})\ , \tag{15.91}$$

where $Q(z)$ denotes the product of primes $q<z$ in $\mathcal{Q}$. This implies that k is squarefree and $k\equiv 1(\operatorname{mod} 4)$.

Our first step is to detect $B(b)$ by a semi-linear sieve (which has nothing to do with the sieve weights (15.89)). To this end we take the sequence $\mathcal{A}=(a_n)$ with

$$a_n=\sum_{\substack{(a,n)=1\\ a\equiv 1(4)}} r(f(a,n))\theta(f(u,v))W(a,n) \tag{15.92}$$

for $n \equiv 1 (\bmod 4)$. Opening θ as in (15.89), we see that our sequence is a sum over $k \leqslant K$ of sequences considered in Section 22.9, see (22.122) with $a = 1$, $q = 4$, and $d = 1$. Theorem 22.22 shows that our sequence has level of distribution

$$D = K^{-1} x^{\frac{1}{2}-\varepsilon} . \tag{15.93}$$

More precisely, we have

$$A_d(x) = \sum_{\substack{n \equiv 0(d) \\ n \equiv 1(4)}} a_n = g(d) X V + r_d , \tag{15.94}$$

with the remainder terms satisfying

$$\sum_{d \leqslant D} | r_d | \ll x^{2-\varepsilon} .$$

In the main term $X \asymp x^2$ and the factor V is given by

$$V = \sum_k \lambda_k h(k) .$$

Here, $g(d)$ and $h(k)$ are multiplicative functions which can be explicitly given (see (22.109) and (22.128)) but we need only the following information:

$$g(p) = \frac{1}{p}\left(1 + O\left(\frac{1}{p}\right)\right) , \tag{15.95}$$

$$h(p) = \frac{\rho(p)}{p}\left(1 + \chi_4(p) + O\left(\frac{1}{p}\right)\right) , \tag{15.96}$$

where we recall that $\rho(p)$ denotes the number of roots $\nu \,(\bmod p)$ of the congruence $f(\nu) \equiv 0 \,(\bmod p)$.

Before applying sieve arguments we have to control the factor V in the main term. Since f is irreducible

$$\sum_{p \leqslant z} \frac{\rho(p)}{p} \log p = \log z + O(1) .$$

Since χ_4 is a quadratic character, it is admissible for f in the sense that the twisted L-function (22.127) has no pole at $s = 1$. Therefore, we also have

$$\sum_{p \leqslant z} \chi_4(p) \frac{\rho(p)}{p} \log p = O(1) .$$

Combining these results we get (recall that q runs over primes in $\mathcal{Q}$)

$$\sum_{\substack{q \leqslant z \\ q \equiv 1(4)}} \frac{h(q)}{q} \log q = \log z + O(1) .$$

Therefore, we are dealing with a linear sieve and so

$$V = \sum_k \lambda_k h(k) \asymp \frac{1}{\log K} . \tag{15.97}$$

Coming back to the sieve to detect $B(b)$ we again have a semi-linear sieve problem but we cannot be as casual in applying it. Since $b \asymp x$, $b \equiv 1(4)$, we need to kill the prime factors in b from

$$P(z) = \prod_{\substack{p \in \mathcal{P} \\ p < z}} p, \tag{15.98}$$

with $z \asymp x^{\frac{1}{2}}$. However, the sifting limit $\beta = 1$ does not allow us to go quite so far because the level of distribution D given by (15.93) is too small, just slightly so because we take

$$K = x^{\mu} \tag{15.99}$$

where μ will be a rather small, but absolute, constant. Therefore, we are going to apply the lower bound semi-linear sieve with

$$z = D^{1/s}, \tag{15.100}$$

where s is larger than one, but not by much. By Theorem 11.12 we obtain

$$S(\mathcal{A}, z) = \sum_{(n,P(z))=1} a_n \gg \sqrt{\frac{s-1}{\log x}} \frac{x^2}{\log K} \,. \tag{15.101}$$

Since $S(\mathcal{A}, z)$ contains more than $S(x)$ we need to subtract off the difference $S(\mathcal{A}, z) - S(x) = S_2(\mathcal{A}, z)$, say, which we write as follows:

$$S_2(\mathcal{A}, z) = \sum\sum_{\substack{(a,b)=1 \\ a \equiv b \equiv 1(4) \\ b = cp_1p_2 \\ p_1 > p_2 \geqslant z}} r(f(a,b))\theta(f(a,b))B(c)W(a,b). \tag{15.102}$$

As in earlier sections of this chapter we need an upper bound for this sum and we expect it to be fairly small due to the fact that the primes p_1, p_2 are relatively close together. Our sieve can be rather crude and we can even derive what we need by holding b fixed and applying the sieve to the sum over a. Therefore, we write

$$S_2(\mathcal{A}, z) = \sum_b S_b \,,$$

where b runs over the same numbers as in (15.102) and

$$S_b = \sum_{\substack{(a,b)=1 \\ a \equiv 1(4)}} r(f(a,b))\theta(f(a,b))W(a,b) \,.$$

Let $w = x^{\gamma}$, where $0 < \gamma < \frac{1}{12}$ is an absolute constant ($\gamma = \frac{1}{13}$ suffices). Any number f with $r(f) > 0$ can be written uniquely as

$$f = e^2 f_1 \quad (f_1, P(w)) = 1 \,,$$

where e has only primes which divide $P(w)$. Recall that $P(w)$ denotes the product (15.98) over primes $p \in \mathcal{P}$, $p < w$. Either $e \leqslant w$ or it has a factor e_1 with

$\sqrt{w} \leqslant e_1 < w$. Therefore,

$$S_b \leqslant \sum_{\substack{e \leqslant w \\ e|P(w)^\infty}} \sum_{\substack{(a,b)=1 \ a\equiv 1(4) \\ f(a,b)\equiv 0(e^2) \\ (f(a,b)e^{-2},P(w))=1}} r(f(a,b))\theta(f(a,b))W(a,b)$$

$$+ \sum_{\sqrt{w}\leqslant e<w} \sum_{\substack{(a,b)=1 \\ a\equiv 1(4) \\ f(a,b)\equiv 0(e^2)}} r(f(a,b))\theta(f(a,b))W(a,b) = S_b^* + S_b^{**}\ , \tag{15.103}$$

say. The second sum can be estimated easily as follows:

$$S_b^{**} \ll x^\varepsilon \sum_{\sqrt{w}<e<w} \sum_{\substack{(a,b)=1 \\ f(a,b)\equiv 0(e^2)}} W(a,b) \ll x^{1+\varepsilon} \sum_{\sqrt{w}<e<w} \frac{\rho(e^2)}{e^2} \ll \frac{x^{1+\varepsilon}}{\sqrt{w}}\ . \tag{15.104}$$

Now we proceed to estimate the difficult part, S_b^*. Since $\theta(f(a,b))$ is a sieve of relatively small level it constrains the variable rather mildly so that S_b^* may be regarded as a divisor sum for the cubic polynomial f. We shall see a number of such problems in later chapters, especially Chapter 22. If we were to open $r(f)$ by the convolution (15.88) we would be forced to consider, even after the Dirichlet hyperbola trick, the variable a in arithmetic progressions to moduli of size $x^{\frac{3}{2}}$, whereas a is only of size x. Clearly, such an approach cannot work. To resolve this difficulty we apply Landreau's inequality

$$r(f) \leqslant r(e^2)r(fe^{-2}) \ll r(e^2) \sum_{\substack{\delta|fe^{-2} \\ \delta\leqslant f^{1/4}}} t(\delta)\ ,$$

where $t(\delta)$ is the multiplicative function, given explicitly in Corollary 22.13, is a divisor-like function. Contrary to what one might think, this step is quite delicate; see the remarks following (15.108). The restriction $\delta \leqslant f^{1/4}$ implies $\delta \leqslant \Delta$ with $\Delta \asymp x^{3/4}$, overcoming our problem. However, there is a price for this; we have replaced $r(f)$ by a function whose average value is larger by a logarithm. This would be fatal were we not to have the sieve factor $\theta(f)$, on whose support all divisor functions behave the same.

We get

$$S_b^* \leqslant \sum_{\substack{e\leqslant w \\ e|P(w)^\infty}} r(e^2) \sum_{\substack{\delta\leqslant\Delta \\ (\delta,e)=1}} t(\delta) \sum_{\substack{(a,b)=1 \ a\equiv 1(4) \\ f(a,b)\equiv 0(e^2\delta) \\ (f(a,b)e^{-2},P(w))=1}} \theta(f(a,b))W(a,b)\ .$$

Note that conditions on the inner sum imply $(\delta, P(w)) = 1$, however, we prefer not to record this condition in order to make the forthcoming sum over δ less constrained. We detect the condition $(f(a,b)e^{-2}, P(w)) = 1$ by a semi-linear upper-bound sieve (ξ_ℓ) with $1 \leqslant \ell \leqslant L$ of level $L = w^2 = x^{2\gamma}$, that is, we introduce the factor

$$\sum_{\substack{\ell|fe^{-2} \\ \ell|P(w)}} \xi_\ell$$

and forget this coprimality condition. Since we need only an upper bound we can ignore at will some of the primes in the sifting range and indeed, for simplicity, we restrict ξ_ℓ by $(\ell, \delta e) = 1$. Introducing this and opening $\theta(f)$ we get $S_b^* \leqslant S_b^+$, where

$$S_b^+ = \sum_{\substack{e \leqslant w \\ e | P(w)^\infty \\ (e,b)=1}} r(e^2) \sum_{\substack{\delta \leqslant \Delta \\ (\delta, eb)=1}} t(\delta) \sum_{\substack{k \leqslant K \\ k | Q(\sqrt{K}) \\ (k,b)=1}} \lambda_k \sum_{\substack{\ell \leqslant L \\ \ell | P(w) \\ (\ell, \delta eb)=1}} \xi_\ell \sum_{\substack{(a,b)=1 \\ a \equiv 1 (\mathrm{mod}\, 4) \\ f(a,b) \equiv 0 (e^2 \ell [\delta, k])}} W(a,b) \ .$$

The last sum over a equals

$$\frac{\rho(4e^2)\varphi(b)}{4e^2 b} \frac{\rho(l[\delta,k])}{l[\delta,k]} \int W(u,b) du + O(X^\varepsilon).$$

Hence

$$S_b^+ = \frac{\varphi(b)}{b} \Big(\int W(u,b) du \Big) \sum_{\substack{e \leqslant w \\ e | P(w)^\infty \\ (e,b)=1}} r(e^2) \frac{\rho(4e^2)}{4e^2}$$

(15.105)

$$\cdot \Big(\sum_{\substack{\ell \leqslant L \\ \ell | P(w) \\ (\ell, eb)=1}} \xi_\ell \frac{\rho(\ell)}{\ell} \sum_{\substack{\delta \leqslant \Delta \\ (\delta, be\ell)=1}} t(\delta) \sum_{\substack{k \leqslant K \\ k | Q(\sqrt{K}) \\ (k,b)=1}} \lambda_k \frac{\rho([\delta,k])}{[\delta,k]} \Big) + O(w\Delta K L x^\varepsilon).$$

Recall that $\Delta \asymp x^{3/4}$ so the last term in (15.105) is $\ll x^{\gamma + \frac{3}{4} + \mu + 2\gamma + \varepsilon} < x^{1-\varepsilon}$ which will be seen to be negligible after we have established (15.106). We execute the summation over $\delta \leqslant \Delta$ by using the generating Dirichlet series

$$T(s) = \sum_{(\delta, be\ell)=1} t(\delta) \frac{\rho([\delta,k])}{[\delta,k]} \delta^{-s}.$$

Keep in mind that $(k, be\ell) = 1$ which is clear by the support of the sieve weights λ_k and ξ_ℓ. Actually, all of the numbers b, e, k, ℓ are mutually co-prime. If $\rho(k) = 0$ then $T(s) = 0$, so we can assume that $\rho(k) \neq 0$. In this case we have

$$\begin{aligned} T(s) &= \frac{\rho(k)}{k} \sum_{(\delta, be\ell)=1} t(\delta) \frac{(\delta,k)}{\delta \rho(k)} \rho\Big(\frac{\delta k}{(\delta,k)} \Big) \delta^{-s} \\ &= \frac{\rho(k)}{k} \prod_{p \nmid be\ell k} \Big(\sum_{\alpha=0}^{\infty} t(p^\alpha) \rho(p^\alpha) p^{-\alpha(s+1)} \Big) \prod_{p | k} \Big(1 + \sum_{\alpha=1}^{\infty} t(p^\alpha) \frac{\rho(p^\alpha)}{\rho(p)} p^{1-\alpha(s+1)} \Big) \\ &= \frac{\rho(k)}{k} \prod_{p \nmid be\ell k} L_p(s+1) \prod_{p|k} M_p(s+1) \ , \end{aligned}$$

say. Let

$$L(s) = \sum_{\delta} t(\delta) \rho(\delta) \delta^{-s} = \prod_p L_p(s).$$

Hence

$$T(s) = L(s+1) \frac{\rho(k)}{k} \prod_{p|k} M_p(s+1) \prod_{p | bek\ell} L_p(s+1)^{-1}.$$

Note that because χ_4 is admissible for f the function $L(s)$ is holomorphic in $\mathrm{Re}\, s > \frac{1}{2}$ except for a simple pole at $s = 1$. By complex integration we deduce that

$$\sum_{\substack{\delta \leqslant \Delta \\ (\delta, be\ell)=1}} t(\delta) \frac{\rho([\delta, k])}{[\delta, k]} = \operatorname*{res}_{s=0} s^{-1} \Delta^s T(s) + O\Big(\frac{\rho(k)}{k} \Delta^{-\frac{1}{2}} (bek\Delta)^{\varepsilon} \Big) .$$

Here, we borrow $\zeta(s)$ and return it. Then, writing $\zeta(s)\Delta^s = s^{-1} + \gamma + \log \Delta + \dots$ and $L(s+1)\zeta(s)^{-1} = G(1 + cs + \dots)$ with $G = \prod_p L_p(1)(1 - 1/p) > 0$ we find

$$\operatorname*{res}_{s=0} s^{-1} \Delta^s T(s) = G \frac{\rho(k)}{k} \prod_{p|k} M_p(1) \prod_{p|bek\ell} L_p(1)^{-1}$$
$$\cdot \Big\{ \gamma + \log \Delta + c + \sum_{p|k} \frac{M_p'}{M_p}(1) - \sum_{p|bek\ell} \frac{L_p'}{L_p}(1) \Big\} .$$

The multiplicative functions, say $c_1(k)$ and $c_2(\ell)$, outside of $\{\dots\}$ are

$$c_1(k) = \prod_{p|k} M_p(1) L_p(1)^{-1} = \prod_{p|k} \Big(3 + O\Big(\frac{1}{p}\Big) \Big) ,$$

$$c_2(\ell) = \prod_{p|\ell} L_p(1)^{-1} = \prod_{p|\ell} \Big(1 + O\Big(\frac{1}{p}\Big) \Big) ,$$

while the additive functions $\gamma_1(k)$ and $\gamma_2(\ell)$ inside $\{\dots\}$ are

$$\gamma_1(k) = \sum_{p|k} \Big(\frac{M_p'}{M_p}(1) - \frac{L_p'}{L_p}(1) \Big) , \qquad \gamma_2(\ell) = - \sum_{p|\ell} \frac{L_p'}{L_p}(1) .$$

Using, for example, Theorem 11.12 for sieves of dimension $3/2$ and $1/2$, respectively, we obtain

$$\sum_k \lambda_k \frac{\rho(k)}{k} c_1(k) \ll (\log K)^{-\frac{3}{2}} , \qquad \sum_{\ell} \xi_l \frac{\rho(\ell)}{\ell} c_2(\ell) \ll (\log L)^{-\frac{1}{2}} .$$

Since $\rho(p) \leqslant p$ we have

$$M_p(1) = 1 + \frac{p}{\rho(p)} \sum_{\alpha=1}^{\infty} t(p^\alpha) \rho(p^\alpha) p^{-\alpha} \leqslant \frac{p}{\rho(p)} L_p(1) ,$$
$$M_p'(1) = -p \frac{\log p}{\rho(p)} \sum_{\alpha=1}^{\infty} \alpha t(p^\alpha) \rho(p^\alpha) p^{-\alpha} = \frac{p}{\rho(p)} L_p'(1) ,$$

so $\gamma_1(p) \leqslant 0$ and $\gamma_1(p) \ll \log p$. Therefore, by Corollary 5.4 we obtain

$$\sum_k \lambda_k \frac{\rho(k)}{k} c_1(k) \gamma_1(k) \leqslant c_1 (\log K)^{-\frac{1}{2}} .$$

Moreover, we also have $0 \leqslant \gamma_2(p) \ll p^{-1} \log p$, whence it follows from Corollary 5.3 with $g'(p) = \frac{1}{2}(1 + g(p))$ that

$$\sum_{\ell} \xi_\ell \frac{\rho(\ell)}{\ell} c_2(\ell) \gamma_2(\ell) \leqslant c_2 (\log L)^{-\frac{1}{2}}$$

where c_1, c_2 are positive constants. Recall that $\log\Delta \asymp \log x$ and $\log L \asymp \log x$. Inserting these estimates in (15.105), we find

$$S_b^+ \ll \frac{x(\log x)^{\frac{1}{2}}}{(\log K)^{\frac{3}{2}}} . \tag{15.106}$$

Summing over $b = cp_1p_2 \sim x$ with $p_1 > p_2 > z$, so $c \ll xz^{-2}$ and $z < p_2 \ll x^{\frac{1}{2}}$ we find that $S_2(\mathcal{A}, z)$ is bounded by

$$\Big(\sum_{\substack{c\ll xz^{-2}\\ c\mid Q(\sqrt{K})}} \frac{1}{c}\Big)\Big(\sum_{z<p_2\leqslant x^{\frac{1}{2}}} \frac{1}{p_2}\Big)\frac{x^2}{(\log K)^{\frac{3}{2}}(\log x)^{\frac{1}{2}}} \ll x^2\Big(\frac{\log xz^{-2}}{(\log K)(\log x)}\Big)^{\frac{3}{2}} .$$

For z given by (15.100) with D given by (15.93) we have $z > K^{-1}x^{1-s/2}$. Hence we conclude that

$$S_2(\mathcal{A}, z) \ll x^2\Big(\frac{\log(K^2x^{s-1})}{(\log K)(\log x)}\Big)^{\frac{3}{2}} . \tag{15.107}$$

Recall that $K = x^{\mu}$. We choose $s = 1 + \mu$ so (15.101) becomes

$$S(\mathcal{A}, z) \gg \mu^{-\frac{1}{2}}x^2(\log x)^{-\frac{3}{2}} ,$$

while (15.107) becomes

$$S_2(\mathcal{A}, z) \ll x^2(\log x)^{-\frac{3}{2}} .$$

Recall that $S(x) = S(\mathcal{A}, z) - S_2(\mathcal{A}, z)$. Choosing μ sufficiently small we conclude that

$$S(x) \gg x^2(\log x)^{-\frac{3}{2}} . \tag{15.108}$$

Remarks. Our application of Corollary 22.13 for $r(f)$ is quite delicate. Indeed, if the multiplicative function (22.37) were a bit worse than the divisor-like function $r(d)$ we would come up with an extra factor (depending on μ) in the upper bound for $S_2(\mathcal{A}, z)$ which exceeds that in the lower bound for $S(\mathcal{A}, z)$. In such a scenario we would not be able to conclude the positive lower bound (15.108) for $S(x)$ by choosing μ small.

Arguments similar to those which gave us the lower bound (15.108) also yield an upper bound of the same order, so we have

$$S(x) \asymp x^2(\log x)^{-\frac{3}{2}}. \tag{15.109}$$

Actually, the proof of the upper bound is considerably simpler because one only needs an estimate for $S(\mathcal{A}, z)$; the contribution of $S_2(\mathcal{A}, z)$ being negative can be ignored.

To complete the proof of Theorem 15.11 we apply Cauchy's inequality getting

$$S(x)^2 \leqslant S^{(2)}(x) \sum\sum_{\substack{(a,b)=1\\ a\equiv b\equiv 1(4)}} r(f(a,b))B(b)W(a,b), \tag{15.110}$$

where $S^{(2)}(x)$ is the same as $S(x)$, but with the sieve weights $\theta(f)$ (see (15.89)) replaced by $\theta(f)^2$. Because $\theta(f)^2$ is also an upper-bound sieve (now of level K^2 rather than K) the result (15.109) holds for $S^{(2)}(x)$. Hence, (15.110) implies (15.87) by (15.109).

CHAPTER 16

Asymptotic Sieve and the Parity Principle

Les Barricades Mystérieuses

16.1. Distribution Functions

Although the original goal of the sieve was the study of sums over prime numbers, this for a long time seemed futile and historically one settled for being able to deal only with sums over almost-primes.

An almost-prime (of order k) is an integer having at most k distinct prime factors (so primes are also almost-primes of order seventeen and 57 is an almost-prime of order two, which is not a prime). These can be counted with various weights. The most direct sieve methods count them with weight one if all of their prime factors are $\geqslant z$ and weight zero otherwise. In this case the number $n \leqslant x$ so counted has at most $k = [\log x/\log z]$ prime factors. A more efficient approach is given by the combinatorial weights introduced in Chapter 25.

However, more natural than any of these are the weights carried by the von Mangoldt functions. Theorem 3.5 gives an example of such a weighted sieve. In contrast to the other weights it gives remarkably strong and elegant results when $D = D(x)$, the level of distribution, is close to x (the maximal level possible), but deteriorates rapidly as the level decreases (see [**50**]).

From these natural weights one may create more general ones by taking linear combinations. It was shown by Bombieri [**8**] that the supply of these is sufficiently rich to generate essentially every nice weight. In Section 16.3 we present an approximation argument which elucidates this.

Let $\mathscr{P}_r$ be the set of squarefree integers

$$n = p_1 p_2 \dots p_r$$

which have exactly r prime factors. Therefore,

$$\mathscr{P}(k) = \bigcup_{1 \leqslant r \leqslant k} \mathscr{P}_r$$

is the set of squarefree almost-primes of order k. The von Mangoldt functions $\Lambda_{(k)}(n)$, for n squarefree, have support on $\mathscr{P}(|k|)$. Given a sequence $\mathcal{A} = (a_n)$ of non-negative real numbers we shall study sums of type

$$\sum_{\substack{n \leqslant x \\ n \in \mathscr{P}_r}} a_n G^*(n) \tag{16.1}$$

where $G^*(n) = G^*_r(n)$ is a nice weight function. More specifically, we consider functions

$$G^*(n) = G\left(\frac{\log p_1}{\log n}, \dots, \frac{\log p_r}{\log n}\right) \tag{16.2}$$

for $n = p_1 p_2 \dots p_r$, where $G(u_1, \dots, u_r)$ is a nice continuous function on the set T_r defined by

$$\begin{aligned} 0 < u_1 < u_2 < \dots < u_r \leqslant 1 \ , \\ u_1 + u_2 + \dots + u_r = 1 \ . \end{aligned} \tag{16.3}$$

One should expect that the sum (16.1) is asymptotically proportional to the volume of $G(u_1, \dots, u_r)$ with respect to a certain measure which perhaps depends very mildly on the sequence $\mathcal{A}$. Bombieri showed that it depends only on the distribution of $\mathcal{A}$ on primes.

The appropriate measure on T_r is defined by

$$d\mu_r = \frac{du_1 \dots du_{r-1}}{u_1 \dots u_r} \quad \text{if } r \geqslant 1 \ , \tag{16.4}$$

where for $r = 1$ this means the discrete measure at the point $u_1 = 1$.

For the simplest sequences such as $a_n = 1$ this would be the measure in question. However, one can find sequences, albeit artificial ones, for which the measure $d\mu_r$ must be altered by a distribution function. We say that the sequence $\mathcal{A} = \big(a_n\big)$ has a distribution function on $\mathscr{P}_r$,

$$\gamma_x(\mathbf{u}) = \gamma_x(u_1, \dots, u_r)$$

if for every $G \in \mathcal{C}_0(T_r)$ we have

$$\sum_{\substack{n \leqslant x \\ n \in \mathscr{P}_r}} a_n \, G^*(n) \sim \left(\int_{T_r} G(\mathbf{u}) \gamma_x(\mathbf{u}) d\mu_r \right) \frac{HA(x)}{\log x} \ ,$$

as $x \to \infty$. Note that $\gamma_x(\mathbf{u})$, if it exists, is defined only up to $o(1)$ as $x \to \infty$.

In the case $r = 1$ this definition means

$$\sum_{p \leqslant x} a_p \sim \delta_x \frac{HA(x)}{\log x} \tag{16.5}$$

as $x \to \infty$ where $\delta_x \sim \gamma_x(1)$ which would be the Prime Number Theorem for the sequence $\mathcal{A}$.

Within the sieve axioms this Prime Number Theorem cannot be established with a reasonable choice of δ_x. Of course, we could also define δ_x as the function which equalizes (16.5), but this would tell us nothing.

16.2. The Basic Theorem

A main thrust of Bombieri's sieve is its demonstration that the distribution of $\mathcal{A}$ on each $\mathscr{P}_r$ is determined by that for $r = 1$, or, indeed, by that for any individual value of r. Specifically, we have

THEOREM 16.1. *Every sequence $\mathcal{A} = \big(a_n\big)$ satisfying the conditions of Theorem 3.5 with*

$$D = D(x) = x \exp\Big(-(\log x)^{\frac{1}{4}} \Big)$$

has a distribution function $\gamma_x(\mathbf{u})$ on $\mathscr{P}_r$ for every r and it is given by

$$\begin{aligned} \gamma_x(\mathbf{u}) &= \delta_x \qquad && \textit{if } r \textit{ is odd,} \\ \gamma_x(\mathbf{u}) &= 2 - \delta_x \qquad && \textit{if } r \textit{ is even.} \end{aligned} \tag{16.6}$$

The beauty of this theorem is that $\gamma_x(\mathbf{u})$ does not depend on $\mathbf{u}$, but only on x and the parity of r. It also shows that

$$0 \leqslant \delta_x \leqslant 2 \ , \tag{16.7}$$

$$0 \leqslant \gamma_x(\mathbf{u}) \leqslant 2 \ . \tag{16.8}$$

The fact that $\delta_x \leqslant 2$ is not obvious from the definition (16.5), but was known in many cases as a fundamental signpost of sieve theory.

By modifying the two examples from Section 12.3 we see that δ_x can take any constant value between 0 and 2. To take the value $\delta_x = \delta$ we choose

$$a_n = 1 + (1-\delta)\lambda(n) \ .$$

In fact, we can arrange for δ_x to have no limit, for example,

$$a_n = 1 + \alpha(n)\lambda(n)$$

where $\alpha(n)$ is any function satisfying

$$|\alpha(n)| \leqslant 1 \ , \qquad |\alpha'(n)| \leqslant \frac{c}{n \log n} \ ,$$

such as $\alpha(n) = \cos(c \log\log n)$ with c a positive constant.

In this example the good level of distribution follows from the Prime Number Theorem for $\lambda(n)$; the factor $\alpha(n)$ can be treated by partial summation. On the other hand, for this sequence we have

$$\sum_{p \leqslant x} a_p = \frac{x}{\log x}\left\{1 - \alpha(x) + o(1)\right\} \ ,$$

also by partial summation.

REMARK. As with the asymptotic formula in Theorem 3.5, Bombieri proves Theorem 16.1 under the less stringent assumption that the level of distribution $D(x) = x^{1-\varepsilon}$ be admissible for every $\varepsilon > 0$.

16.3. Proof of the Basic Theorem

In Theorem 3.5 we gave asymptotic formulae for almost-primes weighted by the von Mangoldt functions $\Lambda_{(k)}(n)$. We deduce Theorem 16.1 from this result by approximating $G^*(n)$ by linear combinations of the $\Lambda_{(k)}(n)$. Since Theorem 3.5 did not apply for vectors $(k) = (1, \dots, 1)$ we must not use these vectors in the decomposition. It turns out that we are able to avoid them except for the basic vector $\Lambda(n) = \Lambda_{(1)}(n)$. This is why we introduce the axiom (16.5).

Rather than $G^*(n)$ it is convenient to approximate $G^*(n)(\log n)^N$ for sufficiently large N and then remove this factor by partial summation. We shall show that $G^*(n)\big(\log n\big)^N$ is well approximated by the linear combination

$$\sum_{|k|=N} c_{(k)}\Lambda_{(k)}(n) + c\Lambda(n)\big(\log n\big)^{N-1}$$

for a suitable choice of the coefficients $c_{(k)}, c$, with $c_{(1,\dots,1)} = 0$. If we used only those Λ_k with k scalar, we would fail to cover all functions G^*.

We begin with several lemmas establishing properties of the approximating functions $\Lambda_{(k)}L^a$. (Recall that L is the function $L(n) = \log n$.)

LEMMA 16.2. *The function $\Lambda_{(k)}L^a$ can be expressed as a linear combination of $\Lambda_{(h)}$ with $|h| = |k| + a$ and $\max h_\nu \geqslant \max k_\nu$.*

PROOF. Because multiplication by L is a derivation in the ring of arithmetic functions, it suffices to prove the lemma for scalar k. For $a=0$ there is nothing to prove. For $a=1$ we have

$$\Lambda_k L = \Lambda_{k+1} - \Lambda * \Lambda_k \ .$$

Hence, multiplying through by L^a we obtain

$$\Lambda_k L^{a+1} = \Lambda_{k+1} L^a - \sum_{b=0}^{a} \binom{a}{b} (\Lambda L^b) * \Lambda_k L^{a-b} \ ,$$

giving the result by induction on a. □

As a corollary we obtain

LEMMA 16.3. *The function*

$$\Lambda L^{a_1-1} * \cdots * \Lambda L^{a_r-1} \ ,$$

with $a_1, \ldots, a_r \geqslant 1$*, can be expressed as a linear combination of the functions* $\Lambda_{(h)}$ *with* $|h| = a_1 + \cdots + a_r$*.*

The elimination of the vectors $(1, \ldots, 1)$ will be achieved by the following result.

LEMMA 16.4. *For the vector* $(k) = (1, \ldots, 1)$ *of dimension* $N \geqslant 1$ *we have*

$$\Lambda_{(1,\ldots,1)} = \sum_{\substack{|h|=N \\ (h)\neq(k)}} \gamma_{(h)} \Lambda_{(h)} + \frac{1}{(N-1)!} \Lambda L^{N-1} \ . \tag{16.9}$$

PROOF. For dimension $N=1$ the result is obvious. For dimension $N=2$ we have

$$\Lambda * \Lambda = -\Lambda_2 + \Lambda L$$

giving the result. For $(k) = (1, \ldots, 1)$ of dimension $N \geqslant 3$ we get by the above that

$$\begin{aligned} \Lambda_{(k)} &= \underbrace{\Lambda * \cdots * \Lambda}_{N-1 \text{ times}} * \Lambda L - \Lambda_2 * \underbrace{\Lambda * \cdots * \Lambda}_{N-1 \text{ times}} \\ &= \frac{1}{N-1} \Lambda_{(k')} L - \Lambda_{(k'')} \ , \end{aligned}$$

where $(k') = (1, \ldots, 1)$ has dimension $N-1$ and$(k'') = (2, 1, \ldots, 1)$ has dimension N. Hence, the result follows by induction using Lemma 16.2. □

Now we are ready to approximate G^* by combinations of von Mangoldt functions.

LEMMA 16.5. *Let* $\big(u_1 \ldots u_r\big)^{-1} G\big(u_1, \ldots, u_r\big)$ *be continuous on the closure of* T_r*. Then, for any* $\varepsilon > 0$ *there exist* N *and coefficients* $\gamma_{(A)}$ *such that*

$$\Big| G^* L^N - \sum_{(A)} \gamma_{(A)} \left(\Lambda L^{a_1-1} * \cdots * \Lambda L^{a_r-1}\right) L^{N-|A|} \Big| \leqslant \varepsilon \, (\underbrace{\Lambda * \cdots * \Lambda}_{r\text{–times}}) \, L^{N-r} \tag{16.10}$$

for all $n \in \mathscr{P}_r$*. Here, the sum runs over vectors* $(A) = (a_1, \ldots, a_r)$ *with* $|A| \leqslant N$*,* $a_r \geqslant \cdots \geqslant a_1 \geqslant 1$*.*

PROOF. Consider the simplex

$$U_r = \{(u_1, \dots, u_r), u_i > 0\ , \quad u_1 + \dots + u_r = 1\}\ ,$$

and for any function F on T_r let $F^\sharp$ be the pull-back of F to U_r which is symmetric in $u_1, \dots, u_r$. If

$$F = \big(u_1 \dots u_r\big)^{-1} G\big(u_1, \dots, u_r\big)\ ,$$

then $F^\sharp$ is continuous in the closure of U_r and symmetric in $u_1, \dots, u_r$ (invariant under the symmetric group S_r. By the Weierstrass approximation theorem there exists a polynomial $P(u_1, \dots, u_r)$, symmetric in $u_1, \dots, u_r$, such that $\big|F^\sharp - P\big| \leqslant \varepsilon$ on U_r. Therefore,

$$\big|G^\sharp(u_1, \dots, u_r) - u_1 \dots u_r\, P(u_1, \dots, u_r)\big| \leqslant \varepsilon u_1 \dots u_r$$

on U_r. We have

$$u_1 \dots u_r\, P(u_1, \dots, u_r) = \sum \gamma_{(A)} \sum_{\sigma \in S_r} u_{\sigma(1)}^{a_1} \cdots u_{\sigma(r)}^{a_r}$$

because P is symmetric, where

$$(A) = (a_1, \dots, a_r), \qquad a_r \geqslant \dots \geqslant a_1 \geqslant 1, \qquad |A| \leqslant r + \deg P = N,$$

say. Hence

$$\big|G^\sharp(u_1, \dots, u_r) - \sum_{(A)} \gamma_{(A)} \sum_{\sigma \in S_r} u_{\sigma(1)}^{a_1} \cdots u_{\sigma(r)}^{a_r}\big| \leqslant \varepsilon u_1 \cdots u_r\ . \tag{16.11}$$

If $n \in \mathscr{P}_r$, $n = p_1 \dots p_r$, put $u_i = \log p_i / \log n$, so $(u_1, \dots, u_r) \in U_r$ and we have

$$\Lambda L^{a_1 - 1} * \dots * \Lambda L^{a_r - 1} = \big(\log n\big)^{|A|} \sum_{\sigma \in S_r} u_{\sigma(1)}^{a_1} \cdots u_{\sigma(r)}^{a_r}\ .$$

Now, Lemma 16.5 follows from (16.11). □

It will be convenient to use the inequality (16.10) for all n. It extends trivially to all n by setting $G^*(n) = 0$ for $n \notin \mathscr{P}_r$. Then, by Lemmas 16.3, 16.4, 16.5 we obtain:

LEMMA 16.6. *Let $(u_1 \dots u_r)^{-1} G(u_1, \dots, u_r)$ be continuous on the closure of T_r. Then, for each ε, there exist N and coefficients c and $c_{(k)}$ such that, for all n,*

$$\Big|G^* L^N - c\Lambda L^{N-1} - \sum_{\substack{|k| = N \\ (k) \neq (1, \dots, 1)}} c_{(k)} \Lambda_{(k)}\Big| \leqslant \varepsilon \Lambda_r L^{N-r}\ .$$

From Lemma 16.6 we first deduce a crude bound for

$$T_N(x) = \sum_{n \leqslant x} a_n G^*(n) \big(\log n\big)^N\ .$$

LEMMA 16.7. *Suppose the conditions of Theorem 3.5 hold. Then, for $N \geqslant 1$ we have*

$$T_N(x) \ll A(x) \big(\log x\big)^{N-1}\ , \tag{16.12}$$

where the implied constant depends on N.

PROOF. This follows from Lemma 16.6 and Theorem 3.5 applied to $\Lambda, \Lambda_{(k)}, \Lambda_r$. In the case of Λ, Theorem 3.5 does not apply so we replace ΛL by Λ_2, by positivity, and then apply it. □

LEMMA 16.8. *Suppose $\mathcal{A} = (a_n)$ is not very sparse, precisely,*

$$\int_1^x A(t)\frac{dt}{t} \ll A(x)\log x \ . \tag{16.13}$$

For any $N \geqslant 1$ we have

$$T_0(x)\big(\log x\big)^N = T_N(x) + O\left(A(x)\big(\log x\big)^{N-1}\right) .$$

PROOF. By partial summation

$$T_0(x) = \frac{T_N(x)}{(\log x)^N} + \int_2^x \frac{T_N(t)}{(\log t)^{N+1}}\frac{dt}{t} \ .$$

The result follows from (16.13) and the bound $T_N(t) \ll A(t)(\log t)^{N-1}$. □

To complete the proof of Theorem 16.1 notice that in Lemma 16.6 we have

$$\sum_{n\leqslant x} a_n G^*(n)\big(\log n\big)^N \sim \big(\log x\big)^N \sum_{n\leqslant x} a_n G^*(n) \ , \tag{16.14}$$

$$\sum_{n\leqslant x} a_n\Lambda(n)\big(\log n\big)^{N-1} \sim \delta_x \frac{HA(x)}{\log x}\big(\log x\big)^N \ , \tag{16.15}$$

$$\sum_{n\leqslant x} a_n\Lambda_{(k)}(n) \sim \frac{(k)!}{(|k|-1)!}\frac{HA(x)}{\log x}\big(\log x\big)^N \tag{16.16}$$

for each $(k) \neq (1,\dots,1)$, with $|k| = N$. For the error term $\varepsilon\Lambda_r L^{N-r}$ in Lemma 16.6 use any crude bound (Brun–Titchmarsh type estimate) or (16.15) with $\delta \leqslant 2$ if $r = 1$ and (16.16) if $r \geqslant 2$. This yields (divide by $\big(\log x\big)^N$ and let ε go to 0):

$$\sum_{n\leqslant x} a_n G^*(n) = \Bigg(c\delta_x + \sum_{\substack{|k|=N\\ k\neq(1,\dots,1)}} c_{(k)}\frac{(k)!}{(|k|-1)!} + o(1)\Bigg) H\frac{A(x)}{\log x} \ , \tag{16.17}$$

as $x \to \infty$. Now it remains to compute the constants c and

$$d = \sum c_{(k)}\frac{(k)!}{(|k|-1)!} \ .$$

Since they are independent of the sequence a_n we apply (16.17) twice, for $a_n = 1$ and $a_n = 1 + \lambda(n)$. The first choice yields by the Prime Number Theorem

$$\sum_{n\leqslant x} G^*(n) \sim \sum_{\substack{p_1<\dots<p_r\\ p_1\cdots p_r\leqslant x}} G\left(\frac{\log p_1}{\log x},\dots,\frac{\log p_r}{\log x}\right) \sim \left(\int_{T_r} G^* d\mu_r\right)\frac{x}{\log x}$$

and $\delta_x = 1$, so

$$c + d = \int_{T_r} G^* d\mu_r \ .$$

The second choice yields, again by the Prime Number Theorem, $\delta_x = 0$, so

$$d = (1 + (-1)^r)\int_{T_r} G^* d\mu_r \quad \text{and} \quad c = -(-1)^r\int_{T_r} G^* d\mu_r \ .$$

Finally, inserting these constants to (16.17), we complete the proof of Theorem 16.1.

16.4. The Parity Phenomenon

Theorem 16.1 treats integers having precisely r prime factors. We are really interested in almost-primes of order r, that is, the integers in the set

$$\mathscr{P}(r) = \bigcup_{1\leqslant s\leqslant r} \mathscr{P}_s \,.$$

Now, for given functions $G_1, \dots, G_r$ on $T_1, \dots, T_r$ respectively, such that each of $(u_1 \dots u_s)^{-1} G_s(u_1, \dots, u_s)$ is continuous on the closure of T_s, we let

$$G^+ = \sum_{s=1}^{r} \int_{T_s} G_s \, d\mu_s \,,$$

$$G^- = \sum_{s=1}^{r} (-1)^s \int_{T_s} G_s \, d\mu_s \,.$$

Now let $G^*(n)$ be the function on $\mathscr{P}(r)$ which is induced by the functions G_s^* on $\mathscr{P}_s$, that is,

$$G^*(n) = G_s\Big(\frac{\log p_1}{\log n}, \dots, \frac{\log p_s}{\log n}\Big) \,,$$

if $n = p_1 \dots p_s$. For convenience we set $G^*(n) = 0$ for $n \notin \mathscr{P}(r)$. From Theorem 16.1 we obtain

THEOREM 16.9. *Let $\mathcal{A} = (a_n)$ satisfy the conditions of Theorem 16.1 for each s, $1 \leqslant s \leqslant r$. Then*

$$\sum_{n\leqslant x} a_n G^*(n) = \big\{G^+ + (1-\delta_x) G^- + o(1)\big\} \frac{HA(x)}{\log x} \,. \tag{16.18}$$

Using $0 \leqslant \delta_x \leqslant 2$ we get immediately lower and upper bounds:

COROLLARY 16.10. *Assume $\mathcal{A} = (a_n)$ satisfies the conditions of Theorem 16.1. Then we have*

$$\sum_{n\leqslant x} a_n G^*(n) \geqslant \big(G^+ - |G^-| + o(1)\big) \frac{HA(x)}{\log x} \,, \tag{16.19}$$

$$\sum_{n\leqslant x} a_n G^*(n) \leqslant \big(G^+ + |G^-| + o(1)\big) \frac{HA(x)}{\log x} \,. \tag{16.20}$$

Since δ_x can be anything between 0 and 2 these inequalities are sharp. Hence the only chance to get the asymptotic formula using exclusively the sieve axioms is for those weight functions $G^*(n)$ having $G^- = 0$.

One underlying goal of the sieve is to show that $\sum a_n G^*(n)$ is not zero, hence the existence of a_n over $\mathscr{P}(r)$. The requirement that $\sum a_n G^*(n)$ be non-zero (or, equivalently, be positive) comes down, in practical terms, to giving a lower bound of the right order of magnitude. From Corollary 16.10 we can accomplish this only if $G^+ > |G^-|$. This means that G_s cannot vanish identically for all s of a given parity. Translating this observation back to our sequence $\mathcal{A} = a_n$, we conclude the following:

PARITY BARRICADE. *For a general sequence $\mathcal{A} = (a_n)$ one can establish using only the sieve axioms (the basic ones given here in terms of the congruence sums) the existence neither of integers with an even number of prime factors nor of integers with an odd number of prime factors.*

For example, one cannot be sure that a_n lives on integers with 13 prime factors but can be sure that it lives on integers having either two or three prime factors. This parity obstruction is the fundamental obstacle in reaching primes. By Theorem 16.9, if we could show the sequence lives on integers with 2007 prime factors, then it lives on primes as well.

The results of this section represent a vast generalization of earlier ideas concerning the limitations of the linear sieve. The original hope that the sieve might produce primes persisted a long time and this hope accidentally stimulated great effort. The inability to produce a positive lower bound seemed in stark contrast to the early success of the upper bound which succeeded in numerous situations to give results of the apparently correct order of magnitude. It was only after many years of experimental work with sieve weights that the upper and lower bounds simultaneously approached critical levels whereby further improvements in either would yield striking results (too much so to accept without suspicion). At this point, in the most favourable situation, the upper bound became twice the expected value and the lower bound just missed positivity. To illustrate the excitement which sieve theory provoked in the 1970's let us point out that improving the factor two for primes in progressions would lead to elimination of the possibility of exceptional characters, a boon for arithmeticians interested in the class numbers of imaginary quadratic fields.

The duality between the upper and lower bounds which had now become clear was reinforced by the examples (12.29) of Selberg showing that each of these bounds is actually attained by special sequences satisfying the sieve axioms. Disappointing though this could have been, it didn't stop the quest for improvements in sieve results, rather refocusing these in more promising directions.

The work of Bombieri which we have presented here reveals how widely this parity barricade spreads over the multiplicative structure of the integers. Understanding this universality has helped to find ways to break it in certain situations. Certainly one needs to find and exploit additional properties of the particular problem. The lesson of this experience is that, no matter how this is done, information coming from the congruence sums $A_d(x)$ is not enough to finish the job. On the other hand, as we have said, it is enough to take you from 2007 prime factors down to one, skipping every second stop along the way.

16.5. The Dichotomy in Action

In this section we are going to show how one can manipulate different instances of Theorem 16.9 to break the parity barrier for specific sequences. For this purpose we have chosen the Goldbach equation

$$p + q = m \ . \tag{16.21}$$

Our goal is to give the number of solutions in primes p, q for almost all even numbers m. Of course, this can be treated by the circle method á la Vinogradov, but we intend to take a different path using the tools of this chapter to shed further light on the parity phenomenon.

Generalizing slightly, we see that this goal can be accomplished by the asymptotic evaluation of the sum

$$W(x) = \sum_{p+q \leqslant x} w(p+q) \log p \tag{16.22}$$

with quite general weights $w(m)$, supported on positive even integers and satisfying

$$0 \leqslant w(m) \leqslant 1. \tag{16.23}$$

By the heuristic arguments in Section 3.1 we expect that

$$\sum_{p+q=m} \log p \sim H^* c(m) \frac{m}{\log m} \tag{16.24}$$

where

$$H^* = 2 \prod_{p>2} \Big(1 - \frac{1}{(p-1)^2}\Big) \tag{16.25}$$

is the usual twin prime constant, and

$$c(m) = \prod_{\substack{p \mid m \\ p>2}} \frac{p-1}{p-2} . \tag{16.26}$$

Hence, we expect the following asymptotic formula holds true:

$$W(x) \sim H^* X^* (\log x)^{-1} \tag{16.27}$$

with

$$X^* = \sum_{m \leqslant x} w(m)\, m\, c(m) . \tag{16.28}$$

We shall succeed if the $w(m)$ are not too sparsely supported. We assume that

$$\sum_{m \leqslant x} w(m) \gg x(\log x)^{-a} \tag{16.29}$$

for some constant $a \geqslant 3$, so that

$$X = \sum_{m \leqslant x} w(m)m \gg x^2 (\log x)^{-2a} . \tag{16.30}$$

We begin by attempting to apply Theorem 16.9 to the sequence $\mathcal{A} = (a_n)$ with

$$a_n = \sum_{p+n \leqslant x} w(p+n) \log p . \tag{16.31}$$

For any odd integer d we have

$$\begin{aligned} A_d(x) = \sum_{\substack{n \leqslant x \\ n \equiv 0 \,(\mathrm{mod}\ d)}} a_n &= \sum_{m \leqslant x} w(m) \sum_{\substack{p<m \\ p \equiv m \,(\mathrm{mod}\ d)}} \log p \\ &= \frac{1}{\varphi(d)} \sum_{\substack{m \leqslant x \\ (m,d)=1}} w(m)m + r_d(\mathcal{A}, x) \end{aligned}$$

say, where by Theorem 22.6,

$$\sum_{\substack{d \leqslant D \\ d\,\mathrm{odd}}} |r_d(\mathcal{A}, x)| \ll x^2 (\log x)^{-A} , \tag{16.32}$$

with

$$D = D(x) = x(\log x)^{-2A} \tag{16.33}$$

for any $A \geqslant 3$, the implied constant depending only on A. Note that, by the Prime Number Theorem,

$$A(x) = A_1(x) = X\{1 + O((\log x)^{-A})\}$$

so, in what follows, we can exchange $A(x)$ with X at will.

To simplify matters we assume that, for any d,

$$\sum_{\substack{m \leqslant x \\ (m,d)=1}} w(m)m = \eta(d)X + O(X(\log x)^{-A}) \ , \tag{16.34}$$

where $\eta(d)$ is a multiplicative function with

$$0 \leqslant \eta(p) \leqslant 1, \qquad \eta(p) = 1 + O(1/p) \ . \tag{16.35}$$

This assumption is really quite mild and it could have been avoided if we had at our disposal sieve axioms slightly more general than those we have actually employed. Because $w(m)$ is supported on even integers we have $\eta(2) = 0$.

By the above arguments we conclude that

$$A_d(x) = g(d)A(x) + r_d(x) \ ,$$

where

$$g(d) = \eta(d)/\varphi(d) \tag{16.36}$$

and

$$\sum_{d \leqslant D} |r_d(x)| \ll A(x)(\log x)^{-A} \ . \tag{16.37}$$

This shows that our sequence $\mathcal{A} = (a_n)$ has level of distribution $D = D(x)$ given by (16.33). Therefore, Theorem 16.9 is applicable but the result is not yet satisfactory because $\delta_x(\mathcal{A})$ has not yet been determined.

We want to show that the sequence has the expected density of primes, that is

$$\delta_x(\mathcal{A}) = 1 \ . \tag{16.38}$$

This particular value translates into the statement that the sequence $\mathcal{A}$ obeys the dichotomy which we mentioned in Section 3.2, there in the context of the sequence of all natural numbers. According to Theorem 16.9 we can determine $\delta_x(\mathcal{A})$ by testing $\mathcal{A} = (a_n)$ against any convenient weight factors $G^*(n)$ of fixed parity. We choose $G^*(n)$ induced by $G(u,v) \geqslant 0$ which is a smooth function compactly supported on $0 < u < v < 1$ with

$$G^+ = \int G(u, 1-u)u^{-1}du > 0 \ . \tag{16.39}$$

Therefore, we need to show that

$$\sum\sum\sum_{p+qr \leqslant x} G\Big(\frac{\log q}{\log qr}, \frac{\log r}{\log qr}\Big)\, w(p+qr)\log p \sim G^+ \frac{HX}{\log x} \ , \tag{16.40}$$

where X is given by (16.30) and

$$H = \prod_p \Big(1 - g(p)\Big)\Big(1 - \frac{1}{p}\Big)^{-1} = \prod_p \Big(1 - \frac{\eta(p)}{p-1}\Big)\Big(1 - \frac{1}{p}\Big)^{-1} . \tag{16.41}$$

Using (16.34) it is easy to show that

$$HX \sim H^* X^* \ .$$

Clearly, we can replace here $\log qr$ by $\log x$ and $\log p$ by the von Mangoldt function without altering the asymptotic value so (16.40) is equivalent to

$$\sum\sum\sum_{\ell+qr\leqslant x} G\Big(\frac{\log q}{\log x}, \frac{\log r}{\log x}\Big)\, w(\ell+qr)\Lambda(\ell) \sim G^+ \frac{HX}{\log x}\,. \tag{16.42}$$

What we have accomplished here is that the original prime variable $n = q$ is replaced by the product of two primes $n = qr$ and we can choose the size of q at will. As it happens we shall choose q to be relatively small.

Let $\mathcal{B} = (b_\ell)$ be the sequence

$$b_\ell = \sum\sum_{\ell+qr\leqslant x} G\Big(\frac{\log q}{\log x}, \frac{\log r}{\log x}\Big)\, w(\ell+qr) \tag{16.43}$$

so that (16.42) can be stated as

$$\sum_{\ell\leqslant x} b_\ell \Lambda(\ell) \sim G^+ HX(\log x)^{-1}. \tag{16.44}$$

For the proof of (16.44) we could appeal a second time to Theorem 16.9, but now for the sequence $\mathcal{B}$ rather than $\mathcal{A}$. Again by Theorem 22.6 the sequence $\mathcal{B}$ has level of distribution given by (16.33) so the expected asymptotic (16.44) amounts to showing that

$$\delta_x(\mathcal{B}) = 1\,. \tag{16.45}$$

What one would accomplish by this argument is that the original prime variable $\ell = p$ can be replaced by the product of two primes $\ell = ps$ with p of size to be chosen at will. Having applied two instances of this test we transform the original equation (16.21) in primes p, q into the equation

$$ps + qr = m \tag{16.46}$$

in primes p, s, q, r. The fact that we can arbitrarily fix the sizes of p, q opens many possibilities for solving the equation (16.46); in particular, it fits well with the dispersion method. In the next chapter we shall develop various combinatorial identities for sums over primes which play a similar role, but those arguments are not as flexible as the dichotomy testing employed here. On the other hand, the combinatorial identities offer the possibility of success even when the level of distribution is not as favourable.

Rather than determining the value of the density (16.45) for $\mathcal{B}$ along the lines of (16.46), we are going to prove (16.44) more quickly by expressing $\Lambda(\ell)$ in terms of the Möbius function:

$$\Lambda(\ell) = -\sum_{d\mid\ell} \mu(d)\log d\,. \tag{16.47}$$

Let $\Lambda^\sharp(\ell)$ denote the partial convolution of (16.47) restricted to $d \leqslant D$ and $\Lambda^\infty(\ell)$ the remaining part. Accordingly, we decompose $S(x) = S^\sharp(x) + S^\infty(x)$, where

$$S^\sharp(x) = \sum_{\ell\leqslant x} b_\ell \Lambda^\sharp(\ell) = -\sum_{d\leqslant D} \mu(d)(\log d)B_d(x)$$

with

$$B_d(x) = \sum_{\substack{\ell \leqslant x \\ \ell \equiv 0 \,(\mathrm{mod}\, d)}} b_\ell = \sum_{m \leqslant x} w(m) \sum_{\substack{qr<m \\ qr \equiv -m \,(\mathrm{mod}\, d)}} \sum G\Big(\frac{\log q}{\log x}, \frac{\log r}{\log x}\Big)$$

$$= \frac{1}{\varphi(d)} \sum_{\substack{qr<m\leqslant x \\ (m,d)=1}} \sum\sum w(m)\, G\Big(\frac{\log q}{\log x}, \frac{\log r}{\log x}\Big) + r_d(\mathcal{B}, x)\ ,$$

say, where the remainder terms $r_d(\mathcal{B}, x)$ satisfy (16.32) by Theorem 22.6. Hence, the leading part $S^\sharp(x)$ is equal to

$$-\sum_{qr<x}\sum G\Big(\frac{\log q}{\log x}, \frac{\log r}{\log x}\Big) \sum_{qr<m\leqslant x} w(m) \sum_{\substack{d\leqslant D \\ (d,m)=1}} \frac{\mu(d)}{\varphi(d)} \log d + O\left(x^2 (\log x)^{-A}\right)\ .$$

The innermost sum over the moduli $d \leqslant D$ is equal to $H^* c(m) + O((\log D)^{-1})$ by Corollary A.12. Hence, we arrive at

$$S^\sharp(x) \sim H^* \sum_{qr<m\leqslant x}\sum\sum w(m)\, c(m)\, G\Big(\frac{\log q}{\log x}, \frac{\log r}{\log x}\Big)\ .$$

Next, we can reverse an earlier step by replacing $\log x$ by $\log qr$ without altering the asymptotic. Then we have, by the Prime Number Theorem,

$$\sum_{qr<m}\sum G\Big(\frac{\log q}{\log qr}, \frac{\log r}{\log qr}\Big) \sim G^+ \frac{m}{\log m}$$

where G^+ is the constant given by the integral (16.39). Finally, we obtain

$$S^\sharp(x) \sim G^+ H^* X^* (\log x)^{-1} \sim G^+ H X (\log x)^{-1} \tag{16.48}$$

which is the same asymptotic value (16.44) as is expected for the full sum $S(x)$.

To finish the job it remains to show that $S^\infty(x) = o(X/\log x)$. In fact, we are going to show the stronger bound

$$S^\infty(x) \ll X(\log x)^{-A}\ . \tag{16.49}$$

To simplify the arguments we are going to further restrict (as we may) our test function $G(u,v)$ to have separated variables

$$G(u,v) = f(u)h(v)$$

say, where $f(u)$ is supported on $\frac{1}{9} < u < \frac{1}{8}$. Then we estimate as follows. We have

$$S^\infty(x) = -\sum_{cd+qr=m\leqslant x}\sum\sum\sum\sum \mu(d)(\log d) f\Big(\frac{\log q}{\log x}\Big) h\Big(\frac{\log r}{\log x}\Big) w(m)$$

$$\ll (\log x) \sum_{c\leqslant C} \sum_{m\leqslant x} \sum_{r\leqslant R} \Big| \sum_{\substack{cd+qr=m \\ D<d\leqslant x/c \\ Q<q\leqslant 2Q}} \sum \mu(d)(\log d) f\Big(\frac{\log q}{\log x}\Big) \Big|$$

where $C = xD^{-1}$, $R = xQ^{-1}$ and $x^{1/9} < Q < x^{1/8}$, after splitting the summation over q into dyadic segments. By Cauchy's inequality,

$$|S^\infty(x)|^2 \ll CRx(\log x)^2 \sum_{c\leqslant C} \sum_{r\leqslant R} \sum_{m} \Big|\sum_{d}\sum_{q}\Big|^2\ .$$

Note that here we dropped the restriction $m \leqslant x$ in the outer summation. This gives

$$|S^\infty(x)|^2 \ll CRx(\log x)^2 \sum_{c\leqslant C}\sum_{r\leqslant R}\sum\sum\sum\sum_{c(d_1-d_2)=(q_1-q_2)r} \beta_{d_1}\beta_{d_2}\gamma_{q_1}\gamma_{q_2}$$

where $\beta_d = \mu(d)\log d$ restricted by $D < d \leqslant x/c$ and $\gamma_q = f(\log q/\log x)$ restricted by $Q < q \leqslant 2Q$. Because r is free of coefficients we interpret the equation in the range of summation as a congruence, obtaining

$$|S^\infty(x)|^2 \ll CQRx(\log x)^2 \sum_{c}\sum_{1\leqslant q\leqslant Q}\Big|\sum\sum_{\substack{c(d_1-d_2)\equiv 0 \,(\mathrm{mod}\ q)\\ |d_1-d_2|\leqslant qR}} \beta_{d_1}\beta_{d_2}\Big| + CRx^3(\log x)^2$$

where the last term comes from the contribution of the diagonal terms $d_1 = d_2$, $q = q_1 - q_2 = 0$.

To estimate this we use the following result.

PROPOSITION 16.11. *Let $A \geqslant 1$ and $x \geqslant 2$. We have*

$$\sum_{q\leqslant Q}\max_{y\leqslant x}\Big|\sum\sum_{\substack{m_1,m_2\leqslant x\\ |m_1-m_2|\leqslant y\\ m_1\equiv m_2\,(\mathrm{mod}\ q)}} \mu(m_1)\mu(m_2)\Big| \ll x^2(\log x)^{-A}\ ,$$

with $Q = x(\log x)^{-B}$ where B and the implied constant depend only on A.

PROOF. Proposition 16.11 is a version of the Barban–Davenport–Halberstam Theorem 9.14 for the Möbius function. The proof of this modified version goes along the lines to be presented in the proof of Proposition 20.9. □

By Proposition 16.11 (with A replaced by any A') we obtain

$$|S^\infty(x)|^2 \ll CQRx^3(\log x)^{-A'} = x^4(\log x)^{2A-A'}\ .$$

This proves (16.49). Adding (16.49) to (16.48) we derive (16.44). This proves (16.45) and (16.38). The latter implies, by Theorem 16.9, that (16.40) holds for any test function G. In particular, we conclude:

THEOREM 16.12. *Let $0 \leqslant w(m) \leqslant 1$ be such that* (16.29) *and* (16.34) *hold. Then*

$$\sum_{p+q\leqslant x} w(p+q)\log p \sim H^*X^*(\log x)^{-1}\ , \tag{16.50}$$

where H^* and X^* are given by (16.25) *and* (16.28).

REMARK. As we have already mentioned, the condition (16.34) for the weights $w(m)$ is a mild one in practice and could be avoided entirely. To this end one needs a slight generalization of the sieve theory wherein the congruence sums $A_d(x)$ are assumed to satisfy (5.15) in place of (5.4). We propose for enthusiastic readers to develop such a theory as a learning exercise.

CHAPTER 17

Combinatorial Identities

Interlude

17.1. Introduction

Following very little development up until the twentieth century the sieve of Eratosthenes has during the past hundred years grown substantially and in two somewhat distinct directions, beginning in the one case with the work of Brun and in the other with the work of Vinogradov. Although it is the direction initiated by Brun to which the words "sieve methods" are usually applied and which constitute the main theme of this book, a very brief discussion of the latter area is certainly in order. Indeed, as the subject has developed in more recent years one sees that these two streams are reapproaching one another.

The methods of Vinogradov and his successors for the estimation of sums over primes begin with a decomposition of the von Mangoldt function $\Lambda(n)$ (or sometimes instead the Möbius function $\mu(n)$) judiciously as a sum of a limited number of other sums. Let $f(n)$ be one of those arithmetic functions for which we intend to study the sum

$$S = \sum_{n \leqslant x} a_n f(n)$$

with coefficients in our destination sequence $\mathcal{A} = (a_n)$. Suppose that $f(n) = f_1(n) + \cdots + f_J(n)$, where the $f_j(n)$ are some arithmetic functions which one can get a grasp on by various means. Hence, our sum now inherits a decomposition

$$S = \sum_{j=1}^{J} S_j, \qquad S_j = \sum_{n \leqslant x} a_n f_j(n).$$

The hope is that, by a clever choice of this decomposition, the resulting sums are more easily dealt with because then one can use different techniques on the different constituents, no one of which applies to all. Although a very simple idea, it leads to deep consequences.

We intend to study the same sum over primes

$$S = \sum_{n \leqslant x} a_n \Lambda(n)$$

as in Bombieri's sieve. Using a combination of the elementary identities

$$\sum_{d|n} \Lambda(d) = \log n, \qquad \sum_{d|n} \mu(d) = 0 \text{ if } n > 1,$$

one rearranges each of the sums S_j into one (or possibly both) of the following shapes:

Type I – *linear forms*

$$\sum_{d\leqslant D} \lambda_d a_{\ell d}, \qquad |\lambda_d| \leqslant 1,$$

Type II – *bilinear forms*

$$\sum_{m\leqslant M} \sum_{n\leqslant N} \alpha_m \, \beta_n \, a_{mn}, \qquad |\alpha_m| \leqslant 1, \ |\beta_n| \leqslant 1.$$

The structure of these forms, truly rooted in the concept of inclusion-exclusion, allows us to remove from consideration all of the terms having complicated, practically unknown, coefficients.

In practice, the main term in our sum S comes from sums over ℓ of one or more of the linear forms. Such a sum of linear forms is itself a bilinear form which is special in that one of the two sequences of coefficients is identically one. We must evaluate this asymptotically and show that the other subsums are small. The special linear forms are ready for application of harmonic analysis. On the other hand, the bounds for the bilinear forms do not depend on the nature of the coefficients α_m, β_n, just the fact that they are bounded. One can treat the bilinear forms by various methods from operator theory (our favourite one being Cauchy's inequality) which smooth out the unknown coefficients and eventually open the way for another application of harmonic analysis.

Note the similarity of all of this to the more recent sieve results described earlier in this book. As compared to those sieve results the current method is more of a gamble. When it works it gives (usually) the asymptotics. Hence, one expects it to be less likely to work.

17.2. Sample Combinatorial Identities

We illustrate with just one of the many identities of this type. This particular one was discovered and applied originally by Linnik (see for example [**126**]) and has since been used successfully by Bombieri, Friedlander and Iwaniec [**11**], Heath-Brown [**84**] and others.

PROPOSITION 17.1. *For each integer $n > 1$,*

$$\frac{\Lambda(n)}{\log n} = \sum_{j=1}^{\infty} \frac{(-1)^{j-1}}{j} \, t_j(n) \tag{17.1}$$

where $t_j(n)$ denotes the number of ways of writing n as the product of j integers each being strictly greater than one and with the order of the factors being distinguished (that is different orders all being counted).

The arithmetic functions t_j remind one of the ordinary divisor functions τ_j, the only difference being that in our case we insist that none of the factors be equal to one. Hence it is easy to see that τ_k can be expressed in terms of the functions t_j and that the reverse is also true, namely

$$t_j(n) = \sum_{k=0}^{j} (-1)^{j-k} \binom{j}{k} \, \tau_k(n).$$

Thus, Linnik's identity reduces the study of $\Lambda(n)$ over a given sequence to the study over the same sequence of the various divisor functions $\tau_k(n)$. In the case where

k is large (so there are many factors) it is possible in practice to arrange bilinear forms of the above type with great flexibility in the choice of M and N. It is then the smallest few values of k which present the limits to the quality of the results. But, still, the estimation of the bilinear forms very often constitutes the bulk of the work.

The proof of Linnik's identity is quite simple. We let $\zeta = \zeta(s)$ denote the Riemann zeta-function. We then have for each $j \geqslant 1$,

$$(\zeta(s) - 1)^j = \sum_{n\geqslant 1} t_j(n) n^{-s},$$

so we deduce the result by comparing coefficients in the identity

$$\log \zeta = \log(1 + (\zeta - 1)) = \sum_{j=1}^{\infty} \frac{(-1)^{j-1}}{j} (\zeta - 1)^j,$$

where on the left-hand side we use the Euler product formula for ζ and on the right side we use the Dirichlet series.

Similarly, one can show that, for each integer $n \geqslant 1$,

$$\mu(n) = \sum_{j=0}^{\infty} (-1)^j \, t_j(n) \, . \tag{17.2}$$

In 1981 D. R. Heath-Brown [**84**] developed a formula for $\Lambda(n)$ which is reminiscent to that of Linnik. Let

$$M(s) = \sum_{m \leqslant z} \mu(m) m^{-s}.$$

Then, the Dirichlet series for $\zeta(s)M(s)$ has coefficients $a_1 = 1$ and $a_n = 0$ for $1 < n \leqslant z$. Consider the identity

$$\frac{\zeta'}{\zeta}(s)(1 - \zeta(s)M(s))^K = \frac{\zeta'}{\zeta}(s) + \sum_{1\leqslant k\leqslant K} (-1)^k \binom{K}{k} \zeta'(s)\zeta^{k-1}(s)M^k(s).$$

Comparing the Dirichlet coefficients on both sides we obtain (the left-hand side has vanishing coefficients for $n \leqslant 2z^K$):

PROPOSITION 17.2. *Let $K \geqslant 1, z \geqslant 1$. Then for any $n \leqslant 2z^K$ we have*

$$\Lambda(n) = - \sum_{1\leqslant k\leqslant K} (-1)^k \binom{K}{k} \sum_{\substack{m_1\ldots m_k n_1 \ldots n_k = n \\ m_1,\ldots,m_k \leqslant z}} \cdots \sum \mu(m_1) \ldots \mu(m_k) \log n_k. \tag{17.3}$$

Each of the identities (17.1) of Linnik and (17.3) of Heath-Brown has a few advantages and disadvantages compared to the other. For instance, the expansion (17.1) has non-negative arithmetic functions $t_j(n)$, a property which can be utilized in conjunction with the sieve, whereas the expansion (17.3) has a number of terms K which can be chosen at will. These and other features help determine the choice of identity for specific applications.

Similarly to Proposition 17.2 one can show that for $n \leqslant z^K$,

$$\mu(n) = - \sum_{1\leqslant k\leqslant K} (-1)^k \binom{K}{k} \sum_{\substack{m_1\ldots m_k n_1 \ldots n_k = n \\ m_1,\ldots,m_k \leqslant z}} \cdots \sum \mu(m_1) \ldots \mu(m_k). \tag{17.4}$$

Nowadays, the most frequently used formula for $\Lambda(n)$ is that which is due to R. C. Vaughan [**157**]. We derive it from

$$\Lambda(n)=\sum_{b|n}\mu(b)\log\frac{n}{b}.$$

Here we keep the terms with $b\leqslant y$ and transform the remaining sum as follows:

$$\sum_{\substack{b|n\\ b>y}}\mu(b)\log\frac{n}{b}=\sum\sum_{\substack{bc|n\\ b>y}}\mu(b)\Lambda(c).$$

Next, we keep the terms with $c>z$ and transform the remaining sum as follows:

$$\sum\sum_{\substack{bc|n\\ b>y,c\leqslant z}}\mu(b)\Lambda(c)=\sum\sum_{\substack{bc|n\\ c\leqslant z}}\mu(b)\Lambda(c)-\sum\sum_{\substack{bc|n\\ b\leqslant y,c\leqslant z}}\mu(b)\Lambda(c).$$

Here the complete sum over all b dividing n/c vanishes unless $c=n$, which is not possible if $n>z$. Adding up the above expressions one obtains

PROPOSITION 17.3. *Let $y,z\geqslant 1$. Then for any $n>z$ we have*

$$\Lambda(n)=\sum_{\substack{b|n\\ b\leqslant y}}\mu(b)\log\frac{n}{b}-\sum\sum_{\substack{bc|n\\ b\leqslant y,c\leqslant z}}\mu(b)\Lambda(c)+\sum\sum_{\substack{bc|n\\ b>y,c>z}}\mu(b)\Lambda(c). \tag{17.5}$$

For some applications the precise coefficients in (17.5) are not important, only their size and range. From (17.5) we derive the following result.

COROLLARY 17.4. *Let $y>1$. For $n>y^3$ we have*

$$\Lambda(n)=\Big(\sum_{\substack{b|n\\ b\leqslant y}}\mu(b)\Big)\log n-\sum\sum_{\substack{bc=n\\ b,c>y}}\big(\beta(c)-\mu(b)\gamma(c)\big)\ , \tag{17.6}$$

where the coefficients are given by

$$\beta(c)=\sum_{\substack{ad=c\\ a,\,d\leqslant y}}\mu(a)\Lambda(d),\quad \gamma(c)=\sum_{\substack{ad=c\\ d>y}}\Lambda(d)\ ,$$

so $|\beta(c)|+\gamma(c)\leqslant\log c$.

PROOF. Note that, choosing $y=z$, we can rewrite (17.5) in the form

$$\Lambda(n)=\sum_{\substack{bc=n\\ b\leqslant y}}\mu(b)\log c-\sum_{\substack{bc=n\\ b\leqslant y}}\Big(\sum_{ad=b}\mu(a)\Lambda(d)\Big)$$
$$-\sum_{\substack{bc=n\\ b,c>y}}\Big(\sum_{\substack{ad=b\\ a,d\leqslant y}}\mu(a)\Lambda(d)\Big)+\sum_{\substack{bc=n\\ b,c>y}}\mu(b)\Big(\sum_{\substack{ad=c\\ d>y}}\Lambda(d)\Big)\ .$$

Here, in the third sum, we have introduced the redundant condition $c>y$. Indeed, we have $c=n/b=n/ad\geqslant n/y^2>y$. Using the formula $\mu*\Lambda=-\mu\log$ we find that the second sum adds nicely to the first one giving the first sum in (17.6). This completes the proof. □

We next derive a formula similar to (17.5) for the Möbius function.

PROPOSITION 17.5. *Let $y, z \geqslant 1$. Then, for any $m > \max(y,z)$ we have*

$$\mu(m) = -\sum_{\substack{bc|m\\ b\leqslant y, c\leqslant z}}\sum \mu(b)\mu(c) + \sum_{\substack{bc|m\\ b>y, c>z}}\sum \mu(b)\mu(c). \tag{17.7}$$

PROOF. We start from

$$\mu(m) = \sum_{bc|m}\sum \mu(b)\mu(c)$$

and split the summation into four ranges according to $b \leqslant y, b > y, c \leqslant z, c > z$. Then we transform the two middle sums into

$$\sum_{\substack{bc|m\\ b\leqslant y, c>z}}\sum \mu(b)\mu(c) = -\sum_{\substack{bc|m\\ b\leqslant y, c\leqslant z}}\sum \mu(b)\mu(c),$$

$$\sum_{\substack{bc|m\\ b>y, c\leqslant z}}\sum \mu(b)\mu(c) = -\sum_{\substack{bc|m\\ b\leqslant y, c\leqslant z}}\sum \mu(b)\mu(c).$$

Adding up the above expressions one obtains (17.7). □

Vaughan's identity (17.5) is not as flexible for arranging bilinear forms as are those of Linnik and Heath-Brown, but it has proved to be sufficient for many basic applications and it is the simplest of all.

17.3. An Identity for Composite Numbers

O. Ramaré showed us a cute formula of his which singles out the composite numbers. Put

$$\gamma_m(x) = \Big(1 + \sum_{p|m,\, p\leqslant\sqrt{x}} 1\Big)^{-1}. \tag{17.8}$$

Then, obviously for n squarefree with $\sqrt{x} < n \leqslant x$ we find

$$\sum_{\substack{pm=n\\ p\leqslant\sqrt{x}}} \gamma_m(x) = \begin{cases} 1 & \text{if } n \text{ is composite,}\\ 0 & \text{if } n \text{ is prime.}\end{cases}$$

Hence, for any sequence $\mathcal{A} = (a_n)$ supported on squarefree numbers we obtain

$$\sum_{\sqrt{x}<n\leqslant x} a_n = \sum_{\sqrt{x}<p\leqslant x} a_p + \sum_{\substack{\sqrt{x}<pm\leqslant x\\ p\leqslant\sqrt{x}}}\sum a_{pm}\gamma_m(x)\ . \tag{17.9}$$

If we restrict the sum on the left by $(n, P(z)) = 1$ we no longer require $\mathcal{A}$ to be supported on squarefree numbers and we can introduce some minor but convenient summation restrictions, getting:

PROPOSITION 17.6. *For any sequence of complex numbers $\mathcal{A} = (a_n)$ and $2 \leqslant z \leqslant \sqrt{x}$ we have*

$$S(\mathcal{A}, z) = \sum_{p\leqslant x} a_p + \sum_{\substack{pm\leqslant x\\ z\leqslant p\leqslant\sqrt{x}\\ (m,P(z))=1}} a_{pm}\gamma_m(x) + E(\sqrt{x}) + E(x,z)\ , \tag{17.10}$$

where

$$|E(y)| \leqslant \sum_{n \leqslant y} |a_n|, \qquad |E(x,z)| \leqslant \sum\sum_{\substack{mp^2 \leqslant x \\ p \geqslant z}} |a_{mp^2}| \ .$$

EXAMPLE 17.1. We apply the proposition to an arithmetic progression $a \pmod q$ with $(a,q)=1$, $q \leqslant \sqrt{x}$ (where we have used it already in Section 9.8 in the proof of the Bombieri–Vinogradov theorem). Suppose $z \leqslant x^{1/6}$ and $q \leqslant x^{1/2}$. In this range we have

$$|E(\sqrt{x})| + |E(x,z)| \ll \frac{x}{qz} \ . \tag{17.11}$$

Indeed, splitting the sum $E(x,z)$ according to $m \leqslant M$ and $m > M$ we can majorize the left-hand side of (17.11) as follows:

$$\begin{aligned} &\frac{\sqrt{x}}{q} + 1 + \sum_{m \leqslant M} \Big(\frac{2}{q}\sqrt{\frac{x}{m}} + 2\Big) + \sum_{z \leqslant p < \sqrt{x/m}} \Big(\frac{x}{qp^2} + 1\Big) \\ &\ll \frac{\sqrt{xM}}{q} + M + \frac{x}{qz} + \sqrt{\frac{x}{M}} \ll \frac{x}{qz}, \end{aligned}$$

on choosing $M = x^{1/3}$.

If we choose z relatively small then the sifting function $S(\mathcal{A}, z)$ can be evaluated asymptotically by purely sieve-theoretic methods (in fact, no positivity of the sequence $\mathcal{A}$ is required) and the formula (17.10) provides a useful expression for the sum over primes

$$S(x) = \sum_{p \leqslant x} a_p \ .$$

17.4. A Further Identity

We are interested not merely in the similarity between these methods (of bilinear forms) to the sieve methods, but also in the implementation of identities of this type within the sieve mechanism. Therefore, we require certain features which, for example, allow us to take advantage of positivity and harmlessly discard small, yet hard to handle, contributions. In this section we develop such an identity which, in the following chapter, will be incorporated into the sieve. This extra usefulness will come with a price. The identity we need is less elegant than those presented above.

For $f : \mathbb{N} \to \mathbb{C}$, an arithmetic function, we split (our notation here is not conventional, but is perhaps quite descriptive)

$$f(n \leqslant z) = \begin{cases} f(n) & \text{if } n \leqslant z, \\ 0 & \text{if } n > z, \end{cases}$$

$$f(n > z) = \begin{cases} 0 & \text{if } n \leqslant z, \\ f(n) & \text{if } n > z. \end{cases}$$

We have

$$f(n > z) = \sum_{bc|n} \mu(b) f(c > z).$$

Split the Möbius function $\mu(b) = \mu(b \leqslant y) + \mu(b > y)$, getting

$$f(n > z) = \sum_{bc|n} \mu(b \leqslant y) f(c > z) + \sum_{bc|n} \mu(b > y) f(c > z).$$

In the first sum insert $f(c > z) = f(c) - f(c \leqslant z)$ getting the identity (a generalized variant of that of Vaughan),

$$f(n > z) = \sum_{b|n} \mu(b \leqslant y) F(n/b) - \sum_{bc|n} \mu(b \leqslant y) f(c \leqslant z) + \sum_{bc|n} \mu(b > y) f(c > z),$$

where $F = f * 1$. To gain more flexibility we split the last sum, say $f(n; y, z)$, into three subsums

$$f_1(n; y, z) = \sum_{bc|n} \mu(b > sy) f(c > sz),$$

$$f_2(n; y, z) = \sum_{bc|n} \mu(sy \geqslant b > y) f(c > z),$$

$$f_3(n; y, z) = \sum_{bc|n} \mu(b > sy) f(sz \geqslant c > z).$$

Naturally, by $f(w \geqslant c > z)$ we mean the function which is equal to $f(c)$ if $w \geqslant c > z$ and vanishes otherwise. Here s can be any number > 1. We choose s to be the unique integer power of 2 in the interval

$$\frac{\Delta\sqrt{x}}{8\delta\sqrt{D}} < s \leqslant \frac{\Delta\sqrt{x}}{4\delta\sqrt{D}},$$

where δ, $\Delta \geqslant 1$ are real numbers, relatively small, whose role is to provide room to accommodate extra sieve variables (in practice, we need both δ and Δ to gain a bigger playground).

Denote

$$F(n; y) = \sum_{b|n} \mu(b \leqslant y) F(n/b),$$

$$F(n; y, z) = \sum_{bc|n} \mu(b \leqslant y) f(c \leqslant z).$$

Therefore, we have

$$f(n > z) = F(n; y) - F(n; y, z) + \sum_{j=1}^{3} f_j(n; y, z). \tag{17.12}$$

We apply the integral operator

$$I_h(Y, Z) = \int_Y^{eY} \int_Z^{eZ} h(y, z)(yz)^{-1} dy\, dz$$

to obtain an identity with smoothed parts (of course, other smoothing devices are available). Note that if the function $h = h(y, z)$ is independent of one, or both, variables the integration over such a variable does nothing. When we apply this operator to (17.12) we denote the results by changing y, z to Y, Z as appropriate. Therefore, in what follows the distinction between small and capital letters must be judiciously observed. For example, for the truncated function $f(n > z)$ this operator gives the logarithmic step-continuous function

$$f(n > Z) = \begin{cases} f(n) & \text{if } n > eZ, \\ f(n)\log \frac{n}{Z} & \text{if } Z < n \leqslant eZ, \\ 0 & \text{if } n \leqslant Z. \end{cases}$$

We introduce this integral operator over the variables y, z in order to facilitate their separation (without loss apart from absolutely bounded factors) in various forthcoming multiple sums.

We shall apply the above considerations for $f = \Lambda$ in which case $F = L$ with $L(n) = \log n$. Thus we have from (17.12),

$$\Lambda(n > Z) = L(n; Y) - L(n; Y, Z) + \sum_{j=1}^{3} \Lambda_j(n; Y, Z). \tag{17.13}$$

More explicitly, here we have

$$L(n; Y) = \sum_{b|n} \Big(\int_Y^{eY} \mu(b \leqslant y) \frac{dy}{y} \Big) \log \frac{n}{b} \ ,$$

$$L(n; Y, Z) = \sum_{bc|n} \Big(\int_Y^{eY} \mu(b \leqslant y) \frac{dy}{y} \Big) \Big(\int_Z^{eZ} \Lambda(c \leqslant z) \frac{dz}{z} \Big) \ ,$$

and similarly for $\Lambda_j(n; Y, Z)$.

We shall take $n \leqslant x$ and make the choices

$$Y = Z = \Delta^{-1} \sqrt{D} \ . \tag{17.14}$$

Note that when y changes over the segment $Y < y \leqslant eY$ then sy satisfies the inequalities $(8\delta)^{-1}\sqrt{x} < sy < \delta^{-1}\sqrt{x}$. The same holds for the variable z.

The obvious fact that, for n squarefree, the left-hand side of (17.13) is supported on primes is lost when we look instead on the right-hand side (and this is the intention of the formula). However, we would not mind, actually we desire, to retain part of this information. To this end we introduce an upper-bound sieve $\{\lambda_\nu, \ \nu \leqslant \Delta\}$, for example the upper-bound beta-sieve of level Δ, so that $|\lambda_\nu| \leqslant 1$ and

$$\theta_n = \sum_{\nu|n} \lambda_\nu \geqslant 0 \ , \quad \text{for all} \quad n \geqslant 1 \ .$$

Clearly, $\Lambda(n > Z) = \theta_n \, \Lambda(n > Z)$, for squarefree n since $Z > \Delta$. Therefore, for a sequence $\mathcal{A} = \big(a_n\big)$ of non-negative real numbers, we have

$$S(x, Z) = \sum_{n \leqslant x} a_n \Lambda(n > Z) = \sum_{n \leqslant x} a_n \theta_n \, \Lambda(n > Z) \ .$$

Hence the decomposition (17.13) yields

$$S(x, Z) = T(x; Y) - T(x; Y, Z) + \sum_{j=1}^{3} S_j(x; Y, Z) \ ,$$

where Y and Z are given by (17.14) and

$$T(x;Y) = \sum_{n\leqslant x} a_n\theta_n L(n;Y)\ ,$$

$$(17.15)\qquad T(x;Y,Z) = \sum_{n\leqslant x} a_n\theta_n L(n;Y,Z)\ ,$$

$$S_j(x;Y,Z) = \sum_{n\leqslant x} a_n\theta_n \Lambda_j(n;Y,Z), \quad j=1,2,3\ .$$

Moreover, the relevant sum over primes

$$(17.16)\qquad S(x) = \sum_{p\leqslant x} a_p \log p$$

is closely approximated by $S(x;Z)$ if we make a harmless assumption that the set is not too sparse. Specifically, we have established the following decomposition.

PROPOSITION 17.7. *Assume that the sequence $\mathcal{A}=(a_n)$ of non-negative numbers satisfies*

$$A(x) \gg A(\sqrt{x})(\log x)^2\ .$$

Then, we have the following decomposition of the sum over primes into smoothed sums:

$$(17.17)\qquad S(x) = T(x;Y) - T(x;Y,Z) + \sum_{j=1}^{3} S_j(x;Y,Z) + O\big(A(x)(\log x)^{-1}\big)$$

where Y and Z are given by (17.14) *and the multilinear forms by* (17.15).

Our identity (17.17) may appear to be more complicated than the combinatorial identities we have stated earlier in this chapter. This is because we have built into it features which are needed for a specific application we have in mind (the proof of Theorem 18.7 which in turn is needed for the problem of primes $p=a^2+b^4$). The simpler looking earlier identities are not really ready for most applications. One often needs to separate the variables of summation and sometimes to smooth them out. There are a number of ways to do this, usually losing a logarithmic factor, but this may not be tolerable when estimating the contribution of the "boundary terms". It would be pointless for us to undergo complicated preparations without having a specific target problem. The reader, being confronted with her or his own problem, might have to invent somewhat different preparations.

However, to gain some comfort with the expressions for the above multilinear forms we decode for illustration the last of these and make some comments. Take

$$S_3(x;Y,Z) = \int_Y^{eY}\int_Z^{eZ} \sum_{n\leqslant x} a_n \Big(\sum_{\substack{\nu|n\\ \nu\leqslant\Delta}} \lambda_\nu\Big)\Big(\sum_{\substack{bc|n\\ b>sy\\ z<c\leqslant sz}} \mu(b)\Lambda(c)\Big)\frac{dy}{y}\frac{dz}{z}\ .$$

Recall that $Y=Z=\Delta^{-1}\sqrt{D}$ and s is uniquely determined as a power of 2 by $1 < 8s\delta\Delta\sqrt{D}/\Delta\sqrt{x} \leqslant 2$. All three parameters δ, Δ, D are at our disposal. Actually, D is set as the level of distribution of the sequence $\mathcal{A}=(a_n)$, so that the special forms $T(x;Y)$ and $T(x;Y,Z)$ can be evaluated asymptotically by an appeal to the congruence sums $A_d(x)$. The parameter Δ reduces $Y=Z$ slightly below $\sqrt{D}$ to make room for the sieve weights λ_ν. Usually $\Delta=D^\eta$ with η a small positive constant suffices while the sieve factors $\theta_n = (\lambda * 1)_n$ effectively reduce the summation to almost-primes, hence avoiding the collateral damage of logarithmic losses due to divisor type factors coming from generous estimations applied along the way. But, what is the parameter δ doing here? It does regulate the size of s. If s is small, both sums $S_2(x; Y, Z)$ and $S_3(x; Y, Z)$ are squeezed into "boundary terms", so a crude treatment works. Sometimes, however, we prefer s quite large so that $S_1(x; Y, Z)$ can be formatted to a bilinear form over more accessible ranges.

CHAPTER 18

Asymptotic Sieve for Primes

Post-Parity Dances

18.1. Combinatorial Transformations

In many places we have been making combinatorial transformations of the arithmetic sequences under consideration. Frequently these can be described in terms of a transformation of a single term; in such a case we may make use of identities for $\Lambda(n)$ such as those described in the previous chapter. Sometimes, however, one cannot conveniently describe the operation on a local level and we need to transform the sequence as a whole. An example of this is the Buchstab identity which we have encountered several times. In this section we elaborate some transformations of Buchstab type leading to a combinatorial identity which, after making a few natural restrictions and reasonable hypotheses, emerges as an approximate formula (see [**36**]) for a general sum over primes. We call such a formula an "asymptotic identity". This will then be applied in the following chapter.

Let $\mathcal{C} = \{c_n\}$ be a sequence of complex numbers such that

$$\sum_1^{\infty} |c_n| < \infty.$$

We shall examine the subsequence over integers free of small prime factors. More precisely, let $z \geqslant 2$, $P(z)$ be the product of all primes $< z$ and

$$S(\mathcal{C}, z) = \sum_{(n,P(z))=1} c_n.$$

Our goal is to express $S(\mathcal{C}, z)$ in terms that can be handled in practice. As usual we introduce, for any $d \mid P(z)$, the subsequence

$$\mathcal{C}_d = \{c_n : n \equiv 0 \,(\mathrm{mod}\, d)\},$$

pretending to have some knowledge about the corresponding partial sum

$$|\mathcal{C}_d| = \sum_{n \equiv 0(d)} c_n,$$

at least when d is not too large. We shall be more specific later. The smaller the sifting level z, the easier the problem. One can reduce the sifting level with the aid of Buchstab's identity. Applying it twice we get

$$S(\mathcal{C}, z) = S(\mathcal{C}, w) - \sum_{w \leqslant p < z} S(\mathcal{C}_p, w) + \sum\sum_{w \leqslant p < q < z} S(\mathcal{C}_{pq}, p),$$

where w is at our disposal subject to $w < z$. Next we express $S(\mathcal{C}, w)$, as well as each $S(\mathcal{C}_p, w)$, in terms of $|\mathcal{C}_d|$ by means of Legendre's identity obtaining

$$S(\mathcal{C}, w) - \sum_{w \leqslant p < z} S(\mathcal{C}_p, w) = \sum_{d|P(z)}^{w} \mu(d)|\mathcal{C}_d|,$$

where $\sum^w$ restricts d to integers having at most one prime factor $\geqslant w$. Since we think of w as being very small the right-hand side will have relatively few terms.

Now we consider the double sum $\sum\sum S(\mathcal{C}_{pq}, p)$ over primes $w \leqslant p < q < z$. If p is large, $p \geqslant y$ say, we shall not be able to do anything about $S(\mathcal{C}_{pq}, p)$ (except to drop it or to give a crude estimate in some cases). Therefore, we leave these terms alone. However, we can make something manageable out of the sum over p in the range $w \leqslant p < y$. This requires modifying the condition $p < q$ (and others that will occur) so that p becomes a free variable. To accomplish this we further partition the range $w \leqslant p < y$ into short segments by appropriately selected points $w = y_K < \cdots < y_1 < y_0 = y$. Then we split the resulting short sums into

$$\sum_{\substack{w \leqslant p < q < z \\ y_{k+1} \leqslant p < y_k}}\sum S(\mathcal{C}_{pq}, p) = \sum_{y_{k+1} \leqslant p < q < y_k}\sum S(\mathcal{C}_{pq}, p) + \sum_{y_{k+1} \leqslant p < y_k \leqslant q < z}\sum S(\mathcal{C}_{pq}, p).$$

Finally, for the second of these we split $S(\mathcal{C}_{pq}, p)$ by Buchstab's identity as follows:

$$S(\mathcal{C}_{pq}, p) = S(\mathcal{C}_{pq}, y_k) + \sum_{p \leqslant r < y_k} S(\mathcal{C}_{pqr}, r).$$

Gathering together the above sums we obtain the following identity.

LEMMA 18.1. *Let $w = y_K < \cdots < y_1 < y_0 = y < z$. We have*

(18.1)
$$\begin{aligned} S(\mathcal{C}, z) - \sum_{y \leqslant p < q < z}\sum S(\mathcal{C}_{pq}, p) &= \sum_{d|P(z)}^{w} \mu(d)|\mathcal{C}_d| + \sum_{0 \leqslant k < K}\ \sum_{y_{k+1} \leqslant p < y_k \leqslant q < z}\sum S(\mathcal{C}_{pq}, y_k) \\ &+ \sum_{0 \leqslant k < K}\Big\{\sum_{y_{k+1} \leqslant p < q < y_k}\sum S(\mathcal{C}_{pq}, p) + \sum\sum_{y_{k+1} \leqslant p \leqslant r < y_k \leqslant q < z}\sum S(\mathcal{C}_{pqr} r)\Big\}. \end{aligned}$$

We wish to estimate the various sums occurring on the right-hand side of (18.1). For some of these it suffices to have a crude bound of the type

$$\sum_{n \equiv 0(d)} |c_n| \leqslant \gamma(d)X, \tag{18.2}$$

where $X > 0$ and γ is non-negative, and also submultiplicative in the sense that $\gamma(d_1 d_2) \leqslant \gamma(d_1)\gamma(d_2)$ for all $d_1, d_2 \geqslant 1$.

In the first sum on the right of (18.1) we split $\sum_d^w = \sum_{d<D}^w + \sum_{d \geqslant D}^w$ and we estimate the latter using (18.2) as follows:

$$\begin{aligned} \Big|\sum_{d \geqslant D}^{w}\Big| &\leqslant X \sum_{D \leqslant d|P(z)}^{w} \gamma(d) \leqslant XD^{-\varepsilon}\Big(1 + \sum_{w \leqslant p < z} \gamma(p)p^{\varepsilon}\Big) \prod_{p<w} (1 + \gamma(p)p^{\varepsilon}) \\ &\leqslant X\big(z/D\big)^{\varepsilon} \prod_{w \leqslant p < z} (1 + \gamma(p)) \prod_{p<w} (1 + 2\gamma(p)) \leqslant 2^{-\log(D/z)/\log w} G^2(z)X, \end{aligned}$$

on choosing $\varepsilon = \log 2/\log w$. Here

$$G(z) = \prod_{p<z}\big(1+\gamma(p)\big). \tag{18.3}$$

We also use (18.2) to bound the last two sums in (18.1) as follows:

$$\Big|\sum\sum_{y_{k+1}\leqslant p<q<y_k} S(\mathcal{C}_{pq},p)\Big| \leqslant X\Big(\sum_{y_{k+1}\leqslant p<y_k}\gamma(p)\Big)^2$$

and

$$\Big|\sum\sum\sum_{y_{k+1}\leqslant p\leqslant r<y_k\leqslant q<z} S(\mathcal{C}_{pqr},r)\Big| \leqslant X\Big(\sum_{y_{k+1}\leqslant q<z}\gamma(q)\Big)\Big(\sum_{y_{k+1}\leqslant p<y_k}\gamma(p)\Big)^2.$$

Since for every k we have

$$1+\sum_{y_{k+1}\leqslant q<z}\gamma(q)\leqslant G(z),$$

we conclude, by summing over k, that the two sums in question are bounded by $\Delta X G^2(z)$, where

$$\Delta = \max_k \sum_{y_{k+1}\leqslant p<y_k}\gamma(p).$$

To estimate Δ we require a bound for $\gamma(p)$. We assume that, for some constant $c>0$ and all primes p, we have

$$\gamma(p)\leqslant cp^{-1}. \tag{18.4}$$

This gives $G(z)\ll(\log z)^c$ and

$$\sum_{y_{k+1}\leqslant p<y_k}\gamma(p)\leqslant c\sum_{y_{k+1}\leqslant m<y_k} m^{-1}\leqslant c\big(w^{-1}+\log(y_k/y_{k+1})\big).$$

To get a good estimate we specialize the y_k for $0\leqslant k\leqslant K$ by the choice

$$y_k = y(w/y)^{k/K}. \tag{18.5}$$

Thus, for $3\leqslant K\leqslant w$ we have

$$\Delta\leqslant c\left(w^{-1}+K^{-1}\log(y/w)\right)\leqslant cK^{-1}\log y\ .$$

Combining the above estimates we get

Lemma 18.2. *Let $3\leqslant K\leqslant w<y<z<D$. Assume that* (18.2) *holds with $\gamma(d)$ satisfying* (18.4). *Then we have*

$$S(\mathcal{C},z)-\sum\sum_{y\leqslant p<q<z}S(\mathcal{C}_{pq},p)=\sum_{\substack{d\mid P(z)\\ d<D}}^{w}\mu(d)|\mathcal{C}_d|+\sum_{0\leqslant k<K}\ \sum\sum_{y_{k+1}\leqslant p<y_k<q<z}S(\mathcal{C}_{pq},y_k)$$
$$+\,\theta X G(z)^2\left(2^{-\frac{\log D/z}{\log w}}+cK^{-1}\log y\right), \tag{18.6}$$

where y_k are given by (18.5) *and $|\theta|\leqslant 1$.*

In practice, we shall estimate the first sum on the right side of (18.6) by special bilinear forms

$$\mathcal{R}(D) = \sum_{d<D} \lambda_d \sum_m c_{dm}, \tag{18.7}$$

and the other sums by general bilinear forms

$$\mathcal{R}(w,y) = \sum_{w\leqslant n<y} \beta_n \sum_{(m,n)=1} \alpha_m c_{mn}. \tag{18.8}$$

A particularly interesting situation arises when c_n are supported on $n \leqslant x$ and are bounded by $\tau(n)$, the number of divisors of n. Then (18.2) holds with $\gamma(d) = \tau(d)d^{-1}$ and $X \ll x\log x$. Choose

$$K = (\log x)^8, \quad w = x^{(\log\log x)^{-3}}, \quad \text{and } D = zx^{(\log\log x)^{-1}}.$$

Suppose that for any coefficients λ_d supported on squarefree numbers with $|\lambda_d| \leqslant 1$ we have

$$\mathcal{R}(D) \ll x(\log x)^{-2}, \tag{18.9}$$

and for any coefficients $|\alpha_m| \leqslant \omega(m)$, $|\beta_n| \leqslant 1$, β_n supported on primes, we have

$$\mathcal{R}(w,y) \ll x(\log x)^{-10}. \tag{18.10}$$

Then, all of the sums on the right-hand side of (18.6) can be successfully bounded, giving

LEMMA 18.3. *Suppose $\mathcal{C} = \{c_n\}$, $1 \leqslant n \leqslant x$, is a sequence of complex numbers bounded by $\tau(n)$ and satisfying (18.9) and (18.10). Then*

$$S(\mathcal{C},z) = \sum_{y\leqslant p<q<z} S(\mathcal{C}_{pq},p) + O\big(x(\log x)^{-2}\big). \tag{18.11}$$

If we can take y near $x^{\frac13}$, say $y = x^{\frac13-\varepsilon}$ with $\varepsilon\log x \geqslant 1$, then

$$\sum_{y\leqslant p<q<z} S(\mathcal{C}_{pq},p) \ll \sum_{y\leqslant p<x^{1/3}} \frac{x}{p}\left(\log\frac{x}{p}\right)^{-1} \ll \frac{\varepsilon x}{\log x}.$$

Similarly, if we can take z near $x^{\frac12}$, say $z = x^{\frac12-\varepsilon}$, then

$$S(\mathcal{C},z) = \sum_{z\leqslant p\leqslant x} c_p + O\Big(\sum_{z\leqslant p<x^{1/2}} \frac{x}{p}\left(\log\frac{x}{p}\right)^{-1}\Big) = \sum_{p\leqslant x} c_p + O\left(\frac{\varepsilon x}{\log x}\right).$$

We obtain

THEOREM 18.4. *Suppose $\mathcal{C} = \{c_n\}, 1 \leqslant n \leqslant x$, is a sequence of complex numbers bounded by $\tau(n)$ such that (18.9) holds with $D = x^{\frac12-\varepsilon}$ and (18.10) holds with $y = x^{\frac13-\varepsilon}$ and $w = x^{(\log\log x)^{-3}}$, where $\varepsilon\log x \geqslant 1$. Then*

$$\sum_{p\leqslant x} c_p \ll \varepsilon\pi(x), \tag{18.12}$$

the implied constant being absolute.

18.2. Coupled Exponents of Distribution

The results of the previous section deal with a situation wherein we have just enough information, although just barely so, to deduce asymptotic results for our sums over primes (or, in our application, uniform distribution of the roots of quadratic congruences). It is worthwhile to elaborate further to draw results which could be useful in less fortunate situations when the hypotheses (18.9) and (18.10) cannot be verified in such wide ranges. In this case we may still hope for results which give the existence (and right order of magnitude) for the primes being sought or, in our application, the density everywhere of the roots.

Suppose that (18.9) holds for $D = x^{\theta-\varepsilon}$ and (18.10) holds for $y = x^{\vartheta-\varepsilon}$ with any $\varepsilon > 0$, the implied constants depending on ε. We call θ, ϑ the first and the second exponents of distribution of the sequence $\mathcal{C} = \{c_n\}$. The first of these is familiar to us; $D = x^{\theta-\varepsilon}$ is what we have been calling the level of distribution. The introduction of the second exponent to the sieve axioms gives us the chance to find primes, not necessarily the full density in the relevant sequence, but at least of some positive proportion which will depend on the exponents θ, ϑ. We shall explain the idea in the following context.

Suppose $c_n = a_n - b_n$ where $0 \leqslant a_n,\ b_n \ll \tau(n)$. Think of $\mathcal{A} = \{a_n\}$ as the sequence of our primary interest and $\mathcal{B} = \{b_n\}$ as a known one which is expected to approximate $\mathcal{A}$ strongly so that θ, ϑ are as large as possible. Most likely $\mathcal{B}$ will be a sequence of the type

$$b_n = \delta \varrho(n), \tag{18.13}$$

where $\delta > 0$ is constant and $\varrho(n)$ is a non-negative multiplicative function satisfying the following distribution law over primes:

$$\sum_{p \leqslant t} \varrho(p) \sim \pi(t). \tag{18.14}$$

This yields

$$S(\mathcal{B}, z) \sim \delta c(\theta) \pi(x) \tag{18.15}$$

and

$$\sum_{y \leqslant p \leqslant q < z} S(\mathcal{B}_{pq}, p) \sim \delta c(\theta, \vartheta) \pi(x), \tag{18.16}$$

where

$$c(\theta) = 1 + \log \frac{1-\theta}{\theta}$$

and

$$c(\theta, \vartheta) = \int\!\!\int_{\substack{\vartheta < u < v < \theta \\ 2u+v<1}} \left\{1 + \log^+ \frac{1-2u-v}{u}\right\} \frac{du\, dv}{uv(1-u-v)}$$

in the range $\frac{1}{5} \leqslant \vartheta \leqslant \frac{1}{3} \leqslant \theta \leqslant \frac{1}{2}$. The expressions for $c(\theta)$ and $c(\theta, \vartheta)$ in other ranges are more complicated. Hence, by (18.11) we get

$$S(\mathcal{A}, z) > \delta\big(c(\theta) - c(\theta, \vartheta) + o(1)\big)\pi(x), \tag{18.17}$$

after dropping $S(\mathcal{A}_{pq}, p)$, which we may do since these terms are non-negative. Suppose further that the first exponent of distribution is $\theta = \frac{1}{2}$. Then we have

$$S(\mathcal{A}, z) = \sum_{p \leqslant x} a_p + o\big(\pi(x)\big).$$

Combining the last two estimates we conclude that

$$\sum_{p \leqslant x} a_p > \delta\big(1 - c(\tfrac{1}{2}, \vartheta) + o(1)\big)\pi(x). \tag{18.18}$$

For $\vartheta = \frac{1}{5}$ it has been computed by E. LeBlanc that $c\left(\frac{1}{2}, \frac{1}{5}\right) = .769664\ldots$. Hence, if $\mathcal{A} = \{a_n\}$ has the exponents of distribution $\left(\frac{1}{2}, \frac{1}{5}\right)$, then the contribution to $\mathcal{A}$ coming from primes exceeds 23% of its expected value. The value of ϑ for which $c\left(\frac{1}{2}, \vartheta\right) = 1$, computed by S. Blight, is $\vartheta = 0.18409770\ldots$. It would be interesting to know in general:

QUESTION. Which pairs of exponents (θ, ϑ) are prime-producing?

This question is attractive for the development of sieve theory, regardless of applications.

Various identities for sums over primes have been crafted with the specific feature in mind of having terms which can be dropped by positivity, yet preserving the chance for a positive lower bound from the remaining terms. Since the remaining terms can offer more flexibility in handling the relevant bilinear forms and exponential sums, one may succeed in unearthing primes from more restricted domains. G. Harman is an architect of the most sophisticated and powerful formulas of this kind. Among many applications we mention Harman's result [**82**] about the fractional part of numbers αp. He proved that if α is irrational then $||\alpha p|| < p^{-3/10}$ for infinitely many primes; in other words, the approximation

$$\left|\alpha - \frac{a}{p}\right| < p^{-13/10}$$

holds for infinitely many fractions a/p with prime denominator. Another impressive application concerns his work with R. Baker and J. Pintz [**3**] on primes in short intervals, coming close to the bound given by the Riemann Hypothesis. This remarkable achievement is again due in great measure to new arrangements in combinatorial sieve theory.

18.3. Finessing Bombieri's Sieve

Bombieri's sieve (see Chapter 3) produces an asymptotic for sums

$$S_k(x) = \sum_{n \leqslant x} a_n \, \Lambda_k(n)$$

with $k \geqslant 2$ for quite general sequences $\mathcal{A} = \big(a_n\big)$ which have the level of distribution $D = D(x)$ close to x. It does, however, fail for the prime sum

$$S(x) = \sum_{n \leqslant x} a_n \, \Lambda(n) \tag{18.19}$$

(the case $k = 1$), and this is by reason of the genetic deficiency of the classical axioms to pass the parity barrier, as explained in Chapter 16.

When the level of distribution $D = D(x)$ is close to x, the asymptotic evaluation of $S(x)$ by Bombieri's argument essentially reduces to the estimation of

$$M_d(x) = \sum_{\substack{m \leqslant x \\ m \equiv 0(d)}} a_m \, \mu(m) \tag{18.20}$$

for small d. One needs to show that there is a considerable cancellation of terms in (18.20) due to the unbiased sign change of $\mu(m)$. In this format the sieve for primes may bounce through the Möbius function back to itself. However, one can break the resonance by a simple device which is reminiscent of the inclusion-exclusion procedure.

We begin by the observation that the following product of Dirichlet series

$$\Big[1 - \zeta(s) \sum_{m \leqslant z^2} \mu(m) m^{-s}\Big]\Big[\sum_{\ell \leqslant xz^{-2}} \Lambda(\ell)\ell^{-s} + \frac{\zeta'}{\zeta}(s)\Big]$$

has no terms n^{-s} with $n \leqslant x$. Hence we deduce that

$$\Lambda(n) = \sum_{\substack{\ell m = n \\ m \leqslant z^2}} \mu(m) \log \ell + \sum_{\substack{k\ell m = n \\ m \leqslant z^2, \ \ell \leqslant xz^{-2}}} \mu(m)\Lambda(\ell)$$

if $xz^{-2} < n \leqslant x$, and the right-hand side vanishes for $n \leqslant xz^{-2}$. Hence, for any complex numbers c_n we have

$$\sum_{xz^{-2} < n \leqslant x} c_n \, \Lambda(n) = \sum_{\substack{\ell m \leqslant x \\ m \leqslant z^2}} c_{\ell m} \, \mu(m) \log \ell + \sum_{\substack{k\ell m \leqslant x \\ m \leqslant z^2, \ \ell \leqslant xz^{-2}}} c_{k\ell m} \, \mu(m) \, \Lambda(\ell) \, .$$

All terms with $z < m \leqslant z^2$ contribute at most

$$2 \sum_{\ell} (\log \ell) \Big| \sum_{\substack{\ell m \leqslant x \\ z < m \leqslant z^2}} c_{\ell m} \, \mu(m) \Big|$$

while these with $m \leqslant z$ contribute at most

$$\sum_{d \leqslant z} \left|C_d'(x)\right| + (\log x) \sum_{d \leqslant xz^{-1}} \left|C_d(x)\right| \, ,$$

where

$$C_d'(x) = \sum_{\substack{n \leqslant x \\ n \equiv 0(d)}} c_n \log n \, , \quad C_d(x) = \sum_{\substack{n \leqslant x \\ n \equiv 0(d)}} c_n$$

(write $\log n/m = \log n - \log m$ to find the factor $\log x$). We conclude

THEOREM 18.5. *Let $1 < z \leqslant \sqrt{x}$. For any complex numbers c_n we have*

$$\begin{aligned} \Big| \sum_{xz^{-2} < n \leqslant x} c_n \, \Lambda(n) \Big| \leqslant \sum_{d \leqslant z} \left|C_d'(x)\right| + (\log x) \sum_{d \leqslant xz^{-1}} \left|C_d(x)\right| \\ + 2(\log x) \sum_{\ell} \Big| \sum_{\substack{\ell m \leqslant x \\ z < m \leqslant z^2}} c_{\ell m} \, \mu(m) \Big| \, . \end{aligned} \tag{18.21}$$

REMARKS. If $D = D(x)$ is the level of distribution of $\mathcal{C} = (c_n)$ then a good choice of z is $z = xD^{-1}$, because none of the outer variables of summation on the right side will exceed D.

Now, let $\mathcal{A} = (a_n)$ be a sequence of non-negative real numbers having the density function g and let $\mathcal{B} = (b_n)$ be its normalized model in the sense of Section 1.4. We apply Theorem 18.5 to $c_n = a_n - b_n$. In this case

$$\sum_{n \leqslant x} c_n \Lambda_n = S(x) - HA(x)x^{-1} \sum_{n \leqslant x} h(n)\, n\, \Lambda(n)$$
$$= S(x) - HA(x)\left\{1 + O\big((\log x)^{-1}\big)\right\}$$

by the prime number theorem. On the other hand, assume that $\mathcal{A}$ has level of distribution D in the (slightly different than usual) sense that

$$R(x; D) \ll A(x)(\log x)^{-2} \tag{18.22}$$

where $R(x; D)$ was defined in (1.17). We find that the terms having $C'_d(x)$, $C_d(x)$ on the right side of (18.21) are negligible (in the first case, after partial summation; see (1.18) (1.19)). This leaves us with the last sum

$$\sum_{\ell} \Big| \sum_{\substack{\ell m \leqslant x \\ z < m \leqslant z^2}} \mu(m)\big(a_{\ell m} - b_{\ell m}\big) \Big| .$$

Here the terms $b_{\ell m} = h(\ell m)\, \ell m H$ are negligible because, by the Prime Number Theorem,

$$\sum_{\substack{m \leqslant x/\ell \\ z < m \leqslant z^2}} h(m)m \ll \frac{x}{\ell} \exp(-\sqrt{\log z}) .$$

Hence, we conclude the following:

THEOREM 18.6. *Let $\mathcal{A} = (a_n)$ satisfy (18.22) for some $D = D(x)$ in the range $\sqrt{x} < D(x) \leqslant x(\log x)^{-20}$. Put $z = xD^{-1}$. Then,*

$$\sum_{n \leqslant x} a_n \Lambda(n) = HA(x)\Big\{1 + O\Big(\frac{1}{\log x}\Big)\Big\} + 2\theta T(x, D) \log x \tag{18.23}$$

where $|\theta| \leqslant 1$ and

$$T(x, D) = \sum_{\ell \leqslant D} \Big| \sum_{\substack{\ell m \leqslant x \\ z < m \leqslant z^2}} a_{\ell m}\, \mu(m) \Big| . \tag{18.24}$$

REMARKS. The sum $T(x, D)$ is a bilinear form for which one has a chance to prove a good bound because of cancellation in the inner sum over m. When $D = D(x)$ is close to x, the task is easier because $z = xD^{-1}$ is quite small. Note that, in the bilinear form $T(x, D)$, the inner sum over m is weighted precisely by the Möbius function, which is more pleasing than in some other expressions, wherein that function occurred in a corrupted form.

18.4. Sifting for Primes in Lacunary Sequences

Yes, the bilinear form (18.24) in Theorem 18.6 is pleasing, but in practice very restrictive; particularly when the sequence $\mathcal{A} = (a_n)$ is lacunary, it proves to be difficult to estimate it. Similarly, Theorem 18.4 does not exhibit thin sets of primes. In this section we are going to prove a result which can succeed for lacunary sequences by demanding stronger inputs from bilinear forms.

As usual, we consider a sequence of real non-negative numbers

$$\mathcal{A} = (a_n) \tag{18.25}$$

with the purpose of showing an asymptotic formula for

$$S(x) = \sum_{n \leqslant x} a_n \Lambda(n). \tag{18.26}$$

The sequence $\mathcal{A}$ can be quite thin although not arbitrarily so. Setting

$$A(x) = \sum_{n \leqslant x} a_n, \tag{18.27}$$

we require that $A(x)$ be slightly larger than $A(\sqrt{x})$; precisely,

$$A(x) \gg A(\sqrt{x})(\log x)^2 . \tag{18.28}$$

As always, $A_d(x)$ are the congruence sums

$$A_d(x) = \sum_{\substack{n \leqslant x \\ n \equiv 0 (\operatorname{mod} d)}} a_n = g(d)A(x) + r_d(x)$$

about which we assume that

$$A_d(x) \ll d^{-1}\tau(d)^8 A(x) , \tag{18.29}$$

uniformly in $d \leqslant x^{1/3}$. Of course, this bound is crude and we need more precise formulas if we hope to obtain asymptotic results. As usual, we assume that g is multiplicative and satisfies

$$0 \leqslant g(p) < 1, \quad g(p) \ll p^{-1}, \tag{18.30}$$

for all primes. Moreover, some regularity in the distribution of $g(p)$ is required. We express this by assuming that for $y \geqslant 2$,

$$\sum_{p \leqslant y} g(p) = \log\log y + c + O\left((\log y)^{-20}\right) \tag{18.31}$$

with some constant c, so that we are dealing with a linear sieve problem. We are also going to assume that

$$\sum_{\substack{d \leqslant y \\ (d,\nu)=1}} \mu(d)g(d) \ll \sigma_\nu (\log y)^{-6} \tag{18.32}$$

uniformly in $\nu \geqslant 1$, $y \geqslant 2$, where

$$\sigma_\nu = \prod_{p|\nu}\Big(1 + \frac{1}{\sqrt{p}}\Big) .$$

Actually, (18.32) follows from (18.31) but in practice it is easier to verify it directly in an individual case.

Let

$$H(x) = -\sum_{d \leqslant x} \mu(d)g(d) \log d . \tag{18.33}$$

The hypothesis (18.31) implies (see Lemma A.9) that

$$H(x) = H + O\left((\log x)^{-1}\right) , \tag{18.34}$$

where H is the positive constant

$$H = -\sum_{d} \mu(d)g(d)\log d = \prod_{p}(1-g(p))\Big(1-\frac{1}{p}\Big)^{-1}. \tag{18.35}$$

We assume that the remainder terms satisfy

$$\sum_{d\leqslant D} \mu^2(d)|r_d(t)| \leqslant A(x)(\log x)^{-2^{22}} \tag{18.36}$$

for all $t \leqslant x$, with some $D = D(x)$ such that

$$x^{\frac{2}{3}} < D(x) < x\ . \tag{18.37}$$

The assumption (18.36) is quite realistic since $D(x)$ is not necessarily required to be near x. As we have seen before, such a choice of $D(x)$ necessitates some additional hypothesis by means of which one would be able to avoid serious problems with the extremely large moduli.

The estimate for bilinear forms that we assume is the following:

$$\sum_{m}\Big|\sum_{\substack{N<n\leqslant 2N\\ mn\leqslant x}} \gamma(n)\mu(mn)a_{mn}\Big| \leqslant A(x)(\log x)^{-2^{22}}\ , \tag{18.38}$$

for every N with

$$\Delta^{-1}\sqrt{D} < N < \delta^{-1}\sqrt{x}\ , \tag{18.39}$$

for some $\delta = \delta(x) \geqslant 2$ and $\Delta = \Delta(x) \geqslant 2$, and where the coefficients are given by

$$\gamma(n) = \gamma(n, C) = \sum_{\substack{d|n\\ d\leqslant C}} \mu(d)\ . \tag{18.40}$$

This is required for every C with

$$1 \leqslant C \leqslant xD^{-1}\ . \tag{18.41}$$

We remark that, since C is relatively small, the presence of the Möbius function in coefficients $\gamma(n, C)$ does not play much of a role.

Finally, for simplicity we assume that the sequence $\mathcal{A}$ is supported on squarefree numbers, that is,

$$a_n = 0 \qquad \text{if } \mu(n) = 0\ . \tag{18.42}$$

Our main result is

THEOREM 18.7. *Assuming the hypotheses* (18.28)–(18.32), (18.36), (18.38) *and* (18.42), *we have*

$$\sum_{p\leqslant x} a_p \log p = HA(x)\left\{1 + O\left(\frac{\log \delta(x)}{\log \Delta(x)}\right)\right\}\ , \tag{18.43}$$

where the implied constant depends only on the function g.

In practice (18.38) can be established in the range (18.39) for $\delta = (\log x)^{\alpha}$ and $\Delta = x^{\eta}$ with some positive constants α, η. For these choices the error term in (18.43) becomes $O(\log\log x/\log x)$ and this cannot be improved by refining our arguments. If $\log \delta(x) \gg \log \Delta(x)$ then the error term exceeds the main term and (18.43) follows from the upper bound sieve. Therefore, for the proof we may assume that $x \geqslant \Delta(x)^A$, $\Delta(x) \geqslant \delta(x)^A$, and $\delta(x) \geqslant A$ for any fixed positive constant

A. Note also that it suffices to prove the theorem with $\Delta(x)$ in the inequality (18.39) replaced by $\Delta(x)^A$ (and similarly for $\delta(x)$).

Note that the restriction $N < \delta^{-1}\sqrt{x}$ in (18.39) makes (18.38) realistic whereas requiring (18.38) for $N = \sqrt{x}$ would not be, for reasons similar to the prohibition of $D(x) = x$ in (18.36). Were we to make this seemingly slight extension in the range of (18.38), we would be able to give a simpler proof, but of a result with virtually no application. Also, our stipulation of the lower bound restriction $N > \Delta^{-1}\sqrt{D}$ in (18.39) is essential; indeed by narrowing this slightly to $N > \sqrt{D}$ we would not be able to break the parity problem.

It is worth mentioning that the source of cancellation in the bilinear form in (18.38) comes from the sign changes of the Möbius function $\mu(mn)$ in the inner sum so there is no need to express this sum in terms of the remainder $r_{mn}(x)$, as in Section 12.7. There is also the Möbius function $\mu(d)$ in the coefficient $\gamma(n, C)$. Here d must be quite a bit smaller than N to ensure that $\mu(d)$ does not completely neutralize $\mu(n)$. By (18.39)–(18.41), we know that $d < C < x\Delta N D^{-3/2}$ so our hypothesis (18.38) can be realistic only if D is somewhat larger than $x^{2/3}$.

The assumption that the sequence $\mathcal{A} = (a_n)$ is supported on squarefree numbers is not a serious problem, but it turns out to be a technical inconvenience in some applications. This restriction can be dispensed with at the expense of making some of the axioms more involved. This has been delivered in [**53**], Theorem 3. These theorems provide the sieve input for the proof of the infinitude of primes $p = a^4 + b^2$ in [**54**]. We come to this subject in Chapter 21.

— Technical Reductions

What we actually need for the proof of Theorem 18.7 are the following hypotheses:

$$\sum_{d \leqslant D} \mu^2(d)\tau_5(d)|r_d(t)| \ll A(x)(\log x)^{-3} \tag{18.44}$$

and

$$\sum_m \tau_5(m) \Big| \sum_{\substack{N<n\leqslant 2N \\ mn\leqslant x}} \gamma(n)\mu(mn)a_{mn} \Big| \ll A(x)(\log x)^{-3}\ , \tag{18.45}$$

in the same notation and ranges as in (18.36) and (18.38). Here the implied constant depends only on g.

In this subsection we derive (18.44) and (18.45) from (18.36) and (18.38) respectively, using the crude bound (18.29). As soon as (18.44) and (18.45) have been established, the hypotheses (18.36), (18.38) and (18.29) will no longer be required.

We begin with a lemma.

Lemma 18.8. *Fix $k \geqslant 2$. Any squarefree integer $n \geqslant 1$ has a divisor $d \leqslant n^{1/k}$ such that*

$$\tau(n) \leqslant (2\tau(d))^k\ .$$

This is just a special case of Landreau's inequalities (see Section 22.5) but in this case the proof is so simple that we give it here.

Proof. Write $n = p_1 \dots p_r$ with $p_1 < \cdots < p_r$ and put $d = p_1 \dots p_{[r/k]}$; thus $d \leqslant n^{1/k}$ and

$$2\tau(d) = 2^{1+[r/k]} \geqslant 2^{r/k} = \tau(n)^{1/k}\ .$$

□

We use the lemma for $k = 3$. First we estimate the following sums:

$$\sum_{d\leqslant x} \tau_5(d)^{4/3} A_d(x) \leqslant \sum_{d\leqslant x} \tau_{10}(d) A_d(x) = \sum_{n\leqslant x} \tau_{11}(n) a_n$$
$$\leqslant \sum_{n\leqslant x} \tau(n)^4 a_n \leqslant \sum_{d\leqslant x^{1/3}} (2\tau(d))^{12} A_d(x)$$

by the lemma. Then, by (18.29) this is bounded by $O\left(A(x)V(x)\right)$ where

$$V(x) = \sideset{}{^\flat}\sum_{d\leqslant x} d^{-1}\tau(d)^{20} \ll (\log x)^{2^{20}}.$$

Here, as usual, $\sum^\flat$ indicates that the summation is restricted to squarefree integers. Therefore,

$$\sum_{d\leqslant x} \tau_5(d)^{4/3} A_d(x) \ll A(x)(\log x)^{2^{20}} . \tag{18.46}$$

Similarly, but using (18.29),

$$\sideset{}{^\flat}\sum_{d\leqslant x} \tau_5(d)^{4/3} g(d) \leqslant \prod_{p\leqslant x} \left(1 + 11g(p)\right) \ll (\log x)^{11} . \tag{18.47}$$

Combining (18.46) and (18.47) we infer that

$$\sideset{}{^\flat}\sum_{d\leqslant x} \tau_5(d)^{4/3} \left|r_d(t)\right| \ll A(x)(\log x)^{2^{20}} . \tag{18.48}$$

Finally, (18.44) follows from (18.36) and (18.48) by Hölder's inequality. The derivation of (18.45) from (18.38) is similar.

We are now ready to proceed to the proof of Theorem 18.7. We decompose our sum over primes in accordance with the combinatorial identity in Proposition 17.7 and then are faced with the asymptotic evaluation of one sum and the estimation of several others. We treat these in turn.

— **Evaluation of $T(x;Y)$**

We are able to evaluate the sum $T(x;y)$ for individual x, y. We split this by means of $\log(n/b) = \log n - \log b$, getting

$$T(x;y) = T_1(x;y) + T_2(x;y).$$

We have

$$\begin{aligned} T_1(x;y) &= \sum_{n\leqslant x} a_n\theta_n \log n \sum_{\substack{b\mid n\\ b\leqslant y}} \mu(b) \\ &= \sum_{b\leqslant y} \mu(b) \sum_{\substack{n\leqslant x\\ n\equiv 0 (\mathrm{mod}\, b)}} a_n\theta_n \log n. \end{aligned} \tag{18.49}$$

Here the inner sum is, by partial summation,

$$V_b(x)\log x - \int_1^x V_b(t)\frac{dt}{t} \tag{18.50}$$

where

$$V_b(x) = \sum_{\substack{n\leqslant x\\ n\equiv 0\,(\mathrm{mod}\,b)}} a_n \sum_{\nu|n} \lambda_\nu \tag{18.51}$$
$$= \sum_\nu \lambda_\nu A_{[\nu,b]}(x) = \sum_\nu \lambda_\nu \left\{g([\nu,b])A(x) + r_{[\nu,b]}(x)\right\}.$$

We have $g([\nu,b]) = g(\nu)g(b/(\nu,b))$ and

$$\sum_{b\leqslant y} \mu(b) g\Big(\frac{b}{(\nu,b)}\Big) = \sum_{d|\nu} \mu(d) \sum_{\substack{b\leqslant y/d\\ (b,\nu)=1}} \mu(b)g(b).$$

This last inner sum is, by (18.32), bounded by $O\left(\sigma_\nu (\log x)^{-6}\right)$. By this and (18.50) we deduce that the contribution to (18.49) coming from the main term in (18.51) is

$$\ll A(x)(\log x)^{-5} \sum_\nu |\lambda_\nu| g(\nu)\tau(\nu)\sigma_\nu \ll A(x)(\log x)^{-3}.$$

The remainder terms in (18.51) make a contribution to (18.49) which is

$$\ll (\log x) \sum_\nu \sum_b |\lambda_\nu \mu(b) r_{[\nu,b]}(x)|$$
$$\ll (\log x) \sum_{d<\Delta y}^{\flat} \tau_3(d)|r_d(x)| \ll A(x)(\log x)^{-2}$$

by (18.44), since $\Delta y < D$. Thus, we obtain the bound $T_1(x;y) \ll A(x)(\log x)^{-2}$.

We now consider the sum

$$T_2(x;y) = -\sum_{b\leqslant y} \mu(b) \log b \sum_\nu \lambda_\nu A_{[\nu,b]}(x).$$

We insert the approximation (1.2) for $A_{[\nu,b]}(x)$. The remainder terms give a contribution which may be estimated in the same way as those for $T_1(x;y)$ giving the same bound $\ll A(x)(\log x)^{-2}$. Here, however, the main term provides also the main term for the total sum $S(x)$. This is

$$-A(x) \sum_\nu \lambda_\nu g(\nu) \sum_{b\leqslant y} \mu(b) g\Big(\frac{b}{(\nu,b)}\Big) \log b\,. \tag{18.52}$$

We extend the summation to all b making an error $\ll A(x)(\log x)^{-3}$ by (18.32). Then, we evaluate the complete inner sum as

$$\sum_{\eta|\nu} \mu(\eta) \sum_{(b,\nu)=1} \mu(b)g(b) \log \eta b = \sum_{\eta|\nu} \mu(\eta) \sum_{(b,\nu)=1} \mu(b)g(b) \log b$$

and this is equal to $-H$ if $\nu = 1$ and equal to zero otherwise. Thus the sum (18.52) and also $T_2(x,y)$ are given by $HA(x) + O\left(A(x)(\log x)^{-2}\right)$.

Combining this with the bound for $T_1(x,y)$ we obtain

$$T(x,y) = HA(x) + O\left(A(x)(\log x)^{-2}\right). \tag{18.53}$$

— **Estimation of $T(x;Y,Z)$**

It again suffices to estimate the contribution for given y, z and here this is given by the sum

$$T(x;y,z)=\sum_{b\leqslant y}\mu(b)\sum_{c\leqslant z}\Lambda(c)\sum_{\nu}\lambda_\nu A_{[\nu,bc]}(x).$$

We insert the approximation (1.2) for $A_{[\nu,bc]}(x)$. For the contribution coming from the main terms, arguing as with $T_1(x;y)$, we are now led to estimate the sum

$$\sum_{b\leqslant y}\mu(b)\sideset{}{^\flat}\sum_{\substack{c\leqslant z\\ c\nmid b}}\Lambda(c)\,g\Big(\frac{bc}{(\nu,bc)}\Big)=\sum_{b\leqslant y}\mu(b)\,g\Big(\frac{b}{(\nu,b)}\Big)\sideset{}{^\flat}\sum_{\substack{c\leqslant z\\ c\nmid b}}\Lambda(c)\,g\Big(\frac{c}{(\nu,c)}\Big)$$

and by (18.32), (18.30) this is $\ll\tau(\nu)\sigma_\nu(\log y)^{-6}(\log\nu z)$. Summing over ν, as with $T_1(x;y)$ we get a contribution $\ll A(x)(\log x)^{-3}$.

The contribution coming from the remainder terms is bounded by

$$\sum_{\nu}|\lambda_\nu|\sideset{}{^\flat}\sum_{b\leqslant y}\sideset{}{^\flat}\sum_{\substack{c\leqslant z\\ c\nmid b}}\Lambda(c)|r_{[\nu,bc]}(x)|\leqslant(\log yz)\sum_{\nu}\sideset{}{^\flat}\sum_{d\leqslant yz}|\,r_{[\nu,d]}(x)\,|\ll A(x)(\log x)^{-2}$$

by (18.44) since $yz\leqslant e^2\Delta^{-2}D<\Delta^{-1}D$. From the two estimates we conclude that

$$T(x;y,z)\ll A(x)(\log x)^{-2}\,. \tag{18.54}$$

— **Estimation of $S_1(x;Y,Z)$**

As before this may be estimated for individual y and z. We have

$$S_1(x;y,z)=\sum\sum\sum_{\substack{bcd\leqslant x\\ b>sy,c>sz}}\mu(b)\Lambda(c)\theta_{bcd}a_{bcd}\,.$$

Note that the conditions of summation imply that c is located in a short interval (in the logarithmic scale), namely

$$\delta^{-1}\sqrt{x}<c<\delta\sqrt{x}\,, \tag{18.55}$$

and d is relatively small, namely

$$d<\delta^2\,. \tag{18.56}$$

Therefore, there is enough room for b out of which to create a nice variable. By dropping $\mu(b)$ we deduce that

$$\begin{aligned}|S_1(x;y,z)|&\leqslant\sum_{c}\Lambda(c)\sum_{d}\sum_{\substack{n\leqslant x\\ n\equiv 0\,(\mathrm{mod}\,cd)}}\theta_n a_n\\ &=\sum_{c}\Lambda(c)\sum_{d}\sum_{\nu}\lambda_\nu A_{[cd,\nu]}(x)\,.\end{aligned} \tag{18.57}$$

Of course, we have lost the possibility of detecting cancellation due to the sign changes of $\mu(b)$, but we have gained the congruence sums which can be evaluated asymptotically due to (18.44). Note that $[cd,\nu]\leqslant cd\nu<\delta^3\Delta\sqrt{x}<D$. We insert (1.2) and estimate the resulting remainder by (18.44) getting

$$|S_1(x;y,z)|\leqslant A(x)L(x)M(x)+O\big(A(x)(\log x)^{-2}\big) \tag{18.58}$$

where $L(x)$, $M(x)$ come from the main terms, namely

$$L(x) = \sum_{c}^{\flat} \Lambda(c) g(c),$$

$$M(x) = \sum_{d}^{\flat} \sum_{\nu} \lambda_\nu g([d, \nu]) \, .$$

Here c, d run over the segments (18.55) and (18.56) respectively. By (18.30) we infer that

$$L(x) \ll \log \delta \, . \tag{18.59}$$

For the estimation of $M(x)$ we write $g([d, \nu]) = g(d) g(\nu/(\nu, d))$ and note that

$$\sum_{\nu} \lambda_\nu g(\nu/(\nu, d)) \geqslant 0 \, .$$

The latter is a property of any upper-bound sieve, and it holds for any multiplicative function $f(\nu)$ with $0 \leqslant f \leqslant 1$ in place of $g(\nu/(\nu, d))$. This positivity property allows us to introduce the factor $(\delta^2/d)^\varepsilon$ into $M(x)$ and extend the summation to all d (Rankin's trick) getting

$$\begin{aligned} M(x) &\leqslant \delta^{2\varepsilon} \sum_{\nu} \lambda_\nu g(\nu) \sum_{d}^{\flat} g(d/(\nu, d)) d^{-\varepsilon} \\ &= \delta^{2\varepsilon} \prod_{p} (1 + g(p) p^{-\varepsilon}) \sum_{\nu} \lambda_\nu g(\nu) h(\nu), \end{aligned}$$

where $h(\nu)$ is the multiplicative function with

$$h(p) = (1 + p^{-\varepsilon})(1 + g(p) p^{-\varepsilon})^{-1} < 2(1 + g(p))^{-1} \, .$$

Note that $g(p)h(p) < 1$, so the sieve theory applies giving

$$\sum_{\nu} \lambda_\nu g(\nu) h(\nu) \ll \prod_{p < \Delta} (1 - g(p) h(p)) = \prod_{p < \Delta} (1 - g(p))(1 + g(p) p^{-\varepsilon})^{-1} \, .$$

Hence,

$$M(x) \ll \delta^{2\varepsilon} \prod_{p \geqslant \Delta} (1 + g(p) p^{-\varepsilon}) \prod_{p < \Delta} (1 - g(p)) \, .$$

We choose $\varepsilon = (\log \Delta)^{-1}$ getting by (18.31),

$$M(x) \ll \prod_{p < \Delta} (1 - g(p)) \ll (\log \Delta)^{-1} \, . \tag{18.60}$$

Combining (18.58), (18.59) and (18.60), we conclude that

$$S_1(x; y, z) \ll A(x) \frac{\log \delta}{\log \Delta} \, . \tag{18.61}$$

We remark that it is for this bound alone that we introduced the λ_ν. Without doing so we would lose a logarithm, and this very tight bound cannot afford such a loss.

— **Estimation of $S_2(x; Y, Z)$**

In this section and the next we finally encounter the sums where (due to contamination of the variables coming from the support of θ) it is necessary to perform the integrated estimate. Whenever a_n occurs we assume $n \leqslant x$ without writing this condition every time. We have

$$S_2(x; y, z) = \sum_e \sum_{y<b\leqslant sy} \sum_{c>z} \mu(b)\Lambda(c)\theta_{ebc}a_{ebc},$$

so that

$$\begin{aligned}|S_2(x; y, z)| &\leqslant (\log x) \sum_k \Big| \sum_{y<b\leqslant sy} \mu(b)\theta_{bk}a_{bk} \Big| \\ &\leqslant (\log x) \sum_{\nu_1} \sum_{\nu_2} |\lambda_{\nu_1\nu_2}| \sum_k \Big| \sum_{\frac{y}{\nu_2}<b\leqslant\frac{sy}{\nu_2}} \mu(b)a_{\nu_1\nu_2 bk} \Big| \ .\end{aligned}$$

In order to remove ν_2 from the range of the inner summation we take advantage of the integration over y (still not integrating over z):

$$\begin{aligned}S_2(x; Y, z) &\leqslant (\log x) \sum_{\nu_1} \sum_{\nu_2} |\lambda_{\nu_1\nu_2}| \sum_k \int_{Y/\Delta}^{eY} \Big| \sum_{y<b\leqslant sy} \mu(b)a_{\nu_1\nu_2 bk} \Big| \frac{dy}{y} \\ &\leqslant (\log x) \int_{Y/\Delta}^{eY} \sum_\ell \tau_3(\ell) \Big| \sum_{y<b\leqslant sy} \mu(b)a_{b\ell} \Big| \frac{dy}{y} \ .\end{aligned} \tag{18.62}$$

By (18.45) we conclude that

$$S_2(x; Y, z) \ll A(x)(\log x)^{-1} \ . \tag{18.63}$$

— **Estimation of $S_3(x; Y, Z)$**

Recall that

$$S_3(x; y, z) = \sum_e \sum_{b>sy} \mu(b) \sum_{z<c\leqslant sz} \Lambda(c)\theta_{eb}a_{ebc} \ .$$

Here we have, for the first time, taken advantage of the fact that $\theta_{ebc} = \theta_{eb}$ since c is prime. In essence this sum is of the same type as that in the previous section but with b and c interchanged. Thus, by analogy to the previous argument, we put c into the inner summation because its range is appropriate for (18.45). However, the fact that c is weighted by Λ rather than μ necessitates splitting it further in order to obtain a factor with sign changes from which we may hope to detect cancellation. We write

$$\mu(c)\Lambda(c) = \sum_{m|c} \mu(m) \log \frac{m}{C} = \Lambda^+(c) - \Lambda^-(c)$$

(recall c is squarefree) with $C = xD^{-1}$ where

$$\Lambda^+(c) = \sum_{m|c} \mu(m) \log^+ \frac{m}{C} \ ,$$

$$\Lambda^-(c) = \sum_{m|c} \mu(m) \log^+ \frac{C}{m} \ .$$

Note that the definition of $\Lambda^-(c)$ is close to $\gamma(c, C)$ in (18.40); precisely, we have

$$\Lambda^-(c) = \int_1^C \gamma(c, t)t^{-1}dt \ .$$

This gives a contribution to $S_3(x;y,z)$ which, by virtue of (18.45) applied $\log\log s$ times, satisfies

$$|S^-(x;y,z)| \leqslant \sum_{\ell} \tau_4(\ell) \Big| \sum_{z<c\leqslant sz} \mu(c)\Lambda^-(c)a_{c\ell} \Big| \ll A(x)(\log x)^{-1} . \tag{18.64}$$

The contribution to $S_3(x;y,z)$ from $\Lambda^+(c)$ is arranged as follows:

$$\begin{aligned} S^+(x;y,z) &= \sum_e \sum_{b>sy} \mu(b)\theta_{eb} \sum\sum_{z<\ell m\leqslant sz} \mu(\ell)\big(\log^+ \tfrac{m}{C}\big) a_{eb\ell m} \\ &= \int_C^x \Big(\sum_{b>sy} \mu(b) \sum_{ebz<x} \theta_{eb} \sum_{\substack{\ell t<sz \\ eb\ell t<x}} \mu(\ell) \sum_{\substack{m>t \\ z<\ell m\leqslant sz}} a_{eb\ell m} \Big) \frac{dt}{t} . \end{aligned}$$

Put $d = eb\ell$, so $d < D$ and d is squarefree because of (18.42). Here the inner sum is equal to $A_d(\min\{x, ebsz\}) - A_d(\max\{dt, ebz\})$. After an application of (1.2) this inner sum becomes

$$g(d) \sum_{\substack{dt<n\leqslant x \\ ebz<n\leqslant ebsz}} a_n + r_d(\min\{x, ebsz\}) - r_d(\max\{dt, ebz\}) . \tag{18.65}$$

We first treat the contribution to $S^+(x;y,z)$ coming from the main term in (18.65). Recall that $d = eb\ell$. We get a significant cancellation from the summation of $\mu(b)$. Note that b ranges over the interval $b_{\min} < b < b_{\max}$, where $b_{\min} = \max\{sy, n/esz\}$ and $b_{\max} = \min\{n/ez, n/e\ell t\}$. This yields

$$\sum_{\substack{(b,e\ell)=1 \\ b_{\min}<b<b_{\max}}} \mu(b)\theta_{eb} g(eb\ell) = g(e\ell) \sum_{\nu_1} \mu(\nu_1) g(\nu_1) \sum_{\nu_2|e} \lambda_{\nu_1\nu_2} \sum_{\substack{(b,e\ell\nu_1)=1 \\ b_{\min}<b\nu_1<b_{\max}}} \mu(b)g(b)$$

and here, by (18.32), the inner sum is $O\left(\sigma_{e\ell\nu_1}(\log x)^{-6}\right)$. Summing over all the other variables trivially we find that the main term in (18.65) contributes to the sum $S^+(x;y,z)$ an amount $S^*(x;y,z)$ which satisfies

$$S^*(x;y,z) \ll A(x)(\log x)^{-1} . \tag{18.66}$$

The remainder terms in (18.65) contribute an amount $S'(x;y,z)$ satisfying

$$|S'(x;y,z)| \leqslant \int_C^x \sideset{}{^\flat}\sum_{d<x/t} \sum_{\substack{k|d \\ k<x/z}} \tau_4(k) \left| r_d\left(\min\{x, ksz\}\right) - r_d\left(\max\{dt, kz\}\right) \right| \frac{dt}{t} .$$

Change the variable of integration t into t/d, integrate in z over $Z < z < eZ$ and change z into z/k, to get

$$|S'(x;y,Z)| \leqslant \int_Z^x \int_C^x \sideset{}{^\flat}\sum_{d<D} \tau_5(d) \left| r_d\left(\min\{x, sz\}\right) - r_d\left(\max\{t, z\}\right) \right| \frac{dt}{t}\,\frac{dz}{z} .$$

Now we can apply (18.44), obtaining

$$S'(x;y,Z) \ll A(x)(\log x)^{-1} . \tag{18.67}$$

Collecting (18.64), (18.66) and (18.67) we conclude that

$$S_3(x;y,Z) \ll A(x)(\log x)^{-1} . \tag{18.68}$$

Combining in Proposition 17.7 the estimates (18.53), (18.54), (18.61), (18.63) and (18.68), we complete the proof of Theorem 18.7.

CHAPTER 19

Equidistribution of Quadratic Roots

Jitterbug

19.1. Equidistribution Modulo One

An important area of number theory is concerned with the distribution modulo one of sequences of real numbers, those which are arithmetically interesting.

The study of this subject was initiated by H. Weyl who proved [**159**] the equidistribution of polynomial values $f(n)$ where f is non-constant with real coefficients, the leading one irrational. He introduced a criterion which reduced the problem to finding bounds of the type

$$W_k(x) = \sum_{n \leqslant x} e\big(kf(n)\big) = o(x)$$

for every $k \neq 0$ and subsequently established such bounds.

We begin this chapter with a simple example of how the sieve can be combined with such bounds to treat a more general problem. In the following sections we shall then consider a question of equidistribution requiring tools more advanced, both in the sieve theory and in the harmonic analysis.

THEOREM 19.1 (Daboussi-Delange). *For any real-valued multiplicative function $\gamma(n)$ with $|\gamma(n)| \leqslant 1$ and any $0 \leqslant \alpha < \beta \leqslant 1$, we have*

$$\sum_{\substack{n \leqslant x \\ \alpha < \{f(n)\} \leqslant \beta}} \gamma(n) = (\beta - \alpha) \sum_{n \leqslant x} \gamma(n) + o(x) \ .$$

This is not a verbatim statement of their result but it is essentially what they prove and we follow their argument closely; see [**26**].

PROOF. By the Weyl criterion the problem reduces to the bound

$$\sum_{n \leqslant x} \gamma(n) e\big(f(n)\big) = o(x) \ ,$$

where we have suppressed the frequency k by changing the polynomial $kf(X)$ to $f(X)$.

Writing the sum in the form

$$\sum\sum_{\substack{\ell m \leqslant x \\ (\ell, m) = 1}} \gamma(\ell)\gamma(m) e\big(f(\ell m)\big) = \sum_{\ell \leqslant L} \sum_{m \leqslant x/\ell} \quad + \quad O\left(xL^{-1}\right) ,$$

where ℓ runs through squarefull numbers and m through squarefree numbers, we see the proof can be reduced to the case that γ is supported on squarefree numbers (change again the polynomial, this time $f(\ell X)$ to $f(X)$).

Now we can decompose the sum as follows:

$$S = \sum\sum_{\substack{mn\leqslant x\\ m|P(z)\\ (n,P(z))=1}} \gamma(m)\gamma(n)e\big(f(mn)\big).$$

By Cauchy's inequality, we have

$$\begin{aligned}|S|^2 &\leqslant \Big(\sum_{\substack{n\leqslant x\\ (n,P(z))=1}} 1\Big) \sum_{\substack{n\leqslant x\\ (n,P(z))=1}} \Big|\sum_{\substack{m\leqslant x/n\\ m|P(z)}} \gamma(m)e\big(f(mn)\big)\Big|^2\\ &\ll \frac{x}{\log z}\sum\sum_{\substack{m_1|P(z)\\ m_2|P(z)}}\Big|\sum_{\substack{n\leqslant x/m_1,\, x/m_2\\ (n,P(z))=1}} e\big(f(m_1n)-f(m_2n)\big)\Big|,\end{aligned}$$

where the first factor has been bounded using a simple upper-bound sieve. The diagonal terms $m_1 = m_2$ make a contribution $O(x^2/\log z)$. For those terms with $m_1 \neq m_2$, we relax the condition $(m, P(z)) = 1$ in the inner sum by Möbius inversion and then use the Weyl bound $o(x)$, applied to the polynomials $f(dm_1X) - f(dm_2X)$ with $d \mid P(z)$, getting a contribution $\varepsilon(x,z)x^2$ where, for every fixed z, $\varepsilon(x,z) \to 0$ as $x \to \infty$. Taking z sufficiently large, we complete the proof. □

An important specific case is exhibited by taking $\gamma(n)$ to be the Möbius function $\mu(n)$. In this case, the sum has been studied much more deeply by Vinogradov, Davenport and others using the methods of bilinear forms which we introduced in Chapter 17.

19.2. Equidistribution Modulo Primes

Let f be an irreducible polynomial of degree $\geqslant 2$ with integer coefficients. For a given prime p the congruence

$$f(\nu) \equiv 0 \,(\operatorname{mod} p)$$

has at most deg f solutions by a well-known theorem of Lagrange. An interesting question is posed by the distribution of these as p varies. Specifically, if we normalize by considering the fractional parts $\{\nu/p\}$, one wonders how many fall into a given segment. Without any reason to guess otherwise, one might expect uniform distribution, namely that the asymptotic formula

$$\#\Big\{(p,\nu) : p \leqslant x, f(\nu) \equiv 0(\operatorname{mod} p),\, \alpha \leqslant \Big\{\frac{\nu}{p}\Big\} < \beta\Big\} \sim (\beta-\alpha)\pi(x) \tag{19.1}$$

holds for any $0 \leqslant \alpha < \beta \leqslant 1$, or equivalently,

$$\lim_{x\to\infty}\frac{1}{\pi(x)}\sum_{p\leqslant x}\sum_{f(\nu)\equiv 0(p)} F\Big(\frac{\nu}{p}\Big) = \int_0^1 F(t)dt$$

for any continuous function F of period 1. The Weyl criterion asserts, in this instance, that (19.1) is equivalent to

$$\sum_{p\leqslant x} \varrho(p) \sim \pi(x) \tag{19.2}$$

and

$$\sum_{p\leqslant x}\varrho_h(p)=o\big(\pi(x)\big),\tag{19.3}$$

for each fixed $h\neq 0$. Here, we denote $\varrho(n)=\varrho_0(n)$ and

$$\varrho_h(n)=\sum_{f(\nu)\equiv 0(\operatorname{mod} n)} e\Big(\frac{h\nu}{n}\Big).\tag{19.4}$$

Note that (19.2) holds by virtue of the Prime Ideal Theorem and that this confirms that the normalization in (19.1) is the correct one.

In this chapter we consider the polynomial $f(X)=X^2+1$ and prove the uniform equidistribution of these roots.

THEOREM 19.2. *For each fixed $h\neq 0$ we have*

$$\sum_{p\leqslant x}\ \sum_{\nu^2+1\equiv 0(\operatorname{mod} p)} e\Big(\frac{h\nu}{p}\Big)=o\big(\pi(x)\big),\tag{19.5}$$

and hence (19.1) *holds.*

The theorem holds for a general irreducible quadratic polynomial. The proof can be found in [**36**] for the definite case and in [**155**] for the indefinite case.

Of course, we are going to apply a sieve capable of detecting primes, in this case Theorem 18.4. This leads us to the problem of estimating special linear forms

$$\mathcal{L}_d(M)=\sum_{M<m\leqslant 2M}\varrho_h(dm),\tag{19.6}$$

as well as general bilinear forms of the type

$$\mathcal{B}(M,N)=\sum\sum_{(m,n)=1}\alpha_m\beta_n\varrho_h(mn).\tag{19.7}$$

Here α_m,β_n are arbitrary but bounded complex numbers with $M<m\leqslant 2M$, $N<n\leqslant 2N$.

It is natural to expect the following bounds:

$$\begin{aligned}\mathcal{L}_d(M)&\ll M^{\frac12}(hdM)^{\varepsilon},\\ \mathcal{B}(M,N)&\ll \|\alpha\|\,\|\beta\|(M+N)^{\frac12}(hMN)^{\varepsilon},\end{aligned}$$

where $\|\alpha\|$ denotes the ℓ_2-norm of $\alpha=(\alpha_m)$, the implied constant depending only on ε. These seem out of reach of current methods. However, we shall establish the following results which, although weaker, are sufficient for interesting applications.

PROPOSITION 19.3. *Suppose $1\leqslant h\ll dM$. Then*

$$\mathcal{L}_d(M)\ll (h,d)^{\frac{1}{20}}d^{\frac{1}{20}}M^{\frac{19}{20}+\varepsilon},\tag{19.8}$$

the implied constant depending on ε.

This will be derived from a somewhat stronger result which bounds a smoothed version $\mathcal{L}_d^*(M)$ of $\mathcal{L}_d(M)$, see (19.23).

PROPOSITION 19.4. *Suppose β_n are supported on primes. For $1\leqslant h\ll MN$ we have*

$$\mathcal{B}(M,N)\ll \|\alpha\varrho\|\,\|\beta\|\big(M^{\frac12}+N^{\frac34}M^{\frac38+\varepsilon}\big),\tag{19.9}$$

the implied constant depending on ε.

We shall prove Proposition 19.4 by bounding $\mathcal{B}(M,N)$ in terms of the linear forms $\mathcal{L}_d^*(M)$ with $d = n_1 n_2$, $N < n_1, n_2 \leqslant 2N$ and h replaced by $(n_2 - n_1)h$. The restriction to n_1, n_2 primes simplifies much of the argument, yet it is just this type that is needed for our application. The harder part of the job is the bound for the linear forms $\mathcal{L}_d^*(M)$ which will be achieved in the next section by relating this sum to a certain Poincaré series and then giving a bound for that series.

From these propositions we can quickly prove the main result. Theorem 18.4 is applicable for the sequence $c_n = \varrho_h(n)$, $1 \leqslant n \leqslant x$. The hypothesis (18.9) follows from Proposition 19.3 and (18.10) follows from Proposition 19.4. Therefore we have

$$\sum_{p \leqslant x} \varrho_h(p) \ll \varepsilon \pi(x).$$

By taking ε arbitrarily small, we conclude the proof of Theorem 19.2.

19.3. The Poincaré Series

The roots of the quadratic congruence

$$\nu^2 + 1 \equiv 0 \,(\mathrm{mod}\, n) \tag{19.10}$$

correspond to primitive representations of the modulus n as the sum of two squares. Precisely,

$$n = \gamma^2 + \delta^2 \quad \text{with } (\gamma, \delta) = 1$$

gives

$$\nu \equiv -\delta\bar{\gamma} \,(\mathrm{mod}\, n) \quad \text{with } \gamma\bar{\gamma} \equiv 1 \,(\mathrm{mod}\, n) .$$

This is a two-to-one correspondence; see the hint after (20.8). Next, we expand the pairs γ, δ to elements of the modular group

$$\sigma = \begin{pmatrix} \alpha & \beta \\ \gamma & \delta \end{pmatrix} \in \Gamma = SL_2(\mathbb{Z}) .$$

Here α is determined modulo γ, namely $\alpha\delta \equiv 1 \,(\mathrm{mod}\, \gamma)$ so σ is determined up to a translation $\begin{pmatrix} 1 & \ell \\ & 1 \end{pmatrix}$ from the left. Note that for $z = i$ we have

$$\sigma z = \frac{\alpha z + \beta}{\gamma z + \delta} = \frac{\alpha}{\gamma} - \frac{\delta}{\gamma(\gamma^2 + \delta^2)} + \frac{i}{\gamma^2 + \delta^2} ,$$

that is,

$$\mathrm{Im}\, \sigma z = \frac{1}{n} , \qquad \mathrm{Re}\, \sigma z = \frac{\alpha}{\gamma} - \frac{\delta}{\gamma n} .$$

On the other hand, we have modulo one,

$$\frac{\nu}{n} \equiv -\frac{\delta\bar{\gamma}}{n} \equiv \frac{\delta\bar{n}}{\gamma} - \frac{\delta}{\gamma n} \equiv \frac{\bar{\delta}}{\gamma} - \frac{\delta}{\gamma n} \equiv \frac{\alpha}{\gamma} - \frac{\delta}{\gamma n} ,$$

that is,

$$\frac{\nu}{n} \equiv \mathrm{Re}\, \sigma z \,\,(\mathrm{mod}\, 1), \qquad \text{for } z = i .$$

By the above transformations we conclude that the Weyl sums can be written in the following way:

$$\varrho_h(n) = \frac{1}{2} \sum_{\substack{\sigma \in \Gamma_\infty \backslash \Gamma \\ (\mathrm{Im}\, \sigma z)^{-1} = n}} e(h \,\mathrm{Re}\, \sigma z) , \qquad \text{for } z = i .$$

Given a positive integer d we decompose the Weyl sum further into cosets with respect to the congruence subgroup $\Gamma_0(d)$ as follows:

$$\varrho_h(n) = \frac{1}{2} \sum_{\tau \in \Gamma_0(d) \backslash \Gamma} \sum_{\substack{\sigma \in \Gamma_\infty \backslash \Gamma_0(d) \\ (\operatorname{Im} \sigma\tau z)^{-1} = n}} e(h \operatorname{Re} \sigma\tau z) \ .$$

We shall use this formula for $n \equiv 0 \,(\operatorname{mod} d)$. Note that for $z = i$ the condition $(\operatorname{Im} \sigma\tau z)^{-1} \equiv 0 \,(\operatorname{mod} d)$ is equivalent to $(\operatorname{Im} \tau z)^{-1} \equiv 0 \,(\operatorname{mod} d)$. Therefore,

$$\varrho_h(n) = \frac{1}{2} \sum_{\tau}^{\diamond} \sum_{\substack{\sigma \in \Gamma_\infty \backslash \Gamma_0(d) \\ (\operatorname{Im} \sigma\tau z)^{-1} = n}} e(h \operatorname{Re} \sigma\tau z) \tag{19.11}$$

if $n \equiv 0 \,(\operatorname{mod} d)$, and the right-hand side vanishes otherwise. Here, the symbol $\diamond$ indicates that the outer summation runs over

$$\tau \in \Gamma_0(d) \setminus \Gamma \quad \text{with} \ (\operatorname{Im} \tau z)^{-1} \equiv 0 \,(\operatorname{mod}\ d) \ .$$

One can easily characterize all such τ's. To this end, note that all the coset representatives of $\Gamma_0(d) \setminus \Gamma$ are given by (cf. Proposition 2.5 of [**106**]),

$$\tau = \begin{pmatrix} * & * \\ u & v \end{pmatrix} \quad \text{with } v|d, \ u \operatorname{mod}(v, d/v) \ .$$

Then, for $z = i$ the condition $(\operatorname{Im} \tau z)^{-1} = u^2 + v^2 \equiv 0 \,(\operatorname{mod}\ d)$ means $v = 1$ and

$$\tau = \begin{pmatrix} * & * \\ u & 1 \end{pmatrix} \quad \text{with } u^2 + 1 \equiv 0 \,(\operatorname{mod}\ d) \ . \tag{19.12}$$

Therefore, the number of such τ's is exactly equal to $\rho(d)$ which is relatively small.

Now we are ready to evaluate the sum of Weyl sums

$$\mathcal{G}_d(N) = \sum_{n \equiv 0\,(d)} \rho_h(n) G(n) \ ,$$

where G is a smooth function supported in a dyadic segment $[N, 2N]$. For notational simplicity, we first consider a more carefully crafted sum

$$\mathcal{F}_d(Y) = \sum_{n \equiv 0\,(d)} \rho_h(n) F(2\pi h/n) \ ,$$

where F is a smooth function supported in a dyadic segment $[Y, 2Y]$. Naturally, these sums translate into one another by taking $G(t) = F(2\pi h/t)$ and $NY \asymp h$. By (19.11) we obtain

$$\mathcal{F}_d(Y) = \frac{1}{2} \sum_{\tau}^{\diamond} P(\tau z) \ , \quad \text{for } z = i \ , \tag{19.13}$$

where the symbol $\diamond$ indicates that the sum runs over τ satisfying (19.12) and where $P(z)$ is a kind of Poincaré series for the group $\Gamma_0(d)$ of level d, namely

$$P(z) = \sum_{\sigma \in \Gamma_\infty \backslash \Gamma_0(d)} F(2\pi h \operatorname{Im} \sigma z) e(h \operatorname{Re} \sigma z).$$

We shall estimate $P(z)$ for each $z \in \mathcal{H}$ uniformly in d by using the spectral theorem. Since $P(z)$ is smooth and square integrable the spectral theorem gives

$$P(z) = \sum_j \langle P, u_j \rangle u_j(z) + \sum_{\mathfrak{a}} \frac{1}{4\pi} \int_{-\infty}^{\infty} \langle P, E_{\mathfrak{a}}(*, \tfrac{1}{2} + it) \rangle E_{\mathfrak{a}}(z, \tfrac{1}{2} + it) dt,$$

where $\{u_j(z)\}$ is an orthonormal system of Maass cusp forms together with the constant function for the zero eigenvalue and $\{E_{\mathfrak{a}}(z,s)\}$ is the eigenpacket of the Eisenstein series; cf. [**93**]. The spectral series converges absolutely and uniformly on compacta. By the Cauchy-Schwarz inequality we obtain

$$|P(z)|^2 \leqslant K(z)R, \tag{19.14}$$

where

$$K(z) = \sum_j h(t_j)|u_j(z)|^2 + \sum_{\mathfrak{a}} \frac{1}{4\pi} \int_{-\infty}^{\infty} h(t)|E_{\mathfrak{a}}(z, \tfrac{1}{2} + it)|^2 dt$$

and

$$R = \sum_j h(t_j)^{-1}|\langle P, u_j\rangle|^2 + \sum_{\mathfrak{a}} \frac{1}{4\pi} \int_{-\infty}^{\infty} h(t)^{-1}|\langle P, E_{\mathfrak{a}}(*, \frac{1}{2} + it)\rangle|^2 dt,$$

for any $h(t) \asymp (1+t^4)^{-1}$. Here, we have introduced the factor $h(t)$ for convergence reasons.

To estimate $K(z)$ observe that it is the spectral decomposition of the automorphic kernel

$$K(z) = \sum_{\gamma \in \Gamma_0(d)} k\big(u(\gamma z, z)\big),$$

where $k(u)$ is the Harish-Chandra/Selberg transform of $h(t)$ (cf. [**93**]), and $u(z,w)$ the hyperbolic distance function given by $u(z,w) = |z-w|^2/4\mathrm{Im}z\,\mathrm{Im}w$. We can make $k(u)$ non-negative by an appropriate choice of $h(t)$; for example,

$$h(t) = (1+t^2)^{-1} - (4+t^2)^{-1} = 3(1+t^2)^{-1}(4+t^2)^{-1}$$

does the job because its Fourier transform is positive and decreasing on $\mathbb{R}^+$. Given τ in the modular group change z into τz, which amounts to conjugating the group $\Gamma_0(d)$ into $\tau^{-1}\Gamma_0(d)\tau$ in the automorphic kernel $K(z)$. Then, extending the summation to the full modular group we can only increase $K(\tau z)$, showing that

$$K(\tau z) \ll y + y^{-1}, \tag{19.15}$$

where the implied constant is absolute. To see this, return to the spectral expansion of the enlarged automorphic kernel with respect to the full modular group and estimate the cusp forms and Eisenstein series trivially.

To estimate R we first compute the inner products of the Poincaré series $P(z)$ against the relevant eigenforms. Let the Fourier expansion of $u_j(z)$ be

$$u_j(z) = y^{\frac{1}{2}} \sum_{n \neq 0} \lambda_j(n) K_{it_j}(2\pi|n|y)e(nx),$$

where $K_\nu(y)$ is the Bessel–Macdonald function ([**66**], 8.40). Then, by the unfolding method we find that (cf. (16.20) of [**109**])

$$\langle P, u_j\rangle = (2\pi h)^{\frac{1}{2}} \overline{\lambda}_j(h)\hat{F}(t_j), \tag{19.16}$$

where

$$\hat{F}(t) = \int_0^{\infty} F(y)K_{it}(y)y^{-\frac{3}{2}}dy.$$

The same formula is valid for the inner product against the Eisenstein series (cf. (16.23) of [**109**]) .

The integral transform $\hat{F}(t)$ decays rapidly. More precisely, suppose $F(y)$ is supported in the dyadic interval $[Y, 2Y]$ and its derivatives satisfy $|F^{(j)}| \leqslant Y^{-j}$ for $0 \leqslant j \leqslant 4$. Then, integrating by parts four times, we derive the bound

$$\hat{F}(t) \ll h(t)(\mathrm{ch}\pi t)^{-\frac{1}{2}}(Y^{it} + Y^{-it} + L)Y^{-\frac{1}{2}}, \tag{19.17}$$

where $L = \log(6 + Y + Y^{-1})$ and the implied constant is absolute. Actually (19.17) combines within it three different bounds obtained separately in the ranges $0 < it \leqslant \frac{1}{2}$, $0 \leqslant t < \frac{1}{2}$, $\frac{1}{2} \leqslant t < \infty$, by methods similar to those applied in [**31**], Lemma 7.1. We point out that the expression $Y^{it} + Y^{-it}$ in (19.17) reflects clearly the difference in the behaviour of $\hat{F}(t)$, depending on whether t is real or purely imaginary. In particular, when t is imaginary this majorizes the contribution from any exceptional eigenvalues. We have,

$$h(t_j)^{-1}|\langle P, u_j\rangle|^2 \ll h|\lambda_j(h)|^2 g(t_j)Y^{-1},$$

where

$$g(t) = h(t)(\mathrm{ch}\pi t)^{-1}(Y^{2it} + Y^{-2it} + L^2). \tag{19.18}$$

The same bound holds true for the inner product against the Eisenstein series. These yield

$$R \ll hQY^{-1} \tag{19.19}$$

where

$$Q = \sum_j g(t_j)|\lambda_j(h)|^2 + \text{Eisenstein part}.$$

Our goal is to show that the spectral sum Q is essentially bounded; precisely we shall obtain (19.22). Such sums are estimated in various papers and books. For example, we may get (19.22) directly from (16.56) and (16.58) of [**109**]; nevertheless, we give here a full proof from scratch because it is quite quick and should convince the reader that the exceptional spectrum is acceptable for our goal. To this end, we appeal to Kuznetsov's formula [**121**] for the group $\Gamma_0(d)$ giving (see Proposition 9.2 of [**31**] or Theorem 16.3 of [**109**]),

$$Q = g_0 + \sum_{c \equiv 0(d)} c^{-1} S(h, h; c)\tilde{g}\left(\frac{4\pi h}{c}\right), \tag{19.20}$$

where $S(h, h; c)$ are Kloosterman sums, and where

$$g_0 = \frac{1}{\pi^2}\int_{-\infty}^{\infty} \mathrm{sh}(\pi t)g(t)t\, dt\ ,$$

$$\tilde{g}(x) = \frac{2i}{\pi}\int_{-\infty}^{\infty} J_{2it}(x)g(t)t\, dt\ ,$$

with $J_\nu(y)$ the Bessel function (see [**66**], 8.40). Clearly,

$$g_0 \ll L^2. \tag{19.21}$$

In order to estimate $\tilde{g}(x)$ we change the real variable t into the complex one $s = \frac{1}{2} + it$ and then move the integration to the vertical line $s = \sigma + it$ with $\frac{1}{2} < \sigma < 1$, by Cauchy's theorem getting

$$\tilde{g}(x) = \frac{2}{\pi i}\int_{(\sigma)} J_{2s-1}(x)g\big((s - \tfrac{1}{2})i\big)\big(s - \tfrac{1}{2}\big)ds.$$

Hence, by the simple bounds (see [**66**], 8.411.6)

$$J_{2s-1}(x) \ll e^{\pi|s|}x^{2\sigma-1},$$
$$g\big((s-\tfrac{1}{2})i\big) \ll |s|^{-4}e^{-\pi|s|}\left(Y^{2\sigma-1}+Y^{1-2\sigma}+L^2\right),$$

we derive

$$\tilde{g}(x) \ll x^{2\sigma-1}\left(Y^{2\sigma-1}+Y^{1-2\sigma}+L^2\right).$$

Inserting these estimates into (19.20) we obtain

$$Q \ll L^2 + h^{2\sigma-1}\left(Y^{2\sigma-1}+Y^{1-2\sigma}+L^2\right)\sum_{c\equiv 0(d)} c^{-2\sigma}|S(h,h;c)|,$$

for any $3/4 < \sigma < 1$. Next, by Weil's bound, $|S(h,h;c)| \leqslant (h,c)^{\frac{1}{2}}c^{\frac{1}{2}}\tau(c)$, we estimate the sum of Kloosterman sums by

$$\sum_{c\equiv 0(d)} (h,c)^{\frac{1}{2}}c^{\frac{1}{2}-2\sigma}\tau(c) \ll \left(\sigma-\tfrac{3}{4}\right)^{-2}d^{\frac{1}{2}-2\sigma}(h,d)^{\frac{1}{2}}\tau(hd).$$

Choosing $\sigma = \frac{3}{4}+\frac{1}{2L}$, we get

$$Q \ll L^2 + L^2(Y+Y^{-1})^{\frac{1}{2}}h^{\frac{1}{2}}d^{-1}(h,d)^{\frac{1}{2}}\tau(hd). \tag{19.22}$$

Finally, combining (19.3), (19.15), (19.19), and (19.22), we conclude:

PROPOSITION 19.5. *Suppose F is supported on $[Y,2Y]$ and has derivatives $|F^{(j)}| \leqslant Y^{-j}$, $0 \leqslant j \leqslant 4$. Then the Poincaré series in (19.13), generated by F, is bounded by*

$$P(\tau z) \ll (y+y^{-1})^{\frac{1}{2}}\Big(\frac{h}{Y}\Big)^{\frac{1}{2}}\left[1+(Y+Y^{-1})^{\frac{1}{2}}h^{\frac{1}{2}}d^{-1}(h,d)^{\frac{1}{2}}\tau(hd)\right]^{\frac{1}{2}}\log(Y+Y^{-1}),$$

uniformly for $z = x+iy \in \mathcal{H}$ and τ in the modular group, the implied constant being absolute.

19.4. Estimation of Linear Forms

In this section we prove Proposition 19.3. By (19.13) and Proposition 19.5 for $z = i$ (so $y = 1$), we obtain:

PROPOSITION 19.6. *Suppose G is supported on $[N,2N]$ and has derivatives $G^{(j)} \ll N^{-j}$, $0 \leqslant j \leqslant 4$. Let $1 \leqslant h \ll N$. Then*

$$\sum_{n\equiv 0(d)} \varrho_h(n)G(n) \ll \tau(d)N^{\frac{1}{2}}\left[1+\tau(hd)(h,d)^{\frac{1}{2}}d^{-1}N^{\frac{1}{2}}\right]^{\frac{1}{2}}\log 2N,$$

where the implied constant depends only on D.

PROOF. Put $G(n) = F(2\pi h/n)$, $z = i$, and sum over the relevant τ. □

By Proposition 19.6 we obtain:

LEMMA 19.7. *Let $1 \leqslant h \ll dM$. Then*

$$\mathcal{L}_d^*(M) = \sum_m \varrho_h(dm)g(m) \ll \tau(hd)^{\frac{3}{2}}(h,d)^{\frac{1}{4}}d^{\frac{1}{4}}M^{\frac{3}{4}}\log 2M, \tag{19.23}$$

where g is any function supported on $[M,2M]$ with derivatives bounded by $g^{(j)} \ll M^{-j}$, $0 \leqslant j \leqslant 4$.

Take g such that $g(m) = \Delta^4$ in $M+\Delta M \leqslant m \leqslant 2M-\Delta M$ and $0 \leqslant g(m) \leqslant \Delta^4$ in the remaining segments. We have

$$\mathcal{L}_d(M) = \Delta^{-4}\mathcal{L}_d^*(M) + O\big((\Delta M + 1)M^{\varepsilon}\big),$$

where the error term bounds the contribution from the segments $M < m < M+\Delta M$ and $2M - \Delta M < m < 2M$, using the trivial estimate $\varrho_h(dm) \ll M^{\varepsilon}$. Choosing Δ optimally we get

$$\mathcal{L}_d(M) \ll |\mathcal{L}_d^*(M)|^{\frac{1}{5}} M^{\frac{4}{5}+\varepsilon} + M^{\varepsilon}.$$

This yields Proposition 19.3 by (19.23).

REMARKS. The above transition from the smooth function $g(m)$ in (19.23) to the sharp characteristic function is wasteful, yet it loses nothing in the range of uniformity for d.

19.5. Estimation of Bilinear Forms

In this section we prove Proposition 19.4. First we arrange the bilinear form as follows:

$$\mathcal{B}(M,N) = \sum_m \alpha_m \sum_{\delta^2+1\equiv 0(m)} \sum_{(n,m)=1} \beta_n \sum_{\substack{\nu^2+1\equiv 0(mn)\\ \nu\equiv\delta(m)}} e\Big(\frac{\nu h}{mn}\Big).$$

Then, by Cauchy's inequality, we obtain

$$|\mathcal{B}(M,N)|^2 \leqslant \|\alpha\varrho\|^2 \sum_m g(m) \sum_{\delta^2+1\equiv 0(m)} \Big|\sum_n \sum_\nu\Big|^2,$$

where g is any function which majorizes the characteristic function of the interval $[M, 2M]$. In fact, we choose g to be smooth and as in (19.23), except that it is now supported on $[M/2, 4M]$. Squaring out the inner sums and changing the order of summation we obtain

$$|\mathcal{B}(M,N)|^2 \leqslant \|\alpha\varrho\|^2 \sum_{n_1} \sum_{n_2} \beta_{n_1}\bar{\beta}_{n_2}\mathcal{B}_{n_1n_2}(M),$$

where

$$\mathcal{B}_{n_1n_2}(M) = \sum_{(m,n_1n_2)=1} g(m) \sum\sum_{\substack{\nu_j^2+1\equiv 0(mn_j)\\ \nu_1\equiv\nu_2(m)}} e\Big(\nu_1\frac{h}{mn_1} - \nu_2\frac{h}{mn_2}\Big).$$

We shall estimate each $\mathcal{B}_{n_1n_2}(M)$ separately. For $n_1 = n_2$ we shall use the trivial bound $\mathcal{B}_{nn}(M) \ll M$. If $n_1 \neq n_2$, since we assumed they are primes, we have $(n_1, n_2) = 1$. In this case m, n_1, n_2 are pairwise coprime so we can write

$$\mathcal{B}_{n_1n_2}(M) = \sum_{\delta^2+1\equiv 0(n_1n_2)} \sum_{(m,n_1n_2)=1} g(m) \sum_{\substack{\nu^2+1\equiv 0(mn_1n_2)\\ \nu\equiv\delta(n_1n_2)}} e\Big(\nu(n_2-n_1)\frac{h}{mn_1n_2}\Big).$$

Here, the condition $(m, n_1n_2) = 1$ can be removed at the cost of two error terms bounded trivially by

$$4 \sum_{m\equiv 0(n_j)} g(m)\varrho(m) \ll N^{-1}M$$

for j = 1,2, again taking advantage of the fact that n_1, n_2 are primes. The resulting sum is now recognized to be exactly $\mathcal{L}^*_{n_1 n_2}(M)$ with h replaced by $(n_2 - n_1)h$. Hence, provided $N^2 \ll M$, (19.23) is applicable giving

$$\mathcal{B}_{n_1 n_2}(M) \ll N^{-1}M + N^{\frac{1}{2}} M^{\frac{3}{4}+\varepsilon}.$$

Summing over all n_1, n_2 we infer that

$$\mathcal{B}^2(M,N) \ll \|\alpha\varrho\|^2 \|\beta\|^2 \left(M + N^{\frac{3}{2}} M^{\frac{3}{4}+\varepsilon}\right) ,$$

which concludes the proof of Proposition 19.4 in the case $N^2 \ll M$. In the opposite case where this last condition does not hold, the proposition is trivial.

CHAPTER 20

Marching over Gaussian Primes

Polonaise

20.1. Introduction

Having spent the previous chapter introducing the set of ideas which finally allow one to wring primes from the sieve we want to illustrate how this is done in an attractive specific case. Our first choice would be to present our own work [**54**] on primes $p = X^2 + Y^4$, but it is simply not possible to give a complete account without filling many more pages than we are willing to spend in this book. Similarly, an account of the arguments used by Heath-Brown in his subsequent paper [**90**] on primes $p = X^3 + 2Y^3$ would also take an inordinate amount of space. We have chosen to present an account of the theorem proved first in [**45**], on the primes $p = \ell^2 + m^2$ where ℓ is prime. Moreover, we introduce some simplification by concentrating on just this specific problem, whereas the result in [**45**] was somewhat more general. Even so, the job is not an easy one.

In Section 20.2 we develop a large sieve inequality for the roots of a quadratic congruence, which we need here and have also applied in a number of other works including [**54**].

In Section 20.3 we consider the sequence $\mathcal{A} = (a_n)$ where

$$a_n = \sum_{\ell^2+m^2=n} \gamma_\ell \tag{20.1}$$

with γ_ℓ a rather general sequence. Here, using the results of Section 20.2 we show that $\mathcal{A}$ has a level of distribution which is essentially optimal. Although our main goal is to capture primes, we cannot resist the opportunity, when presented with a sequence having such good distribution, to illustrate the application of Bombieri's asymptotic sieve. As a result we have

THEOREM 20.1. *Let γ_ℓ be arbitrary numbers with $0 \leqslant \gamma_\ell \leqslant 1$ and supported on primes. Then, for any $k \geqslant 2$ we have*

$$\sum_{\ell^2+m^2\leqslant x} \gamma_\ell \Lambda_k(\ell^2 + m^2) = kHX(\log x)^{k-1}\{1 + O((\log x)^{-\frac{1}{4}})\} + O(x(\log x)^{-A})$$

with

$$H = \prod_p \Big(1 - \frac{\chi_4(p)}{p-1}\Big), \qquad X = \sum_{\ell\leqslant\sqrt{x}} \gamma_\ell \sqrt{x - \ell^2}\ , \tag{20.2}$$

where χ_4 is the non-principal character of modulus four.

Note that the result is trivial if $X \ll x(\log x)^{-A}$. Otherwise, one can use Proposition 20.7 with $D = x(\log x)^{-4A-4}$, showing that Theorem 3.5 is applicable and the theorem follows. As a special case, we derive the following (obviously Theorem 20.1 can be applied with coefficients $\gamma_\ell \ll \log \ell$ by a proper normalization):

COROLLARY 20.2. *We have for $k \geqslant 2$,*

$$\sum_{\ell^2+m^2\leqslant x} \Lambda(\ell)\Lambda_k(\ell^2+m^2) = \frac{\pi k}{4} Hx(\log x)^{k-1}\left(1+O(\log x)^{-\frac{1}{4}}\right). \tag{20.3}$$

As we have seen in Bombieri's asymptotic sieve the assumption $k \geqslant 2$ is necessary due to the parity problem. In Chapter 18 we showed how to introduce a bilinear remainder $T(x, D)$ which circumvents the parity problem when $k = 1$. In the fourth section of this chapter (see Proposition 20.8) we successfully estimate this remainder for the sequence $\gamma_\ell = \log \ell$ at primes, so that by Theorem 18.6 we deduce Corollary 20.2 for the case $k = 1$, that is for primes!

THEOREM 20.3. *As $x \to \infty$ we have*

$$\sum\sum_{\ell^2+m^2\leqslant x} \Lambda(\ell)\Lambda(\ell^2+m^2) \sim \frac{\pi}{4} Hx\ .$$

The Gaussian primes $\ell + im$ with free co-ordinates ℓ, m appear so densely that a well-known conjecture of Erdös asserts that one can walk on them to infinity taking absolutely bounded steps. Of course, this is not possible if one steps only on the primes in Theorem 20.3 which lie vertically above the rational primes.

20.2. Large Sieve for Quadratic Roots

In this section we consider the distribution of the roots of the congruence

$$\nu^2 + 1 \equiv 0 \,(\mathrm{mod}\, d) \tag{20.4}$$

as d varies. We study these via the Weyl harmonics

$$\rho_h(d) = \sum_{\nu^2+1\equiv 0(d)} e\Big(\frac{\nu h}{d}\Big)\ . \tag{20.5}$$

We shall prove

PROPOSITION 20.4. *For any complex numbers α_h we have*

$$\sum_{d\leqslant D}\Big|\sum_{|h|\leqslant H} \alpha_h \rho_h(d)\Big| \leqslant 50 D^{\frac{1}{2}}(D+H)^{\frac{1}{2}}\|\alpha\|\ .$$

This proposition follows immediately from the following inequality:

$$\sum_{d\leqslant D}\ \sum_{\nu^2+1\equiv 0(d)}\Big|\sum_{n\leqslant N} \alpha_n e\Big(\frac{\nu n}{d}\Big)\Big| \leqslant 50 D^{\frac{1}{2}}(D+N)^{\frac{1}{2}}\|\alpha\|\ . \tag{20.6}$$

In turn, (20.6) is a consequence of the large sieve for the additive characters $e(\nu n/d)$. Specifically, we shall show

PROPOSITION 20.5. *For any complex numbers α_n we have*

$$\sum_{D<d\leqslant 2D}\ \sum_{\nu^2+1\equiv 0(d)}\Big|\sum_{n\leqslant N} \alpha_n e\Big(\frac{\nu n}{d}\Big)\Big|^2 \leqslant 200(D+N)\|\alpha\|^2.$$

From this, (20.6) follows by dyadic subdivision followed by Cauchy's inequality.

We shall deduce Proposition 20.5 from the classical large sieve inequality of Davenport and Halberstam for additive characters. To this end we need a well-spacing of the rational points ν/d modulo one. Thinking of these as typical rational points we can see that these are spaced at a distance D^{-2} but, taking advantage of their properties as roots of the congruence, we can improve this to D^{-1}, best possible since there are about D of them.

PROOF. The argument begins with the observation that there is a one-to-one correspondence between the roots ν of (20.4) and the representations

$$d = r^2 + s^2, \ (r,s) = 1, \ -s < r \leqslant s, \tag{20.7}$$

given by

$$\nu \equiv r\bar{s} \ (\operatorname{mod} d). \tag{20.8}$$

Hint: If d is the power of a prime $p \equiv 1 \,(\operatorname{mod} 4)$ we have two solutions of (20.4) and two representations of (20.7) and obviously (20.8) gives the correspondence. If $p \equiv 3 \,(\operatorname{mod} 4)$ we have no solution for either. For $d = 2^\alpha$ the argument is trivial. For other integers the claim follows by multiplicativity.

Hence, we have

$$\frac{\nu}{d} \equiv \frac{r\bar{s}}{d} \equiv -\frac{r\bar{d}}{s} + \frac{r}{ds} \equiv -\frac{\bar{r}}{s} + \frac{r}{ds} \ (\operatorname{mod} 1) \ .$$

The last term is bounded by

$$\frac{|r|}{ds} = \frac{|r|}{(r^2+s^2)s} < \frac{1}{2s^2} \ .$$

Let d run over the segment $8D < d \leqslant 9D$. Consider the roots with $s > 1$ and r having fixed sign. The corresponding fractions $\nu/d \neq \nu_1/d_1$ are spaced by

$$\left\| \frac{\nu}{d} - \frac{\nu_1}{d_1} \right\| > \frac{1}{ss_1} - \max\left\{ \frac{1}{2s^2}, \frac{1}{2s_1^2} \right\} > \frac{1}{2ss_1} > \frac{1}{18D}$$

because $2\sqrt{D} < s, s_1 < 3\sqrt{D}$. Hence, by the large sieve,

$$\sum_{8D<d\leqslant 9D} \ \sum_{\nu^2+1\equiv 0(d)} \Big| \sum_{n\leqslant N} \alpha_n e\Big(\frac{\nu n}{d}\Big) \Big|^2 \leqslant 2(18D+N)\|\alpha\|^2 \ ,$$

where the factor two comes from considering the integers r having the other sign. Adding up over a few of these intervals for d, we complete the proof of Proposition 20.5. □

Proposition 20.5 generalizes easily to give:

COROLLARY 20.6. *Let $q \geqslant 1$. For any complex numbers α_n we have*

$$\sum_{\substack{D<d\leqslant 2D \\ (d,q)=1}} \ \sum_{\nu^2+1\equiv 0(d)} \Big| \sum_{n\leqslant N} \alpha_n e\Big(\frac{\nu n\bar{q}}{d}\Big) \Big|^2 \leqslant 200(qD+N+q)\|\alpha\|^2. \tag{20.9}$$

This follows by splitting the summation over n into progressions $\alpha \,(\operatorname{mod} q)$ and applying Proposition 20.5 to each of these separately.

20.3. Level of Absolute Distribution

We consider the sequence $\mathcal{A} = (a_n)$ with

$$a_n = \sum_{l^2+m^2=n} \gamma_\ell \, . \tag{20.10}$$

We write

$$A_d(x) = \sum_{\substack{n\leqslant x\\ n\equiv 0(d)}} a_n = \sum_{\substack{\ell^2+m^2\leqslant x\\ \ell^2+m^2\equiv 0(d)}} \gamma_\ell \, .$$

We define the remainder terms $r_d(x)$ by

$$A_d(x) = \frac{\rho(d)}{d} X + r_d(x) \tag{20.11}$$

where $\rho(d) = \rho_0(d)$ is the number of roots of (20.4) and X is given by (20.2).

Our bound for the remainder is

PROPOSITION 20.7. *Let γ_ℓ be complex numbers, $|\gamma_\ell| \leqslant 1$, and supported on primes $1 \leqslant \ell \leqslant \sqrt{x}$. Then, for $D \leqslant x$ we have*

$$R(x;D) = \sup_{x'\leqslant x} \sum_{d\leqslant D} |r_d(x')| \ll D^{\frac14} x^{\frac34} \log x, \tag{20.12}$$

where the implied constant is absolute.

PROOF. For notational simplicity, we treat only the case $x' = x$; the general case is similar. First we remove from $A_d(x)$ the terms with $(\ell, d) \neq 1$ getting

$$A_d^*(x) = \sum_{\substack{\ell^2+m^2\leqslant x\\ \ell^2+m^2\equiv 0(d)\\ (\ell,d)\neq 1}} \gamma_\ell.$$

Since ℓ is prime, $(\ell, d) \neq 1$ means that $\ell | d$ and so we can easily compare $A_d^*(x)$ with the original sum $A_d(x)$, namely

$$|A_d(x) - A_d^*(x)| \leqslant \sum_{\ell | d} |\gamma_\ell| \sum_{\substack{m^2+1\leqslant x\ell^{-2}\\ m^2+1\equiv 0(d/\ell)}} 1,$$

and so

$$\sum_{d\leqslant D} |A_d(x) - A_d^*(x)| \leqslant \sum_{\ell} |\gamma_\ell| \sum_{m^2+1\leqslant x\ell^{-2}} \tau(m^2+1) \ll x^{\frac12} (\log x)^2 \, ,$$

which is much better than what is needed.

Next, we shall evaluate $A_d^*(x)$ by summing over m and using Fourier analysis. As usual, we find it easier to do this for a smooth approximant:

$$A_d^*(f) = \sum_{\substack{\ell^2+m^2\equiv 0(d)\\ (\ell,d)=1}} \gamma_\ell f(\ell^2+m^2).$$

Here f is a smooth function such that

$$\operatorname{supp} f = [1,\, x], \quad 0 \leqslant f(u) \leqslant 1 \;,$$
$$f(u) = 1, \qquad \text{if } \; y \leqslant u \leqslant x - y \;,$$
$$f^{(j)}(u) \ll y^{-j}, \quad \text{for } j = 0, 1, 2 \;,$$

where the parameter y will be chosen later, subject to $\sqrt{x} \leqslant y \leqslant \frac{1}{2}x$. We estimate what is lost in the process of smoothing as follows:

$$\begin{aligned}\sum_{d \leqslant D} |A_d^*(x) - A_d^*(f)| &\leqslant \Big(\sum_{\ell^2+m^2 \leqslant y} + \sum_{x-y \leqslant \ell^2+m^2 \leqslant x} \Big) \tau(\ell^2 + m^2) \\ &\ll y \log x \;,\end{aligned} \tag{20.13}$$

which will suffice provided y is not too close to x.

We now proceed to evaluate $A_d^*(f)$.

$$A_d^*(f) = \sum_{\nu^2+1 \equiv 0(d)} \; \sum_{(\ell,d)=1} \gamma_\ell \sum_{m \equiv \nu\ell(d)} f(\ell^2 + m^2).$$

We remove the condition $(\ell, d) = 1$, replacing $A_d^*(f)$, say by $A_d(f)$, and making the same error $O(y \log x)$ as when we imposed it. By Poisson's formula

$$A_d(f) = \frac{1}{d} \sum_{\ell} \gamma_\ell \sum_{k \in \mathbb{Z}} \rho_{k\ell}(d) F_\ell\Big(\frac{k}{d}\Big),$$

where $F_\ell(v)$ is the Fourier transform

$$F_\ell(v) = \int f(\ell^2 + t^2) e(-vt) dt.$$

The zero frequency $k = 0$ yields the main term

$$\begin{aligned} M_d(f) &= \frac{\rho(d)}{d} \sum_{\ell} \gamma_\ell F_\ell(0) = \frac{\rho(d)}{d} \Big\{ \sum_{\ell} \gamma_\ell \sqrt{x - \ell^2} + O(y) \Big\}, \\ &= \frac{\rho(d)}{d} X + O\Big(\frac{\rho(d)}{d} y \Big). \end{aligned} \tag{20.14}$$

The sum over the frequencies $k \neq 0$ gives the remainder term $r_d(f)$. This will be estimated using the large sieve. To enable the use of Proposition 20.4 we need to separate the variable d from k and ℓ, and we accomplish this by the change of variable $t = w\sqrt{x}/k$ in the Fourier integral, getting

$$F_\ell\Big(\frac{k}{d}\Big) = \frac{2\sqrt{x}}{k} \int_0^\infty f\Big(\ell^2 + \frac{xw^2}{k^2}\Big) \cos\Big(\frac{2\pi w \sqrt{x}}{d}\Big) dw. \tag{20.15}$$

This expression is useful for small $|k|$. However, if we estimate using this expression, the sum over k diverges, hence we give an alternative expression. Integrating (20.15) by parts twice, we get

$$F_\ell\Big(\frac{k}{d}\Big) = \frac{\sqrt{x} d^2}{\pi^2 k^3} \int_0^\infty \Big(f' + \frac{2w^2 x}{k^2} f'' \Big) \Big(\ell^2 + \frac{xw^2}{k^2} \Big) \cos\Big(\frac{2\pi w \sqrt{x}}{d}\Big) dw. \tag{20.16}$$

Put

$$R(f, D) = \sum_{D < d \leqslant 2D} |r_d(f)|.$$

We have

$$R(f,D) \leqslant \frac{1}{D} \sum_{D<d\leqslant 2D} \Big| \sum\sum_{k\ell\neq 0} \gamma_\ell F_\ell\Big(\frac{k}{d}\Big)\rho_{k\ell}(d)\Big| \leqslant \sum_K R_K(f,D) ,$$

where

$$R_K(f,D) = \frac{1}{D} \sum_{D<d\leqslant 2D} \Big| \sum_{K\leqslant |k|<2K} \sum_\ell \gamma_\ell F_\ell\Big(\frac{k}{d}\Big)\rho_{k\ell}(d)\Big|$$

and K runs over powers of 2.

For $K \leqslant D\sqrt{x}y^{-1}$ we use (20.15) getting, by Proposition 20.4,

$$\begin{aligned} R_K(f,D) &\leqslant \frac{2\sqrt{x}}{DK} \int_0^{2K} \sum_{D<d\leqslant 2D} \Big| \sum_{K\leqslant |k|<2K} \frac{K}{k}\gamma_\ell f\Big(\ell^2 + \frac{xw^2}{k^2}\Big)\rho_{k\ell}(d)\Big| dw \\ &\ll \frac{\sqrt{x}}{D} D^{\frac12}(D+K\sqrt{x})^{\frac12}(K\sqrt{x})^{\frac12}\log x \ll x^{\frac54}\Big(\frac{K}{y}\Big)^{\frac12}\log x . \end{aligned}$$

For $K > D\sqrt{x}y^{-1}$ we use (20.16) getting, again by Proposition 20.4,

$$\begin{aligned} R_K(f,D) \leqslant \frac{4\sqrt{x}D}{\pi^2 K^3} \int_0^{2K} &\sum_{D<d\leqslant 2D} \Big| \sum_{K\leqslant |k|<2K} \frac{K^3}{k^3} \\ &\cdot \sum_\ell \gamma_\ell \Big(f' + \frac{2w^2x}{k^2}f''\Big)\Big(\ell^2 + \frac{xw^2}{k^2}\Big)\rho_{k\ell}(d)\Big| dw \\ &\ll \frac{x^{\frac32}D}{y^2K^2} D^{\frac12}(D+K\sqrt{x})^{\frac12}(K\sqrt{x})^{\frac12}\log x \ll \frac{x^2D^{\frac32}}{y^2K}\log x. \end{aligned}$$

Summing as K runs over powers of 2, we conclude that

$$R(f,D) \ll \frac{x}{y}(Dx)^{\frac12}\log x. \tag{20.17}$$

Combining (20.13), (20.14), (20.17), we deduce that

$$R(x,D) \ll y\log x + \frac{x}{y}(Dx)^{\frac12}\log x.$$

Choosing $y = x^{\frac34}D^{\frac14}$, we complete the proof of Proposition 20.7 in the case that $x' = x$ in (20.12). The general case is similar. □

20.4. Level of Bilinear Distribution

Let $\mathcal{A} = (a_n)$ be as in (20.10), but now with the specific choice

$$\gamma_\ell = \begin{cases} \log \ell, & \text{if } \ell \text{ prime,} \\ 0, & \text{else.} \end{cases} \tag{20.18}$$

We shall estimate bilinear forms of the following type (see Theorem 18.6):

$$T(x,D) = \sum_{n\leqslant D} \Big| \sum_{\substack{mn\leqslant x \\ z<m\leqslant z^2}} a_{mn}\mu(m)\Big| . \tag{20.19}$$

PROPOSITION 20.8. *Let* $D = x^{1-\delta}$, $z = x^{\delta}$, *with* δ *a small positive number. Then we have*

$$T(x, D) \ll x(\log x)^{-A}$$

for any $A > 0$.

Note that it suffices to prove the bound

$$B(M, N) \ll MN(\log MN)^{-A} \tag{20.20}$$

for $M = N^{\delta}$, where

$$B(M, N) = \sum_{n\sim N}\Big|\sum_{m\sim M} a_{mn}\mu(m)\Big|$$

and $m \sim M$ means $M < m \leqslant M'$ for some unspecified M', $M < M' \leqslant 2M$. To see this we split the summation in $T(x, D)$ into subintervals of type

$$M < m \leqslant (1+\theta)M\ ,$$
$$N < n \leqslant (1+\theta)N\ ,$$

for $\theta = (\log x)^{-C}$ with a suitable constant $C = C(A)$ and replace the condition $mn \leqslant x$ by $MN \leqslant x$. In this way we cover exactly the whole range except for those satisfying $(1+\theta)^{-2}x < mn < (1+\theta)^2 x$, for which the trivial bound suffices.

Next, we can restrict the sum $B(M, N)$ by the condition $(m, n) = 1$. To do this we transfer the common factor $d = (m, n)$ from m to n if $d < N^{\frac{\delta}{2}}$ and estimate trivially if $d \geqslant N^{\frac{\delta}{2}}$.

Therefore, it suffices to prove that

$$B^*(M, N) \ll MN(\log MN)^{-A} \tag{20.21}$$

where

$$B^*(M, N) = \sum_{n\sim N}\Big|\sum_{\substack{m\sim M\\(m,n)=1}} a_{mn}\mu(m)\Big|\ .$$

Typically, for bilinear forms of this nature, one applies Cauchy's inequality and interchanges the order of summation. However, in our case a_{mn} has multiplicity which would become more difficult to treat after application of Cauchy's inequality. Our next step is to express the variables in terms of Gaussian integers so that there is no multiplicity, after which Cauchy's inequality can be applied without leading to such complications.

In the following $\mathfrak{a}$, $\mathfrak{b}$, $\mathfrak{m}$, $\mathfrak{n}$ denote Gaussian integers. Then we can write

$$a_n = \sum_{N\mathfrak{a}=n} \gamma(\mathfrak{a})$$

with

$$\gamma(\mathfrak{a}) = \gamma_\ell \text{ and } \ell = \mathrm{Re}\,\mathfrak{a}. \tag{20.22}$$

Hence,

$$B^*(M, N) \leqslant \sum_{\mathfrak{n}}\Big|\sum_{(\mathfrak{m},\mathfrak{n})=1}^{\pi} \gamma(\mathfrak{m}\mathfrak{n})\mu(\mathfrak{m})\Big|$$

where $N(\mathfrak{m}) \sim M$, $N(\mathfrak{n}) \sim N$ and $\mu(\mathfrak{m})$ is the Möbius function for the Gaussian field. Here $\sum_{\mathfrak{m}}^{\pi}$ restricts the summation to primary integers $\mathfrak{m}$, which means, not divisible by any rational integer greater than one, or equivalently, integers $\mathfrak{m}$ satisfying $(\mathfrak{m}, \overline{\mathfrak{m}}) \mid (1+i)$.

We next remove the coprimality condition using the same argument by which we inserted it, but now in the Gaussian domain. Therefore, it suffices to show that

$$\mathcal{B}(M,N) \ll MN(\log MN)^{-A}$$

where

$$\mathcal{B}(M,N) = \sum_{\mathfrak{n}\sim N} \Big| \sum_{\mathfrak{m}\sim M}^{\pi} \gamma(\mathfrak{m}\mathfrak{n})\mu(\mathfrak{m}) \Big|$$

and $\mathfrak{n} \sim N$ means $N(\mathfrak{n}) \sim N$. By Cauchy's inequality,

$$\mathcal{B}^2(M,N) \ll N\mathcal{C}(M,N) \ ,$$

where

$$\mathcal{C}(M,N) = \sum_{\mathfrak{n}\sim N} \Big| \sum_{\mathfrak{m}\sim M}^{\pi} \gamma(\mathfrak{m}\mathfrak{n})\mu(\mathfrak{m}) \Big|^2 .$$

Now we need to show that

$$\mathcal{C}(M,N) \leqslant NM^2(\log MN)^{-A} \ . \tag{20.23}$$

Interchanging the order of summation we have

$$\mathcal{C}(M,N) = \sum_{\mathfrak{m}_1}^{\pi} \sum_{\mathfrak{m}_2}^{\pi} \mu(\mathfrak{m}_1)\mu(\mathfrak{m}_2)\mathcal{D}(\mathfrak{m}_1,\mathfrak{m}_2) \tag{20.24}$$

where

$$\mathcal{D}(\mathfrak{m}_1,\mathfrak{m}_2) = \sum_{\mathfrak{n}} \gamma(\mathfrak{n}\mathfrak{m}_1)\overline{\gamma}(\mathfrak{n}\mathfrak{m}_2) \ .$$

Put $\ell_1 = \operatorname{Re} \mathfrak{n}\mathfrak{m}_1, \ell_2 = \operatorname{Re} \mathfrak{n}\mathfrak{m}_2$, that is,

$$\begin{aligned} \mathfrak{n}\mathfrak{m}_1 + \overline{\mathfrak{n}}\,\overline{\mathfrak{m}}_1 &= 2\ell_1 \ , \\ \mathfrak{n}\mathfrak{m}_2 + \overline{\mathfrak{n}}\,\overline{\mathfrak{m}}_2 &= 2\ell_2 \ . \end{aligned} \tag{20.25}$$

Define $\Delta = \frac{i}{2}(\mathfrak{m}_1\overline{\mathfrak{m}}_2 - \overline{\mathfrak{m}}_1\mathfrak{m}_2) = \operatorname{Im} \overline{\mathfrak{m}}_1\mathfrak{m}_2$.

The contribution to $C(M,N)$ of terms with $\Delta = 0$ is

$$\mathcal{C}_0(M,N) \ll N \sum^{\pi} \sum_{\operatorname{Im} \overline{\mathfrak{m}}_1\mathfrak{m}_2 = 0}^{\pi} 1 \ll NM \log MN \ . \tag{20.26}$$

Now let $\Delta \neq 0$. Then, the system (20.25) gives

$$\overline{\mathfrak{n}} = i(\ell_1\mathfrak{m}_2 - \ell_2\mathfrak{m}_1)/\Delta \ .$$

The conditions for $\mathfrak{n}$ now translate into

$$\begin{aligned} &\ell_1\mathfrak{m}_2 \equiv \ell_2\mathfrak{m}_1 (\operatorname{mod} \Delta) \ , \\ &|\ell_1\mathfrak{m}_2 - \ell_2\mathfrak{m}_1|^2 \sim \Delta^2 N \ , \end{aligned}$$

that is,

$$\mathcal{D}(\mathfrak{m}_1,\mathfrak{m}_2) = \sum\sum_{\substack{\ell_1,\ell_2 \leqslant 2\sqrt{MN} \\ \ell_1\mathfrak{m}_2 \equiv \ell_2\mathfrak{m}_1 (\operatorname{mod} \Delta) \\ |\ell_1\mathfrak{m}_2 - \ell_2\mathfrak{m}_1|^2 \sim \Delta^2 N}} \gamma_{\ell_1}\overline{\gamma}_{\ell_2} \ . \tag{20.27}$$

Note that the restrictions $\ell_1,\ \ell_2 \leqslant 2\sqrt{MN}$ are redundant (see (20.25)).

Now we require a theorem for primes in arithmetic progression similar to, but more complicated than, the Barban–Davenport–Halberstam theorem which was given in Corollary 9.15. We show the following:

PROPOSITION 20.9. *Let* γ_l *be as in* (20.18). *Then, for* $Q = x(\log x)^{-B(A)}$,

$$\sum_{q \leqslant Q} \max_{a,\mathfrak{a},y} \Big| \sum_{\substack{\ell_1,\ell_2 \leqslant x \\ |\ell_1 - \mathfrak{a}\ell_2| \leqslant y \\ \ell_1 \equiv a\ell_2 (q)}} \sum \gamma_{\ell_1}\overline{\gamma}_{\ell_2} - \frac{1}{\varphi(q)} \sum_{\substack{\ell_1,\ell_2 \leqslant x \\ |\ell_1 - \mathfrak{a}\ell_2| \leqslant y}} \sum \gamma_{\ell_1}\overline{\gamma}_{\ell_2} \Big| \ll x^2(\log x)^{-A}, \tag{20.28}$$

where a *runs over rational integers relatively prime to* q, $\mathfrak{a}$ *over complex numbers, and* y *over real numbers.*

PROOF. We can assume that for each q we have $(a, q) = 1$ because, if $\ell_1 > Q$ this follows since ℓ_1 is prime and, if $\ell_1 \leqslant Q$ the trivial bound gives the result.

We would like to remove the condition $|\ell_1 - \mathfrak{a}\ell_2| \leqslant y$ so as to make the summations over ℓ_1, ℓ_2 independent. To this end we split the summation over ℓ_1, ℓ_2 into small boxes $\mathcal{L}_{12}$ given by

$$L_1 < \ell_1 \leqslant (1+\theta)L_1, \qquad L_2 < \ell_2 \leqslant (1+\theta)L_2,$$

where L_1, L_2 run over the numbers $(1+\theta)^j \leqslant x$. We shall choose θ later. We obtain, for given q, a, $\mathfrak{a}$, y,

$$\sum_{\substack{\ell_1,\ell_2 \\ \ell_1 \equiv a\ell_2 \\ |\ell_1 - \mathfrak{a}\ell_2| \leqslant y}} \sum - \frac{1}{\varphi(q)} \sum_{\substack{\ell_1,\ell_2 \\ |\ell_1 - \mathfrak{a}\ell_2| \leqslant y}} \sum$$

$$= \sum_{\substack{L_1,L_2 \\ |L_1 - \mathfrak{a}L_2| \leqslant y}} \sum \Big(\sum_{\substack{\ell_1,\ell_2 \leqslant x \\ \ell_1 \sim L_1, \ell_2 \sim L_2 \\ \ell_1 \equiv a\ell_2 (q)}} \sum \gamma_{\ell_1}\overline{\gamma}_{\ell_2} - \frac{1}{\varphi(q)} \sum_{\substack{\ell_1,\ell_2 \leqslant x \\ \ell_1 \sim L_1, \ell_2 \sim L_2}} \sum \gamma_{\ell_1}\overline{\gamma}_{\ell_2} \Big)$$

$$+ O\Big(\sum_{\substack{\ell_1,\ell_2 \leqslant x \\ y - \theta x \leqslant |\ell_1 - \mathfrak{a}\ell_2| \leqslant y + \theta x \\ \ell_1 \equiv a\ell_2 (q)}} \sum 1 + \frac{1}{\varphi(q)} \sum_{\substack{\ell_1,\ell_2 \leqslant x \\ y - \theta x \leqslant |\ell_1 - \mathfrak{a}\ell_2| \leqslant y + \theta x}} \sum 1 \Big).$$

Here, the error term takes care of the terms not covered precisely by the subdivision. Fixing ℓ_2 and summing trivially over ℓ_1 we find that this error term is bounded by $(1 + \theta x/\varphi(q))x$. Then summing over q this is $O((Q + \theta x)x \log x)$.

For the main part of the sum we treat each individual box $\mathcal{L}_{12}$. In this box we write

$$\sum_{\substack{\ell_1,\ell_2 \\ \ell_1 \equiv a\ell_2 (q)}} \sum - \frac{1}{\varphi(q)} \sum_{\ell_1,\ell_2} \sum = \sum_{\alpha_1 \equiv a\alpha_2 (q)}^{\star} \Big(\sum_{\ell_1 \equiv \alpha_1} - \frac{1}{\varphi(q)} \sum_{\ell_1} \Big) \Big(\sum_{\ell_2 \equiv \alpha_2} - \frac{1}{\varphi(q)} \sum_{\ell_2} \Big).$$

Now, applying Cauchy's inequality and Corrollary 9.15, we get for each box the bound $O(x^2(\log x)^{-3A-4})$. Therefore, our original sum is bounded by

$$\theta^{-2}x^2(\log x)^{-2-3A} + (Q + \theta x)x \log x.$$

Choosing $\theta = (\log x)^{-A-1}$, we complete the proof of Proposition 20.9. □

Now we return to (20.27). Inserting this into (20.24) we see that the sum $\mathcal{C}(M, N)$ is similar to that considered in Proposition 20.9. To make it ready for the application we split $\mathcal{D}(\mathfrak{m}_1, \mathfrak{m}_2)$ into classes $\ell_1 \equiv a\ell_2 \,(\bmod \Delta)$ where a ranges over reduced residue classes modulo Δ such that

$$a\mathfrak{m}_2 \equiv \mathfrak{m}_1 (\bmod \Delta). \tag{20.29}$$

From Proposition 20.9 we obtain

$$C(M,N) = \sum^{\pi}\sum^{\pi}_{\mathfrak{m}_1,\mathfrak{m}_2\sim M} \mu(\mathfrak{m}_1)\mu(\mathfrak{m}_2)\frac{\eta(\Delta)}{\varphi(\Delta)} \sum_{\substack{\ell_1,\ell_2 \\ |\ell_1\mathfrak{m}_2-\ell_2\mathfrak{m}_1|\sim|\Delta|\sqrt{N}}}\sum \gamma_{\ell_1}\gamma_{\ell_2} + O(NM^2(\log MN)^{-A}). \tag{20.30}$$

Here $\eta(\Delta)$ denotes the number of rational residue classes $a \bmod \Delta, (a,\Delta) = 1$, satisfying (20.29). We shall first simplify the main term by executing the summation over ℓ_1, ℓ_2 using the Prime Number Theorem. We have

$$\sum_{\ell_1}\sum_{\ell_2}\gamma_{\ell_1}\gamma_{l_2} = \iint_{|\ell_1\mathfrak{m}_2-\ell_2\mathfrak{m}_1|\sim|\Delta|\sqrt{N}} d\ell_1 d\ell_2 + O(MN(\log MN)^{-A}).$$

Recall that the region of integration $|\ell_1\mathfrak{m}_2 - \ell_2\mathfrak{m}_1| \sim |\Delta|\sqrt{N}$ is determined by the system (20.25) where $\mathfrak{n} = u + iv$ is a complex number in the annulus $N < |\mathfrak{n}| \leqslant N'$. Therefore, the area of this region is easily computed by a linear change of variables $(\ell_1,\ell_2) \to (u,v)$ whose Jacobian is $|\Delta|$ getting

$$\int\int dl_1 dl_2 = \pi|\Delta|(N'-N)\ .$$

Hence,

$$C(M,N) = \pi(N'-N)S_M + O(NM^2(\log MN)^{-A}), \tag{20.31}$$

where

$$S_M = \sum^{\pi}\sum^{\pi}_{\mathfrak{m}_1,\mathfrak{m}_2\sim M} \mu(\mathfrak{m}_1)\mu(\mathfrak{m}_2)\frac{\eta(\Delta)|\Delta|}{\varphi(\Delta)}. \tag{20.32}$$

It remains to estimate S_M. We begin by extracting the greatest common divisor of $\mathfrak{m}_1$ and $\mathfrak{m}_2$. We get (note the change of variables)

$$S_M = \sum_{\mathfrak{d}}^{\pi} \mu^2(\mathfrak{d}) \sum^{\pi}\sum^{\pi}_{\substack{(\mathfrak{m}_1,\mathfrak{m}_2)=1 \\ (\mathfrak{m}_1\mathfrak{m}_2,\mathfrak{d})=1 \\ \mathfrak{m}_1,\mathfrak{m}_2\sim M/d}} \mu(\mathfrak{m}_1)\mu(\mathfrak{m}_2)\frac{\eta(d\Delta)|d\Delta|}{\varphi(d\Delta)},$$

where $d = N(\mathfrak{d})$ and $\Delta = \operatorname{Im}\overline{\mathfrak{m}}_1\mathfrak{m}_2$. Here, we have

$$\eta(d\Delta) = \sum^{\star}_{\substack{\alpha(\operatorname{mod} d\Delta) \\ \alpha\mathfrak{m}_2\equiv\mathfrak{m}_1(\operatorname{mod}\overline{\mathfrak{d}}\Delta)}} 1.$$

Note that $(\mathfrak{m}_1\mathfrak{m}_2,\overline{\mathfrak{d}}\Delta) = 1$, so the congruence determines α uniquely modulo $\overline{\mathfrak{d}}\Delta$ and α can be taken to be rational because Δ is rational and $\mathfrak{m}_1,\mathfrak{m}_2,\mathfrak{d}$ are primitive. Such a solution can be extended to reduced residue classes modulo $d\Delta$, and the number of these is given by

$$\eta(d\Delta) = d\prod_{p|d,p\nmid\Delta}(1-\frac{1}{p}). \tag{20.33}$$

Hence

$$\frac{\eta(d\Delta)d|\Delta|}{\varphi(d\Delta)} = d\prod_{p|\Delta}(1-\frac{1}{p})^{-1} = \frac{d|\Delta|}{\varphi(\Delta)},$$

and

$$S_M = \sum_{\mathfrak{d}}^{\pi} \mu^2(\mathfrak{d}) d \sum\sum_{\substack{(\mathfrak{m}_1,\mathfrak{m}_2)=1 \\ (\mathfrak{m}_1\mathfrak{m}_2,\mathfrak{d})=1 \\ \mathfrak{m}_1,\mathfrak{m}_2 \sim M/d}}^{\pi\quad\pi} \mu(\mathfrak{m}_1)\mu(\mathfrak{m}_2)\frac{|\Delta|}{\varphi(\Delta)}.$$

We have

$$\frac{|\Delta|}{\varphi(\Delta)} = \sum_{h|\Delta}^{\flat} \frac{1}{\varphi(h)},$$

so

$$S_M = \sum_{|\mathfrak{d}|<2M}^{\pi} \mu^2(\mathfrak{d}) d \sum_{h\leqslant 2M}^{\flat} \frac{1}{\varphi(h)} S_M(\mathfrak{d}, h),$$

where

$$S_M(\mathfrak{d}, h) = \sum\sum_{\substack{(\mathfrak{m}_1,\mathfrak{m}_2)=1 \\ (\mathfrak{m}_1\mathfrak{m}_2,\mathfrak{d})=1 \\ \mathfrak{m}_1\overline{\mathfrak{m}}_2 \equiv \overline{\mathfrak{m}}_1\mathfrak{m}_2 (2h) \\ \mathfrak{m}_1,\mathfrak{m}_2 \sim M/d}}^{\pi\quad\pi} \mu(\mathfrak{m}_1)\mu(\mathfrak{m}_2).$$

Detecting the congruence $\mathfrak{m}_1\overline{\mathfrak{m}}_2 \equiv \overline{\mathfrak{m}}_1\mathfrak{m}_2 (\mathrm{mod}\, 2h)$ and using the orthogonality of the characters of the group $G = \mathbb{Z}[i]/2h\mathbb{Z}[i]$, we get

$$S_M(\mathfrak{d}, h) = \frac{1}{|G|} \sum_{\chi\in\hat{G}} \sum\sum_{\substack{(\mathfrak{m}_1,\mathfrak{m}_2)=1 \\ (\mathfrak{m}_1\mathfrak{m}_2,\mathfrak{d})=1 \\ \mathfrak{m}_1,\mathfrak{m}_2 \sim M/d}}^{\pi\quad\pi} \mu(\mathfrak{m}_1)\mu(\mathfrak{m}_2)\psi(\mathfrak{m}_1)\overline{\psi}(\mathfrak{m}_2),$$

where $\psi(\mathfrak{a}) = \chi(\mathfrak{a})\overline{\chi}(\overline{\mathfrak{a}})$. To estimate this we appeal to the Siegel–Walfisz theorem in the form

PROPOSITION 20.10. *For any character ψ on ideals we have*

$$\sum_{N\mathfrak{m}\leqslant x} \mu(\mathfrak{m})\psi(\mathfrak{m}) \leqslant x(\log x)^{-A} \ ,$$

for any $A > 0$, where the implied constant depends only on A.

Applying this bound, handling easily the coprimality condition $(\mathfrak{m}_1\mathfrak{m}_2, \mathfrak{d}) = 1$, then summing over h, and $\mathfrak{d}$ we obtain the bound

$$S_M \leqslant M^2(\log M)^{-A}.$$

Combining this with (20.31) we obtain (20.23), completing the proof of Proposition 20.8.

CHAPTER 21

Primes Represented by Polynomials

Ballet de Prôtos

21.1. Introduction

The representation of primes by the values of polynomials with integer coefficients has been a main focus of number theory from very early days and has given motivation to much of the development of sieve methods. The celebrated theorem of Dirichlet on primes in arithmetic progressions answers the question for linear polynomials, although, as we have seen, important problems remain open about uniformity in the coefficients. For polynomials of higher degree, we think we know the answer but cannot justify it with proofs.

Let $f(X) \in \mathbb{Z}[X]$ be an irreducible polynomial of degree $r \geqslant 1$ with positive leading coefficient. These conditions are not sufficient because the values of $f(X)$ could all be divisible by a fixed prime, for example,

$$f(X) = X^2 - X + 2 = X(X-1) + 2$$

is always even. In similar fashion, all values of $f(X) = X^p - X + p$ are divisible by p. To prevent this from happening we assume that $\omega(p)$, the number of solutions of $f(X) \equiv 0 \,(\mathrm{mod}\, p)$, satisfies

$$\omega(p) < p \tag{21.1}$$

for all primes. Since, by Lagrange's theorem, $\omega(p) \leqslant r$, verification of (21.1) is only needed for $p \leqslant r$.

Under these conditions it is expected that $f(X)$ represents infinitely many primes and

$$\Big| \{n \leqslant x;\ f(n) \text{ prime}\} \Big| \sim \frac{Hx}{r \log x} \tag{21.2}$$

where

$$H = \prod_p \Big(1 - \frac{\omega(p)}{p}\Big)\Big(1 - \frac{1}{p}\Big)^{-1}.$$

This asymptotic (21.2) was conjectured by Bateman and Horn [**5**], having been derived in certain cases by Hardy and Littlewood using heuristic arguments based on the circle method. Another more straightforward heuristic based on the randomness of the Möbius function leads to the same formula in a more general context; see (3.5)–(3.7).

Sieve methods give a start to the problem but so far fail to solve it. One gets, by either the Λ^2-sieve or the beta-sieve, fairly good upper bounds $(2+\varepsilon)Hx/\log x$ which are out by a factor of $2r$; see Theorem 7.10 or Theorem 11.13. This is understandable because we have a linear sieve problem, so the sieving limit is 2 and the exponent of distribution is $1/r$.

There are also successes for sieve lower bounds, but unfortunately not for primes. The best result known in full generality asserts (see Theorem 25.4) that $f(X)$ represents infinitely many integers having at most $r+1$ distinct prime factors. Similar statements and conjectures apply to systems of polynomials.

Up to now we have been talking about polynomials in a single variable. There is much more room if we allow for additional variables. Thus, for example, every prime is a sum of four squares by Lagrange's theorem since every positive integer is. There are even examples (cf. [**138**], p. 191) of polynomials whose positive range is precisely the set of prime numbers. There are natural problems about primes for polynomials in several variables coming from algebraic number theory, those primes represented by norm forms. In this case the problem is solved by the prime ideal theorem.

After polynomials in one variable the obvious place to look is two variables and here we find both natural questions and considerable history. A major line of development which begins with Fermat, then Euler, then Gauss and Dirichlet, gives the prime values of $x^2 + y^2$ and then of any binary quadratic form $\varphi(x, y) = ax^2 + bxy + cy^2$.

In this chapter we are going to present and discuss some more recent successes with prime values of polynomials in two variables which represent highlights of modern sieve theory.

21.2. Prime Values of Quadratic Polynomials in Two Variables

The case of a quadratic polynomial in two variables,

$$F(X, Y) = aX^2 + bXY + cY^2 + \alpha X + \beta Y + \gamma \in \mathbb{Z}[X, Y] ,$$

is resolved in general, and by sieve methods. Obvious necessary conditions are that

$$(a, b, c, \alpha, \beta, \gamma) = 1 ,$$

$$F(X, Y) \text{ is irreducible in } \mathbb{Q}[X, Y] ,$$

$$F(X, Y) \text{ represents arbitrarily large odd numbers} .$$

Of course, we want $F(X, Y)$ to depend essentially on two variables; for example, $F(X, Y) = (X + Y)^2 + 1$ is too tough for us. This condition may be expressed by the requirement that

$$\frac{\partial F}{\partial X}, \frac{\partial F}{\partial Y} \text{ are linearly independent.}$$

We introduce two discriminants:

$$\Delta = b^2 - 4ac ,$$

$$D = a\beta^2 - b\alpha\beta + c\alpha^2 - \Delta\gamma .$$

THEOREM 21.1. *For $F(X, Y)$ satisfying the above conditions and with either $D = 0$ or Δ a square, we have*

$$\big|\{p \leqslant x, p = F(m, n)\}\big| \asymp \frac{x}{\log x} .$$

Because of the condition on the discriminants this case reduces, in one way or another, to classical analytic results about primes in arithmetic progressions or binary quadratic forms, due to Dirichlet, Hecke, Weber. In particular, one requires a zero-free region for the L-functions attached to Dirichlet characters and also to class group characters.

THEOREM 21.2. *For $F(X, Y)$ satisfying the above conditions, with $D \neq 0$ and Δ not a square, we have*

$$\left|\{p \leqslant x, p = F(m,n)\}\right| \asymp \frac{x}{(\log x)^{3/2}} .$$

The special case of $F(X,Y) = X^2 + Y^2 + b$ is contained in Theorem 14.8. The general case can be proved similarly using the semi-linear sieve and the Bombieri–Vinogradov theorem. Here, to resolve issues involving the class group one requires ergodic ideas of Linnik about ambiguous classes (see [**102**]).

Unexpectedly, Theorem 21.2 has been recently applied by P. Sarnak [**141**] to a problem of geometric flavour which goes back to Apollonius. Given three tangent circles they determine two different ones tangent to them. Recall that if the radius of a circle is r then its curvature is $a = \pm 1/r$ when viewed from the outside/inside. The relation amongst the curvatures of the four tangent circles is given by Descartes' equation

$$2(a_1^2 + a_2^2 + a_3^2 + a_4^2) - (a_1 + a_2 + a_3 + a_4)^2 = 0 .$$

Now we begin with four mutually tangent circles, three of which are in the interior of the largest one and exterior to each other. Next, we place four circles, one in each of the four newly-created lunar-like regions, so that they are tangent to the three circles that bound the lune. There are now twelve new lunes. We repeat the process indefinitely, obtaining a dense packing of the initial large circle by tangent circles of various generations. An interesting fact is that, if the initial four curvatures are integers, then so also are all the subsequent curvatures in the packing. This property follows nicely by applying inversions and the Descartes relation. There are fascinating problems concerning these "integral packings". Very appealing is the conjecture by J. Lagarias which asserts that for any given integral Apollonian packing, the set of curvatures that is achieved consists of all sufficiently large integers. Sarnak succeeded to prove the weaker but still very striking conclusion that such a set must contain infinitely many prime numbers. To this end he showed (as a by-product of other deep investigations) that all sufficiently large values of a certain quadratic polynomial in two variables are in the set. He then shows that the conditions of Theorem 21.2 are satisfied and concludes that "Apollonian primes" occur in abundance.

In very recent work, J. Cho and H. Kim apply Theorem 21.2 to count the split primes p of $\mathbb{Q}(\sqrt{-2})$ for which $27p + 256$ is a norm and thereby verify the existence of certain infinite families of Galois extensions. Note that $F(X,Y) = 108X^2 + 6Y^2 + 28X + 4Y - 7 = n$ if and only if $27n + 256 = (54X + 7)^2 + 2(9Y + 3)^2$ and that we then have $n \equiv 1 \text{ or } 3 \pmod 8$.

21.3. Primes Captured by $X^2 + Y^4$

One of the things that makes non-linear polynomials so difficult is the lacunarity of the sequence. Adding an extra variable helps to ameliorate this and in the cases we have treated so far the sequences have been almost dense. The first case successfully treated, of a lacunary polynomial, was X^2+Y^4. The seemingly simpler example $X^2 + Y^3$ seems harder due to lack of structure. In fact, any polynomial $F(X,Y) = \varphi(X,Y^2)$ where φ is a positive definite binary quadratic form, possesses the requisite structure while $\varphi(X,Y^2) + 1$ does not.

THEOREM 21.3. *We have*

$$\sum\sum_{a^2+b^4\leqslant x} \Lambda(a^2 + b^4) = \gamma\, x^{\frac{3}{4}} \left\{1 + O\Big(\frac{\log\log x}{\log x}\Big)\right\} , \tag{21.3}$$

where a, b run over positive integers, and γ is a positive constant, namely

$$\gamma = \frac{4}{\pi}\int_0^1 (1-t^4)^{\frac{1}{2}}\, dt \;.$$

From a geometric viewpoint this theorem counts the Gaussian primes which lie directly above the squares. Theorem 21.3 follows from the asymptotic sieve for primes and we have already prepared much of the groundwork in previous chapters. Theorem 18.7 forms the basis for the work. (In our original paper [**53**] we actually used the theorem in a form slightly modified in some technical aspects.) Although quite general, Theorem 18.7 requires the verification of two key conditions which are in practice quite difficult and one may say that they form the heart of the matter. These two conditions are those which deal with the linear and bilinear forms.

The first of these, (18.36), concerns the absolute level of distribution. Here, already, the lacunarity of the sequence

$$a_n = \big|\{(a,b) = 1;\; a^2 + b^4 = n\}\big|$$

limits the level to nothing better than exponent $3/4$. This optimal level is achieved in Theorem 22.14, Section 22.6. That result in turn is powered by the large sieve inequality for quadratic roots given in Proposition 20.4.

Because there is no hope to go beyond $3/4$ for this sequence $\mathcal{A}$, we are required to estimate the bilinear sum in the second condition (18.38) in a very broad range. As a result, that part of the work, given in [**54**], is long and difficult. After much deliberation we have decided it is not reasonable to include it in this book. For those wishing to see the structure of the argument used in verifying the bilinear form bound (18.38), but without going through the many details, we recommend the synopsis "In studying ... achieve two goals at once" on pages 948–951 of [**54**].

However, we would like to present here a number of ingredients and byproducts which seem to be of interest. All of the results we highlight deal with the Jacobi symbol in one form or another. Perhaps that is not so surprising in view of the nature of the polynomial $X^2+(Y^2)^2$. We require the distribution of complex sequences in residue classes to large moduli, very similar to that in the Barban–Davenport–Halberstam Theorem 9.14, but now involving twists by the Jacobi symbol. From several of these results (see Proposition 14.1 of [**54**]), we extract the following.

THEOREM 21.4. *Let $\varepsilon > 0$, $R \geqslant S^{\varepsilon}$, $S \geqslant R^{\varepsilon}$, and $\alpha_{r,s}$ with $(r,s) = 1$ be complex numbers supported in the box $R < r \leqslant 2R$, $S < s \leqslant 2S$. Then,*

$$\sum_{\substack{(RS)^{4\varepsilon}\leqslant d\leqslant (RS)^{1-4\varepsilon}\\ d\,\mathrm{odd}}} \;\sum_{a(\mathrm{mod}\, d)} \Bigg|\sum\sum_{s\equiv ar(\mathrm{mod}\, d)} \alpha_{r,s}\Big(\frac{r}{d}\Big)\Bigg|^2 \ll (RS)^{1-\varepsilon}\|\alpha\|^2$$

where $\left(\frac{r}{d}\right)$ is the Jacobi symbol, the implied constant depending only on ε.

Many of our arguments enter the Gaussian domain. We say that $z \in \mathbb{Z}[i]$ is primary, primitive if $z \equiv 1(\mathrm{mod}\, 2(1+i))$, $(z, \bar{z}) = 1$. For two Gaussian integers $z = a + bi$, $w = c + di$ which are primary, primitive we define the Dirichlet symbol by

$$\left(\frac{z}{w}\right) = \left(\frac{\mathrm{Re}\, w z}{|w|^2}\right) = \left(\frac{ac - bd}{c^2 + d^2}\right) .$$

This symbol satisfies the elegant reciprocity law

$$\left(\frac{w}{z}\right) = \left(\frac{z}{w}\right) .$$

From Proposition 21.3 of [**54**] we find:

THEOREM 21.5. *Let α_w, β_z be complex numbers with $|\alpha_w| \leqslant 1$, $|\beta_z| \leqslant 1$ which are supported on primary, primitive integers in the discs $|w|^2 \leqslant M$, $|z|^2 \leqslant N$. We have*

$$\sum_w \sum_z \alpha_w \beta_z \left(\frac{z}{w}\right) \ll (M+N)^{\frac{1}{12}} (MN)^{\frac{11}{12}+\varepsilon} ,$$

for any $\varepsilon > 0$, the implied constant depending only on ε.

Our final result in this section deals with a sum over primes. Like many of the results of this book it is established by means of sieve methods.

The Legendre symbol (p/q) provides a beautiful connection between a pair of primes and has long fascinated number theorists. It is interesting that, to primes $p \equiv 1\,(\mathrm{mod}\, 4)$ we can associate in a natural fashion a symbol which is internal to the prime itself. This is given by the Jacobi symbol

$$\sigma_p = \left(\frac{s}{r}\right) ,$$

where $p = r^2 + s^2$ with $r, s > 0$ and r odd, which we call the "spin" of p.

Life being what it is, one expects that the spin takes each sign about half of the time. A special case of (26.2) in [**54**] confirms this:

THEOREM 21.6. *We have*

$$\sum_{\substack{p \leqslant x \\ p \equiv 1 (\mathrm{mod}\, 4)}} \sigma_p \ll x^{\frac{76}{77}} .$$

One of the ingredients used in the proof of Theorem 21.6 is the verification of a sieve axiom made possible by the bilinear form estimate given in Theorem 21.5.

21.4. Primes Represented by $X^3 + 2Y^3$

In this section we give a brief overview of the work of Heath-Brown [**90**] concerning primes represented by the cubic form $X^3 + 2Y^3$. It would be nice to give a complete proof or even an extensive description of the ideas but, as in the case of the polynomial $X^2 + Y^4$, we can only show some highlights. As this is a book on sieve methods we again concentrate mainly on the sieve aspects of the work even though, as for $X^2 + Y^4$, it is the bilinear form estimate that is the most difficult part.

We begin with an extract from his main result. In fact, more is proved and the result is stated in a slightly different form; see also [**92**].

THEOREM 21.7. *There are infinitely many primes of the form* $a^3 + 2b^3$ *with positive integers* a, b. *Moreover,*

$$\sum\sum_{x^{\frac{1}{3}} < a,\, b < (1+\delta)x^{\frac{1}{3}}} \Lambda(a^3 + 2b^3) \sim \mathfrak{S}\, \delta^2 x^{\frac{2}{3}} ,$$

where $\mathfrak{S}$ *is given by the product*

$$\mathfrak{S} = \prod_p \Big(1 + \frac{1 - \rho(p)}{p}\Big)$$

with $\rho(p)$ *being the number of roots of the congruence* $\nu^3 \equiv 2 \,(\bmod p)$. *Here,* $\delta = (\log\log x)^{-\eta}$ *with any sufficiently small constant* $\eta > 0$.

Naturally enough, the proof involves an application of the linear sieve to the sequence $\mathcal{A} = (a_n)$ where a_n denotes the number of representations of n in the form $a^3 + 2b^3$ with the variables a, b in appropriate ranges. One shows that this has the maximum possible level of distribution, that is, $D(x) = x^{\frac{2}{3}}(\log x)^{-A}$. The arguments, given in Heath-Brown's Lemma 3.2, are similar but somewhat different than those provided here in Section 22.8 which give $D(x) = x^{\frac{2}{3}-\varepsilon}$. The seemingly small discrepancy between $(\log x)^{-A}$ and $x^{-\varepsilon}$ is crucial for his work.

This level of distribution is required in the asymptotic identity for sums over primes which is developed in somewhat the same spirit as that in Lemma 18.2. The proofs of these, in both cases, use the Buchstab identity and the Fundamental Lemma. The Fundamental Lemma is required in order that the resulting pieces in the decomposition all retain the correct order of magnitude. In contrast, if one attempted to use the exact combinatorial identities of Chapter 17 one would lose this feature. This feature is very important because subsequent estimates for the bilinear forms do not apply completely to all pieces. Fortunately, they barely miss doing so and the missing parts, because we have kept the correct order of magnitude, need only be estimated crudely by the upper-bound sieve. If the level of distribution were somewhat smaller the missing parts of the bilinear forms would have positive measure and the argument would fail.

The estimates for the bilinear forms constitute the heart of the work. The idea goes by considering the full norm form

$$N(a, b, c) = a^3 + 2b^3 + 4c^3 - 6abc$$

with the side condition $c = 0$. This way one can exploit the multiplicative structure of the bilinear forms.

A very similar idea was applied to the polynomial $a^2 + b^4$ which was considered in the previous section. There one works with the norm form $N(a, c) = a^2 + c^2$ with the side condition $c = b^2$. On the one hand, the form $a^3 + 2b^3$ represents a thinner sequence than the polynomial $a^2 + b^4$; on the other hand, the homogeneity of the form helps in the arguments. When working with $N(a, b, c)$ at some point one needs to capture the zero value of one out of the three variables. On the other hand, when working with $N(a, c)$ one needs to capture the square value of one out of the two variables. These tasks require very different realizations.

The main arguments in the work of Heath-Brown are performed in the ring of integers of the cubic field $K = \mathbb{Q}(\rho)$ with $\rho = 2^{\frac{1}{3}} > 0$, which is

$$\mathbb{Z}[\rho] = \{a + b\rho + c\rho^2;\ a, b, c \in \mathbb{Z}\}.$$

The norm of $\beta = a + b\rho + c\rho^2$ is given by

$$N\beta = a^3 + 2b^3 + 4c^3 - 6abc\ .$$

The cubic extension $K/\mathbb{Q}$ has class number one, that is, $\mathbb{Z}[\rho]$ has unique factorization. The fundamental unit is

$$\varepsilon_0 = 1 + \rho + \rho^2\ .$$

The rational primes which are norms have positive density by the Prime Ideal Theorem so they are plentiful. The problem is to capture those $p = N\beta = a + b\rho$, that is, with β in the lower rank sub-module

$$M[\rho] = \mathbb{Z} + \rho\mathbb{Z} \subset \mathbb{Z} + \rho\mathbb{Z} + \rho^2\mathbb{Z}$$

which consists of $\beta \in \mathbb{Z}[\rho]$ with $c = 0$.

For counting purposes it is necessary to restrict the arguments to finite subsets of $M[\rho]$. Let $\eta > 0$ be a small number and $x \geqslant 2$ a large number. We consider the subset

$$\mathcal{A} = \{a + b\rho;\ (a, b) = 1,\ x^{\frac{1}{3}} < a, b \leqslant (1 + \eta)x^{\frac{1}{3}}\}.$$

Note that no two elements of $\mathcal{A}$ are associates provided η is sufficiently small and so the set

$$\mathfrak{A} = \{(\alpha); \alpha \in \mathcal{A}\}$$

contains distinct ideals. Here (α) denotes the principal ideal generated by α. The elements of $\mathfrak{A}$ enjoy a few attractive features inherited from the arithmetic of the field K which are convenient when performing the inclusion-exclusion arguments of the combinatorial sieve. For example, it turns out that the elements of $\mathfrak{A}$ are divisible only by primes of first degree and cannot be divisible by two primes of the same norm.

The axioms of the asymptotic sieve for primes require estimates for bilinear forms of type

$$\sum\sum_{\substack{\mathfrak{a},\mathfrak{b}\subset\mathbb{Z}[\rho]\\ \mathfrak{a}\mathfrak{b}\in\mathfrak{A}}} d_{\mathfrak{a}} f_{\mathfrak{b}}$$

for some special coefficients $d_{\mathfrak{a}}$, $f_{\mathfrak{b}}$, but they are so complicated that in practice one can consider them as general functions. The key point is that $d_{\mathfrak{a}}$, $f_{\mathfrak{b}}$ are independent and one of these, say $f_{\mathfrak{b}}$, is well distributed over residue classes to relatively small moduli. Writing $\mathfrak{a} = (\alpha)$, $\mathfrak{b} = (\beta)$ with $\alpha = a_1 + b_1\rho + c_1\rho^2$, $\beta = a_2 + b_2\rho + c_2\rho^2$, one needs to perform summation on the surface

$$a_1c_2 + b_1b_2 + c_1a_2 = 0$$

with relevant coefficients. This connects the vectors $\hat{\alpha} = (a_1, b_1, c_1)$, $\hat{\beta} = (a_2, b_2, c_2)$ quite inseparably, so a non-trivial bound for the bilinear form with general coefficients has a chance to be true.

This brief description is a paraphrase of what is happening behind the scenes in Heath-Brown's work, just to show how challenging the problem is. Before the final attack on the bilinear forms is launched, Heath-Brown needs many results for sums over the ring $\mathbb{Z}[\rho]$ and for sums restricted to the sub-module $M[\rho]$. As an example of the first, in his Lemma 13.1 he derives a new stronger version of the large sieve type estimate in three dimensions. As an example of the latter, in his Lemma 3.2 he establishes the optimal exponent of distribution $2/3$ for the sequence $\mathcal{A}$.

Here, we describe a special case of Lemma 3.10 of [**90**], a part of the argument in which the structure of the form $X^3 + 2Y^3$ is crucially used.

Let $\mathcal{C} \subset \mathbb{R}^3$ be a cube with edges parallel to the axes and of length $\sim V^{\frac{1}{3}}$ with $\log V \asymp \log x$. Suppose $f_{(\beta)}$ is well distributed over residue classes to relatively small moduli. Assume (among other estimates of the same flavour) that

$$\sum_{\substack{\hat{\beta} \in \mathcal{C} \\ \beta \equiv \nu \,(\mathrm{mod}\, q)}} f_{(\beta)} \ll V \exp(-\lambda\sqrt{\log x})$$

for every $\nu \in \mathbb{Z}[\rho]$ and every $q \leqslant (\log x)^A$ for every $A \geqslant 1$, where $\lambda > 0$ and the implied constant depend only on A. Notice the absence of a main term, because $f_{(\beta)}$ is a function which emerges from combinatorial sieve identities but with the expected main term subtracted out. Let $d_{\mathfrak{a}}$ be complex numbers, absolutely bounded and supported on ideals $\mathfrak{a}$ which have no small prime divisors. (In applications $d_{\mathfrak{a}}$ takes on the values 0, ± 1 on products of primes in segments created by a multiple use of Buchstab's formula.) Then,

$$\sum_{\substack{\mathfrak{ab} \in \mathfrak{A} \\ V < N\mathfrak{b} \leqslant 2V}} d_{\mathfrak{a}} f_{\mathfrak{b}} \ll x^{\frac{2}{3}} (\log x)^{-A}$$

for every A, provided $x^{\frac{1}{3}+\varepsilon} \ll V \ll x^{\frac{1}{2}-\varepsilon}$.

Because in practice the coefficients $d_{\mathfrak{a}}$, $f_{\mathfrak{b}}$ are of a similar nature, one can interchange them. But $N\mathfrak{a} \asymp U$ with $UV = x$ so one can also cover the range when one of the two is between $x^{\frac{1}{2}+\varepsilon}$ and $x^{\frac{2}{3}-\varepsilon}$.

In order to match the ranges from the other results, those about the level of distribution of $\mathcal{A}$, one still needs to cover the three ranges: between $x^{\frac{1}{3}-\varepsilon}$ and $x^{\frac{1}{3}+\varepsilon}$, between $x^{\frac{1}{2}-\varepsilon}$ and $x^{\frac{1}{2}+\varepsilon}$, between $x^{\frac{2}{3}-\varepsilon}$ and $x^{\frac{2}{3}+\varepsilon}$. These, however, are quite short ranges in the logarithmic scale, so one can estimate this contribution crudely by any upper-bound sieve because the coefficients $d_{\mathfrak{a}}$, $f_{\mathfrak{b}}$ are supported on almost primes. This technique for handling boundary contributions has been used in sieve theory in a number of problems, for example, in Section 14.4, in Section 18.1 and in Bombieri's paper [**8**] on the asymptotic sieve.

Finally we remark that, as in the case of primes of the form $X^2 + Y^4$, our description here barely scratches the surface of the many complexities that need to be overcome.

21.5. A Binary Polynomial of Higher Degree

As we have mentioned earlier, the polynomial $X^2 + Y^3$ with density $x^{\frac{5}{6}}$ would seem a natural target for arguments like those of the previous two sections. Perhaps even more interesting would be the polynomial $27X^2 + 4Y^3$ which is no doubt of equivalent difficulty but whose prime values would produce elliptic curves with only one place of bad reduction (which would be of multiplicative type).

However, neither of the methods for polynomials $X^2 + Y^4$ nor $X^3 + 2Y^3$ apply here because the requisite structure is missing. In the case of $X^2 + Y^4$ the proof takes place within the Gaussian domain. This suggests the possibility of studying $X^2 + Y^6$ (or $27X^2 + 4Y^6$, thereby unearthing some of the previous elliptic curves).

Let us consider the sequence $\mathcal{A} = (a_n)$ with

$$a_n = \left|\{(a, b) = 1;\, a^2 + b^6 = n\}\right| .$$

Note that $A(x) \sim c_1 x^{2/3}$ with c_1 a positive constant and we shall establish, as a special case of Theorem 22.14, the optimal level of absolute distribution

$$D(x) = x^{2/3}(\log x)^{-A} . \tag{21.4}$$

Clearly this is, by necessity, weaker than the exponent $\frac{3}{4}$ attained for $X^2 + Y^4$. Therefore, were we to attempt to use the asymptotic sieve for primes we would need to establish the bilinear form estimate in a wider range than before. Since this is a sparser sequence of similar shape, serious difficulties should be expected.

In Section 17.2 we introduced a number of combinatorial identities such as that of Linnik which related primes to divisor functions. As a result, sums

$$\sum_{n\leqslant x} a_n\, \tau_k(n) \tag{21.5}$$

are the components for the target sum

$$S(x) = \sum_{n\leqslant x} a_n\, \Lambda(n) .$$

For $k = 1$, the sum (21.5) is just $A(x) \sim c_1 x^{2/3}$. For $k = 2$, one can show that

$$\sum_{n\leqslant x} a_n\, \tau(n) \sim c_2 x^{2/3} \log x$$

using the Dirichlet hyperbola method. Because of the level of distribution (21.4), one just misses reaching the asymptotic formula (21.5) for $k = 3$. However, using more precise arguments at the borderline of Dirichlet's techniques, one crosses that border, obtaining (see [**56**])

$$\sum_{\substack{a^2+b^6\leqslant x\\ (a,b)=1}} \tau_3(a^2 + b^6) \sim c_3 x^{2/3}(\log x)^2 . \tag{21.6}$$

The next step toward primes would be an asymptotic formula for (21.5) with $k = 4$, which would be a nice problem to work on. In the current state, for $k = 4$ one just barely succeeds for the polynomial $a^2 + b^4$ and the relevant asymptotic formula has been derived by V. Tipu [**153**].

Similarly, one can approach primes of the polynomial $X^2 + 1$ by establishing asymptotics for the sums of $\tau_k(n^2 + 1)$ and measure the progress by viewing how large a value of k we can treat. For $k = 2$ the Dirichlet hyperbola method yields (see Section 2.1),

$$\sum_{n\leqslant x} \tau(n^2 + 1) \sim \frac{3}{\pi}\, x \log x , \tag{21.7}$$

however, the case $\tau_3(n^2+1)$ remains an outstanding challenge. Perhaps, in this case, the spectral theory of automorphic forms offers a chance. One may also consider the divisor sum (21.7) taken over primes. This is a much more difficult problem. Using the Elliott–Halberstam conjecture, an upper-bound sieve and Proposition 19.3, one can establish the asymptotic formula

$$\sum_{p\leqslant x} \tau(p^2 + 1) \sim cx , \qquad c = 2\prod_p \Big(1 + \frac{1}{p(p-1)} + \frac{\chi_4(p)}{p-1}\Big) . \tag{21.8}$$

CHAPTER 22

Level of Distribution of Arithmetic Sequences

Harmonic Variations II

22.1. Introduction

Modern sieve theory is enhanced by the implementation of harmonic analysis. The most common way of employing this is to improve the level of distribution of the sequence $\mathcal{A} = (a_n)$ to be treated by the sieve. This amounts to employing harmonic analysis in the asymptotic evaluation of the congruence sums

$$A_d(x) = \sum_{\substack{n \leqslant x \\ n \equiv 0 \,(\mathrm{mod}\ d)}} a_n \ .$$

These sums, as we have seen all along, are our basic starting point and we could not have come this far in a book on sieve methods without already broaching the topic. For some sequences (such as polynomials in one variable) achieving an optimal level of absolute distribution takes very little work. Moreover, for linear sequences (short intervals and arithmetic progressions), we have come a good deal beyond this in Chapter 13 by applying arguments from the theory of exponential sums. In this chapter we further exemplify such analytic techniques, as well as more advanced ones, in treating a number of other basic arithmetic sequences.

22.2. Primes in Arithmetic Progressions

The sets of primes in the arithmetic progressions $qm + a$, with $(a, q) = 1$, as m varies, form the building blocks of many constructions in analytic number theory, especially in conjunction with sieve methods. For example, consider the sequence $\mathcal{A} = (a_n)$ the characteristic function of the set $n = \alpha p + \beta$ which generalizes, in particular, both the Goldbach and twin prime problems. For this sequence our basic congruence sum $A_d(x)$ counts the primes up to x in the progression $qm + a$ with $q = d$ and $a = -\bar{\alpha}\beta$. To this end one needs first of all a good level of distribution for primes $p \leqslant x$ in progressions to very large moduli q.

The prime number theorem for progressions

$$\psi(x; q, a) = \frac{x}{\varphi(q)} \left\{1 + O\big((\log x)^{-A}\big)\right\} \tag{22.1}$$

for $(a, q) = 1$, is only known to hold in a very limited range $q < (\log x)^A$ with any fixed A (Siegel–Walfisz theorem), which is much less than needed. The GRH offers much more, namely

$$\psi(x; q, a) = \frac{x}{\varphi(q)} + O\big(\sqrt{x}(\log x)^2\big) \ , \tag{22.2}$$

where the implied constant is absolute. This shows that (22.1) holds uniformly with $q < \sqrt{x}(\log x)^{-A-3}$. Even this does not represent the last word since it is expected that the error term in (22.2) can be improved to

$$\psi(x; q, a) = \frac{x}{\varphi(q)} + O\big(q^{-\frac{1}{2}} x^{\frac{1}{2}+\varepsilon}\big) \tag{22.3}$$

if $(a, q) = 1$ and $q < x^{1-\varepsilon}$ for any $\varepsilon > 0$, the implied constant depending only on ε. This was conjectured by H. Montgomery.

Looking through the explicit formula one can say that results of the strength of (22.3) require the Riemann Hypothesis and a good understanding of the vertical distribution of low-lying zeros of L-functions. Improving (22.2) in the q-aspect in the range $q \leqslant x^{\theta}$, even for some $\theta > \frac{1}{2}$, seems to be completely beyond current technology.

When used in connection with the sieve one needs the asymptotic formula averaged over many progressions. As we are able to prove stronger results about such averages, this is a handy feature. We have already seen (and proved) the basic result of this kind, the Bombieri–Vinogradov Theorem of Section 9.8, which reaches the same range of moduli as the GRH bound (22.2). The error term is much weaker, saving only powers of the logarithm; nevertheless, in practice this weakness does not usually matter.

The great usefulness of the Bombieri–Vinogradov theorem and the fact that it goes so far beyond the range of results for individual moduli naturally suggest the question as to how far could it go. As we showed in Section 9.9 one cannot go too far with such predictions, the asymptotic formula (22.1) fails for $q > x(\log x)^{-B}$ for every B (and in fact even sooner, see [**48**], [**49**]). Nevertheless, it would be of great value to prove the Elliott-Halberstam Conjecture:

CONJECTURE. *We have*

$$\sum_{q \leqslant Q} \max_{(a,q)=1} \left| \psi(x; q, a) - \frac{x}{\varphi(q)} \right| \ll x\,(\log x)^{-A}\ , \tag{22.4}$$

where $Q = x^{\theta}$ *with any* $\theta < 1$ *and any* $A > 0$, *the implied constant depending only on* θ *and* A.

Weaker versions of this conjecture have already been used in Section 7.14, Section 14.7 and Section 14.8 to derive remarkable consequences.

In the form stated in (22.4) there has been no progress in going beyond the Bombieri–Vinogradov level of $Q = \sqrt{x}(\log x)^{-B}$. However, for many problems, including the twin prime and Goldbach problems which motivated much of the work on sieves in the first place, one can afford to dispense with the maximum over the reduced classes a in (22.4) and deal with individual a, taking the sum over $q \leqslant Q$, $(q, a) = 1$. In the twin prime type of problem (shifted primes), one can even allow the implied constant to depend on a. Once one makes these concessions, a small improvement is possible; see [**12**].

THEOREM 22.1. *Let* $a \neq 0$. *We have*

$$\sum_{\substack{q \leqslant x^{\frac{1}{2}+\delta(x)} \\ (q,a)=1}} \left| \pi(x; q, a) - \frac{\pi(x)}{\varphi(q)} \right| \ll x \left(\delta(x) + \frac{\log\log x}{\log x} \right)^2 \tag{22.5}$$

for any $\delta(x) \geqslant 0$, *the implied constant depending only on* a.

Theorem 22.1 goes beyond the level of the Riemann Hypothesis but, since it does not go far beyond (the result becomes trivial as soon as $\delta(x) \gg 1$), it is mostly of theoretical interest.

Bounding the sum in (22.5) is equivalent to bounding the sum

$$E(x;q,a) = \sum_{\substack{q\leqslant Q\\(q,a)=1}} \lambda_q\Big(\psi(x;q,a) - \frac{x}{\varphi(q)}\Big)$$

for arbitrary coefficients $|\lambda_q| \leqslant 1$. With the development of flexible forms of the sieve remainder, such as that presented in Section 12.7, it became clear that bounding the sum $E(x;q,a)$ with specific coefficients λ_q sufficed for many applications. Moreover, far greater levels Q could be reached for the relevant coefficients.

We record a few results which exceed the range of moduli covered by the Bombieri–Vinogradov theorem and the Generalized Riemann Hypothesis. These are based on works of Bombieri, Fouvry and the authors. For example, we have (a special case of [**11**], Theorem 8, with exponent of bilinear distribution 29/56):

THEOREM 22.2. *Let $a \neq 0$, $\varepsilon > 0$, $A \geqslant 1$. Then for any complex numbers γ_q, δ_r with $|\gamma_q| \leqslant 1$, $|\delta_r| \leqslant 1$ and $x \geqslant 2$, we have*

$$\sum\sum_{\substack{q\leqslant Q,\, r\leqslant R\\(qr,a)=1}} \gamma_q\,\delta_r\left(\psi(x;qr,a) - \frac{x}{\varphi(qr)}\right) \ll x(\log x)^{-A}\ , \tag{22.6}$$

where $Q = x^{9/28-\varepsilon}$, $R = x^{11/56}$ and the implied constant depends on ε, a, A.

In applications we encounter not just the primes but also the almost-primes. We define for $z_1,\dots,z_r \geqslant 2$ the set (with multiplicities)

$$\mathcal{P}(z_1,\dots,z_r) = \left\{p^{(r)} = p_1\dots p_r;\ p_1 \geqslant z_1,\dots,p_r \geqslant z_r\right\}\ .$$

Let $\pi_r(x;q,a)$ be the number of $p^{(r)} \in \mathcal{P}(z_1,\dots,z_r)$ with $p^{(r)} \leqslant x$, $p^{(r)} \equiv a\,(\mathrm{mod}\, q)$ and $\pi_r(x;q)$ the number of $p^{(r)}$ with $p^{(r)} \leqslant x$, $(p^{(r)},q) = 1$. Thus we have

$$\pi_r(x;q) = \sum_{a(\mathrm{mod}\, q)}^{\star} \pi_r(x;q,a)\ .$$

Our Theorem 9.16 implies

THEOREM 22.3. *We have, for any $A > 0$ and some $B(A) > 0$,*

$$\sum_{q<Q} \max_{(a,q)=1} \left|\pi_r(x;q,a) - \frac{1}{\varphi(q)}\,\pi_r(x;q)\right| \ll x(\log x)^{-A}$$

with $Q = x^{\frac12}(\log x)^{-B(A)}$, where the implied constant depends on r and A.

Recall (see Section 12.7) that the arithmetic function λ is called well-factorable of level Q if, for any $Q_1 \geqslant 1$, $Q_2 \geqslant 1$, $Q_1Q_2 = Q$ there exists λ', λ'' supported on $q_1 \leqslant Q_1$, $q_2 \leqslant Q_2$, respectively, and bounded by one, such that $\lambda = \lambda' * \lambda''$.

THEOREM 22.4. *Let $r \geqslant 1$. For any fixed a, b, $ab \neq 0$, $(a,b) = 1$ and for any well-factorable function λ of level $Q \leqslant x^{4/7-\varepsilon}$, we have*

$$\sum_{(q,ab)=1} \lambda_q\left(\pi_r(x;q;a\bar b) - \frac{1}{\varphi(q)}\pi_r(x;q)\right) \ll x(\log x)^{-A}\ , \tag{22.7}$$

where $\bar{b}$ is the multiplicative inverse of b modulo q, A is any positive constant, and the implied constant depends only on ε, a, b, r and A.

In the case of primes (that is $r = 1$) and $b = 1$ this is Theorem 10 of [**11**]. The argument involves the decomposition of sums over primes into sums of divisor-type functions which are then recombined into bilinear forms. For general r the same argument works. In the case of general b the only novelty required is to transfer the factor $\bar{b}$ in the residue class $a\bar{b}$ into a factor b in one of the components in the bilinear form.

Inspecting the arguments of [**11**] in detail one can see that the method works for a and b fixed, but not necessarily small. One can accomplish essentially the same result (22.7) for $0 \leqslant |ab| \leqslant x^{\delta}$ with δ a small, positive absolute number, at the cost of reducing slightly the exponent of distribution $4/7$. For example, we obtain

THEOREM 22.5. *Let $r \geqslant 1$, $\delta > 0$, $A > 0$. Then (22.7) holds for all $(a, b) = 1$ with $0 < |ab| \leqslant x^{\delta}$ and for any well-factorable function λ_q of level $Q = x^{4/7-c\delta}$ with c some positive absolute number and the implied constant in $\ll$ depending only on r, δ, A.*

REMARKS. Although the residue class $\bar{b}$ changes with the modulus q, because it does so in this specific arithmetic fashion ($\bar{b}$ stands for the multiplicative inverse of b modulo q), the arguments of [**11**] are able to tolerate this variation. However, were b to change with q in a more general fashion the arguments would break down for intrinsic reasons. For example, imagine that q runs over the squarefree products $q = q_1 \dots q_r$ while the residue class is $h_j\bar{b} \pmod{q_j}$ with fixed but distinct numbers h_j for $1 \leqslant j \leqslant r$. In this situation the present arguments of [**11**] fail, which is unfortunate, especially in view of the work [**64**] of GPY on small gaps between primes where one requires the residue class to vary in such a fashion; see Section 7.13. Should we be allowed to vary the residue class arbitrarily in estimates of type (22.6) and (22.7), then it would follow along the arguments of GPY that $p_{n+1} - p_n$ is bounded by an absolute constant infinitely often.

Here we propose a conjecture which asks less and still would produce bounded gaps between primes.

CONJECTURE. *Let $C \leqslant x^{\delta} < A, B \leqslant x^{1/4+\delta}$ where $\delta > 0$ is a sufficiently small absolute constant. Then we have*

$$\sum_{a \leqslant A} \sum_{b \leqslant B} \sum_{c \leqslant C} \mu(abc)\mu(c) \sum_{\substack{n \leqslant x,\, n \equiv \gamma(c) \\ n \equiv \alpha(a),\, n \equiv \beta(b)}} \Lambda(n) \ll x(\log x)^{-J}, \tag{22.8}$$

where α, β, γ are reduced residue classes each of which may depend on its respective modulus a, b, c and *J is arbitrary, the implied constant depending only on J.*

Notice there is no need to subtract from the inner sum the main term $x/\varphi(abc)$ because of the cancellation due to the Möbius function. We actually do not require this conjecture for arbitrary classes α, β, γ of the moduli a, b, c; it suffices that they are roots of a particular polynomial $P(X)$:

$$P(\alpha) \equiv 0 \pmod{a}, \quad P(\beta) \equiv 0 \pmod{b}, \quad P(\gamma) \equiv 0 \pmod{c} .$$

In the work of [**64**] the polynomial P would be a shift of the polynomial Q given in (7.162). The similarity of (22.8) to (22.6) raises hope that (22.8) may be within reach.

REMARKS. Since c occurring in (22.8) is relatively small, the triple sum is essentially a bilinear form in the variables a and b with natural coefficients, namely the Möbius function. The variables a and b can be assumed to be essentially the same size and the ensuing expression does not have a form as flexible as the well-factorable coefficients in the linear sieve. But this defect can be overcome when the dimension is large, as is the case in GPY. The idea is to sacrifice somewhat in the main term in order to obtain the flexibility. As we showed in Section 7.10, the Selberg sieve yields approximately the same result when we restrict the support of the weights $\{\rho_d;\ d \leqslant \sqrt{D}\}$ by $d|P(z)$ with z considerably smaller than $\sqrt{D}$; for example, $z = D^{(\log k)/2k}$ sacrifices the main term by only a factor $1 + O(1/k)$. Since, for k large, such moduli d have only small prime divisors, this opens the possibility of arranging the remainder in a well-factorable shape. This possibility was communicated to us by J. Pintz and details have now been presented in [**134**]. Hence, it seems that the problem of the residue class moving with the modulus remains the main obstacle to achieving bounded gaps along the lines of GPY.

22.3. Absolute Level for Additive Convolutions

In this section we consider a sequence $\mathcal{A} = (a_n)$ of type

$$a_n = \sum_{\substack{\ell+m=n \\ (\ell,m)=1}} \alpha_\ell \beta_m \tag{22.9}$$

with α_ℓ, β_m being quite general sequences. Our goal is to show, under very mild restrictions, that $\mathcal{A}$, restricted by $n \leqslant x$, has level of distribution

$$D = D(x) = x(\log x)^{-A}\ , \tag{22.10}$$

essentially as large as could be hoped for. We have already taken full advantage of this feature in Section 16.5 when exemplifying the nature of the parity problem.

We require only that

$$|\alpha_\ell| \leqslant \tau_r(\ell), \quad |\beta_m| \leqslant \tau_r(m)\ , \tag{22.11}$$

for some $r \geqslant 1$, and that one of them satisfy a weak Siegel–Walfisz condition, say

$$\sum_{\substack{m\leqslant y \\ m\equiv a \,(\mathrm{mod}\ q) \\ m\equiv 0 \,(\mathrm{mod}\ t)}} \beta_m = \frac{1}{\varphi(q)} \sum_{\substack{m\leqslant y \\ (m,q)=1 \\ m\equiv 0 \,(\mathrm{mod}\ t)}} \beta_m + O\big(x(\log x)^{-B}\big) \tag{22.12}$$

for any a, q, t with $(at, q) = 1$, all $2 \leqslant y \leqslant x$ and some $B > 0$, where the implied constant depends only on r and B. Note that this condition is trivial if $q \gg (\log x)^{B+r^2}$.

THEOREM 22.6. *For $d \geqslant 1$, $x \geqslant 2$ put*

$$r_d(x) = \sum_{\substack{n\leqslant x \\ n\equiv 0 \,(\mathrm{mod}\ d)}} a_n - \frac{1}{\varphi(d)} \sum_{\substack{n\leqslant x \\ (n,d)=1}} a_n\ .$$

Under the assumption (22.12) *we have*

$$\sum_{d\leqslant D}|r_d(x)| \ll \bigl(D + x(\log x)^{-B}\bigr)^{\frac{1}{2}} x^{\frac{3}{2}} (\log x)^{3r^2/2} ,$$

where the implied constant depends only on r *and* B.

Proof. We begin with the crude bound

$$\sum_{\ell+m=n} \tau_r(\ell)\tau_r(m) \leqslant \sum_{1\leqslant \ell<n} \tau_r^2(\ell) \ll n(\log n)^{r^2-1} ,$$

from which we derive

$$\sum\sum_{\substack{x-y<\ell+m\leqslant x+y\\ \ell+m\equiv 0\,(\mathrm{mod}\ d)}} \tau_r(\ell)\tau_r(m) \ll \bigl(d^{-1}y+1\bigr)x(\log x)^{r^2-1} . \tag{22.13}$$

We first establish Theorem 22.6 for the slightly modified remainder term

$$r_d(L,M) = \sum\sum_{\substack{\ell\leqslant L,\, m\leqslant M\\ (\ell,m)=1\\ \ell+m\equiv 0\,(\mathrm{mod}\ d)}} \alpha_\ell\beta_m - \frac{1}{\varphi(d)} \sum\sum_{\substack{\ell\leqslant L,\, m\leqslant M\\ (\ell,m)=1\\ (\ell m,d)=1}} \alpha_\ell\beta_m$$

with L, $M \leqslant x$. We retain the condition $(\ell m, d) = 1$ but remove the other coprimality $(\ell, m) = 1$ using the Möbius device, obtaining

$$r_d(L,M) = \sum_{(t,d)=1} \mu(t) r_{dt}(L,M)$$

where

$$r_{dt}(L,M) = \sum\sum_{\substack{\ell\leqslant L/t,\, m\leqslant M/t\\ (\ell m,d)=1\\ \ell+m\equiv 0\,(\mathrm{mod}\ d)}} \alpha_{\ell t}\beta_{mt} - \frac{1}{\varphi(d)} \sum\sum_{\substack{\ell\leqslant L/t,\, m\leqslant M/t\\ (\ell m,d)=1}} \alpha_{\ell t}\beta_{mt} .$$

Estimating trivially as in (22.13), we find

$$r_{dt}(L,M) \ll \tau_r^2(t)\Bigl(\frac{x}{dt}+1\Bigr)\frac{x}{t}(\log x)^{r^2-1} .$$

We use this rough bound for $t \geqslant T$, say, getting

$$\sum_{T<t\leqslant x} |r_{dt}(L,M)| \ll \Bigl(\frac{x}{dT}+1\Bigr)x(\log x)^{2r^2-1} . \tag{22.14}$$

For each $1 \leqslant t \leqslant T$ we estimate $r_{dt}(L,M)$ more precisely using the large sieve inequality (9.52). To this end we detect the congruence $\ell \equiv -m\,(\mathrm{mod}\ d)$ by multiplicative characters, arriving at

$$r_{dt}(L,M) = \frac{1}{\varphi(d)} \sum_{\substack{\chi\,(\mathrm{mod}\ d)\\ \chi\neq\chi_0}} \Bigl(\sum_{\ell\leqslant L/t} \chi(\ell)\alpha_{\ell t}\Bigr)\Bigl(\sum_{m\leqslant M/t} \bar\chi(-m)\beta_{mt}\Bigr) .$$

After summing over d we bound this just as was (9.77), arriving at the same bound (9.76), which in the current case becomes

$$\tau_r^2(t)t^{-1}\bigl(D + x(\log x)^{-B}\bigr)x(\log x)^{r^2-1} .$$

Summing this over $t \leqslant T$ and combining with the bound (22.14) for $t > T$, we derive

$$\sum_{d \leqslant D} |r_d(L,M)| \ll \big(D + x(\log x)^{-B}\big) x (\log x)^{2r^2-1} , \tag{22.15}$$

by choosing $T = (\log x)^B$.

Now we are ready to complete the proof of Theorem 22.6. We split the summation over ℓ into short intervals $L < \ell \leqslant (1+\eta)L$ with $0 < \eta < 1$ to be chosen. The number of such partial sums is $O(\eta^{-1} \log x)$. For each of these sums the condition $\ell + m \leqslant x$ becomes $L + m \leqslant x$ except for the last few terms $(1-\eta)x < \ell + m < (1+\eta)x$ whose contribution was estimated in (22.13). Using (22.15) for all of the partial sums but the last one and using the crude bound (22.13) for the last sum, we arrive at

$$\sum_{d \leqslant D} |r_d(x)| \ll \eta^{-1}\big(D + x(\log x)^{-B}\big) x (\log x)^{2r^2} + (\eta x + D) x (\log x)^{r^2} .$$

Finally, choosing the optimal value of η we complete the proof of Theorem 22.6. □

22.4. Bilinear Level for Quadratic Polynomials

Let $F(X) \in \mathbb{Z}[X]$ be an irreducible quadratic. Questions about the prime values of F are very attractive. On the one hand, such problems are beyond our current abilities. On the other hand, they seem not too far beyond, and the theory of automorphic functions seems to offer possibilities for attacking them. In this section we are going to use that theory to establish a strong statement about the level of distribution which will allow us in Chapter 25 to deduce that $F(X)$ represents infinitely many numbers with at most two prime factors, the best we can do to date. For simplicity we consider the polynomial

$$F(X) = X^2 + 1 .$$

We choose the sequence $\mathcal{A} = \big(a_n\big)$ with

$$a_n = f(a) \quad \text{if} \quad n = a^2 + 1$$

and $a_n = 0$ otherwise, where $f(v)$ is a non-negative, smooth function, supported on a dyadic type segment $v \asymp \sqrt{x}$, whose derivatives satisfy

$$f^{(j)} \ll x^{-j/2} , \quad j \geqslant 0 ,$$

the implied constant depending on j. Naturally, one expects that

$$A_d(x) = \sum_{a^2+1 \equiv 0 \,(\mathrm{mod}\, d)} f(a)$$

is well approximated by $\rho(d) d^{-1} \hat{f}(0)$, where $\rho(d)$ is the number of solutions to the congruence

$$\nu^2 + 1 \equiv 0 \,(\mathrm{mod}\, d)$$

and

$$\hat{f}(u) = \int f(v) e(-uv) dv$$

is the Fourier transform of f. We assume that

$$\hat{f}(0) = \int f(v) dv \asymp \sqrt{x} .$$

Accordingly, we define the remainder term

$$r_d(x) = A_d(x) - \frac{\rho(d)}{d}\,\hat{f}(0)\ . \tag{22.16}$$

We consider the bilinear form

$$\mathcal{B}(M,N) = \sum\sum_{(m,n)=1} \alpha_m \beta_n r_{mn}(x) \tag{22.17}$$

where α_m, β_n are real numbers supported on dyadic type segments $m \asymp M$, $n \asymp N$, respectively, with $|\alpha_m| \leqslant 1$, $|\beta_n| \leqslant 1$. We also assume that

$$\beta_n = 0\ , \quad \text{unless } \big(n, P(x^{3\varepsilon})\big) = 1,\ n \text{ squarefree.} \tag{22.18}$$

THEOREM 22.7. *For bilinear forms* (22.17) *with*

$$M \ll x^{\frac{1}{2}-3\varepsilon} \quad \text{and} \quad M^7 N^{10} < x^{4-16\varepsilon}\ , \tag{22.19}$$

we have

$$\mathcal{B}(M,N) \ll x^{\frac{1}{2}-\varepsilon}\ . \tag{22.20}$$

PROOF. If $M < N^2$ then $M^7N^{10} > (MN)^8$, so (22.19) implies $MN \ll x^{\frac{1}{2}-2\varepsilon}$. In this case (22.20) follows trivially since $r_{mn}(x) \ll \tau(mn)$. Therefore, we can assume $N^2 \leqslant M$.

We begin with the inequality

$$\big|\mathcal{B}(M,N)\big| \leqslant \sum_m \sum_{\substack{\nu \,(\mathrm{mod}\, m)\\ \nu^2+1\equiv 0(m)}} \Big| \sum_{(n,m)=1} \beta_n \Big\{ \sum_{\substack{a^2+1\equiv 0(n)\\ a\equiv \nu(m)}} f(a) - \frac{\rho(n)}{mn}\,\hat{f}(0)\Big\}\Big|\ .$$

Hence, by Cauchy's inequality we get

$$\mathcal{B}(M,N) \ll M^{\frac{1}{2}} \mathcal{T}(M,N)^{\frac{1}{2}} \tag{22.21}$$

where

$$\mathcal{T}(M,N) = \sum_m \alpha(m) \sum_{\substack{\nu \,(\mathrm{mod}\, m)\\ \nu^2+1\equiv 0(m)}} \Big(\sum_{(n_1,m)=1} \beta_{n_1} \Big\{ \sum_{\substack{a_1^2+1\equiv 0(n_1)\\ a_1\equiv \nu(m)}} f(a_1) - \frac{\rho(n_1)}{mn_1}\,\hat{f}(0)\Big\}\Big)$$
$$\Big(\sum_{(n_2,m)=1} \beta_{n_2} \Big\{ \sum_{\substack{a_2^2+1\equiv 0(n_2)\\ a_2\equiv \nu(m)}} f(a_2) - \frac{\rho(n_2)}{mn_2}\,\hat{f}(0)\Big\}\Big)\ .$$

Here $\alpha(m)$ is a smooth majorant of the characteristic function of the segment $m \asymp M$.

First, we estimate crudely $\mathcal{T}_0(M,N)$, which is the contribution from terms with $(n_1,n_2) \neq 1$. Because of (22.18) the common factor $\ell = (n_1,n_2)$ satisfies $\ell \geqslant x^{3\varepsilon}$. Despite the fact that such pairs occur rarely, care is needed in the estimation of their contribution.

We shall treat the sum $\mathcal{T}_{00}(M,N)$ of terms involving products $f(a_1)f(a_2)$. The other three sums involving the products $f(a_1)\hat{f}(0)$, $\hat{f}(0)f(a_2)$, and $\hat{f}(0)\hat{f}(0)$ can be treated along the same lines, but more easily, and give the same bounds.

We have $a_1 \equiv a_2 \,(\mathrm{mod}\, m)$ and write the quadratic congruences as equations:

$$a_1^2 + 1 = b_1 k_1 \ell m, \qquad a_2^2 + 1 = b_2 k_2 \ell m,$$

where $n_1 = k_1\ell$, $n_2 = k_2\ell$, $(k_1, k_2) = 1$. Clearly, $\mathcal{T}_{00}(M, N)$ is bounded by the number of solutions to the above system in $a_1, a_2, b_1, b_2, k_1, k_2, \ell, m$.

Given a_1, b_1, k_1, ℓ, m we first estimate the number of values a_2, b_2, k_2. To this end note that $a_2^2 \equiv a_1^2 \pmod{\ell m}$ and $a_2 \equiv a_1 \pmod{m}$. Hence the number of choices of a_2 modulo ℓm is bounded by $\tau(\ell)$ which is bounded in terms of ε, because $\ell \mid n_1$ and hence has no small prime factors. Therefore, the number of a_2 is bounded by $1 + \sqrt{x}\,/\ell m$. Given a_2 the factor b_2k_2 is determined and the number of ways of decomposing this product is bounded in terms of ε, because $k_2 \mid n_2$. We have thus shown that $\mathcal{T}_{00}(M, N)$ is bounded by $\big(1+\sqrt{x}/x^{3\varepsilon}M\big)$ times the number of solutions to

$$a_1^2 + 1 = b_1k_1\ell m\ .$$

Considering b_1k_1 and ℓm as two separate variables, each with multiplicity bounded in terms of ε, we find that the number of solutions to this last equation is bounded by

$$\sum_{a_1} \tau(a_1^2 + 1) f(a_1) \ll \sqrt{x}\, \log x\ .$$

Finally, we get

$$\mathcal{T}_{00}(M, N) \ll \Big(1 + M^{-1}x^{\frac{1}{2}-3\varepsilon}\Big)\, x^{\frac{1}{2}} \log x\ . \tag{22.22}$$

Similarly, we get the same bound for the other three contributions, and hence (22.22) holds for $\mathcal{T}_0(M, N)$.

Now we proceed to the essential part of $\mathcal{T}(M, N)$, namely $\mathcal{T}^*(M, N)$ which sums over pairs $(n_1, n_2) = 1$. In this case we develop the Fourier expansion

$$\begin{aligned}\sum_{\substack{a_1^2+1\equiv 0(n_1)\\ a_1\equiv\nu(m)}} f(a_1) &= \sum_{\substack{\nu_1(mn_1)\\ \nu_1^2+1\equiv 0(n_1)\\ \nu_1\equiv\nu(m)}}\ \sum_{a_1\equiv\nu_1(mn_1)} f(a_1)\\ &= \frac{1}{mn_1}\sum_{h_1}\sum_{\nu_1} e\Big(\frac{\nu_1 h_1}{mn_1}\Big)\hat{f}\Big(\frac{h_1}{mn_1}\Big)\ .\end{aligned}$$

Here $h_1 = 0$ gives the main term $\hat{f}(0)\rho(n_1)/mn_1$. Hence,

$$\mathcal{T}^*(M, N) = \sum\sum_{(n_1,n_2)=1} \frac{\beta_{n_1}\beta_{n_2}}{n_1n_2} \sum\sum_{h_1h_2\neq 0} \mathcal{T}(n_1n_2; h_1, h_2)$$

where

$$\mathcal{T}(n_1, n_2; h_1, h_2) = \sum_m \frac{\alpha(m)}{m^2} \sum_{\nu(mn_1n_2)} e\Big(\frac{\nu(h_1n_2 + h_2n_1)}{mn_1n_2}\Big)\hat{f}\Big(\frac{h_1}{mn_1}\Big)\hat{f}\Big(\frac{h_2}{mn_2}\Big)$$

since m, n_1, n_2 are pairwise prime. Note that the Fourier transform satisfies

$$\hat{f}(u) \ll \sqrt{x}\big(1 + |u|\sqrt{x}\big)^{-A}$$

for every $A \geqslant 0$. Hence, the Fourier series can be truncated to

$$|h_1|, |h_2| \leqslant H = MNx^{\frac{1}{2}(\varepsilon-1)}\ , \tag{22.23}$$

with the remaining terms contributing a negligible amount.

Now we estimate $\mathcal{T}(n_1, n_2; h_1, h_2)$ using Lemma 19.7 with $d = n_1n_2$ and $h = h_1n_2 + h_2n_1$. Note that $d \asymp N^2$, $h \ll HN$ and $h \neq 0$, because $(n_1, n_2) = 1$ and

$|h_1|, |h_2| \ll Nx^{-\varepsilon/2}$, so Lemma 19.7 is applicable. Since we have reduced to the case $N^2 \leqslant M$, we get

$$\mathcal{T}(n_1, n_2; h_1, h_2) \ll \tau(hn_1n_2)^{\frac{3}{2}}(h, n_1n_2)^{\frac{1}{4}} N^{\frac{1}{2}} M^{-\frac{5}{4}} x \log x \ .$$

Here $\tau(hn_1n_2) \ll \tau(h) \ll x^\varepsilon$ and $(h, n_1n_2) = (h_1, n_1)(h_2, n_2)$. Summing over h_1, h_2, n_1, n_2 we get by (22.23),

$$\mathcal{T}^*(M, N) \ll N^{\frac{5}{2}} M^{\frac{3}{4}} x^\varepsilon \ . \tag{22.24}$$

Adding (22.24) to (22.22) we get by (22.21) that

$$\mathcal{B}(M, N) \ll \left(M + x^{\frac{1}{2} - 3\varepsilon}\right)^{\frac{1}{2}} x^{\frac{1}{4}} (\log x)^{\frac{1}{2}} + N^{\frac{5}{4}} M^{\frac{7}{8}} x^\varepsilon \ .$$

This simplifies to (22.20), subject to (22.19). □

22.5. Truncations of Divisor Functions

In the remaining sections of this chapter we are going to study the level of distribution of some further polynomial sequences which are quite lacunary. Far from being able to find primes in very thin sequences, current technology does not even permit us to succeed in the detection of squarefree integers.

Even for the very simplest counting functions, the divisor sums

$$D_r(x) = \sum_{n \leqslant x} \tau_r(n) a_n \ ,$$

an asymptotic formula is often beyond reach. For applications we shall need estimates for these divisor sums, for example, in connection with bounds for the sieve remainder, and for such purposes a sufficiently good upper bound will suffice. More specifically, the trivial upper bound

$$D_r(x) \ll A(x) x^\varepsilon \ , \tag{22.25}$$

will not be good enough because the loss of x^ε from the expected result is simply too large, swamping the gains of a power of the logarithm which we may have elsewhere saved. On the other hand, the loss of a not too large power of the logarithm is frequently good enough to do the job.

In dealing with this problem there is a very useful technical device. The following is a special case of a basic inequality of B. Landreau.

PROPOSITION 22.8. *Let $k \geqslant 2$ be an integer. For all $n \geqslant 1$ we have*

$$\tau(n) \leqslant C \sum_{\substack{d|n \\ d \leqslant n^{1/k}}} \left(2^{\omega(d)} \tau(d)\right)^k \tag{22.26}$$

where C is a constant depending on k.

We shall need this for $k = 4$, and we prove it in Corollary 22.11 with an explicit constant:

$$\tau(n) \leqslant \sum_{\substack{d|n \\ d \leqslant n^{1/4}}} (2\tau(d))^8 \ . \tag{22.27}$$

The proof of Proposition 22.8 is given in [**123**], and indeed the result there is much more general. See also the first result of this kind in [**161**] .

We require some more refined inequalities for the divisor functions. The first of these sharpens Lemma 6.3 of [**55**]. A result optimal for the method can be found in [**110**].

PROPOSITION 22.9. *If $n \geqslant 1$ is a squarefree integer, then*

$$\tau(n) \leqslant 4^9 \sum_{\substack{d|n \\ d \leqslant n^{1/4}}} \tau(d) \ . \tag{22.28}$$

PROOF. This proceeds in the same fashion as in Lemma 6.3 of [**55**] but with a different choice of parameter. We have

$$\tau_4(n) \leqslant 4 \sum_{\substack{d|n \\ d \leqslant n^{1/4}}} \tau_3\Big(\frac{n}{d}\Big)$$

and, for any positive multiplicative function f,

$$\frac{\tau_4(n)}{\tau_3(n)} \leqslant 4 \sum_{\substack{d|n \\ d \leqslant n^{1/4}}} \frac{1}{\tau_3(d)} \leqslant 4\Big(\sum_{\substack{d|n \\ d \leqslant n^{1/4}}} f(d)\Big)^{1/9} \Big(\sum_{d|n} \frac{1}{\tau_3(d)^{9/8} f(d)^{1/8}}\Big)^{8/9}$$

by Hölder's inequality. Hence

$$4^9 \sum_{\substack{d|n \\ d \leqslant n^{1/4}}} f(d) \geqslant \Big(\frac{\tau_4(n)}{\tau_3(n)}\Big)^9 \Big(\sum_{d|n} \frac{1}{\tau_3(d)^{9/8} f(d)^{1/8}}\Big)^{-8} .$$

Here the right-hand side is a multiplicative function of n and if we choose $f(p) = 2$ we see that on the primes it exceeds 2; that is,

$$\Big(\frac{4}{3}\Big)^9 (1 + 3^{-9/8} 2^{-1/8})^{-8} > 2 ,$$

so that it exceeds $\tau(n)$ for all squarefree n. □

We shall need some bounds for higher divisor functions and also at integers not necessarily squarefree. To obtain these we use a simpler method which, however, gives weaker results.

PROPOSITION 22.10. *Let $k \geqslant 2$ and $r \geqslant 2$. Any $n \geqslant 1$ has a divisor $d \mid n$, $d \leqslant n^{1/k}$ such that*

$$\tau_r(n) \leqslant \big(2\tau(d)\big)^{(r-1)k(\log k)/(\log 2)} . \tag{22.29}$$

PROOF. We first consider the ordinary divisor function, that is, we take $r = 2$. Write $n = \ell m$ where ℓ collects those primes occurring with exponent $\geqslant k$ and m is made up of those having exponent $< k$. Thus $(\ell, m) = 1$. Put

$$d_\ell = \prod_{p^\alpha \| \ell} p^{[\alpha/k]}$$

so that $d_\ell \leqslant \ell^{1/k}$ and

$$\tau(d_\ell)^k = \prod_{p^\alpha \| \ell} \Big(\Big[\frac{\alpha}{k}\Big] + 1\Big)^k \geqslant \prod_{p^\alpha \| \ell} (\alpha + 1) = \tau(\ell) .$$

Write $m = p_1^{\alpha_1} \dots p_r^{\alpha_r}$ with $p_1 < \dots < p_r$. Setting $d_m = p_1 \dots p_{[r/k]}$ we have $d_m < m^{1/k}$ and

$$(2\tau(d_m))^k \geqslant 2^r \geqslant \prod_j (\alpha_j + 1)^{(\log 2)/(\log k)} = \tau(m)^{(\log 2)/(\log k)} .$$

Take $d = d_\ell d_m$ completing the proof in this case.

For the case of general r we note that

$$\tau_r(p^\alpha) = \binom{\alpha + r - 1}{r - 1} \leqslant (\alpha + 1)^{r-1} .$$

Hence $\tau_r(n) \leqslant \tau(n)^{r-1}$ and the result follows from the case $r = 2$. □

COROLLARY 22.11. *Let $k \geqslant 2$, $r \geqslant 2$, $n \geqslant 1$. We have*

$$\tau_r(n) \leqslant \sum_{\substack{d|n \\ d \leqslant n^{1/k}}} \big(2\tau(d)\big)^{(r-1)k(\log k)/(\log 2)} . \tag{22.30}$$

We next use the previous two propositions to prove a bound valid for general positive integers which is quite strong on the squarefree part of the integer without caring so much about the contribution coming from the higher powers.

PROPOSITION 22.12. *For any $n \geqslant 1$ we have*

$$\tau(n) \leqslant 2^{18} 3^8 \sum_{\substack{d|n \\ d \leqslant n^{1/4}}} f(d) ,$$

where $f(d)$ is the multiplicative function given by $f(p^\alpha) = 2\alpha^{13}$ for $\alpha \geqslant 1$.

PROOF. We write $n = \ell m^2$ where ℓ is squarefree. Then $\tau(n) \leqslant \tau(\ell)\tau(m^2)$ and

$$\tau(\ell) \leqslant 4^9 \sum_{\substack{a|\ell \\ a \leqslant \ell^{1/4}}} \tau(a)$$

by Proposition 22.9. We have $\tau(m^2) \leqslant \tau(m)^{\log 3/\log 2}$ and m has a divisor $b \leqslant m^{1/4}$ such that $\tau(m) \leqslant (2\tau(b))^8$ by Proposition 22.10. Hence

$$\tau(m) \leqslant 4^9 3^8 \sum_{\substack{a|\ell \\ a \leqslant \ell^{1/4}}} \tau(a)\tau(b)^{8 \log 3/\log 2}$$

for some $b \mid m$, $b \leqslant m^{1/4}$. Putting $d = ab^2$ we have $d \mid n$, $d \leqslant n^{1/4}$ and the multiplicity of d is bounded by the multiplicative function which for prime powers p^α is

$$2\Big(\Big[\frac{\alpha}{2}\Big] + 1\Big)^{8(\log 3)/(\log 2)} \leqslant 2\alpha^{13} .$$

□

Using the same arguments as in the proof of Proposition 22.12 we obtain

COROLLARY 22.13. *The statement of Proposition 22.12 holds with $\tau(n)$ replaced by $r(n)$ and the multiplicative function $f(d)$ given by*

$$f(p^\alpha) = (1 + \chi_4(p)^\alpha)\alpha^{13} \quad \text{for} \quad \alpha \geqslant 1 . \tag{22.31}$$

These inequalities for divisor-like functions are handy in many situations. For example, returning to the question of the divisor sums $D_r(x)$ we can improve the simple bound (22.25). Indeed, using Corollary 22.11 we have

$$D_r(x) \leqslant \sum_{d \leqslant x^{1/k}} \big(2\tau(d)\big)^{(r-1)(\log k)/(\log 2)} A_d(x) \ .$$

Hence, taking any $k > 1/\theta$ where θ is the exponent of distribution of $\mathcal{A} = (a_n)$, we infer the bound

$$D_r(x) \ll X(\log x)^A$$

for some $A = A(r, \theta, \kappa)$, which is satisfactory in certain situations.

22.6. Gaussian Sequences

The sequence $\mathcal{A}$ considered in Section 22.4 is one of the most basic examples of a Gaussian sequence. By a Gaussian sequence we mean a sequence $\mathcal{A} = \big(a_n\big)$ supported on integers n which can be written as the sum of two squares, that is, on norms of Gaussian integers. Although the problem of obtaining primes in that sequence remains a formidable obstacle, the attainment of the optimal level (one half) of absolute distribution is not.

In this section and the next we are going to consider some other basic Gaussian sequences, including some which are quite lacunary, but for which the attainment of this optimal level requires more delicate arguments.

The method we present in this section works for quite general sequences

$$a_n = \sum_{\substack{\ell^2 + m^2 = n \\ (\ell, m) = 1,\, q | m}} \gamma_\ell \tag{22.32}$$

where q is a fixed positive integer, ℓ, m run over positive integers, and γ_ℓ are any complex numbers satisfying $|\gamma_\ell| \leqslant 1$. The problem is more attractive when γ_ℓ is a lacunary sequence, but a certain spacing property is helpful. Therefore, to control the spacing we fix a positive integer r and assume that γ_ℓ is supported on r-th powers prime to q:

$$\gamma_\ell = 0 \qquad \text{if } \ell \neq k^r \text{ or } (\ell, q) \neq 1\ , \tag{22.33}$$

$$|\gamma_\ell| \leqslant 1 \qquad \text{if } \ell = k^r \text{ and } (\ell, q) = 1\ . \tag{22.34}$$

In this case we have

$$\sum_{n \leqslant x} a_n = \frac{1}{q} X + O\Big(x^{\frac{1}{2r}} \log x\Big) \tag{22.35}$$

where

$$X = \sum_{\ell < \sqrt{x}} \gamma_\ell \frac{\varphi(\ell)}{\ell} \sqrt{x - \ell^2}\ . \tag{22.36}$$

Since $X \ll x^{\frac{1}{2} + \frac{1}{2r}}$, the best level of absolute distribution one can hope for is

$$D(x) = x^{\frac{1}{2} + \frac{1}{2r}} \big(\log x\big)^{-A}\ , \tag{22.37}$$

and we are going to achieve this.

For any d not too large with $(d,q)=1$ we expect that $A_d(x)$ is well approximated by

$$M_d(x)=\frac{\rho(d)}{qd}\sum_{\substack{\ell<\sqrt{x}\\(\ell,d)=1}}\gamma_\ell\,\frac{\varphi(\ell)}{\ell}\,\sqrt{x-\ell^2} \tag{22.38}$$

where $\rho(d)$ is the number of solutions of the congruence

$$\alpha^2+1\equiv 0\,(\mathrm{mod}\,d)\ . \tag{22.39}$$

THEOREM 22.14. *Let γ_ℓ be supported on r-th powers and $|\gamma_\ell|\leqslant 1$. Then, for*

$$x^{\frac{1}{2}}\leqslant D\leqslant x^{\frac{r+1}{2r}} \tag{22.40}$$

we have

$$\sum_{d\leqslant D}\big|A_d(x)-M_d(x)\big|\ll D^{\frac{1}{4}}x^{\frac{3(r+1)}{8r}}\big(\log x\big)^{130} \tag{22.41}$$

where the implied constant depends only on r.

Particular cases of this theorem are applied elsewhere in the book. Examples are the sequence (22.32) with γ_ℓ the characteristic function on primes in Chapter 20, on squares in Chapter 21, for both of which sequences the corresponding sum over primes has been successfully evaluated, and on cubes, also in Chapter 21, for which the goal of finding primes remains open.

The main ingredients are the large sieve inequality for the roots of the congruence (22.39), which we already proved in Proposition 20.4, as well as a special case of Corollary 22.11.

PROOF. We proceed to the proof of the theorem. As usual, before applying harmonic analysis it helps to introduce some smoothing factors. Let $f(t)$ be a smooth function on $[0,\infty]$ such that

$$f(t)=1 \qquad \text{if} \quad 0\leqslant t\leqslant (1-\eta)x\ ,$$

$$f(t)=0 \qquad \text{if} \quad t\geqslant x\ ,$$

$$f^{(j)}(t)\ll(\eta x)^{-j} \quad \text{for} \quad j=0,1,2,$$

where η will be chosen later with $x^{-1/4r}\leqslant\eta\leqslant 1$.

We replace $A_d(x), M_d(x)$ by their smooth counterparts

$$A_d(f)=\sum_{n\equiv 0(d)}a_n\,f(n) \tag{22.42}$$

and

$$M_d(f)=\frac{\rho(d)}{qd}\sum_{(\ell,d)=1}\gamma_\ell\,\frac{\varphi(\ell d)}{\ell d}\int_0^\infty f(\ell^2+t^2)\,dt\ . \tag{22.43}$$

We estimate the corrections resulting from this modification by elementary arguments as follows:

$$\sum_{d\leqslant D}\big|A_d(x)-A_d(f)\big|\leqslant\sideset{}{'}\sum_{\substack{(1-\eta)x<\ell^2+m^2\leqslant x\\(\ell,m)=1}}|\gamma_\ell|\tau(\ell^2+m^2)+O\big(\sqrt{x}\,\log x\big)\ .$$

Here $\sum'$ means that the terms with a value of ℓ which is nearest to $\sqrt{x}$ are omitted. For the remaining points we apply (22.27), which is the case $k = 4$ of Corollary 22.11, getting the bound

$$\sum_{\ell<\sqrt{x}}{}' |\gamma_\ell| \sum_{\substack{d\leqslant x^{\frac14}\\(d,\ell)=1}} \tau(d)^8 \sum_{\substack{(1-\eta)x<\ell^2+m^2\leqslant x\\ \ell^2+m^2\equiv 0(d)}} 1 .$$

Note that m runs over an interval of length $O\left(\eta x/\sqrt{x-\ell^2}\right)$. Splitting into residue classes $m \equiv \alpha\ell (\operatorname{mod} d)$ with α running over the roots of (22.39) we estimate the above sum by

$$\ll \eta x \Big(\sum_d\Big)\Big(\sum_\ell{}'\Big) + x^{\frac14+\frac{1}{2r}}(\log x)^{256},$$

where the first sum is

$$\sum_{d\leqslant x^{\frac14}} \tau^8(d)\rho(d)d^{-1} \ll (\log x)^{256} ,$$

and the second sum is

$$\begin{aligned}\sum_{\ell<\sqrt{x}}{}' |\gamma_\ell|(x-\ell^2)^{-\frac12} &\leqslant \sum_{k<x^{\frac{1}{2r}}}{}' (x-k^{2r})^{-\frac12}\\ &\ll x^{\frac{1-2r}{4r}} \sum_{k<x^{\frac{1}{2r}}}{}' \left(x^{\frac{1}{2r}}-k\right)^{-\frac12} \ll x^{\frac{1-r}{2r}} .\end{aligned}$$

Hence, we obtain

$$\sum_{d\leqslant D} \left|A_d(x) - A_d(f)\right| \ll \eta x^{\frac{r+1}{2r}} (\log x)^{256} . \tag{22.44}$$

An easier variant of the same argument shows that

$$\sum_{d\leqslant D} \left|M_d(x) - M_d(f)\right| \ll \eta x^{\frac{r+1}{2r}} \log x . \tag{22.45}$$

Next, we decompose $A_d(f)$ as follows (expand m to all $\mathbb{Z}$):

$$\begin{aligned}A_d(f) &= \tfrac12 \sum\sum_{\substack{\ell^2+m^2\equiv 0(d)\\(\ell,m)=1,\, q|m}} \gamma_\ell\, f(\ell^2+m^2)\\ &= \tfrac12 \sum_{\alpha^2+1\equiv 0(d)} \sum_\ell \gamma_\ell \sum_{\substack{(m,\ell)=1\\ m\equiv\alpha\bar{q}\ell(d)}} f(\ell^2+m^2q^2)\\ &= \tfrac12 \sum_{\alpha^2+1\equiv 0(d)} \sum_a \mu(a) \sum_\ell \gamma_{a\ell} \sum_{m\equiv\alpha\bar{q}\ell(d/(a,d))} f\left(a^2(\ell^2+m^2q^2)\right) .\end{aligned}$$

To the inner sum we apply Poisson's formula

$$\sum_m = \frac{(a,d)}{d} \sum_{h\in\mathbb{Z}} e\Big(\alpha h\bar{q}\ell\frac{(a,d)}{d}\Big) F_{a\ell}\Big(\frac{h(a,d)}{d}\Big) ,$$

where $F_{a\ell}(v)$ is the Fourier integral

$$\begin{aligned} F_{a\ell}(v) &= \int_{-\infty}^{\infty} f\big(a^2(\ell^2+t^2q^2)\big)e(-vt)dt \\ &= 2\int_0^{\infty} f\big(a^2(\ell^2+t^2q^2)\big)\cos(2\pi vt)dt\ . \end{aligned} \tag{22.46}$$

Integrating (22.46) by parts twice we get an alternative expression

$$F_{a\ell}(v) = \Big(\frac{aq}{\pi v}\Big)^2 \int_0^{\infty} (f' + 2a^2t^2q^2 f'')\big(a^2(\ell^2+t^2q^2)\big)\cos(2\pi vt)dt\ . \tag{22.47}$$

The zero frequency $h = 0$ yields exactly $M_d(f)$, so we have

$$\big|A_d(f) - M_d(f)\big| \leqslant \frac{1}{d}\sideset{}{^\flat}\sum_a \sum_{\substack{bc=d \\ b|a}} \rho(b)b \sum_{\substack{\alpha(\mathrm{mod}\, c) \\ \alpha^2+1\equiv 0(c)}} \big|W_a(c,\alpha)\big|$$

where

$$W_a(c,\alpha) = \sum_{h>0}\sum_{\ell} \gamma_{a\ell}\, e\Big(\frac{\alpha h\bar{q}\ell}{c}\Big)F_{a\ell}\Big(\frac{h}{c}\Big) \tag{22.48}$$

and $\sum^\flat$ denotes a sum over squarefree integers. Summing over the moduli d in a dyadic segment we get

$$\sum_{\substack{D<d\leqslant 2D \\ (d,q)=1}} \big|A_d(f) - M_d(f)\big| \leqslant \frac{1}{D}\sideset{}{^\flat}\sum_a \sum_{b|a} \rho(b)bV_a(D/b)\ , \tag{22.49}$$

where

$$V_a(C) = \sum_{\substack{C<c\leqslant 2C \\ (c,q)=1}} \sum_{\alpha^2+1\equiv 0(c)} \big|W_a(c,\alpha)\big|\ . \tag{22.50}$$

Next we split the outer summation in (22.48) into dyadic segments $H \leqslant h < 2H$ and we shall treat these partial sums separately. By (22.50) we obtain

$$V_a(C) \leqslant \sum_H V_a(C,H) \tag{22.51}$$

where

$$V_a(C,H) = \sum_{\substack{C<c\leqslant 2C \\ (c,q)=1}} \sum_{\alpha^2+1\equiv 0(c)} \big|W_a(H;c,\alpha)\big| \tag{22.52}$$

and

$$W_a(H;c,\alpha) = \sum_{H\leqslant h<2H}\sum_{\ell} \gamma_{a\ell}\, e\Big(\frac{\alpha h\bar{q}\ell}{c}\Big)F_{a\ell}\Big(\frac{h}{c}\Big)\ . \tag{22.53}$$

We wish to separate the modulus c from the variables h, ℓ in the Fourier integral $F_{a\ell}(h/c)$. This can be easily achieved by changing t into tH/h in the integrals (22.46) or (22.47), and then holding t fixed. Note that the trivial integration over t in (22.46) or (22.47) gives

$$F_{a\ell}(v) \ll \frac{\sqrt{x}}{a}\ , \quad F_{a\ell}(v) \ll \frac{\sqrt{x}}{a}\Big(\frac{a}{\eta v\sqrt{x}}\Big)^2\ ,$$

respectively, because $f \ll 1$, $f' \ll (\eta x)^{-1}$, $f'' \ll (\eta x)^{-2}$. Hence, the Fourier integral at $v = h/c$ satisfies

$$F_{a\ell}(h/c) \ll \frac{\sqrt{x}}{a} \min\left\{1, \left(\frac{aC}{\eta H\sqrt{x}}\right)^2\right\} . \tag{22.54}$$

By (22.52), (22.53) we obtain

$$V_a(C,H) \ll U_a(C,H)\frac{\sqrt{x}}{a} \min\left\{1, \left(\frac{aC}{\eta H\sqrt{x}}\right)^2\right\} , \tag{22.55}$$

where

$$U_a(C,H) = \sum_{\substack{C<c\leqslant 2C \\ (c,q)=1}} \sum_{\alpha^2+1\equiv 0(c)} \Big| \sum_{H\leqslant h<2H} \sum_{\ell} \gamma_{a\ell}\, \xi_{h\ell}\, e\Big(\frac{\alpha h\bar{q}\ell}{c}\Big)\Big| \tag{22.56}$$

with some coefficients $\xi_{h\ell}$ which do not depend on c, α and which satisfy $|\xi_{h\ell}| \leqslant 1$.

Now we are ready to apply the large sieve for quadratic roots, Corollary 20.6, which yields

$$U_a(C,H) \ll C^{\frac{1}{2}}\big(C + H\sqrt{x/a}\big)^{\frac{1}{2}} E^{\frac{1}{2}} \tag{22.57}$$

where

$$E = \sum_n \Big(\sum_{\substack{h\ell=n \\ H\leqslant h<2H}} |\gamma_{a\ell}|\Big)^2 \ll Ha^{-1}x^{\frac{1}{2r}} \log x .$$

This estimate for E is not immediate. Because a is squarefree and $a\ell$ is an r-th power it follows that $\ell = a^{r-1}m^r$ with $m \leqslant a^{-1}x^{\frac{1}{2r}} = M$, say. Therefore, we see that E is bounded by the number of solutions of

$$h_1 m_1^r = h_2 m_2^r$$

with $H \leqslant h_1, h_2 < 2H$ and $m_1, m_2 \leqslant M$. The solutions are given explicitly by $m_1 = st_1$, $m_2 = st_2$ with $(t_1,t_2) = 1$, $st_1, st_2 \leqslant M$ and $h_1 = kt_2^r$, $h_2 = kt_1^r$ with $k \leqslant 4H(t_1^r + t_2^r)^{-1}$. Hence,

$$E \leqslant 8HM \sum\sum_{t_1,t_2\leqslant M} (t_1^r + t_2^r)^{-1}(t_1+t_2)^{-1} \leqslant 16HM \sum_{t\leqslant M} t^{-r} .$$

The worst case is $r = 1$, yielding $E \ll Ha^{-1}x^{\frac{1}{2r}}\log x$, and (22.57) gives

$$U_a(C,H) \ll C^{\frac{1}{2}}\big(C + H\sqrt{x}\,/\,a\big)^{\frac{1}{2}} H^{\frac{1}{2}} a^{-\frac{1}{2}} x^{\frac{1}{4r}} (\log x)^{\frac{1}{2}} . \tag{22.58}$$

Introducing (22.58) to (22.55) we see that the series (22.51) over the dyadic endpoints H converges and the largest contribution is at

$$H \asymp aC/\eta\sqrt{x} . \tag{22.59}$$

Therefore, we obtain

$$V_a(C) \ll (\eta a)^{-1} C^{\frac{3}{2}} x^{\frac{r+1}{4r}} (\log x)^{\frac{3}{2}} . \tag{22.60}$$

Inserting this into (22.49) we arrive at

$$\sum_{\substack{D<d\leqslant 2D \\ (d,q)=1}} |A_d(f) - M_d(f)| \ll \eta^{-1} D^{\frac{1}{2}} x^{\frac{r+1}{4r}} (\log x)^{\frac{5}{2}} . \tag{22.61}$$

This bound remains the same for the sum over all $d \leqslant D$. Finally, combining this with (22.44), (22.45) we conclude (22.41) by choosing

$$\eta = D^{\frac{1}{4}} x^{-\frac{r+1}{8r}} (\log x)^{\frac{5}{4}-128} . \tag{22.62}$$

This completes the proof of the theorem. □

At the level of generality for which Theorem 22.14 holds one cannot hope to give a completely explicit formula for the main term. However, we can proceed rather far for special coefficients. We have $M_d(x) = (\rho(d)/qd)X_d$, where

$$X_d = \sum_{\substack{\ell<\sqrt{x} \\ (\ell,d)=1}} \gamma_\ell \frac{\varphi(\ell)}{\ell} \sqrt{x-\ell^2} . \tag{22.63}$$

By Mellin inversion we find for $0 < y < x$,

$$\sqrt{x-y} = \frac{\sqrt{x}}{2\pi i} \int_{(\sigma)} \mathcal{B}(s) \Big(\frac{x}{y}\Big)^s ds$$

where $\sigma > 0$ and

$$\mathcal{B}(s) = \int_0^1 \sqrt{1-y}\, y^{s-1} dy = B\left(s, \tfrac{3}{2}\right) = \frac{\Gamma(s)\Gamma\left(\frac{3}{2}\right)}{\Gamma\left(s+\frac{3}{2}\right)} .$$

Moreover, the contour integral vanishes if $y \geqslant x$. Therefore,

$$X_d = \frac{\sqrt{x}}{2\pi i} \int_{(\sigma)} \mathcal{B}(s) x^s Z(2s) ds ,$$

where $\sigma > \frac{1}{2}$ and

$$Z(s) = \sum_{(\ell,dq)=1} \gamma_\ell \frac{\varphi(\ell)}{\ell} \ell^{-s} .$$

To go further, we now specialize to the important case of γ_ℓ, the characteristic function of r-th powers. Here $Z(s)$ is given by

$$\begin{aligned} Z(s) &= \sum_{(k,dq)=1} \prod_{p|k} \Big(1-\frac{1}{p}\Big) k^{-rs} = \prod_{p\nmid dq} \Big(1-\frac{1}{p^{rs}}\Big)^{-1} \Big(1-\frac{1}{p^{rs+1}}\Big) \\ &= \frac{\zeta(rs)}{\zeta(rs+1)} \prod_{p|dq} \Big(1-\frac{1}{p^{rs}}\Big)\Big(1-\frac{1}{p^{rs+1}}\Big)^{-1} . \end{aligned}$$

Moving the contour to $\sigma = \varepsilon$ we encounter a simple pole of $Z(2s)$ at $s = \frac{1}{2r}$ with residue

$$\operatorname*{res}_{s=\frac{1}{2r}} Z(2s) = \frac{1}{2r\zeta(2)} h(d)h(q)$$

where

$$h(q) = \prod_{p|q} \Big(1+\frac{1}{p}\Big)^{-1} . \tag{22.64}$$

Because $\mathcal{B}(s) \ll |s|^{-\frac{3}{2}}$ on the line $s = \varepsilon$, the integral converges absolutely so that

$$X_d = x^{\frac{r+1}{2r}} \mathcal{B}\Big(\frac{1}{2r}\Big) \operatorname*{res}_{s=\frac{1}{2r}} Z(2s) + O\left(\tau(d) x^{\frac{1}{2}+\varepsilon}\right) .$$

Gathering the above formulas we obtain:

PROPOSITION 22.15. *Let γ_ℓ be the characteristic function of r-th powers prime to q. We have,*

$$M_d(x) = g(d)\frac{h(q)}{q}X + O\left(\tau(d)\rho(d)d^{-1}x^{\frac{1}{2}+\varepsilon}\right) , \tag{22.65}$$

where

$$g(d) = \frac{\rho(d)}{d}\prod_{p|d}\left(1+\frac{1}{p}\right)^{-1} \tag{22.66}$$

and

$$X = \frac{\Gamma(\frac{1}{2r})\Gamma(\frac{3}{2})}{2\zeta(2)\Gamma(\frac{1}{2r}+\frac{3}{2})}\ x^{\frac{1}{2}+\frac{1}{2r}} . \tag{22.67}$$

22.7. Sum of Two Biquadrates

In this section we are going to consider a sequence still thinner than those in the previous section. Let $\mathcal{A} = (a_n)$ where a_n denotes the number of representations

$$a^4 + b^4 = n \tag{22.68}$$

in coprime integers a, b. Note that $A(x) \asymp x^{1/2}$ so that the best we can hope for in the level of absolute distribution is

$$D = D(x) = x^{\frac{1}{2}}(\log x)^{-B} \tag{22.69}$$

for some B, and we are going to attain this.

Arguments very similar to those we give here would work for sums of two k-th powers or even more generally for binary forms of degree k leading to $D(x) = x^{2/k}(\log x)^{-B}$. The fact that one can get the best level of distribution for such thin sequences shows the tremendous benefit conferred by the homogeneity of the polynomial.

The expected main term for the congruence sum $A_d(x)$ is $g(d)X$ where the density function $g(d)$ is given by (see (22.89))

$$g(d) = \frac{\rho(d)}{d}\prod_{p|d}\left(1+\frac{1}{p}\right)^{-1} , \tag{22.70}$$

with $\rho(d)$ being the number of roots $\omega \pmod d$ of the congruence

$$\omega^4 + 1 \equiv 0 \pmod d . \tag{22.71}$$

Moreover, X should approximate to

$$A(x) = \sum_{n\leqslant x} a_n = \sum\sum_{\substack{a^4+b^4\leqslant x\\(a,b)=1}} 1 . \tag{22.72}$$

We choose X to be $\zeta(2)^{-1}$ times the Lebesgue measure of the set $u^4 + v^4 \leqslant x$, that is

$$X = 6\pi^{-2}\gamma\sqrt{x} \tag{22.73}$$

where γ is the measure of the set $u^4 + v^4 \leqslant 1$. Thus we have

$$\gamma = 4\int_0^1 (1-u^4)^{\frac{1}{4}}du = \int_0^1 (1-t)^{\frac{1}{4}}t^{-\frac{3}{4}}dt = \frac{1}{2\sqrt{\pi}}\,\Gamma\left(\frac{1}{4}\right)^2 . \tag{22.74}$$

THEOREM 22.16. *Let $r_d(x) = A_d(x) - g(d)X$ where $g(d)$ is given by (22.70) and X by (22.73). Then, for $2 \leqslant D \leqslant \sqrt{x}$ we have*

$$\sum_{d \leqslant D} |r_d(x)| \ll D^{\frac{1}{4}} x^{\frac{3}{8}} (\log x)^{256} , \tag{22.75}$$

where the implied constant is absolute.

REMARKS. The result shows that

$$D = D(x) = \sqrt{x} (\log x)^{-B} \tag{22.76}$$

with B sufficiently large is the level of absolute distribution. As we have already mentioned, this is best possible apart from the value of B.

In some applications it is helpful to utilize (22.75) together with a rough estimation for $A_d(x)$ for individual moduli d. We shall give a quick estimation which has right order of magnitude for any $d \ll \sqrt{x}$. We shall also use this in the proof of Corollary 22.18.

PROPOSITION 22.17. *For any d we have*

$$A_d(x) \leqslant 2\rho(d)\big(1 + 4\sqrt{x} \,/\, d\big) . \tag{22.77}$$

PROOF. We split the solutions of $a^4 + b^4 \equiv 0 \,(\bmod\, d)$ with $(a, b) = 1$, $|a|, |b| \leqslant x^{\frac{1}{4}}$ into classes $\mathcal{C}(\omega)$ defined by the congruence

$$a = \omega b \,(\bmod\, d) \tag{22.78}$$

where ω runs over the roots of (22.71). Let (a, b) be a solution in the class $\mathcal{C}(\omega)$ having the largest value of $|a|$. Suppose (a', b') is another solution in $\mathcal{C}(\omega)$, so $ab' \equiv a'b \,(\bmod\, d)$. We interpret this congruence as the equation

$$ab' - a'b = dh , \tag{22.79}$$

where $|h| \leqslant 2\sqrt{x}\, d^{-1}$. Given h, the pair (a', b') is determined by $a' \equiv \bar{b}\, dh \,(\bmod\, a)$. Since $|a'| \leqslant |a|$ we have $|\mathcal{C}(\omega)| \leqslant 2\big(1 + 4\sqrt{x}\, d^{-1}\big)$ and (22.77) follows. □

Combining (22.77) with the inequalities $g(d) \leqslant \rho(d)d^{-1}$, $X \leqslant 4\sqrt{x}$, we obtain

$$|r_d(x)| \leqslant 2\rho(d)(1 + 6\sqrt{x} \,/ d) . \tag{22.80}$$

Note that $\rho(d)$ is a multiplicative function with

$$\rho(p^\alpha) = \begin{cases} 4 & \text{if } p \equiv 1 \,(\bmod\, 8), \\ 0 & \text{if } p \not\equiv 1 \,(\bmod\, 8), \end{cases} \tag{22.81}$$

for $p > 2$, $\alpha > 0$ and $\rho(2) = 1$, $\rho(2^\alpha) = 0$ if $\alpha > 1$. Hence, for $A \geqslant 1$, $D \geqslant 2$ we have

$$\sum_{d \leqslant D} A^{\Omega(d)} \rho(d) d^{-1} \ll (\log D)^A . \tag{22.82}$$

By (22.80), (22.82) and (22.75) we derive, using Cauchy's inequality, the following.

COROLLARY 22.18. *For $2 \leqslant D \leqslant \sqrt{x}$ we have*

$$\sum_{d \leqslant D} A^{\Omega(d)} |r_d(x)| \ll D^{\frac{1}{8}} x^{\frac{7}{16}} (\log x)^{A+128} , \tag{22.83}$$

where the implied constant depends only on A.

In the proof of Theorem 22.16 we shall encounter the sequence $\tilde{\mathcal{A}} = (\tilde{a}_n)$ where $\tilde{a}_n$ is the number of all representations (22.68) with no coprimality restriction. We only need a crude bound for

$$A^{(r)}(y) = \sum_{m \leqslant y} \tau_r(m)\, \tilde{a}_m \,, \qquad r = 2, 5 \,.$$

Pulling out the common factor of the variables a, b we derive

$$A^{(5)}(y) \leqslant \sum_{c^4 \leqslant y} \tau_5(c^4) \sum_{m \leqslant yc^{-4}} \tau_5(m) a_m \,.$$

By the version of Landreau's inequality found in Corollary 22.11, we have

$$\tau_5(m) \leqslant \sum_{\substack{d|m \\ d \leqslant \sqrt{m}}} \big(2\tau(d)\big)^8 \,.$$

Hence, by (22.77) we get

$$\sum_{m \leqslant z} \tau_5(m) \tilde{a}_m \ll \sqrt{z} \sum_{d \leqslant \sqrt{z}} \tau^8(d) d^{-1} \ll \sqrt{z} \big(\log 2z\big)^{256}$$

and

$$A^{(5)}(y) \ll \sqrt{y} \big(\log 2y\big)^{256} \,. \tag{22.84}$$

Similarly,

$$A^{(2)}(y) \ll \sqrt{y}\, \log y \,. \tag{22.85}$$

As has happened frequently before, we find it convenient to deduce Theorem 22.16 from the corresponding result for a smoothed version of the sum.

Let f be a smooth function supported on $[0, 2x]$ with

$$f^{(j)} \ll (\eta x)^{-j} \,, \quad j = 0, 1, \ldots, 8 \tag{22.86}$$

where $0 < \eta \leqslant 1$. We consider the sum

$$A_d(f) = \sum_{n \equiv 0 \, (\mathrm{mod}\, d)} a_n\, f(n)$$

for any d. We have

$$A_d(f) = \sideset{}{^*}\sum\sideset{}{^*}\sum_{\substack{\alpha, \beta \, (\mathrm{mod}\, d) \\ \alpha^4 + \beta^4 \equiv 0 \, (\mathrm{mod}\, d)}} \sum\sum_{\substack{(a,b)=1 \\ a, b \equiv \alpha, \beta \, (\mathrm{mod}\, d)}} f(a^4 + b^4) \,,$$

where, as usual, $\sum^*$ restricts the variable of summation to reduced residue classes. Remove the condition $(a, b) = 1$ by the Möbius formula getting

$$\begin{aligned} A_d(f) &= \sum_{(c,d)=1} \mu(c) \sideset{}{^*}\sum\sideset{}{^*}\sum_{\substack{\alpha, \beta \, (\mathrm{mod}\, d) \\ \alpha^4 + \beta^4 \equiv 0 \, (\mathrm{mod}\, d)}} \sum\sum_{a, b \equiv \alpha, \beta \, (\mathrm{mod}\, d)} f\big((a^4 + b^4)\, c\big) \\ &= A_d^{\sharp}(f) + A_d^{\infty}(f) \,, \end{aligned} \tag{22.87}$$

say, where $A_d^{\sharp}(f)$ is the sum restricted by $c \leqslant C$ and $A_d^{\infty}(f)$ is the sum of the remaining terms. Here $C > 0$ will be chosen later.

We find by (22.84)

$$\sum_d |A_d^\infty(f)| \leqslant \sum_{c>C} \sum_{1\leqslant n\leqslant 2xc^{-4}} \tau(n)\tilde{a}_n \ll C^{-1}x^{\frac{1}{2}} \log x \ . \tag{22.88}$$

Next we evaluate $A_d^\sharp(f)$ by Poisson's formula which gives

$$A_d^\sharp(f) = \sum_{\substack{(c,d)=1\\ c\leqslant C}} \mu(c)(cd)^{-2} \sum\sum_{k,\,\ell\in\mathbb{Z}} G(k,\ell;d)F(k/cd,\ell/cd) \ ,$$

where

$$G(k,\ell;d) = \sideset{}{^*}\sum\sideset{}{^*}\sum_{\alpha^4+\beta^4\equiv 0 \,(\mathrm{mod}\, d)} e\left(\frac{\alpha k+\beta\ell}{d}\right)$$

and $F(u,v)$ is the Fourier transform of $f(a,b) = f(a^4+b^4)$.

The zero frequencies yield the main term

$$M_d(f) = \sum_{\substack{c\leqslant C\\ (c,d)=1}} \mu(c)(cd)^{-2} G(0,0;d)F(0,0) \ .$$

Here we have $G(0,0;d) = \rho(d)\varphi(d)$ and

$$F(0,0) = \iint f(a^4+b^4)da\, db = I(f) \ ,$$

say. We also have

$$\sum_{\substack{c\leqslant C\\ (c,d)=1}} \mu(c)c^{-2} = \frac{6}{\pi^2}\prod_{p|d}\left(1-\frac{1}{p^2}\right)^{-1} + O\left(C^{-1}\right) \ .$$

Hence

$$M_d(f) = g(d)\frac{6}{\pi^2}I(f) + O\left(g(d)\sqrt{x}\,C^{-1}\right) \tag{22.89}$$

where

$$g(d) = \rho(d)d^{-1}\prod_{p|d}\left(1+\frac{1}{p}\right)^{-1} \ . \tag{22.90}$$

For k, ℓ not both zero we write

$$G(k,\ell;d) = \sum_{\omega^4+1\equiv 0(d)} \sideset{}{^\star}\sum_{\alpha(d)} e\big(\alpha(k+\omega\ell)/d\big)$$

where the innermost sum is the Ramanujan sum. Hence,

$$\big|G(k,\ell;d)\big| \leqslant \sum_{\omega^4+1\equiv 0(d)} (d, k+\omega\ell) \ .$$

If $q|d$ and $q|(k+\omega\ell)$ then $k^4+\ell^4\equiv 0 \,(\mathrm{mod}\, q)$. This implies $(d,k+\omega\ell)\big|(d,k^4+\ell^4)$ so

$$\big|G(k,\ell;d)\big| \leqslant \rho(d)(d,k^4+\ell^4) \ . \tag{22.91}$$

The Fourier transform satisfies the bound

$$F(u,v) \ll \sqrt{x}\big(1+|u|\eta x^{\frac{1}{4}}\big)^{-4}\big(1+|v|\eta x^{\frac{1}{4}}\big)^{-4}$$

which follows on applying partial integration four times in each variable to the formula

$$F(u,v)=4\int_0^\infty\int_0^\infty f(a^4+b^4)\cos(2\pi au)\cos(2\pi bv)da\,db\ .$$

Note that the boundary terms in the partial integrations vanish even when $f^{(j)}(t)$ does not vanish at $t=0$. Hence,

$$F(u,v)\ll\sqrt{x}\left(1+(u^4+v^4)\eta^4x\right)^{-1}\ . \tag{22.92}$$

By the above estimates we see that the error term satisfies the bound

$$\left|A_d^\sharp(f)-M_d(f)\right|\ll\rho(d)d^{-2}\sqrt{x}\sum_{c\leqslant C}c^{-2}\sum_{m=1}^\infty(d,m)\left(1+mx\eta^4c^{-4}d^{-4}\right)^{-1}\tilde{a}_m\ .$$

Hence,

$$\sum_{D<d\leqslant 2D}\left|A_d^\sharp(f)-M_d(f)\right|\ll\frac{\sqrt{x}}{D}\sum_{c\leqslant C}c^{-2}\sum_{m=1}^\infty\tau_5(m)\left(1+mx\eta^4c^{-4}D^{-4}\right)^{-1}\tilde{a}_m$$

because

$$\sum_{D<d\leqslant 2D}\frac{\rho(d)}{d}(d,m)\leqslant\sum_{\delta|m}\rho(\delta)\sum_{D/\delta<d\leqslant 2D/\delta}\frac{\rho(d)}{d}\ll(1*\rho)(m)\ll\tau_5(m)\ .$$

Now, by (22.84) we get

$$\sum_{D<d\leqslant 2D}\left|A_d^\sharp(f)-M_d(f)\right|\ll\frac{\sqrt{x}}{D}\sum_{c\leqslant C}c^{-2}\frac{c^2D^2}{\eta^2\sqrt{x}}(\log x)^{256}\ll\frac{CD}{\eta^2}(\log x)^{256}\ .$$

Replacing here $M_d(f)$ by (22.89) and adding (22.88) we have

$$\begin{aligned}\sum_{D<d\leqslant 2D}\left|A_d(f)-g(d)\frac{6}{\pi^2}I(f)\right|&\ll C^{-1}\sqrt{x}\,\log x+\eta^{-2}CD(\log x)^{256}\\&\ll\eta^{-1}D^{\frac12}x^{\frac14}(\log x)^{256}\end{aligned}$$

on choosing $C=\eta x^{\frac14}D^{-\frac12}$. Hence, we conclude the following version of (22.75).

PROPOSITION 22.19. *If f satisfies (22.86) with $0<\eta\leqslant 1$ then*

$$\sum_{d\leqslant D}\left|A_d(f)-g(d)\frac{6}{\pi^2}I(f)\right|\ll\eta^{-1}D^{\frac12}x^{\frac14}(\log x)^{256}\ , \tag{22.93}$$

where the implied constant is absolute.

We are now ready to give the proof of Theorem 22.16. We derive (22.75) by applying (22.93) for two smooth functions $0\leqslant f^-,\ f^+\leqslant 1$ supported on the interval $[0,(1+2\eta)x]$ with $f^\pm(y)=1$ if $0\leqslant y\leqslant(1\pm\eta)x$, respectively. Hence, $f^-\leqslant\chi\leqslant f^+$ where χ is the characteristic function of the segment $[0,x]$ and so $A_d(f^-)\leqslant A_d(x)\leqslant A_d(f^+)$. The corresponding integrals $I(f^-),I(f^+)$ are equal to

$$\iint_{a^4+b^4\leqslant x}da\,db=\gamma\sqrt{x}\ ,$$

up to an error term which does not exceed the measure of the set $\{(1-\eta)x<a^4+b^4\leqslant(1+2\eta)x\}$, which is $O(\eta\sqrt{x})$.

Combining these two estimates for f^-, f^+ we get

$$\sum_{d\leqslant D} \left|A_d(x) - g(d)X\right| \ll \eta\sqrt{x}\,\log x + \eta^{-1}D^{\frac{1}{2}}x^{\frac{1}{4}}\left(\log x\right)^{256}$$
$$\ll D^{\frac{1}{4}}x^{\frac{3}{8}}\left(\log x\right)^{256},$$

by choosing $\eta = D^{\frac{1}{4}}x^{-\frac{1}{8}}$. This completes the proof of Theorem 22.16.

As a quick application of Theorem 22.16 we are able to evaluate asymptotically the divisor sum for our sequence $\mathcal{A}$.

THEOREM 22.20. *We have, for $\mathcal{A}$ given by* (22.68),

$$\sum_{n\leqslant x} \tau(n)a_n = G\sqrt{x}\,\log x + O\big(\sqrt{x}\,\log\log x\big) \tag{22.94}$$

where G is the positive constant given in (22.96)–(22.99).

PROOF. We write

$$\tau(n) = 2\sum_{\substack{d|n\\ d<\sqrt{n}}} 1 + O(1)\ .$$

Hence, our sum equals

$$2\sum_{d<\sqrt{x}} \big(A_d(x) - A_d(d^2)\big) + O(\sqrt{x})\ .$$

By (22.77) we have

$$\sum_{d<\sqrt{x}} A_d(d^2) \leqslant 10\sum_{d<\sqrt{x}} \rho(d) \ll \sqrt{x}$$

and

$$\sum_{D<d<\sqrt{x}} A_d(x) \leqslant 10\sqrt{x}\sum_{D<d<\sqrt{x}} \rho(d)d^{-1} \ll \sqrt{x}\,\log\log x$$

where $D = D(x)$ is given by (22.76). Now we are left with

$$2\sum_{d\leqslant D} A_d(x) = 2X\sum_{d\leqslant D} g(d) + O(\sqrt{x})$$

by (22.75). By Theorem A.5 we obtain

$$\sum_{d\leqslant D} g(d) = G\log D + O(1)\ , \tag{22.95}$$

where

$$G = \prod_p \big(1 + g(p) + g(p^2) + \cdots\big)\Big(1 - \frac{1}{p}\Big)\ . \tag{22.96}$$

Here we have $g(2) = 1/3$, $g(2^\alpha) = 0$ if $\alpha > 1$ and for any $p > 2$,

$$g(p^\alpha) = \rho(p)\Big(1 + \frac{1}{p}\Big)^{-1} p^{-\alpha}\ , \quad \alpha = 1, 2, \ldots \tag{22.97}$$

Hence,

$$G = \frac{2}{3}\prod_{p>2}\left(1+\frac{\rho(p)}{p}\left(1+\frac{1}{p}\right)^{-1}\left(1-\frac{1}{p}\right)^{-1}\right)\left(1-\frac{1}{p}\right)$$
$$(22.98)\qquad = \frac{2}{3}\prod_{p>2}\left(1+\frac{\rho(p)-1}{p+1}-\frac{1}{p(p+1)}\right).$$

We have for $p > 2$,

$$(22.99)\qquad \rho(p) = \sum_{\chi\,(\mathrm{mod}\,8)} \chi(p) = 1 + \sum_{\substack{\chi\,(\mathrm{mod}\,8)\\ \chi\neq\chi_0}} \chi(p)\,,$$

so the product converges to a positive constant. This completes the proof of Theorem 22.20. □

A somewhat different proof of Theorem 22.20 (in a more general form) was given by S. Daniel [**27**].

22.8. Binary Forms

In the previous sections we have considered the level of distribution of binary quadratic and quartic forms. In fact, these were diagonal forms (Gaussian sequences), but of course one could apply similar arguments to more general forms.

Now we consider irreducible binary forms

$$(22.100)\qquad f(X,Y) = c_0X^r + c_1X^{r-1}Y + \cdots + c_rY^r$$

with integer coefficients $c_0, c_1, \ldots, c_r$, where $c_0 \neq 0$. Although we are no longer in the realm of Gaussian sequences we continue to benefit from the homogeneity of $f(X,Y)$ which allows us again to obtain what is essentially the largest possible level of distribution.

Motivated by a specific application in Section 15.4, we introduce some restrictions on the variables. In fact, we consider the form $f(X,dY)$, where d is a given number and we seek estimates which are uniform in d while the dependence on the coefficients $c_0, c_1, \ldots, c_r$ is not of our concern. We consider the sequence $\mathcal{A} = (a_n)$ with

$$(22.101)\qquad a_n = \sum_{\substack{(m,dn)=1\ m\equiv a(q)\\ f(m,dn)\equiv 0(k)}} W(m,dn)$$

for $dn \equiv b(\mathrm{mod}\,q)$. Here we take $W(u,v)$ to be a smooth function with compact support in a box

$$(22.102)\qquad u \asymp x,\ v \asymp x$$

and with partial derivatives satisfying

$$(22.103)\qquad \frac{\partial^{i+j}}{\partial^i u\,\partial^j v}W(u,v) \ll x^{-i-j}, \qquad 0 \leqslant i,j \leqslant 2.$$

Of course, with a little more work (as in previous sections) we could succeed with less smooth test functions $W(u,v)$, but we do not go for that here.

The restriction

$$(22.104)\qquad (m,dn) \equiv (a,b)\ (\mathrm{mod}\,q)$$

is needed for our applications, but it is not serious in our arguments because we do not seek uniformity in q. We assume that

$$(abdk, q) = 1 \tag{22.105}$$

and

$$(d, k) = 1. \tag{22.106}$$

The fact that we only need the zero class modulo k for $f(m, dn)$ is critical to the argument which exploits the homogeneity of $f(X, Y)$.

Our goal is to evaluate the sum

$$N_k(W) = \sum_{n \equiv b\bar{d}(q)} a_n \tag{22.107}$$

with an error term which is useful for k as large as possible. For technical reasons we achieve this goal only on average over k (still uniformly for individual d's). The expected main term is

$$M_k(W) = g(d)h(k)j(q)\zeta(2)^{-1}\hat{W}(0,0) \tag{22.108}$$

where

$$g(d) = \frac{1}{d}\prod_{p|d}\left(1 - \frac{p}{(p-1)(p+1)}\right), \tag{22.109}$$

$$h(k) = \frac{\rho(k)}{k}\prod_{p|k}(1 + \frac{1}{p})^{-1}, \tag{22.110}$$

$$j(q) = \prod_{p|q}(p-1)^{-1}(p+1)^{-1}, \tag{22.111}$$

$\zeta(2) = \pi^2/6$, and $\hat{W}$ is the Fourier transform of W, so

$$\hat{W}(0,0) = \int\!\!\int W(u,v)\,du\,dv\,. \tag{22.112}$$

Moreover, $\rho(k)$ denotes the number of roots $\omega \pmod k$ of

$$f(\omega, 1) \equiv 0 \pmod k\,. \tag{22.113}$$

THEOREM 22.21. *Let $(d, q) = 1$. For $K \geqslant 1$ and $x \geqslant 1$ we have*

$$\sum_{\substack{k \leqslant K \\ (k,dq)=1}} |N_k(W) - M_k(W)| \ll \big(K + x + x\sqrt{K/d}\big)(dqK)^{\varepsilon} \tag{22.114}$$

with any $\varepsilon > 0$, the implied constant depending only on ε and the coefficients of $f(X, Y)$.

REMARK. The main term $M_k(W)$ has the order of magnitude x^2/dk so our estimate (22.114) is meaningful for

$$d \ll x^{1-\varepsilon}, \; dK \ll x^{2-\varepsilon}. \tag{22.115}$$

PROOF. We begin as with the sum for two biquadrates in the previous section by writing

$$N_k(W) = \sum_{\substack{\alpha,\beta\,(\mathrm{mod}\,kq)\\ \alpha,\beta\equiv a,b\,(\mathrm{mod}\,q)\\ f(\alpha,\beta)\equiv 0\,(\mathrm{mod}\,k)}}^{*}\sum^{*} \sum_{\substack{(m,dn)=1\\ m,dn\equiv\alpha,\beta\,(\mathrm{mod}\,kq)}}\sum W(m,dn).$$

Now we remove the coprimality condition $(m, dn) = 1$ by the Möbius formula getting

$$N_k(W) = \sum_{(c,kq)=1} \mu(c) N_{ck}(W) \tag{22.116}$$

where

$$N_{ck}(W) = \sum_{\alpha}^{\star}\sum_{\beta}^{\star} \sum_{m,n\equiv\alpha\bar{c},\beta\overline{cd'}\,(\mathrm{mod}\,kq)}\sum W(cm, cd'n)\ , \tag{22.117}$$

$d' = d/(c,d)$ and the outer summation is over the reduced residue classes $\alpha, \beta(\mathrm{mod}\,kq)$ as before.

For large c we use a trivial bound

$$|N_{ck}(W)| \leqslant \sum_{\substack{m\ n\\ f(m,d'n)\equiv 0(k)}}\sum |W(cm, cd'n)|.$$

Hence,

$$\sum_{k\leqslant K} |N_{ck}(W)| \leqslant \sum_m\sum_n |W(cm, cd'n)|\tau(f(m, d'n))$$
$$\ll x^{\varepsilon} \sum_m\sum_n |W(cm, cd'n)| \ll \frac{x^{2+\varepsilon}}{c^2 d'}.$$

Summing over $c > C$ we conclude that

$$\sum_{c>C}\sum_{k\leqslant K} |N_{ck}(W)| \ll (dC)^{-1}x^{2+\varepsilon}. \tag{22.118}$$

For small c we evaluate the inner sum over m, n in (22.117) by Poisson's formula getting

$$N_{ck}(W) = \frac{1}{d'(ckq)^2}\sum_u\sum_v \hat{W}\Big(\frac{u}{ckq}, \frac{v}{d'ckq}\Big)G(u, v; kq) \tag{22.119}$$

where u, v run over integers, $\hat{W}$ is the Fourier transform of W and $G(u, v; kq)$ is the corresponding complete exponential sum

$$G(u, v; kq) = \sum_{\substack{\alpha,\beta\,(\mathrm{mod}\,kq)\\ \alpha,\beta\equiv a,b\,(\mathrm{mod}\,q)\\ f(\alpha,\beta)\equiv 0\,(\mathrm{mod}\,k)}}^{*}\sum^{*} e\Big(\frac{\alpha\bar{c}u + \beta\overline{d'c}v}{kq}\Big)\ .$$

Note that this sum depends on other parameters which we do not display.

Changing variables and using the homogeneity of $f(X, Y)$ we can simplify this as follows:

$$G(u, v; kq) = e\Big(\frac{a\overline{ck}u + b\overline{d'ck}v}{q}\Big) \sum_{f(\omega,1)\equiv 0(k)} R(d'u\omega + v; k)$$

where

$$R(h;k) = \sum\nolimits^{\star}_{\gamma \,(\mathrm{mod}\, k)} e\left(\frac{\gamma h}{k}\right)$$

is the Ramanujan sum. Hence for $u = v = 0$ we get

$$G(0,0;kq) = \rho(k)\varphi(k). \tag{22.120}$$

For the other frequencies we estimate the Ramanujan sum by $r = (h,k)$ with $h = d'u\omega + v$. Let $\delta = (u,v,k)$, $u_1 = u/\delta$, $v_1 = v/\delta$, $k_1 = k/\delta$, so that we have $r = \delta(d'u_1\omega + v_1, k_1) = \delta r_1$, say. Considering the congruence $f(\omega,1) \equiv 0 \,(\mathrm{mod}\, k_1)$, one sees that

$$f(-v_1, d'u_1) \equiv 0 \,(\mathrm{mod}\, k_1)\ .$$

Therefore, r_1 is a common divisor of k_1 and $f(-v_1, d'u_1)$ and hence, r is a common divisor of k and $f(-v, d'u)$, showing that

$$|G(u,v;kq)| \leqslant \big(f(-v,d'u),k\big)\rho(k)\ . \tag{22.121}$$

Inserting (22.120) and (22.121) into (22.119), we obtain

$$\begin{aligned}\Big|N_{ck}(W) &- \frac{\rho(k)\varphi(k)}{d'(ckq)^2}\hat{W}(0,0)\Big| \\ &\leqslant \frac{\rho(k)}{d'(ckq)^2} \sum\sum_{(u,v)\neq(0,0)} \Big|\hat{W}\Big(\frac{u}{ckq}, \frac{v}{d'ckq}\Big)\Big|\big(f(-v,d'u),k\big)\Big|.\end{aligned}$$

By (22.103) we find that the Fourier transform $\hat{W}(u,v)$ decays rapidly, namely

$$\hat{W}(u,v) \ll x^2(1+|u|x)^{-2}(1+|v|x)^{-2}.$$

Hence,

$$\begin{aligned}N_{ck}(W) &- \frac{\rho(k)\varphi(k)}{d'(ckq)^2}\hat{W}(0,0) \\ &\ll \frac{\rho(k)x^2}{d'(ckq)^2} \sum\sum_{(u,v)\neq(0,0)} \left(1+\frac{|u|x}{ckq}\right)^{-2}\left(1+\frac{|v|x}{d'ckq}\right)^{-2}\big(f(-v,d'u),k\big).\end{aligned}$$

Although this is a relatively simple summation we cannot execute it precisely enough to achieve the desired uniformity with respect to k. To this end, we average over k in a dyadic segment $\frac{1}{2}K < k \leqslant K$, getting

$$\begin{aligned}\sum_{\frac{1}{2}K<k\leqslant K} &\Big|N_{ck}(W) - \frac{\rho(k)\varphi(k)}{d'(ckq)^2}\hat{W}(0,0)\Big| \\ &\ll \frac{x^2}{d'(cq)^2K} \sum\sum_{(u,v)\neq(0,0)} \left(1+\frac{|u|x}{cKq}\right)^{-2}\left(1+\frac{|v|x}{d'cKq}\right)^{-2}(|v|+d|u|)^{\varepsilon} \\ &\ll \frac{x^2}{d'(cq)^2K}\left(1+\frac{cKq}{x}\right)\frac{d'cKq}{x}(cdqK)^{\varepsilon} = \frac{1}{c}(x+cK)(cdqK)^{\varepsilon}.\end{aligned}$$

This last bound also holds for the sum over the whole range $1 \leqslant k \leqslant K$.

It remains to sum over $1 \leqslant c \leqslant C$, $(c, kq) = 1$. In the main term we encounter the sum

$$\sum_{\substack{c \leqslant C \\ (c,kq)=1}} \mu(c)(c,d)c^{-2} = \sum_{(c,kq)=1} \mu(c)(c,d)c^{-2} + O(\tau(d)C^{-1})$$

$$= \frac{1}{\zeta(2)} \prod_{p|kq} (1 - \frac{1}{p^2})^{-1} \prod_{p|d} \Big(1 - \frac{p}{(p-1)(p+1)}\Big) + O\Big(\frac{\tau(d)}{C}\Big) .$$

The sum over $c \leqslant C$ of the error terms is $\ll (x + CK)(CdqK)^{\varepsilon}$.

Gathering the above estimates we arrive at

$$\sum_{\substack{k \leqslant K \\ (k,dq)=1}} |N_k(W) - M_k(W)| \ll (dC)^{-1}x^{2+\varepsilon} + (x + CK)(CdqK)^{\varepsilon}.$$

Finally, choosing $C = 1 + x/\sqrt{dK}$ we obtain (22.114). □

22.9. Binary Cubic Forms with Twists

In this section, partly for application in Section 15.4, we generalize the results of the previous section, but only for cubic forms. We consider the sequence $\mathcal{A} = (a_n)$ with

$$a_n = \sum_{\substack{(m,dn)=1,\ m \equiv a \,(\mathrm{mod}\, q) \\ f(m,dn) \equiv 0 \,(\mathrm{mod}\, k)}} r(f(m,dn))W(m,dn) \tag{22.122}$$

where $r(f)$ is a special arithmetic function. The sequence in the previous section had the trivial twist $r(f) = 1$. Here we take

$$r(f) = (1 * \chi)(f) = \sum_{t|f} \chi(t) , \tag{22.123}$$

where χ is a character of modulus q, not necessarily primitive.

Our goal is to evaluate the sum

$$N_k(W) = \sum_{n \equiv b\bar{d} \,(\mathrm{mod}\, q)} a_n . \tag{22.124}$$

To this end we use the results of the previous section. Throughout we assume that $f(X,Y)$ is an irreducible cubic form and the classes a, b modulo q are such that

$$\chi(f(a,b)) = 1 . \tag{22.125}$$

Recall that $\rho(k)$ denotes the number of roots $\omega (\mathrm{mod}\, k)$ of the congruence

$$f(\omega) \equiv 0 \,(\mathrm{mod}\, k) ,$$

where $f(X) = f(X,1)$ is the corresponding cubic polynomial. It is known that the associated zeta-function

$$\zeta_f(s) = \sum_{k=1}^{\infty} \rho(k)k^{-s} \tag{22.126}$$

is analytic in $\operatorname{Re} s > \frac{1}{2}$ with a simple pole at $s = 1$, because $f(X)$ is irreducible. To simplify our arguments we are going to assume that $\chi \,(\operatorname{mod} q)$ is a character such that the L-function

$$L_f(s, \chi) = \sum_{k=1}^{\infty} \rho(k)\chi(k)k^{-s} \tag{22.127}$$

is analytic in $\operatorname{Re} s > \frac{1}{2}$ with no poles. In particular, χ is non-principal; we call χ an admissible character for f.

The expected main term $M_k(W)$ which approximates $N_k(W)$ is a bit more complicated than that in the previous section. We are particularly interested in how this looks in terms of k. Recall that the k-factor of the main term in the previous section is given by the multiplicative function

$$h(k) = \frac{\rho(k)}{k} \prod_{p|k} \Big(1 + \frac{1}{p}\Big)^{-1}. \tag{22.128}$$

Put

$$L(s) = \sum_{l=1}^{\infty} \chi(\ell)h(\ell)\ell^{-s} = \prod_{p} L_p(s), \tag{22.129}$$

where

$$L_p(s) = \sum_{\alpha=0}^{\infty} \chi(p^\alpha)h(p^\alpha)p^{-\alpha s}. \tag{22.130}$$

Now we are ready to predict the main term

$$M_k(W) = g_\chi(d)h_\chi(k)j(q)L(0)2\zeta(2)^{-1}\hat{W}(0, 0) \tag{22.131}$$

where

$$g_\chi(d) = g(d) \prod_{p|d} L_p(0)^{-1} \tag{22.132}$$

and

$$h_\chi(k) = \prod_{p|k} (h(p) + L_p(0) - 1)L_p(0)^{-1}. \tag{22.133}$$

THEOREM 22.22. *Let $f(X, Y)$ be an irreducible cubic form and $\chi \,(\operatorname{mod} q)$ an admissible character. Let d be coprime with the leading coefficient $f(1, 0)$ and with q. For $K \geqslant 1$, $x \geqslant 1$ we have*

$$\sum_{\substack{k \leqslant K \\ (k,dq)=1}}^{\flat} |N_k(W) - M_k(W)| \ll (K + x^{\frac{1}{4}}\sqrt{K/d})x^{\frac{3}{2}}(dK)^{\varepsilon} \tag{22.134}$$

with any $\varepsilon > 0$, the implied constant depending only on ε, q, and the coefficients of $f(X, Y)$.

Note that the main term $M_k(W)$ has size essentially x^2/dk, so the result is meaningful in the range

$$dk < x^{\frac{1}{2}-\varepsilon}. \tag{22.135}$$

PROOF. Note that for $f = f(m, dn)$ in the range of summation we have

$$(f, d) = 1,\ \chi(f) = 1 \text{ and } f \ll x^3.$$

Hence, we can shorten the involution (22.125) by the Dirichlet hyperbola method getting

$$r(f) = \sum_{\substack{t|f\\ t\leqslant y}} \chi(t) + \sum_{\substack{t|f\\ t<|f|/y}} \chi(t)$$

with any $y > 0$, the best size of y being $y \asymp Y = x^{3/2}$. To ease the separation of variables in the forthcoming arguments we integrate the above expression with respect to y against some smooth function $w(y/Y)y^{-1}$, where $w(\xi)$ is compactly supported on $\mathbb{R}^+$ with

$$\int w(\xi)\xi^{-1}d\xi = 1\,. \tag{22.136}$$

We get

$$\begin{aligned} r(f) &= \sum_{t|f} \chi(t) \int \big(w(y/Y) + w(|f|/yY)\big)y^{-1}dy \\ &= \int \sum_{\substack{t|f\\ t\leqslant y}} \chi(t)\big(w(y/Y) + w(|f|/yY)\big)y^{-1}dy\,. \end{aligned}$$

Now Theorem 22.21 is applicable with k replaced by $[k, t]$, K replaced by KY and the test function $W(u, v)$ by

$$\big[w(y/Y) + w(f(u, v)/yY)\big]W(u, v).$$

Hence, the bound (22.114) becomes (22.134), but the main term (22.108) has to be modified accordingly. Now it should be

$$g(d)j(q)\zeta(2)^{-1} \int \sum_{\substack{t\leqslant y\\ (t,d)=1}} \chi(t)h([k, t])\Big[w(\tfrac{y}{Y})\hat{W}(0, 0) + \hat{W}_y(0, 0)\Big]\frac{dy}{y}$$

where

$$W_y(u, v) = w(f(u, v)/yY)W(u, v)$$

and $\hat{W}_y$ is the Fourier transform of W_y, so

$$\hat{W}_y(0, 0) = \iint w(f(u, v)/yY)W(u, v)du\,dv.$$

The integration will simplify a lot if we execute the summation over $t \leqslant y$ first. Expanding to all $(t, d) = 1$ we obtain the infinite series

$$S(d, k) = \sum_{(t,d)=1} \chi(t)h\big([k, t]\big) \tag{22.137}$$

and the tail is estimated by $O\big(k^{-1}y^{-\frac{1}{2}}(dky)^{\varepsilon}\big)$ using our assumption that the L-function (22.129) is holomorphic in $\operatorname{Re} s > \frac{1}{2}$. This tail alters the main term by at most $O\big(g(d)k^{-1}Y^{-\frac{1}{2}}x^2(dkY)^{\varepsilon}\big)$. Summing over $k \leqslant K$ we find that the total alteration is $O\big(d^{-1}x^{\frac{5}{4}}(dKx)^{\varepsilon}\big)$, which is clearly absorbed by the right-hand side of (22.134).

This completes the proof of (22.135) with the main term $M_k(W)$ given by

$$M_k(W) = g(d)j(q)S(d,k)2\zeta(2)^{-1}\hat{W}(0,0) \tag{22.138}$$

because of (22.136).

It remains to compute the series (22.137) for $(k, dq) = 1$, k squarefree. We get

$$\begin{aligned} S(d,k) &= \sum_{\delta|k} h(k/\delta) \sum_{\substack{(t,dk/\delta)=1 \\ t\equiv 0(\delta)}} \chi(t)h(t) \\ &= \sum_{\delta|k} h(k/\delta) \prod_{p|\delta} (L_p(0)-1) \prod_{p\nmid dk} L_p(0) \\ &= L(0) \prod_{p|d} L_p(0)^{-1} \prod_{p|k} \big(h(p)+L(p)-1\big) L_p(0)^{-1}. \end{aligned}$$

This shows that (22.138) agrees with the main term (22.132), completing the proof of Theorem 22.22. □

22.10. Level of Distribution Along Orbits

One problem that arises when the sifting sequence $\mathcal{A}$ is parametrized by several variables is that we would already run into trouble just counting the points. In which order should we do this? The result might depend on that order. So far we have been counting the points in accordance with their placement in Euclidean or hyperbolic domains. In order to consider more general problems, Bourgain, Gamburd and Sarnak [**13**] have been led to revise this method of counting. In this section we highlight some of the novelties of their work. Full details at this degree of generality are beyond the scope of the book and we recommend to the reader the original papers.

First we have to choose the variables of summation. For this we take a subgroup Λ of $\mathbb{Z}^n$ or of $SL_n(\mathbb{Z})$ and consider the orbit

$$\mathcal{O} = b + \Lambda$$

in the case of $\mathbb{Z}^n$ with b a fixed vector in $\mathbb{Z}^n$, or

$$\mathcal{O} = b\Lambda$$

in the case of $SL_n(\mathbb{Z})$ where now b may be a fixed vector in $\mathbb{Z}^n$ or else a fixed matrix in $SL_n(\mathbb{Z})$. Of course, we need lots of points of summation so we don't want Λ to be very thin. What we should mean by this is not obvious. The proper notion, as BGS found, is to request that Λ be Zariski dense.

Given a polynomial $f \in \mathbb{Z}[x_1, \dots, x_n]$, the question is to estimate the number of points in the orbit $\mathcal{O}$ for which $f(x)$ is prime or almost-prime, that is $f(x)$ has at most, say, r prime factors. For example, if $\Lambda = \mathbb{Z}$, $\mathcal{O} = \mathbb{Z}$, and

$$f(x) = \prod_{h\in\mathcal{H}} (x-h)$$

where $\mathcal{H} = \{h_1, \dots, h_r\}$ is an admissible set (see Section 7.13), we are asking for the simultaneous primality of the $x - h_j$. The Hardy–Littlewood prime r-tuple conjecture asserts that there are infinitely many of these.

Some very nice results in the literature can be formulated within the above framework. For example, if $\Lambda \subset \mathbb{Z}^3$ has $\operatorname{rank}(\Lambda) = 2$ and $f(x) = x_1x_2x_3$ then we are asking for primes p_1, p_2 such that the linear combination $\alpha p_1 + \beta p_2$ is also prime with specific coefficients α, β depending on Λ. This is a slight variation of the three-prime theorem of Vinogradov. If $\Lambda \subset \mathbb{Z}^4$ has $\text{rank}(\Lambda) = 2$ and $f(x) = x_1x_2x_3x_4$ then we are asking for primes p_1, p_2 such that two linear combinations $\alpha p_1 + \beta p_2$ and $\gamma p_1 + \delta p_2$ are also prime. The asymptotic formula for the number of these has been established by Green and Tao [**77**].

A proof of the prime r-tuple conjecture is beyond our current means but we have seen that the lower-bound sieve of Brun produces almost-primes, and in plentiful quantities. If the orbit $\mathcal{O}$ is thin then even this goal is not an easy one and, in this context, [**13**] introduced the following notion. We say that $(\mathcal{O}, f)$ "saturates" if, for some fixed r, the set of points x in $\mathcal{O}$ for which $f(x)$ has at most r prime factors is Zariski dense. This means that the set of such points is relatively large. In the case of one variable it merely means that the set of such points is infinite but in more than one variable the notion of Zariski density has greater intricacy. Roughly speaking, it means that none of the variables gets lost. An example which is not Zariski dense would be

$$\mathcal{O} = \Lambda = \left\{ \begin{pmatrix} 1 & 1 \\ 0 & 1 \end{pmatrix}^m ; m \in \mathbb{Z} \right\} \subset SL_2(\mathbb{Z}) .$$

In this case $f(x_1, x_2, x_3, x_4) = f(1, m, 0, 1)$ so it is really a one-variable problem.

The smallest $r = r_0$ for which $(\mathcal{O}, f)$ saturates will be called the saturation number. The r-tuple conjecture predicts that it is just the number of factors.

Asking for primes in values of polynomials is only feasible when there are no local obstructions. For example, $f(x) = x^2 - x + 2$, although irreducible, represents only even integers. Therefore, we must assume that $(\mathcal{O}, f)$ is primitive. This means that for every q there exists x in $\mathcal{O}$ with $\big(f(x), q\big) = 1$. Under certain mild conditions on Λ this primitivity can be verified by examining a single value of q, judiciously chosen.

In [**13**] there are a number of very general conjectures. In the case of SL_2 there is the following.

CONJECTURE. Let $\Lambda \subset SL_2(\mathbb{Z})$ have Zariski closure SL_2, $b = [b_1, b_2] \in \mathbb{Z}^2$ and $\mathcal{O} = b\Lambda$, the corresponding orbit. Let $\pi(\mathcal{O})$ be the set of points $x = [x_1, x_2] \in \mathcal{O}$ with prime co-ordinates. Then the Zariski closure of $\pi(\mathcal{O})$ is the full Zariski closure of $\mathcal{O}$ if and only if $(\mathcal{O}, f)$ is primitive for $f(x) = x_1x_2$.

They view this conjecture as an $SL_2(\mathbb{Z})$ analogue of the Dirichlet Prime Number Theorem.

Another conjecture, which actually implies the previous one, asserts that, for $\mathcal{O} = \Lambda \subset SL_2(\mathbb{Z})$ Zariski dense, and for f a polynomial in four variables with t irreducible factors and of course with $(\mathcal{O}, f)$ primitive, the saturation number is $r_0(\mathcal{O}, f) = t$.

This conjecture has been proven in a few interesting cases. For example, it stimulated the work [**20**] wherein one finds the following result. Take

$$\mathcal{O} = \Lambda = \left\{ \begin{pmatrix} x_1 & x_2 \\ x_3 & x_4 \end{pmatrix} \in SL_2(\mathbb{Z}) \right\}$$

and $f(x) = x_1^2 + x_2^2 + x_3^2 + x_4^2 - 2$. Then the conjecture holds. In the more difficult and more interesting case where $f(x) = x_1^2 + x_2^2 + x_3^2 + x_4^2$ it is proven conditionally in [**20**] and in Section 14.7.

Sarnak has proven the conjecture for some Λ which are thin. An excellent example is the problem of integral Apollonian packing; see the presentation in Section 22.2. In this case Λ has infinite index in $SL_2(\mathbb{Z})$, thus very thin in that sense, yet is still Zariski dense.

We emphasize the special case for subgroups $\Lambda \subset SL_2(\mathbb{Z})$ of one of their theorems. Subject to certain conditions they prove that, if $\mathcal{O} = \Lambda$, $(\mathcal{O}, f)$ is primitive as usual and Λ is Zariski dense in SL_2, then $(\mathcal{O}, f)$ saturates. The precise conditions are of a geometric flavour which we do not state but, roughly speaking, amount to the resulting density function $g(d)$ in the sieve application being of the form $g(p) \sim \kappa/p$ where κ is a positive integer. To prove such a theorem they apply the lower-bound Brun sieve, for example, Theorem 6.9 will do. As we know, this requires a study of the congruence sums and, for these, a positive exponent of distribution. The great novelty is in their construction and evaluation of these congruence sums. Counting the points in the orbits in such generality creates a challenging issue. Their approach is through a translation of the problem to the language of graphs followed by the application of a spectral theory relevant to that setting.

Let $\Lambda \subset SL_n(\mathbb{Z})$. Since we are interested in a lower bound we can reduce Λ to a slightly smaller subgroup Γ which has convenient properties. For example, we can require Γ to be a free group on k generators (with $k = 2$ in our case) which we denote by $A_1, \dots, A_k$. Now, $\mathcal{O} = b\,\Gamma$ for some fixed b in $SL_n(\mathbb{Z})$. Since Γ acts transitively we can identify Γ with $\mathcal{O}$. In this way, $\mathcal{O}$ becomes a $2k$-regular graph on joining x to xA_j and to xA_j^{-1} for each j. This makes $\mathcal{O}$ the Cayley graph of Γ with respect to the generating set

$$S = \{A_1, A_1^{-1}, \dots, A_k, A_k^{-1}\} \,. \tag{22.139}$$

We denote the graph by $\mathcal{G}(\Gamma, S)$. For $x,\, y \in \mathcal{O}$ let $w(x, y)$ be the distance in the graph between x and y. This will be the yardstick used to count the points in the congruence sums. We choose a large number L and consider the ball

$$\mathcal{B} = \{x \in \mathcal{O};\ w(b, x) \leqslant L\} \,,$$

centred at some fixed b. Now, our sifting sequence $\mathcal{A} = (a_n)$ is given by

$$a_n = \{x \in \mathcal{B};\ |f(x)| = n\} \,.$$

One shows that the number of points of the orbit in the ball $\mathcal{B}$ is given by

$$X = \sum_n a_n = \sum_{x \in \mathcal{B}} 1 = \frac{k(2k-1)^L - 1}{k-1} \;;$$

thus $X = 2 \cdot 3^L - 1$ in our case of $k = 2$. More generally, for d squarefree we need to evaluate

$$A_d = \sum_{n \equiv 0\,(d)} a_n = \sum_{\substack{x \in \mathcal{B} \\ f(x) \equiv 0\,(d)}} 1 \,.$$

To this end we consider $\Gamma_d \subset GL_n(\mathbb{Z}/d\mathbb{Z})$, the reduction of Γ modulo d, and correspondingly $\mathcal{O}_d$, the reduction of $\mathcal{O}$ modulo d. Let $\Gamma(d) \subset \Gamma$ be the kernel of this reduction so that $\Gamma_d = \Gamma/\Gamma(d)$. Then $\Gamma(d)$ is a congruence subgroup of Γ of finite index. Every point $x \in \mathcal{O}$ can be written as $\rho\delta$ with $\rho \in \mathcal{O}_d$ and $\delta \in \Gamma(d)$.

However, we only count those points with $f(x) \equiv 0 \,(\bmod d)$ which means we restrict ρ to the set

$$\mathcal{O}_d^{(f)} = \{\rho \in \mathcal{O}_d;\ f(x) \equiv 0 \,(\bmod d)\} .$$

Having described this decomposition, we write

$$A_d = \sum_{\rho \in \mathcal{O}_d^{(f)}} A(d,\rho) \qquad \text{with} \qquad A(d,\rho) = \sum_{\substack{\delta \in \Gamma(d) \\ \rho\delta \in \mathcal{B}}} 1 .$$

We shall evaluate $A(d,\rho)$ for every ρ separately. To this end we are going to perform harmonic analysis and expand the characteristic function of the ball $\mathcal{B}$. As in Section 13.2 (Euclidean) and in Section 19.3 (hyperbolic), we need an orthonormal basis in the space of functions which is appropriate to the problem at hand. In this case that amounts to $\Gamma(d)$-periodic functions on $\mathcal{O}$. Recall that the $2k$-regular Cayley graph $\mathcal{G}_d = \big(\mathcal{O}/\Gamma(d), S\big)$ is of size $|\mathcal{O}_d| = |\Gamma_d|$. Using the appropriate invariant, self-adjoint operator of the graph we obtain the spectral expansion of $A(d,\rho)$. As expected, the largest eigenvalue $\lambda_0 = 2k$ with constant eigenfunction gives the main term. The other eigenvalues are spaced below the largest by an amount independent of d. This crucial feature is known as the "gap principle" and a graph having this property is known as an "expander graph". The size of the gap determines the quality of the remainder. From the expander graph property one obtains

$$A(d,\rho) = \frac{X}{|\mathcal{O}_d|} + O\big(X^{1-\eta}\big)$$

for some $\eta = \eta(\Gamma) > 0$. Summing over $\rho \in \mathcal{O}_d^{(f)}$ we find

$$A_d = g(d)X + O\big(d^C X^{1-\eta}\big)$$

for some $C = C(\Gamma, f) > 0$, where $g(d) = |\mathcal{O}_d^{(f)}|/|\mathcal{O}_d|$. The density function $g(d)$ is multiplicative, however, proving this property requires considerable skill. In fact, the above presentation has been somewhat simplified in this regard since one needs to move further down to a group of larger level Nd where N is fixed and depends on f (to resolve issues related to the content of f, along with other technicalities concerning the factorization of congruence subgroups).

Because $g(p) \sim \kappa p^{-1}$ it follows that $\mathcal{A}$ has level of distribution

$$D = X^{\vartheta - \varepsilon}, \quad \vartheta = \eta/(C+1) .$$

The level of distribution part is even harder to deal with than the density function. The proof that the relevant graphs are expanders constitutes the main part of the arguments in the paper [**13**]. Actually, one requires this property for the family of Cayley graphs $\mathcal{G}\big(\Gamma/\Gamma(d), S\big)$ obtained by reduction modulo d and we need to know that they are absolute expanders (that is, there is a positive bound for the spectral gap independent of d). In the paper [**13**] they conjecture such a bound for $\Lambda \subset GL_m(\mathbb{Z})$ and they succeed for $m = 2$. Precisely, they prove in their Theorem 1.7 the following result.

Let Λ be a subgroup of $SL_2(\mathbb{Z})$ which is Zariski dense in SL_2 and let S be a finite symmetric set of generators for Λ. Then, for d squarefree, the family of Cayley graphs $\mathcal{G}\big(\Lambda/\Lambda(d), S\big)$ is an absolute expander family.

We still need to relate the level of distribution D to the size of $n = |f(x)|$ for $x \in \mathcal{B}$. We write $x = bA_{\alpha_1} \cdots A_{\alpha_r}$ where A_{α_q} are in the set (22.139) of generators of Λ. For $x \in \mathcal{B}$ this means that such a representation can be achieved with $r \leqslant L$. Hence, the ℓ_2-norm of x satisfies $\|x\| \leqslant c^{L+1}$, where

$$c = \max\{\|b\|, \|A_1\|, \|A_1^{-1}\|, \ldots, \|A_k\|, \|A_k^{-1}\|\}$$

is a constant depending on Λ and the choice of the central point b. Since f is a polynomial we get

$$n = |f(x)| \ll c^{L \deg f} \ll X^{(\deg f)\,(\log c)/\log(2k-1)} .$$

Now, by Brun's lower-bound sieve, this completes the proof of the theorem. The fact that the lower bound for almost-primes did not lose a fixed power of X suffices to show that the set is Zariski dense.

22.11. Future Prospects

To a great extent, important parts of sieve theory are now fairly well understood and are reaching a state of maturity. There remain of course important motivating questions, such as the twin prime conjecture, that seem hopelessly out of reach of the sieve alone but it is plausible that future developments will see a role for the sieve in their eventual solution.

Exciting recent advances have come, not so much from new choices for the sieve weights, but rather from inventive use of different parts of mathematics for the estimation of congruence sums and bilinear forms. We have tried in this chapter to give a sample of such developments. The reader will note the concentration, in this recent progress, on polynomials in several variables. For such polynomials, much more so for algebraic varieties, it is not even easy to guess which of them should contain primes, and, if so, how many should we expect and in what way should we count them. As we saw in the previous section, the paper [**13**] of Bourgain, Gamburd, and Sarnak provides a very promising framework to address these complex issues.

Congruence sums are treated by harmonic analysis and a good error term comes from the fact that, in the spectral expansion, the main term is well isolated from the other terms due to gaps in the spectrum. For example, when we apply Poisson summation, the zero frequency yields the main term while the other frequencies (non-zero integers) contribute to the error term. When we applied, in Chapter 19, the spectral expansion of the hyperbolic Laplacian for congruence groups, the lowest eigenvalue gives the main term, while the next eigenvalue, being at an absolute constant distance away (due to Selberg), guarantees that the error term is relatively small. For varieties of much more complex structure there is no spectral theory but the structure provided by a group action can, as we have just seen, lead to successful conclusions by similar gap principles.

There are other techniques for counting points on varieties; one of the oldest is the circle method which is not so heavily based on a group structure. The demand from sieve theory for good evaluations of congruence sums has inspired a great number of developments in harmonic analysis. This trend will certainly continue and hopefully both subjects will thereby flourish.

CHAPTER 23

Primes in Short Intervals

Zeta Aria

23.1. Introduction

Right from its beginnings, the sieve was very useful in providing upper bounds for the number of primes in very short intervals, but providing lower bounds proved to be stubborn for a long time. Analytic methods, especially on the Riemann Hypothesis, were able to give such results; nevertheless, it must have been a surprise to some when Hoheisel succeeded in proving unconditionally that the expected asymptotic formula

$$\pi(x) - \pi(x-y) \sim \frac{y}{\log x} \tag{23.1}$$

as $x \to \infty$, holds uniformly for $x^\theta \leqslant y < x$ with some absolute constant $\theta < 1$. This was based on existing results about the distribution of zeros of the Riemann zeta-function which say there are not many zeros to the right of the central line $\operatorname{Re} s = \frac{1}{2}$ (zero-density estimates) and there are none that are very close to the boundary line $\operatorname{Re} s = 1$.

A particularly important ingredient in Hoheisel's [**96**] work is the Littlewood zero-free region

$$\sigma > 1 - \eta \frac{\log\log(|t|+3)}{\log(|t|+3)} \tag{23.2}$$

for $s = \sigma + it$, where η is a (rather small) positive absolute constant. Today, we have the zero-free region (23.2) with $\eta = \eta(t) \to \infty$ as $|t| \to \infty$, which is a consequence of a broader region due to Vinogradov [**158**]. Following this, the exponent θ in Hoheisel's original method depends only on the exponent c in the zero-density estimate

$$\Big|\{\rho = \beta + i\gamma;\, \zeta(\rho) = 0,\, \beta \geqslant \alpha, |\gamma| \leqslant T\}\Big| \ll T^{c(1-\alpha)}(\log T)^A\ .$$

Consequently, every θ with $1 - 1/c < \theta < 1$ is good.

Significant progress in the theory of the zeta-function led to many improvements in the zero-density estimates, notably due to Ingham, Montgomery and Huxley, leading to the exponent $c = 12/5$ so that we now know (23.1) for every $\theta > 7/12$. Needless to say, the Riemann Hypothesis gives (23.1) with every $\theta > 1/2$.

Although the exponent $7/12$ of Huxley for the asymptotic formula (23.1) has not been reduced, there has been substantial progress in establishing lower bounds

$$\pi(x) - \pi(x-y) \gg \frac{y}{\log x} \tag{23.3}$$

for considerably shorter intervals (see [**108**] for the first such result). This breakthrough has been made possible by combining sieve methods with the same analytic ingredients and with their subsequent refinements. At present the record belongs to Baker, Harman and Pintz [**3**] who established (23.3) for $\theta = 0.525$, not so far from the consequence of the Riemann Hypothesis. Their method uses many powerful tools and arguments, both analytic and combinatorial, and these are extremely complicated.

In this chapter our goal is to show the means by which the sieve can break the barrier of 7/12 for the lower bound (23.3) without appealing to any zero-density estimates. The point we make is that the sieve provides great flexibility for combinatorial arrangements which allow one to avoid deeper considerations of the zeros of $\zeta(s)$. All we need is the zero-free region (23.2) with $\eta(t) \to \infty$ which is necessary to crash the parity barrier of the linear sieve. We have chosen to work with

$$y = x^{\theta}, \ \tfrac{11}{20} < \theta < \tfrac{7}{12}\ . \tag{23.4}$$

The restriction of 11/20 comes from Proposition 23.4 which is a refined version of the basic Lemma 1 of [**91**]. This proposition provides a pretty good level of distribution for our sequence $\mathcal{A} = \big(f(n)\big)$, namely (23.9). Note that if θ is close to 7/12 then the exponent of distribution $\vartheta = \frac{2}{5}(6\theta - 1)$ is near one. Therefore, applying the lower-bound linear sieve directly, we just miss prime numbers. We need only a little further saving in order to succeed. To this end we appeal to the terms of the combinatorial sieve formulas which are usually ignored by positivity. But now we can evaluate asymptotically some parts of these terms which fall into the range of Proposition 23.4. At this point we break the parity barrier which, in the general case, prevents us from getting primes using the linear sieve.

In order to succeed with every $11/20 < \theta < 7/12$ we would need to combine the sieve inequalities with combinatorial identities of the type introduced in Chapter 17 and thereby achieve a further decomposition of the relevant sums over primes. However, we want to employ as little as possible without using combinatorial decompositions other than those which are intrinsic within the sieve structure alone. Therefore, we only show the arguments for θ close to 7/12, because the reader may see clearly that the lower bound (23.3) emerges quickly without any numerical computation. In earlier versions of this chapter we dealt with $\theta = 4/7$ and with $\theta = 5/9$ but decided in the end that, although much simpler than works in the literature, these were still a little too intricate for what we had hoped to achieve in this book. Specifically, we verify by numerical computations that the present version works for

$$\theta = 0.58 = \tfrac{7}{12} - \tfrac{1}{300}.$$

It is also interesting to study the distribution of primes in almost all short intervals rather than every short interval. One should expect to succeed with considerably shorter intervals and indeed Selberg [**143**] showed how to succeed in this by employing zeros of the zeta-function for the problem. Specifically, he gave the relation $\theta = 1 - 2/c$ between the constant c in the zero-density bound and the exponent θ for intervals of length $y = x^{\theta+\varepsilon}$ such that the asymptotic formula (23.1) holds for almost all x. From Huxley's density theorem this gives the result with $\theta = \frac{1}{6}$.

As in the Hoheisel problem further progress became possible by giving up on the asymptotic formula and using the sieve to show the existence of primes in almost all short intervals. The first such result, already quite strong, was $\theta = \frac{1}{10}$, established by Harman. To do this he developed [**81**] new sieve identities for which he has found many applications which are presented in his book [**83**]. The current record is $\theta = \frac{1}{20}$, which is due to Jia [**114**]. For almost-primes one does much better; see Section 6.10.

23.2. The Sieve Argument

For technical convenience we consider integers in the segment $x - y < n < x$ weighted by a smooth function $f(n)$. We assume that $f(u) \geqslant 0$, $f(u)$ is supported in $x - y < u < x$, and its derivatives satisfy

$$f^{(j)}(u) \ll T^{-j} , \quad j = 0, 1, 2, \dots \tag{23.5}$$

where

$$T = x/y = x^{1-\theta} \tag{23.6}$$

and the implied constant depends only on j. We also assume that

$$y_f = \int f(u)du \asymp y . \tag{23.7}$$

Our goal is to show that

$$S = \sum_p f(p) \gg y\big(\log x\big)^{-1} . \tag{23.8}$$

To accomplish this we apply the linear sieve to the sequence $\mathcal{A} = \big(a_n\big)$ with $a_n = f(n)$. Note that $S = S(\mathcal{A}, \sqrt{x})$. However, as we have seen in Chapter 16, the sieve cannot succeed with $S(\mathcal{A}, z)$ for z as large as $\sqrt{x}$.

We shall prove in Lemma 23.8 that our sequence $\mathcal{A}$ has bilinear level of distribution

$$D = D(x) = x^{\vartheta}(\log x)^{-B} , \qquad \vartheta = \frac{2}{5}(6\theta - 1) . \tag{23.9}$$

Hence, for $z = \sqrt{D}$ the linear sieve yields, for any x large in terms of ε,

$$S(\mathcal{A}, \sqrt{D}) \geqslant S_2(\mathcal{A}, \sqrt{D}) + O(\varepsilon y/\log x)$$

where

$$S_2(\mathcal{A}, \sqrt{D}) = \sum\sum_{(D/p_1)^{\frac{1}{3}} \leqslant p_2 < p_1 < \sqrt{D}} S(\mathcal{A}_{p_1 p_2}, p_2) ;$$

see Theorem 12.19. Since $\sqrt{D}$ is somewhat smaller than $\sqrt{x}$ this suggests to use the Buchstab identity

$$S(\mathcal{A}, \sqrt{x}) = S(\mathcal{A}, \sqrt{D}) - \sum_{\sqrt{D} \leqslant p < \sqrt{x}} S(\mathcal{A}_p, p) .$$

Adding these formulas we derive the following inequality:

$$\begin{aligned}\sum_p f(p) \geqslant &\sum_{D^{\frac{1}{6}} < p_2 < \sqrt{T}} \ \sum_{D p_2^{-3} < p_1 < \sqrt{D}} \ \sum_q f(p_1 p_2 q) \\ &- \sum_{\sqrt{D} < p < \sqrt{x}} \ \sum_q f(pq) + O(\varepsilon y/\log x) ,\end{aligned}$$

where q, p, p_1, p_2 run over prime numbers. Here the triple sum over q, p_1, p_2, say S''', is just a small part of $S_2(\mathcal{A}, \sqrt{D})$. We shall evaluate S''' asymptotically while the double sum over q, p, say S'', will be estimated from above by the sieve.

Note that in the range of S''' we have $q \ll \sqrt{D}$; therefore Case 3 of Proposition 23.7 applies, giving

$$S''' = \sum_{p_1} \sum_{p_2} \frac{\big(1+o(1)\big)y_f}{p_1 p_2 \log(x/p_1p_2)} \sim c_3(\theta)\,\frac{y_f}{\log x}$$

where

$$\begin{aligned} c_3(\theta) &= \int_{\vartheta/6}^{(1-\theta)/2} \int_{\vartheta-3\beta}^{\vartheta/2} \frac{d\alpha\, d\beta}{\alpha\beta(1-\alpha-\beta)} \\ &\geqslant c_3\Big(\frac{7}{12}\Big) = \int_{1/6}^{5/24} \frac{1}{\beta(1-\beta)} \log\frac{2\beta}{(1-2\beta)(1-3\beta)}\, d\beta\ . \end{aligned} \tag{23.10}$$

To estimate S'' we apply the sieve to the subsequence $\mathcal{A}_p$ for every $\sqrt{D} < p < \sqrt{x}$. Clearly, $\mathcal{A}_p$ has the level of distribution $p^{-1}y(\log x)^{-B}$, but we shall prove in Lemma 23.9 that a larger function

$$D_p(x) = p^{-1}\sqrt{xy}\,(\log x)^{-B}$$

is the level of distribution on average over p in the segment $T < p < \sqrt{x}$. Since $T < \sqrt{D}$ we obtain by the upper-bound sieve

$$S'' \leqslant \sum_{\sqrt{D}<p<\sqrt{x}} \frac{\big(2+o(1)\big)y_f}{p\,\log(\sqrt{xy}/p)} \sim c_2(\theta)\,\frac{y_f}{\log x}$$

where

$$\begin{aligned} c_2(\theta) = 4\int_{\vartheta/2}^{1/2} \frac{d\gamma}{\gamma(1+\theta-2\gamma)} &= \frac{4}{1+\theta}\log\Big(1+\frac{(1+\theta)(1-\vartheta)}{\theta\vartheta}\Big) \\ &\leqslant \frac{4(1-\vartheta)}{\theta\vartheta} = \frac{2(7-12\theta)}{\theta(6\theta-1)}\ . \end{aligned} \tag{23.11}$$

We need to show that $c_3(\theta) > c_2(\theta)$. It is clear that any θ close to $7/12$ is fine, because $c_3(7/12) > 0$ while $c_2(7/12) = 0$. One can show that $c_2(\theta) < c_3(7/12)$ for $\theta = 0.58$, so we conclude

THEOREM 23.1. *The lower bound* (23.3) *holds for every* y *with* $x^{0.58} \leqslant y \leqslant x$ *and every* x *sufficiently large.*

23.3. Estimates for Dirichlet Polynomials

Let (a_n) be a sequence of complex numbers. Our goal is to pick up the subsequence of (a_n) in a short interval. In Section 13.4 we considered such a problem and got interesting results by using exponential sums; see Proposition 13.7. These are non-trivial even for very short intervals. However, the multiplicative harmonics n^{it}, where t is a free real variable, sometimes work better than the additive characters in the exponential sums, specifically for intervals which are not so very short. To this end one needs estimates for the relevant Dirichlet polynomials

$$D(s) = \sum_{1\leqslant n\leqslant N} a_n n^{-s} \tag{23.12}$$

at points s on the imaginary line. In this section we provide quite general results which we use later in conjunction with sieve methods.

By Cauchy's inequality

$$|D(it)|^2 \leqslant GN \tag{23.13}$$

where

$$G = \sum_{1 \leqslant n \leqslant N} |a_n|^2 . \tag{23.14}$$

Of course, the bound (23.13) cannot be improved in general at every given point, but due to an "almost orthogonality" of the harmonics n^{it}, we can expect very strong estimates for almost all points. For example, the classical mean-value theorem asserts that

$$\int_0^T |D(it)|^2 dt = (T + O(N))G . \tag{23.15}$$

Hence, we can think of G as the mean value of $|D(it)|^2$.

We need to know how often $|D(it)|^2$ takes large values relative to G in a set of points $0 \leqslant t_1 < \cdots < t_R \leqslant T$ which are well-spaced, that is, satisfying

$$t_{r+1} - t_r \geqslant 1 \quad \text{for } 1 \leqslant r < R. \tag{23.16}$$

Here we collect basic results of this kind (see Chapter 9 of [**109**] for proofs and for more information).

Proposition 23.2. *For any Dirichlet polynomial $D(s)$ of length $N \geqslant 2$ and for any well-spaced points $0 \leqslant t_r \leqslant T$ with $T \geqslant 1$ we have*

$$\sum_{r=1}^{R} |D(it_r)|^2 \ll G\left[N + \min(T, R^{\frac{2}{3}}T^{\frac{1}{3}}N^{\frac{1}{3}})\right] (\log NT)^6 \tag{23.17}$$

where the implied constant is absolute.

Corollary 23.3. *Let $V > 0$. The number of well-spaced points for which*

$$|D(it_r)| \geqslant V \tag{23.18}$$

satisfies

$$R \ll \left[GNV^{-2} + \min(GTV^{-2}, G^3TNV^{-6})\right] (\log NT)^{18} , \tag{23.19}$$

where the implied constant is absolute.

We are interested in three Dirichlet polynomials

$$L(s) = \sum_{\ell} \lambda_\ell \, \ell^{-s} , \tag{23.20}$$

$$M(s) = \sum_{m} \alpha_m \, m^{-s} , \tag{23.21}$$

$$N(s) = \sum_{n} \beta_n \, n^{-s} , \tag{23.22}$$

with coefficients $\lambda_\ell, \alpha_m, \beta_n$ which are bounded and supported in dyadic type intervals

$$\ell \asymp L , \quad m \asymp M , \quad n \asymp N , \quad LMN = X , \quad \text{say} .$$

We assume that one of these polynomials has somewhat regular coefficients. Specifically, we postulate that $L(s)$ has a non-trivial approximation by $g(s)L^{1-s}$ on the line $\mathrm{Re}\, s = \frac{1}{2}$, where $g(s)$ is holomorphic with

$$g(s) \ll |s|^{-2}\,, \quad \text{if } \mathrm{Re}\, s = \tfrac{1}{2}\,. \tag{23.23}$$

REGULARITY CONDITION. Let $A \geqslant 1$ and $L \geqslant 2$. For s with $\mathrm{Re}\, s = \frac{1}{2}$ and $|s| \leqslant L^A$ we have

$$L(s) = g(s)L^{1-s} + O\left(L^{\frac{1}{2}} (\log L)^{-A}\right)\,, \tag{23.24}$$

the implied constant depending only on A.

EXAMPLE I. If $\lambda_\ell = h(\ell/L)$, where $h(u)$ is a smooth function, compactly supported on $\mathbb{R}^+$, then (23.24) holds with

$$g(s) = \int h(u)u^{-s}du\,. \tag{23.25}$$

Actually, in this case we have (23.24) with a better error term $O\big(L^{\frac{1}{2}-\delta}\big)$ where $\delta = \delta(A) > 0$, which follows by van der Corput's estimate for exponential sums (cf. [**109**]).

EXAMPLE II. If $\lambda_\ell = h(\ell/L)\Lambda(\ell)/\log L$, then we get (23.24) by the zero-free region (23.2) with $\eta = \eta(t) \to \infty$.

Aside from the Regularity Condition we require nothing special about our polynomials $L(s), M(s), N(s)$. Put

$$L^\flat = L(s) - g(s)L^{1-s}\,. \tag{23.26}$$

Our goal is to estimate

$$Z(L, M, N) = \sum_r \big|L^\flat(s_r)M(s_r)N(s_r)\big| \tag{23.27}$$

where the summation runs over the well-spaced points $s_r = \frac{1}{2} + it_r$ with $0 \leqslant t_1 < \cdots < t_R \leqslant T$. Estimating $M(s_r)N(s_r)$ trivially we get by the Regularity Condition (23.24)

$$L^\flat(s_r)M(s_r)N(s_r) \ll X^{\frac{1}{2}} (\log L)^{-A}\,.$$

But we wish to have essentially the same bound for the whole sum $Z(L, M, N)$. In other words, we need to save R on top of $(\log L)^A$.

PROPOSITION 23.4. *Let $L, M, N \geqslant 2$ with $MN \ll L^A$ for some $A \geqslant 20$ and*

$$1 \leqslant T \ll X^{\frac{9}{20}} (\log X)^{-A}, \tag{23.28}$$

where $X = LMN$, the implied constants depending only on A. Then we have

$$Z(L, M, N) \ll X^{\frac{1}{2}} (\log X)^{20-A} \tag{23.29}$$

in each of the following three cases:

CASE 1: $L(s)$ is as in Example I and

$$M,\, N \ll XT^{-\frac{6}{5}} (\log X)^{-3A}\,; \tag{23.30}$$

CASE 2: $L(s)$ is as in Example I and

$$T \ll M \ll XT^{-1}(\log X)^{-3A}, \quad MN \ll XT^{-\frac{1}{2}} (\log X)^{-3A}\,; \tag{23.31}$$

CASE 3: $L(s)$ satisfies the Regularity Condition (23.24) *and*

$$L \ll T^{\frac{1}{2}}, \quad M,\, N \ll XT^{-\frac{6}{5}}(\log X)^{-3A}. \tag{23.32}$$

REMARK. In applications we need Case 3 only for special $L(s)$ such as in Example II.

We begin the proof of Proposition 23.4 by dividing the set of well-spaced points $s_r = \frac{1}{2} + it_r$ into disjoint classes according to the size of $|L^\flat(s_r)|$, getting

$$Z(L, M, N) = \sum_V V\, Z_V(L, M, N)$$

where

$$Z_V(L, M, N) = \sum_r |M(s_r)N(s_r)| \tag{23.33}$$

and s_r runs over the points for which

$$V < |L^\flat(s_r)| \leqslant 2V\ . \tag{23.34}$$

Note that all the points with $V < 1$ contribute at most

$$\begin{aligned}\sum_{0<V<1} V\, Z_V(L, M, N) &\ll (M+T)^{\frac{1}{2}}(N+T)^{\frac{1}{2}}(\log X)^6\\ &= (MN + MT + NT + T^2)^{\frac{1}{2}}(\log X)^6\\ &\ll X^{\frac{1}{2}}(\log X)^{6-A}\end{aligned}$$

by Cauchy's inequality, by (23.17) for $M(s), N(s)$ in place of $D(s)$, and by the conditions

$$M, N \ll XT^{-1}(\log X)^{-2A}\ , \tag{23.35}$$

which follow in all three cases. The result is better than (23.29) so we are left with

$$1 \leqslant V \ll L^{\frac{1}{2}}(\log L)^{-A}\ , \tag{23.36}$$

by the regularity condition (23.24). There are $O(\log L)$ such dyadic values of V. For each of these we need to show that

$$Z_V^2(M, N) \ll V^{-2}X(\log X)^{38-2A}\ . \tag{23.37}$$

Applying Cauchy's inequality we derive by (23.17) that

$$\begin{aligned}Z_V^2(M, N) &\ll \big[M + \Delta(T, R, M)\big]\big[N + \Delta(T, R, N)\big](\log X)^6\\ &= \big[MN + M\Delta(T, R, N) + N\Delta(T, R, M) + \Delta(T, R, M)\Delta(T, R, N)\big](\log X)^6\end{aligned}$$

where

$$\Delta(T, R, N) = \min\Big(T, R^{\frac{2}{3}}T^{\frac{1}{3}}N^{\frac{1}{3}}\Big)\ . \tag{23.38}$$

For the leading term MN (and only for this one) we use (23.36), getting the bound

$$V^2MN \ll X(\log X)^{-2A}\ , \tag{23.39}$$

which is sufficient for (23.37). For the two mixed terms we need to show that

$$M\Delta(R, T, N) \ll V^{-2}X(\log X)^{32-2A}\ , \tag{23.40}$$

$$N\Delta(R, T.M) \ll V^{-2}X(\log X)^{32-2A}\ , \tag{23.41}$$

while for the last term we need

$$\Delta(T,R,M)\Delta(T,R,N) \ll V^{-2}X(\log X)^{32-2A}\ . \tag{23.42}$$

We no longer require (23.36), but instead we need to know how often $|L^\flat(s)|$ takes large values on the set of well-spaced points $s_r = \frac{1}{2} + it_r$. To this end we apply Corollary 23.3.

LEMMA 23.5. *The number of well-spaced points* $s_r = \frac{1}{2} + it_r$ *with* $0 \leqslant t_1 < \cdots < t_R \leqslant T$ *for which* (23.34) *holds satisfies*

$$R \ll \left[L^2V^{-4} + \min\left(TV^{-4}, L^2TV^{-12}\right)\right](\log LT)^{27}\ . \tag{23.43}$$

PROOF. Since $L^\flat(s) = L(s) - g(s)L^{1-s}$ we have either $|L(s)| > \frac{1}{2}V$, or else $|g(s)|\sqrt{L} > \frac{1}{2}V$ for every point $s = s_r$. For the first class of points (23.43) follows from (23.19) applied to the polynomial $D(s) = L^2(s)$. For the points of the second class we have

$$V^4 < 16L^2|g(s_r)|^4 \ll L^2|s_r|^{-8}$$

by (23.23). Summing this bound over the points s_r we get $RV^4 \ll L^2$, which is included in the leading term of (23.43). □

Lemma 23.5 holds for general polynomials $L(s)$ with bounded coefficients λ_ℓ. However, if λ_ℓ is a nice smooth function in ℓ, then we have a bound for R which does not depend on the length L of the polynomial $L(s)$.

LEMMA 23.6. *Let* $\lambda_\ell = h(\ell/L)$ *be as in Example I. Then the number of well-spaced points* $s_r = \frac{1}{2} + it_r$ *with* $0 \leqslant t_1 < \cdots < t_R \leqslant T$ *for which* (23.34) *holds satisfies*

$$R \ll TV^{-4}(\log 2T)^3\ . \tag{23.44}$$

PROOF. By Mellin inversion (23.25) yields

$$h(u) = \frac{1}{2\pi i}\int_{(\sigma)} g(1-z)u^{-z}dz\ .$$

Thus, by contour integration we have

$$L^\flat(s) = \frac{1}{2\pi i}\int_{(0)} g(1-z)\zeta(s+z)L^z dz\ ,$$

where $\zeta(s)$ is the Riemann zeta-function (the main term $g(s)L^{1-s}$, which was subtracted from $L(s)$, comes here as the residue at $z = 1-s$). Hence, by (23.23),

$$L^\flat(s) \ll \int_{-\infty}^{\infty} |\zeta(s+iu)|(u^2+1)^{-1}du\ ,$$

and by Hölder's inequality,

$$\begin{aligned}|L^\flat(s_r)|^4 &\ll \int_{-\infty}^{\infty} |\zeta(s_r+iu)|^4(u^2+1)^{-1}du\\ &= \int_{-\infty}^{\infty} |\zeta(\tfrac{1}{2}+iu)|^4\left((u-t_r)^2+1\right)^{-1}du\ .\end{aligned}$$

Summing over the well-spaced points $0 \leqslant t_1 < \cdots < t_R \leqslant T$, we get

$$\sum_r \left((u-t_r)^2+1\right)^{-1} \ll \left(1+|u|/T\right)^{-2}\ .$$

Hence,

$$\sum_r \left|L^\flat(s_r)\right|^4 \ll \int_{-\infty}^{\infty} \left|\zeta(\tfrac{1}{2}+iu)\right|^4 \left(1+|u|/T\right)^{-2} du \ll T(\log 2T)^3$$

by the fourth power-moment estimate for the Riemann zeta-function on the critical line. This implies (23.44). □

Now we are ready to derive (23.40), (23.41), (23.42) from (23.43) and (23.44).

PROOF. We shall often use the convexity inequality $\min(a,b) \leqslant a^\alpha b^\beta$ with $\alpha+\beta=1$, where all numbers are positive.

In each of the three cases we have

$$R \ll TV^{-4}(\log X)^{27} \tag{23.45}$$

by (23.43) and (23.44) respectively. Combining (23.45) with (23.43) we derive by the convexity inequality (with exponents $\frac{3}{4}$, $\frac{1}{4}$) the following:

$$\begin{aligned} R &\ll \left(TV^{-4}\right)^{\frac{3}{4}}\left(L^2V^{-4}+L^2TV^{-12}\right)^{\frac{1}{4}}(\log X)^{27} \\ &\ll L^{\frac{1}{2}}\left(T^{\frac{3}{4}}V^{-4}+TV^{-6}\right)(\log X)^{27}\ . \end{aligned}$$

Hence

$$\begin{aligned} M\Delta(T,R,N) &= M\min\left(T, R^{\frac{2}{3}}T^{\frac{1}{3}}N^{\frac{1}{3}}\right) \\ &\ll M\min\left[T, (X/M)^{\frac{1}{3}}\left(T^{\frac{5}{6}}V^{-\frac{8}{3}}+TV^{-4}\right)\right](\log X)^{18} \\ &\leqslant \left[MT^{\frac{1}{4}}\left((X/M)^{\frac{1}{3}}T^{\frac{5}{6}}V^{-\frac{8}{3}}\right)^{\frac{3}{4}} + MT^{\frac{1}{2}}\left((X/M)^{\frac{1}{3}}TV^{-4}\right)^{\frac{1}{2}}\right](\log X)^{18} \\ &= V^{-2}\left(M^{\frac{3}{4}}T^{\frac{7}{8}}X^{\frac{1}{4}} + M^{\frac{5}{6}}TX^{\frac{1}{6}}\right)(\log X)^{18}\ . \end{aligned}$$

This estimate implies (23.40), provided M satisfies (23.32). Similarly, we get (23.41), provided N satisfies (23.32). Therefore, to complete Case 1 and Case 3 it remains to establish (23.42). As before, using the convexity inequality (but now with exponents $\frac{7}{8}$, $\frac{1}{8}$, rather than $\frac{3}{4}$, $\frac{1}{4}$), we get

$$R \ll L^{\frac{1}{4}}\left(T^{\frac{7}{8}}V^{-4}+TV^{-6}\right)(\log X)^{27}\ .$$

Note that $\min(a,b)\min(a,c) \leqslant \min(a^2, bc)$, hence

$$\begin{aligned} \Delta(T,R,M)\Delta(T,R,N) &\leqslant \Delta(T,R,\sqrt{MN})^2 = \min\left[T, R^{\frac{2}{3}}T^{\frac{1}{3}}(MN)^{\frac{1}{6}}\right]^2 \\ &\ll \min\left[T, X^{\frac{1}{6}}\left(T^{\frac{11}{12}}V^{-\frac{8}{3}}+TV^{-\frac{10}{3}}\right)\right]^2(\log X)^{36} \\ &\leqslant V^{-2}\left[T^{\frac{5}{4}}\left(X^{\frac{1}{6}}T^{\frac{11}{12}}\right)^{\frac{3}{4}} + T^{\frac{7}{5}}\left(X^{\frac{1}{6}}T\right)^{\frac{3}{5}}\right](\log X)^{36} \\ &= V^{-2}\left[T^{\frac{31}{16}}X^{\frac{1}{8}} + T^2X^{\frac{1}{10}}\right](\log X)^{36}\ . \end{aligned}$$

This yields (23.42) by (23.28) and the proof of Proposition 23.4 is complete in Case 1 and Case 3. Case 2 is straightforward. In this case we apply Hölder's inequality (rather than Cauchy's) obtaining

$$\begin{aligned} Z_V^2(L,M,N) &\ll R^{\frac{1}{2}}\Big(\sum_r |M(s_r)|^2\Big)\Big(\sum_r |N(s_r)|^4\Big)^{\frac{1}{2}} \\ &\ll (TV^{-4})^{\frac{1}{2}}(T+M)(T+N^2)^{\frac{1}{2}}(\log X)^{23} \end{aligned}$$

by (23.45) and (23.17). This yields (23.37), subject to the conditions (23.31), completing the proof of Proposition 23.4 in every case. □

23.4. Triple Products in Short Intervals

Recall from Section 23.2 that we are going to apply the sieve to the sequence $\mathcal{A} = \big(a_n\big)$ with

$$a_n = f(n) \tag{23.46}$$

where $f(v)$ is a non-negative smooth function supported on the interval $x - y < v < x$ with derivatives satisfying (23.5), and the mean value being

$$\int f(v)dv = y_f \asymp y\ .$$

Let α_m, β_n be any complex numbers which are bounded and supported on dyadic type intervals $m \asymp M$, $n \asymp N$, respectively. We consider the triple sum

$$\mathcal{A}(L,M,N) = \sum_\ell \sum_m \sum_n \lambda_\ell \alpha_m \beta_n f(\ell mn) \tag{23.47}$$

for various special sequences (λ_ℓ). Because of the support of α_m, β_n and $f(v)$, we may and we do assume that the coefficients λ_ℓ in (23.47) are supported on a dyadic type interval $\ell \asymp L$ where

$$LMN = x\ . \tag{23.48}$$

Moreover, if needed, we can attach to λ_ℓ a redundant factor $h(\ell/L)$, where $h(u)$ is a smooth function, compactly supported on $\mathbb{R}^+$, with $h(u) = 1$ on a sufficiently large segment.

First, for any $d \asymp MN$ we evaluate the sum

$$\mathcal{A}_d(\lambda) = \sum_\ell \lambda_\ell f(\ell d)\ . \tag{23.49}$$

We have

$$\mathcal{A}_d(\lambda) = \frac{1}{2\pi i}\int_{(\sigma)} F(s)L(s)d^{-s}ds$$

where $F(s)$ is the Mellin transform of $f(v)$,

$$F(s) = \int f(v)v^{s-1}dv\ . \tag{23.50}$$

Subtracting $g(s)L^{1-s}$ from $L(s)$ we obtain

$$\mathcal{A}_d(\lambda) = \frac{1}{2\pi i}\int_{(\sigma)} F(s)g(s)L^{1-s}d^{-s}ds + \mathcal{A}_d^\flat(\lambda)\ , \tag{23.51}$$

say, where

$$\mathcal{A}_d^\flat(\lambda) = \frac{1}{2\pi i}\int_{(\sigma)} F(s)L^\flat(s)d^{-s}ds\ .$$

Opening $F(s)$ by (23.50) we see that the integral in (23.51) (the expected main term) equals y_f/d, because

$$\int f(v)\Big(\frac{1}{2\pi i}\int_{(\sigma)} g(s)(v/dL)^{s-1}ds\Big)dv$$
$$= \int f(v)h(v/dL)dv = \int f(v)dv = y_f .$$

Hence,

$$\mathcal{A}(L,M,N) = y_f\Big(\sum_m \alpha_m m^{-1}\Big)\Big(\sum_n \beta_n n^{-1}\Big) + \mathcal{A}^\flat(L,M,N) ,$$

where the last term

$$\mathcal{A}^\flat(L,M,N) = \frac{1}{2\pi i}\int_{(\sigma)} F(s)L^\flat(s)M(s)N(s)ds$$

is to be considered an error term.

We estimate $\mathcal{A}^\flat(L,M,N)$ using Proposition 23.4. We choose the line $\sigma = \frac{1}{2}$ on which we have

$$F(s) \ll \frac{y}{\sqrt{x}}\Big(1+\frac{|s|}{T}\Big)^{-2}$$

with

$$T = x/y . \tag{23.52}$$

Hence,

$$\mathcal{A}^\flat(L,M,N) \ll \frac{y}{\sqrt{x}}\int_{-LT^2}^{LT^2} \big|L^\flat MN(\tfrac{1}{2}+it)\big|\Big(1+\frac{|t|}{T}\Big)^{-2}dt + \frac{y}{L}$$

by the trivial estimation $\big|L^\flat MN(\frac{1}{2}+it)\big| \ll \sqrt{x}$, applied for $|t| > LT^2$. Next we split this into integrals over intervals of length T getting

$$\mathcal{A}^\flat(L,M,N) \ll \frac{y}{\sqrt{x}}\sum_{|k|<LT}(1+|k|)^{-2}\int_0^T \big|L^\flat MN(\tfrac{1}{2}+it+ikT)\big|dt + \frac{y}{L} .$$

For every k we estimate the integral by a sum of type (23.27) for polynomials with coefficients $\lambda_\ell \ell^{ikT}$, $\alpha_m m^{ikT}$, $\beta_n n^{ikT}$ in place of $\lambda_\ell, \alpha_m, \beta_n$. If the original coefficients satisfy the conditions of Proposition 23.4, then so do the shifted ones. Therefore, by Proposition 23.4, we conclude

PROPOSITION 23.7. *Let $x \geqslant 3$ and y satisfy*

$$x^{\frac{11}{20}}(\log x)^{2A} \leqslant y \leqslant x , \tag{23.53}$$

where A is a large constant. Then we have

$$\sum_\ell\sum_m\sum_n \lambda_\ell\alpha_m\beta_n f(\ell mn) = y_f\Big(\sum_m \alpha_m m^{-1}\Big)\Big(\sum_n \beta_n n^{-1}\Big) + O\big(y(\log x)^{20-A}\big) , \tag{23.54}$$

in every one of the three cases in Proposition 23.4, where $T = xy^{-1}$.

23.5. Level of Bilinear Distribution

For the coefficients $\lambda_\ell = 1$ we have

$$A_d = \sum_{n\equiv 0\,(\mathrm{mod}\,d)} a_n = d^{-1}y_f + r_d\,, \tag{23.55}$$

say, and Proposition 23.7 (see also Proposition 23.4, Case 1) yields

LEMMA 23.8. *For any complex numbers α_m, β_n with $|\alpha_m| \leqslant 1$, $|\beta_n| \leqslant 1$ we have*

$$\sum\sum_{m,n\leqslant\sqrt{D}} \alpha_m\beta_n r_{mn} \ll y(\log x)^{-A} \tag{23.56}$$

where

$$D = D(x) = \left(xT^{-\frac{6}{5}}\right)^2(\log x)^{-B}\,, \tag{23.57}$$

A is any positive number and $B = B(A)$ depends only on A.

In other words, Lemma 23.8 says that the sequence $\mathcal{A} = \big(f(n)\big)$ has bilinear level of distribution D.

Proposition 23.4 (see Proposition 23.4, Case 2) yields

LEMMA 23.9. *For any complex numbers α_m, β_n with $|\alpha_m| \leqslant 1$, $|\beta_n| \leqslant 1$ we have*

$$\sum_{M<m\leqslant 2M}\ \sum_{N<n\leqslant 2N} \alpha_m\beta_n r_{mn}(x) \ll y(\log x)^{-A}\,, \tag{23.58}$$

subject to the following conditions:

$$T \ll M \ll xT^{-1}(\log x)^{-B}, \quad MN \ll xT^{-\frac{1}{2}}(\log x)^{-B}\,, \tag{23.59}$$

where A is any positive number and $B = B(A)$ depends only on A.

CHAPTER 24

The Least Prime in an Arithmetic Progression

Phantom of the Opera

24.1. Introduction

Because of the parity barrier one cannot produce primes exclusively within the basic sieve axioms. Nevertheless, it is possible to find primes in special sets by combining the sieve mechanism with analytic estimates that take advantage of the special properties of the involved sequence. The results of such a sifting can be quite impressive, especially when the target set is thin. We have seen how this works for capturing primes $p = a^2 + b^4$ in Chapter 22 and for studying short gaps between primes in Chapters 6 and 23.

In this chapter we address the question as to how far we can go, in terms of the modulus q, before we encounter the first prime $p \equiv a \pmod q$. In 1943 Ju.V. Linnik [**125**] showed that

$$p_{\min(q,a)} \ll q^L \tag{24.1}$$

for any $q > 1$, $(a, q) = 1$, where L and the implied constant are absolute. Linnik's original ideas set the theory of Dirichlet's L-functions at its most profound level. In some arguments Linnik employs sieve methods to save logarithmic factors, which may seem to be a small gain but are critical to the outcome and could not be achieved by traditional complex function theory alone. Along these lines Linnik established a number of very subtle properties of the zeros of L-functions which are treasures in their own right.

Our goal is to prove (24.1) by applying sieve methods more directly to avoid zeros of L-functions in most places. Eventually, the zeros do appear in our presentation, but only in classical arguments. Avoiding the zeros of L-functions completely might be a nice challenge, however, we do not see at present how to do so with reasonable clarity. More importantly, we hope the reader would like to see how the sieve and the complex variable methods complement each other, offering great shortcuts for solving difficult problems.

As we have said, one cannot succeed by the sieve without external inputs. In our arguments this will be the classical zero-free region for L-functions, which asserts that $L(s, \chi) \neq 0$ for any $\chi \pmod q$ and $s = \sigma + it$ with

$$\sigma > 1 - c/q \log(|t| + 2)\ , \tag{24.2}$$

where c is a positive, absolute constant, except for at most one simple, real zero of $L(s, \chi)$ in the case that χ is real. For a better understanding of the elements of the proof, we intentionally arrange the arguments so that our appeal to the zero-free region is the last thing to be done (we even compromise the strength of some intermediate results for this purpose of clarity).

The exceptional zero cannot be eliminated by present means, and its possible existence is perhaps the most frustrating situation in analytic number theory. All proofs of Linnik's theorem take care of this case separately. The extra help that has been used in such a scenario is the Deuring–Heilbronn phenomenon (the repulsion property of the exceptional zero). This says that the closer the exceptional real zero is to the point $s = 1$, the further that all the other zeros of all L-functions are pushed away from this point. A quantitative version of this phenomenon was established by Linnik, and it is the highlight of his work on the least prime in an arithmetic progression. There are several interesting approaches to this phenomenon; some employ ideas of Selberg's sieve method (see [**68**]), others apply the power-sum method of Turan (see [**9**]).

To demonstrate the new potential of the sieve, we skip the exotic territory of repulsion entirely. As a matter of fact, the exceptional real character simplifies rather than complicates our arguments. In this case we apply the sieve directly to a sequence especially designed so that the relevant sieve has a very small dimension. Therefore, it is not surprising that we can obtain prime numbers in such a sequence without facing the parity barrier. Any sublinear sieve would do the job, but we take advantage that the dimension is arbitrarily small so that a crude Brun sieve does so quickly. No need for the zero-free region, nor the Prime Number Theorem, in this case. Of course, the final estimate does not match the correct (expected) order of magnitude because the sifting sequence is thin, yet it is positive, giving (24.1). Through this exceptional case the reader can see that sometimes the sieve works better when applied to a thinner subsequence.

24.2. The Exceptional Case

First, we treat the case when we think that the exceptional character does exist. We do not really assume this to be the case, that so we end up with unconditional results. These, however, will be useful for the final conclusion only when we must consider the exceptional character.

Let $\chi \pmod q$ be a real, non-principal character. Put

$$\lambda(n) = \sum_{d|n} \chi(d) \; . \tag{24.3}$$

We shall apply the sieve to the sequence $\mathcal{A} = \big(\lambda(n)a_n\big)$, where a_n is the characteristic function of the arithmetic progression $n \equiv a \pmod q$ with $(a, q) = 1$. Clearly, we must assume that

$$\chi(a) = 1 \; , \tag{24.4}$$

or else there is nothing in $\mathcal{A}$. We need to evaluate the sums

$$A_d(x) = \sum_{\substack{n \leqslant x \\ n \equiv 0 (d)}} \lambda(n)a_n = \sum_{\substack{n \leqslant x/d \\ n \equiv a\bar{d}(q)}} \lambda(dn)$$

for $(d, q) = 1$. The function $\lambda(n)$ is multiplicative; precisely we have,

$$\lambda(dn) = \sum_{\delta | (d,n)} \mu(\delta)\chi(\delta)\lambda(d/\delta)\lambda(n/\delta) \; . \tag{24.5}$$

Hence, we get

$$A_d(x) = \sum_{\delta|d} \mu(\delta)\chi(\delta)\lambda(d/\delta)A(x/\delta d; q, a\overline{\delta d})$$

where

$$A(y;q,\alpha) = \sum_{\substack{m\leqslant y\\ m\equiv\alpha(q)}} \lambda(m)\ .$$

We write

$$\lambda(m) = \sum_{k\ell=m} \chi(k) = \bigl(1+\chi(m)\bigr)\sum_{\substack{k\ell=m\\ k<\ell}} \chi(k) + \chi(\sqrt{m})$$

for $(m,q) = 1$, where the last term $\chi(\sqrt{m})$ vanishes, unless $\sqrt{m}$ is an integer (a traditional convention), that is, if $m = k^2$. This gives

$$A(y;q,\alpha) = \bigl(1+\chi(\alpha)\bigr)A^*(y;q,\alpha) + O\bigl(\sqrt{y}\bigr)$$

where

$$A^*(y;q,\alpha) = \sum_{\substack{k\ell\leqslant y, k<\ell\\ k\ell\equiv\alpha \,(\mathrm{mod}\, q)}} \chi(k) = \frac{1}{q}\sum_{k<\sqrt{y}} \chi(k)\Bigl(\frac{y}{k}-k\Bigr) + O\bigl(\sqrt{y}\bigr)\ .$$

Hence,

$$A^*(y;q,\alpha) = L(1,\chi)\frac{y}{q} + O\bigl(\sqrt{y}\bigr),$$

$$A(y;q,\alpha) = \bigl(1+\chi(\alpha)\bigr)L(1,\chi)\frac{y}{q} + O\bigl(\sqrt{y}\bigr),$$

where the implied constant is absolute. Here we have $\alpha = a\overline{\delta d}$, $\chi(\alpha) = \chi(a)\chi(d/\delta) = \chi(d/\delta) = 1$, or else $\lambda(d/\delta) = 0$. Therefore,

$$A_d(x) = 2L(1,\chi)\frac{\nu(d)x}{dq} + O\bigl(\tau_3(d)\sqrt{x/d}\bigr) \tag{24.6}$$

where

$$\nu(d) = \sum_{\delta|d} \mu(\delta)\frac{\chi(\delta)}{\delta}\lambda\Bigl(\frac{d}{\delta}\Bigr)\ . \tag{24.7}$$

This is multiplicative with

$$\nu(p) = 1 + \chi(p)\Bigl(1-\frac{1}{p}\Bigr)\ . \tag{24.8}$$

We write (24.6) in the familiar sieve format

$$A_d(x) = g(d)X + r_d(x)\ , \tag{24.9}$$

where $g(d) = \nu(d)/d$ stands for the sifting density function,

$$X = 2L(1,\chi)\frac{x}{q} \tag{24.10}$$

and $r_d(x)$ is the remainder term, $r_d(x) \ll \tau_3(d)\sqrt{x/d}$. Hence, the remainder of level $D \geqslant 2$ satisfies

$$R(x,D) = \sum_{d<D} |r_d(x)| \ll \sqrt{xD}(\log D)^2\ , \tag{24.11}$$

where the implied constant is absolute.

We seek primes in $\mathcal{A}$ so we wish to estimate $S(\mathcal{A},\sqrt{x})$. Hypothetically, it is possible to get a lower bound for $S(\mathcal{A},\sqrt{x})$ because we think that $\chi(p)=-1$ very often for the exceptional character, so the density function at such primes is very small, $g(p)=p^{-2}$, and we have a sieve problem of low dimension. However, at this moment we do not want to play with the exceptional character so we proceed by the Buchstab formula

$$S(\mathcal{A},\sqrt{x})=S(\mathcal{A},z)-\sum_{z\leqslant p<\sqrt{x}}S(\mathcal{A}_p,p)\ . \tag{24.12}$$

By the Fundamental Lemma (see Theorem 6.9), we have

$$S(\mathcal{A},z)=XV(z)\left\{1+O(e^{-s})\right\}+O\big(\sqrt{xD}(\log D)^2\big)\ , \tag{24.13}$$

where $s=\log D/\log z\geqslant 2$ and

$$V(z)=\prod_{\substack{p<z\\ p\nmid q}}\Big(1-\frac{1}{p}\Big)\Big(1-\frac{\chi(p)}{p}\Big)\ . \tag{24.14}$$

To simplify (24.13) a bit we replace e^{-s} by s^{-1} and we choose

$$D=x/q^3(\log x)^8\ , \tag{24.15}$$

so for $x\geqslant q^8$ the error term is negligible giving

$$S(\mathcal{A},z)=XV(z)\Big\{1+O\Big(\frac{\log z}{\log x}\Big)\Big\}\ .$$

On the other hand, by the upper-bound sieve of level D/p we get

$$S(\mathcal{A}_p,p)\ll g(p)XV(p)+\sqrt{xD/p}(\log D)^2$$

for all $p\leqslant\sqrt{x}$. Adding these estimates we arrive at

$$S(\mathcal{A},\sqrt{x})=XV(z)\Big\{1+O\Big(\frac{\log z}{\log x}+\delta(z,x)\Big)\Big\}\ , \tag{24.16}$$

where

$$\delta(z,x)=\sum_{z\leqslant p<x}(1+\chi(p))p^{-1}\ . \tag{24.17}$$

Hence, we conclude the following (choose $z=q^2$):

PROPOSITION 24.1. *Let $\chi\ (\mathrm{mod}\, q)$ be a real, non-principal character and $\chi(a)=1$. Then, for $x\geqslant q^8$ we have*

$$\pi(x;q,a)=L(1,\chi)V(z)\frac{x}{q}\Big\{1+O\Big(\frac{\log z}{\log x}+\delta(z,x)\Big)\Big\} \tag{24.18}$$

with $z=q^2$, where $V(z)$ is given by (24.14), $\delta(z,x)$ is given by (24.17), and the implied constant is absolute.

It is easy to estimate $\delta(z,x)$ in terms of $L(1,\chi)$, or in terms of any real zero of $L(s,\chi)$. We have

$$\Delta(z,x)=\sum_{z\leqslant n<x}\lambda(n)n^{-1}=L(1,\chi)\big[\log\frac{x}{z}+O(z^{-1/4}))\big]$$

if $x \geqslant z \geqslant q^2$, where the implied constant is absolute, (cf. (22.109) of [**109**]). Hence, by the obvious inequality $\delta(z,x)\Delta(1,z) \leqslant \Delta(z,xz)$, we get

$$\delta(z,x)\Delta(1,z) \leqslant L(1,\chi)\big[\log\frac{x}{z} + O(z^{-1/4})\big] .$$

Using the trivial bound $\Delta(1,z) \geqslant 1$, we find

$$\delta(z,x) < L(1,\chi)\big[\log\frac{x}{z} + O(z^{-1/4})\big] . \tag{24.19}$$

We can also estimate $\Delta(1,z)$ from below as follows:

$$\begin{aligned}\Delta(1,z) &> z^{\beta-1}\sum_{1\leqslant n<z}\lambda(n)n^{-\beta}\Big(1-\frac{n}{z}\Big)\\ &= \frac{1}{2\pi i}\int_{(1)}\zeta(s+\beta)L(s+\beta,\chi)z^{s+\beta-1}\frac{ds}{s(s+1)} ,\end{aligned}$$

where β is any number < 1. We choose β to be a real zero of $L(s,\chi)$, so we have only a simple pole at $s = 1-\beta$ with residue $L(1,\chi)/(1-\beta)(2-\beta)$. We move the integration to the line $\sigma = \frac{1}{2} - \beta$ where the integral is bounded by $O\left(q^{1/4}z^{-1/2}\right)$. We obtain

$$\Delta(1,z) > \frac{L(1,\chi)}{(1-\beta)(2-\beta)} + O\big(q^{\frac14}z^{-\frac12}\big) .$$

Here, the error term is much smaller than the main term provided $z \geqslant q^2$ and q is large, so $\Delta(1,z) \geqslant L(1,\chi)(1-\beta)^{-1}$. Combining the two inequalities, we obtain

$$\delta(z,x) < (1-\beta)\big[\log x + O(\log z)\big] \tag{24.20}$$

if $x > z \geqslant q^2$, where β is any real zero of $L(s,\chi)$ and the implied constant is absolute. Note that q being large is not required for (24.20).

COROLLARY 24.2. *Let $\chi \,(\mathrm{mod}\, q)$ be a real, non-principal character, β any real zero of $L(s,\chi)$ and let $\chi(a) = 1$. Then for x with*

$$q^A \leqslant x \leqslant e^{1/A(1-\beta)} ,$$

with $A \geqslant 8$ being any large constant, we have

$$\pi(x;q,a) > \frac{xL^*(1,\chi)}{2\varphi(q)\log q} ,$$

where $L^*(1,\chi)$ denotes the L-function at $s = 1$ with all the local factors at $p < q^2$ omitted.

24.3. A Parity-Preserving Sieve Inequality

Next, we are going to apply the sieve to the sequence $\mathcal{A} = \big(a_n\big)$, the characteristic function of the arithmetic progression $n \equiv a \,(\mathrm{mod}\, q)$. Our goal is to estimate

$$S(\mathcal{A},z) = \sum_{\substack{n\leqslant x\\ (n,P(z))=1}} a_n$$

for $z = \sqrt{x}$. For $(d,q) = 1$ we have

$$A_d(x) = \frac{x}{dq} + O(1) ,$$

so we have a problem for the linear sieve. For the level of distribution of $\mathcal{A}$ we take

$$D = \frac{x}{q}(\log x)^{-4/3}. \tag{24.21}$$

Recall that the lower-bound linear sieve works only in the range $z \leqslant \sqrt{D}$. We take $z = \sqrt{D}$ getting (see (12.10))

$$S(\mathcal{A}, \sqrt{D}) = S^-(\mathcal{A}, \sqrt{D}) + \sum_{n \text{ even}} S_n(\mathcal{A}, \sqrt{D})$$

where S^- and S_n are defined in (6.27), (6.28). Usually one discards all of the sums $S_n(\mathcal{A}, z)$ getting the lower bound $S(\mathcal{A}, z) \geqslant S^-(\mathcal{A}, z)$, but this would not suffice for us. Indeed, we have

$$S^-(\mathcal{A}, \sqrt{D}) = XV^-(D, \sqrt{D}) + O(D)$$

where $O(D)$ is the bound for the remainder term. By Theorem 11.12,

$$V^-(D, \sqrt{2}) \geqslant \left\{f(2) + O(\log D)^{-\frac{1}{6}}\right\} V(\sqrt{D})$$

with $f(2) = 0$, so this part is negligible. We actually have, however,

$$S(\mathcal{A}, \sqrt{D}) \geqslant \sum_{n \text{ even}} S_n(\mathcal{A}, \sqrt{D}) + O(D) \tag{24.22}$$

and shall gain enough for our purpose by only exploiting the single additional sum

$$S_4(\mathcal{A}, \sqrt{D}) = \sideset{}{'}\sum_{p_4<p_3<p_2<p_1<\sqrt{D}} S(\mathcal{A}_{p_1p_2p_3p_4}, p_4).$$

Here, $\sum'$ restricts the summation by the conditions $p_1p_2^3 < D$, $p_1p_2p_3p_4^3 \geqslant D$. Dropping still more terms we deduce that

$$S(\mathcal{A}, \sqrt{D}) \geqslant \frac{1}{24} \sideset{}{''}\sum_{\substack{p_4p_3p_2p_1\, p\leqslant x \\ p_4p_3p_2p_1p\equiv a(q)}} 1 + O(D) \tag{24.23}$$

where $\sum''$ indicates that the prime variables p_r run independently over the segment

$$D^{\frac{1}{6}} < p_r < D^{\frac{1}{5}}, \quad r = 1, 2, 3, 4. \tag{24.24}$$

REMARK. We have estimated essentially a sum over primes by a sum over products of five primes. Note the same parity! A closer look at the other sums $S_n(\mathcal{A}, \sqrt{D})$, which we discarded, reveals that they are essentially over products of $n+1$ primes (if D is close to x), so the parity is odd throughout all these sums. Therefore, the formula (24.22) does not break the parity barrier (see the analysis in Section 16.4).

We have chosen to work with products of five primes rather than three for technical advantage (products of any larger fixed odd number of primes would also succeed).

Before applying characters to detect the congruence $p_4p_3p_2p_1p \equiv a \pmod{q}$ we exploit the positivity at this point and partition the sum into suitable blocks so that the separation of variables will not become an issue later. It is essential that we can do this now and without much loss because the forthcoming arguments will be so delicate that anything like partial summation will inflict irreparable damage (certainly, losing a logarithmic factor will kill the arguments, even losing a relatively large constant factor is not acceptable). To this end, we fix a smooth function $f(u)$ supported on $\left[\frac{1}{2},1\right]$ with $0 \leqslant f(u) \leqslant 1$ and put

$$\Psi_{\mathfrak{X}}(x;q,a) = \sum_{\substack{pp_1p_2p_3p_4 \equiv a(q) \\ x_r < p_r \leqslant 2x_r}} \cdots \sum f(pp_1p_2x_3x_4/x)\log p ,$$

where $\mathfrak{X} = \left[x_1,x_2,x_3,x_4\right]$ runs over the vectors of dyadic partition points of the segment $\left[D^{\frac{1}{6}}, D^{\frac{1}{5}}\right]$. Then (24.23) yields

$$S(\mathcal{A},\sqrt{D}) \geqslant \frac{1}{24\log x}\sum_{\mathfrak{X}} \psi_{\mathfrak{X}}(x;q,a) + O(D) . \tag{24.25}$$

Notice that we did not partition the range of p and we included the variables p_1, p_2 along with p in the argument of the smoothing function f. These seemingly technical devices will pay nice dividends in our estimations of the relevant character sums.

Note that on the left-hand side of (24.25) we have

$$S(\mathcal{A},\sqrt{D}) = S(\mathcal{A},\sqrt{x}) + O\left(\frac{x}{\varphi(q)}\,\frac{\log q}{(\log x)^2}\right) , \tag{24.26}$$

by the Brun–Titchmarsh bound.

24.4. Estimation of $\psi_{\mathfrak{X}}(x;q,a)$

Applying the orthogonality of characters we write

$$\psi_{\mathfrak{X}}(x;q,a) = \frac{1}{\varphi(q)} \sum_{\chi\,(\mathrm{mod}\,q)} \overline{\chi}(a)\psi_{\mathfrak{X}}(x,\chi) ,$$

where

$$\psi_{\mathfrak{X}}(x,\chi) = \sum_{\substack{pp_1p_2p_3p_4 \\ x_r < p_r \leqslant 2x_r}} \cdots \sum \chi(pp_1p_2p_3p_4) f(pp_1p_2x_3x_4/x)\log p . \tag{24.27}$$

The principal character $\chi_0\,(\mathrm{mod}\,q)$ gives the main term $\psi_{\mathfrak{X}}(x,\chi_0)/\varphi(q)$. We also put aside the contribution of the real character $\chi_1\,(\mathrm{mod}\,q)$, say, which is given by $\chi_1(a)\psi_{\mathfrak{X}}(x,\chi_1)/\varphi(q)$. Let $\Delta_{\mathfrak{X}}(x,q,a)/\varphi(q)$ denote the contribution of the other characters so we have

$$\psi_{\mathfrak{X}}(x,q,a) = \frac{1}{\varphi(q)}\left\{\psi_{\mathfrak{X}}(x,\chi_0) + \chi_1(a)\psi_{\mathfrak{X}}(x,\chi_1) + \Delta_{\mathfrak{X}}(x;q,a)\right\} . \tag{24.28}$$

For any $\chi\,(\mathrm{mod}\,q)$ we have

$$\psi_{\mathfrak{X}}(x,\chi) = \psi_{\mathfrak{Y}}(y,\chi)\Big(\sum_{x_3<p_3\leqslant 2x_3}\chi(p_3)\Big)\Big(\sum_{x_4<p_4\leqslant 2x_4}\chi(p_4)\Big)$$

where $\mathfrak{Y} = [x_1,x_2]$, $y = x/x_3x_4$ and

$$\psi_{\mathfrak{Y}}(y,\chi) = \sum_{\substack{p \; p_1 \; p_2 \\ x_r < p_r \leqslant 2x_r}}\sum\sum \chi(pp_1p_2) f(pp_1p_2/y)\log p . \tag{24.29}$$

We estimate the error term $\Delta_{\mathfrak{X}}(x;q,a)$ in the following fashion which resembles the circle method for ternary additive problems (here, we have a multiplicative analogue):

$$\left|\Delta_{\mathfrak{X}}(x;q,a)\right| \leqslant \max_{\chi\neq\chi_0,\chi_1} \left|\psi_{\mathfrak{Y}}(y,\chi)\right| \Big(\sum_{\chi}\Big|\sum_{p_3}\chi(p_3)\Big|^2\Big)^{\frac{1}{2}} \Big(\sum_{\chi}\Big|\sum_{p_4}\chi(p_4)\Big|^2\Big)^{\frac{1}{2}} .$$

It is easy to see that for any $X \geqslant q^2$,

$$\sum_{\chi \,(\mathrm{mod}\, q)} \Big| \sum_{X<p\leqslant 2X} \chi(p)\Big|^2 \ll \Big(\frac{X}{\log X}\Big)^2 \tag{24.30}$$

where the implied constant is absolute. To this end we square out and estimate the resulting sum over primes $p_1 \equiv p_2 \,(\mathrm{mod}\, q)$ by the Brun–Titchmarsh theorem.

We now need a modest but non-trivial estimate of $\psi_{\mathfrak{Y}}(y,\chi)$ for any $\chi \neq \chi_0, \chi_1$ (which is like asking for a non-trivial estimate of the corresponding exponential sum in the circle method at any point of the minor arcs). For this we have

PROPOSITION 24.3. *Let $\chi \,(\mathrm{mod}\, q)$ be a non-trivial character and*

$$\delta_\chi = \min_{\rho}\{1-\beta : |\gamma| \leqslant \log q\} \tag{24.31}$$

where $\rho = \beta + i\gamma$ denote zeros of $L(s,\chi)$. For $x_1, x_2 \geqslant q$ and $q^2x_1x_2 \leqslant y^{\frac{3}{4}}$ we have

$$\psi_{\mathfrak{Y}}(y,\chi) \ll \Big\{\frac{1}{\delta_\chi \log y} + \frac{1}{\log q}\Big\} \frac{y}{(\log x_1)(\log x_2)} , \tag{24.32}$$

where the implied constant is absolute.

REMARKS. Estimating trivially, one obtains $\psi_{\mathfrak{Y}}(y,\chi) \ll y/(\log x_1)(\log x_2)$, so the expression in braces $\{\dots\}$ is the saving factor (if q, y are sufficiently large). The proof of (24.32) does not require the zero-free region nor even the Prime Number Theorem. However, the arguments are more analytic than sieve-theoretic in nature, so we postpone the proof to the last section.

Combining (24.30) and (24.32) we obtain

$$\Delta_{\mathfrak{X}}(x;q,a) \ll \Big\{\frac{1}{\delta(q)\log x} + \frac{1}{\log q}\Big\}\psi_{\mathfrak{X}}(x) , \tag{24.33}$$

where $\delta(q) = \min\{\delta_\chi : \chi \neq \chi_0, \chi_1\}$ and the implied constant is absolute (in fact, we get (24.33) with $x/(\log x_1)\dots(\log x_4)$ in place of $\psi_{\mathfrak{X}}(x)$, but these have the same order of magnitude without using the PNT). Here we can write $\psi_{\mathfrak{X}}(x)$ rather than $\psi_{\mathfrak{X}}(x,\chi_0)$ because they coincide.

For estimating $\psi_{\mathfrak{X}}(x,\chi_1)$ we have two options. First, using (24.32) we get immediately

$$\psi_{\mathfrak{X}}(x,\chi_1) \ll \Big\{\frac{1}{\delta_{\chi_1}\log x} + \frac{1}{\log q}\Big\}\psi_{\mathfrak{X}}(x) . \tag{24.34}$$

Recall that δ_{χ_1} is defined by (24.31). This bound is satisfactory if $\delta_{\chi_1}\log q$ is larger than some positive absolute constant. In case this cannot be established we have a second option. We replace each $\chi(p), \chi(p_1), \chi(p_2), \chi(p_3), \chi(p_4)$ by -1 and obtain

$$\psi_{\mathfrak{X}}(x,\chi_1) = -\big\{1 + O\big(\delta(z,x)\big)\big\}\,\psi_{\mathfrak{X}}(x) \tag{24.35}$$

with $z = q^2$, where $\delta(z, x)$ is defined by (24.17) for the character χ_1. This bound is satisfactory if $\delta(z, x)$ is sufficiently small, and indeed it is small by (24.20) provided that $L(s, \chi_1)$ has a real zero β_1 with $(1 - \beta_1) \log q$ sufficiently small. Therefore, either (24.34) or (24.35) is useful.

24.5. Conclusion

We are now ready to derive Linnik's bound (24.1) from the assorted results in Sections 24.2, 24.3, 24.4. Note that we have not yet used the zero-free region (24.2), but of course it accompanied our thinking while we were compiling the ingredients for the final punch. Now it is time to use (24.2). We only take it for the non-trivial characters $\chi \,(\mathrm{mod}\, q)$ in rectangles of height $|t| \leqslant \log q$.

Suppose $\chi_1 \,(\mathrm{mod}\, q)$ is a non-principal real character such that $L(s, \chi_1)$ has a real zero β_1 with

$$(1 - \beta_1) \log q \leqslant A^{-2} , \tag{24.36}$$

where $A \geqslant 8$ is a large constant. If $\chi_1(a) = 1$ then Corollary 24.2 yields (24.1) with $L = A$. If $\chi_1(a) = -1$ then (24.28), (24.33), (24.35) and (24.20) yield

$$\psi_{\mathfrak{X}}(x; q, a) = \frac{2}{\varphi(q)} \psi_{\mathfrak{X}}(x) \left\{1 + O\left(A^{-1}\right)\right\}$$

for $q^A \leqslant x \leqslant q^{1/A(1-\beta_1)}$. Summing over $\mathfrak{X}$ we get from (24.25) and (24.26)

$$\pi(x; q, a) \gg \frac{x}{\varphi(q) \log x} . \tag{24.37}$$

This yields (24.1) with $L = A$.

Now we can assume that no $L(s, \chi)$ with $\chi \neq \chi_0$ vanishes in the rectangle

$$(1 - \sigma) \log q \leqslant A^{-2} , \quad |t| \leqslant \log q .$$

Hence (24.33) and (24.34) hold with $\delta(q) \geqslant 1/A^2 \log q$ and $\delta_{\chi_1} \geqslant 1/A^2 \log q$. Then (24.28) gives

$$\psi_{\mathfrak{X}}(x, q, a) = \frac{1}{\varphi(q)} \psi_{\mathfrak{X}}(x) \left\{1 + O\Big(A^2 \frac{\log q}{\log x} + \frac{1}{\log q}\Big)\right\} \gg \frac{1}{\varphi(q)} \psi_{\mathfrak{X}}(x)$$

if $x \geqslant q^{A^2 B}$, where B is a large constant. Summing over $\mathfrak{X}$ we get from (24.25) and (24.26),

$$\pi(x; q, a) \gg \frac{x}{\varphi(q) \log x} , \quad \text{if } x \geqslant q^{A^2 B} . \tag{24.38}$$

This yields (24.1) with $L = A^2 B$.

24.6. Character Sums over Triple-Primes

In this section we prove Proposition 24.3. Obviously, we can assume that $\chi \,(\mathrm{mod}\ q)$ is primitive. We start by the "explicit" formula

$$\sum_{n} \chi(n) \Lambda(n) f\left(\frac{n}{y}\right) = -\sum_{\beta \geqslant \frac{1}{2}} \hat{f}(\rho) y^{\rho} + O\left(\sqrt{y} (\log q)^2\right) \tag{24.39}$$

where $\hat{f}(x)$ is the Mellin transform of $f(u)$, $\rho = \beta + i\gamma$ runs over the zeros of $L(s,\chi)$ and the implied constant is absolute. This gives

$$\psi_{\mathfrak{Y}}(y,\chi) = -\sum_{\beta \geqslant \frac{1}{2}} \hat{f}(\rho) y^{\rho} \Big(\sum_{p_1} \chi(p_1) p_1^{-\rho}\Big)\Big(\sum_{p_2} \chi(p_2) p_2^{-\rho}\Big) + O\left(\sqrt{y}(\log q)^2\right) .$$

Note that the error term absorbs the contribution of prime powers p^2, $p^3, \ldots$, which are missing in $\psi_{\mathfrak{Y}}(y,\chi)$. We have

$$\hat{f}(\rho) \ll |\rho|^{-3} \tag{24.40}$$

and hence

$$\sum_{|\gamma| > T} |\hat{f}(\rho)| \ll T^{-2} \log q . \tag{24.41}$$

We choose $T = \log q$ and estimate the terms in $\psi_{\mathfrak{Y}}(y,\chi)$ with $|\gamma| > T$ trivially, getting

$$\psi_{\mathfrak{Y}}(y,\chi) = -\sum\nolimits^{\rho} \hat{f}(\rho) \left(\frac{y}{x_1 x_2}\right)^{\rho} \left(\sum_{p_1} \chi(p_1) \left(\frac{x_1}{p_1}\right)^{\rho}\right)\left(\sum_{p_2} \chi(p_2) \left(\frac{x_2}{p_2}\right)^{\rho}\right)$$
$$+ O\left(y/(\log x_1)(\log x_2)\log q\right) .$$

Here $\sum^{\rho}$ denotes summation over the zeros $\rho = \beta + i\gamma$ of $L(s,\chi)$, restricted by $\beta \geqslant \frac{1}{2}$, $|\gamma| \leqslant \log q$. We have

$$\left|\left(\frac{y}{x_1 x_2}\right)^{\rho}\right| \leqslant \frac{y}{x_1 x_2}\left(\frac{q^2 x_1 x_2}{y}\right)^{\delta_\chi} q^{2(\beta-1)} \leqslant (x_1 x_2)^{-1} y^{1-\delta_\chi/4} q^{2(\beta-1)} .$$

Hence,

$$\psi_{\mathfrak{Y}}(y,\chi) \ll T_\chi(x_1,x_2) y / x_1 x_2 \delta_\chi \log y \;+\; y/(\log x_1)(\log x_2)\log q$$

where

$$T_\chi(x_1,x_2) = \sum\nolimits^{\rho} q^{2(\beta-1)} \Big| \sum_{x_1 < p_1 \leqslant 2x_1} \chi(p_1)\Big(\frac{x_1}{p_1}\Big)^{\rho}\Big| \Big| \sum_{x_2 < p_2 \leqslant 2x_2} \chi(p_2)\Big(\frac{x_2}{p_2}\Big)^{\rho}\Big| . \tag{24.42}$$

Note that we have ignored the factor (24.40), because it does not help (the problem occurs with bounded zeros). We also estimate $y^{-\delta_\chi/4}$ by $4/\delta_\chi \log y$ (working with $y^{-\delta_\chi/4}$ would lead to a better Linnik constant which we do not compute anyway). Later, we also replace $q^{2(\beta-1)}$ by $1/(1-\beta)\log q$.

To complete the proof of (24.32) it remains to show that

$$T_\chi(x_1,x_2) \ll x_1 x_2/(\log x_1)(\log x_2) \tag{24.43}$$

where the implied constant is absolute. If we knew that

$$\sum\nolimits^{\rho} q^{2(\beta-1)} \ll 1 , \tag{24.44}$$

then (24.43) would follow by a trivial estimation of the sums over p_1, p_2. The above estimate (24.44) is true, it is a kind of log-free density estimate for the zeros of $L(s,\chi)$. However, we can avoid a proof of (24.44) (which would be very long) by gaining a bit from the cancellation in the sums over p_1, p_2. When the variation of $\rho = \beta + i\gamma$ with respect to γ exceeds $(\log q)^{-1}$ we do have a change in the argument of $(x/p)^{\rho}$. This observation should explain why we did not want to separate p from p_1, p_2 in the smoothing function $f(pp_1p_2/y)$. Moreover, it is worth mentioning that for this purpose we use two prime variables p_1, p_2, rather than one, because we can (and we must) apply the duality principle.

LEMMA 24.4. *For $x \geqslant q$ and any complex numbers c_p we have*

$$\sum\nolimits^{\rho} q^{2(\beta-1)} \Big| \sum_{x<p\leqslant 2x} c_p \Big(\frac{x}{p}\Big)^{\rho} \Big|^2 \ll \frac{x}{\log x} \sum_p |c_p|^2 , \tag{24.45}$$

where the implied constant is absolute.

Now (24.45) yields (23.34) by Cauchy's inequality.

For the proof of Lemma 24.4 it suffices to show that, for any complex numbers a_ρ,

$$\sum_{x<p\leqslant 2x} \frac{\log p}{p} \Big| \sum\nolimits^{\rho} a_\rho \Big(\frac{x}{p}\Big)^{\rho} q^{\beta-1} \Big|^2 \ll \sum_\rho |a_\rho|^2 . \tag{24.46}$$

First we smooth the outer summation by introducing a factor $h(p/x)$, then we square out and execute the summation over p, getting

$$\sum_p \frac{\log p}{p} h\Big(\frac{p}{x}\Big) \Big(\frac{x}{p}\Big)^{\rho_1+\bar\rho_2} \ll \big(1 + |\gamma_1 - \gamma_2| \log x\big)^{-2} . \tag{24.47}$$

This follows by partial summation using $\psi(y) = y + O\big(y(\log y)^{-4}\big)$ and the fact that the Fourier transform of $h(u)$ satisfies $\hat h(v) \ll (1+|v|)^{-2}$. Hence, the left side of (24.46) is estimated by

$$\sum\nolimits^{\rho_1} \sum\nolimits^{\rho_2} |a_{\rho_1} a_{\rho_2}| \, \big(1 + |\gamma_1 - \gamma_2| \log q\big)^{-2} q^{\beta_1+\beta_2-2} \leqslant \sum\nolimits^{\rho} A(\gamma) |a_\rho|^2$$

where

$$A(t) = \sum_\rho \big(1 + |\gamma - t| \log q\big)^{-2} \big(1 + (1-\beta)\log q\big)^{-1} .$$

Here we employed $q^{\beta_1+\beta_2-2} \leqslant q^{\beta_1-1} \leqslant \big(1 + (1-\beta_1)\log q\big)^{-1}$. We also dropped the restrictions on ρ in $A(t)$. The use of the Prime Number Theorem in the proof of (24.47) is not necessary; one can apply an upper-bound sieve to (24.46) before squaring out.

To complete the proof of (24.46) (and also (24.45) by duality) it remains to show that $A(t)$ is absolutely bounded for $|t| \leqslant q$.

LEMMA 24.5. *For $|t| \leqslant q$ we have*

$$A(t) \leqslant 2 + O\big(1/\log q\big) \tag{24.48}$$

where the implied constant is absolute.

PROOF. For any complex number s we have

$$\frac{L'}{L}(s,\chi) = -\frac{1}{2}\log\frac{q}{\pi} - \frac{1}{2}\psi\Big(\frac{s_\chi}{2}\Big) + B(\chi) + \sum_\rho \Big(\frac{1}{s-\rho} + \frac{1}{\rho}\Big)$$

where the summation runs over all zeros $\rho = \beta + i\gamma$ with $\beta > 0$, $B(\chi)$ is a constant with

$$\operatorname{Re} B(\chi) = -\sum_\rho \operatorname{Re} \frac{1}{\rho}$$

while

$$s_\chi = s + \tfrac{1}{2}\big(1+\chi(-1)\big)\ , \quad \psi(s) = \Gamma'(s)/\Gamma(s)\ ;$$

see (5.24) of [**109**]. Hence, for $s = \sigma + it$ with $1 < \sigma \leqslant 2$ we have

$$\sum_\rho \operatorname{Re}\frac{1}{s-\rho} = \frac{1}{2}\log q|s| + \operatorname{Re}\frac{L'}{L}(s,\chi) + O(1)\ ,$$

$$\operatorname{Re}\frac{1}{s-\rho} = \frac{\sigma-\beta}{(\sigma-\beta)^2+(\gamma-t)^2} \geqslant \frac{1}{\sigma-\beta}\left(1+\frac{|\gamma-t|}{\sigma-1}\right)^{-2}\ ,$$

$$\left|\frac{L'}{L}(s,\chi)\right| \leqslant -\frac{\zeta'}{\zeta}(\sigma) = \frac{1}{\sigma-1} + O(1)\ ,$$

and

$$\sum_\rho \frac{\sigma-1}{\sigma-\beta}\left(1+\frac{|\gamma-t|}{\sigma-1}\right)^{-2} \leqslant 1 + \tfrac{1}{2}(\sigma-1)\big(\log q|s| + O(1)\big)\ .$$

For $\sigma = 1+1/\log q$ the left side is equal to $A(t)$. This proves (24.48), and completes the proof of Linnik's estimate (24.1). □

The determination of the Linnik constant L along our sieve argument is possible, but not very encouraging. A proof of Linnik's theorem with many sieve ingredients and avoiding traditional lines has also been given in [**39**] and, as here, the constant there has not been computed. Analytic methods of proving (24.1) also do not lead easily to an explicit value of L; nevertheless, Heath-Brown [**88**] developed many interesting analytic refinements (in the zero-free region he takes into consideration the possibility of several zeros near the boundary) producing the constant $L = 5.5$, a remarkable improvement of the previously known exponents.

24.7. Ghosts in the Sieve Opera House

The arguments which give Linnik's theorem under the assumption of exceptional characters rely for their great importance on the fact that we are able to complete the proof in the opposite case.

In this section we show that this method of using the exceptional characters is much more general, applying to many interesting sequences for which, unfortunately, the complementary case has not yet been resolved. In fact, in combination with the sieve, the results obtained can even go beyond those found on assumption of the Riemann Hypothesis. In particular, we can find, in wide ranges of the modulus, primes $p \leqslant q^{2-\delta}$ in the arithmetic progression $p \equiv a \,(\mathrm{mod}\, q)$ and we can find in wide ranges primes in short intervals $[x - x^{\frac{1}{2}-\delta},\, x]$.

No wonder then that we welcome the exceptional characters into the arsenal of tools in analytic number theory, even though we perceive them as ghosts in the house who will eventually disappear forever, just as soon as a conflict with the legitimate room-mates is found.

In the presentation we do not make any reference to the exceptional characters so that the intermediate statements are unconditional and give interesting illustrations of the sieve at work. However, the final conclusions are only meaningful when the characters are indeed exceptional.

We target the primes in the sequence $\mathcal{A} = (a_n)$ with $a_n \geqslant 0$, which may be quite lacunary. We are going to reduce the problem of finding primes in $\mathcal{A}$ to finding primes in another sequence $\mathcal{B} = (b_n)$ with $b_n \geqslant 0$, similar to $\mathcal{A}$ but simpler in many respects, in particular being relatively larger. The chief tool in the transfer from $\mathcal{A}$ to $\mathcal{B}$ will be a real primitive character χ of modulus Δ. This will play the role of $\chi \,(\mathrm{mod}\, q)$ in Section 24.2 but we have changed the notation for the modulus to avoid confusion in some applications. In fact, for estimating the least prime $p \equiv a \,(\mathrm{mod}\, q)$ we now prefer to employ $\chi \,(\mathrm{mod}\, \Delta)$ with $(\Delta, q) = 1$ rather than with $\Delta = q$; therefore one will hear different tunes in this entr'acte.

Of course, we require some relations between $\mathcal{A}$ and $\mathcal{B}$. The essence of these relations is that the sequences share similar asymptotics for their congruence sums twisted by divisor-like functions. Specifically, we need the following divisor-like functions:

$$\lambda = \chi * 1$$

and

$$\lambda' = \chi * \log = \chi * 1 * \Lambda = \lambda * \Lambda \,.$$

Note that λ has the same structure as the divisor function $\tau = 1 * 1$. One may think of $\lambda(n)$ as the Hecke eigenvalue for an Eisenstein series (of weight one and central character χ), so one can see the product formula (24.5) in this context. Moreover, λ' has a similar divisor-like structure when we see it from the expression $\lambda' = \chi * \log$, but it also has another expression $\lambda' = \lambda * \Lambda$ which will allow us to single out the target primes. This having been done, the von Mangoldt function Λ in the terms left behind will be replaced by an upper-bound sieve of relatively small level, so λ' will be majorized by a function like

$$\lambda'' = \lambda * 1 = \chi * 1 * 1 = \chi * \tau$$

whose structure is similar to that of the divisor function $\tau_3 = 1 * 1 * 1$.

The fact that we only require functions like τ_k with $k \leqslant 3$ means that we only require exponents of distribution of $\mathcal{A}$, $\mathcal{B}$, as large as $\frac{2}{3} + \delta$ (and only for special remainders), a fact which is responsible for the method's success beyond the territory of RH. Note the contrast with the Linnik identity (17.3) which contains, not only higher rank divisor-like functions, but also requires the consideration of quite general bilinear forms to overcome the parity barrier. There will be no issue of parity here because, when χ is exceptional, the density functions $g(d) = \lambda(d)\tau_r(d)d^{-1}$ have arbitrarily small dimension in the sense of sieve theory.

In what follows we shall limit the ranges of our sums by various parameters, three of which x, y, z, are assumed to satisfy

$$\Delta^8 \leqslant z \leqslant y \leqslant x^{1/8} \;; \tag{24.49}$$

we shall make more specific choices later.

Because we work with the von Mangoldt function we need a crude bound for the contribution from higher powers of primes; the following estimates suffice:

$$\sum_{\substack{p^\alpha \leqslant x \\ p<z}}\sum a_{p^\alpha} \ll A(x)(\log x)^{-2} \,, \qquad \sum_{\substack{p^\alpha \leqslant x \\ p<z}}\sum b_{p^\alpha} \ll B(x)(\log x)^{-2} \,, \tag{24.50}$$

where

$$A(x) = \sum_{n \leqslant x} a_n \,, \qquad B(x) = \sum_{n \leqslant x} b_n \,. \tag{24.51}$$

In order to effect the reduction from $\mathcal{A}$ to $\mathcal{B}$ we make the following:

HYPOTHESIS A-B. For each of the functions $f = 1,\ \lambda',\ \lambda''$ we have uniformly for $d \leqslant y$,

$$\frac{1}{A(x)} \sum_{\substack{n\leqslant x\\ n\equiv 0\,(\mathrm{mod}\ d)}} f(n)a_n = \frac{1}{B(x)} \sum_{\substack{n\leqslant x\\ n\equiv 0\,(\mathrm{mod}\ d)}} f(n)b_n + O\Big(\frac{\tau_{3r}(d)}{d}(\log x)^{-5r}\Big) .$$

REMARKS. In practice, the verification of Hypothesis A-B requires an asymptotic evaluation of each side of the equation. The resulting main terms may not be simple to formulate in this generality, for example, involving polynomials in logarithms; nevertheless, it will not be hard to see that they agree up to the error term $O\big(\tau_{3r}(d)d^{-1}(\log x)^{-5r}\big)$, where r is a positive integer. Actually, inspecting our arguments one could see that we only need good error terms on average over $d \leqslant y$ where y will be the level of the sieve to be applied. Here is how one can be convinced that Hypothesis A-B is realistic. We think of having a family of sequences $\mathcal{A}^{(\ell)} = (a_n^{(\ell)})$, $0 \leqslant \ell < L$, each of which is a slight mutant of $\mathcal{A} = (a_n)$, suggesting that the twisted sums $\sum f(n)a_n^{(\ell)}$ over $n \leqslant x$, $n \equiv 0(\mathrm{mod}\, d)$ are asymptotically equal. We then choose the model sequence $\mathcal{B} = (b_n)$ as the average of the $\mathcal{A}^{(\ell)}$. If the family $\mathcal{A}^{(\ell)}$, $0 \leqslant \ell < L$ is complete, then $\mathcal{B}$ is not only larger than $\mathcal{A}$ but, above all, it is more uniformly distributed; hence it is presumably easier to search for primes in $\mathcal{B}$ than in $\mathcal{A}$. These remarks will be clearer when we consider the following example.

EXAMPLE (Arithmetic Progression). Let $(a, q) = 1$. Take

$$\begin{aligned} a_n &= 1 \quad \text{if} \quad n \equiv a\,(\mathrm{mod}\ q)\ , \\ b_n &= 1 \quad \text{if} \quad (n, q) = 1\ , \end{aligned}$$

and $a_n = b_n = 0$ otherwise. Then we have

$$A(x) = \frac{x}{q} + O(1)\ , \qquad B(x) = \frac{\varphi(q)}{q}x + O\big(\tau(q)\big)\ .$$

Let $(\Delta, q) = 1$, $\Delta \leqslant x^{1/80000}$, $y = x^{1/10000}$, $x \geqslant q^{119/60}$. Then Hypothesis A-B holds by arguments similar to those in the proof of Theorem 5 of [**51**]. Note that $119/60 < 2$. Perhaps it is worth mentioning that the verification of Hypothesis A-B for $f = \lambda''$ in this example relies on the Riemann Hypothesis for a variety over a finite field, as does the proof of Theorem 5 of [**51**].

For some problems Hypothesis A-B turns out to be too idealistic to hold. It fails, not for intrinsic reasons, but rather due to the difficulty in creating a single model sequence $\mathcal{B} = (b_n)$ for which the main terms agree to the necessary precision with those of the target sequence $\mathcal{A} = (a_n)$, for each of $f = 1,\ \lambda',\ \lambda''$ and all $d \leqslant y$. Therefore, we propose to work with the following less demanding hypothesis which unfortunately is also less appealing in appearance.

HYPOTHESIS AB. Let $\xi = (\xi_d)$ be an upper-bound sieve supported on $d \leqslant y$ and $\theta = 1 * \xi = (\theta_n)$. For $f = \lambda'$ we have,

$$\frac{1}{A(x)} \sum_{n\leqslant x} f(n)\theta_n a_n = \frac{1}{B(x)} \sum_{n\leqslant x} f(n)\theta_n b_n + O\Big(\frac{1}{\log x}\Big)\ ,$$

and for $f = \lambda'' - 1$,

$$\frac{1}{A(x)} \sum_{n\leqslant x} f(n)\theta_n a_n \ll \frac{1}{B(x)} \sum_{n\leqslant x} f(n)\theta_n b_n + \frac{1}{(\log x)^2}\ .$$

EXAMPLE (Short Interval). Take

$$a_n = 1 \quad \text{if} \quad x - u < n \leqslant x \ ,$$
$$b_n = 1 \quad \text{if} \quad x' - u' < n \leqslant x' \ ,$$

and $a_n = b_n = 0$ otherwise. Then we have

$$A(x) = u + O(1) \ , \qquad B(x) = u' + O(1) \ .$$

Let $\Delta \leqslant x^{1/2400}$, $y = x^{1/300}$, $x^{149/300} \leqslant u, u' \leqslant x \exp(-\sqrt{\log x})$ and $\frac{1}{2}x < x' \leqslant x$. Then, Hypothesis AB holds by exponential sum arguments similar to those given in [**2**] for $a_n = \tau_3(n)$. Note that $149/300 < \frac{1}{2}$.

Hypothesis AB completely suffices to compare the sum over primes in $\mathcal{A}$, precisely the sum

$$S_{\mathcal{A}}(x) = \sum_{n \leqslant x} a_n \Lambda(n) \ , \tag{24.52}$$

with that for the model sequence $\mathcal{B}$,

$$S_{\mathcal{B}}(x) = \sum_{n \leqslant x} b_n \Lambda(n) \ . \tag{24.53}$$

Now we are ready to begin the arguments. Throughout, $\xi = (\xi_d)$ is an upper-bound sieve supported on $d < y$, $d \mid P(z)$ with $|\xi(d)| \leqslant 1$. Hence, $\theta = 1 * \xi = (\theta_n)$ with

$$0 \leqslant \theta_n = \sum_{d|n} \xi_d \leqslant \tau(n) \ . \tag{24.54}$$

According to (24.49) we assume that the sieve level is $y \leqslant x^{1/8}$ and the sifting range is $z = y^{1/t} \geqslant \Delta^8$ with $t \geqslant 1$. Later on we shall assume that the sifting variable $t = (\log y)/(\log z)$ is sufficiently large in terms of the dimension of the applied sieve. Choosing (ξ_d) to be Brun's sieve (see (6.53)) we have $\xi_d = \mu(d)$ for $d < z$ whence $\theta_n = 0$ if $1 < n < z$. Occasionally this property offers a technical advantage, although it is nowhere essential.

We start from the sum

$$S'_{\mathcal{A}}(x) = \sum_{n \leqslant x} \lambda'(n) \theta_n a_n$$

and relate it to

$$S'_{\mathcal{B}}(x) = \sum_{n \leqslant x} \lambda'(n) \theta_n b_n$$

by Hypothesis AB for $f = \lambda'$, obtaining

$$S'_{\mathcal{A}}(x) = \frac{A(x)}{B(x)} S'_{\mathcal{B}}(x) + O\Big(\frac{A(x)}{\log x}\Big) \ .$$

On the other hand, using $\lambda' = \lambda * \Lambda$ we write $S'_{\mathcal{A}}(x)$ as a double sum:

$$S'_{\mathcal{A}}(x) = \sum\sum_{mn \leqslant x} \lambda(m) \Lambda(n) \theta_{mn} a_{mn} \ .$$

Here, we single out the contribution of $m = 1$, which comes within a small error of the target sum $S_{\mathcal{A}}(x)$; precisely we have

$$\sum_{n \leqslant x} \Lambda(n) \theta_n a_n = S_{\mathcal{A}}(x) + \sum_{p^\alpha \leqslant x} \xi_p (\log p) a_{p^\alpha} = S_{\mathcal{A}}(x) + O\big(A(x)/\log x\big)$$

by (24.50) (observe by the way that $\xi_p \leqslant 0$). In the remaining sum with $m \neq 1$ we estimate $\Lambda(n)$ by $\log x$ so that

$$S_{\mathcal{A}}(x) + O\big(A(x)(\log x)^{-1}\big) \leqslant S'_{\mathcal{A}}(x) \leqslant S_{\mathcal{A}}(x) + T_{\mathcal{A}}(x)\,\log x$$

where

$$T_{\mathcal{A}}(x) = \sum\sum_{\substack{mn \leqslant x \\ m \neq 1}} \lambda(m)\theta_{mn} a_{mn}\ .$$

We write this as the single-variable sum

$$T_{\mathcal{A}}(x) = \sum_{n \leqslant x} \big(\lambda''(n) - 1\big)\theta_n a_n\ .$$

In similar fashion,

$$S_{\mathcal{B}}(x) + O\big(B(x)(\log x)^{-1}\big) \leqslant S'_{\mathcal{B}}(x) \leqslant S_{\mathcal{B}}(x) + T_{\mathcal{B}}(x)\,\log x\ ,$$

where

$$T_{\mathcal{B}}(x) = \sum_{n \leqslant x} \big(\lambda''(n) - 1\big)\theta_n b_n\ . \tag{24.55}$$

Next we relate $T_{\mathcal{A}}(x)$ to $T_{\mathcal{B}}(x)$ in the same fashion that we related $S'_{\mathcal{A}}(x)$ to $S'_{\mathcal{B}}(x)$. This time we apply Hypothesis AB for $f = \lambda'' - 1$, obtaining

$$T_{\mathcal{A}}(x) \ll \frac{A(x)}{B(x)}\,T_{\mathcal{B}}(x) + \frac{A(x)}{(\log x)^2}\ .$$

Combining the above results we arrive at the following inequality:

$$\Big|S_{\mathcal{A}}(x) - \frac{A(x)}{B(x)}\,S_{\mathcal{B}}(x)\Big| \ll \frac{A(x)}{B(x)}\,T_{\mathcal{B}}(x)\log x + \frac{A(x)}{\log x}\ . \tag{24.56}$$

We are now finished with the original sequence $\mathcal{A}$ and it remains to estimate $T_{\mathcal{B}}(x)$ for the model sequence $\mathcal{B}$.

REMARKS. Even though we might for convenience assume very strong properties of the b_n (because $\mathcal{B} = (b_n)$ can be chosen to be nice, as in Section 1.4), it is pleasing to show that one can manage with very modest requirements. The reason we can do so is that we only require a good estimation of $T_{\mathcal{B}}(x)$, not an asymptotic formula. In consequence we are able to succeed using elementary arguments (albeit with a lot of help from sieve technology) whereas, obtaining an asymptotic formula by analytic methods would inevitably bring onto the scene both $L(1,\chi)$ and $L'(1,\chi)$, and estimating the latter in terms of the former hurts us a bit (in an earlier version of this entr'acte we were able to do just that, although not without pain).

We assume that

$$b_n \leqslant \rho(n)\ , \tag{24.57}$$

where $\rho(n)$ is a completely multiplicative function with $0 \leqslant \rho(p) \leqslant r$ for some positive integer r. We shall allow the implied constants to depend on r, but not on anything else. Our goal is the upper bound

$$T_{\mathcal{B}}(x)\log x \ll x\big(L(1,\chi)\log x + e^{-t}\big)\Big(\frac{\log x}{\log z}\Big)^{3r+1}\ . \tag{24.58}$$

We begin the proof by splitting

$$T_{\mathcal{B}}(x) = T(x, z) + T(x \mid z) \ , \tag{24.59}$$

where $T(x, z)$ is the contribution from those n in (24.55) which are free of prime divisors $\geqslant z$ and $T(x \mid z)$ is the remaining sum.

First, we are going to estimate $T(x, z)$. Our goal is the bound

$$T(x, z) \ll \frac{x}{\log z} e^{-t} \ . \tag{24.60}$$

We have

$$T(x, z) \leqslant \sum_{\substack{1<n\leqslant x \\ p|n \Rightarrow p<z}} \lambda''(n)\rho(n)\theta_n = T''(x, z) \ ,$$

say. It clearly suffices to prove (24.60) for $T^\flat(x, z)$ which denotes the subsum of $T''(x, z)$ restricted to squarefree n (note that θ_n depends only on the squarefree kernel of n). Moreover, we can restrict $T^\flat(x, z)$ to $z \leqslant n \leqslant x$ because $\theta_n = 0$ for $1 < n < z$, $n \mid P(z)$. This nice feature of the sieve we chose is a bonus but is not essential because the contribution from small n would in any case be negligible. Extracting from n its largest prime factor we get

$$T^\flat(x, z) \leqslant 3r \sum\sum_{\substack{z\leqslant pm\leqslant x \\ p<z,\, m|P(p)}} \tau_3(m)\tau_r(m)\theta_{pm} \ .$$

Note that here $\theta_{pm} = \theta_m - \theta_m^-$, where θ_m^- stands for the corresponding lower-bound sieve convolution (see (6.45)). Moreover, p cannot be very small. Indeed, $pm \leqslant p^{1+\nu(m)} \leqslant p^{\tau(m)}$, so $p \geqslant z^{1/\tau(m)}$. Applying these observations we obtain

$$\begin{aligned} T^\flat(x, z) &\leqslant 3r \sum_{\substack{m|P(z) \\ m\leqslant x}} \tau_3(m)\tau_r(m)(\theta_m - \theta_m^-) \sum_{z^{1/\tau(m)}\leqslant p\leqslant x/m} 1 \\ &\ll \frac{x}{\log z} \sum_{m|P(z)} m^{-1}\tau(m)\tau_3(m)\tau_r(m)(\theta_m - \theta_m^-) \ . \end{aligned}$$

The last sum over all $m \mid P(z)$ is $O(e^{-t})$ by the Fundamental Lemma applied with dimension $\kappa = 6r$ for both sieves θ_m, θ_m^-. Note that for $m \neq 1$ we cannot expect any cancellation since both θ_m and $-\theta_m^-$ are then non-negative. Nevertheless, this shows that $T^\flat(x, z)$ satisfies (24.60), hence so do $T''(x, z)$ and $T(x, z)$.

Next, we are going to estimate the sum $T(x \mid z)$ which runs over numbers having a prime divisor $p \geqslant z$, that is,

$$T(x \mid z) = \sum\sum_{\substack{pn\leqslant x,\, p\geqslant z \\ p'|n \Rightarrow p'\leqslant p}} (\lambda''(pn) - 1)\theta_n\rho(pn)$$

because $\theta_{pn} = \theta_n$ for $p \geqslant z$. The contribution of terms with $p \mid n$ is trivially bounded by $O\big(z^{-1}x(\log x)^{6r}\big)$ which is negligible. If $p \nmid n$ we write $\lambda''(pn) - 1 = \lambda''(p)\big(\lambda''(n) - 1\big) + \lambda(p)$ and accordingly we have

$$T(x \mid z) \leqslant T''(x \mid z) + T^*(x \mid z) + O\big(z^{-1}x(\log x)^{6r}\big) \ ,$$

where

$$T''(x \mid z) = \sum_{n\leqslant x} \big(\lambda''(n) - 1\big)\, \theta_n\, \rho(n) \sum_{z\leqslant p\leqslant x/n} \lambda''(p)\rho(p) \ ,$$

$$T^*(x \mid z) = \sum_{n\leqslant x} \theta_n\, \rho(n) \sum_{z\leqslant p\leqslant x/n} \lambda(p)\rho(p) \ .$$

First we estimate $T^*(x \mid z)$. In the inner sum we use the inequality $\rho(p) \leqslant r$ and then we forget that p runs over primes, receiving the bound

$$r \sum_{z<\ell\leqslant x/n} \lambda(\ell) \ll \frac{x}{n} L(1,\chi) \ .$$

Next, the outer sum is treated via the upper-bound sieve of dimension r as follows:

$$\sum_{n\leqslant x} \theta_n \rho(n) n^{-1} \ll \sum_{m\leqslant x}{}^{\flat}\, \theta_m \tau_r(m) m^{-1}$$
$$\leqslant \prod_{z\leqslant p\leqslant x} \Big(1 + \frac{2r}{p}\Big) \sum_{m|P(z)} \theta_m \tau_r(m) m^{-1} \ll \Big(\frac{\log x}{\log z}\Big)^{2r} \ .$$

In the above, as usual, the superscript $\flat$ restricts the summation to squarefree numbers. Hence, we deduce

$$T^*(x \mid z) \ll x \Big(\frac{\log x}{\log z}\Big)^{2r} L(1,\chi) \ . \tag{24.61}$$

It remains to estimate $T''(x \mid z)$, which is the key part of $T_{\mathcal{B}}(x)$. In the inner sum $\lambda''(p)\rho(p) \leqslant 3r$ so that

$$T''(x \mid z) \ll \frac{x}{\log z} \sum_{n\leqslant x} (\lambda''(n) - 1)\theta_n \rho(n) n^{-1} \ .$$

Applying Rankin's trick, we arrive at

$$T''(x \mid z) \ll \frac{x^s}{\log z} Z(s) \tag{24.62}$$

for any $s > 1$, where

$$Z(s) = \sum_{1}^{\infty} \big(\lambda''(n) - 1\big)\, \theta_n \rho(n) n^{-s} \ .$$

Opening the sieve convolution θ_n and changing the order of summation we get

$$Z(s) = \sum_{d|P(z)} \xi_d \Big(\sum_{n\equiv 0\,(d)} \lambda''(n) \frac{\rho(n)}{n^s} - \frac{\rho(d)}{d^s} L(s,\rho) \Big) \ ,$$

where

$$L(s,\rho) = \sum_{1}^{\infty} \rho(n) n^{-s} \ .$$

Put

$$L(s,\lambda''\rho) = \sum_{1}^{\infty} \lambda''(n)\rho(n) n^{-s} = \prod_p \ell_p(s) \ ,$$

where

$$\ell_p(s) = \sum_{\alpha=0}^{\infty} \lambda''(p^\alpha)\rho(p^\alpha) p^{-\alpha s}$$

are the local factors of $L(s,\lambda''\rho)$. Hence, for d squarefree we have

$$\sum_{n\equiv 0\,(d)} \lambda''(n)\rho(n)n^{-s} = L(s,\lambda''\rho)\prod_{p|d}\Big(1-\frac{1}{\ell_p(s)}\Big)$$

and $Z(s)$ can be written as

$$Z(s) = L(s,\lambda''\rho)\sum_d \xi_d \prod_{p|d}\Big(1-\frac{1}{\ell_p(s)}\Big) - L(s,\rho)\sum_d \xi_d \frac{\rho(d)}{d^s}\ .$$

Here, the density functions attached to the sieve weights ξ_d satisfy (5.38) with $\kappa = 3r$ so, by the Fundamental Lemma, we obtain

$$Z(s) = \big(1+O(e^{-t})\big)L^*(s,\lambda''\rho) - \big(1+O(e^{-t})\big)L^*(s,\rho)\ ,$$

where $t = (\log y)/(\log z)$ and the superscript $*$ means that the local factors of the L-functions at primes $p < z$ are to be omitted. It remains to evaluate these incomplete L-functions.

Since ρ is completely multiplicative we have $\lambda''\rho = (\lambda * 1)\rho = \lambda\rho * \rho$ and $L^*(s,\lambda''\rho) = L^*(s,\lambda\rho)L^*(s,\rho)$, hence

$$Z(s) = \big[\big(1+O(e^{-t})\big)L^*(s,\lambda\rho) - 1\big]L^*(s,\rho)\ . \tag{24.63}$$

Using

$$\sum_{z<p\leqslant u}\frac{1}{p} = \log\frac{\log u}{\log z} + O\Big(\frac{1}{\log z}\Big)\ ,$$

it follows by partial summation that

$$\sum_{p>z} p^{-s} = \int_{(s-1)\log z}^{\infty} e^{-v}\frac{dv}{v} + O\big(z^{1-s}(\log z)^{-1}\big)\ .$$

Assuming $(s-1)\log z \leqslant 1$, this gives

$$\sum_{p>z} p^{-s} \leqslant 1 - \log\big((s-1)\log z\big) + O(1/\log z)\ .$$

Hence, we derive

$$L^*(s,\rho) \ll \exp\Big(\sum_{p>z}\rho(p)p^{-s}\Big) \ll \big((s-1)\log z\big)^{-r}\ .$$

Similarly, we find that

$$L^*(s,\lambda\rho) \ll \big((s-1)\log z\big)^{-2r}\ .$$

But we can do better by using (see (24.19)):

$$\sum_{z<p\leqslant u}\frac{\lambda(p)}{p} < L(1,\chi)\Big(\log\frac{u}{z} + O(z^{-1/4}\Big)\ .$$

Hence, it follows by partial summation that

$$\sum_{p>z}\lambda(p)p^{-s} < L(1,\chi)\Big(\frac{z^{1-s}}{s-1} + O\big(z^{-1/4}\big)\Big) \ll \frac{L(1,\chi)}{s-1}\ .$$

We have

$$L^*(s,\lambda\rho) = \exp\Big(\sum_{p>z}\lambda(p)\rho(p)p^{-s} + O(z^{-1})\Big)\ .$$

This, together with the previous bounds, implies

$$L^*(s,\lambda\rho) = 1 + O\Big(\frac{L(1,\chi)}{s-1}\big((s-1)\log z\big)^{-2r}\Big) .$$

Next, employing these bounds in (24.63) we obtain

$$Z(s) \ll \Big(\frac{L(1,\chi)}{s-1} + e^{-t}\Big)\big((s-1)\log z\big)^{-3r} .$$

Now, taking $s = 1 + 1/\log x$ we conclude from (24.62) that

$$T''(x\,|\,z) \ll \frac{x}{\log z}\big(L(1,\chi)\log x + e^{-t}\big)\Big(\frac{\log x}{\log z}\Big)^{3r} , \tag{24.64}$$

where $t = (\log y)/(\log z)$ is the sifting variable. Note that the bounds (24.60) and (24.61) are stronger than (24.64). This completes the proof of (24.58).

Combining (24.58) with (24.56) for $z = \Delta^8$, we conclude:

Proposition 24.6. *Let $\mathcal{A} = (a_n)$, $\mathcal{B} = (b_n)$ be sequences of non-negative reals and let $A(x)$, $B(x)$, $S_{\mathcal{A}}(x)$, $S_{\mathcal{B}}(x)$ be defined by* (24.51), (24.52) *and* (24.53). *Suppose $\mathcal{A}$, $\mathcal{B}$ satisfy Hypothesis AB where $\Delta^8 \leqslant y \leqslant x^{1/8}$, that they satisfy* (24.50), *and also $b_n \leqslant \rho(n)$ for a totally multiplicative function $\rho(n)$ with $\rho(p) \leqslant r$. Then we have*

$$S_{\mathcal{A}}(x) - \frac{A(x)}{B(x)}\,S_{\mathcal{B}}(x) \ll x\,\frac{A(x)}{B(x)}\,\big(L(1,\chi)\log x + e^{-t}\big)\Big(\frac{\log x}{\log \Delta}\Big)^{4r} + \frac{A(x)}{\log x} ,$$

where $t = \log y\,/8\log\Delta$ and the implied constant depends only on r.

To clarify this result we add a few further conditions and we make specific choices of the relevant parameters. Let $0 < \delta \leqslant 1/8$ be an admissible exponent of distribution for all three sequences $\mathcal{A}_f = \big(f(n)a_n\big)$ with $f = 1,\ \lambda',\ \lambda''$. This ensures that Hypothesis AB holds with our sieve of level $y \leqslant x^\delta$ for every $x \geqslant \Delta^c$ with any number c sufficiently large in terms of δ. Suppose the model sequence $\mathcal{B} = (b_n)$ is bounded (the case $r = 1$) and it has a lot of primes, that is,

$$S_{\mathcal{B}}(x) \asymp x . \tag{24.65}$$

Then, on choosing $z = \Delta^8$ and $y = \Delta^{\delta c}$, we find that, for any x in the segment $\Delta^c \leqslant x \leqslant \Delta^{8c}$, Proposition 24.6 simplifies to

$$S_{\mathcal{A}}(x) = \frac{A(x)}{B(x)}\,S_{\mathcal{B}}(x)\big[1 + O\big(\eta(c,\Delta)\big)\big] ,$$

where

$$\eta(c,\Delta) = c^5 L(1,\chi)\log\Delta + c^4 e^{-\delta c/8} + (c\log\Delta)^{-1} .$$

Finally, if $\chi \,(\mathrm{mod}\,\Delta)$ runs over exceptional characters, by which we mean

$$L(1,\chi)\log\Delta \to 0 \quad \text{as } \Delta \to \infty , \tag{24.66}$$

then we can let $c \to \infty$ slowly so that $\eta(c,\Delta) \to 0$, obtaining the asymptotic formula

$$S_{\mathcal{A}}(x) \sim \frac{A(x)}{B(x)}\,S_{\mathcal{B}}(x) \tag{24.67}$$

in the range $\Delta^c \leqslant x \leqslant \Delta^{8c}$.

In practice, the condition (24.65) requires our model $\mathcal{B} = (b_n)$ to have positive density, a property which may not always be convenient. The reason for this deficiency lies in our crude working assumption (24.57) which ignores the possibility of

$\mathcal{B}$ having short support. However, this feature can be recovered as follows: Suppose we have several models, say $\mathcal{B}^{(j)} = (b_n^{(j)})$ for $j = 1, 2, \ldots, J$ having disjoint supports on $n \leqslant x$, each of which has the same mean value $\mathcal{B}^{(j)} = \mathcal{B}(x)$ and satisfies Hypothesis AB. Then, by (24.56), summing over j we obtain Proposition 24.6 with $S_{\mathcal{B}}(x)$ replaced by $S(x)\,/J$ where

$$S(x) = \sum_j S_{\mathcal{B}^{(j)}}(x) \ .$$

Hence, the conclusion (24.67) becomes $JS_{\mathcal{A}}(x) \sim A(x)B(x)^{-1}S(x)$, subject to the condition (24.65) being replaced by $S(x) \asymp x$. Since we may achieve $S_{\mathcal{B}^{(j)}}(x) \sim S_{\mathcal{B}}(x)$ for every j we have $S(x) \sim JS_{\mathcal{B}}(x)$ so the asymptotic formula (24.67) remains true under the weaker condition $S(x) \asymp x$.

REMARKS. The comparative asymptotic formula (24.67) does not require the PNT. Our first example (arithmetic progression) depends only on a single model $\mathcal{B}$ whereas the second example (short interval) needs several models $\mathcal{B}^{(j)}$ which are shifted intervals covering the segment $\frac{1}{2}x < n \leqslant x$. Heath-Brown [**85**] employed exceptional characters for prime twins and the authors [**55**] employed them for primes of the form $x^2 + y^6$ but the results of this section do not quickly cover those examples.

CHAPTER 25

Almost-Prime Sieve

Chance Music

25.1. Introduction

Let $P(z)$ be the product of all primes $p < z$ and

$$S(\mathcal{A}, z) = \sum_{\substack{n \leqslant x \\ (n, P(z))=1}} a_n \,. \tag{25.1}$$

Let $D = D(x)$ be the level of distribution of $\mathcal{A} = \big(a_n\big)$ and put $s = \log D / \log z$. The most basic sieve lower bound

$$S(\mathcal{A}, z) \gg XV(z) \tag{25.2}$$

holds provided s is sufficiently large in terms of the sieve dimension κ. Note that the numbers a_n counted by $S(\mathcal{A}, z)$ are supported on integers having at most $r = [s \log x / \log D]$ prime divisors, so the lower bound (25.2) ensures us the existence of such numbers, in fact, by letting $x \to \infty$ we get infinitely many such numbers in $\mathcal{A} = \big(a_n\big)$. Numbers having at most r distinct prime factors are called almost-primes of order r.

For example, consider the integers in the short interval $x - x^\theta < n \leqslant x$ with $0 < \theta \leqslant 1$. This has level of distribution $D = x^{\theta - \varepsilon}$, so applying the linear sieve we obtain

$$S(\mathcal{A}, z) \gg \frac{x^\theta}{\log x} \,,$$

for $z = x^{\theta/2 - \varepsilon}$, because the sifting limit is $\beta = 2$. This shows that our short interval contains almost-primes of order $r = [2/\theta]$ if $2/\theta \notin \mathbb{Z}$ and $r = 2/\theta + 1$ if $2/\theta \in \mathbb{Z}$.

Another example is the characteristic function of an irreducible polynomial $h(X) \in \mathbb{Z}[X]$, $\deg h = g \geqslant 1$. The straightforward application of the linear sieve in this case produces values of $h(X)$ which are almost-primes of order $r = 2g+1$. Since the exponent of distribution in this case is $1/g$ and $\kappa = 1$ this is quite analogous to the example of the short interval.

One expects that these results are out from the truth, approximately by a factor of two. To see that an improvement of that size is the most we can expect, consider the example of the following generalized Selberg sequence (compare with Section 12.3) of integers $n = \ell m$ with

$$\ell = p_1 p_2 \dots p_{r-1},\ p_1, \dots, p_{r-1} \asymp x^{1/r} \,,$$

$$m \asymp x^{1/r} \,,\ \Omega(m) \equiv 0 \,(\bmod 2) \,,$$

so m runs through integers having an even number of prime factors and, as usual, the p_j run through primes. It follows easily from the Prime Number Theorem that the sequence $\mathcal{A}$ has exponent of distribution $1/r$ (fix the p_j and observe the cancellation in the sum over m). On the other hand, it is obvious that every number in $\mathcal{A}$ contains at least $r+1$ prime factors.

There is a central, motivating conjecture in this direction. Let $\mathcal{A} = (a_n)$ be a sequence of sieve dimension $\kappa = 1$ having exponent of absolute distribution θ. If r is an integer with $r > 1/\theta$ then, for all sufficiently large x,

$$\sum_{\substack{n\leqslant x\\ \nu(n)\leqslant r}} a_n \gg \frac{X}{\log x}\ . \tag{25.3}$$

The best result in the direction of the conjecture, which improves on earlier work of Richert [**140**] and of Laborde [**122**], is due to Greaves [**73**], [**76**] and asserts that if the condition $r > 1/\theta$ is replaced by the stronger one $r > 1/\theta + 0.124$, then (25.3) holds.

The previous example shows that the conjecture is best possible; this "sifting limit" for the almost-prime linear sieve cannot be any better.

To come somewhat closer to this barrier than we have so far it is useful to note that the above arguments (the lower bound for $S(\mathcal{A}, z)$) give us almost-primes with the extra property of having no small prime factors. If one is willing to give up this nice property and is mainly interested in lowering the value of r, then there are possibilities other than just estimating the sifting function. One of these is to modify the sifting function $S(\mathcal{A}, z)$ by attaching to a_n weights $\mathcal{W} = \big(w(n)\big)$ which are no longer always positive. Thus we consider the weighted sum

$$S(\mathcal{A}\cdot\mathcal{W}, z) = \sum_{\substack{n\leqslant x\\ (n,P(z))=1}} a_n\, w(n)\ . \tag{25.4}$$

If we succeed with these weights to prove that

$$S(\mathcal{A}\cdot\mathcal{W}, z) \gg XV(z)\ , \tag{25.5}$$

then we know that some of the weights $w(n)$ are positive. The idea will be to create weights $w(n)$ with positive support on almost-primes of order r with r as small as possible. Of course, one could choose $w(n) = 1$ for all $\big(n, P(z)\big) = 1$ and get almost-primes as before, but it turns out that one can do better with different choices. There is a long history in the development of such weights, some of them being quite complicated. Generally speaking, the theory of the weighted sieve is still at an experimental stage.

In this chapter we give an account of a weighted sieve of Richert which provides a combination of power and elegant simplicity.

25.2. Evaluation of the Weighted Sum

We assume that the weights are of the type

$$w(n) = 1 - \frac{1}{\eta} \sum_{\substack{p\mid n\\ z\leqslant p<y}} w_p \tag{25.6}$$

where η is a positive constant and $0 \leqslant w_p \leqslant 1$ are free for the time being. Thus, our weights w_p depend solely on the prime factors of n. Other choices are possible but we do not consider them.

For the weights (25.6) we have

$$S(\mathcal{A}\cdot\mathcal{W},z)=S(\mathcal{A},z)-\frac{1}{\eta}\sum_{z\leqslant p<y}w_p\,S(\mathcal{A}_p,z)\ ,$$

so the problem reduces to giving a lower bound for $S(\mathcal{A},z)$ and upper bounds for the $S(\mathcal{A}_p,z)$. Of course, for the latter sums we need bounds only on average over p.

Now we assume that our original sequence $\mathcal{A}=(a_n)$ is giving rise to a linear sieve problem, that is, (1.2) holds with a density function $g(p)$ satisfying (5.38) with $\kappa=1$. We also assume that the remainder terms $r_d(x)$ satisfy

$$R(x,\,D|N)=\sum_{d\leqslant D}\Big|\sum_{n\leqslant N}\alpha_n\,r_{dn}(x)\Big|\ll X\big(\log x\big)^{-3} \tag{25.7}$$

for any complex coefficients $|\alpha_n|\leqslant 1$. When $N=1$ this is just the absolute remainder $R(x,D)$, however, in many applications we can obtain stronger results by allowing N to be somewhat larger.

To take advantage of $R(x,\,D|N)$ we apply the linear sieve with well-factorable remainder. Let ε be any small positive number and x sufficiently large in terms of ε. By Theorem 12.19 we get

$$S(\mathcal{A},z)\geqslant XV(z)\{f(s)-\varepsilon\} \tag{25.8}$$

where $s=\log(DN)/\log z$. Similarly, by Theorem 12.20, provided $y\leqslant D$, we get

$$\sum_{z\leqslant p<y}w_p\,S(\mathcal{A}_p,z)\leqslant XV(z)\Big\{\sum_{z\leqslant p<y}g(p)\,w_p\,F(s_p)+\varepsilon\Big\} \tag{25.9}$$

where $s_p=\log(DN/p)/\log z$. Observe that the restriction $y\leqslant D$ is necessary in order to incorporate p inside the single factor d of dn in (25.7). Note also that the resulting multiplicity of the coefficients is bounded by the number of prime factors, hence absolutely bounded by $\log x$. The saving of $(\log x)^2$ in (25.7) as compared to the expected main term $X\big(\log x\big)^{-1}$ is sufficient to accommodate this multiplicity.

Adding these estimates we obtain

$$S(\mathcal{A}\cdot\mathcal{W},z)\geqslant XV(z)\left\{W(z,y)-\left(1+\tfrac{1}{\eta}\right)\varepsilon\right\} \tag{25.10}$$

where

$$W(z,y)=f(s)-\frac{1}{\eta}\sum_{z\leqslant p<y}g(p)\,w_p\,F(s_p)\ . \tag{25.11}$$

To simplify we restrict the range $z\leqslant p<y$ by the conditions

$$(DN)^{\frac14}\leqslant z\leqslant(DN)^{\frac12}\leqslant y\leqslant D\ . \tag{25.12}$$

In this range $2\leqslant s\leqslant 4$ and $0<s_p\leqslant 3$, so the functions f,F are elementary, given by $f(s)=2e^{\gamma}s^{-1}\log(s-1)$, $F(s_p)=2e^{\gamma}s_p^{-1}$. Therefore,

$$W(z,y)=\frac{2e^{\gamma}}{s}\Big\{\log(s-1)-\frac{s}{\eta}\sum_{z\leqslant p<y}g(p)\,w_p/s_p\Big\}\ . \tag{25.13}$$

25.3. Richert's Weights

The first weights of the type (25.6) are due to Kuhn [**120**] who took $w_p = 1$. Following Richert [**140**] we choose

$$w_p = 1 - \frac{\log p}{\log y} . \tag{25.14}$$

Hence,

$$\begin{aligned} s \sum_{z \leqslant p < y} g(p) w_p / s_p &= \sum_{z \leqslant p < y} g(p) \Big(1 - \frac{\log p}{\log y}\Big) \frac{\log(DN)}{\log(DN/p)} \\ &= \int_z^y \Big(1 - \frac{\log t}{\log y}\Big) \frac{\log(DN)}{\log(DN/t)} \frac{dt}{t \log t} + O\Big(\frac{1}{\log z}\Big) , \end{aligned}$$

by (5.47). Put $DN = z^s = y^u$ so the conditions (25.12) become $2 \leqslant s \leqslant 4$ and

$$u \geqslant 1 + \log N / \log D , \tag{25.15}$$

while the above integral becomes

$$\int_{1/s}^{1/u} \frac{1 - \alpha u}{1 - \alpha} \frac{d\alpha}{\alpha} = \log \frac{s-1}{u-1} - u \log \frac{u(s-1)}{s(u-1)} .$$

Hence,

$$W(z, y) = \frac{2e^\gamma}{s} \left(\log(s-1) - \frac{1}{\eta} \log \frac{s-1}{u-1} + \frac{u}{\eta} \log \frac{u(s-1)}{s(u-1)} \right) .$$

Now we choose $s = 4$ (that is, $z = (DN)^{\frac{1}{4}}$), which is the best we can do in our restricted range (25.12), getting

$$W(z, y) = \frac{e^\gamma}{2} (\log 3) \Big(1 - \frac{\delta(u)}{\eta}\Big) \tag{25.16}$$

where

$$\delta(u) = \frac{1}{\log 3} \left(\log \frac{3}{u-1} - u \log \frac{3u}{4(u-1)} \right) . \tag{25.17}$$

We wish (25.16) to be positive so that (25.5) holds. This is the case for η sufficiently large, precisely we require

$$\eta > \delta(u) . \tag{25.18}$$

On the other hand, we wish to have η not too large so that $w(n) \leqslant 0$ for every $n \leqslant x$, $\big(n, P(z)\big) = 1$ having more than r distinct prime factors. If $\nu(n) \geqslant r + 1$, we have

$$\begin{aligned} w(n) &\leqslant 1 - \frac{1}{\eta} \sum_{p \mid n} \Big(1 - \frac{\log p}{\log y}\Big) \leqslant 1 - \frac{1}{\eta} \Big(\nu(n) - \frac{\log n}{\log y}\Big) \\ &\leqslant 1 - \frac{1}{\eta} \Big(r + 1 - \frac{\log x}{\log y}\Big) . \end{aligned}$$

Hence,

$$\eta = r + 1 - \frac{\log x}{\log y} \tag{25.19}$$

is a good choice, provided this is positive, a condition which will be easily verified for our forthcoming choice of the parameter u.

For η given by (25.19), the condition (25.18) becomes

$$\frac{\log x}{\log DN} < \frac{r+1-\delta(u)}{u} = \Delta_r(u)\ , \quad \text{say}\ . \tag{25.20}$$

Recall that u must also satisfy the condition (25.15) which does not depend on r. We would like the requirement (25.20) to be as weak as possible so as to obtain a stronger final result. We find by differentiation that the maximum of $\Delta_r(u)$ is attained at

$$u = 1 + 3^{-r}\ . \tag{25.21}$$

We put $\delta_r = \delta\big(1+3^{-r}\big)$ which is

$$\delta_r = 1 - r\,3^{-r} - \frac{\big(1+3^{-r}\big)}{\log 3}\Big(\log\frac{3}{4}\big(1+3^{-r}\big)\Big) \tag{25.22}$$

and $\Delta_r = \Delta_r\big(1+3^{-r}\big)$ which in turn is

$$\Delta_r = r + \frac{1}{\log 3}\log\Big(\frac{3}{4}\big(1+3^{-r}\big)\Big)\ . \tag{25.23}$$

It is easy to check that $\delta_r \geqslant 0$ and

$$r - 1 < \Delta_r \leqslant r\ . \tag{25.24}$$

For u given by (25.21) the conditions (25.15), (25.20) in practice become

$$D \geqslant N^{3^r}\ , \quad DN \geqslant x^{1/\Delta_r+\varepsilon}\ . \tag{25.25}$$

The first condition in (25.25) holds for relatively small N (it is trivial for $N=1$). However, if we have (25.7) for the remainder $R(x,\, D|N)$ which allows large values of N, then this condition becomes so severe that it destroys our reasoning behind the optimization of (25.20), and it pays to make a different choice of u. We simply take $y = D$, that is,

$$u = \frac{\log DN}{\log D} \tag{25.26}$$

rather than (25.21), so the condition (25.15) is automatically satisfied while the condition (25.20) in practice becomes

$$\frac{\log x}{\log D} \leqslant r + 1 - \delta\Big(\frac{\log DN}{\log D}\Big) - \varepsilon\ . \tag{25.27}$$

Having made the two choices of u, namely (25.21) or (25.26), we conclude two results about almost-primes in $\mathcal{A} = \big(a_n\big)$.

THEOREM 25.1. *Let $\mathcal{A} = \big(a_n\big)$ be a sequence of non-negative numbers which satisfies the linear sieve conditions* (1.2) *and* (5.38) *with $\kappa = 1$. Suppose the remainder $R(x,\, D|N)$ satisfies* (25.7) *with D, N satisfying* (25.27) *(recall that Δ_r is given by* (25.23)*. Then we have*

$$\sum_{\substack{n\leqslant x\\ (n,P(z))=1\\ \nu(n)\leqslant r}} a_n \asymp XV(x)\ , \tag{25.28}$$

where $z = (DN)^{\frac{1}{4}}$ and the implied constant depends on r and ε.

Because almost-primes of order $r = 2$ are the best one can hope to obtain by the general weighted sieve it is worthwhile to be more specific in this case. We have:

COROLLARY 25.2. *Let $\mathcal{A} = (a_n)$ be a sequence of non-negative numbers which satisfies the linear sieve conditions* (1.2) *and* (5.38) *with $\kappa = 1$. Suppose the remainder $R(x, D|N)$ satisfies* (25.7) *with*

$$D \geqslant N^9, \qquad DN \geqslant x^{6/11}.$$

Then, we have

$$\sum_{\substack{n \leqslant x \\ (n, P(x^{1/8}))=1 \\ \nu(n) \leqslant 2}} a_n \asymp XV(x) . \tag{25.29}$$

THEOREM 25.3. *Let $\mathcal{A} = (a_n)$ be a sequence of non-negative numbers which satisfies the linear sieve conditions* (1.2) *and* (5.38) *with $\kappa = 1$. Suppose the remainder $R(x, D|N)$ satisfies* (25.7) *with D, N satisfying* (25.27) *(recall that $\delta(u)$ is given by* (25.17)*). Then we have*

$$\sum_{\substack{n \leqslant x \\ (n, P(z))=1 \\ \nu(n) \leqslant r}} a_n \asymp XV(x) , \tag{25.30}$$

where $z = (DN)^{\frac{1}{4}}$ and the implied constant depends on r and ε.

The work of Greaves envisions an almost-prime sieve with progress beyond the experimental level. In particular, rather than simply appending weights onto an existing sieve, Greaves' idea is to rebuild the sieve ab initio. He develops a form of the beta-sieve with constraints modified so as to implement more efficiently the goal of producing almost-primes. His approach gives significant quantitative improvements and approaches the limit of what one can hope to achieve. Because, although conceptually interesting, it is long and complicated and already well-presented in Greaves' book [**76**] and his survey article [**75**], we shall confine ourselves to applications of the method of Richert developed above. It would be interesting to see what could be attained by building the bilinear form of the remainder into the Greaves construction.

25.4. Almost-Prime Values of Polynomials

Our first two applications of the weighted sieve, due to Richert [**140**], concern the sequence of values of a polynomial of arbitrary degree in a single variable.

THEOREM 25.4. *Let $h(X) \in \mathbb{Z}[X]$ be irreducible of degree $g \geqslant 1$ and with positive leading coefficient. Moreover, suppose that not all the values of $h(X)$ are divisible by any fixed prime. This means that the number of roots of $h(X) \equiv 0 (\operatorname{mod} p)$ is strictly less than p for all $p \leqslant g$. Then, for any Y sufficiently large we have*

$$\left|\left\{m \leqslant Y;\ \nu\big(h(m)\big) \leqslant g+1\right\}\right| \gg Y\big(\log Y\big)^{-1} .$$

PROOF. We apply Theorem 25.1 for the sequence which is the characteristic function of the values of $h(X)$ with $x \asymp Y^g$. For this sequence we have no improvement on the level of absolute distribution, so we consider the remainder $R(x, D|N)$

with $N = 1$. But for that we have best possible absolute level $D = Y^{1-\varepsilon}$. The condition (25.25) of Theorem 25.1 is satisfied with $r = g+1$, because $\Delta_{g+1} > g$. □

In similar fashion we establish

THEOREM 25.5. *Let $h(X) \in \mathbb{Z}[X]$ be irreducible of degree $g \geqslant 1$ and with positive leading coefficient. Moreover, suppose that the number of roots of $h(X) \equiv 0 \,(\bmod\, p)$ prime to p is strictly less than $p-1$ for all $p \leqslant g+1$. Then, for any Y sufficiently large we have*

$$\left|\{p \leqslant Y;\ \nu\big(h(p)\big) \leqslant 2g+1\}\right| \gg Y\big(\log Y\big)^{-2} .$$

PROOF. We apply Theorem 25.1 to the characteristic function of the values of $h(X)$ at primes. By the Bombieri–Vinogradov theorem we can take $D = Y^{\frac{1}{2}-\varepsilon}$ and the result follows with $r = 2g+1$ because $\Delta_{2g+1} > 2g$. □

Our next example is the polynomial in two variables $F(X, Y) = X^4 + Y^4$. The sequence $\mathcal{A} = (a_n)$ with a_n being the number of integral representations

$$n = a^4 + b^4 \tag{25.31}$$

has level of distribution $D = D(x) = \sqrt{x}(\log x)^{-A}$ by Theorem 22.16 and it satisfies the linear sieve axioms. Therefore, a straightforward application of the lower-bound linear sieve (Theorem 12.5) yields infinitely many numbers of the form (25.31) which have at most five distinct prime factors and one narrowly misses getting four. However, by the weighted sieve we can achieve three with plenty of room to spare.

THEOREM 25.6. *For the sequence* (25.31) *we have*

$$\sum_{\substack{n \leqslant x \\ (n, P(x^{1/9}))=1 \\ \nu(n) \leqslant 3}} a_n \asymp \sqrt{x}(\log x)^{-1} .$$

PROOF. This follows from Theorem 25.1 because $\Delta_3 > 2$, (see (25.24)). □

Actually, Δ_3 is closer to three, $\Delta_3 = \log 21/\log 3 = 2.77\ldots$, so we are not so far from reaching almost-primes of order two. This goal makes a nice challenge.

25.5. Further Applications

Now we are going to give some examples where we can profit from the bilinear version of the sieve. In this case it pays to use Theorem 25.3, although Theorem 25.1 also gives some interesting results. We will consider only the case of almost-primes of order two.

Our first example is the characteristic function of the integers in the short interval $x - x^\theta < n \leqslant x$ with $\frac{1}{3} < \theta < \frac{3}{5}$. In this case the remainder $R(x, D|N)$ is admissible for

$$D = x^{\theta-\varepsilon} , \quad N = x^{(3\theta-1)/4} , \tag{25.32}$$

by Proposition 13.7 (take $(\alpha, \beta) = \left(\frac{1}{2}, \frac{3}{2}\right)$ in (13.53)). Therefore, the condition (25.27) becomes

$$\frac{1}{\theta} < 3 - \delta\Big(\frac{7\theta-1}{4\theta}\Big) . \tag{25.33}$$

For $\theta = 6/13$ we have $\delta(29/24) = 1.18919\ldots$, so (25.33) holds giving

THEOREM 25.7. *For $x^{6/13} \leqslant y \leqslant x$ with x sufficiently large, we have*

$$\left|\{n;\ x-y<n\leqslant x,\ \nu(n)\leqslant 2\}\right| \gg y\left(\log x\right)^{-1} .$$

For our next application we replace the short interval by the arithmetic progression $m \equiv a \,(\mathrm{mod}\, q)$ where $(a,q)=1$, $m \leqslant x$ with $q \leqslant x^{1-\theta}$, $\frac{1}{3}<\theta<\frac{3}{5}$. In this case the remainder $R(x,\ D|N)$ is admissible for the same levels (25.32), but now by Proposition 13.3. Therefore, Theorem 25.3 yields

THEOREM 25.8. *Let $(a,q)=1$. For $x \geqslant q^{13/7}$ and q sufficiently large we have*

$$\left|\{m\leqslant x;\ m\equiv a \,(\mathrm{mod}\, q),\ \nu(m)\leqslant 2\}\right| \gg x/\varphi(q)\log x .$$

In particular, there is an $m \equiv a\,(\mathrm{mod}\, q)$, $m \leqslant q^{\frac{13}{7}}$, m an almost-prime of order two.

Our next example is the characteristic function of the values of a quadratic polynomial, specifically m^2+1, with integers $1 \leqslant m \leqslant x^{\frac{1}{2}}$. In this case the remainder $R(x,\ D|N)$ is admissible for

$$D = x^{\frac{1}{2}-\varepsilon}\ , \quad N = x^{\frac{1}{20}}\ , \tag{25.34}$$

by Theorem 22.7.

Now, for $r=2$, either Theorem 25.1 or Theorem 25.3 applies. Indeed, the conditions (25.25) hold because

$$\Delta_2 = 2-\log(\tfrac{6}{5})/\log 3 > \tfrac{20}{11}\ ,$$

while the condition (25.27) holds because $\delta(7/6)<1$. Therefore, either theorem yields

THEOREM 25.9. *For $x \geqslant 2$ we have*

$$\left|\{m\leqslant \sqrt{x};\ \nu(m^2+1)\leqslant 2\}\right| \gg \sqrt{x}/\log x .$$

Consequently, the polynomial X^2+1 represents infinitely often numbers having at most two prime factors. If two, these are necessarily distinct because $p^2=m^2+1$ is not possible.

25.6. Twin Almost-Primes

Recall the hypothetical formula for twin primes, the number of $p \leqslant x$ such $p-2$ is prime:

$$\pi_2(x) \sim Bx(\log x)^{-2}\ ,$$

where B is the twin prime constant, a special case of (3.6):

$$B = 2\prod_{p>2}(1-(p-1)^{-2})\ .$$

Using the well-factorable level of distribution given by Theorem 22.4 and the linear sieve in the form of Theorem 12.20, we immediately obtain the upper bound

$$\pi_2(x) \leqslant (\tfrac{7}{2}+\varepsilon)Bx(\log x)^{-2} \tag{25.35}$$

for all x sufficiently large in terms of ε.

In this section we treat the twin-prime type lower bound problem in a straightforward fashion (without weights) to re-establish, in particular, the well-known result of Chen:

THEOREM 25.10. *There are infinitely many primes p such that $p-2$ is almost-prime of order two.*

Actually, Chen's weighted sieve is capable of producing primes p such that $p-2$ has at most two prime divisors each of which is larger than $p^{1/8}$. We shall take a step closer to genuine twin primes by requiring these two prime divisors to be larger than $p^{3/11}$. We shall also consider the more general statement replacing $p-2$ by $ap-b$. Such results have nice applications, for example, to the Artin conjecture for primitive roots (and related questions involving Euclidean rings and elliptic curves) where it is important that $\frac{3}{11} > \frac{1}{4}$ (see [**78**], [**87**] and [**24**]). In fact, there are many other applications due, for example, to the connection of $2p+1$ or $(p-1)/2$ with Fermat's little theorem. Our precise result is the following.

THEOREM 25.11. *Let $a \geqslant 1$, $c \geqslant 1$ and $b \neq 0$ be pairwise coprime integers with $2|abc$. Then, for all x sufficiently large in terms of abc, we have*

$$|\{p \leqslant x;\ ap+b = cp''\}| \geqslant \frac{W(abc)}{31c} \frac{Bx}{(\log x)^2}\,, \tag{25.36}$$

where p'' denotes a number having at most two prime factors, each one larger than $x^{3/11}$, and

$$W(d) = \prod_{\substack{p|d\\ p>2}} \Big(1 - \frac{1}{p-1}\Big)^{-1}.$$

REMARK. Actually, the result holds uniformly for $|abc| \leqslant (\log x)^A$ where A is any positive constant provided x is sufficiently large in terms of A. We do not elaborate on this since, in the theorem to follow this one, we further relax the constraints on the size of the coefficients, although at the cost of weakening somewhat the exponent $3/11$.

Let $\pi_2(x,z)$ denote the number of primes $p \leqslant x$ such that $p-2 = p_1p_2$ with $p_1 \geqslant p_2 \geqslant z$. Theorem 25.11 yields:

COROLLARY 25.12. *For all x sufficiently large*

$$\pi_2(x) + \pi_2(x, x^{\frac{3}{11}}) > \tfrac{1}{31} Bx(\log x)^{-2}. \tag{25.37}$$

Of course, upper bounds of the same order of magnitude for these follow at once from Brun's sieve.

PROOF. To begin the proof of the theorem we let $\mathcal{A} = (a_n)$ be the sequence of numbers $n = (ap+b)/c$ with $p \leqslant x$ and

$$ap + b \equiv c \,(\mathrm{mod}\ 2c)\ .$$

Note that this congruence condition for p means the same as

$$p \equiv (c-b)\bar{a} \,(\mathrm{mod}\ 2c/(2,a))\ ,$$

where the residue class is coprime with the modulus. Note also that the elements of $\mathcal{A}$ are prime to $2ab$, therefore our sifting set of primes will be all primes which do not divide $2ab$. The sieve density function $g(d)$ at primes is given by $g(p) = 1/(p-1)$ if $p \nmid c$ and $g(p) = 1/p$ if $p \mid c$.

By Theorem 22.4 our sequence has well-factorable level of distribution $D = x^{4/7-\varepsilon}$. Therefore, applying the linear sieve inequality Theorem 12.19 we get

$$S(\mathcal{A}, z) > \Big(f\Big(\frac{4s}{7}\Big) - \varepsilon\Big)|\mathcal{A}|V(z)(2,c)\frac{\varphi(c)}{c}W(abc)$$

for $z = x^{1/s}$ where

$$|\mathcal{A}| = \pi\Big(x; \frac{2c}{(2,a)}, (c-b)\bar{a}\Big) \sim \frac{1}{\varphi\left(2c/(2,a)\right)}\frac{x}{\log x}$$

and

$$V(z) = \prod_{2<p<z}\Big(1 - \frac{1}{p-1}\Big) = e^{-\gamma}B(\log z)^{-1}.$$

Assuming $3 < s \leqslant 4$, the sifting function $S(\mathcal{A}, z)$ catches numbers $(ap+b)/c$ having at most three prime divisors, each of which is $\geqslant z$. We just do not want those with exactly three prime divisors. Thus we need to subtract from $S(\mathcal{A}, z)$ the number of prime solutions to $ap + b = cp_1p_2p_3$. We interpret this equation as $p = (cp_1p_2p_3 - b)a^{-1}$ and let $\mathcal{B} = (b_m)$ be the characteristic function of the sequence of numbers

$$m = (cp_1p_2p_3 - b)a^{-1} \leqslant x\ ,$$

formed as p_1, p_2, p_3 run through primes satisfying

$$p_1 \geqslant p_2 \geqslant p_3 \geqslant z, \quad cp_1p_2p_3 - b \equiv a\,(\mathrm{mod}\ 2a)\ .$$

This means that $p_1p_2p_3 \leqslant (ax+b)c^{-1}$ and

$$p_1p_2p_3 \equiv (a+b)\bar{c}\,(\mathrm{mod}\ 2a/(2,c))\ .$$

Note that the elements of $\mathcal{B}$ are prime to $2bc$, therefore the sifting set of primes will be all primes which do not divide $2bc$. The sieve density function $g(d)$ at odd primes is given by $g(p) = 1/(p-1)$ if $p \nmid a$ and $g(p) = 1/p$ if $p|a$. After this switch the number of unwanted solutions is estimated by $S(\mathcal{B}, z)$. Note that the reversed sequence $\mathcal{B}$ has fewer elements than $\mathcal{A}$ since z is near $x^{1/3}$ (more precisely $|\mathcal{B}| < \frac{1}{12}|\mathcal{A}|$ for $s = 11/3$), yet it has the same well-factorable level of distribution as $\mathcal{A}$, that is, $D = x^{4/7-\varepsilon}$, due to the great generality of Theorem 22.4. Therefore, applying the linear sieve inequality of Theorem 12.20, we get

$$S(\mathcal{B}, z) < \Big(F\Big(\frac{4s}{7}\Big) + \varepsilon\Big)|\mathcal{B}|V(z)(2,a)\frac{\varphi(a)}{a}W(abc).$$

It remains to estimate the cardinality $|\mathcal{B}|$. By the Prime Number Theorem,

$$|\mathcal{B}| \sim \frac{1}{6}\frac{a}{c\varphi(2a/(2,c))}\sum_{\substack{p_2,p_3\geqslant z\\ p_2p_3\leqslant x}}\sum\frac{x}{p_2p_3}\Big(\log\frac{2x}{p_2p_3}\Big)^{-1} \sim \frac{aG(s)}{c\varphi(2a/(2,c))}\frac{x}{\log x}$$

where

$$G(s) = \frac{1}{6}\iint\limits_{\substack{\alpha,\beta>1/s\\ \alpha+\beta<1-1/s}}\frac{d\alpha d\beta}{\alpha\beta(1-\alpha-\beta)} = \frac{s}{6}\iint\limits_{\substack{\alpha,\beta>1\\ \alpha+\beta<s-1}}\frac{d\alpha d\beta}{\alpha\beta(s-\alpha-\beta)}.$$

Since $\alpha\beta(s-\alpha-\beta) \geqslant s-2$ and the area of the triangular domain of integration is $\frac{1}{2}(s-3)^2$, we get easily a quite good upper bound

$$G(s) < \frac{s(s-3)^2}{12(s-2)}\ . \tag{25.38}$$

Combining the above estimates, we find that the left-hand side of (25.36) is bounded below by

$$\begin{aligned} S(\mathcal{A},z)-S(\mathcal{B},z) &> \Big\{f\Big(\frac{4s}{7}\Big)-F\Big(\frac{4s}{7}\Big)G(s)+o(1)\Big\}\,\frac{W(abc)}{c}\,\frac{x}{\log x}\,\frac{e^{-\gamma}B}{\log z}\\ &\sim \frac{7}{2}\Big\{\log\Big(\frac{4s}{7}-1\Big)-G(s)\Big\}\,\frac{W(abc)}{c}\,\frac{Bx}{(\log x)^2}\ . \end{aligned}$$

Here we used, in addition to Mertens' formula, the identity

$$\prod_{\substack{p|c\\ p>2}}\Big(1-\frac{1}{p}\Big)=(2,c)\,\frac{\varphi(c)}{c}=\frac{1}{c}\,\varphi\Big(\frac{2c}{(2,a)}\Big)$$

and the same identity with a and c interchanged. For $s=11/3$ we compute that the above constant is greater than $1/31$. This completes the proof of Theorem 25.11. □

In some problems one needs numbers $ap+b$ which have at most two prime divisors and where a and b are not so small. To this end, we apply Theorem 22.5 and the weighted sieve.

THEOREM 25.13. *Let η be a sufficiently small positive absolute constant. Let $A>0$ and $0<a|b|\leqslant x^{\eta}$, $1\leqslant c\leqslant(\log x)^A$ with a, b, c, pairwise coprime integers such that $2|abc$. Then, for all x sufficiently large in terms of A we have*

$$|\{p\leqslant x;\ ap+b=cp''\}|\gg\frac{W(abc)}{c}\,\frac{x}{(\log x)^2}\ , \tag{25.39}$$

where p'' denotes a number having at most two prime factors, each larger than $x^{1/8}$.

PROOF. According to Theorem 22.5 our sequence $\mathcal{A}=(a_n)$ has well-factorable level of distribution $x^{4/7-2\eta}$. Thus, Theorem 25.1 applies with $r=2$, giving the result. □

We remark that we needed a direct application of the weighted sieve to compensate for the fact that the switching technique is not available due to the size of a. Given GRH it would be, but in some other examples the switching technique is not applicable at all. For example, we can apply Theorem 22.4 and the weighted sieve of Theorem 25.1 to prove:

THEOREM 25.14. *Let $\mathbf{u}=[u_1,\dots,u_r]$ be positive reals with $u_1+\ldots+u_r=1$. Let $\pi_2(x;\mathbf{u})$ denote the number of products $p^{(r)}=p_1\dots p_r$ with $x^{u_j}<p_j\leqslant 2x^{u_j}$ such that*

$$p^{(r)}+2=p''\ ,$$

where p'' denotes a number having at most two prime factors, each one larger than $x^{1/8}$. Then, for all x sufficiently large we have

$$\pi_2(x;\mathbf{u})\asymp x(\log x)^{-r-1}\ ,$$

where the implied constant depends only on $\mathbf{u}$.

APPENDIX A

Mean Values of Arithmetic Functions

Backstage

A.1. Simple Estimates

In sieve theory one encounters a variety of sums of multiplicative (and other arithmetic) functions over special integers having prime divisors in the sifting range. In this chapter we deliver a few elementary estimates for the mean value

$$\mathcal{M}_f(x) = \sum_{m \leqslant x} f(m) \ , \tag{A.1}$$

where m runs over all positive integers.

Throughout this section f is a non-negative multiplicative function supported on squarefree numbers. If P is the sifting range the additional summation condition $m|P$ can be implemented by assuming that f is supported on the divisors of P.

Together with f, we consider another multiplicative function,

$$g(m) = f(m)m^{-1} \ . \tag{A.2}$$

The following estimates need no explanation:

$$\mathcal{M}_f(x) \leqslant x\mathcal{M}_g(x) \leqslant x \prod_{p \leqslant x} \Big(1 + \frac{f(p)}{p}\Big) \ . \tag{A.3}$$

If $f(p) \geqslant 1$ one can do slightly better. Write $f = 1 * h$ with $h = \mu * f$ by Möbius inversion. Since $h(p) = f(p) - 1 \geqslant 0$ we infer

$$\mathcal{M}_f(x) = \sum_{mn \leqslant x} h(m) \leqslant x\mathcal{M}_h(x);$$

hence by (A.3) we get

$$\mathcal{M}_f(x) \leqslant x \prod_{p \leqslant x} \Big(1 + \frac{f(p) - 1}{p}\Big) \ . \tag{A.4}$$

These two results can be combined. Suppose $f(p) \geqslant 1$ for $p|P$. Then we have

$$\mathcal{M}_f(x) \leqslant x \prod_{\substack{p \leqslant x \\ p \nmid P}} \Big(1 + \frac{f(p)}{p}\Big) \prod_{\substack{p \leqslant x \\ p | P}} \Big(1 + \frac{f(p) - 1}{p}\Big) \ . \tag{A.5}$$

Next, in place of the assumption $f(p) \geqslant 1$, suppose that

$$\sum_{y < p \leqslant x} g(p) \log p \leqslant a \log \frac{x}{y} + b \tag{A.6}$$

for all $2 \leqslant y \leqslant x$ where $a \geqslant 0, b \geqslant 1$ are constants. By partial summation we infer

$$\sum_{p \leqslant x} f(p) \log p \leqslant cx \ , \tag{A.7}$$

where $c = a + b$. Then, using Tchebyshev's ideas we obtain

$$\sum_{m \leqslant x} f(m) \log m = \sum_{np \leqslant x} f(np) \log p \leqslant cx\mathcal{M}_g(x).$$

Hence, again by partial summation, $\mathcal{M}_f(x) \leqslant 1 + cL(x)\mathcal{M}_g(x)$, where

$$L(x) = \int_2^x (\log t)^{-1} dt < 2x(\log x)^{-1} \ .$$

Therefore,

$$\mathcal{M}_f(x) < \frac{3cx}{\log x}\mathcal{M}_g(x) \leqslant \frac{3cx}{\log x} \prod_{p \leqslant x} \Big(1 + \frac{f(p)}{p}\Big) \ . \tag{A.8}$$

This bound is as strong as (A.4) apart from a constant factor. Moreover, in order for it to hold we do not require the assumption $f(p) \geqslant 1$ but rather (A.6), which is less demanding in practice.

Let $z \geqslant 2$ and now let $P(z)$ be the product of all primes $p < z$. We are next going to consider the restricted sum

$$\mathcal{M}_f(x, z) = \sum_{\substack{m \leqslant x \\ m | P(z)}} f(m) \ . \tag{A.9}$$

In other words, we sum over all integers free of large prime factors. This extra restriction does not ruin the condition (A.6) so (A.8) becomes

$$\mathcal{M}_f(x) \leqslant \frac{3cx}{\log x} \prod_{p < z} \Big(1 + \frac{f(p)}{p}\Big) \ . \tag{A.10}$$

However, if x is much larger than z in the logarithmic scale, that is, if $s = \log x / \log z$ is large, we can do better than (A.10).

To this end we use "Rankin's trick", that is, we proceed as follows. Let $\varepsilon \geqslant 0$ so we have

$$\mathcal{M}_f(x, z) - \mathcal{M}_f(\sqrt{x}, z) \leqslant \sum_{\substack{m \leqslant x \\ m | P(z)}} f(m) \Big(\frac{m}{\sqrt{x}}\Big)^\varepsilon \ .$$

The multiplicative function $f(m)m^\varepsilon$, when restricted to the divisors of $P(z)$, satisfies (A.6) with a, b replaced by az^ε, bz^ε respectively. Hence, (A.10) yields

$$\mathcal{M}_f(x, z) - \mathcal{M}_f(\sqrt{x}, z) \leqslant \Big(\frac{z}{\sqrt{x}}\Big)^\varepsilon \frac{3cx}{\log x} \prod_{p < z} \Big(1 + \frac{f(p)}{p} p^\varepsilon\Big) \ .$$

We choose $\varepsilon = 2 \log C / \log z$ where $C \geqslant 1$ is a constant. Then, we have $(z/\sqrt{x})^\varepsilon = C^{2-s}$ and, for every $p < z$,

$$1 + \frac{f(p)}{p} p^\varepsilon = 1 + \frac{f(p)}{p} \Big(1 + O\Big(\frac{\log p}{\log z}\Big)\Big) \ ,$$

where the implied constant depends only on C. Hence,

$$\prod_{p<z}\Big(1+\frac{f(p)}{p}p^{\varepsilon}\Big) \ll \prod_{p<z}\Big(1+\frac{f(p)}{p}\Big)$$

by (A.6). We obtain

$$\mathcal{M}_f(x,z)-\mathcal{M}_f(\sqrt{x},z) \ll C^{-s}\frac{x}{\log x}\prod_{p<z}\Big(1+\frac{f(p)}{p}\Big)\,,$$

where the implied constant depends only on a, b, and C. Applying (A.10) for $\mathcal{M}_f(\sqrt{x})$, we conclude:

THEOREM A.1. *Let f be a non-negative multiplicative function, supported on squarefree numbers and satisfying*

$$\sum_{w\leqslant p<z} f(p)\frac{\log p}{p} \leqslant a\log\frac{z}{w}+b\,,$$

for all w with $2\leqslant w<z$, where $a\geqslant 0$, $b\geqslant 1$ are constants. Let $C\geqslant 1$. Then, for $z\geqslant C^2$ and $x\geqslant 2$ we have

$$\mathcal{M}_f(x,z) \ll C^{-s}\frac{x}{\log x}\prod_{p<z}\Big(1+\frac{f(p)}{p}\Big)\,, \tag{A.11}$$

where $s=\log x/\log z$ and the implied constant depends only on a, b and C.

We shall also need a bound for a similar sum with mixed restrictions, specifically

$$\mathcal{M}_f(x,y,z) = \sum\sum_{\substack{mn\leqslant x\\ m|P(z),\, m\geqslant y\\ (n,P(z))=1}} f(mn)\,. \tag{A.12}$$

Note that

$$\mathcal{M}_f(x,y,z) \leqslant \sum_{\substack{n\leqslant xy^{-1}\\ (n,P(z))=1}} f(n)\mathcal{M}_f\Big(\frac{x}{n},z\Big)\,.$$

By (A.11) we get

$$\mathcal{M}_f\Big(\frac{x}{n},z\Big) \ll C^{-t}\frac{x}{n\log y}\prod_{p<z}\Big(1+\frac{f(p)}{p}\Big)\,,$$

where $t=\log y/\log z$. Moreover, we have

$$\sum_{\substack{n\leqslant xy^{-1}\\ (n,P(z))=1}} \frac{f(n)}{n} \leqslant \prod_{z\leqslant p\leqslant xy^{-1}}\Big(1+\frac{f(p)}{p}\Big)\,.$$

Hence, we conclude:

COROLLARY A.2. *Let f satisfy the conditions of Theorem A.1. Let $C\geqslant 1$. Then, for $z\geqslant C^2$ and $x\geqslant y\geqslant 2$ we have*

$$\mathcal{M}_f(x,y,z) \ll C^{-t}\frac{x}{\log y}\prod_{p<z}\Big(1+\frac{f(p)}{p}\Big)\,, \tag{A.13}$$

where $t=\log y/\log z$ and the implied constant depends only on a, b and C.

REMARK. In the derivation of (A.13) we tacitly assumed that $z \leqslant x$, but this is not necessary because $\mathcal{M}_f(x, y, z)$ does not change when z increases beyond x.

The Rankin trick also works nicely to produce a lower bound for the sum $\mathcal{M}_g(x)$.

THEOREM A.3. *Let $g(m)$ be a multiplicative function supported on squarefree numbers and satisfying* (A.6) *with some $0 \leqslant a < 1$, $b \geqslant 1$. Then we have*

$$\mathcal{M}_g(x) \gg \prod_{p \leqslant x} \left(1 + g(p)\right) ,$$

where the implied constant depends only on a and b.

PROOF. For technical reasons we assume $a > 0$; the result for $a = 0$ is obvious. Put

$$\mathcal{N}_g(x) = \sum_{\substack{m > x \\ m | P(x)}} g(m)$$

so that

$$\mathcal{M}_g(x) + \mathcal{N}_g(x) = \prod_{p \leqslant x} \left(1 + g(p)\right) = W_g(x), \text{ say} .$$

We need an upper bound for $\mathcal{N}_g(x)$. By Rankin's trick,

$$\mathcal{N}_g(x) \leqslant x^{-\varepsilon} \sum_{m | P(x)} g(m) m^{\varepsilon} = x^{-\varepsilon} \prod_{p \leqslant x} \left(1 + g(p) p^{\varepsilon}\right) .$$

Hence,

$$\begin{aligned} \frac{\mathcal{N}_g(x)}{W_g(x)} &\leqslant x^{-\varepsilon} \prod_{p \leqslant x} \Big(1 + \frac{g(p)}{1 + g(p)} (p^{\varepsilon} - 1)\Big) \\ &\leqslant x^{-\varepsilon} \exp\Big(\sum_{p \leqslant x} \frac{g(p)}{1 + g(p)} (p^{\varepsilon} - 1)\Big) . \end{aligned}$$

Let $S(x, y)$ be the sum on the left-hand side of (A.6). By (A.6), using partial summation we obtain

$$\begin{aligned} \sum_{p \leqslant x} g(p)(p^{\varepsilon} - 1) &= - \int_1^x \frac{(y^{\varepsilon} - 1)}{\log y} \, d\left(S(x, y)\right) \\ &= \varepsilon \, S(x, 1) + \int_1^x S(x, y) \, d\left(\frac{y^{\varepsilon} - 1}{\log y}\right) \\ &\leqslant \varepsilon \left(a \log x + b\right) + \int_1^x \left(a \log \frac{x}{y} + b\right) d\left(\frac{y^{\varepsilon} - 1}{\log y}\right) \\ &= a \int_0^1 \frac{x^{\varepsilon u} - 1}{u} du + b \, \frac{x^{\varepsilon} - 1}{\log x} \leqslant \left(a + \frac{b}{\log x}\right)\left(x^{\varepsilon} - 1\right) . \end{aligned}$$

Choosing $\varepsilon = \alpha / \log x$ we get

$$\begin{aligned} \frac{\mathcal{N}_g(x)}{W_g(x)} &\leqslant \exp\Big(a(e^{\alpha} - 1) - \alpha + O\Big(\frac{b}{\log x}\Big)\Big) \\ &= a e^{(1-a)} \Big(1 + O\Big(\frac{b}{\log x}\Big)\Big) \end{aligned}$$

if we choose $\alpha = -\log a$, which is positive since $a < 1$. Hence

$$\mathcal{M}_g(x) \geqslant \Big(1 - ae^{1-a} + O\Big(\frac{1}{\log x}\Big)\Big) W_g(x) \ ,$$

completing the proof. □

THEOREM A.4. *Let g satisfy the conditions of Theorem A.3 and suppose that $f(p) = g(p)p$ satisfies*

$$\sum_{p \leqslant y} f(p) \log p \gg y \tag{A.14}$$

for all sufficiently large y. Then, for all $x \geqslant 2$,

$$\mathcal{M}_f(x) \gg \frac{x}{\log x} \prod_{p \leqslant x} \Big(1 + \frac{f(p)}{p}\Big) .$$

PROOF. We have

$$\begin{aligned} \log x \sum_{n \leqslant x} f(n) \geqslant \sum_{n \leqslant x} f(n) \log n &= \sum\sum_{mp \leqslant x} f(mp) \log p \\ &\geqslant \sum_{m \leqslant \sqrt{x}} f(m) \sum_{p \leqslant x/m} f(p) \log p \end{aligned}$$

so, using (A.14) and Theorem A.3, this is seen to be

$$\gg x \sum_{m \leqslant \sqrt{x}} \frac{f(m)}{m} \gg x \prod_{p \leqslant \sqrt{x}} \Big(1 + \frac{f(p)}{p}\Big) \asymp \prod_{p \leqslant x} \Big(1 + \frac{f(p)}{p}\Big) .$$

□

A.2. Asymptotic Formulas for Full Sums

In order to establish an asymptotic formula for $\mathcal{M}_g(x)$ we require a condition to be placed on the distribution of g at primes, which is somewhat stronger than (A.6), namely that

$$\sum_{p \leqslant x} g(p) \log p = k \log x + \delta(x) \tag{A.15}$$

where $\delta(x)$ is bounded for all $x \geqslant 2$. We do not need any specific property of $\delta(x)$ except that $\delta(x) \ll 1$; nevertheless, having introduced this we are able to state a nice exact formula, see (A.22). For many functions that we meet in practice this condition can be established by elementary methods (it is weaker than the Prime Number Theorem). With some applications in mind we allow $g(p)$ to be negative but not very much so. Precisely, we assume that (A.15) holds with $k > -\frac{1}{2}$. Moreover, we need two minor estimates:

$$\prod_{w \leqslant p < z} (1 + |g(p)|) \ll \Big(\frac{\log z}{\log w}\Big)^{|k|} \qquad \text{if } z > w \geqslant 2 \tag{A.16}$$

and

$$\sum_p g(p)^2 \log p < \infty. \tag{A.17}$$

Our arguments are modeled on E. Wirsing [**160**]; they begin by consideration of the logarithmically smoothed mean value

$$\sum_{m\leqslant x} g(m)\log\frac{x}{m} = \int_1^x \mathcal{M}_g(t)t^{-1}dt.$$

As before we have (Tchebyshev's ideas)

$$\begin{aligned}\sum_{m\leqslant x} g(m)\log m &= \sum_{np\leqslant x} g(np)\log p\\ &= \sum_{n\leqslant x} g(n)\sum_{p\leqslant x/n} g(p)\log p - \sum_{np^2\leqslant x} g(np)g(p)\log p\\ &= \sum_{n\leqslant x} g(n)\left\{k\log\frac{x}{n}+\delta\left(\frac{x}{n}\right)\right\} - \sum_{np^2\leqslant x} g(np)g(p)\log p.\end{aligned}$$

Hence,

$$(k+1)\sum_{m\leqslant x} g(m)\log m = k\mathcal{M}_g(x)\log x + \Delta_g(x)\ , \tag{A.18}$$

where

$$\Delta_g(x) = \sum_{n\leqslant x} g(n)\delta\left(\frac{x}{n}\right) - \sum_{np^2\leqslant x} g(np)g(p)\log p.$$

For any $x\geqslant 2$ we have by (A.16),

$$\Delta_g(x) \ll \mathcal{M}_{|g|}(x)\Big(1+\sum_p g(p)^2\log p\Big) \ll (\log x)^{|k|}. \tag{A.19}$$

We write (A.18) as follows:

$$\mathcal{M}_g(x)\log x - (k+1)\int_1^x \mathcal{M}_g(t)t^{-1}dt = \Delta_g(x).$$

Here, we extract the contribution of the integral over $1\leqslant t\leqslant 2$, which is $(k+1)\log 2$, and putting this on the right side we get for $x\geqslant 2$,

$$\mathcal{M}_g(x)\log x - (k+1)\int_2^x \mathcal{M}_g(t)t^{-1}dt = \Delta_g^*(x) \tag{A.20}$$

where $\Delta_g^*(x) = \Delta_g(x) + (k+1)\log 2$. Next we divide by $x(\log x)^{k+2}$ and integrate:

$$\begin{aligned}\int_2^x \mathcal{M}_g(t)t^{-1}(\log t)^{-k-1}dt - (k+1)\int_2^x t^{-1}(\log t)^{-\kappa-2}\int_2^t \mathcal{M}_g(u)u^{-1}du\\ = \int_2^x \Delta_g^*(t)t^{-1}(\log t)^{-k-2}dt.\end{aligned}$$

Changing the order of integration we find that the left-hand side is equal to

$$(\log x)^{-k-1}\int_2^x \mathcal{M}_g(t)t^{-1}dt.$$

Combining this result with (A.20) we arrive at the following identity:

$$\mathcal{M}_g(x) = -(\log x)^k\int_2^x \Delta_g^*(t)d(\log t)^{-k-1} + \Delta_g^*(x)(\log x)^{-1}. \tag{A.21}$$

Since $k > -\frac{1}{2}$, the above integral converges by virtue of (A.19). Extending the integration to infinity we obtain another identity:

$$\mathcal{M}_g(x) = \{c_g + \gamma_g(x)\}(\log x)^k \tag{A.22}$$

for $x \geqslant 2$, where c_g is the constant

$$c_g = -\int_2^\infty \Delta_g^*(t)d(\log t)^{-k-1} \tag{A.23}$$

and $\gamma_g(x)$ is the error term given by

$$\gamma_g(x) = \int_x^\infty (\Delta_g(t) - \Delta_g(x))d(\log t)^{-k-1}.$$

By (A.19) we have

$$\gamma_g(x) \ll (\log x)^{|k|-k-1}.$$

Here the exponent is negative, so (A.22) becomes an asymptotic formula for $\mathcal{M}_g(x)$. However, the constant c_g as given by (A.23) is not appealing. We shall compute c_g in another way.

Consider the zeta-function formed by g:

$$\zeta_g(s) = \sum_{m=1}^\infty g(m)m^{-s}.$$

The series converges absolutely for $s > 0$. We compute by partial summation, using (A.22), that

$$\begin{aligned}\zeta_g(s) &= \int_1^\infty x^{-s}d\mathcal{M}_g(x) = -\int_1^\infty \mathcal{M}_g(x)dx^{-s}\\ &= -\int_0^\infty \mathcal{M}_g(e^t)de^{-st} = -\int_0^\infty (c_g + O(t^{-\varepsilon}))t^k de^{-st}\\ &= (c_g + O(s^\varepsilon))s^{-k}\Gamma(k+1) \qquad \text{as } s \to 0+ .\end{aligned}$$

Comparing this with the k-th power of the Riemann zeta-function, we find

$$\zeta(s+1)^{-k}\zeta_g(s) \sim c_g\Gamma(k+1).$$

On the other hand, we have the product over primes:

$$\zeta(s+1)^{-k}\zeta_g(s) = \prod_p (1 - p^{-s-1})^k(1 + g(p)p^{-s})$$

which converges absolutely for $s > 0$ and has a limit as $s \to 0$ by virtue of assumption (A.15). Hence, the constant c_g is given by

$$c_g = \frac{1}{\Gamma(k+1)}\prod_p \Big(1 - \frac{1}{p}\Big)^k (1 + g(p)). \tag{A.24}$$

Thus, we have established the following result.

THEOREM A.5. *Suppose g is a multiplicative function supported on squarefree numbers which satisfies* (A.15) *with $k > -\frac{1}{2}$ and $\delta(x)$ bounded. Assume that* (A.16) *and* (A.17) *also hold. Then, the mean value of g satisfies the asymptotic formula*

$$\mathcal{M}_g(x) = c_g(\log x)^k + O((\log x)^{|k|-1}) \tag{A.25}$$

if $x \geqslant 2$, where c_g is given by (A.24). *The implied constant depends only on k and the constants implied in* (A.15)–(A.17).

For use in Section 7.13 we needed the following corollary of Theorem A.5. Let Δ be a positive integer and suppose $g(m)$ is a multiplicative function, supported on squarefree numbers, such that

$$g(p) = \frac{\kappa}{p} + O\left(\frac{1}{p^2}\right), \qquad \kappa \geqslant 1\,. \tag{A.26}$$

Then, the conditions of Theorem A.5 are satisfied, giving

$$\sum_{\substack{m\leqslant x\\(m,\Delta)=1}} g(m) = \frac{c}{\Gamma(\kappa+1)}(\log x)^\kappa + O\Big((\log x)^{\kappa-1}\Big) \tag{A.27}$$

where

$$c = \prod_{p\nmid\Delta}\Big(1-\frac{1}{p}\Big)^\kappa(1+g(p))\prod_{p\mid\Delta}\Big(1-\frac{1}{p}\Big)^\kappa, \tag{A.28}$$

and the implied constant depends only on Δ and on the one in (A.26). Hence, by partial summation we derive the following:

COROLLARY A.6. *Let κ and ℓ be positive integers and assume g satisfies* (A.26). *Denote,*

$$S_\ell(x) = \sideset{}{^\flat}\sum_{\substack{m\leqslant x\\(m,\Delta)=1}} g(m)\left(\log\frac{x}{m}\right)^\ell .$$

Then, for $x \geqslant 2$,

$$S_\ell(x) = c\,\frac{\ell!}{(\ell+\kappa)!}(\log x)^{\ell+\kappa}\Big(1+O\Big(\frac{1}{\log x}\Big)\Big), \tag{A.29}$$

where the implied constant depends only on k, ℓ, Δ and on the one in (A.26).

A.3. Asymptotic Formulas for Restricted Sums

Sieve theory demands asymptotics for the mean values of multiplicative functions over integers free from large prime divisors. In this section we derive the required asymptotics from the results for the unrestricted mean values given in Theorem A.5.

Let $P(z)$ be the product of all primes $< z$ and

$$\mathcal{M}_g(x,z) = \sum_{\substack{m\leqslant x\\m\mid P(z)}} g(m). \tag{A.30}$$

Throughout this section, we assume that g satisfies the hypotheses of Theorem A.5 and, to simplify notation, we do not carry the subscript g in $\mathcal{M}_g$ or c_g. For $z \geqslant x \geqslant 2$ we have

$$\mathcal{M}(x,z) = \mathcal{M}(x) = c(\log x)^k + O((\log x)^{|k|-1}). \tag{A.31}$$

For $2 \leqslant z < x$ we have the following recurrence formula:

$$\mathcal{M}(x,z) = \mathcal{M}(x) - \sum_{z\leqslant p<x} g(p)\mathcal{M}(x/p,p). \tag{A.32}$$

To prove this, write every m in $\mathcal{M}(x)$ which is not in $\mathcal{M}(x,z)$ as $m = np$ where p is the largest prime divisor of m with $z \leqslant p < x$. Such a factorization is unique and it is also determined by the condition $n|P(p)$. By this recurrence formula, we can extend (A.31) to any level $z = x^{1/s}$ with $s > 0$ by induction on [s].

We postulate the following formula:

$$\mathcal{M}(x,z) = cm(s)(\log x)^k + O((\log x)^{|k|-1}) , \tag{A.33}$$

for all $x, z \geqslant 2$, where $m(s)$ is a suitable continuous function of $s = \log x/\log z$, to be determined. We know this holds with

$$m(s) = 1, \qquad \text{if} \quad 0 < s \leqslant 1. \tag{A.34}$$

Suppose $s > 1$. Inserting (A.33) into (A.31) we get

$$\mathcal{M}(x,z) = c(\log x)^k - c \sum_{z \leqslant p < x/2} g(p)m(s_p)\Big(\log \frac{x}{p}\Big)^k$$
$$+ O\left((\log x)^{|k|-1} + \sum_{z \leqslant p < x} |g(p)|\Big(\log \frac{2x}{p}\Big)^{|k|-1}\right) ,$$

where $s_p = \log x/\log p - 1$. Hence, by (A.15) and (A.16), using partial summation we derive

$$\mathcal{M}(x,z) = c\left(1 + \int_1^s m(t-1)dt^{-k}\right)(\log x)^k + O\big((\log x)^{|k|-1}\big) .$$

Therefore, (A.33) holds with $m(s)$ given by

$$m(s) = 1 + \int_1^s m(t-1)dt^{-k}, \qquad \text{if} \quad s \geqslant 1. \tag{A.35}$$

Differentiating (A.35) we get

$$s^{k+1}m'(s) = -k(s-1)^k m(s-1). \tag{A.36}$$

Putting $\mathfrak{m}(s) = s^k m(s)$ this equation becomes

$$s\mathfrak{m}'(s) = k\mathfrak{m}(s) - k\mathfrak{m}(s-1), \qquad \text{if} \quad s > 1. \tag{A.37}$$

The asymptotic behaviour of this function will be developed in Section B.2. As a result of this we have

$$\mathfrak{m}(s) = \mathfrak{m}(\infty) + O(s^{-s}), \tag{A.38}$$

where

$$\mathfrak{m}(\infty) = e^{\gamma k}\Gamma(k+1). \tag{A.39}$$

We shall state the asymptotic formula (A.33) in a form harmonized with sieve theory, that is, we express the main term by the product

$$W(z) = \prod_{p<z} (1 + g(p)). \tag{A.40}$$

We have trivially

$$W(z) = \prod_{p<z} \Big(1 - \frac{1}{p}\Big)^{-k} \prod_{p<z} \Big(1 - \frac{1}{p}\Big)^k (1 + g(p)) ,$$

and hence, by Mertens' formula (2.21) and the hypothesis (A.15), we find that

$$W(z) = e^{\gamma k}\Gamma(k+1)c_g(\log z)^k\{1 + O((\log z)^{-1})\} , \tag{A.41}$$

where c_g is the constant (A.24).

Combining (A.41) with (A.33) we conclude the following result.

THEOREM A.7. *Let g be a multiplicative function which satisfies* (A.15) *with $k > -\frac{1}{2}$ and $\delta(x)$ bounded. Assume that* (A.16) *and* (A.17) *also hold. Let $P(z)$ be the product of all primes $< z$. Then*

$$\mathcal{M}_g(x,z) = W(z)\{\mathfrak{f}(s) + O((\log x)^{-\eta})\}\ , \tag{A.42}$$

where $W(z)$ is the product (A.40), *$s = \log x/\log z$, $\mathfrak{f}(s)$ is the continuous solution to the differential-difference problem:*

$$\begin{aligned} s^{-k}\mathfrak{f}(s) &= e^{-\gamma k}\Gamma(k+1)^{-1}, \quad \text{if } 0 < s \leqslant 1,\\ s\mathfrak{f}'(s) &= k\mathfrak{f}(s) - k\mathfrak{f}(s-1), \quad \text{if } s > 1, \end{aligned} \tag{A.43}$$

and $\eta = 1 + k - |k|$. The implied constant depends on s.

REMARK. The function $\mathfrak{f}(s) - 1$ changes sign in every interval of length one; see Corollary B.9.

A.4. The Linear Case

In this section g is a non-negative multiplicative function supported on square-free numbers such that

$$\sum_{p \leqslant x} g(p)\log p = \log x + O(1)\ , \tag{A.44}$$

$$\sum_{p} g(p)^2 \log p < \infty\ . \tag{A.45}$$

In this case, Theorem A.5 with $k = 1$ yields

$$\sum_{m \leqslant x} g(m) = c\log x + O(1)\ , \tag{A.46}$$

where

$$c = \prod_p \Big(1 - \frac{1}{p}\Big)(1 + g(p))\ . \tag{A.47}$$

However, in some applications (see, for example, Section 18.4) we shall need to have asymptotic formulae more precise than (A.46), namely

$$\sum_{m \leqslant x} g(m) = c\log x + b + O((\log x)^{-A}) \tag{A.48}$$

for all $x \geqslant 2$ where b is a constant. This could be derived by refining the proof of Theorem A.5, subject to the correspondingly stronger hypothesis

$$\sum_{p \leqslant x} g(p)\log p = \log x + a + O((\log x)^{-A}). \tag{A.49}$$

While (A.49) is just as deep as the Prime Number Theorem, the asymptotic formula (A.48) can in practice be verified directly by simple arguments in many cases. For example, we often have

$$f(m) = mg(m) = \sum_{d|m} u(d)$$

where u is an arithmetic function such that $u(d) \ll (\log 2d)^{-A-1}$. In this case, (A.48) can be derived rather easily with the constants

$$c = \sum_d \frac{u(d)}{d} \quad \text{and} \quad b = \sum_d \frac{u(d)}{d}(\gamma - \log d).$$

In order to keep our exposition of sieve theory at an elementary level we shall accept (A.48) as a hypothesis, rather than (A.49), whenever such a strong asymptotic formula is needed. From (A.48) we deduce the following:

THEOREM A.8. *Suppose g is a multiplicative function supported on squarefree numbers with $0 \leqslant g(p) < 1$ and which satisfies (A.48) with some $A \geqslant 2$. Then, for any $q \geqslant 1$, we have*

$$\sum_{\substack{m \leqslant x \\ (m,q)=1}} g(m) = \gamma_q \{c \log x + c\delta_q + b + O(\tau(q)(\log x)^{-A})\}\ , \tag{A.50}$$

where

$$\gamma_q = \prod_{p|q}(1+g(p))^{-1} \quad \textit{and} \quad \delta_q = \sum_{p|q} \frac{g(p)}{1+g(p)} \log p, \tag{A.51}$$

the implied constant depending on g and A.

PROOF. Note that $g(p) \ll (\log p)^{-A}$ by (A.48). Denote

$$P(s) = \prod_{p|q}(1+g(p)p^{-s})^{-1} = \sum_n \alpha_n n^{-s},$$

$$Q(s) = \prod_{p|q}(1-g(p)p^{-s})^{-1} = \sum_n |\alpha_n| n^{-s}.$$

Observe that

$$\sum_{(m,q)=1} g(m)m^{-s} = \prod_{p\nmid q}(1+g(p)p^{-s}) = P(s)\sum_m g(m)m^{-s}.$$

Hence, the sum in (A.50) is

$$\begin{aligned}\sum\sum_{mn \leqslant x} \alpha_n g(m) &= \sum_{n \leqslant x} \alpha_n \Big\{c \log \frac{x}{n} + b + O\Big(\big(\log \frac{2x}{n}\big)^{-A}\Big)\Big\} \\ &= \sum_n \alpha_n \Big(c \log \frac{x}{n} + b\Big) + O\Big((1 + \sum_n |\alpha_n|(\log n)^j)(\log x)^{-A}\Big)\ ,\end{aligned}$$

with $j = A+1$. Here, we have

$$\sum_n \alpha_n = P(0) = \gamma_q\ ,$$

$$-\sum_n \alpha_n \log n = P'(0) = P(0)\delta_q = \gamma_q \delta_q\ ,$$

and

$$\sum_n |\alpha_n|(\log n)^j = (-1)^j Q^{(j)}(0) \ll Q(0)\omega(q)^j \ll \gamma_q \tau(q). \tag{A.52}$$

The latter estimate in (A.52) is obvious while the former is derived as follows. We have

$$R(s) = -\frac{Q'}{Q}(s) = \sum_{p|q}\sum_{\ell=1}^{\infty} g(p)^\ell p^{-s\ell}\log p.$$

Differentiating the equation $-Q'(s) = Q(s)R(s)$ repeatedly, $j-1$ times, we get the recurrence formula

$$-Q^{(j)}(0) = \sum_{0\leqslant k<j}\binom{j-1}{k} Q^{(k)}(0)R^{(j-1-k)}(0).$$

Hence,

$$|Q^{(j)}(0)| \leqslant |Q^{(j-1)}(0)| \sum_{p|q}\sum_{\ell=1}^{\infty} g(p)^\ell (1+2\ell\log p)^j.$$

Since $g(p) < 1$ and $g(p) \ll (\log p)^{-j}$ the last sum is bounded uniformly in p. This shows that $|Q^{(j)}(0)| \ll |Q^{(j-1)}(0)|\omega(q) \ll Q(0)\omega(q)^j$ by induction. This completes the proof of Theorem A.8. □

A.5. A Main Term Computation

In this section we provide asymptotic formulas for sums of multiplicative functions which are needed for the evaluation of the main term in various linear sieve problems. Throughout, we assume that $g(d)$ is a multiplicative function which satisfies the following conditions:

(A.53) $$0 \leqslant g(p) < 1, \quad \text{for all } p\ ,$$

(A.54) $$\sum_{p\leqslant x} g(p)\log p = \log x + c_g + O\big((\log x)^{-A}\big)\ ,$$

for all $x \geqslant 2$, where c_g is a constant depending on g, $A \geqslant 0$ is any number and the implied constant depends only on g and A.

Clearly (A.54) holds for $g(p) = p^{-1}$ by the Prime Number Theorem. In many of our arguments it would be sufficient to use (A.54) with $A = 0$, a condition which is elementary in practice. However, in one instance we genuinely need (A.54) for positive A so we cannot avoid a Prime Number Theorem in that case. To simplify our arguments we assume that (A.54) holds with any $A \geqslant 0$.

We begin by investigating the multiplicative function $\Delta(a)$ defined by

(A.55) $$\Delta(p^\alpha) = \Delta(p) = 1 - g(p)p\ , \quad \text{if } \alpha \geqslant 1\ .$$

We also consider

(A.56) $$\delta(a) = \Delta(a)a^{-1}\ .$$

By (A.54) we get

(A.57) $$\sum_{p^\alpha > z} \delta(p^\alpha) \ll (\log z)^{-A}\ , \quad \text{if } z \geqslant 2\ .$$

Hence, by partial summation

(A.58) $$\sum_{z<p^\alpha\leqslant y} \Delta(p^\alpha) \ll y(\log z)^{-A}\ .$$

LEMMA A.9. *For $y \geqslant 2$ we have*

$$\sum_{a\leqslant y} \Delta(a) \ll y(\log y)^{-A} . \tag{A.59}$$

PROOF. We use the following Buchstab type formula:

$$\sum_{a\leqslant y} \Delta(a) = 1 + \sum_{1\leqslant m\leqslant y} \Delta(m) \sum_{q_m<p^\alpha\leqslant y/m} \Delta(p^\alpha) ,$$

where q_m is the largest prime power divisor of m. Let r be the number of distinct prime factors of m, so $q_m > m^{1/r}$. Hence, $m \leqslant y^{r/(r+1)}$ and

$$\sum_{q_m<p^\alpha\leqslant y/m} \Delta(p^\alpha) \ll \frac{y}{m}\Big(\log \frac{y}{m}\Big)^{-A}$$
$$\ll \frac{y}{m}\Big(\frac{r+1}{\log y}\Big)^{A} \ll \frac{\tau(m)y}{m(\log y)^A} .$$

Next we have the trivial bound

$$\sum_{1\leqslant m\leqslant z} |\delta(m)|\tau(m) \leqslant \prod_{p<z}(1+2|\delta(p)|+\dots) \ll (\log z)^4 .$$

Putting together these results, we obtain (A.59). □

COROLLARY A.10. *For $z \geqslant 2$ we have*

$$\sum_{a>z} \delta(a) \ll (\log z)^{-A} . \tag{A.60}$$

PROOF. This follows from (A.59) by partial summation. □

Next we consider, for $x \geqslant 1$, the sum

$$G_\ell(x) = \sum_{d\leqslant x} \mu(d)g(d)\Big(\log \frac{x}{d}\Big)^\ell . \tag{A.61}$$

For any $d \geqslant 1$ we have

$$\mu(d)g(d) = \sum_{ab=d} \delta(a)\mu(b)b^{-1} . \tag{A.62}$$

Hence, by (A.60) and by the corresponding result for $\mu(b)b^{-1}$, namely

$$\sum_{b>z} \mu(b)b^{-1} \ll (\log z)^{-A} , \tag{A.63}$$

we find that

$$\sum_{d>z} \mu(d)g(d) \ll (\log z)^{-A} . \tag{A.64}$$

This yields, by partial summation,

$$G_\ell(x) = \sum_{d=1}^{\infty} \mu(d)g(d)\Big(\log \frac{x}{d}\Big)^\ell + O\big((\log 2x)^{-A}\big) .$$

Hence, we obtain the asymptotic expansion

$$G_\ell(x) = \sum_{0\leqslant k\leqslant \ell} \gamma_\ell(k)(\log x)^k + O\big((\log 2x)^{-A}\big) , \tag{A.65}$$

where $\gamma_\ell(k)$ are the constants given by the convergent series

$$\gamma_\ell(k) = \binom{\ell}{k} \sum_{d=1}^{\infty} \mu(d)g(d)(-\log d)^{\ell-k} . \tag{A.66}$$

For $k = \ell$ we have

$$\gamma_\ell(\ell) = \sum_{d} \mu(d)g(d) = \Big(\sum_{a} \delta(a)\Big)\Big(\sum_{b} \mu(b)b^{-1}\Big) = 0 .$$

For $k = \ell - 1$ we have

$$\begin{aligned}
\gamma_\ell(\ell-1) &= -\ell \sum_{d} \mu(d)g(d)\log d \\
&= -\ell \sum_{a}\sum_{b} \delta(a)\mu(b)b^{-1}\log(ab) \\
&= -\ell\Big(\sum_{a} \delta(a)\Big)\Big(\sum_{b} \mu(b)b^{-1}\log b\Big) \\
&= \ell \sum_{a} \delta(a) = \ell \prod_{p}(1 + \delta(p) + \delta(p^2) + \dots) = \ell H ,
\end{aligned}$$

where

$$H = \prod_{p}(1 - g(p))\Big(1 - \frac{1}{p}\Big)^{-1} . \tag{A.67}$$

We do not need any further terms, so we conclude the above arguments, getting:

LEMMA A.11. *For $\ell \geqslant 0$ and $x \geqslant 1$, we have*

$$\sum_{d \leqslant x} \mu(d)g(d)\Big(\log \frac{x}{d}\Big)^{\ell} = \ell H (\log x)^{\ell-1} + O\big((\log 2x)^{\ell-2}\big) , \tag{A.68}$$

where the implied constant depends on g and ℓ.

COROLLARY A.12. *For $x \geqslant 1$,*

$$-\sum_{d \leqslant x} \mu(d)g(d)\log d = H + O\big((\log 2x)^{-1}\big) . \tag{A.69}$$

COROLLARY A.13. *For $q \geqslant 1$, $\ell \geqslant 1$, $x \geqslant 1$,*

$$\sum_{\substack{d \leqslant x \\ (d,q)=1}} \mu(d)g(d)\Big(\log \frac{x}{d}\Big)^{\ell} = \prod_{p|q}(1 - g(p))^{-1} \ell H (\log x)^{\ell-1} + O\big(\omega(2q)(\log 2x)^{\ell-2}\big), \tag{A.70}$$

where the implied constant depends only on g and ℓ.

PROOF. Let $\tilde{g}(d)$ denote the multiplicative function with $g(p^\alpha) = g(p)^\alpha$ for all $\alpha \geqslant 0$. Then,

$$\begin{aligned}\sum_{(m,q)=1} \mu(m)g(m)m^{-s} &= \prod_{p\nmid q}(1-g(p)p^{-s})\\ &= \prod_{p\mid q}(1-g(p)p^{-s})^{-1}\prod_{p}(1-g(p)p^{-s})\\ &= \Big(\sum_{d\mid q^\infty} \tilde{g}(d)d^{-s}\Big)\sum_m \mu(m)g(m)m^{-s}\ .\end{aligned}$$

Hence, the sum (A.70) is equal to

$$\begin{aligned}&\sum_{d\mid q^\infty} \tilde{g}(d) \sum_{dm\leqslant x} \mu(m)g(m)(\log x/dm)^\ell\\ &\quad= \sum_{\substack{d\mid q^\infty\\ d\leqslant x}} \tilde{g}(d)\left\{\ell H\left(\log x/d\right)^{\ell-1} + O\left(\left(\log 2x/d\right)^{\ell-2}\right)\right\}\\ &\quad= \sum_{d\mid q^\infty} \tilde{g}(d)\left\{\ell H\left(\log x\right)^{\ell-1} + O\left(\left(\log 2d\right)\left(\log 2x\right)^{\ell-2}\right)\right\}\\ &\quad= Q(0)\ell H\left(\log x\right)^{\ell-1} + O\left(\omega(2q)\left(\log 2x\right)^{\ell-2}\right)\end{aligned}$$

by (A.68) and (A.52), where

$$Q(0) = \prod_{p\mid q}(1-g(p))^{-1}\ ,$$

completing the proof. □

A.6. Evaluation of $P_{(k)}(x)$

Our goal in this section is the evaluation of the sums

$$P_{(k)}(x) = \sum_{d\leqslant x} g(d)\Lambda_{(k)}(d)$$

involving the vector von Mangoldt functions which arose in Chapter 3 in connection with the the asymptotic sieve. As usual, $g(d)$ is a multiplicative function satisfying

$$g(d) \leqslant \tau_r(d)d^{-1}, \tag{A.71}$$

$$0 \leqslant g(p) < 1, \tag{A.72}$$

and, in this section, we assume the two-sided (but weak) linear sieve condition:

$$\sum_{p\leqslant x} g(p)\log p = \log x + O(1)\ . \tag{A.73}$$

LEMMA A.14. *For $x \geqslant 1$, we have*

$$P_{(k)}(x) = c\big((k)\big)\left(\log x\right)^k + O\big((\log 2x)^{k-1}\big) \tag{A.74}$$

where $(k) = (k_1, \ldots, k_r)$, $|k| = k_1 + k_2 + \cdots + k_r$ and

$$c\big((k)\big) = k_1!\ldots k_r!/(k_1+\cdots+k_r)!\ .$$

PROOF. First we establish (A.74) for completely multiplicative functions and for scalars. We have

$$P_1(x) = \sum_{d \leqslant x} g(d)\Lambda(d) = \log x + O(1)$$

by (A.73) and (A.71). Hence, by the recurrence formula $\Lambda_k = \Lambda_{k-1}L + \Lambda_{k-1} * \Lambda$ with $k \geqslant 2$ (see (3.15)) and, using the induction hypothesis, we obtain

$$\begin{aligned} P_k(x) &= \sum_{d \leqslant x} g(d)\Lambda_k(d) \\ &= \sum_{d \leqslant x} g(d)\Lambda_{k-1}(d)\big(\log d + P_1(x/d)\big) \\ &= P_{k-1}(x)\big(\log x + O(1)\big) \ . \end{aligned}$$

Hence,

$$P_k(x) = \big(\log x\big)^k + O\big((\log 2x)^{k-1}\big) \ . \tag{A.75}$$

Note that (A.75) holds for $k = 0$ without the error term.

Next, we generalize (A.75) by partial summation, obtaining

$$\sum_{d \leqslant x} g(d)\Lambda_k(d)\Big(\log \frac{x}{d}\Big)^{\ell} = \frac{k\,!\,\ell\,!}{(k+\ell)!}\big(\log x\big)^{k+\ell} + O\big((\log 2x)^{k+\ell-1}\big) \ . \tag{A.76}$$

Finally, we generalize (A.75) to vectors by induction. Suppose $(k) = \big(k_1, (k')\big)$. We have

$$\begin{aligned} P_{(k)}(x) &= \sum\sum_{mn \leqslant x} g(m)g(n)\Lambda_{k_1}(m)\Lambda_{(k')}(n) \\ &= \sum_{m \leqslant x} g(m)\Lambda_{k_1}(m)P_{(k')}(x/m) \\ &= \sum_{m \leqslant x} g(m)\Lambda_{k_1}(m)\left\{c\big((k')\big)\Big(\log \frac{x}{m}\Big)^{|k'|} + O\left(\big(\log 2x\big)^{|k'-1|}\right)\right\} \\ &= c\big((k')\big)\,\frac{k_1!k'!}{(k_1+k')!}\big(\log x\big)^{|k|} + O\left(\big(\log 2x\big)^{|k-1|}\right) \ . \end{aligned}$$

This completes the proof of (A.74) for totally multiplicative functions $g(d)$.

If $g(d)$ is not totally multiplicative then it can be made so by a suitable modification at prime powers, that is, at numbers d which are not squarefree. However, the contribution to $P_{(k)}(x)$ of such terms is negligible. Indeed, we have

$$\begin{aligned} \sum_{\substack{d \leqslant x \\ \mu(d)=0}} g(d)\Lambda_{(k)}(d) &\leqslant \sum_{\substack{mp^{\alpha} \leqslant x \\ \alpha \geqslant 2, p \nmid m}} g(mp^{\alpha})\Lambda_{|k|}(mp^{\alpha}) \\ &\ll \sum_{m \leqslant x}\sum_{\substack{p^{\alpha} \leqslant x \\ \alpha \geqslant 2}} (mp^{\alpha})^{-1} \sum_{0 \leqslant j < |k|} \binom{k}{j} \Lambda_j(m)\Lambda_{|k|-j}(p^{\alpha}) \ . \end{aligned}$$

Here $j = |k|$ is omitted, because $\Lambda_0(p^{\alpha}) = 0$. Using (A.75), we see that this is bounded by $O\big((\log 2x)^{|k|-1}\big)$, completing the proof of Lemma A.14. □

A.7. Evaluation of $h_{(k)}(x)$

Let $g(d)$ be a multiplicative function satisfying (A.53), (A.54) and

$$g(d) \leqslant \tau_r(d)d^{-1} .$$

Our goal here is the evaluation of the sum (3.40) which appears in the main term of Bombieri's sieve; see Proposition 3.4. Recall that this sum is given by

$$h_{(k)}(y) = \sum_{d<y} \mu_{(k')}(d)g(d)\Big(\log\frac{y}{d}\Big)^{\ell} ,$$

where $(k) = \big(\ell, (k')\big)$ with $\ell \geqslant 2$ and $\mu_{(k')} = \mu * \Lambda_{(k')}$. Therefore,

$$\begin{aligned} h_{(k)}(y) &= \sum_{mn\leqslant y} \mu(m)g(mn)\Lambda_{(k')}(n)\Big(\log\frac{y}{mn}\Big)^{\ell} \\ &= \mathop{\sum\sum}_{\substack{dn\leqslant y\\ d\mid n}} \mu(d)g(dn)\Lambda_{(k')}(n) \sum_{\substack{m\leqslant y/dn\\ (m,n)=1}} \mu(m)g(m)\Big(\log\frac{y}{dmn}\Big)^{\ell} . \end{aligned}$$

The inner sum was evaluated in Corollary A.13 giving

$$\begin{aligned} h_{(k)}(y) = &\mathop{\sum\sum}_{\substack{dn\leqslant y\\ d\mid n}} \mu(d)g(dn)\Lambda_{(k')}(n)\prod_{p\mid n}\big(1-g(p)\big)^{-1}\ell H\Big(\log\frac{y}{dn}\Big)^{\ell-1} \\ &+ O\Big(\sum_{n\leqslant y} n^{-1}\Lambda_{|k'|}(n)\big(\log 2y\big)^{\ell-2}\Big) . \end{aligned}$$

Here, the error term is $O\big((\log 2y)^{|k|-2}\big)$ by (A.74). In the main terms, the contribution from $d \neq 1$ is negligible. Indeed, we estimate this contribution crudely as follows:

$$\mathop{\sum\sum}_{\substack{dn\leqslant y\\ d\mid n,\, d\neq 1}} \ll \sum_{\substack{mp^{\alpha}\leqslant y\\ \alpha\geqslant 2,\, p\nmid m}} g(m)g(p^{\alpha})\Lambda_{|k'|}(mp^{\alpha-1})(\log y)^{\ell-1} \ll (\log 2y)^{k-2} ,$$

by the same arguments as used at the end of the previous section.

Therefore we are left with

$$h_{(k)}(y) = \ell H \sum_{n\leqslant y} h(n)\Lambda_{(k')}(n)\Big(\log\frac{y}{n}\Big)^{\ell-1} + O\big((\log 2y)^{|k|-2}\big) ,$$

where

$$h(n) = g(n)\prod_{p\mid n}\big(1-g(p)\big)^{-1} .$$

This function satisfies the conditions which were imposed on g, so Lemma A.14 is applicable, giving

$$\sum_{n\leqslant x} h(n)\Lambda_{(k')}(n) = c\big((k')\big)\big(\log x\big)^{|k'|} + O\big((\log 2x)^{|k'|-1}\big) .$$

Hence, we infer by partial summation that

$$h_{(k)}(y) = \ell H c\big((k')\big)\,\frac{k'!(\ell-1)!}{(k'+\ell-1)!}\,(\log y)^{|k|-1} + O\big((\log 2y)^{|k|-2}\big) .$$

This proves Proposition 3.4.

A.8. Congruence Sums

Our goal in this section is to evaluate asymptotically the sum

$$\mathcal{M}_f(x,q) = \sum_{\substack{m\leqslant x\\ m\equiv 0(q)}} f(m) \tag{A.77}$$

for q as large as possible in terms of x. We shall take advantage of the residue class being zero modulo q as well as on f being multiplicative. We have in mind applications to linear sieve problems, so naturally we can, for practical purposes, assume that

$$\mathcal{M}_f(x) = \sum_{m\leqslant x} f(m) = c_f x + O(x^\theta)\ , \tag{A.78}$$

with some constants $c_f > 0$ and $0 < \theta < 1$. This shows that the corresponding zeta function

$$Z(s) = \sum_m f(m)m^{-s} = \prod_p \Big(\sum_{\alpha\geqslant 0} f(p^\alpha)p^{-\alpha s}\Big) \tag{A.79}$$

is analytic in $\operatorname{Re} s > \theta$ with only a simple pole at $s = 1$. Hence, comparing $Z(s)$ with $\zeta(s)$ we find that the constant c_f in the leading term in (A.78) must be

$$c_f = \prod_p \Big(1 - \frac{1}{p}\Big)\Big(\sum_{\alpha\geqslant 0} f(p^\alpha)p^{-\alpha}\Big)\ . \tag{A.80}$$

We are going to derive a formula for $\mathcal{M}_f(x,q)$ from (A.78). This would be easy if f were completely multiplicative. Indeed, in that case we have

$$\begin{aligned}\mathcal{M}_f(x,q) &= f(q)\mathcal{M}_f(x/q)\\ &= c_f\, f(q)q^{-1}x + O\big(f(q)(x/q)^\theta\big)\ .\end{aligned} \tag{A.81}$$

This result is meaningful for q up to $o(x)$.

However, we do need results of similar quality for multiplicative functions f which are not completely multiplicative. If f is supported on squarefree numbers one can infer the desired formulas from Theorem A.8, but we do not want such a restriction. Instead, we impose some conditions which control the values $f(m)$ at higher prime powers. Specifically, we assume that

$$0 \leqslant f(m) \leqslant \tau_r(m) \tag{A.82}$$

and

$$|f^*(m)| \leqslant \tau_r(m) \tag{A.83}$$

where $f^*(m)$ are the coefficients of $Z^*(s) = Z(s)^{-1}$,

$$Z^*(s) = \sum_m f^*(m)m^{-s} = \prod_p \Big(\sum_{\alpha\geqslant 0} f(p^\alpha)p^{-\alpha s}\Big)^{-1}\ . \tag{A.84}$$

For example, our conditions are clearly satisfied if the zeta function $Z(s)$ is given by an Euler product of degree r:

$$Z(s) = \prod_p \big(1-\alpha_1(p)p^{-s}\big)^{-1}\dots\big(1-\alpha_r(p)p^{-s}\big)^{-1}\ , \tag{A.85}$$

with

$$|\alpha_j(p)| \leqslant 1 \ , \quad 1 \leqslant j \leqslant r \ . \tag{A.86}$$

Given $q \geqslant 1$ we write

$$\sum_{m \equiv 0(q)} f(m) m^{-s} = Z(s) Z_q(s) Z_q^*(s) \ ,$$

where

$$Z_q(s) = \prod_{p^\nu \| q} \Big(\sum_{\alpha \geqslant \nu} f(p^\alpha) p^{-\alpha s} \Big) = \sum_n f_q(n) n^{-s} \ ,$$

$$Z_q^*(s) = \prod_{p|q} \Big(\sum_{\alpha \geqslant 0} f(p^\alpha) p^{-\alpha s} \Big)^{-1} = \sum_n f_q^*(n) n^{-s} \ .$$

Now we evaluate the sum (A.77) as follows:

$$\begin{aligned} \mathcal{M}_f(x, q) &= \sum\sum\sum_{\ell n m \leqslant x} f_q^*(\ell) f_q(n) f(m) \\ &= \sum_\ell \sum_n f_q^*(\ell) f_q(n) \left\{ c_f \frac{x}{\ell n} + O\Big(\Big(\frac{x}{\ell n} \Big)^\theta \Big) \right\} \\ &= c_f x \, Z_q^*(1) Z_q(1) + O\big(x^\theta Z_q^+(\theta) Z_q(\theta) \big) \end{aligned}$$

where

$$Z_q^+(s) = \sum_\ell \big| f_q^*(\ell) \big| \ell^{-s} \ .$$

Applying (A.83) we get

$$Z_q^+(s) \leqslant \prod_{p|q} \big(1 - p^{-s} \big)^{-r} \ . \tag{A.87}$$

Similarly, applying (A.83) and the sub-multiplicativity of τ_r, which means that

$$\tau_r(ab) \leqslant \tau_r(a) \tau_r(b) \ ,$$

we derive

$$Z_q(s) \leqslant q^{-s} \tau_r(q) \prod_{p|q} \big(1 - p^{-s} \big)^{-r} \ . \tag{A.88}$$

We can do better at prime places as follows:

$$\begin{aligned} Z_p(s) &= f(p) p^{-s} + \sum_{\alpha \geqslant 2} f(p^\alpha) p^{-\alpha s} \\ &\leqslant f(p) p^{-s} + \big(1 - p^{-s} \big)^{-r} - 1 - r p^{-s} \\ &\leqslant p^{-s} \big(1 - p^{-s} \big)^{-r} \left\{ f(p) + O\big(p^{-s} \big) \right\} \ . \end{aligned}$$

Hence, we conclude that

$$Z_q^*(s) Z_q(s) \ll \tilde{f}(q) q^{-s} \ , \tag{A.89}$$

where

$$\tilde{f}(q) = \sum_{abc^2 = q} f(a) b^{-\varepsilon} c^{\varepsilon} \ , \tag{A.90}$$

with any fixed $\varepsilon > 0$, and an implied constant depending only on ε and r. Moreover, we have $c_f Z_q^*(1)Z_q(1) = \hat{f}(q)$ with

$$\hat{f}(q) = \prod_{p^\nu \| q}\Big(1-\frac{1}{p}\Big)\Big(\sum_{\alpha\geqslant\nu} f(p^\alpha)p^{-\alpha}\Big)\,. \tag{A.91}$$

Hence, we have established the following formula.

LEMMA A.15. *Let $f(m)$ be a multiplicative function satisfying* (A.78), (A.82) *and* (A.83). *Then, for any $q \geqslant 1$ we have*

$$\mathcal{M}_f(x,q) = \hat{f}(q)x + O\big(\tilde{f}(q)(x/q)^\theta\big) \tag{A.92}$$

where $\hat{f}$, $\tilde{f}$ are the multiplicative functions given by (A.91), (A.90) *respectively, with ε a small positive constant. The implied constant depends on ε, r and the constant implied in* (A.78).

APPENDIX B

Differential-Difference Equations

Back to the Future

B.1. Adjoint Equations

Differential equations with delayed argument occur in mathematical models of real time processes. In number theory one encounters solutions to differential-difference equations in the asymptotics for the average value of a multiplicative function taken over integers having prime divisors of a certain type in selected segments. All we need in sieve theory are the linear equations of type

$$sQ'(s) + aQ(s) + bQ(s-1) = 0, \quad \text{if } s > \beta, \tag{B.1}$$

where a, b are constant real coefficients and $\beta \geqslant 1$. For example, the beta-sieve theory makes use of this for $a = \kappa$ and $b = \pm\kappa$. Different coefficients are relevant to other sieves (see Section 7.9).

Equivalently, we can write (B.1) as

$$(s^a Q(s))' = -bs^{a-1}Q(s-1), \quad \text{if } s > \beta. \tag{B.2}$$

Hence, given $Q(s)$ in the initial segment $\beta - 1 < s \leqslant \beta$, it extends as a continuous function uniquely to all $s > \beta$ by repeated integration. If we suppose $Q(s)$ is smooth inside the initial segment then it is smooth everywhere except for the points $\beta + n$ with $n = 0, 1, 2, \ldots$ and at those it has continuous derivatives of order $\leqslant n$. In general $Q(s)$ is not smooth at these points so we are not able to write a simple formula for it.

A standard technique for solving the differential-difference equation (B.1) applies the Laplace transform to obtain an ordinary differential equation, and after the latter is solved, one gets $Q(s)$ by the inverse Laplace transform (in the complex plane). We shall apply an alternative technique employing the adjoint equation which, aside from being more elementary, quickly reveals a number of properties of the solution which are required for the development of the beta-sieve.

The adjoint to (B.1) is the following equation with advanced argument:

$$(sq(s))' = aq(s) + bq(s+1), \qquad \text{if } s > 0. \tag{B.3}$$

Equivalently, we write this as

$$\left(s^{1-a}q(s)\right)' = bs^{-a}q(s+1), \qquad \text{if } s > 0. \tag{B.4}$$

An equation with advanced argument as above usually has a nice solution $q(s) \not\equiv 0$ of $\mathcal{C}^\infty$-class. There is a reasonable explanation. Think of s as a time variable. The equation (B.1) describes the process $Q(s)$ at the moment s which happens after elapse of a finite time from the initial period. To the contrary (B.4) describes

$q(s)$ at the present moment from which the beginning of the process is distanced infinitely far away towards the future. That is why $q(s)$ is smooth at any s. In fact $q(s)$ is holomorphic in $\mathrm{Re}\, s > 0$ and consequently it will be possible to give a simple expression for it.

To a pair of functions $Q(s), q(s)$ we associate the "inner product"

$$\langle Q, q\rangle = sQ(s)q(s) - b\int_{s-1}^{s} Q(x)q(x+1)dx\ , \tag{B.5}$$

which is defined for $s \geqslant \beta$. The key property of this product is that it is constant for adjoint functions. Indeed, differentiating (B.5) we get

$$sQ'(s)q(s) + Q(s)(sq(s))' - bQ(s)q(s+1) + bQ(s-1)q(s) = 0$$

by invoking (B.1) and (B.3).

Using (B.2) and (B.4) we can also write the inner product by partial integration as follows:

$$\langle Q, q\rangle = (s-1)Q(s-1)q(s-1) + \int_{s-1}^{s} x^{1-a}q(x)dx^{a}Q(x). \tag{B.6}$$

The inner product is instrumental for studying the local properties of $Q(s)$ whose behaviour is often erratic and is difficult to grasp by the methods of complex analysis. Since the inner product is constant one can view this as an integral representation for $Q(s)$ over a unit segment. Concerning the kernel function $q(s)$, think of this as a given simple function having regular asymptotic behaviour.

For instance, if $a + b = n + 1$ is a positive integer, then $q(s)$ is a polynomial of degree n,

$$q(s) = \sum_{0\leqslant \ell\leqslant n} \binom{n}{\ell} a_\ell\, s^{n-\ell}\ ,$$

where $a_0 = 1$ and a_ℓ with $\ell > 0$ are determined by the recurrence formula

$$\ell a_\ell + b\sum_{0\leqslant j<\ell} \binom{\ell}{j} a_j = 0.$$

In particular, for $a + b = 1, 2, 3$, we get

$$q(s) = 1, \quad s - b, \quad s^2 - 2bs + b(b - \tfrac{1}{2})\ , \tag{B.7}$$

respectively. Hence, for $a = b = \kappa = \frac{1}{2}, 1, \frac{3}{2}$, we have

$$q(s) = 1, \quad s - 1, \quad s^2 - 3s + \tfrac{3}{2}. \tag{B.8}$$

In general, $q(s)$ can be given by the contour integral

$$q(s) = \frac{\Gamma(a+b)}{2\pi i}\int_{\mathcal{C}} z^{-a-b}\exp\Big(sz + b\int_0^z (1 - e^u)u^{-1}du\Big)dz\ , \tag{B.9}$$

where $\mathcal{C}$ is a contour enveloping the negative real axis in the positive direction and the power function is defined using $-\pi < \arg z \leqslant \pi$. Indeed, the function being integrated has exponential decay as $z \to \infty$ along $\mathcal{C}$, thus we can integrate by parts and verify (B.3) as follows:

$$\begin{aligned} sq'(s) &= \frac{\Gamma(a+b)}{2\pi i}\int_{\mathcal{C}} z^{1-a-b}\exp\left(b\int_0^z (1-e^u)u^{-1}du\right)de^{sz} \\ &= -\frac{\Gamma(a+b)}{2\pi i}\int_{\mathcal{C}} e^{sz}dz^{-1-a-b}\exp\left(b\int_0^z (1-e^u)u^{-1}du\right)dz \\ &= \frac{\Gamma(a+b)}{2\pi i}\int_{\mathcal{C}} z^{-a-b}(be^z+a-1)\exp\left(sz+b\int_0^z (1-e^u)u^{-1}du\right)dz \\ &= (a-1)q(s)+bq(s+1). \end{aligned}$$

It is possible and also convenient to transform $q(s)$ into an integral over the positive reals, simply by compressing $\mathcal{C}$ to the negative real axis and then changing the variable z into $-z$. This succeeds if $a+b<1$, otherwise there is a problem of convergence. We overcome that problem by expanding

$$R(z) = \exp\left(b\int_0^z (1-e^u)u^{-1}du\right)$$

into a Taylor series, say

$$R(z) = \sum_{0\leqslant \ell<N} R^{(\ell)}(0)\frac{z^\ell}{\ell!} + R_N(z)\ ,$$

and integrating explicitly everything but the remainder term $R_N(z)$. Using the formula

$$\frac{1}{2\pi i}\int_{\mathcal{C}} e^{sz}z^{-\nu}dz = \frac{s^{\nu+1}}{\Gamma(\nu)}\ ,$$

we obtain

$$q(s) = \sum_{0\leqslant \ell<N}\binom{a+b-1}{\ell}R^{(\ell)}(0)s^{a+b-1-\ell} + \frac{\Gamma(a+b)}{2\pi i}\int_{\mathcal{C}} e^{sz}R_N(z)z^{-a-b}dz\ ,$$

where

$$\binom{a+b-1}{\ell} = \frac{\Gamma(a+b)}{\ell!\,\Gamma(a+b-\ell)}\ .$$

Here, $R_N(z) \ll |z|^N$ as $z\to 0$ so if $N > a+b-1$ one can shrink the contour of integration into two copies of the half-line $(-\infty, 0)$ having opposite orientations. When approaching from below we reach the value

$$z^{-a-b} = |z|^{-a-b}e^{-\pi i(a+b)}$$

while from above we reach the complex conjugate of this value. Having done this, the contour integral becomes

$$\frac{1}{2\pi i}\int_{\mathcal{C}} e^{sz}R_N(z)z^{-a-b}dz = \pi^{-1}\sin(\pi(a+b))\int_0^\infty e^{-sz}R_N(-z)z^{-a-b}dz.$$

Finally, using the functional equation for the gamma function,

$$\Gamma(a+b)\Gamma(1-a-b)\sin\pi(a+b) = \pi\ ,$$

we conclude the following result.

LEMMA B.1. *If $N > a+b-1$ we have*

$$q(s) = \sum_{0\leqslant \ell<N} \binom{a+b-1}{\ell} R^{(\ell)}(0) s^{a+b-1-\ell} + \frac{1}{\Gamma(1-a-b)} \int_0^\infty e^{-sz} R_N(-z) z^{-a-b} dz.$$

We shall call $q(s) = q(s;a,b)$ given by (B.9) the standard solution to (B.3). It follows from Lemma B.1 that $q(s;a,b)$ is holomorphic in Re $s > 0$ and it is an entire function of either variable, a or b.

If $a+b<1$, the formula of Lemma B.1 yields

$$q(s) = \frac{1}{\Gamma(1-a-b)} \int_0^\infty \exp\left(-sz + b\int_0^z (1-e^{-u})u^{-1}du\right) z^{-a-b}dz \tag{B.10}$$

and if $a+b<2$ it yields

$$q(s) = s^{a+b-1} + \frac{1}{\Gamma(1-a-b)} \int_0^\infty e^{-sz} \left\{\exp\left(b\int_0^z (1-e^{-u})u^{-1}du\right) - 1\right\} z^{-a-b}dz. \tag{B.11}$$

In general, it follows by Lemma B.1 that

$$q(s) \sim s^{a+b-1}, \quad \text{as } s\to\infty. \tag{B.12}$$

LEMMA B.2. *If $a+b\leqslant 1$ and $a<1$, then*

$$s^{1-a}q(s) \sim e^{b\gamma}\frac{\Gamma(1-a)}{\Gamma(1-a-b)}, \quad \textit{as } s\to 0+. \tag{B.13}$$

PROOF. If $a+b=1$, then $q(s)=1$ so the result is obvious. Let $a+b<1$. We have by (8.367.12) of [**66**],

$$\gamma = \int_0^1 (1-e^{-u})u^{-1}du - \int_1^\infty e^{-u}u^{-1}du \tag{B.14}$$

and

$$\mathcal{E}(z) = \int_z^\infty e^{-u}u^{-1}du = \int_0^z (1-e^{-u})u^{-1}du - \log z - \gamma. \tag{B.15}$$

Hence, by (B.10) we get

$$\begin{aligned} e^{-b\gamma}\Gamma(1-a-b)q(s) &= \int_0^\infty \exp\left(-sz + b\mathcal{E}(z)\right) z^{-a}dz \\ &= \Gamma(1-a)s^{a-1} + \int_0^\infty e^{-sz}(e^{b\mathcal{E}(z)}-1)z^{-a}dz. \end{aligned}$$

The last integral is bounded; hence we obtain the asymptotic (B.13). □

B.2. Zeros of the Adjoint Function

We have seen that the zeros of $q(s)$ for $(a,b)=(\kappa,\kappa)$ play a role in the beta-sieve theory. In general, the distribution of zeros of $q(s)$ is a fascinating subject on its own merit. Here, we present only a few elementary results.

LEMMA B.3. *If $b \leqslant 0$ or $a+b \leqslant 1$, then $q(s)$ is positive for all $s>0$.*

PROOF. By (B.12) it follows that $q(s)$ is positive for all sufficiently large s. If $b=0$, then $q(s)=s^{a-1}>0$ for all $s>0$. Let $b<0$ and suppose $q(s)$ has a zero, say α is the largest one. Then $q'(\alpha)\geqslant 0$, $q(\alpha+1)>0$ and $\alpha q'(\alpha)=bq(\alpha+1)<0$ which is a contradiction. If $a+b<1$, then $q(s)$ is positive by (B.10) while for $a+b=1$ we have $q(s)=1$. This completes the proof. □

LEMMA B.4. *If $a+b\geqslant 1$, then the number of real, positive zeros of $q(s)$ is fewer than $a+b$.*

PROOF. If $a+b$ is a positive integer, then $q(s)$ is a polynomial of degree $a+b-1$ so the assertion is true. Now suppose $n<a+b<n+1$ where n is a positive integer. Differentiating the equation (B.3) or the integral formula (B.9) n times we find that $\Gamma(a+b)^{-1}\Gamma(a+b-n)q^{(n)}(s)$ is the standard solution to (B.3) with the coefficient a reduced to $a-n$, and therefore $q^{(n)}(s)$ is positive by virtue of the previous observation. This implies $q(s)$ has at most n real positive zeros. □

LEMMA B.5. *If $b>0$ and $a+b>1$, then $q(s)$ has a real, positive zero.*

PROOF. First we show that if $q'(s)$ has a positive zero, say $\alpha'>0$, then $q(s)$ has a zero $\alpha>\alpha'$. Indeed, assuming α' is the largest zero of $q'(s)$ we derive $q(\alpha'+1)>q(\alpha')$ and

$$0=\alpha' q'(\alpha')=(a-1)q(\alpha')+bq(\alpha'+1)>(a+b-1)q(\alpha'),$$

hence $q(\alpha')$ is negative so $q(s)$ must have a zero $\alpha>\alpha'$. Now by this property we can reduce the proof by repeated differentiation to the case $1<a+b\leqslant 2$. If $a+b=2$, then $q(s)=s-b$ so $\alpha=b$ is the zero of $q(s)$. If $1<a+b<2$, then by (B.11) it follows that $q(s)$ is negative as $s\to 0+$ while it is positive as $s\to\infty$, so $q(s)$ has a real, positive zero. □

Arguing as above one can show that the largest zero α of $q(s)$ is a continuous function of the coefficients a,b for $b>0$ and $a+b>1$. By playing with the equation (B.3) we show the following.

LEMMA B.6. *Let $b>0$ and $a+b>1$ and let α be largest zero of $q(s)$. Then α is simple and it is located in the following segments:*

$$\begin{aligned} &0<\alpha<b && \text{if } 1<a+b<2,\\ &b<\alpha<b+\left(\tfrac12 b\,(a+b-2)\right)^{\frac12} && \text{if } 2<a+b<3,\\ &\alpha>b+\left(\tfrac12 b\,(a+b-2)\right)^{\frac12} && \text{if } a+b>3.\end{aligned}\tag{B.16}$$

PROOF. From the equation (B.3) it is clear that, if α is a multiple zero of $q(s)$, then $\alpha+1$ is a zero of $q(s)$. Hence, the largest real zero of $q(s)$ is simple.

Suppose $1 < a+b < 2$. By (B.11), $q''(s) < 0 < q'(s)$ for $s > 0$, $q(s) < 0$ for small s and $q(s) > 0$ for large s, so $q(s)$ has exactly one zero $\alpha > 0$. By (B.3) we have

$$sq'(s) = (a-1)q(s) + bq(s+1) < (a+b-1)q(s) + bq'(s) \ . \tag{B.17}$$

Taking $s = \alpha$ we get $\alpha < b$, proving the first part of (B.16).

Next, suppose $2 < a+b < 3$. Let α' be the largest zero of $q'(s)$. This exists by Lemma B.5 and satisfies $\alpha' < \alpha$ by the arguments used in the proof of Lemma B.5. Since $q''(s) > 0$ we get by Lemma B.3 the inequality (B.17) but in the reverse direction, so that $\alpha > b$. Next, in the direction of (B.17) we obtain

$$sq'(s) < (a+b-1)q(s) + bq'(s) + \tfrac{1}{2}bq''(s)$$

because $q'''(s) < 0$. Moreover, applying (B.17) for $q'(s)$ we find that

$$sq''(s) < (a+b-2)q'(s) + bq''(s) \ .$$

Combining the two previous inequalities we derive, for all $s > b$,

$$\big((s-b)^2 - \tfrac{1}{2}b(a+b-2)\big)q'(s) < (a+b-1)(s-b)q(s) \ . \tag{B.18}$$

Taking $s = \alpha$ we obtain the upper bound in the second part of (B.16).

Now, suppose $a+b > 3$. The largest zeros of $q(s)$, $q'(s)$, $q''(s)$, satisfy $\alpha > \alpha' > \alpha''$ and we find that the inequality (B.18) holds in the reverse direction. Taking $s = \alpha$, this yields the third part of (B.16). □

We are now going to justify some results stated without proof in Section 11.19. Note that the inequalities (B.16) in the case $a = b = \kappa$ become those in (11.197).

To prove (11.195) we use the formula (B.11) at $s = \alpha$, getting

$$\alpha^\kappa = -\frac{1}{\Gamma(1-2\kappa)} \int_0^\infty e^{-y} y^{-\kappa} r(y/\alpha)\, dy$$

where

$$r(z) = z^{-\kappa} \exp\Big\{\kappa \int_0^z (1-e^{-u})u^{-1} du - 1\Big\} = e^{\kappa\gamma + \kappa\mathcal{E}(z)} - z^\kappa \ ,$$

by (B.14) and (B.15). Since $\alpha \to 0$ as $\kappa \to \frac{1}{2}^+$ we have $r(y/\alpha) \sim r(\infty) = e^{\kappa\gamma}$ giving

$$\alpha^\kappa \sim -e^{\kappa\gamma}\frac{\Gamma(1-\kappa)}{\Gamma(1-2\kappa)} \sim e^{\kappa\gamma}\Gamma(\tfrac{1}{2})(2\kappa - 1) \ ,$$

which yields (11.195).

The proof of (11.198) is easy. Recall that $A = A(\kappa)$ is given by (11.57) with $p(s)$ given by (11.42). Since $sp(s) \sim 1$ as $s \to \infty$, the differential-difference equation (11.41) can be written as the integral equation (see also (11.65)):

$$sp(s) + \kappa \int_s^{s+1} p(u)\, du = 1 \ .$$

Since $p(u)$ is decreasing we infer from the above equation that $(s+\kappa)p(s) > 1$. Hence, using $p(u) > (s+1+\kappa)^{-1}$ to estimate the above integral we arrive at two estimates

$$\frac{1}{s+\kappa} < p(s) < \frac{s+1}{s(s+1+\kappa)} \ . \tag{B.19}$$

For $s = \beta$ this yields (11.198).

The proof of (11.196) is not short (see [**104**]), and we are going to address the relevant question in the next section.

B.3. The Largest Zeros of the Adjoint Polynomial

Throughout, we assume that $b > 0$ and $a + b > 1$ so we know that the solution $q(s)$ of the adjoint equation (B.3) has at least one positive zero. Let α be the largest such. Our goal is to give a glimpse of how large α is and how far distanced the other zeros are from α. To this end we could employ the integral representation in Lemma B.1 and apply analytic estimates. That such a brute force method actually succeeds can be "predicted" since it turns out that α runs miles ahead of the other zeros; see the proof of Theorem 6 of [**104**]. In this section we present direct arguments exploiting exclusively the differential-difference equation (B.3); a great deal, although not everything, can be achieved along such lines. Of course, as we know by now, the practical aspect of this investigation is diminished by the existence of sieve constructions other than the beta-sieve which yield superior results (when κ is relatively large and s is relatively small). Therefore, our presentation is driven mainly by curiosity. One interesting observation will be that, if $a \leqslant 1$, then all the zeros of $q(s)$ are real.

To make things easier, we assume that $n = a + b - 1$ is a positive integer so that $q(s)$ is a monic polynomial of degree n. We write

$$q(s) = (s - \alpha_1) \dots (s - \alpha_n) \,,$$

where $\alpha_1, \dots, \alpha_n$ are the (complex) roots of $q(s)$. Hence, the differential-difference equation (B.3) becomes

$$\text{(B.20)} \qquad \sum_j \frac{s}{s - \alpha_j} = a - 1 + b \prod_j \Big(1 + \frac{1}{s - \alpha_j}\Big) \,.$$

We already know that α, the largest real zero of $q(s)$, is positive and simple. Comparing the residues of (B.20) at $s = \alpha$ we obtain our first relation among the zeros:

$$\text{(B.21)} \qquad \alpha = b \prod_{\alpha_j \neq \alpha} \Big(1 + \frac{1}{\alpha - \alpha_j}\Big) \,.$$

Next, we write (B.20) in the following fashion:

$$\sum_{\alpha_j \neq \alpha} \frac{s}{s - \alpha_j} = a - 1 + b \prod_{\alpha_j \neq \alpha} \Big(1 + \frac{1}{s - \alpha_j}\Big) + \frac{1}{s - \alpha} \Big[b \prod_{\alpha_j \neq \alpha} \Big(1 + \frac{1}{s - \alpha_j}\Big) - s \Big] \,.$$

Using (B.21) we can evaluate the above expression at $s = \alpha$ as follows:

$$\begin{aligned} \sum_{\alpha_j \neq \alpha} \frac{\alpha}{\alpha - \alpha_j} &= a - 1 + \alpha + \Big[b \prod_{\alpha_j \neq \alpha} \Big(1 + \frac{1}{s - \alpha_j}\Big) - s \Big]'_{s=\alpha} \\ &= a - 2 + \alpha + \alpha \sum_{\alpha_j \neq \alpha} \Big(\frac{1}{\alpha - \alpha_j + 1} - \frac{1}{\alpha - \alpha_j} \Big) \,. \end{aligned}$$

Hence, we obtain a second relation among the zeros:

$$\sum_{\alpha_j \neq \alpha} \Big(\frac{1}{\alpha - \alpha_j} + \frac{1}{(\alpha - \alpha_j)(\alpha - \alpha_j + 1)} \Big) = \frac{a-2}{\alpha} + 1 \,. \tag{B.22}$$

Next, inspired by these two relations, we put

$$\delta = \frac{2}{\alpha} + \sum_{\alpha_j \neq \alpha} \Big[\log\Big(1 + \frac{1}{\alpha - \alpha_j}\Big) - \frac{1}{\alpha - \alpha_j} - \frac{1}{(\alpha - \alpha_j)(\alpha - \alpha_j + 1)} \Big] \tag{B.23}$$

with the expectation that δ is quite small. Now (B.21) becomes

$$\log \frac{\alpha}{b} = \frac{a}{\alpha} + 1 + \delta \,. \tag{B.24}$$

This formula tells us that α is well-approximated by the root of the simple transcendental equation

$$\log \frac{x}{b} = \frac{a}{x} + 1 \,. \tag{B.25}$$

To estimate δ we require a third relation among the zeros, namely

$$\sum_{\alpha_j \neq \alpha} \frac{1}{\alpha - \alpha_j - 1} = \frac{a-1}{\alpha - 1} + 1 \,, \tag{B.26}$$

which follows from (B.20) at $s = \alpha - 1$. Note that $\alpha - 1$ is not a zero of $q(s)$ since otherwise (B.21) would imply that $\alpha = 0$. In fact, the following is true.

LEMMA B.7. *The real zero γ of $q(s)$ closest to α, if it exists, satisfies $\gamma < \alpha - 1$.*

PROOF. First we consider the case $q(0) \neq 0$ so $\gamma \neq 0$. Then γ is also the real zero of $sq(s)$ closest to α. Since $sq(s) > 0$ for all $s > \alpha$ and $sq(s) < 0$ for $\gamma < s < \alpha$ it follows that $q(s)$ is decreasing at $s = \gamma$. By (B.19) we get

$$bq(s+1) = \big(sq(s)\big)' < 0 \quad \text{at } s = \gamma \,,$$

whence $\gamma + 1 < \alpha$.

Now, we consider similarly the case $q(0) = 0$. We have $q(1) = 0$ so $\alpha \geqslant 1$. If $\gamma \geqslant \alpha - 1$ then $\gamma > 0$. Since $q(s) > 0$ for all $s > \alpha$ and $q(s) < 0$ for $\gamma < s < \alpha$ it follows that $q(s)$ is decreasing at $s = \gamma$. By (B.19) we obtain

$$bq(s+1) = sq(s)' < 0 \quad \text{at } s = \gamma \,,$$

whence $\gamma + 1 < \alpha$. □

Returning to the relation (B.26), we tacitly assumed (for peace of mind) that $\alpha \neq 1$ so that the right-hand side makes sense. Note that $\alpha = 1$ implies $(a-1)q(0) = 0$, but $q(0) \neq 0$ so $a = 1$.

Subtracting (B.26) from (B.22), we find our fourth relation between α and the other zeros:

$$\sum_{\alpha_j \neq \alpha} \frac{1}{(\alpha - \alpha_j - 1)(\alpha - \alpha_j)(\alpha - \alpha_j + 1)} = \frac{a + \alpha - 2}{2\alpha(\alpha - 1)} \,. \tag{B.27}$$

It would have been helpful if all the zeros $\alpha_1, \ldots, \alpha_n$ were real, but this is not always true. However, we do have the following result.

LEMMA B.8. *If $a \leqslant 1$ then all the zeros of $q(s)$ are real and are greater than or equal to one.*

PROOF. Suppose the assertion is false. In that case there is a zero of $q(s+1)$, say η, such that $\eta \neq 0$ and η is not a zero of $q(s)$. We can choose η uniquely from the set of numbers $\alpha_j - 1$ by specifying first that it have the largest imaginary part and then, in case of a draw, the smallest real part. By (B.20) we have

$$\sum_j \frac{\eta}{\eta - \alpha_j} = a - 1 \ .$$

We can rearrange this equation into the following shape:

$$\eta\Big(\sum_j \frac{1}{|\eta - \alpha_j|^2} - \frac{a-1}{|\eta|^2}\Big) = \sum_j \frac{\alpha_j}{|\eta - \alpha_j|^2} \ .$$

Since $a \leqslant 1$ this shows that ηc with some $c \geqslant 1$ lies in the convex hull of the points $\alpha_1, \dots, \alpha_n$, which obviously contradicts our choice of η. □

REMARKS. If $a \leqslant 1$ is an integer then the proof of the absence of complex zeros of $q(s)$ can be derived from the equation

$$s^a P'(s) = b(s+1)^{a-1} P(s+1) \ ,$$

where $P(s) = s^{1-a} q(s)$; see (B.4). This shows that the complex zeros of the polynomial $P'(s)$ are shifted by one from the complex zeros of $P(s)$. However, by Gauss' theorem, all zeros of the derivative lie in the convex hull of the zeros of the polynomial. This implies that all the zeros of $P(s)$ are real.

Now we come back to (B.27). Suppose all the zeros of $q(s)$ are real, so all the terms in (B.27) are positive. Taking only the one term at the zero γ closest to α we find that

$$\text{(B.28)} \qquad \alpha - \gamma > \Big(\frac{2\alpha(\alpha-1)}{a+\alpha-2}\Big)^{\frac{1}{3}} \ .$$

Then, inserting this lower bound back into (B.27), we obtain

$$\text{(B.29)} \qquad \sum_{\alpha_j \neq \alpha} \frac{1}{(\alpha - \alpha_j - 1)(\alpha - \alpha_j + 1)} < \Big(\frac{a+\alpha-2}{2\alpha(\alpha-1)}\Big)^{\frac{2}{3}} \ .$$

Hence, we deduce by (B.23) that

$$\text{(B.30)} \qquad \delta \ll \Big(\frac{a+\alpha}{\alpha^2}\Big)^{\frac{2}{3}} \ .$$

REMARKS. Although the last three estimates (B.28), (B.29), and (B.30), hold under the assumption of all zeros of $q(s)$ being real, nevertheless, one can dispense with this condition and establish results of similar quality using analytic arguments. In the most important case when $q(s)$ is given by (B.3) with $a = b = \kappa$, the relevant analysis is carried out in [**104**]. Recall that, in this case, the largest zero $\beta = \alpha + 1$ of $q(s-1)$ is just the sifting limit of the beta-sieve for dimension κ. One can show that

$$\delta \ll \kappa^{-\frac{2}{3}} \ ,$$

which agrees with (B.30). Hence, it follows from (B.24) that

$$\alpha = c\kappa + O\big(\kappa^{\frac{1}{3}}\big) \ ,$$

where $c = 3.591\ldots$ is the same constant appearing in (6.68). Moreover, it turns out that the distance from α to the nearest zero γ is growing to infinity as $\kappa \to \infty$; specifically we have

$$\alpha - \gamma \gg \kappa^{\frac{1}{3}} ,$$

a bound which agrees with (B.28).

B.4. An Example

For use in Section A.3 we develop, by the method of the adjoint equation, some results on the asymptotic behaviour of $m(s)$, defined as the continuous solution of the equation

$$s^{k+1} m'(s) = -k(s-1)^k m(s-1) , \tag{B.31}$$

with initial condition

$$m(s) = 1, \qquad \text{if } \ 0 < s \leqslant 1. \tag{B.32}$$

Putting $\mathfrak{m}(s) = s^k m(s)$, the equation becomes

$$s\mathfrak{m}'(s) = k\mathfrak{m}(s) - k\mathfrak{m}(s-1), \qquad \text{if } \ s > 1. \tag{B.33}$$

Note that a constant function satisfies (B.33) but our initial values are not constant, namely

$$\mathfrak{m}(s) = s^{-k}, \qquad \text{if } 0 < s < 1. \tag{B.34}$$

Writing (B.34) as

$$s\mathfrak{m}'(s) = k \int_{s-1}^{s} \mathfrak{m}'(t)dt ,$$

it follows from Lemma 11.2 that $\mathfrak{m}'(s) \ll s^{-s}$ and so

$$\mathfrak{m}(s) = \mathfrak{m}(\infty) + O(s^{-s}) , \tag{B.35}$$

where $\mathfrak{m}(\infty)$ is a constant.

The adjoint to (B.33) is the equation

$$(s\mathfrak{p}(s))' = -k\mathfrak{p}(s) + k\mathfrak{p}(s+1) , \tag{B.36}$$

whose standard solution is given by

$$\mathfrak{p}(s) = \int_0^\infty \exp\left(-sz + k\int_0^z (1-e^{-u})u^{-1}du\right) dz . \tag{B.37}$$

Therefore, $\mathfrak{p}(s)$ is positive for all $s > 0$. Moreover, by (B.12) and (B.13),

$$\mathfrak{p}(s) \sim s^{-1} \qquad \text{as } s \to \infty \tag{B.38}$$

and

$$\mathfrak{p}(s) \sim e^{\gamma k}\Gamma(k+1)s^{-k-1} \qquad \text{as } s \to 0. \tag{B.39}$$

Having these asymptotics we can compute the constant $\mathfrak{m}(\infty)$ by examining the inner product

$$\begin{aligned}\langle \mathfrak{m}, \mathfrak{p}\rangle &= s\mathfrak{m}(s)\mathfrak{p}(s) - k\int_{s-1}^{s} \mathfrak{m}(x)\mathfrak{p}(x+1)dx \\ &= (s-1)\mathfrak{m}(s-1)\mathfrak{p}(s-1) + \int_{s-1}^{s} x^{k+1}\mathfrak{p}(x)dx^{-k}\mathfrak{m}(x).\end{aligned}$$

Letting s tend to infinity from the first expression we obtain $\langle \mathfrak{m}, \mathfrak{p} \rangle = \mathfrak{m}(\infty)$ by the conditions (B.35) and (B.38). On the other hand, letting s tend to one, from the second expression we obtain $\langle \mathfrak{m}, \mathfrak{p} \rangle = e^{\gamma k}\Gamma(k+1)$, by the initial condition (B.34) and the asymptotic (B.39). Therefore,

$$\mathfrak{m}(\infty) = e^{\gamma k}\Gamma(k+1). \tag{B.40}$$

Since both $\mathfrak{m}(s)$ and the constant function $\mathfrak{m}(\infty)$ satisfy (B.33), we deduce by subtracting inner products of these against $\mathfrak{p}(s)$ the following integral formula:

$$s(\mathfrak{m}(s) - \mathfrak{m}(\infty)) = k \int_{s-1}^{s} (\mathfrak{m}(x) - \mathfrak{m}(\infty))\mathfrak{p}(x+1)dx \tag{B.41}$$

for $s \geqslant 1$. Since $\mathfrak{p}(s)$ is positive this tells us:

COROLLARY B.9. *If k is negative then $\mathfrak{m}(s) - \mathfrak{m}(\infty)$ changes sign in every interval of length 1.*

Bibliography

1. N. C. Ankeny and H. Onishi, *The general sieve*, Acta Arith. **10** (1964/1965), 31–62.
2. F. V. Atkinson, *A divisor problem*, Quart. J. Math. **12** (1941), 193–200.
3. R. C. Baker, G. Harman, and J. Pintz, *The difference between consecutive primes II*, Proc. London Math. Soc. **83** (2001), 532–562.
4. M. B. Barban and P. P. Vehov, *An extremal problem*, Trudy Moskov. Mat. Obšč. **18** (1968), 83–90.
5. P. T. Bateman and R. A. Horn, *A heuristic asymptotic formula concerning the distribution of prime numbers*, Math. Comp. **16** (1962), 363–367.
6. V. Blomer, *Uniform bounds for Fourier coefficients of theta-series with arithmetic applications*, Acta Arith. **114** (2004), 1–21.
7. V. Blomer and J. Brüdern, *A three squares theorem with almost primes*, Bull. London Math. Soc. **37** (2005), 507–513.
8. E. Bombieri, *The asymptotic sieve*, Rend. Accad. Naz. XL Ser. V **1/2** (1975/76), 243–269.
9. ______, *Le Grand Crible dans la Théorie Analytique des Nombres*, Astérisque **18** (1987), 103 pp.
10. E. Bombieri and H. Davenport, *On the large sieve method*, Number Theory and Analysis, Plenum, (New York), 1969, pp. 9–22.
11. E. Bombieri, J. B. Friedlander, and H. Iwaniec, *Primes in arithmetic progressions to large moduli*, Acta Math. **156** (1986), 203–251.
12. ______, *Primes in arithmetic progressions to large moduli III*, J. Amer. Math. Soc. **2** (1989), 215–224.
13. J. Bourgain, A. Gamburd, and P. Sarnak, *Sieving and expanders*, C. R. Math. Acad. Sci. Paris **343** (2006), 155–159.
14. R. P. Brent, *Irregularities in the distribution of primes and twin primes*, Math. Comp. **29** (1975), 43–56.
15. J. Brüdern and É. Fouvry, *Le crible à vecteurs*, Compos. Math. **102** (1996), 337–355.
16. V. Brun, *Uber das Goldbachsche Gesetz und die Anzahl der Primzahllpaare*, Archiv for Math. og Naturvid **B 34** (1915), 19 pages.
17. ______, *Le crible d'Eratosthene et le théorème de Goldbach*, Skr. Norske Vid.-Akad. Kristiana (1920), no. 3, 36.
18. ______, *Das Sieb des Erathosthenes*, 5. Skand. Mat. Kongr., Helsingfors (1922), 197–203.
19. A. A. Buchstab, *New improvements in the method of the sieve of Eratosthenes*, Mat. Sbornik **4** (1938), 375–387.
20. F. Châtelet, *Points rationnels sur certaines surfaces cubiques*, Les Tendances Géom. en Algèbre et Théorie des Nombres, CNRS, (Paris), 1966, pp. 67–75.
21. R. Chelluri, *Equidistribution of the Roots of Quadratic Congruences*, Ph.D. thesis, Rutgers, 2004.
22. J.-R. Chen, *On the representation of a larger even integer as the sum of a prime and the product of at most two primes*, Sci. Sinica **16** (1973), 157–176.
23. A. C. Cojocaru, *Reductions of an elliptic curve with almost prime orders*, Acta Arith. **119** (2005), 265–289.
24. A. C. Cojocaru and M. R. Murty, *An Introduction to Sieve Methods and their Applications*, London Math. Soc. Student Texts, vol. 66, Cambridge Univ. Press, (Cambridge), 2006.
25. J. B. Conrey and H. Iwaniec, *The cubic moment of central values of automorphic L-functions*, Ann. of Math. **151** (2000), 1175–1216.
26. H. Daboussi and H. Delange, *Quelques propriétés des fonctions multiplicatives de module au plus égal à* 1, C. R. Acad. Sci. Paris Sér. A **278** (1974), 657–660.

27. S. Daniel, *On the divisor-sum problem for binary forms*, J. Reine Angew. Math. **507** (1999), 107–129.
28. H. Davenport, *Multiplicative Number Theory*, third ed., Grad. Texts in Math., vol. 74, Springer-Verlag, (New York), 2000, Revised by H. L. Montgomery.
29. H. Davenport and H. Halberstam, *The values of a trigonometrical polynomial at well spaced points*, Mathematika **13** (1966), 91–96.
30. N. G. de Bruijn, *The asymptotic behaviour of a function occurring in the theory of primes*, J. Indian Math. Soc. (N.S.) **15** (1951), 25–32.
31. J.-M. Deshouillers and H. Iwaniec, *Kloosterman sums and Fourier coefficients of cusp forms*, Invent. Math. **70** (1982/83), 219–288.
32. H. Diamond, H. Halberstam, and H.-E. Richert, *Combinatorial sieves of dimension exceeding one*, J. Number Theory **28** (1988), 306–346.
33. H. G. Diamond and H. Halberstam, *A Higher-Dimensional Sieve Method*, Cambridge Tracts in Mathematics, vol. 177, Cambridge Univ. Press, (Cambridge), 2008.
34. W. Duke, *Hyperbolic distribution problems and half-integral weight Maass forms*, Invent. Math. **92** (1988), no. 1, 73–90.
35. ______, *On ternary quadratic forms*, J. Number Theory **110** (2005), 37–43.
36. W. Duke, J. B. Friedlander, and H. Iwaniec, *Equidistribution of roots of a quadratic congruence to prime moduli*, Ann. of Math. **141** (1995), 423–441.
37. ______, *Bilinear forms with Kloosterman fractions*, Invent. Math. **128** (1997), 23–43.
38. W. Duke and R. Schulze-Pillot, *Representation of integers by positive ternary quadratic forms and equidistribution of lattice points on ellipsoids*, Invent. Math. **99** (1990), 49–57.
39. P. D. T. A. Elliott, *The least prime primitive root and Linnik's theorem*, Number Theory for the Millennium, I (Urbana, IL, 2000), A K Peters, (Natick MA), 2002, pp. 393–418.
40. P. Erdös, *On integers of the form $2^k + p$ and some related problems*, Summa Brasil. Math. **2** (1950), 113–123.
41. T. Estermann, *Einige Sätze über quadratfreie Zahlen*, Math. Ann. **105** (1931), 653–662.
42. L. Euler, *De tabule numerorum primorum*, Novi Commentarii Acad. Scient. Petropol. **19** (1775), 132–183.
43. K. Ford, *On Bombieri's asymptotic sieve*, Trans. Amer. Math. Soc. **357** (2005), 1663–1674.
44. É. Fouvry, *Autour du théorème de Bombieri-Vinogradov*, Acta Math. **152** (1984), 219–244.
45. É. Fouvry and H. Iwaniec, *Gaussian primes*, Acta Arith. **79** (1997), 249–287.
46. J. B. Friedlander, *Sifting short intervals II*, Math. Proc. Cambridge Philos. Soc. **92** (1982), 381–384.
47. ______, *Moments of sifted sequences*, Math. Ann. **267** (1984), 101–106.
48. J. B. Friedlander and A. Granville, *Limitations to the equi-distribution of primes I*, Ann. of Math. **129** (1989), 363–382.
49. J. B. Friedlander, A. Granville, A. Hildebrand, and H. Maier, *Oscillation theorems for primes in arithmetic progressions and for sifting functions*, J. Amer. Math. Soc. **4** (1991), 25–86.
50. J. B. Friedlander and H. Iwaniec, *On Bombieri's asymptotic sieve*, Ann. Sc. Norm. Sup. (Pisa) **5** (1978), 719–756.
51. ______, *Incomplete Kloosterman sums and a divisor problem*, Ann. of Math. **121** (1985), 319–350, appendix B. J. Birch and E. Bombieri.
52. ______, *The Brun-Titchmarsh theorem*, Analytic Number Theory (Kyoto, 1996), London Math. Soc. Lecture Note Ser., vol. 247, Cambridge Univ. Press, (Cambridge), 1997, pp. 85–93.
53. ______, *Asymptotic sieve for primes*, Ann. of Math. **148** (1998), 1041–1065.
54. ______, *The polynomial $X^2 + Y^4$ captures its primes*, Ann. of Math. **148** (1998), 945–1040.
55. ______, *The illusory sieve*, Int. J. Number Theory **1** (2005), 459–494.
56. ______, *A polynomial divisor problem*, J. Reine Angew. Math. **601** (2006), 109–137.
57. ______, *Hyperbolic prime number theorem*, Acta Math. **202** (2009), 1–19.
58. ______, *Ternary quadratic forms with rational zeros*, J. Théorie des Nombres de Bordeaux (to appear).
59. P. X. Gallagher, *The large sieve*, Mathematika **14** (1967), 14–20.
60. ______, *Bombieri's mean value theorem*, Mathematika **15** (1968), 1–6.
61. ______, *A larger sieve*, Acta Arith. **18** (1971), 77–81.
62. A. O. Gelfond and Ju. V. Linnik, *Elementary Methods in Analytic Number Theory*, Rand McNally, (Chicago), 1965.

63. D. A. Goldston, S. W. Graham, J. Pintz, and C. Y. Yıldırım, *Small gaps between products of two primes*, Proc. Lond. Math. Soc. **98** (2009), 741–774.
64. D. A. Goldston, J. Pintz, and C. Y. Yıldırım, *Primes in tuples I*, Ann. of Math. **170** (2009), 819–862.
65. D. A. Goldston and C. Y. Yıldırım, *Higher correlations of divisor sums related to primes III. Small gaps between primes*, Proc. Lond. Math. Soc. **95** (2007), 653–686.
66. I. S. Gradshteyn and I. M. Ryzhik, *Table of Integrals, Series, and Products*, 4th ed., Academic Press, (New York), 1965.
67. S. Graham, *An asymptotic estimate related to Selberg's sieve*, J. Number Theory **10** (1978), 83–94.
68. ______, *On Linnik's constant*, Acta Arith. **39** (1981), 163–179.
69. S. W. Graham and G. Kolesnik, *Van der Corput's Method of Exponential Sums*, London Math. Soc. Lecture Notes, vol. 126, Cambridge Univ. Press, (Cambridge), 1991.
70. A. Granville, *Smooth numbers: computational number theory and beyond*, Algorithmic Number Theory: Lattices, Number Fields, Curves and Cryptography, Math. Sci. Res. Inst. Publ., Cambridge Univ. Press, (Cambridge), 2008, pp. 267–323.
71. A. Granville and K. Soundararajan, *The number of unsieved integers up to x*, Acta Arith. **115** (2004), 305–328.
72. G. Greaves, *On the representation of a number in the form $x^2+y^2+p^2+q^2$ where p, q are odd primes*, Acta Arith. **29** (1976), 257–274.
73. ______, *Weighted sieves*, Séminaire de Théorie des Nombres, 1979–1980, Exp. 29, Univ. Bordeaux I, (Talence), 1980.
74. ______, *A weighted sieve of Brun's type*, Acta Arith. **40** (1981/82), 297–332.
75. ______, *A comparison of some weighted sieves*, Elementary and Analytic Theory of Numbers (Warsaw, 1982), Banach Center Publ., vol. 17, PWN, (Warsaw), 1985, pp. 143–153.
76. ______, *Sieves in Number Theory*, Ergeb. der Math. und ihrer Grenzgeb., vol. 43, Springer-Verlag, (Berlin), 2001.
77. B. Green and T. Tao, *An inverse theorem for the Gowers $U^3(G)$ norm*, Proc. Edin. Math. Soc. **51** (2008), 73–153.
78. R. Gupta and M. R. Murty, *A remark on Artin's conjecture*, Invent. Math. **78** (1984), 127–130.
79. H. Halberstam and H.-E. Richert, *Sieve Methods*, London Math. Soc. Monographs, vol. 4, Academic Press, (London), 1974.
80. G. H. Hardy and J. E. Littlewood, *Some problems of 'partitio numerorum' III. On the expression of a number as a sum of primes*, Acta Math. **44** (1922), 1–70.
81. G. Harman, *Primes in short intervals*, Math. Zeit. **180** (1982), 335–348.
82. ______, *On the distribution of αp modulo one*, J. London Math. Soc. **27** (1983), 9–18.
83. ______, *Prime-Detecting Sieves*, London Math. Soc. Monographs, vol. 33, Princeton University Press, (Princeton), 2007.
84. D. R. Heath-Brown, *Prime numbers in short intervals and a generalized Vaughan identity*, Canad. J. Math. **34** (1982), 1365–1377.
85. ______, *Prime twins and Siegel zeros*, Proc. London Math. Soc. **47** (1983), 193–224.
86. ______, *The square sieve and consecutive square-free numbers*, Math. Ann. **266** (1984), no. 3, 251–259.
87. ______, *Artin's conjecture for primitive roots*, Quart. J. Math. (Oxford) **37** (1986), 27–38.
88. ______, *Zero-free regions for Dirichlet L-functions, and the least prime in an arithmetic progression*, Proc. London Math. Soc. **64** (1992), 265–338.
89. ______, *Almost-prime k-tuples*, Mathematika **44** (1997), 245–266.
90. ______, *Primes represented by x^3+2y^3*, Acta Math. **186** (2001), 1–84.
91. D. R. Heath-Brown and H. Iwaniec, *On the difference between consecutive primes*, Invent. Math. **55** (1979), 49–69.
92. D. R. Heath-Brown and B. Z. Moroz, *On the representation of primes by cubic polynomials in two variables*, Proc. London Math. Soc. **88** (2004), 289–312.
93. D. A. Hejhal, *The Selberg Trace Formula for* PSL(2, **R**). *II*, Lecture Notes in Math., vol. 1001, Springer-Verlag, (Berlin), 1983.
94. A. Hildebrand, *On the number of positive integers $\leqslant x$ and free of prime factors $> y$*, J. Number Theory **22** (1986), 289–307.

95. A. Hildebrand and G. Tenenbaum, *Integers without large prime factors*, J. Théorie des Nombres de Bordeaux **5** (1993), 411–484.
96. G. Hoheisel, *Primzahlprobleme in der Analysis*, Sitzungsber. Pruess. Akad. Wiss. (1930), 580–588.
97. R. Holowinsky, *A sieve method for shifted convolution sums*, Duke Math. J. **146** (2009), 401–448.
98. R. Holowinsky and K. Soundararajan, *Mass equidistribution of Hecke eigenforms*, Ann. of Math. (to appear).
99. M. N. Huxley, *The Distribution of Prime Numbers*, Clarendon Press, (Oxford), 1972.
100. A. E. Ingham, *Some asymptotic formulae in the theory of numbers*, J. London Math. Soc **2** (1927), 202–208.
101. H. Iwaniec, *On the error term in the linear sieve*, Acta Arith. **19** (1971), 1–30.
102. ———, *Primes represented by quadratic polynomials in two variables*, Acta Arith. **24** (1973/74), 435–459.
103. ———, *A new form of the error term in the linear sieve*, Acta Arith. **37** (1980), 307–320.
104. ———, *Rosser's sieve*, Acta Arith. **36** (1980), 171–202.
105. ———, *Fourier coefficients of modular forms of half-integral weight*, Invent. Math. **87** (1987), 385–401.
106. ———, *Topics in Classical Automorphic Forms*, Grad. Studies in Math., vol. 17, Amer. Math. Soc., (Providence), 1997.
107. H. Iwaniec and J. Jiménez Urroz, *Orders of CM elliptic curves modulo p with at most two primes*, Ann. Sc. Norm. Sup. (Pisa) (to appear).
108. H. Iwaniec and M. Jutila, *Primes in short intervals*, Ark. Mat. **17** (1979), 167–176.
109. H. Iwaniec and E. Kowalski, *Analytic Number Theory*, Colloq. Publications, vol. 53, Amer. Math. Soc., (Providence), 2004.
110. H. Iwaniec and R. Munshi, *Cubic polynomials and quadratic forms*, J. London Math. Soc. **81** (2010), 45–64.
111. H. Iwaniec and J. Pomykała, *Sums and differences of quartic norms*, Mathematika **40** (1993), 233–245.
112. H. Iwaniec and P. Sarnak, *The non-vanishing of central values of automorphic L-functions and Landau-Siegel zeros*, Israel J. Math. **120** (2000), 155–177.
113. H. Iwaniec, J. van de Lune, and H. J. J. te Riele, *The limits of Buchstab's iteration sieve*, Indag. Math. **42** (1980), 409–417.
114. C. Jia, *Almost all short intervals containing prime numbers*, Acta Arith. **76** (1996), 21–84.
115. D. Johnson, *Mean values of Hecke L-functions*, J. Reine Angew. Math. **305** (1979), 195–205.
116. W. B. Jurkat and H.-E. Richert, *An improvement of Selberg's sieve method I*, Acta Arith. **11** (1965), 217–240.
117. A. A. Karatsuba, *Analogues of Kloosterman sums*, Izv. Ross. Akad. Nauk, Ser. Mat. **59** (1995), 93–102.
118. I. Kobayashi, *A note on the Selberg sieve and the large sieve*, Proc. Japan Acad. **49** (1973), 1–5.
119. N. Koblitz, *Almost primality of group orders of elliptic curves defined over small finite fields*, Experiment. Math. **10** (2001), 553–558.
120. P. Kuhn, *Neue Abschätzungen auf Grund der Viggo Brunschen Siebmethode*, Tolfte Skand. Matematikerkong., Lund, 1953, Lunds Univ. Mat. Inst., (Lund), 1954, pp. 160–168.
121. N. V. Kuznetsov, *The Petersson conjecture for cusp forms of weight zero and the Linnik conjecture. Sums of Kloosterman sums*, Mat. Sbornik **111(153)** (1980), 334–383.
122. M. Laborde, *Buchstab's sifting weights*, Mathematika **26** (1979), 250–257.
123. B. Landreau, *Majorations de fonctions arithmétiques en moyenne sur des ensembles de faible densité*, Séminaire de Théorie des Nombres, 1987–1988, Exp. 13, Univ. Bordeaux I, (Talence), 1988.
124. Ju. V. Linnik, *The large sieve*, Dokl. Akad. Nauk. SSSR **30** (1941), 292–294.
125. ———, *On the least prime in an arithmetic progression I. The basic theorem*, Mat. Sbornik **15(57)** (1944), 139–178.
126. ———, *The Dispersion Method in Binary Additive Problems*, Amer. Math. Soc., (Providence), 1963.
127. F. Mertens, *Ein Beitrag zur analytischen Zahlenthheorie*, J. Reine Angew. Math. (1874), 46–62.

128. H. L. Montgomery, *A note on the large sieve*, J. London Math. Soc. **43** (1968), 93–98.
129. ———, *Topics in Multiplicative Number Theory*, Lecture Notes in Math. v. 227, Springer-Verlag, (Berlin), 1971.
130. H. L. Montgomery and R. C. Vaughan, *The large sieve*, Mathematika **20** (1973), 119–134.
131. ———, *Hilbert's inequality*, J. London Math. Soc. **8** (1974), 73–82.
132. ———, *Multiplicative Number Theory. I. Classical Theory*, Studies in Advanced Math., vol. 97, Cambridge Univ. Press, (Cambridge), 2007.
133. Y. Motohashi, *On some improvements of the Brun-Titchmarsh theorem*, J. Math. Soc. Japan **26** (1974), 306–323.
134. Y. Motohashi and J. Pintz, *A smoothed GPY sieve*, Bull. London Math. Soc. **40** (2008), 298–310.
135. M. Nair and G. Tenenbaum, *Short sums of certain arithmetic functions*, Acta Math. **180** (1998), 119–144.
136. O. Ramaré, *Arithmetical Aspects of the Large Sieve Inequality*, Harish-Chandra Research Inst. Lecture Notes, Hindustan Book Agency, (New Delhi), 2009.
137. A. Rényi, *On the large sieve of Ju. V. Linnik*, Compos. Math. **8** (1950), 68–75.
138. P. Ribenboim, *The New Book of Prime Number Records*, Springer-Verlag, (New York), 1996.
139. G. Ricci, *Sui grandi divisori primi delle coppie di interi in posti corrispondenti di due progressioni aritmetiche. Applicazione del metodo di Brun*, Ann. Mat. Pura Appl. **11** (1933), 91–110.
140. H.-E. Richert, *Selberg's sieve with weights*, Mathematika **16** (1969), 1–22.
141. P. Sarnak, *Integral Apollonian packings*, MAA Lecture, January 2009.
142. L. G. Schnirelmann, *Über additive Eigenschaften von Zahlen*, Math. Ann. **107** (1933), 649–690.
143. A. Selberg, *On the normal density of primes in small intervals, and the difference between consecutive primes*, Arch. Math. Naturvid. **47** (1943), 87–105.
144. ———, *On the zeros of Riemann's zeta-function*, Collected papers. Vol. I, Springer-Verlag, (Berlin), 1989, pp. 85–155.
145. ———, *Lectures on sieves*, Collected papers. Vol. II, Springer-Verlag, (Berlin), 1991, pp. 65–247.
146. J-P. Serre, *A Course in Arithmetic*, Graduate Texts in Math., No. 7, Springer-Verlag, (New York), 1973.
147. ———, *Spécialisation des éléments de* $\mathrm{Br}_2(\mathbf{Q}(T_1, \cdots, T_n))$, C. R. Acad. Sci. Paris Sér. I Math. **311** (1990), 397–402.
148. D. Shanks and J. W. Wrench, Jr., *Brun's constant*, Math. Comp. **28** (1974), 293–299; corrigenda, ibid. 28 (1974), 1183.
149. R. A. Smith, *The circle problem in an arithmetic progression*, Canad. Math. Bull. **11** (1968), 175–184.
150. ———, *On* $\sum r(n)r(n+a)$, Proc. Nat. Inst. Sci. India (A) **34** (1968), 132–137.
151. K. Soundararajan, *Small gaps between prime numbers: the work of Goldston-Pintz-Yıldırım*, Bull. Amer. Math. Soc. **44** (2007), 1–18.
152. V. Tartakovskiĭ, *Sur quelques sommes du type de Viggo Brun*, Dokl. Akad. Nauk SSSR (N.S.) (1939), 126–129.
153. V. S. Tipu, *Polynomial Divisor Problems*, Ph.D. thesis, Toronto, 2008.
154. E. C. Titchmarsh, *The Theory of the Riemann Zeta-Function*, 2nd ed., Clarendon Press, (Oxford), 1986, ed. D. R. Heath-Brown.
155. A. Tóth, *Roots of quadratic congruences*, Inter. Math. Res. Notices **14** (2000), 719–739.
156. J. H. van Lint and H.-E. Richert, *On primes in arithmetic progressions*, Acta Arith. **11** (1965), 209–216.
157. R. C. Vaughan, *Sommes trigonométriques sur les nombres premiers*, C. R. Acad. Sci. Paris Sér. A-B **285** (1977), 981–983.
158. I. M. Vinogradov, *The method of trigonometrical sums in the theory of numbers*, Dover, (Mineola), 2004, Reprint of the 1954 translation by K. F. Roth and A. Davenport.
159. H. Weyl, *Über die Gleichverteilung von Zahlen mod Eins*, Math. Ann. **77** (1916), 313–352.
160. E. Wirsing, *Das asymptotische Verhalten von Summen über multiplikative Funktionen*, Math. Ann. **143** (1961), 75–102.
161. D. Wolke, *A new proof of a theorem of van der Corput*, J. London Math. Soc. **5** (1972), 609–612.

Index

Titles in This Series

TITLES IN THIS SERIES